Climate: Present, Past and Future

First published in 1977, the second volume of *Climate: Present, Past and Future* covers parts III and IV of Hubert Lamb's seminal and pioneering study of climatology.

Part III provides a survey of evidence of types of climates over the last million years, and of methods of dating that evidence. Through the earlier stages of the Earth's development the book traces what is known of the various geographies presented by the drifting continents and indicates what can be learnt about climatic regimes and the causes of climatic change. From the last ice age to the present our knowledge of the succession of climates is summarized, indicating prevailing temperatures, rainfalls, wind and ocean current patterns where possible.

Part IV considers events during the fifteen years prior to the book's initial publication, leading on to the problems of estimating the most probable future course of climatic development, and the influence of Man's activities on climate.

Alongside the reissue of volume 1, this *Routledge Revival* will be essential reading for anyone interested in both the causes and workings of climate and in the history of climatology itself.

Climate:
Present, Past and Future

Volume 2
Climatic History and the Future

H. H. Lamb

Routledge
Taylor & Francis Group

First published in 1977
by Methuen & Co. Ltd

This edition first published in 2011 by Routledge
2 Park Square, Milton Park, Abingdon, Oxon, OX14 4RN

Simultaneously published in the USA and Canada
by Routledge
711 Third Avenue, New York, NY 10017

Routledge is an imprint of the Taylor & Francis Group, an informa business

© 1977 H. H. Lamb

Publisher's Note

The publisher has gone to great lengths to ensure the quality of this reprint but points out that some imperfections in the original copies may be apparent.

Disclaimer

The publisher has made every effort to trace copyright holders and welcomes correspondence from those they have been unable to contact.

A Library of Congress record exists under LC Control Number: 77008305

ISBN 13: 978-0-415-67959-6 (set)
ISBN 13: 978-0-415-67951-0 (hbk)
ISBN 13: 978-0-203-80430-8 (ebk)
ISBN 13: 978-0-415-68223-7 (pbk)

CLIMATE

PRESENT, PAST AND FUTURE

CLIMATE

PRESENT, PAST AND FUTURE

Volume 2

Climatic history and the future

H. H. LAMB

METHUEN & CO LTD   LONDON

BARNES & NOBLE BOOKS   NEW YORK

First published in 1977
by Methuen & Co Ltd,
11 New Fetter Lane, London EC4
© *1977 H. H. Lamb*
Printed in Great Britain by
William Clowes & Sons Ltd,
Beccles, Suffolk

ISBN 0 416 11540 3

Library of Congress Cataloging in Publication Data

Lamb, H. H.
  Climatic history and the future.

  (His Climate; v. 2)
  Bibliography: p.
  Includes indexes.
  1. Paleoclimatology. 2. Climatology. 3. Weather
forecasting. I. Title.
QC981.L28 vol. 2   551.6s [551.6] 77-8305

ISBN 0-06-473881-7

Contents

Introduction to Parts III and IV

Part III Climatic history

Appendix to Part III

Part IV The future

Appendix to Part IV

Tables

Part III

Appendix to Part III

Part IV

Figures

Part III

Appendix to Part III

Plates

The plates are between pages 352 and 353

Acknowledgements

The author expresses very sincere thanks to all those who have provided material specially for inclusion in this volume, to those who have permitted the reproduction of their diagrams and data, and to those who have read and advised on various chapters. It is an honour and a pleasure to acknowledge the generosity of so many who are leaders in their fields of science.

My particular gratitude is due to Sir HARRY GODWIN F.R.S., Emeritus Professor of Botany in the University of Cambridge, who first made contact with me in 1959, at a time when almost no meteorologists anywhere in the world were contributing to the interpretation of evidence of the climate of past centuries and millennia. Through the regular Quaternary Discussions organized by Godwin, the writer was led into an awareness of the vast range of work relevant to this theme in other sciences and enabled to meet and hear the expositions of workers in every part of the world.

My gratitude also goes to the University of East Anglia, to the Vice Chancellor, Dr Frank THISTLETHWAITE, and Senate of the University who invited me to establish the Climatic Research Unit and to many colleagues in the university for diverse facilities, advice and assistance and, perhaps most of all, for their tolerance of my inability, while the compilation and writing of this work continued, to give full attention to the day-to-day matters basic to the institution. The compilation was, however, a necessary foundation for the work of the Climatic Research Unit. That gratitude is all the more real because the writing of this interdisciplinary volume could hardly have been done but for my transfer in 1972 to the university, and the setting up of the Climatic Research Unit, which made the necessary contacts with a great diversity of workers and sources of information much easier. Thanks are due therefore to those grant-giving bodies, particularly the Nuffield, Wolfson and Rockefeller Foundations and Shell International, whose vision of the needs and faith in the possibilities of the subject, produced the funds needed to set up the Climatic Research Unit and have sustained it since.

Special thanks are also due to my friend GORDON MANLEY, Emeritus Professor of Environmental Sciences in the University of Lancaster, who read and checked the entire typescript text and made many helpful suggestions, and who also supplied valuable data.

I wish to thank particularly those who gave time to scrutinizing the writing on their own subjects in this book, notably Professor R. G. WEST of the Cambridge University Sub-Department of Quaternary Research on the use of botanical evidence, Dr G. R. COOPE of Birmingham University on the sections about insects, Dr N. J. SHACKLETON of the Cambridge University Palaeotemperatures Laboratory on the oxygen isotope work, and Dr V. R. SWITSUR on radiocarbon dating. Professor F. H. VINE and Dr G. S. BOULTON of the University of East Anglia kindly looked over chapters 15 and 16 respectively for me, and Dr E. S. WOOD of Guildford gave me valued advice on the archaeology.

Many gifts of data and information are acknowledged in the text and tables in the Appendix. Particularly valuable items came from: Professor R. A. BRYSON of the University of Wisconsin; Drs H. H. and H. J. B. BIRKS, N. J. SHACKLETON and R. G. WEST of Cambridge; M. I. BUDYKO, A. V. SHNITNIKOV, K. YA. KONDRATYEV and I. V. MAKSIMOV of Leningrad; Professor H. E. LANDSBERG of the University of Maryland; Professor G. MANLEY of the University of Lancaster; Dr J. M. MITCHELL of the United States Environmental Data Service; Dr A. D. VERNEKAR of the Travelers Research Center, Hartford, Connecticut; Dr J. D. FRIEDMAN of the United States Geological Survey, Washington; Dr L. M. LIBBY of the University of California, Los Angeles; and Dr W. L. GATES of the RAND Corporation, Santa Monica. Dr V. C. LAMARCHE of the Laboratory of Tree-Ring Research, University of Arizona, generously made his longest bristlecone pine tree-ring series available to me before publishing it himself and Professor H. C. FRITTS and Dr J. E. KUTZBACH similarly provided their constructions of atmospheric circulation patterns derived from tree-ring work. I am equally indebted for information and results from tree-ring work to Dr J. M. FLETCHER, whose X-ray work at Oxford on tree rings in English oaks is becoming justly known, and to Dr G. SIRÉN, formerly of the Forestry Research Institute of Finland, Professor B. FRENZEL of Stuttgart (Universität Hohenheim), and my colleague Dr H. T. MÖRTH, who has joined the Climatic Research Unit at Norwich. Some of the meteorological and oceanographic data in the eastern Atlantic and European areas from the seventeenth century to the present decade were kindly made available to me by colleagues in the British, Danish, Dutch and German meteorological services, notably Drs DUMSTREI and FRYDENDAHL of Copenhagen, Drs KRIJNEN and SCHUURMANS of De Bilt and Dr TRENKLE of Freiburg, as well as through the agency of Professor FLOHN of the University of Bonn, and by the Academy and National Library of the Faeroe Islands in Thorshavn. Particularly interesting unpublished research material on the history of droughts in Africa was generously supplied by Ms SHARON NICHOLSON of the Universities of Wisconsin and Bonn and data on the Southern Oscillation by my colleague at Norwich Mr P. B. WRIGHT. Similar generosity from Dr N. J. SHACKLETON of Cambridge, Dr W. DANSGAARD of Copenhagen, Professor A. T. WILSON of the University of Waikato, New Zealand, Dr A. MCINTYRE of the Lamont-Doherty Observatory, New York, Professors J. IMBRIE of Brown University and J. P. KENNETT of Kingston, Rhode Island, Dr W. F. RUDDIMAN of the U.S. Naval Oceanographic Office, Drs J. C. BRIDEN of Leeds University and A. G. SMITH of Cambridge, provided the most recent research results from studies of the ocean bed, of ice-sheets and of palaeogeographies derived from considerations of continental drift. Archaeological and palaeobotanical results were kindly supplied by workers in these fields, notably Dr H. NICHOLS of Boulder, Colorado, Dr B. FREDSKILD of Copenhagen, and Dr P. ASHBEE of the University of East Anglia and Dr G. SINGH for India. Contributions of historical material from Professor J. SANDNES, FRU VORREN and Dr P. A. TALLANTIRE of the University of Trondheim, Norway, from Professor A. HOLMSEN of Oslo and workers in the Nordisk Ødegårdsprojekt must be acknowledged, as well as notable items supplied by Professor J. DE VRIES of the History Department, University of California at Berkeley, Professor LADURIE of Paris, Dr J. MALEY of Montpellier and Drs A. T. and J. GROVE of Cambridge. The help of Professor F. K. HARE of Toronto, Professor J. SANDNES of Trondheim, Frk T. JEXLEV of Copenhagen and Mr MICHAEL HUNT of Anglia Television, Norwich, in securing diagrams and data needed for this book is also thankfully acknowledged.

Other contributions too numerous to specify here took the form of permission, in all cases generously granted, to copy diagrams and pictures and in some cases the special supply of such items for this book: these are gratefully acknowledged to the authors and publishers concerned, as specified under the figures and plates concerned.

I also wish to acknowledge material and computational help unfailingly supplied by all members of the staff of the Climatic Research Unit, including work from their own researches by Mrs B. M. GRAY, Dr P. M. KELLY and Mr J. A. KINGTON. The long text and tables were quickly and beautifully typed by Mrs SUSAN BOLAND, Mrs MARGARET GIBSON (whose accurate work on the tables constituted an unheard-of record) and Miss CHRIS PRING.

Finally, I cannot adequately express by indebtedness to my wife and family for their support, patience and tolerance during the twelve years when so many free days and evenings had to be given up to the writing of these two volumes. If we had foreseen how long it would take, it would have been hard to find the morale to see it through.

Preface

The first volume of this work was a product of the thinking about how climate is generated and, in particular, the role of the general circulation of the atmosphere and the oceans, which the writer had the opportunity to develop over fifteen years in the branches of the Meteorological Office concerned with world climatology and the development of long-range weather forecasting. Its ten main chapters were devoted to the fundamental observations and theory on the production of climate and its variability. The volume included the first publication in book form of a considerable body of forecasting rules – of British, German and Russian origin – applicable to the weather in Europe over a few months ahead, based on understanding of the seasonal development and tendencies in the atmospheric circulation and the influence of solar variation, volcanic activity, ocean surface temperature anomalies, and variations of the Arctic sea ice, and expressed in statistical terms.

The development of a capability of forecasting the prevailing weather character over the seasons ahead and the fluctuations or trend of climate over some years, and even over many decades, ahead is demanded by many practical planning needs of the age. Since Volume I was published the position has moved on in various ways.

The computer revolution in meteorology has made possible the identification of an enormous further number of statistically significant relationships connecting this, that and the other types of summer, winter, spring, autumn, January, etc., weather with various preconditions of the global atmospheric circulation, temperature and rainfall anomalies over many preceding months; but little or no improvement of the monthly and seasonal forecasts has resulted. Presumably, the explanation must lie in lack of recognition of the physical processes and chains of causation which underlay the statistics of the past. Until these processes and the large-scale evolutions of the circulation of the atmosphere and oceans over the months involved are identified, many statistical relationships may be mistaken for independent indications of a particular forecast development (and all taken as giving added weight to the evidence for it) which are in reality only diverse features of a single process.

The successes achieved in Germany some decades ago by the pioneer of seasonal weather forecasting, Professor FRANZ BAUR, leave no doubt that the formulation of statistical rules which express real physical processes in the Earth and its fluid envelope, and in the surrounding universe (e.g. fluctuations of solar output and of tidal forces), is a valid method which has something to contribute to the solution of the problem of advising on future weather probabilities. Moreover, a probability statement is the form of advice on long-term weather prospects best adapted to planners who must assess the risks to their undertakings.

These years have also seen continued expansion of the effort, and the variety of approaches, put into mathematical modelling of the atmosphere and oceans, their dynamics and energy exchanges, etc., and the whole realm of climate. This promises great advances in our theoretical understanding and knowledge of the processes on every scale of time and space involved and of their interactions. The prospects of any specific forecast by these means is thought to be still remote, however, since there are as yet large percentage errors in many features of the present-day climatic average conditions calculated by the models. Nevertheless, the space relationships indicated are sufficiently realistic to make possible illuminating experiments on the consequences of introducing anomalies and changes at this or that place and time. The main practical contribution of such work so far has accordingly been in assessing the probable effects (and the geographical range of the effects) of various human activities and Man-made pollution, including local heat sources.

On the other side, the concern over the vulnerability of an ever increasingly overpopulated world, greedy for ever higher living standards, to climatic fluctuations affecting the total harvests and food supply increases yearly. This concern has been greatly stimulated by the droughts and floods and other weather extremes of recent years, starting with the colder winters in Europe and eastern North America in the 1960s (particularly 1962–3), and the Indian monsoon failures and droughts in the Soviet and Chinese grainlands in that decade and since. The drought which continued over many years in Africa just south of the Sahara brought a large-scale disaster in the Sahel and Ethiopia in 1972 and 1973, which caused mass migration and political upheaval, while the failures of the grain harvests in India and the Soviet Union in 1972 disrupted the world economy in foodstuffs.

The demand for sound advice on future climatic tendencies, particularly in such marginal areas (including also the Arctic fringe and the other arid regions of the Earth) is pressing. But no such advice can be given without both knowledge and understanding. There is great need therefore to pursue research aimed at (*a*) reconstructing the facts of the past record of climate and (*b*) specifying the processes which produce climate and cause its fluctuations and changes. Without (*a*) climatology is in the position of a young science in which little laboratory work has been done to observe the facts of Nature's behaviour. The past record is also needed to test, calibrate and improve the theoretical models now being worked out in the mathematical laboratories. And those laboratories are needed to speed our progress with (*b*), without which the now growing welter of statistical results may only produce confusion.

There is also a pay-off in terms of the new light which this subject can shed on human history and on the problems of interpretation in all those sciences which are concerned with evidence of the past. It might be possible, were it not for the danger of arguing in a circle and the accusation of inventing climatic changes to explain the puzzles of history, to deduce the history of climate from the ups and downs of human affairs, the details of food prices and the like. But we can already glimpse a more satisfying outcome: so to establish the climatic record by firm physical evidence that it throws new light on causes in human history.

This is an exciting time, in which, after decades of neglect of climatology, new discoveries about the record of the past and the processes at work are being made, some of them rather easily. Most meteorologists seem hitherto to have been too readily convinced that the observations of more than about a century ago were too rough and unreliable to be worth investigating. The brilliant theorists and the advanced laboratories of meteorology today need to keep in touch with the quest for the facts of the past climatic sequence in which botanists, historians, isotope physicists and others are collaborating, in order to test and calibrate their models. If there is a danger of this being missed, it is for an honourable – but I believe insufficient – reason: a preference on the part of meteorologists for study of just those recent years in which the physical state and motion of the atmosphere in three dimensions, over the whole globe, can be most accurately and fully specified. The need remains, however, to observe the long-term processes which takes decades, centuries or millennia for their completion: they also have their sharp phases when abrupt shifts take place.

The bulk of this volume is devoted to presenting an outline of the past record of climate and the many types of evidence on which knowledge of it can now be built up. The author is acutely aware of his temerity in describing and discussing, however sketchily, the evidence and methods in such a great variety of fields of learning. It must have been impossible to do justice to them all and to the many advanced workers and work touched on. The excuse is that the subject needs such a base from which to advance. It will soon be impossible for one author or one book to survey it all. This work therefore is designed to serve as a necessary guidebook to the science of climatology and climate development at a time when it seems set for a great leap forward. The subject is fascinating for its diversity and touches nearly every aspect of our lives and environment. The need for advance to contribute to solving the pressing problems of the world-wide economy is very great.

It may be that some apology is needed, not only to various researchers whose work is inadequately presented, but also for a usage of terms that does not conform to current practice in various branches of learning. The shock that is likely, for instance, to geologists in a different use of the words era, period, etc., and to the classicist and mathematician over different use of the word epoch, is explained by the difficulty inevitable in a work where the same stock of English words has to be used to cover intervals of time from the longest units of Earth history down to the shortest-lived atmospheric processes. The standard geological terminology is given in Table 1 in Appendix V.

The same Appendix is designed to provide an assembly of data, largely composed of some of the longest series of weather observation material and parameteorological or proxy data which appear in some way representative of the condition of global climate. These series, some of which are not yet well known and have not been readily accessible, may provide a source from which interrelationships as yet unknown may be detected.

The last two chapters and Appendix VI indicate the present position as regards understanding the effects of Man's activities and providing a forecasting capacity and some of the steps that are being taken, and should be taken, to cater better for the future.

Introduction to
Parts III and IV

Chapter 12

Man's awareness of climatic changes

Man's awareness of the fickleness of climate can be traced back to the dawn of history and may have existed long before that in tales carefully transmitted from generation to generation, which even now survive enshrined in myth and legend. In contrast, the view, regarded as scientific, which was widely taught in the earlier part of this century, that climate was essentially constant apart from random fluctuations from year to year was at variance with the attitudes and experience of most earlier generations. It has also had to be abandoned in face of the significant changes in many parts of the world that occurred between 1900 and 1950 and other changes since.

Among the earliest written reports of *climatic fluctuations* are the inscriptions recording the yearly levels of the Nile flood (see p. 140), some of them from around 3000 B.C. Other inscribed tablets, or steles, quoted by BELL (1971), show a reasoned awareness of the liability of the Nile to fluctuations lasting some years. The following examples are among many from this source:

(1) From the tomb of ANKHTIFI, relating to the earliest known great famine around 2180–2130 B.C., in Upper Egypt, near Luxor: '...All Upper Egypt was dying of hunger, to such a degree that everyone had come to eating his children, but I managed that no one died of hunger in this nome [district]. I made a loan of grain to Upper Egypt....I kept alive the house of Elephantine during these years....'

(2) From the so-called prophecy of NEFERTY composed between about 2000 and 1950 B.C. in Lower Egypt:

'The river of Egypt is empty, men cross over the water on foot . . .. Men search for water on which the ships may sail . . .. The South wind drives away the North wind and the sky has still only one wind . . .. The birds no longer hatch their eggs in the swamps of the delta . . .. Foes are in the East, Asiatics are come down into Egypt . . ..'

(3) In a letter of the same period written by a certain HEKANAKHT to his family while on a business journey:

'The whole land is perished, but you have not been hungry . . .. When I came hither southwards I fixed your rations properly . . . our food is fixed for us in proportion to the flood. So be patient . . ..'

So, too, we have the Biblical report in the First Book of MOSES (Genesis 41:29–57) of JOSEPH's successful management of the economy of Egypt through a sequence of seven fat years of abundance and the seven lean years which followed: a date in the middle of the second millennium B.C. may be assumed. And in the First Book of Kings is reported a drought in Palestine in King AHAB's reign (874–853 B.C.), which according to two references in the New Testament lasted $3\frac{1}{2}$ years.[1] KLEIN (1914) reports that there are many other droughts in Palestine recorded in other ancient Hebrew texts (e.g. the *Mishna, Tosefta, Midrash, Talmud* and others). Certain predictive rules (*Volks-Wetterregeln*) seem also to have been developed for spotting the imminence of a rainless season, depending on knowledge of the usual dates for the first rains of the season, spaced at about 30-day intervals: the seemingly obvious conclusion was drawn, if the first two of these expected monthly rains failed (KLEIN 1914, pp. 38–9).

In numerous other ancient writings and legends we may distinguish knowledge of certain *climatic catastrophes*, which changed the face of the world and from which recovery was either long delayed or never fully established. The commonest of these accounts, usually of the distant past, are those that relate to a great flood, of which NOAH's flood is perhaps the best known example.[2]

Legends of an ancient flood also existed in Babylon, where a fragment of the account in an inscription dated as early as the nineteenth century B.C. has been found. Stratigraphic evidence of water-laid clay at Ur has been held to show a flood layer probably dated somewhat before 3000 B.C. and a layer of mud and river sand of similar age at Nineveh which seems to be associated with a lasting climatic change. Similar deposits at Kish and elsewhere in Mesopotamia have given a range of dates between 4000 and 2400 B.C., which seems to have been a wetter time than since (see Chapter 16, pp. 389, 394).

Flood legends are also found in India, southeast Asia, the East Indies, Melanesia, Polynesia, New Guinea and Australia, as well as among the original peoples of North and South America, in China and in parts of Europe. Some of the legends clearly relate to river

1. LUKE 4:25, and JAMES 5:17.   2. Genesis 6:17 to 8:14.

floods, perhaps to eras of frequent flooding of the river due to a heavier (and more frequent) rainfall regime. Unlike most others, the Chinese legend does not attribute any religious significance to the flood, but tells that it was finally brought under control by building up the river banks. Others among the flood memories may be thought to be connected with the postglacial eustatic rise of world sea level, which drowned what had previously been coastal lowlands and doubtless raised the soil-water level in the lower parts of flat river valleys. This process must have caused much coastal land to be lost in the eastern Mediterranean and Mesopotamia, and on other similar coasts, around 6000 years ago and continued, at times probably still alarmingly, for another 2000 years. Moreover, there is some evidence (in lake levels, etc.) that the monsoon rains were particularly active in, or about, those same millennia.

A folk memory of a climatic disaster of quite another kind may, perhaps, be identified in the old Norse legend of *Ragnarök*, the doom of the gods and end of the world, accompanied by a long and dreadful winter, the *Fimbulvinter*. The snow drives from all quarters with a biting wind; three such winters follow each other and there is no summer in between. As is known from ancient poems from tenth century Denmark, this legend was already old when it was written in SNORRI STURLUSON's prose *Edda* in Iceland about A.D. 1220. Can it, perhaps, have originated as an imaginative tale of the end of an old way of life in Scandinavia when the climate became colder in the early Iron Age, between about 1000 and 500 B.C.? The story is completed by telling of a conflagration (possibly an Icelandic insertion deriving from fear of the volcanoes?), that wars raged over the Earth and wolves attacked, but after some time a new world arose and some of Odin's sons survived in it.

BERGERON *et al.* (1956), writing of this northern legend, suggested that a prehistoric climatic experience such as the climatic deterioration that set in about the beginning of the Iron Age may be the basis of it. They also report that very similar legends exist in the highlands of Persia (Iran) and in the Altai in central Asia: a golden age is said to have been ended by a cold, snowy winter that destroyed everything living. In some Indian mythology a conflagration (possibly volcanic?) and rain in floods for 12 years is spoken of as the end of the world. The Indian concept of successive ages of the world points to continual renewal followed again by deterioration, destruction and renewal once more, a cosmic cycle in which Man plays no part.

Scientific writing on the subject of changes of climate may, perhaps, be said to begin with ARISTOTLE (384–322 B.C.), e.g. in the following passage about changes of climate and environment in Greece, quoted by SHAW (1926) from *The works of Aristotle, Meteorologika*, Book I, chapter 14:

> The same parts of the Earth are not always moist or dry, but they change according as rivers come into existence or dry up. And so the relation of land to sea changes too, and a place does not always remain land or sea throughout all time . . .. But we must suppose these changes to follow some order and cycle.

About a century earlier HERODOTUS (in his *History*, Book II, chapter 13) had shown an awareness that the river Nile was continually silting its bed and building its delta and a recognition that rainfall varies.

Texts from Italy in classical times not only record isolated years with extreme winters or heat and drought in summer, but seem also to trace a gradual climatic change (largely by its effects upon the vegetation). PLINY THE ELDER (A.D. 23–79) stated, in his *Natural History*, Book XVI, 15, that the beech tree (*fagus*), which he regarded as a mountain tree that descended to the plain in northern Italy, formerly grew within the precincts of Rome itself but did so no longer in his day, the climate seeming to be too hot there, as it was also for chestnut trees. THEOPHRASTUS (371–287 B.C.) had reported (*Hist. Plant.* III, 10) that there were beech trees in the neighbourhood of Rome in his lifetime, apparently confirming the change noted by PLINY. There are other reports, too, which seem to show that through the times of the Roman republic and empire Italy's climate, and that of Europe generally, was becoming warmer, a recovery from a period around 500 B.C. which was colder than now. By late Roman times, particularly in the fourth century A.D., it may have been warmer than now. I am indebted to H. WARNER ALLEN (personal communication, 27 March 1960) for the following notes:[1]

> COLUMELLA, Roman writer about A.D. 30–60, in *De Re Rustica* I, 5, warned agriculturalists to be up to date in their methods, as the traditional ways of husbandry might not be adapted to the latest conditions. He supported this remark by quoting from the SASERNAS, writers of books on agriculture at the beginning of the last century B.C. (which are now lost), the statement that in their time the vine and the olive were still slowly working their way northwards up the leg of Italy[2].
>
> COLUMELLA added that 'many notable authorities have held that through the long wasting of time weather and climate change . . . chief among them . . . the great astronomer HIPPARCHUS . . . and the trustworthy writer SASERNA seems to have accepted it. For in his book he concludes that the position of the heavens has changed . . . regions which previously on account of the regular severity of the weather could give no protection to any vine or olive stock planted there, now that the former cold has abated and the weather is warmer, produce olive crops and vintages in the greatest abundance . . . .'

The vine was first cultivated at Rome about 150 B.C. By the time that STRABO wrote his *Geography* (about 10 B.C.) the olive was cultivated in southern France, and its northern limit was about the same as in modern times. Later on there may be evidence of still greater warmth. In the fourth century A.D. the vine was cultivated about Paris and in at least a few

---

1. There is a reference on p. 75 in H. W. ALLEN, *A History of Wine* (London, Faber & Faber, 1961), 304 pp.

2. The Greek colonists had introduced the systematic cultivation of the vine and olive to the south of Italy about 700 B.C., when they founded Cumae (Κύμη) on the coast just west of Naples (40°8′N).

places in southern England. There is some complication about the significance of this,[1] since an earlier prohibition of the Roman republic against planting the vine in the trans-Alpine provinces had been lifted by Probus in A.D. 280. It is known, however, that the ban had not been uniformly enforced. Passages in the writings of AMMIANUS MARCELLINUS (c. 330–392) about Roman campaigns in the 350s in Gaul have been thought to indicate that the grain harvest was ordinarily gathered in rather earlier (in July in the north and in June in the south) than now.

The foregoing comments from ancient Roman authors are largely taken from a collection by ROTHMAN (1848) who was at pains to produce every counter-argument against deducing any change of climate. He was especially impressed by the similarities between the descriptions of the classical writers of around 100 B.C. to A.D. 100 and the position in the eighteenth and nineteenth centuries A.D. But the simplest interpretation is that some general warming of the climate was going on from the times before 100 B.C. until at least A.D. 350–60, and this seems now to be supported by various kinds of fossil evidence to be discussed in this volume.

In China, too, according to CHU KO-CHEN (1973), many early writers noted how the climate had varied with time. Particular interest may be accorded to CHIN LI-HSIANG (A.D. 1252–1303)[2] because he adopted a sceptical attitude to such reports but in the end decided, from the evidence of the pattern of cultivation in China during the Chou, Chin and two Han dynasties, in the period between 900 B.C. and A.D. 220, that the climate has been rather warmer then than in his own times. We may notice in passing here an appearance of broadly anti-phase relationship with the climatic change in Europe over the thousand years or more concerned. We shall look at this again in rather more detail, and in a longer perspective, in a later chapter (see, particularly fig. 16.23, p. 402): for it is certain that both China and Europe were still experiencing a climatic regime warmer than now, in fact the latter stages of the so-called 'Postglacial Climatic Optimum', around 2000 B.C., and to some extent even later than that.

We must suppose that the primitive peoples in many fringe regions who were driven to migrate in search of warmer climates or more dependable pastures by the climatic shifts that brought to an end the postglacial warmest times, or that followed the lesser periods of warmth in late Roman times and in the late Middle Ages, knew quite well that the compulsion was climatic: conditions had changed and were, at least for the time being, intolerable in their former homelands. This must be true, even when they have left no written record (and even if the lust for heroic adventure, plunder, rapine and destruction clouded the

1. There is certainly no implication that the natural range of the vine spread from south of Rome to England in the 480 years between 200 B.C. and A.D. 280. The discussion is about a plant that was introduced for cultivation. It would be reasonable to deduce from SASERNA's writing that until some time between 200 and 100 B.C. weather hazards (perhaps late frosts or summer and early autumn rain and hail?) had made the grape harvests unreliable in central Italy; olive trees (and more doubtfully vines) may also have been killed in some years by severe winter frosts.

2. Cited by CHU KO-CHEN in an earlier article (1961).

minds of the raiders when they descended on the inhabitants of more settled neighbouring lands). We shall see in a later chapter that there are grounds for seeing the Highland troubles of Scotland in the late Middle Ages in this light. Farther north, on the frontier of European civilization, there are unambiguous written records of the changes in the natural environment during those times.

In the first centuries after the Viking settlement of Iceland in A.D. 870–930, and of southwest Greenland from 987 onwards, sea ice was practically never mentioned in the accounts of the voyages; though it had been encountered by the earlier Irish monk-explorers in remote waters, presumably in that area, in the sixth to eighth centuries. It was ice filling one of the northwestern fjords (Arnafjord) 'and all the fjords on the northern side of the island' in the cold spring (in or about A.D. 865) that followed the first Viking overwintering in Iceland that gave the island its name. From that time until A.D. 1200 there were only one or two years each century when ice is reported to have caused any difficulty in sailing about Iceland and southern Greenland.[1] Yet about 1250 the *King's Mirror* (*Konungs Skuggsja*) tells us that the East Greenland ice was formidable, and IVAR BARDARSON or BAARDSON in his *Description of Greenland* reports that by about 1342 the old sailing route along the 65°N parallel to the Greenland coast – and thence following the coast round to the colonies on the western side – had been abandoned because of the increase of the ice; a track going just a day and a night west from Iceland and thence southwest to avoid the ice was laid down instead.[2] The situation worsened further after that, as is recorded in a letter of Pope ALEXANDER VI in 1492, of which the following extract is given in translation by STEFANSSON (1943, p. 157):

> '. . . the church of Garda is situated at the ends of the Earth in Greenland, and the people dwelling there are accustomed to live on dried fish and milk for lack of bread, wine and oil . . . shipping to that country is very infrequent because of the extensive freezing of the waters – no ship having put in to shore, it is believed, for eighty years – or, if voyages happened to be made, it could have been, it is thought, only in the month of August . . . and . . . it is also said that no bishop or priest . . . has been in residence for eighty years or thereabouts . . . .'[3]

In Europe generally the end of the thirteenth century, and the first part of the fourteenth, saw an acute crisis developing in human affairs, and 'the climate played a major

1. The original sources have been very thoroughly surveyed, perhaps first by PETTERSSON (1914). See also KOCH (1945).
2. *Det gamle Gronlands beskrivelse* by IVAR BARDARSON (MS edited by F. JONSSON) (Copenhagen, 1930).
3. The letter here quoted is the most interesting one for our purpose of a number of examples of papal correspondence concerning Greenland, Iceland and Norway in the Middle Ages in the Vatican archives, discussed by STEFANSSON (1943, p. 157), who

believed there must be others there in which the relevant allusions were as yet undiscovered or which might be still uncatalogued.

   JOHANNIS BURCKHARDI, an official chronicler of papal concerns from 1483 to 1506, was also interested in the variations of the years in Rome and kept records of the weather of the twelve days of Christmas each year, because he believed them significant for the twelve months of the year that followed.

part in the mischief seventy or eighty years before the Black Death' (the bubonic plague of 1348–50), according to ZIEGLER (1969). There was a series of disastrous harvests. There were famines in England in 1272, 1277, 1283, 1292 and 1311. And in one, two or three years between 1314 and 1319 almost every country in Europe lost almost the whole harvest: the poor were reduced to eating dogs, cats and even children. There were further disastrous harvest failures in 1332 and 1345–8. In some areas substantial percentages of the population died. But in the retreat of cultivation and change of land use that followed, particularly in connection with the enclosures of land in England for sheep rearing, one finds the enclosing landowners (a ready scapegoat) and the plague blamed for the troubles rather than the climate.

It is from eastern Europe, however, that, apart from Greenland (and in much more detail than for that country), we have the harshest climatic disasters in the late Middle Ages reported: winters of such unaccustomed severity and depth of snow that many people died and a few winters so mild that the people were puzzled and anxious about seed time and harvest.[1] There were summers with continual rains and others with such prolonged heat and drought that, in either case, the crops failed. In the ensuing famines people took in desperation to cannibalism or emigrated to the west. The accounts of those seasons in the chronicles of the monasteries in the areas that are now Poland and on the plains of Russia, from which the relevant reports have been transcribed by BUCHINSKY (1957), make terrible reading. The earliest serious case seems to have been in A.D. 1215, when early frosts destroyed the harvest throughout the district about Novgorod, people ate pine bark and sold their children into slavery for bread, 'many common graves were filled with corpses, but they could not bury them all ... those who remained alive hastened to the sea'.[2] Other bad years came in 1229 and 1230, and in the latter there were many incidents of cannibalism 'over the whole district of Russia with the sole exception of Kiev'. (Reports of a few such incidents in the Middle Ages in western Europe, including the British Isles, also occur, though much more rarely.) After that, there seems to have been some respite: bad years came more seldom, and

1. In the 300 years from A.D. 1200 to 1499 at least twenty-five winters must be ranked as very severe in eastern Europe (those of 1219–20, 1281, 1284, 1314, 1363–4, 1377, 1378, 1383, 1393, 1400, 1403, 1404, 1409, 1417, 1422–3, 1440, 1442, 1443, 1444, 1448, 1468–9, 1476–7, 1480–1, 1495–6, 1497–8 and those of 1420–1 and 1470–1 as very snowy). Five winters were outstandingly mild (those of 1303, 1371, 1405, 1474–5, though followed by a cold, snowy spring, and 1492–3 when January and February brought a false spring in southern Russia – with the grass growing tall and birds nesting – which was also followed by severe frosts in March). Probably the winter of 1452–3, which followed a long mild autumn and was remarkably snowless, should be added to this list of very mild winters.

In these lists the winters are designated by the year in which the January fell. Cases where there is

some doubt as to which of two neighbouring winters is meant have been printed as 1219–20, 1422–3, etc., and because of the old style of calendar it is probable that the winter concerned was the one which spanned the two years (of the modern calendar) mentioned.

2. The records go back to much earlier times. There is an interesting passage in a book published nearly six centuries later which may indicate some lingering folk memory, due to transmitted tales, of this change of climate: Ja. MARKOVIČ, in his *Memoirs of Little Russia* [i.e. Ukraine], published in St Petersburg in 1798, wrote of the climate of the Ukraine that 'it does not possess such a degree of warmth as other lands in the same latitude ... but they say that in olden times the winter was more supportable ...' (cited by I. E. BUCHINSKY, *O klimate proshlogo russkoy ravniny* (Leningrad, Gidrometeoizdat, 1957, 2nd edn).

(with possible exceptions in 1309 and 1332) were less serious when they did come, until the 1360s. It was in the fifteenth century that the calamities were most frequent. From 1421 to 1423 there was famine for 3 years in Russia, again leading to cannibalism and an exodus to the west. In 1436 because of famine and dearth many of the people left Novgorod and went to Germany. In 1438 similar events occurred about Smolensk, and it was reported that 'the wild animals ate people and people ate people and small children'. Further years of famine occurred between 1466 and 1485, but seem to have been less severe than those earlier in the century. The winter of 1565 saw once again widespread deaths and unburied corpses of people and animals, and there were many severe years in the seventeenth century. It is never clear that the reporters were aware of any trend, though the people who emigrated may well have concluded that conditions in their former homelands had taken a persistent turn for the worse. Indeed there seem to have been times in the fifteenth and seventeenth centuries (perhaps also in the early part of the thirteenth century) when certain areas were largely depopulated.

In those parts of Europe where there are glaciers there was sooner or later no doubt about the source of the troubles in the next few centuries, which have come to be known as the Little Ice Age, since the extent of ice on land and sea probably exceeded that at any other time since the last major glaciation. LADURIE (1967, 1971) has assembled a remarkable collection of reports, dated maps and pictures, which document the advances of the glaciers in the Alps and other attendant local disasters. The earliest (briefly) documented advance tells that the Vernagt glacier in the Tyrol made some initial advance quite early in the thirteenth century.[1] Archaeological study (KINZL 1932) and recent radiocarbon dating (OESCHGER and RÖTHLISBERGER 1961) of the walls of a medieval irrigation course, the Oberriederin, led off from a subglacial stream emerging from the Aletsch glacier, high up in the Swiss Alps at a point which is at present covered by the glacier, indicate that the structure was destroyed by advance of the glacier some time between about A.D. 1200 and 1350 – i.e. possibly as early as the first great advance of the Arctic sea ice off east Greenland. It is known from ancient documents that it was out of use by 1385. At Grindelwald there was in 1760 (LADURIE 1971, p. 250) convincing evidence of a glacier advance about A.D. 1280 (see also footnote, p. 447), in the stumps of larch trees that had been overrun by the ice on the slopes of the Eiger and the Fischerhorn and a documentary record of a church (at Burgbiel) that was moved to be out of danger of the glacier and of floods. No more reports of glacier advance are known until the late sixteenth century, when they became numerous. A report (quoted by LADURIE 1971, p. 250) of a horseback journey by one, SEBASTIAN MÜNSTER, from the upper Rhone valley in Switzerland over the Furka pass in August 1546, however, indicates

1. Known as the Vernagtferner. *'Ferner'* (from the old High German word *'Firn'*) is the word commonly in use for 'glacier' in the eastern Alps. The Vernagtferner is a few kilometres north of the Hintereisferner, which has been the object of measurements since 1893 and on which mass-balance studies have been carried out since 1952 by HOINKES (1962, 1964). The flow of the Vernagtferner itself was measured from 1888. Both these glaciers are fed from the ice on the heights of the Ötztaler Alpen, above the Rofental which joins the Ötztal from the southwest, near its head.

that he crossed the Rhone where it emerged from the glacier front which was about 200 m across and 10–15 m high: the glacier was evidently already much farther forward than in this century and in an advancing attitude. Advancing glaciers from side-valleys in the Alps in the following decades frequently crossed the main valley bottoms and formed ice-dammed lakes there. Thus, the Allalin glacier blocked the Saas valley in 1589 and a lake, the Mattmarksee, formed; but the ice barrier gave way in September 1589 and the lake drained in a torrential flood down the valley, a disaster that was many times repeated later, e.g. in 1633, 1680, 1719, 1724, 1733, 1740, 1752, 1755, 1764, 1766, 1772.[1] The ice-dammed Mattmarksee was a normal feature of the Saastal landscape from 1600 until the nineteenth century; the residual lake in 1920, dammed by the moraine formed in recent centuries, had a maximum depth of 4 m and held an estimated 560 000 m^3 of water, to be compared with the reported 29 m depth and $18 \cdot 8 \times 10^6$ m^3 of water in 1834. Similar disasters are recorded in many other Alpine valleys in those centuries, including in the Ötztal and about Chamounix on the French side (LADURIE 1971, p. 250).

Between 1530 and 1575 the inhabitants of Chamounix lamented the fact that the glaciers made their surroundings very cold, but complaints of physical destruction wrought by the advancing glaciers only came later. There must have been some (probably earlier) advance of the glaciers in the upper levels but no immediate menace at that stage of the sixteenth century. But by 1580 three of the great glaciers, probably the Argentière, the Mer de Glace and Les Bossons, were described by a traveller as spreading through rifts in the mountains and descending almost to the valley floor. In 1600, at the time of a tax reform in Savoy, a document records that the glaciers of the Arve and other rivers had ruined much land in the parish of Chamounix, had destroyed twelve houses in the village of Chastelard (which later disappeared altogether) and that another village, Les Bois, had been left uninhabited because of the glaciers. Les Bois and another hamlet, Bonanay, also subsequently disappeared totally. In 1600 the glaciers both on the French and the Italian side of Mont Blanc had advanced so far that the people of the valleys were in panic. Ice-dammed lakes which formed and burst, as in Saastal, were the agency of much of the destruction.

At Grindelwald, where there had been either a glacial stillstand or a retreat (see LADURIE 1971, p. 161) after the advance in the thirteenth century, there is a documentary report that the (lower?) glacier broke through its end-moraine in 1588. By 1600 the glaciers had thrust right forward to their most advanced positions in recent centuries and destroyed some houses and barns. More or less the same advanced positions of the glacier fronts were maintained throughout the seventeenth century. In 1703 there was some renewed increase

---

1. The worst catastrophes were those in 1633, 1680 and 1772, the last one precipitated by two days of heavy rain. On each occasion many houses were carried away.

There seems to have been some resumption of events of the same character in Saasthal in recent years (RÖTHLISBERGER 1974): an ice avalanche on 30 August 1965, when part of the snout of the Allalin glacier broke away, causing heavy damage and loss of life, and two incidents, in 1968 and 1970, when temporary blockage of the meltwater channels under the Gruben glacier caused lakes to form which subsequently broke their dam with heavy erosion of the moraine leading to destruction of part of the village of Saas Balen and burial of the best meadows.

and 'the meadows of the prior, which are still entered in the shepherds' books . . . are in fact buried under the ice'. There was some retreat of the ice after 1720 and a swift withdrawal in 1748 to a point 500 m below its present front. There may have been some readvance of the lower glacier by 1780 and it was still in a forward position through much of the nineteenth century.

There seems to have been a good parallel with the course of events in the far north. Iceland seems to have enjoyed a rather easier time than for more than a hundred years earlier around A.D. 1500 and in the half century that followed, though contact with the old colony in Greenland had been lost. There was anxiety about its fate and several expeditions were sent. One in 1540 reached Greenland and more or less established that the population of the old Norse colony had died out. After about 1550 the sea ice increased beyond its previous limits, and expeditions commonly had to make a wide sweep south of Cape Farewell to get round the ice to reach the west coast of Greenland. Conditions then deteriorated in Iceland too, and the advance of the glaciers and ice sheets is vividly clear from accounts like the following extracts from a parish register in the southeast of the island (EYTHORSSON 1952):

> 1709. *Breiðamörk* (farm). Half King's ownership, half owned by the farmer. Derelict . . . a little woodland, now surrounded by the glacier . . .
> *Fjall* (farm). Owned by the Church. Derelict. Lies northeast of Breiðamörk. Fourteen years ago had farmhouse and buildings, all now come under the glacier . . ..

From a book dated 1712 it is learnt that Fjall had been abandoned in 1700 or earlier, Breiðamörk in 1698. 'There was some grass visible then, but since the glacier has swept over all except the hillock on which the farmhouse at Breiðamörk stood and that is surrounded by ice so that it is no use even for sheep'. Later that hillock also was overrun by the ice.

Documents in Norway indicate (HOEL and WERENSKIØLD 1962) the beginning of a great advance of the glaciers around 1660–70, and in the 1680s the glaciers and glacier rivers destroyed meadows and pastures. In 1676 landslides and torrents destroyed farmland at Buar in Hardanger, below the Folgefonn ice sheet. About 1690 the glacier Tverrbreen in Jostedal could still only be seen high up in the mountains, but not very many years after that it advanced about 200 m in 10 years and was threatening farmhouses in the valley. The glaciers in Nordfjord were advancing until about 1742. The subsequent history of retreat and readvances began about 1750.

Of these same times in Scotland we have numerous accounts which were summarized by TREVELYAN (1942, p. 432) as follows:

> The last half dozen years of William's reign [i.e. the 1690s] had been the 'dear years' of Scottish memory, six consecutive seasons of disastrous weather when the harvest would not ripen. The country had not the means to buy food from abroad, so the people had laid themselves down and died. Many parishes had been reduced to a half or a third of

their inhabitants. This sombre experience, from which the nation was slowly emerging during the years when the Treaty of Union [with England] was under debate, coloured the North Briton's outlook, deepened his superstitions and darkened his political passions, especially in relation to the hated English who had watched the kindly Scots die of hunger and had moved no finger . . . . Fortunately a cycle of fat years . . . followed. [This last is presumably a reference to better years around 1718–19 and in the 1730s, long after the Union.]

WALTON'S (1952) survey of the matter as far as northeast Scotland was concerned led to the statement that famine and food shortages were never very far from the population at the beginning of the eighteenth century. In 1709 the recovery from the seven ill years – which meant the harvest failures in 1693–8 and 1700 – was interrupted by cold unseasonable weather, presumably meaning considerable snowfalls on the hills, continuing till mid June (New Style calendar) with 'great loss of flocks and cattell'. The next important food shortage was in 1740–2, and similar troubles recurred, though perhaps less severely, in 1771–2 and in 1782.

It was the upland parishes in all parts of Scotland that were severely hit, and a leading Scottish figure, ANDREW FLETCHER of Saltoun in Midlothian, in a discourse to the Edinburgh Parliament in 1698, criticized also the well-to-do in the Scottish lowlands who still had their 'delicate morsels' and luxuries 'when so many are already dead, and so many . . . struggling with death . . .'.

From other accounts it appears that in the seventeenth century the valley bottoms in western and central Scotland, in Ayrshire, Fife and elsewhere, were wet and ploughing had spread up the steep hillsides, as the problem of draining the lower ground had not been tackled, and this probably increased the vulnerability of the community to the cold, wet summers in the 1690s and in other bad years as far back as the 1580s. The *Old Statistical Account*, from parish records at Duthil and Rothiemurchus, in Speyside, in the following century, recalls that in the late 1600s

the poorer sort of people frequented the churchyard to pull a mess of nettles, and frequently struggled and fought over it . . . being the earliest spring green which they greedily fed upon . . .. So many families perished from want that, for six miles in a well inhabited tract there was not a smoke remaining. Nursing women were found dead upon the public roads and babes in the agonies of death sucking at their mother's breasts . . .' (quoted from Sir JOHN SINCLAIR's *Statistical Account of Scotland*, Edinburgh 1791–9.)

More details of the weather of those worst seasons and of the incidence of death and disease among people and cattle are given by GRAHAM (1899).

The severe weather of some years in and about the seventeenth century led cultivators in many countries to decisions which surely indicate a dawning awareness of climatic change.

Orange and mandarin cultivation which had been practised for centuries in the Kiangsi province of China was virtually abandoned after a succession of cold waves between 1654 and 1676, when the trees were repeatedly killed by frost and the growers became fearful of attempting to replace them with new trees. Similarly, the northern limit of vine growing in France was permanently shifted somewhat farther south. There was speculation in those times about the cause of abandonment and decay of the medieval vineyards in England (as opposed to the laments over bad seasons on the part of those who persisted in trying to operate them in the fourteenth and fifteenth centuries). JOHN ROSE, Gard'ner to His Majesty (King Charles II) at his Royal Garden in St James's, in his publication *The English Vineyard Vindicated* (London 1666) wrote of the decline of the former English vineyards: 'the strange decay of them among us . . . [noting that the northernmost vineyards in Europe (Germany) extended to our own latitude] must proceed from no other cause than that of our own neglect'. This seems to have been intended to counter suggestions by others that the climate had become less suitable. Nevertheless, when SAMUEL PEPYS in July (August N.S.) 1661 visited the vineyard which the Cecils had planted in 1611 at Hatfield House, in a walled garden in a favourable site on a sheltered, south-facing bank, his only comments were on the coldness of the weather and the size of the gooseberries. There is no record that it ever achieved any success, and Lord CECIL's successor the late Marquis of Salisbury (personal communication, 5 August 1959) described 'the whole thing as a rich man's fantaisie'.

The remarkable turn of the climate of Europe towards greater warmth from soon after the beginning of the eighteenth century, and affecting all seasons of the year in the 1730s, seems to have produced little comment at the time, though by then the temperatures were being observed with thermometers and entered into regularly maintained observation books in a number of places. The splendid sunny, warm and dry summers of 1718 and 1719 (and in some parts of Europe 1717 also) and some other bountiful years later in the century naturally called forth enthusiasm. But such conditions still readily led to anxieties about drought. Doubtless, an easing of climatic conditions was more readily taken for granted, and more liable to pass unnoticed, than when the seasons became severe. The health and population statistics in northern Europe and in Iceland registered the effect of the warmer decades, and it may be more than a coincidence that a more cheerful social and political mood than in the seventeenth century gradually became evident in many parts of Europe.

In the Annals of Dunfermline, once the medieval capital of Scotland, we read that in 1733-4 puritan opposition was temporarily overcome so far as to allow an annual New Year Ball in the Town House and that in 1733 also 'wheat was first grown' in the district. The latter statement was almost certainly fallacious, and shows how far earlier, better days had been forgotten: enough wheat had been grown farther north, in Aberdeen and Strathmore, in the early 1500s to allow an export trade.

The renewed incidence of occasional severe winters in Europe from 1739-40 onwards was probably the main factor in the anxieties about the stability of climate which, as we shall see in the next chapter, led to moves from the 1770s onwards to institute official networks of regular and continuing weather observations. The average level of summer temperatures continued high through the remainder of the eighteenth century and on till about 1810; indeed their level was marginally higher than for 1933–52, the warmest part of the twentieth century, and 0·3 to 0·6°C above the averages for 1850–1950 in most of Europe (VON RUDLOFF 1967, p. 281). The prevailing pattern of the latter half of the eighteenth century was one, however, of rather short warm summers and all the other seasons of the year colder than in the present century.

THOMAS BARKER of Lyndon, Rutland (England), wrote the following shrewd meteorological observation at the end of 1775 in his fine series of weather observations and account of the seasons at Lyndon, which he maintained from 1733 to 1798:

> For a good many years past the seasons have been in general wet, the nature of East winds has been very different from what it formerly was. For several years after the great frost of 1740 there were a great many Northeast winds in spring, but they were generally cold and dry, stopping vegetation; but for the last ten years the East winds have been often very wet; many of the greatest summer floods came by rains out of that quarter, and many times it came rain almost as certainly as the winds turned East.

BARKER was evidently summarizing his impressions of a number of wet summers and other seasons in the east midlands of England between 1760 and 1775.[1]

Climate has continued to vary, and its variations have affected individuals and human society in many ways, some of them matters of extreme anxiety for whole populations, or involving enormous economic stakes, and others trivial or affecting chiefly the colour and mood of life. In 1784, with the situation around the beginning of the eighteenth century and the last decade of the previous century still in mind, there was such great concern over the renewed increase of ice about the coasts of Iceland (as well as the activity of the volcanoes) that the Danish government debated whether to evacuate the Icelanders and resettle them in Europe (KOCH 1945). Such a proposal was probably beyond the resources of the time, and, although the population of Iceland declined by about 25% with the thousands of deaths in the prevailing famine, the community did survive. And, on another level, we learn that it was the colder years in the beginning of the nineteenth century which led to the introduction of certain warm designs of women's underwear (the 'spencer' and the 'bosom friend') in Europe and ultimately put an end to the daring fashions of the years after the French Revolution which 'exposed the person': it was said at the time that it took the aid of the north wind to enforce the return to modesty in women's dress.[2]

---

1. It is interesting to note that if the date mentioned in the passage were changed from 1740 to 1940, it might have been written around 1970 and could then be taken as an apt description of the change in the usual character of easterly wind situations in England just 200 years later, from the 1940s to the 1960s and after.

2. I have this from a footnote (p. 587) in *The Diary of a Country Parson 1758–1802* by JAMES WOODFORDE of Weston Longville, near Norwich (edited by J. BERESFORD) (Oxford University Press, The World's Classics series No. 514, 1949).

Adaptations to the warmer climate of the first half of the present century, particularly the almost unbroken run of mild winters in the British Isles from 1896 to 1939, were presumably unconscious (in view of the prevalent doctrine of the time about constancy of climate), but they were none the less real. There was, it is true, wide public awareness in England of the cessation of the series of skating winters that our Victorian grandparents had quite often enjoyed. Previous fashions for wearing long clothes and thick underwear were given up and unkindly attributed to the unreasonable mental attitudes and inhibitions of our nineteenth century forebears. Their cautious habit of laying in food stocks for the winter in lonely habitations was also increasingly given up, and this necessitated many helicopter relief operations in the snowier seasons in the 1950s and 1960s. The decades of mild winters also coincided with the introduction of plumbing in most homes in Britain and may account for the practice of placing the pipes on an outside wall of the house (commonly the north or east wall, to be out of sight from the front of the house), a practice which has given trouble in all the severer frosts which increased markedly in frequency from 1939 onwards. The twentieth-century fashion among the well-to-do for building their houses on hilltop sites with a view to the south and west may also owe something to the length and warmth of the summers, which culminated between 1933 and 1952. Though there was some analogy with the planning of the handsome eighteenth century terraces at Bath (in another relatively warm period of our climate), previous generations had mostly shunned such sites because of their exposure to the stormy winds and the rain. In the fifteenth and sixteenth centuries houses in England had been built where possible on gentle northern slopes. CAMDEN's *Britannia* (1586) recommended the low-lying site of Oxford as sheltered by woody hills from 'the unhealthy South wind and the violent West wind and admitting only the gentle East and the North wind which drives away pestiferous vapours' (quoted by WILLIAMS FREEMAN 1928).

In recent years the most widely publicized event which has reawakened widespread concern about climatic change is the greatly increasing drought since about 1955 of a whole zone of Africa, the 'Sahel' zone, just south of the Sahara (WINSTANLEY 1973a, 1973b), which has caused untold deaths and hardship and has caused whole populations to migrate south before the advancing desert. The time of relatively high rainfall, averaging double the mean of 1968–72, was from about 1908 to 1956, the peaks nearly coinciding with those of the frequency of the westerly type regime over the British Isles; so the beginnings of the decline can be traced as far back as the 1930s. The debate (DALBY and HARRISON CHURCH 1973) about the trend, and what should be done either towards the artificial rehabilitation of the drought-stricken lands or the resettlement of their peoples elsewhere in Africa, seems likely to call for measures which may tax the skills and resources of the international community.

In several years since about 1960 shortfalls of the harvest in Soviet central Asia due to dry summers and/or lack of snow in the preceding winter have led to grain purchases from western countries on such a massive scale that the former reserves of stored grain surpluses were virtually exhausted by 1973. The severest case had been in 1972, when the price of wheat in western countries (including England) doubled in about 4 months as a direct result of the pressure of Russian demand. Similar harvest failures in China and India since 1960 contributed to the reduction of the world's reserves of stored grain.

In the last decade also in Iceland there is renewed anxiety about symptoms of climatic change and its possible threat to the future, as on so many occasions in past centuries. Renewed increase of the sea ice in the East Greenland Current during the 1960s culminated in April–May 1968 with a period of several weeks when Iceland was half surrounded by the ice, which had not occurred for eighty years. The increased volume of cold water transported by the East Greenland Current also resulted in the herring leaving the area: the consequent drop in the Icelandic herring fishery was held to have provoked two devaluations of that country's currency, totalling 40% (KRISTJANSSON 1969).

In the international concern over these examples of the threat, and actual impact, of climatic variation in recent years on a scale which may upset, and which in some cases has upset, the world's currencies and trading arrangements, the demand arises for scientifically based climatic forecasts.

This book is an attempt to marshal existing knowledge of climate and its behaviour, so as to provide the basis on which forecasting techniques will have to be developed. In a final section of the book we shall examine briefly what forecasting methods, or approaches to a forecasting method, appear possible and reasonable.

Widespread anxiety has also developed in recent years over the possible impact of Man's activities in changing climate either inadvertently, as an unintended side effect of increasing industrial activity particularly the output of carbon dioxide and soon of waste heat, or by deliberate schemes contrived to open up or preserve certain regions for agriculture, industry and increased population.

Little attempt has been made so far to study the economic return which might be expected from investment in climatic research directed towards meeting the need for advice on probable future development of the natural climate or Man's potential or actual effects on it. Cases cited in this chapter already make it clear that the economic losses from climatic shifts and disasters are enormous; indeed, because of their effects on the currency units themselves the losses may be literally incalculable. Data on the economic impact of short-term weather and climatic fluctuations on most kinds of human activity are very scarce, though the matter has been approached by MAUNDER (1970) and in the unpublished report on *Climatic Fluctuation and the Problems of Foresight* (1972) submitted by a working group to the World Meteorological Organization, Geneva. The impact of year-by-year variations on just one localized and relatively small-scale activity indicates the big money values that are involved even in such cases. The total farm incomes from the dairy produce of 750 000 cows in the South Auckland province of New Zealand differs between a wet January and a dry January by about 2 million New Zealand dollars; even if only some small proportion of this could be saved by reliable foresight about the coming season, it is likely to be a considerable sum. Similarly, in the 1950s it was realized that many millions of pounds sterling was being lost yearly by shipping insurance in London alone because the insurance rates had not been adjusted to the increased frequency of North Atlantic hurricanes in those years as compared with the beginning of the century.

Requests for advice on future climate, which have confronted climatology in recent years, include the following:

(1) Estimates of temperature fluctuations and trend over the next 30–50 years for planning the provision of heating oil and nuclear power.

(2) Estimates of rainfall fluctuations and trend over the next 10–50 years for planning the provision of water supply for domestic and industrial use.

(3) (a) Estimates of the probable incidence of severe winters, and snowiness of winters, in the next 1–5 or up to 25 years, for provision of road salt, highway snow-clearing equipment, coal stocks, heating-oil storage and distribution, and winter sports facilities.
(b) Estimation of the probability of a severe winter occurring in the same winter in Europe, eastern U.S.A. and the Far East.

(4) Advice on recent changes in the frequency of heavy, prolonged (24 to 36-hour) rains in lowland districts of England and probability of continuance of the new regime, with a view to siting new towns and suburban building and providing drainage appropriate to the flood risk.

(5) Estimates of trends in wind direction frequencies and storm frequencies affecting roughness of the North Sea and the frequency of sea floods on low-lying coasts and in estuaries (especially those which lead to great cities/ports).

(6) Estimates of future trends in tropical cyclones/hurricanes affecting shipping and coastal defences, warning systems, etc.

(7) Diagnosis and prognosis of sea temperature trends, causing displacements of important fisheries and affecting their yields.

(8) Estimates of future tendency of the frequency of blowing sand affecting activities (building, etc.) in arid countries.

(9) Estimates of trend of summer temperature to guide policy in forest planting near the northern limit and on high ground, growth of exotic crops and fruits (e.g. maize, vines, etc., in England), for planning honey production, and development of tourist facilities, etc.

(10) Estimates of future trend of glaciers for their effects on river flow and hydroelectric power development.

Questions have also been asked about the possible effects on world climate of various new, or increasing, pollutants in the atmosphere as well as questions on the estimated effectiveness of schemes designed to increase the availability of water in Africa and central Asia and any undesirable side effects of such schemes on global climate.

The total sums spent yearly in the whole world on research into the past meteorological record (last 100 years) are estimated to be in the region of 4 million U.S. dollars. About another 8 million U.S. dollars is probably spent on research in the various field sciences uncovering the facts of the climatic record over many thousands, and even some millions, of years past. Very little indeed is being spent at present on establishing the climatic record of the earlier part of the era of meteorological instrument observations and of historical times (the last 1000 to 3000 years) for which descriptive accounts exist. The W.M.O. working group report estimated that at least 10 million U.S. dollars yearly of the spending on the advanced mathematical laboratories of meteorology can be regarded as going towards improving theoretical construction of climate. It seems therefore that the total sums being spent are very small in comparison to the economic and (partly incalculable) human stakes involved in climatic changes and what could be done with better knowledge leading to any forecasting capacity. There is also some danger at the present stage of developing, at considerable expense, the theory (and elegant mathematically complete theoretical models) of climate without adequate knowledge of the observable behaviour of climate, and of the processes involved in climatic changes, over the last 300 to 3000 years. The soundness of the theoretical constructions, and of forecast advice given on the basis of them, is needlessly jeopardized as long as this observational base of the subject is neglected.

Climatology has arrived at a crossroads in recent years, a point where it must either stand still or arrange advances along several different roads simultaneously. Only by the latter decision can it hope to meet the challenge presented by the demand for advice on probable future climatic developments and trend. The many ingenious formulations of theory of climate in terms of global mathematical models are likely to prove fallacious in one or another way unless and until equal attention is paid to establishing the past record of climatic behaviour and assimilating the lessons from it, particularly as regards those processes identifiable in it which take from more than one year up to many decades or centuries for their completion. At present we see meteorological theoreticians developing brilliant models from the global and three-dimensional observation coverage usually of the last 15–20 years and, on the other hand, field scientists who from their analysis of observational data of the last 10^5–10^7 years pronounce with some confidence that the present interglacial is liable to end (see, for example, KUKLA and MATTHEWS 1972) within the next few centuries or at most a few millennia. The theoreticians tend to have no comment on the latter group of scientists' diagnosis other than a tendency to mild disbelief, which is based on the absence of evidence in the material that they have put into the meteorological models. It is, moreover, difficult to bring the two types of mind together – though it must be done. The establishing of the past climatic record and its interpretation in terms of physical and mathematical theory is an exciting interdisciplinary quest, perhaps the broadest challenge in any field of science to interdisciplinary collaboration. And the problems of our times demand that it be pursued. Climatology promises to become the most essential branch of the Earth sciences to human welfare.[1]

1. This conclusion was first put into words, and published, by FLOHN (1970) in the *W.M.O. Bulletin*.

There is an evident need – and it may be an urgent need – to study the potential losses to present-day national and regional economies, and the strains to international trade arrangements, that could arise from a return to the global climatic regime of just 150 to 300 years ago, or from the wider range of climatic variability that seems to have occurred in that period and earlier. Indeed, the question should be considered as to whether the present highly organized international order is any less vulnerable to climatic shifts than were its predecessors, particularly when the growing world population strains every resource.

Part III

Climatic history

Chapter 13

Evidence of past weather and climate

In reconstructing the climatic events and regimes of the past, use can be made of an almost limitless variety of kinds of evidence. This chapter will serve as a guide to the main ones. Our aim here is to indicate the potentialities of each, and to illustrate the contributions to knowledge that have come from them, but also to point to some of the precautions that are necessary and the uncertainties of interpretation that remain.[1] Because nearly all aspects of our environment, and indeed of human affairs, are in some degree or at some points sensitive to weather and climate, the most multifarious traces of past climatic events can be found, given the required insight. Our list of possible sources of evidence therefore cannot be exhaustive. The great variety of the fossil witnesses of the past is important, however, because the often necessarily imaginative interpretations of any one thread of evidence must be treated with reserve until they find support from at least one other, unrelated, field of study and the independent reasoning suggested thereby. Only so can one altogether avoid arguing in a circle, as one may, for example, in conjecturing some climatic event, or sequences of events, in the past to explain a human migration or some change in the composition of the vegetation of a given region and then regarding the hypothesis as proven by its ability to explain all similar happenings around the same time.

---

1. These uncertainties include the margins of error of the physical measurements used, and of any dates, or values of the climatic variables derived therefrom by the methods described. In some places in this chapter it has been thought useful to specify the size of these margins of error. Where this is done, it must be understood that this describes the situation that applies (as indicated by the latest information in the author's hands) in laboratories using the apparatus and techniques available in 1970–2.

The aim of meteorology in this quest for the facts of the past is to identify the true nature of the weather events and the atmospheric circulation processes at work. It is the large-scale atmospheric circulation that chiefly accounts for the distribution of weather phenomena at any time, and through understanding the atmospheric circulation patterns we may be able to proceed from interpretation of the field data at a skeleton (or even a fragmentary) network of points to discern, in outline, the pattern over intermediate points and ideally over the whole Earth. The logic of the patterns and processes arrived at may also provide indications on the one hand of the prevailing energy level of the general atmospheric circulation and hence, by implication, of the global radiation and heat budget at any given time and on the other hand of the range of weather phenomena, temperature and rainfall values, etc., that occurred over any selected span of time.

In making such a diagnosis the meteorologist will be seen to be adopting the geologist's principle of uniformitarianism – that 'the present is the key to the past' – since our theoretical understanding is necessarily founded upon analysis of the phenomena occurring within the period for which we have adequate observational coverage and measurements. This principle is, at bottom, nothing other than a belief in the permanency of those physical laws whose working has been rightly understood.

13.1   Evidence from meteorology and parameteorological phenomena

Observation data

The earliest meteorological instruments were presumably the wind vane and the rain gauge. The former betrays its simple origins by the derivation of the word from the old English 'fane', a flag. The instrument in the form of a weathercock has been in continuous use since the early Middle Ages, and vanes had already been used long before by the ancient Greeks and Romans.[1] Well levels were probably measured in antiquity, as the height of the yearly Nile flood certainly was; and because of the great importance of droughts to the economy, it seems likely that some kind of rain gauge would very early be hit upon. A network of rainfall measurements with gauges is said to have been instituted in Korea in A.D. 1441–2, and a photograph of one of the gauges has been published (CENT. MET. OFFICE 1961). But the earliest reports of rain measurements now readily to hand are from places in England in the last part of the seventeenth century and a continuous record from the Netherlands, starting at Delft, from 1715 (LABRIJN 1945).[2] The first known temperature-measuring instrument

1. HELLMANN (1908) reports that Terentius VARRO refers to vanes constructed in Roman villas to show the wind direction by means of a windrose on the ceiling of a room; also that rainfall measurements made in Palestine in the first century A.D. are preserved in the *Mishna*, a collection of ancient Jewish books other than the Bible. (According to D. DEVAHUTI in *Harsha: a political study* (Oxford – Clarendon Press 1970) rainfall measurements also exist for parts of northern India in the seventh century A.D. The figures and distribution appear similar to today's.) Representations of some late Roman ships of the fourth to fifth century A.D. have been found with an instrument on the mast resembling a rotating cup anemometer.

2. The first series was from near Burnley, Lancashire, 1677–1704, another at Upminster near London, 1697–1715. In France rainfall records for Versailles

was due to GALILEO in 1597 and the fluid-type thermometer in 1611. The barometer was invented by TORRICELLI in 1643, and experimental observations were made with one on the top of the Puy de Dome and in Clermont Ferrand at the foot of the mountain in 1648 and 1649.

The era of regular daily meteorological observations with instruments dawned with the series instituted by the *Accademia del Cimento* in Florence (Firenze) in 1654 and continued till 1670. The Florentine thermometers were distributed for use elsewhere in Europe and were used, for instance, for an observation series in Paris starting in 1658, and probably in Oxford or London from 1659. Soon afterwards the newly formed Royal Society of London encouraged the taking of meteorological observations, if possible at a network of places, and advised on how to draw up a daily register of the observations to make a 'history of the weather' (HOOKE 1663). The Society also concerned itself with the design of instruments: by 1670 barometers for use on land and sea had been invented (the latter, a type with a constriction in the tube, was never used until reinvented about 100 years later), also a thermometer with a scale of degrees reading from 0 at the freezing point of water, each degree being about 2·4°C (originally double that), and a pressure-plate anemometer. All these were due to HOOKE, who had graduated from being ROBERT BOYLE's laboratory assistant to become the Royal Society's Curator of Experiments. Much careful labour is entailed in reducing the observations made with early meteorological instruments to comparability with modern series. One must examine the records for the behaviour of the instruments which often aged rapidly (e.g. see MANLEY 1962, p. 46) and for their exposure (thermometers in the early days of instrument observations were commonly set up in an unheated north room and rain gauges on roofs[1]) and must establish what corrections need to be applied (in the case of barometric pressure these always include correction for the temperature of the instrument and its height above sea level, as well as adjustment of the value to standard gravity at latitude 45°). Furthermore, the measurements have to be converted from a great variety of ancient scales to modern units: conversion data (and some notes on comparison tests of the early instruments) are given by LAMB and JOHNSON (1966, pp. 6, 12–16).[2] The Vienna barometer observations, even about 1800, were described as being made 63·6 *Wiener Füsse* (Vienna feet) above the pavement of the Bäckerstrasse and 101·7 *Wiener Füsse* above the average level of the Danube, which was related to a mark on a

---

start in 1685 and at the Astronomical Observatory in Paris in 1688, but are not continuous to the present day. Continuous records for England begin at Southwick near Oundle in 1726 and Plymouth in 1727.

1. The practices here mentioned seem to have been common in the seventeenth and early eighteenth centuries. Later, thermometers were exposed in various sorts of screen or box out of doors on a north wall. Louvred screens on stands in the open were gradually introduced from the mid nineteenth century onwards.

2. Particular care is needed also over wind direction abbreviations in Latin language texts (which were often used):

S. (= *Septentrionalis*)	stands for	North
S. Or.	,, ,,	North-East
Or. (= *Orientalis*)	,, ,,	East
M. Or.	,, ,,	South-East
M. or Mer. (= *Meridionalis*)	,, ,,	South
M. Occ.	,, ,,	South-West
Occ. (= *Occidentalis*)	,, ,,	West
S. Occ.	,, ,,	North-West

certain bridge. An excellent guide to these problems has been given by VON RUDLOFF (1967). Despite all the difficulties, it has been possible to arrive at a continuous, and apparently homogeneous, series of monthly mean temperature values for typical lowland sites in central England (MANLEY 1959, 1961, 1974) that stand up to all available comparisons from as early as 1680, to which point the individual monthly values are probably within 0·3 to 0·5°C. MANLEY's (1953) account of the instruments and exposures, and the overlapping records for different places used to obtain this long series, is instructive as to the problems and methods of standardization. The series is kept going today as the arithmetic mean of the values for Cambridge (Botanical Gardens), 52°2′N 0°1′E 12 m, Ross on Wye, 51°9′N 2°6′W 68 m and Nelson, Lancashire, 53°8′N 2°2′W 165 m. Such area values, provided the places represent a reasonably homogeneous climatic region, have the virtue that they are likely to be subject to smaller errors than the individual stations' records from which they are composed. For this reason, and because of the impossibility of maintaining a given station in the same place, with its environment unchanged, for hundreds of years,[1] area averages have been adopted in constructing other long records, e.g. England and Wales rainfall from 1727 (NICHOLAS and GLASSPOOLE 1931, see also WALES-SMITH 1971) and eastern United States temperature from 1738 (LANDSBERG et al. 1968).

There are other places in central, northern and western Europe for which observation records already about 250 years long exist, e.g. Berlin beginning 1700 and continuous from 1730 (LENKE 1961). And daily observations of wind and weather, initially without instruments, in or near London are continuous from 1669 in the archives of the Meteorological Office. Observation series, with instrument measurements either already standardized or believed capable of reduction to homogeneity with modern series (see e.g. LAMB and JOHNSON 1966, VON RUDLOFF 1967 for bibliographies), mostly available as monthly values but in many cases with the original daily observation record also extant, for the following places, continuous from the dates shown, mark the gradual extension of the network to cover the whole world:

London (Kew Observatory) rainfall from 1697; 'Eastern U.S.A.' (effectively Philadelphia) from 1737 (Cambridge, Mass., from 1743 and some earlier fragments – Cambridge 1715, Boston, Mass., 1725–6); Uppsala (Sweden) 1739; Zwanenburg (Netherlands) 1743 (some earlier shorter runs overlap: Hoofddorp temperatures from 1706, Delft rainfall from 1715); St Petersburg 1751 (some information from 1726);

---

1. Sites of observing stations and exposures of instruments are liable to change over the years with the inevitable changes in the use of the land, through the growth of towns, erection of walls and buildings and the laying of paved surfaces nearby, and even through the growth of trees. Observation sites on airfields, where the most highly trained observing staffs are commonly stationed, have proved notoriously impermanent. In an effort to minimize the effect of these changes on our perception and assessment of any real climatic trend, the World Meteorological Organization has recommended each country to designate a limited number of observing stations as 'benchmark stations' where every effort will be made to maintain the observations at the same spot indefinitely into the future and where there is a good prospect of avoiding change in the nature of the environment.

Copenhagen 1752; Paris 1753; Basle 1755; Stockholm 1756; Turin 1756 (Padua rainfall from 1725); Milan 1761; Trondheim 1762; Edinburgh 1764 (earlier fragment 1731–6); Prague 1775; Vienna 1775; southern Iceland 1779 (Reykjavik from 1798 but interrupted after 1840); Palermo (Sicily) 1791; Madras 1796; Bombay 1817; Archangel 1813; Gibraltar 1822; Barnaul (Siberia) 1836 (some data for Astrakhan from 1745); Irkutsk 1836; Peking 1841 (rainfall at Seoul, Korea, from 1770 and some monthly details, numbers of raindays, snowdays, etc., from 1626); Sitka (Alaska) 1843 (but not unbroken till 1908 onwards); west Greenland 1831–52 and 1873 onwards; St Louis, Missouri (U.S.A.), 1836; Toronto 1841; Sacramento, California, 1850; St Helena (south Atlantic) 1841; Hobart (Tasmania) 1841; Cape Town 1842; Rio de Janeiro 1851; Mauritius 1853; Auckland (New Zealand) 1854; Santiago de Chile 1861 (Valparaiso rainfall from 1824); Azores 1865 (some record of measurements from 1840); Tokyo 1870; Honolulu 1883; Laurie Island (South Orkneys) 1903; Spitsbergen 1911; Antarctica (Graham Land, Antarctic Peninsula) 1944 (fragments from 1900).

The first mountain stations were established on the Alpine passes St Gotthard 1781, and St Bernard 1817; the first summit stations Ben Nevis (Scotland) 1883–1904, and Sonnblick (Austria) continuous from 1886.

In many cases fragmentary records exist from earlier years, e.g. expeditions in the Canadian Arctic in the 1820s and 1850s, various ships in the Antarctic (e.g. Weddell Sea area 1820s, Ross Sea area 1840–2), the Falkland Islands 1859–68 and 1875–7, the International Polar Year and preliminary expedition bases on islands near 80°N 1879–83 and the drift of Nansen's ship *Fram* in the ice near the North Pole 1893–6.

There is still a need for better indexing and archiving of the early records. The most comprehensive single collection is the monthly values printed in the volumes of *World Weather Records*, of which the first three, giving data to 1950, were published by the Smithsonian Institution, Washington, in the *Smithsonian Miscellaneous Collections*, and the volumes for later years by the U.S. Weather Bureau.

Ships apparently began measuring sea surface temperatures, presumably in water collected by bucket, about 1780 or earlier. Such measurements were instituted by BENJAMIN FRANKLIN (1786) as a means whereby the westbound U.S. mail packet ships could avoid the Gulf Stream, and in 1785 he had even attempted to sample the subsurface water down to 100 foot depth with a cask fitted with valves. The agreement between the independent surveys of temperatures prevailing in the Atlantic Ocean constituted by assembling separately on maps the observations of American and British ships (cf. figs. 8.9 and 10.12) is so good as to suggest that the instruments and observation method were good enough to give reliable mean values (probably within a few tenths of a degree) from the first.

Numerous ships' log books, containing observations made daily both when in harbour and at sea, particularly on board naval ships, are preserved in the Public Records Office in London and possibly in other capitals. Even though for long no instruments were carried, this source adds a wealth of observations from 1680 onwards (and a few in earlier years) in British and Irish havens, as well as elsewhere in Europe, the Mediterranean, the West Indies, the American colonies of those times, along the coasts and on the ocean crossings – occasionally even to the Far East. A sample page from one of these logs is illustrated in Plate I.

There are naturally some problems in the use of archaic words: e.g. mizzle = drizzle. The names applied to different forces of the wind before the adoption of the Beaufort scale (put forward by the British Admiral BEAUFORT in 1806) varied somewhat at different dates but were to some extent standardized. 'Light airs' and gentle 'cat-skins', ruffling only patches of water, were distinguished from 'breezes' (or 'breeses'), which whether soft or moderate blew over the whole face of the sea within sight. Once a breeze became a little brisk it was called a 'gale' (or 'gaile'). The lighter 'gales' were sometimes called 'gentle gales', whereas 'small gales', 'moderate' or 'loom gales' were the best sailing weather in which a ship could carry all her sails. In a 'brisk', 'stiff' or 'strong gale' a ship could just carry her topsails. A 'hard gale' seems to have been about equivalent to the modern description of a gale, with Beaufort force $\geqslant 8$. (The above descriptions rely on *The Four Years' Voyages of Capt. George Roberts* (1726), pp. 108–9, and Daniel Defoe on the *Tempest* of 1703. I am indebted to Mr J. A. KINGTON of the Climatic Research Unit, University of East Anglia, for drawing my attention to them.)

In 1775 the French government, being concerned about the possible effects of the weather of the preceding years on human health and agricultural production, addressed an appeal to the country's doctors which led to the organizing in 1778 of a nation-wide network of observations (SANSON 1955, KINGTON 1970) and for the next decade or so the *Société Royale de Médecine* in Paris was the collecting centre for a good coverage of reports, which are still there. Unhappily the arrangement did not survive the revolution; and it was not until 1821 that the French government again, being concerned at the 'sudden variations in the behaviour of the seasons' to which the climate at that time appeared 'more and more subject', tried (apparently with indifferent success) to organize an observation network, this time using the police posts throughout the country.[1] The Ministry of the Interior in Prussia organized a network in 1817, and increasing efforts seem to have been made from about this time in England and Scotland relying mainly on voluntary observers.

An international network, extending for a time from Godthaab in Greenland to Moscow and Rome was arranged by the *Societas Meteorologica Palatina* (the first meteorological society), and the collected daily observations were published in yearly

---

1. The German poet GOETHE, as a young man, had in fact been responsible for setting up the first government-run weather observations network anywhere, in 1776 in the small state of Sachsen-Weimar, while briefly holding the position of minister of government. Later in life (J. P. ECKERMANN, *Gespräche mit Goethe*, entry for 13 February 1829) he voiced the frustration which many must have felt:

'The subjects of meteorology ... which we daily see working ... presuppose a synthesis – only the cooperating circumstances are so many that Man is not equal to this synthesis and therefore uselessly wearies himself in observations and inquiries.' It is a happy event that we now can bring the devoted work of so many pioneer observers to fruition.

volumes of *Ephemerides* by the Society in Mannheim from 1781 to 1792, when this development of the science also was snuffed out in the troubled times that followed the French revolution. It was not until the first International Meteorological Congress in Vienna in 1873 and the founding of many national meteorological services about that time that this setback to the development of the observation network was made good.

H. W. BRANDES in 1820 used the observations of the Mannheim network to draw the first daily synoptic weather maps, covering Europe for the year 1783. He used lines of equal departure of barometric pressure from the overall mean in place of isobars, but these effectively revealed the travelling low and high pressure systems. Apart from this notable experiment,[1] the early observers never saw their patient labours[2] applied to elucidating the weather processes that interested them.

The longest continuous series of daily weather maps covers the area about the British Isles and North Sea from 1861 (MSS maps only 1861–72, prepared in the United Kingdom Meteorological Office); printed maps appear in various countries' Daily Weather Reports from about 1873. Fig. 13.1 shows, however, the quality of the daily maps that can now be produced for much earlier years.[3] The gradual extension of the coverage of daily synoptic surface weather and atmospheric circulation maps to the whole globe is indicated by the following list of the main chart series available for study:

> *Cartes synoptiques journalières*, 1873–6 (Copenhagen, Danish Met. Inst., 1876–9). These maps cover the north Atlantic and Europe.
> *Bulletin of International Meteorological Observations*, 1877 ff. (Washington, U.S. Signals Office, 1877–84). These maps cover most of the northern hemisphere, excluding the Arctic.
> *Tägliche synoptische Wetterkarten für den Nordatlantischen Ozean und anliegenden Theile der Kontinente*, 1880–1912 (Copenhagen, Danish Met. Inst., and Hamburg, Deutsche Seewarte, 1884–1931).

1. BRANDES's charts were subsequently lost and may not have become widely known until specimens of them were reconstructed from his data tables by H. H. HILDEBRANDSSON towards the end of the nineteenth century.
2. Among the early observers many were doctors of medicine, parsons, and rural administrators of various kinds stationed at outposts, indulging an interest in science in their leisure. Of these, the Moravian missionaries on the Labrador coast in the 1790s and SVEINN PÁLSSON's record in Iceland from 1798 to 1840 deserve special mention. Their meticulous work and steadfastness commonly surpassed that of the official observatories, as is well shown by the observation record maintained at Salem, Massachusetts (HAVENS 1958), by Dr E. A. HOLYOKE from 1754 to a few weeks before his death in 1829 at the age of

100, though he himself did not consider that he had managed to maintain his former standards after his 92nd birthday and recommended that that final portion of his records should not be published.
3. A series of daily weather maps covering the area of the eastern Atlantic and Europe from Iceland to 30°E and from Trondheim to northern Italy, of which fig. 13.1 is a sample, is being constructed by Mr J. A. KINGTON in the Climatic Research Unit, University of East Anglia, Norwich, in a project sponsored by the Research Committee of the Meteorological Office. The series has been completed for the years 1781 to 1785 and is being continued. The sources used have been described by KINGTON in a number of articles. Photographic examples of some of them and a 7-day sequence of the maps in November 1784 are given in KINGTON (1975).

Historical Daily Weather Maps, Northern Hemisphere Sea Level, 1899 ff. (Washington, U.S. Dept of Commerce, Weather Bureau, 1945 ff.).

MEINARDUS, W. and MECKLING, L., *Meteorologischer Atlas: mittlere Isobarenkarten und tägliche synoptische Wetterkarten der höheren südlichen Breiten, Okt. 1901 bis März 1904 (Deutsche Südpolar Expedition*, ERICH VON DRYGALSKI) (Berlin, Reimer, 1911). 4 vols.

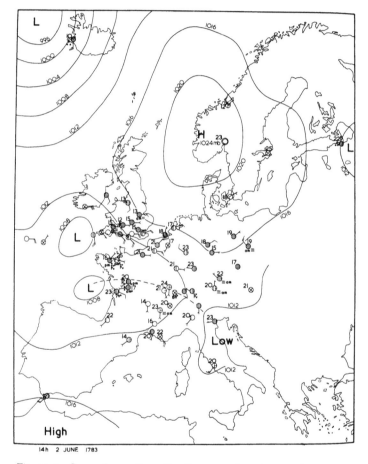

Fig. 13.1   Synoptic weather situation at 14 h, 2 June 1783.
    Isobars indicate barometric pressure reduced to mean sea level and standard gravity.
    Plotted observations according to the usual notations: surface air temperature in °C, wind direction and Beaufort force, cloud cover, present weather (left of circle) and weather of the last six hours (lower right).
    (*Data assembled and chart plotted by* J. A. KINGTON; *see* OLIVER *and* KINGTON *1970.*)

Southern Hemisphere Daily Surface Weather Maps, 1951 ff. (Pretoria, Weather Bureau, unpublished – available on microfilm).

For many studies of the climatic sequence it is sufficient, and handier, to use maps of the mean conditions of a whole month, or season, and averages of these over longer or shorter terms of years. The relationship of monthly mean circulation charts to the temperature, rainfall and other conditions prevailing during the month has been illustrated in Volume I (Chapter 7, figs. 7.16–7.20).

Monthly mean pressure charts for each January and July back to 1750, and certain period averages, have been reconstructed by LAMB and JOHNSON (1966) and tests of reliability applied to the analysis.[1] The area reliably covered is greater for 10-year means than for the individual years, but for both includes most of western and central Europe between 40° and 60°N back to 1750. Mean maps for 40 or more years cover the North Atlantic and Europe reliably back to the same date. At the time of writing, data are being compiled on charts for other months and for earlier years.

Our ability to compile the history of the weather, and to derive the atmospheric circulation regimes that produced that history, can be extended back long before the invention of meteorological instruments by the judicious use of

(1) Weather diaries.
(2) Descriptive accounts of weather, chiefly outstanding events, and the general nature of particular seasons.
(3) Records of parameteorological phenomena (parched or flooded ground, frozen rivers, etc.).

The most useful early weather diaries which have so far been submitted to scientific analysis are:

(a) Observations on 171 days of the year by CLAUDIUS PTOLEMAEUS, presumably at Alexandria, about A.D. 120 (for a list of published editions and the original meteorological analysis see HELLMANN 1916).

1. The procedure adopted was to draw the first isobaric analysis of the monthly mean pressure charts for years in the present century using observation reports only for the same skeleton network of places as were available in the eighteenth century. Next, another set of maps of recent years on which observations had been entered for the places that would have been available in the middle of the nineteenth century was analysed. Only after this were charts with the full modern network of data analysed. The distribution of errors in the analyses produced with the early observation networks was then studied. Guided by this study, it was decided to print isobars on the maps of any period only where the root mean square error appeared to be less than 2·5 mb in winter and less than 1 mb in summer and where there was no evidence of any systematic error due to analyst's bias exceeding 1 mb. It was found that for averages of 40 years or more the analysis over the North Atlantic was valid within these limits back to the eighteenth century.

(b) MERLE's weather diary, 1337–44, kept mainly at Driby, Lincolnshire, in eastern England (printed by SYMONS 1891).[1]

(c) The register of weather observations by W. HALLER, Keeper of the Public Records, in Zurich 1546–76 (printed in extenso by WOLF 1872 etc.; analysed by FLOHN 1949).

(d) The Danish astronomer TYCHO BRAHE's daily weather and wind observations on the island of Hven in the Sound between Denmark and Sweden, 1582–97 (printed in extenso by LA COUR 1876).

(e) Observations given by the Landgrave HERMANN IV of Hessen from 1621 to 1650, chiefly for the area of Kassel or Rotenburg-Fulda (analysis and further details by LENKE 1960).

Many other weather diaries in various countries were listed by HELLMANN (1926, 1927), the earliest of these in central Europe relating to the years 1490–1505, and seem not to have been used yet (some may unhappily have changed hands and been lost since). There are doubtless many others that have never been widely beyond the circle of the writer's family. From time to time previously undiscovered early weather diaries come to light in the catalogues of antiquarian booksellers, but are liable to remain inaccessible to scientific study because they command high prices as a collector's curio.[2]

Besides these obvious sources, there are many reports of weather occurrences and of related conditions such as floods, good and bad harvests, great frosts and snows, to be found here and there in state, local, monastic and manorial papers, chronicles and account books in Europe and somewhat similar types of information in the Far East. The region of the European lowlands near 50°N, from Ireland in the west to Russia in the east, is particularly well covered by such records; indeed it seems likely that every season of any sort of outstanding character since the year A.D. 1100 is known. The region is, moreover, a promising one for meteorological analysis since it lies open to the prevailing W and SW winds and, being always under the influence of the large-scale pattern of warm ridges and

1. This diary was formerly alleged to relate to Oxford, presumably because the Reverend WILLIAM MERLE was said to have been a Fellow of Merton College, Oxford, though his name cannot now be traced on the list (which is extant) of the Fellows of that College (LAWRENCE 1972). 'Oxford' had been entered, albeit in a much more modern script, on the original Latin manuscript. Merle was, in fact, the parish priest of Driby in Lincolnshire near the east coast of England, 250 km north-northeast of Oxford, during the years concerned; and it seems unlikely that a man strict enough about his tasks to keep a weather diary for 7 years would be a continual absentee from the scene of his public duty. Mr E. N. LAWRENCE of the Meteorological Office has kindly investigated this problem for me. There are not many references in the dairy to place, but nine were found to relate to Lincolnshire and three to Oxford, one was to a place just over the Lincolnshire border possibly on the way to Oxford, and there was one other, evidently on a journey in eastern England, in Cambridgeshire. Some of the observations, e.g. of fog with easterly winds in May in 1340 and 1342, also suggest the North Sea fogs of the east coast. The picture which seems to emerge is of a priest normally resident in his parish but having some appointment which necessitated occasional journeys to Oxford and elsewhere.

2. A weather diary kept at Urbino, Italy, from 1591 to 1597 by FREDERICO BONAVENTURA was, however, recently (1971) secured by the library of the Swiss Federal Institute of Technology, Zürich, for scientific use.

cold troughs in the flow of the upper W'lies and the surface weather systems associated with them, affords indications of the prevailing state of the northern hemisphere circumpolar vortex.

The by now numerous published collections of these reports facilitate the work of the analyst (see particularly SHORT 1749, ARAGO 1858, LOWE 1870, HENNIG 1904, NORLIND 1914, SPEERSCHNEIDER 1915, 1927, THORODDSEN 1916–17, VANDERLINDEN 1924, EASTON 1928, BAKER 1932, BRITTON 1937, MÜLLER 1953, BUCHINSKY 1957, WEIKINN 1958–63, ARAKAWA 1954, 1955, 1956, and CHU KO-CHEN 1961). Care must be taken, however, over the many and various kinds of possible unrepresentativeness of the reports available. The great shortcoming of medieval annals and chronicles as sources of weather information is that their main concern was with unusual weather and with extremes and their effects, whereas for our interpretation of the climatic regime then ruling these details are only incidental to discovering what was the normal range of experience of those times. Any scientific use of these reports has to be adapted to this characteristic. The most reliable indications of European surface weather in the early manuscripts to use in analysis of the character of individual seasons appear to be:

(a) *Mildness or severity of the weather prevailing in the winter months:* The effects upon landscape, transport and agriculture are likely to have been reported in all important cases. It is possible to identify confidently persistent spells: mild winters by rains, flooding and thunderstorms even in continental regions, also by early or out-of-season flowering of plants; severe winters by frozen waterways and by many sorts of privation and damage.

(b) *Raininess or drought in the summer months:* Again the effects upon the landscape and upon agriculture are likely to have achieved mention in all important cases. Wet summers produce flooding and ruined crops, though dramatic accounts of individual thunderstorms may occur in otherwise good summers. Dry summers are known by parched ground and dwindling rivers, whilst the grain crops are usually good (except in areas with insufficient ground water, near the arid margin, where crop failures may result); forest fires are also particularly liable to occur. The rain character of a summer is surer of faithful recording than the temperature, since one oppressive heat wave might well be the only recorded reference to temperature in an otherwise poor summer.

In some respects, as will be noticed, it appears safer to rely on such parameteorological evidence as registers the integrated effect of a whole season of obviously well-marked character rather than on sparse reports of the weather itself. In the twentieth century, however, the levels of many rivers and lakes (and the question of whether they freeze or not), which were formerly useful indicators of the weather, are increasingly determined (intentionally or unintentionally) by the activities of Man.

Crude numerical indices may be designed, as defined below, to eliminate as far as possible the effects of any varying attentiveness to reporting (i.e. different frequency of reports, regardless of the character of the weather) in different periods. These indices are applied to the years in groups of not less than ten, thus also eliminating any uncertainty due to calendar usages (see next section) as to the exact year to which a given report was meant to apply.

(a) *Winter mildness/severity index:* The excess number per decade of unmistakably mild or cold winter months (Decembers, Januarys and Februarys only) over months of unmistakably opposite character, excesses of cold months being counted negative. Unremarkable decades score about 0. Extreme decade values of the index in Europe in the last 1000 years range from about +10 to −20.

(b) *Summer wetness/dryness index:* Each high summer month (Julys and Augusts only) with unmistakable evidence of drought counted 0, unremarkable months ½, months with unmistakable evidence of frequent rains and wetness counted 1. Unremarkable decades score about 10. Extreme decade values of the index in Europe in the last 1000 years range from about 4 to 17.

The index values decade by decade for three different longitudes in Europe are tabulated in Appendix V, Table 4, and examples of analysis of these to reveal the course of climatic history in Europe and Asia will be found in fig. 13.3 and fig. 2 in Appendix V.

General trustworthiness of the early data in these collections, and of the index values here derived from them, is suggested by a comparison made between the summer wetness index values for England for the decades between A.D. 1200 and 1350 and figures for the summers and autumns (actually from about June through September) derived by TITOW (1970) from the audited account records of manors owned by the bishopric of Winchester (not known to LAMB when compiling the summer index values): the comparison yielded a correlation coefficient of +0·77 over the fifteen pairs of figures. It seems fair to conclude that our knowledge of the character of individual seasons in England is reasonably complete, at least as regards any important particulars, as far back as this. The more dramatic the weather of any given year, the more reports – indeed the more widespread reports – of it are likely to have been made and correspondingly many should survive to our own day. Thus, it is possible to map the weather over Europe each summer and each winter all through the peculiarly harsh decade of the 1430s. This at once proves groundless BRITTON's (1937) suspicion that most of the reports of severe winters in that decade were due to mistakes and repetitions on the part of the medieval chroniclers: he seems never to have considered the possibility that there could have been so many severe spells or even that there could have been more than one long one, which he took to be in 1433–4, in one decade. The map now compiled for winter 1431–2 is reproduced here as fig. 13.2 and is of a type (persistent anticyclone over Scandinavia) well known to bring bitter weather to most of Europe.

Owing to the strong correlation between the index values here defined and the average temperature and rainfall conditions prevailing in each decade over the period since about

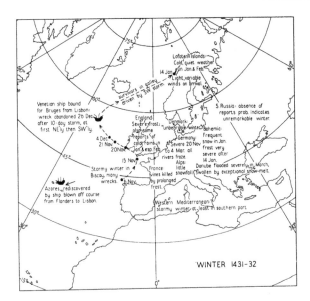

WINTER 1431-32

Fig. 13.2   Map of weather reported in the winter of 1431–2.

1700, for which comparison was possible, it has proved possible to establish regression equations (as follows) with which to derive probable climatic averages of these items for times before the introduction of thermometers and rain gauges (LAMB 1965):[1]

$$T_{DJF} = 3·69 + 0·11m$$ standard error of the result ±0·25°C

$$T_{JA} = 16·40 − 0·062W$$ standard error of the result ±0·36°C

$$T_Y = 0·864\left(\frac{T_{DJF} + T_{JA}}{2}\right) + 0·75$$ standard error (when T_{DJF} and T_{JA} are accurate) ±0·16°C

$$R_{JA} = 6·52W + 29·1$$ standard error of the result ±4·01

$$R_{10} = 7·81T_{DJF} + 66·6$$ standard error of the result ±4·29

$$R_Y = 9·80T_Y + 6·2$$ standard error of the result ±4·65

1. The correlation coefficients which made it possible to derive estimates of the temperature and rainfall averages for different periods from the decade values of the summer wetness and winter mildness indices were:

Rainfall over England and Wales, as % of the 1916–50 averages

			r
(1) In high summer (July and August)	*v.*	Summer wetness index since 1740	$r = +0·91 ± 0·04$
(2) Over the other 10 months of the year	*v.*	Winter temperature since 1740	$r = +0·65 ± 0·12$
(3) Over the whole year	*v.*	Temperature of the whole year since 1740	$r = +0·42 ± 0·18$

Temperature in central England

(1) In winter (December, January and February)	*v.*	Winter mildness index since 1680	$r = +0·87 ± 0·05$
(2) In high summer (July and August)	*v.*	Summer wetness index	$r = −0·23 ± 0·18$
(3) Over the whole year	*v.*	Arithmetic mean of the high summer and winter averages since 1680	$r = +0·89 ± 0·04$

In these equations T_{DJF}, T_{JA}, T_Y stand for the mean temperature (°C) in the lowlands of central England in December, January and February; in July and August; and over the whole year, respectively. R_{JA}, R_{10} and R_Y stand for the mean rainfall in England and Wales in July and August, over the 10 months September to June, and over the whole year, respectively, in each case expressed as a percentage of the 1916–50 averages. m is the average decade value of the winter mildness index, W the average value of the summer wetness index. Average values of the indexes over several successive decades are used, to avoid relying too much on a supposed evenness of (i.e. constant attentativeness to) reporting.

The history of changing values of prevailing temperature and rainfall in Britain derived by this method has been illustrated in Volume 1, figs. 6.6 and 6.7, together with the margins of error of the method and the corrections to which the results appear liable. (The values are tabulated here in Appendix V, Table 3). One is clearly already at the point where independent control provided by altogether other lines of evidence is desirable. This check was obtained in the study here reported by reference to known shifts of the height limits of trees and cultivation of grain and vine crops in central Europe, of the northern limit of thriving vineyards (at times affecting England), and by reasoning about sea temperatures in the Atlantic, all of which suggested an upward adjustment of the temperatures derived from the regression equations by 0·3 to 0·5°C in the winters and summers of the twelfth to fourteenth centuries A.D. Such adjustments probably imply changes in the reporting (e.g. no notice taken of much winter weather that would nowadays be ranked as mild) – and hence there should be changes in the correlation coefficients also – attributable to the somewhat different conditions which the reporters in earlier centuries regarded as normal. In situations in early times where the case is altered by sparseness of the reporting, similar indexes averaged over 100 or 200 years may give meaningful results. Because the frequency of months noted in the Russian chronicles in the Middle Ages was fairly steadily about half that for western and central Europe (where one third of the winter months and 40% of Julys and Augusts figured between A.D. 1100 and 1550), index values obtained for Russia were doubled in the present author's studies for comparison with western and central Europe. But in such cases it may be better to define indices by the *ratios* of frequency of reports of wet to dry summers and mild to cold winter weather.

Analysis of the meteorological observations

Fig. 13.2 illustrates how directly mapping of the European weather reports available for years as early as 1432 may reveal the dominant characteristics of the atmospheric circulation (in this case a large and persistent blocking anticyclone over northern Europe). Various indirect approaches have been developed (LAMB 1963, LAMB et al. 1966, 1970) to derive information about the circulation patterns prevailing in earlier epochs, the results of which will be used in later chapters.

When 50-year averages of the summer wetness and winter mildness indexes are plotted on diagrams showing the sequence of the last 1000 years (tentatively extended to twelve centuries) in different European longitudes near 50°N (fig. 13.3), it is seen that all across Europe there was a predominance of dry summers and mild winters about the 1100s and of wet summers and cold winters in the late 1500s and 1600s. Between these extreme phases of the climatic pattern, from about 1200 to 1500, there was an apparent westward movement of the anomaly concentrations shown on fig. 13.3, such as would correspond to shortening wave length in the upper W'lies. Between 1700 and the warm climatic period in the first half of the twentieth century there was a contrary movement, eastward across Europe, implying renewed lengthening of the prevailing wave length in the upper W'lies. A similar tendency is perhaps discernible in the earliest part of the diagram, before the height of the early medieval warmth. Measurement of the longitude shifts between the cold climate period 1550–1700 and the warmest times in Europe in 1150–1300 and the present century shows that they

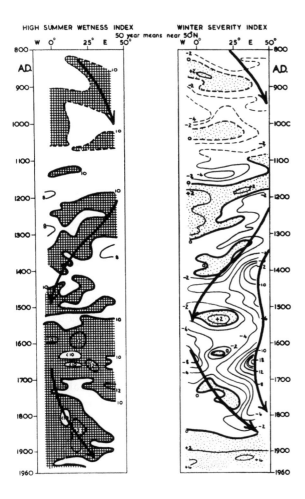

Fig. 13.3 50-year mean values of the LAMB summer and winter indices at different longitudes in Europe near 50°N between England and Russia from A.D. 800 to 1960.

Hatched areas indicate excess of wet over dry months in high summer (July and August).

Stippled areas mark more very mild than severe winter months (December, January and February).

Schematic arrows make it easier to follow the movement of the greatest anomalies during times of change. In the periods of climatic standstill – i.e. when some persistent character of the atmospheric circulation must be presumed (A.D. 1050–1200, 1550–1700, and after 1900) – all longitudes tended to fare alike as regards prevalence of dryness or wetness, warmth or cold.

amounted to 20–30°, and if one assumes that the mean position of the waves in the upper W'lies was anchored throughout near the Rocky Mountains, wavelength changes of about 15–25° are implied. These figures are so great as to suggest, on the basis of the Rossby equation (Volume 1, Chapter 3), that (a) the main flow of the upper W'lies must have been displaced south in the time of colder climate by some 3–5° of latitude across the American and Atlantic sectors (a displacement probably liable to lead to more frequent blocking over Europe) and that (b) at the same time the circulation generally weakened somewhat. The tracks of depressions crossing the Atlantic and the range of influence of frequent cyclonic weather should have been displaced south by a similar amount. These results are illustrated by the maps in a later chapter (figs. 17.8, 17.9). It is further implied that the area of the relatively quiet cold climates of the polar cap underwent a corresponding expansion, allowing the ice on the polar seas to spread.

Derivation of the circulation regimes that probably prevailed in the last ice age and in the warmest postglacial times is also possible, using various types of field data from those times. It depends upon calculation, by means of concepts and equations used in numerical daily weather forecasting, of the cyclonic and anticyclonic development distributions produced by the prevailing upper air temperature fields, themselves derived from the surface temperatures indicated by botanical and other field evidence. The flow of the upper winds, also indicated by the upper air temperature fields, indicates the steering of the cyclones and anticyclones once formed; hence the general course of the surface isobars and winds can be tentatively sketched.

Besides enabling us in these ways to derive a picture of the geographical patterns and morphology of the general wind circulation in all the main climatic eras that have diversified the history of the last 50 000 years, meteorological analysis of the data of the last 1200 years in Europe (and the Far East), and in appropriate extra detail of the last 100–200 years, can give us insight into the range of the year-to-year, decade-to-decade and century-to-century variations and the extent to which their incidence is affected by periodicities (or quasi-periodicities – see Volume 1, Chapter 6). In fig. 13.4(a) the year-by-year variations of temperature in central England between 1850 and 1965 can be followed, and the scales at the left-hand margin indicate the range of variability in terms of standard deviations from the overall means of this period. The dots in this diagram are the values for the individual years and the continuous lines show the 10-year averages for periods centred at any point on the curves. The variations from one decade to another are shown by the dots in fig. 13.4(b) and the smooth curves in this diagram are the 100-year averages. Here the vertical (ordinate) scale of fig. 13.4(a) has been multiplied by $\sqrt{10}$, so that the standard deviation scales printed beside fig. 13.4(a) should apply without adjustment (since the standard deviation of the mean of n observations is $1/\sqrt{n}$ times the standard deviation of the individual observations). By this standard of comparison we see that the departures from the longer-term temperature level presented by some decades, though less than the variations of the individual years, are very significant and it is probable that the differences between the warmest and coldest centuries (regardless of the alternative estimates for the warm centuries in the early Middle Ages) have a still higher level of statistical significance. Fig. 13.4(*c*) extends this comparison to the course of the last 12 000 years, since the end of the last ice age, including several thousand years of the warmest postglacial climate. Estimated 1000-year averages are shown by the continuous lines and the dots plot the average values derived for the last ten centuries. Once again the ordinate scale is multiplied by $\sqrt{10}$, and the enormous range of the changes of prevailing temperature by comparison with the statistical measure of present-day year-to-year variability stands out strongly. The shaded areas on fig. 13.4(*c*) which one has to use instead of point values for plotting the curves indicate the margins of uncertainty of (*a*) the temperature data supplied by palaeobotanical evidence and (*b*) the times to which the temperature data refer (obtained from radiocarbon dating).

Figs. 13.4(*a*) and (*b*) may help us to a realistic appraisal of the year-to-year and longer-term variations which probably went on in other climatic epochs and which should be superposed on the grossly smoothed curves of temperature change over thousands of years, which we derive from the various types of field evidence. The probable range of these variations in other times can be judged by the figures for comparable geographical and climatic situations, where such exist, in the world today. In warm ice-free eras the situation of England as regards the possibilities of temperature differences from one year to another should presumably be more nearly equivalent to places near the subtropical anticyclone zone today. The inter-yearly standard deviation of mean temperature for each month of the year in Britain in the presence of the great ice sheets of the ice ages (though still nearer than other parts of northern Europe to the open waters of the Atlantic Ocean) may be judged comparable with the values for south Greenland or Spitsbergen at the present time, about twice the values shown in fig. 13.4(*a*), and in extreme phases perhaps three times the present values. The ice age situation in central North America and eastern Europe, where immediately beyond the edge of the great ice sheet there was the extensive land surface of a continent, open to summer heating much as now, has no match in the world today: at such points the year-to-year variations of temperature may have been greater than now occur anywhere on Earth, especially in comparison with present-day summers. The nearest approaches to the ice age situation in the world today as regards year-to-year temperature variability are presumably offered by continental situations near the thickest of the Arctic sea ice in summer and places in latitudes 40–50°N in America and Eurasia in winter that are geographically open to warm as well as cold air: standard deviations about three times as great at those here shown for present-day England seem likely to have applied in continental interiors within, perhaps, 500 km south of the edge of the ice sheets.

The question of the greatest rapidity with which major climatic changes occur is an important one on which more precise data should be sought. Although significant climatic changes did occur within the last 1000 years, as represented in figs. 13.4(*a*) and (*b*), and had disastrous effects on human history in some areas (notably Greenland), they may not have been the greatest or most rapid changes of postglacial times. To resolve this question, and to

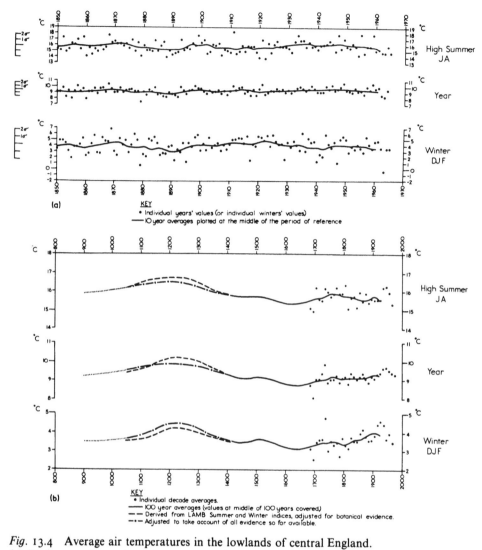

Fig. 13.4 Average air temperatures in the lowlands of central England.
(*a*) Each year 1850–1965 and 10-year running means.
(*b*) Each decade 1680–1960s, and 100-year running means from
  about A.D. 800 (derived as described in the text).
(*c*) Each century from A.D. 1000 and 1000-year running means from
  about 10 000 B.C. (see text).
Note: JA means July and August, DJF means December, January and February.

follow the course of events (the 'ups' and 'downs') within the period of onset of other major
climatic shifts, we shall have to draw on evidence from fields other than meteorology. It will
be illuminating to build up more factual knowledge and more detailed meteorological
analysis of the onset of colder conditions around A.D. 1430 and again around 1560, but it is
doubtful whether these changes were more abrupt than the shifts towards warmth around
A.D. 980 and 1730 or 1900. The great and 'permanent' change that set in between about 1200

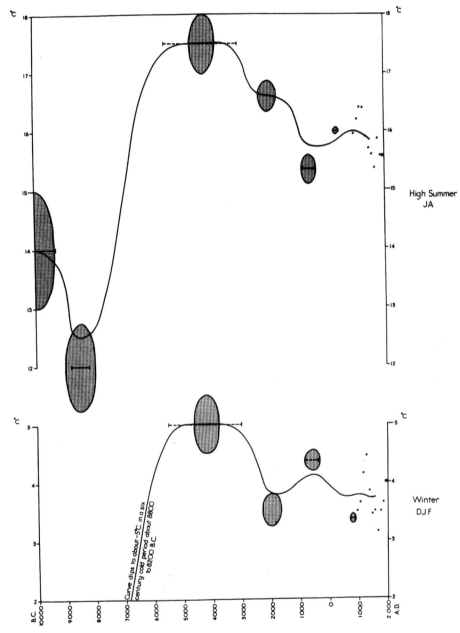

High Summer
JA

Winter
DJF

(c) <u>KEY</u>
——— Course of the supposed (or, last millenium, calculated) 1000-year averages.
• Individual century averages.
⊕ 1900–1965 (66 years)
Ovals shaded: approx. ranges within which estimates of temperature lie and 2σ error
margins of dating (C¹⁴ method) on either side of the most probable date.
⊢——⊣ Supposed duration of the epoch to which the "observed" mean value applies.

and 500 B.C., and ended the warmest postglacial times, cannot be covered by similar data; but dendroclimatic evidence (see later section of this chapter) of its course, including sequences of tree-growth data over any period from 10 to several hundred years long within it, should be sought and analysed.

Most types of dating applicable to such early times, and determinations of age differences by methods other than the use of year-rings in trees or lake varves, etc., entail margins of error that preclude any possibility of establishing the detail and duration of the transitions that here concern us. They can indicate a maximum estimate of the time span within which a given climatic shift took place, but no indication of how few years elapsed before significant new extremes were felt. The point is important because of the indications (e.g. BRYSON *et al.* 1970) that various climatic regimes that ruled at one time or another within the last 50 000 years were separated by transitions that, in some cases, took no more than a few centuries to complete.

Contributions from physical and dynamical meteorology, theoretical 'models', etc.

The invention of the radio-sonde in the 1930s, an instrument carried by balloon for making ascents ('soundings') through the upper atmosphere, measuring pressure, temperature and humidity, and sending the values back to Earth by automatic radio transmitter, while the ground station tracking the balloon by radar reveals the upper winds, made it possible for the first time to survey the atmosphere in three dimensions daily in all weathers. The development of a network of these observations from about 1940 onwards has shown the behaviour of the circumpolar vortex, the jet stream and the waves in the upper W'lies, and has led to those advances in our understanding from which it is now possible to view the atmospheric circulation over a hemisphere (and ultimately over the whole Earth) as a single entity and observe the working of the laws which govern the development and steering of cyclones and anticyclones. Before this, meteorology was ill equipped to contribute to the analysis of the great changes of climate which could be seen to have taken place in the past.

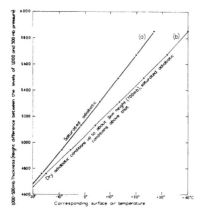

Fig. 13.5   1000 to 500 mb thickness (i.e. depth of the layer of the atmosphere between the levels where the pressure is 1000 and where it is 500 millibars) for different values of surface air temperature (near the 1000 mb level) under conditions of convective equilibrium:

(*a*) With saturation of the air and condensation of moisture occurring in rising air throughout the layers between the 1000 mb and 500 mb pressure levels.

(*b*) With saturation at heights above about 3 km – i.e. dry adiabatic lapse rate of temperature from about 1000 to 700 mb and saturated adiabatic lapse rate between 700 and 500 mb.

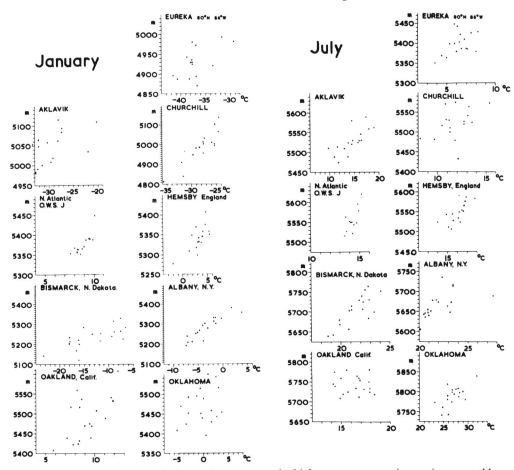

Fig. 13.6 Scatter diagrams of observed 1000–500 mb thickness accompanying various monthly mean surface air temperatures at places representing different special climates at the present day (1945–65). (*a*) In January. (*b*) In July.

Now, however, we can use this understanding to reconstruct, at least in outline, the prevailing patterns of the circulation from the field evidence of surface temperatures and so on, so long as that evidence is firmly enough dated to establish contemporaneity – i.e. to give a (skeleton) synoptic distribution of surface conditions all belonging to a single time and climatic regime.

In the analyses developed by the present writer, we proceed from the evidence of prevailing surface temperatures to derive the probable 1000–500 mb thickness using either

(i) whichever of the curves in fig. 13.5 appears appropriate: (*a*) in oceanic or other moist, cloudy climates, (*b*) in dry, especially subtropical desert, climates;

or (ii) whichever of the plots in figs. 13.6(*a*) and (*b*) of actual association between surface temperature and 1000–500 mb thickness in present-day climates appears most appropriate to cases where neither of the standard lapse rates in fig. 13.5 can be expected to apply – especially where surface inversions are prevalent over polar cold surfaces and continental snow cover in winter.

Northern hemisphere mean thickness maps derived for January and July in different climatic epochs are shown later in figs 16.8 and 16.9: the map for the Januarys about the last glacial maximum appears here below in fig. 13.7. These thickness maps indicate, by their gradients,

(a)

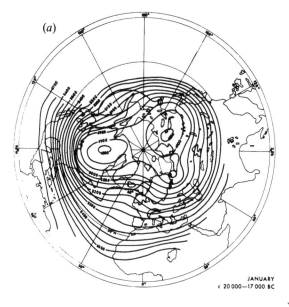

JANUARY
c 20 000—17 000 BC

Fig. 13.7   (*a*) Probable average 1000–500 mb thickness distribution in January and geography of the northern hemisphere about the maximum development of the last ice age (Würm III) around 20 000–17 000 B.C. (derived from prevailing surface temperature values indicated by botanical, marine-biological and oxygen isotope evidence).

(*b*) Prevailing areas of cyclonic and anticyclonic development, derived by computation from fig. 13.7(*a*) (LAMB and WOODRUFFE 1970), and suggested mean isobars of mean sea level pressure in January around 20 000–17 000 B.C.

Cross hatching = strong indications of cyclonic development.

Vertical hatching = weaker indications but cyclonic development predominates.

Stipple = strong indications of anticyclonic development.

Light shading = weaker indications but anticyclonic development predominates.

(*b*)

January 20 000—17 000 BC

the thermal winds flowing around the polar cold centre, i.e. the (broadly W'ly) vector wind component which the 1000–500 mb layer adds to the surface pressure-gradient wind.

SUTCLIFFE's development theory (1947) showed that the horizontal divergence of the thermal wind, div V', through a layer of the atmosphere, is given (subject to certain approximations) by

$$f \operatorname{div} V' = -V' \nabla (f + 2\zeta_0 + \zeta')$$

where f is the Coriolis parameter, V' is the thermal wind through the layer, and ζ_0 and ζ' are the vertical components of vorticity of the (in this case surface) wind at the bottom of the layer and of the thermal wind through the layer respectively. SUTCLIFFE (1939) showed that the horizontal divergence of the thermal wind represents the vertical motion of the atmosphere in this layer, with (in the case of the lower atmosphere or troposphere) upward motion in cyclonic development areas feeding the net divergence in the upper levels and downward motion in anticyclonic development areas fed by net convergence in the upper troposphere. Calculation of div V' is tantamount to calculating the tendency to cyclonic (div V' positive) or anticyclonic (div V' negative) development. The ζ_0 term in the equation is the 'steering term' concerned with the translation (or movement) of individual cyclones and anticyclones and may be omitted in climatological studies designed to indicate the regions of repeated genesis of cyclonic and anticyclonic systems.

Geostrophic wind measurements may be used for V' and computation of ζ'. Values of $f \operatorname{div} V'$ obtained from climatological mean thickness charts in this way are one or two orders of magnitude smaller than those given by the instantaneous synoptic charts used in daily weather forecasting: this is because monthly (or longer-term) mean thickness gradients are weaker and the smoother mean flow gives weaker shears and curvature. Small-scale areas of development of cyclonic or anticyclonic features may therefore be derived which are quite unreal and result only from minor errors in drawing the smooth lines on the mean maps or in reading off values from them: such features must be neglected and so far as possible eliminated. Values of thickness read off the charts were therefore smoothed arithmetically by taking the value h_1 read at any point in a square grid of points 300 km apart and deriving the smoothed value H_1 as

$$H_1 = \frac{4h_1 + \sum h}{8}$$

where $\sum h$ is the sum of the unsmoothed values at the four nearest surrounding points. Since we are concerned with areas of consistent tendency to cyclonic or anticyclonic development, the field of $f \operatorname{div} V'$ obtained from the calculations was mapped as the numbers of points giving positive values among the nine neighbouring grid points in a 3×3 square centred about each point.

Cyclonic development areas calculated in this way on charts of the present day appear to correspond to:

(a) Areas of much frontal activity on the warmer side of monthly mean low pressure regions.
(b) Areas where frontal wave development or the occlusion process repeatedly occur.
(c) Paths of warm front waves and other cyclonic features 'plunging' southeastwards (northern hemisphere case).

Anticyclonic development areas so calculated correspond to:

(a) Rear sides of cyclonic circulations, where outbreaks of polar cold air are frequent.
(b) Main subsidence regions in the subtropical and high latitude anticyclones and surface ridges of high pressure.

The final stage of the reconstruction is to sketch the probable course of the mean sea level isobars indicated by combining these results with the prevalent steering of surface depressions and anticyclones by the main currents of thermal wind – i.e. by the run of the thickness isopleths in zones of strong gradient. The process is illustrated in figs. 13.7(a) and (b) by the thickness isopleths, computed development areas and putative surface isobars sketched for the maximum development stage of the last glaciation over the northern hemisphere.

Other types of theoretical approach not so directly linked with observed data from the epoch under study may be useful in other connexions, notably mathematical 'models' or schemes which, in however simplified form, view the circulation of the atmosphere and the oceans as a whole and, by a series of suitably formulated equations, compute the fluxes and in some cases the patterns whereby the heat, moisture and momentum exchanges should take place under various assumed conditions of albedo, equator–pole temperature difference, etc.[1] Such studies may be used to explore – indeed lend themselves to exploring – the probable effects of alteration of one or more of the environmental influences and hence predict the circulation and climatic regimes that might result, for instance, from a change of the solar constant or from Man's activities (deliberate or otherwise) affecting the transparency of the atmosphere.

---

1. U. RADOK has expressed the view (personal communication, September 1975) that our knowledge of the past record of climate can only be improved to a very limited extent and that therefore most past events, including the twentieth century warming, can never be adequately explained. Hence, it is argued, there should be greater emphasis on a completely new record for the future based on climatic 'indices' which have only lately become observable; processes and causes should be sought in these new observation series and in theoretical models. The analysis of past events, he admits, remains useful for showing what can happen and how often it did happen. This is an extreme standpoint which does less than justice to the reconstructions of past temperature distributions, atmospheric and ocean circulations – notably those of the last ice age maximum illustrated elsewhere in this volume – and is surely over-bold as regards the possibility of detecting and accurately modelling long-term processes on the basis of a few years of modern observation coverage.

A detailed survey of this ever-widening field of research (as more and more theoretical formulations are developed and tried out) would be beyond the scope of this book; but a few general remarks will help to show the relationship of this scientific method to the matters with which we are concerned. In these comments the author relies to a great extent upon judgements and opinions published by ROBINSON (1970), SELLERS (1969) and LORENZ (1968, 1969, 1970), whose involvement with the developments concerned has been much closer than his own.

There are two different kinds of mathematical model used to simulate the climatic problem:

(1) Comprehensive dynamical models developed for numerical daily weather forecasting from the instantaneous state of the atmosphere, notably the successive schemes developed by SMAGORINSKY (e.g. 1963, see also MYAKODA et al. 1969) and by BUSHBY and TIMPSON (1967).

(2) Climatic-regime models which proceed from specification of the *average* state of the boundary conditions imposed on the atmosphere and the fluxes within it to study how any given balance is brought about, following the line pioneered by DEFANT (1921) and ÅNGSTRÖM (1935) – see Volume I, Chapter 3 – treating the atmospheric circulation as a turbulent exchange, or *Austausch*, mechanism. Without going into details of the circulation pattern, the solutions of their equations indicate such items as variations of the distribution of mean temperature by latitude associated with circulation strength.

Models of type (1) proceed from the instantaneous state of the atmosphere, expressed in the equations of motion, and for momentum, heat and moisture transfer, etc., to develop its future state from one to a few days ahead. The initial conditions are specified using observations at a network of points of the order of 300 km apart (which means that disturbances less than perhaps 500 km across, and their effects are ignored). The integrations are done by finite difference methods, using time steps of 10 to 30 minutes. The resulting forecasts have proved useful to about 5 days ahead, and there is some skill beyond this which holds promise for a limited further extension of the forecasts in time. The results still look like real weather patterns much farther ahead than 5 days without, however, showing useful resemblance to the particular developments in any actual case. It must be presumed that at over 5–10 days, and still more at 30–50 days and longer, ahead the effects of items neglected in the equations, and the growth of systems which were at the outset small enough to slip through the meshes of the observation net, become important.[1] Moreover, although the

1. Experiments by LORENZ (1969) with the effects of introducing observational errors into the initial conditions have indicated that errors of various sizes double in the first 3 to 8 days of the computations. It is, of course, the largest-scale systems (which constitute the framework or steering pattern of the following weather development over the hemisphere) which are least affected by the growth of errors arising from observational errors and initially small systems. The computed behaviour of the very large systems, such as planetary waves in the westerlies, may therefore have some relation to reality beyond the time at which the individual weather systems cease to be reliably predicted.

computational programme is already one of the most complex (and costly) so far ever carried out, the interests of simplicity have so far demanded that moisture be treated as precipitated as soon as it is condensed: this is a serious shortcoming where the object is to go on to study the development of climate and climatic changes, since cloud cover has a great effect upon the albedo and therefore on the amount and distribution of the heating of the Earth. Similarly, sea temperatures and sea ice are incorporated only as climatological average values and therefore insensitive to climatic change. The computing capacity required to continue the integrations to represent the circulation and climatic distribution after several hundred years is far beyond our present or foreseeable resources. Efforts are nevertheless proceeding to combine the advanced atmospheric models with models of the ocean because of the known importance (see Volume 1, Chapter 10) of ocean–atmosphere interactions. MANABE and BRYAN (1969) in one of the first such comprehensive attempts (which required 1200 hours computing time) geared the ocean model to the atmosphere in such a way that the processes of 100 years in the ocean were made equivalent to one atmospheric year.

LORENZ has expressed concern about another aspect of this approach, which he fears may defeat our efforts at understanding. The set of equations which represents the behaviour of the atmosphere has many solutions, each representing a possible condition of the atmosphere: climate can be seen – can be mathematically defined – as the statistics of occurrence of these various conditions. It is possible, LORENZ suggests, that different sets of solutions exist almost independently of the initial conditions. In this case, long-term climatic changes might occur (as it were, inconsequentially) without any change of the environmental conditions. The latter presumably do affect the atmosphere; indeed this is implicit in the mathematical formulation, but the effects would become very difficult to establish: an environmental change (such as the addition of carbon dioxide) which ought to cause a rise of temperature by one or two degrees might happen just when the atmosphere was spontaneously going over to a regime which brought a fall of temperature (due to this suggested characteristic of the system of equations, being what is mathematically called 'intransitive' – or 'almost-intransitive' according to a definition given by LORENZ). This pessimism seems at present, however, premature and attributable to its authors' excessive (or sole) reliance on the mathematical model method of attack. If by pursuit of the methods of observational science, we can establish, and assess, enough cases of actual climatic change under the influence of a given event (e.g. solar disturbance, volcanic dust, absence or diminution of the polar ice) in the past, or enough detailed observation of the working of carbon dioxide and other pollutants at the present day, it may still be possible to reach a confident diagnosis and establish the magnitude of the effect.

SELLERS (1969) has proposed an elegantly simple global climatic model of type (2) and explored its implications. Of these the most instructive and convincing are qualitative: revelation of the complex web of global interrelationships that has to be expected when any change in the external controls is imposed. Thus, one of his conclusions is that manipulation of the albedo (e.g. by blackening the ice with soot) south of 70°s should ultimately produce a larger rise of temperature in the Arctic than a similar action directly upon the ice north of 70°N. He also finds that 'a relatively small change in the ability of the atmosphere and oceans to transfer heat poleward could conceivably offset any increase or decrease in the intensity of solar radiation'. SELLERS himself, however, also carefully reviewed the conclusions derived by a number of other workers with global models falling within the same category. These have produced results so different that removal or disappearance of the Arctic ice might be expected to raise the surface temperature in the central Arctic by 30–35°C in winter and 5°C in summer (BUDYKO 1966) or by various other figures ranging from these ones down to his own figure of just 7°C averaged over the year. Some models have suggested that the atmospheric circulation should be intensified when there is no ice covering the Arctic Ocean (and reducing the amount of water vapour put into the atmosphere), others that it should be, on balance, weakened (by the reduced equator–pole thermal contrast).

One must conclude that knowledge and understanding of climatic changes depends, and will continue to depend, greatly on observation. Advance will only be achieved by keeping theoretical constructions closely constrained by observation of what actually occurs and has occurred in the past. Just as we have shown in the first part of this section how well-established analysis of atmospheric processes may be used to complete the meteorological analysis of a skeleton network of field data on this or that climate of the past, the theoretician will, it appears achieve his most reliable interpretations – and eventual advances of theory – by operating within the closest framework of observational data that field studies can establish. And these interpretations must be critically examined for their ability to explain all known aspects and any new facts that come to light.

Numerical models may also give guidance in choosing between various far-flung hypotheses, put forward on the basis of more or less gifted flashes of physical insight, to explain our scanty knowledge and inadequate observation data on some past climatic eras and various proposed causes of specific climatic changes (see also Chapter 20, p. 686).

Reconstruction of the temperatures, the downput of rain and snow, the wind regimes, and so on, that prevailed in bygone climatic eras is an exercise in the assembly and judicious interpretation of data derived from categories of evidence perhaps more miscellaneous than any other scientific problem is concerned with, and contributed by many fields of learning, as shown by the remaining sections of this chapter. All the numerical values of climatic elements which we can use, or derive, for past eras are but estimates, to which it may or may not be possible to assign margins of probable error. Certainty and precision, and many kinds of detail, are not attainable. Nevertheless, the gross features of the climatic sequence, and of each of the main climatic regimes discerned within it, can be stated with some confidence insofar as:

(i) Values of the climatic elements deduced from non-meteorological evidence are soundly based on (a) understanding of the complex of factors that govern the observed biological, geomorphological phenomena and other field observations, and (b) real climatic thresholds affecting the biological and other occurrences registered in the field data.

(ii) The reconstructed climatic distributions, including the atmospheric and ocean circulation modes prevailing, depend straightforwardly on known atmospheric and ocean behaviour and the physical and dynamical laws that determine it.

Corroboration should be sought by independent use of different types of data and different approaches. The circulation patterns reconstructed from data on one climatic element (so far usually temperature) should be expected to explain the apparent distribution in the same climatic era of other elements (precipitation, storminess, cloud cover, etc.) and perhaps to predict the occurrence in the era concerned of phenomena not previously recognized within it.

13.2 Dating methods

For information on past weather and climate to be useful, in particular if there is to be any possibility of interpreting its occurrence in relation to the atmospheric processes and external influences ruling at the same time, it must be securely dated.

Calendars and calendar usages in manuscripts and inscriptions

Astronomical measurements give the period of the Earth's passage round the Sun as 365 days, 5 hours, 49 minutes and 46 seconds (365·2422 days).[1] The modern Christian calendar, called the *Gregorian calendar* since its enactment by a decree of Pope Gregory XIII in March 1582, adopts a 365-day year except in leap years, which have 366 days by the addition of a 29th day to February; every fourth year is a leap year except the century years, which are only leap years when the number of hundreds is divisible by 4 (i.e. 1900 was not a leap year, but 2000 will be). It is suggested that all thousand years except those divisible by 4 should be leap years; if this is adopted, this calendar will keep constant phase to within one day in 20 000 years.

Confusion can arise in the use of many old-style manuscripts in which the months January and February were sometimes ascribed to the previous year or the months were numbered counting from March as the first month. This practice seems to have been very gradually abandoned and certainty can only be obtained from internal checking within each document. Others wrote 'January 1739–40', meaning January 1740. This has been a more important source of confusion in using medieval chronicles, owing to the diversity of

1. These units are, of course, 'mean solar days', marking rotation of the Earth until the same meridian
faces the sun (not rotation relative to a fixed point in
space or distant star, the 'sidereal day', which takes
about 23 hours and 56 minutes).

practices that obtained as regards reckoning when the year began. Countries, towns, courts, monasteries, the Vatican, each had their own style and New Year's Day was variously taken as Christmas Day, 1 January, 1 March, 25 March (Lady Day), and changes of practice were made quite frequently.[1] Hence, inevitably errors of one year have crept into the dating of weather events, and in some cases the same event has been ascribed by later copyists to two successive years. Much, or most, of this confusion can be eliminated by mapping the data.

The Gregorian or 'New Style' calendar was adopted in Italy, France, Spain and Portugal in 1582; in the German Catholic states, Flanders, Austria and the then Spanish Netherlands from 1 January 1583; in Poland in 1586, Hungary in 1587, the German Protestant states and the free Netherlands in 1700; Sweden changed gradually by omitting 11 leap days between 1700 and 1740; Britain and her then American colonies and Ireland changed in 1752, the day after 2 September Old Style being called 14 September New Style; Japan adopted this calendar in 1872, China in 1912, Bulgaria in 1915, Turkey and Soviet Russia in 1917, Jugoslavia and Rumania in 1919, and Greece in February 1923.

Dates found in documents from many countries, whether in Europe in earlier times or in modern Japan, basically using the Gregorian or Julian (see below) calendars are unfortunately liable to be in the more obscure form of the year of some reign or dynasty.

The corrections to be applied to convert dates on the Old Style calendar to the Gregorian (or New Style) are as follows:

After 29 Feb. 1900 O.S.	+13 days	29 Feb. 1100–28 Feb. 1300	+7 days
29 Feb. 1800–28 Feb. 1900	+12 days	29 Feb. 1000–28 Feb. 1100	+6 days
29 Feb. 1700–28 Feb. 1800	+11 days	29 Feb. 900–28 Feb. 1000	+5 days
29 Feb. 1500–28 Feb. 1700	+10 days	29 Feb. 700–28 Feb. 900	+4 days
29 Feb. 1400–28 Feb. 1500	+9 days	29 Feb. 600–28 Feb. 700	+3 days
29 Feb. 1300–28 Feb. 1400	+8 days	29 Feb. 500–28 Feb. 600	+2 days

This Old Style Christian calendar was introduceed in Italy in the sixth century and is believed to have reached England in A.D. 596 with St Augustine, but was not ordered (by the bishops) to be used generally until 816.

The Roman *Julian calendar* was the same as the Christian Old Style one except as regards the numbering of the years and the demarcation of the months in the usage of the Latin language. It was adopted by decree of Julius Caesar from January 46 B.C., the month of July previously counted as the fifth month (Quintilis) being then renamed after him. February had at first 29 days normally and 30 in every fourth year, both figures being reduced by one in the year 8 B.C. when Augustus Caesar added this day to the month then renamed after him. The Julian calendar was then in the shape in which it continued for over fifteen centuries. But misunderstandings of the instructions mean that dates in the early centuries of its history are not reliable to within a day or two.

1. See REGINALD L. POOLE, *Studies in Chronology and History* (Oxford, 1934), Chapter I, 'The Beginning of the Year in the Middle Ages'.

Before Julius Caesar the Roman year had 12 months with alternatively 29 and 30 days plus one extra day (i.e. 355 days) and an extra month of 22 or 23 days added (but not with systematic regularity) at the discretion of the government. Its correspondence with the seasons wandered somewhat accordingly, and to rectify this the year 45 B.C. was given 445 days as an exceptional measure in the process of bringing in the Julian calendar. So before 45 B.C. Roman dates may be misleading (or obscure) to the extent of a sizeable fraction of a year.

In general, however, the chronology of events in ancient Roman and Greek history has been established by cross-checks available with Egyptian and Mesopotamian chronology, the latter including Alexander the Great, and with the astronomical observations recorded, including eclipses, which must tally with astronomical calculation back to those times.

Many early calendars took as a basic unit the lunar month (about 29·5 days), the recurrences of new moon and full moon being the easiest way of keeping track of the progress of the year in readily followed steps. The months were commonly taken as 29 and 30 days alternately. Unfortunately, the year of the Earth's orbital passage about the sun is not perfectly divisible in this way: about 11 days have to be added if the seasons are to be kept at constant dates. The Jewish and Muslim calendars ignore this, and undergo a complete cycle of phase change with respect to the seasons in every 34 of their (lunar) years, which approximate to 33 solar years. The earliest Roman calendar and the calendars used in the Greek states in Classical times, i.e. after about 750 B.C., seem to have been kept in reasonably constant phase with the seasons by *ad hoc* intercalations of an extra month before the beginning of spring, and were presumably seldom in error by more than one to two months and never by as much as a year.

The most consistently maintained ancient calendar appears to have been the Egyptian civil calendar, which had three 4-month seasons making up 12 months of 30 days each and 5 days added at the end of the year to make a 365-day year, which was intended to begin with the Season of Inundations (EDWARDS 1970). Because no leap years were introduced, this calendar became 97 days out in 400 years and just one year out (i.e. had one year too many in about 1500 (actually 1508) years of the true solar calendar (the so-called 'Sothic cycle'). The Egyptians were aware of the error but did not correct it. Instead, the date of the heliacal rising of 'Sothus', meaning the date at which Sirius (or Sothus) the Dog Star, prominent in the desert sky at a declination about 17°s, rose exactly with the sun and therefore became invisible throughout its approximately 11 hours above the horizon at 30°N – or, more practically, the first day on which Sirius could just be seen on the eastern horizon before sunrise – served as a (nearly) fixed point, or marker, of the true solar year. It occurred at a date conveniently preceding the seasonal flood of the Nile.[1]

The beginning of the Egyptian civil year coincided with the heliacal rising of Sirius in A.D. 139, according to the report of a Latin writer (Censorinus) in the following century; it

1. Actually the (true) date of this heliacal rising is affected by the rotation of the Earth's orbit in space and the precession of the equinoxes and is now about two months later (August) than it was 4000–5000 years ago.

follows from the Sothic cycles that the same ideal coincidence of the heliacal rising of Sirius with the beginning of the civil year must also have occurred about 1500, 3000 and 4500 years earlier, the earliest of these being perhaps the period when this Egyptian calendar was instituted. From this knowledge the dates in Egyptian papyrus records and inscriptions on stone can be fixed, back to the reigns of kings and the heights of the Nile flood and other events given in a yearly table on the Palermo Stone, an inscribed stone from about 2400 B.C. The events recorded, including the details on the Nile floods, are normally given according to the year of the king's reign. Confirmatory evidence from a papyrus fragment now in Turin indicates a span of 955 years for the first eight dynasties, suggesting that the first Egyptian dynasty began in 3119 B.C. Further checks on the dates in Egyptian records can be obtained from the observations of the new moon used for the contemporary religious calendars, the date of new moon returning to the same date of the same Egyptian civil month after a 25-year cycle. If the wrong one of these cycles is chosen a dating error of 25, 50 or 75, . . . Egyptian civil years could be the consequence. Errors may mar the ancient record itself when poor visibility affected observation of the tip of the crescent of the waning moon, giving the wrong date for new moon: this would indicate a wrong year within the 25-year lunar cycle. Taking account of these sources of error, and on the other hand of the degree of corroboration shown where duplicate evidence exists, it appears that errors in the modern interpretation of the ancient Egyptian chronology are very small, in general probably less than 10 years back to 1800 B.C. and no more than about 100 years back to 3119 B.C. Hence, Egyptian chronology can be useful in exploring the errors of other methods of dating, including those of radiocarbon dating.[1]

Babylonian dates from a surviving list of kings and their reigns copied by Ptolemy of Alexandria (SACHS 1970), with some fixes given by the recorded dates of eclipses, can be regarded as certain back to 650, and perhaps to 750 B.C. A list of the earlier Assyrian kings ruling in Mesopotamia, with cross-checks obtained from the correspondence they exchanged with Egyptian rulers, continues this chronology back to about 1350 B.C. with, it is thought, a maximum error of 20 years. Before this date our knowledge of the kings in that area may be incomplete. The date of the end of the first dynasty of Babylon, tentatively regarded as 1594 B.C., appears liable to an error of ±50–60 years, or possibly 100–120 years, associated with uncertainty about how the abundant observations made of Venus as a morning star (which performs approximately 56-year cycles) should be aligned with the list of kings known to us. Babylonian dates from a 500-years-long but unanchored 'floating chronology', a list of earlier kings back to the first dynastic period there, in the third millennium B.C., though subject to a similar range of uncertainty, appear to confirm a greater error than this in radiocarbon dating about that time.

Ancient Chinese calendars, from the Shang dynasty, c. 1766–1122 B.C., or earlier, were reckoned in cycles of 60 days, subdivided into 10-day periods, and combined into longer

---

1. More about the reliability of ancient Egyptian dates and their value for checking other dating systems can be found in C. RENFREW, *Before Civilization* (London, Jonathan Cape, 1973).

cycles of 12 solar years, which were fixed by observation of Jupiter (the length of Jupiter's year is 11·86 Earth years). It should therefore be possible to correct recorded dates by the astronomical observations. It became known that the Earth's year was really close to 365 days, but at least as late as the 'Time of Warring States', around 247 B.C., a 360-day year was still in general use, divided into 15-day 'solar-terms' and seventy-two 5-day periods (pentads), each named by the seasons of nature ('Awakening Insects', for instance, being in early March and five days later 'The Oriole sings').[1] The solstices and equinoxes had been determined and were presumably used to correct the calendar. Chinese chronology, reckoned in cycles of 60 years in the earliest times, is usually taken as reliable back to 2357 B.C. Details of the individual reigns of the Emperors are known from 163 B.C.

Hindu chronology is counted from the death of Buddha, supposedly in 483 B.C.; but this datum, from which ancient Indian documents are reckoned, is subject to an uncertainty of about 60 years, due to discrepancies in different sources, and may have been as early as 544 B.C.

Dendrochronology

Yearly growth rings are prominent in the stems of trees that grow in the seasonal climates of middle and higher latitudes. The newest cells, and therefore the youngest ring, are those next the bark. Each tree ring is formed by the succession from larger cells and softer wood produced in spring to smaller cells and harder, denser wood produced in summer and autumn. Microscopic examination of sections from sufficiently sensitive types of tree, particularly if supplemented by X-ray photography for subsequent measurements of differences of optical density, which can be calibrated against physical density of the wood (FLETCHER and HUGHES 1970), can reveal even some detail of the succession of periods more and less favourable for growth within individual seasons.

A climatic event such as an extreme season of some kind or a group of extreme seasons, which can be identified with a year or years of anomalous tree rings, can be dated in favourable circumstances to the exact year. More usually there is some margin of uncertainty due to occasional years in the tree's life which produced either a very thin ring or, on the other hand, a double ring. When different trees in the same locality are examined, the thinnest rings produced by the worst years for growth may appear to be missing altogether in the sections from some trees. In areas where standard tree ring chronologies presenting the

1. These names of the pentads mark a fascinating stage in Man's (in this case the generality of country folk's) observation of the natural world, and they could probably be taken as the average dates at some period in the first millennium B.C. of the phenological events referred to. This might yield information about how the seasonal climate of that time compared with today's. For this purpose, however, one would need to know within rather narrow limits which part of China – a country with a strong temperature gradient from south to north – the names originated in. This possibility is now, however, prejudiced by the primitive misinterpretations of some phenomena, such as bird migration, revealed in another of their pentad names later in March, 'The Falcon Turns into the Dove', the dove being in China a seasonal migrant, disappearing in autumn and returning in spring (CHU KO-CHEN 1961).

prevailing pattern found in many trees have been established it is claimed that the unresolvable uncertainty of dating of any feature rarely amounts to more than a year or two, even when the chronology extends back over several thousand years. Recently, LA MARCHE and HARLAN (1973) have built up a second tree ring chronology in the White Mountains of California, using different samples of wood taken from the same kinds of trees as the master chronology in the same area, and report that over the 5500 years covered by both chronologies the dating agrees with high precision, demonstrating tree ring work as the most accurate method of dating beyond the range of actual written records and in the best cases capable of dating to the exact year. Certainly, features of the ring sequence of a few decades

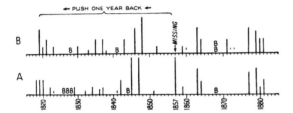

Fig. 13.8   Tree ring plots as used originally by A. E. DOUGLASS and in the University of Arizona Tree Ring Research Laboratory at Tucson.
(*a*) Rings of an Arizona pine A.D. 1815–85.
(*b*) Rings of another tree in the same locality with the 1857 ring missing. (In tree (*a*) the year 1857 is represented by a very narrow ring.
In the plot normal rings are not marked (though they are, of course, counted). Vertical lines indicate narrow rings the longer the line, the narrower the ring. A letter *B* is entered to indicate an unusually broad ring.
  (*This diagram, which follows* GLOCK (*1937*), *was originally published by* F. E. ZEUNER *in* Dating the Past (*London, Methuen, 4th edn, 1958*), *and is reproduced here by courtesy of the author and the publishers.*)

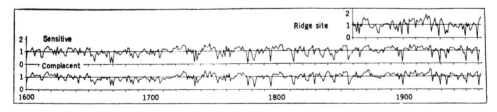

Fig. 13.9   Ring width chronologies representing averages of a number of trees: sixteen trees at an exposed ridge site and nine trees each representing sensitive and complacent sets.
  (*Diagram from* FERGUSON (*1968*) *reproduced by kind permission.*)

spanning a climatic change or some other drastic event could be studied in detail without risk of errors amounting to more than a very few years.

Trees in ever-moist tropical climates with no seasons generally show no growth rings. (This is one of the indications that the trees of the Carboniferous geological era about 300 million years ago which produced the coal measures in central and northwest Europe grew in a moist tropical climate.) H. T. MÖRTH reports (unpublished personal communication 10 February 1972) that trees growing today near the 2000 m level in the Kenya highlands do show growth rings, but the rings are not annual: they are determined by a thin dark ring, only about three cells thick, which is formed whenever, and as often as, a dry spell occurs, with less than 40 mm rain in a month. At such times the growth of these trees ceases: growth is resumed and produces light-coloured wood as soon as the monthly rainfall once more exceeds this. Trees in Java also show growth rings which appear to be annual and are presumably related to differences of growth in the wetter and drier halves of the year.

How a master tree ring chronology is built up has been described by FERGUSON (1968) and is illustrated here by figs. 13.8 and 13.9 and the photographs in Plate I. (The implications of such records as regards year-by-year differences of weather are discussed in a later section. Here we are only concerned with their value for dating the events they register.) Trees of some species and in some sites are more sensitive than others. The most sensitive records come from exposed sites and old, hardwood trees. If the trees are too sensitive, there are many missing rings and double rings where the growth of one season was interrupted, e.g. by some climatic shock or an attack of caterpillars: either peculiarity may introduce an error of one year each time it occurs. (The extra ring, always a narrow ring, formed when growth is resumed later in the same season is called a 'false ring'.) Difficult rings are commoner in hardwood than in softwood trees. On the other hand, if the trees are too complacent (see Plate I) – i.e. show too little variation in ring width from year to year, as is particularly liable to occur in the interior of a forest – the best fit which aligns together the corresponding rings in different trees affected by the same external events is hard to establish (possibly indeterminate) and considerable errors of dating may result. The best method therefore is to use many trees and rather sensitive sites to establish a master chronology for an area small enough to ensure a common experience of environmental stimuli and hazards.

Maximum response to climatic fluctuations is found at the limits of the forest (LAMARCHE and HARLAN 1973). Fallen logs, and timbers taken from buildings, can sometimes be used to extend a chronology by 'cross-dating' their ring series with the older portion of the ring record of trees still living in the area.

The longest tree-ring chronologies in the world have been established for the very long-living trees, the Giant Redwoods (*Sequoia*) and (above all) Bristlecone pine (*Pinus aristata*) which lives to over 4000 years, in the western U.S.A. In the case of bristlecone pine in the White Mountains of east central California the chronology of year rings extends back over 7000 years to before 5000 B.C., based on two or more trees at all times since 4500 B.C. And in these best tree ring series repetition tests showed no error back at least as far as 3535 B.C.

(LA MARCHE and HARLAN 1973). There is also a 'floating chronology' of nearly 500 years of rings in dead specimen indicated by radiocarbon tests as about 9000 years old, which may some day be fixed by cross-dating if specimens are found which bridge the gap to the beginning of the existing continuous ring series of the master chronology.

Other parts of the world for which long tree ring chronologies have been established (see also references in BANNISTER (1963)) are:

Alaska, from *c.* 1000 A.D.
  Based on spruce trees (*Picea*) (FRITTS 1965, GIDDINGS 1941).
Mackenzie River, northwest Canada, from *c.* 1500 A.D.
  Based on spruce trees (*Picea*) (FRITTS 1965).
British Columbia (Yoho National Park), from 1581 A.D.
  Based on *Picea engelmannii* spruce (BRAY and STRUIK 1963).
Finnish Lapland, from 1181 A.D.
  Based on pines (SIRÉN 1961).
Northern Urals, from *c.* 1700 A.D.
  Based on conifers (ADAMENKO 1963).
Novgorod, western Russia, 880–1462 A.D.
  Based on pine logs (THOMPSON 1967).
Samarkand, from 14th century A.D.
  Based on savin beams (RUDAKOV 1968).
Japan, from *c.* 1630 A.D.
  Based on keyaki (*Zelkova serrata*) (OUTI 1964).
Taiwan (Formosa), Taipingshan highlands, from *c.* 1640 A.D.
  Based on Formosan cypress (OUTI 1964).
England, *c.* 160 B.C.–320 A.D. and *c.* 850–1500 A.D.
  Based on oaks (both periods floating chronologies) (SCHOVE and LOWTHER 1957).
Germany, a floating series in the Bronze Age and various complete chronologies from between 500 and 900 A.D. to date. Based on oaks, beeches, larches, pines and spruce (HUBER and HOLDHEIDE 1942, HUBER 1960, HUBER and SIEBENLIST 1963, BREHME 1951, BECKER and SIEBENLIST 1970).
New Zealand, Fjordland, from *c.* 1300 A.D.
  Based on the southern conifer Rimu (*Dacrydium cupressinum*) (WARDLE 1963).

Growth of lichens

A specual method, to be known as lichenometry, has been used for dating the retreat of glaciers and disappearance of perennial snowbeds. Lichen spores reach the newly exposed rock faces, and some species (e.g. *Rhizocarpon geographicum*) are thought to have nearly constant growth rates after an initial flush. Growth is slow and continues over a very long time. An estimate of the time elapsed since the rock became available for colonization by the lichens can be obtained by measuring the size of the largest lichen present. Errors may arise from growth rate changes, if the subsequent climate varied widely enough. Applying this method to lichens at different heights above sea level registering the rise of the snow line since the seventeenth century in southwest Greenland, PITMAN (1973) has derived an interesting temperature curve: mean annual air temperature there around 1950 was $+0 \cdot 1°$C, having apparently increased from a value about $-2°$C in 1860–80 and $-2 \cdot 2°$C between about 1815 and 1837. Higher values, about $-2°$C seem to have prevailed in the middle or earlier part of the eighteenth century, but a value about $-2 \cdot 6°$C was indicated for the time around 1680 and $-3 \cdot 2°$C around 1660. The 95% confidence limits attached to the method indicate, however, that that is about the limit of what can be derived: before 1700 the temperature can hardly be indicated within a most probable range of 1°C and the date also within a wide range. Use of the same method has indicated that there were probably permanent snowbeds in the Cairngorm Mountains in Scotland in the seventeenth and eighteenth centuries.[1]

Varves in lake beds and some other sediments

Mud and clay sediments formed at the bottom of lakes (found also where former lakes have dried up and been buried by a later soil and vegetation), and similarly formed on some parts of the sea bed near land, commonly show yearly layers, called 'varves' (Plate II). These are produced by regularly alternating sequences of colour, grain size, and mineral and organic composition, due to seasonal differences in the relative amounts (and kinds) of organic material and wind- and waterborne mineral particles deposited. These differences are particularly pronounced where the streams feeding a lake are frozen for part of the year and the seasonal melt is followed by a period of rapid or torrential flow. Wherever a thicker or thinner varve is attributable to some climatic event, this can be dated in the chronology represented by a yearly varve series, much as with tree rings.[2] Care must be exercised not to be misled by using parts of the deposit that have been altered by slumping and other movements or by stream scouring.

As a method of dating, the study of varve series was pioneered by DE GEER (1912, 1940 – see also 1954, 1957) and used as a first means of dating the retreat of the edge of the great Scandinavian ice sheet across northern Europe at the end of the last glaciation. The varve sequence shown by the laminated deposits that formed in a succession of the lakes always forming at the edge of the melting ice, which have since disappeared, and none of which

---

1. Today the position is much more marginal. A few snowbeds survive here or there through most summers, but their positions vary and depend on the depth of snow accumulated in winds from this or that direction during the preceding winter or over several winters.

2. Not all varve series represent an annual cycle of deposition. Some in small lakes are produced by much shorter time scale variations in the material deposited. But the conditions producing varve formation have been studied, and at least in *large* lakes it seems that the yearly rhythm predominates (TAUBER 1970).

spanned the whole 10 000–15 000 years, provided the chronology through matching (and assumed teleconnections between) similar features in different former lake basins not too far apart. Most of the individual varves are from a few millimetres to a few centimetres in thickness, though some are thinner and some significantly thicker than this. OLAUSSON (1971) further reports that the lifetime of individual lakes in Sweden and Finland between formation and silting up and becoming choked by bog plants is mostly between 5000 and 15 000 years, figures which he believes should also apply in the other interglacial periods in the Pleistocene. As with tree rings, errors may arise in a chronology based on varve counting, through occasional missing varves in poor years for deposition, occasional double (or even triple) varves in other years and mistakes in cross-correlating varve series in different lake basins. The liability to error is reduced by comparing the varve profiles in many lake basins. The procedure left the Swedish chronology open to suggestions that DE GEER's dates for the earliest (glacial) varves measured were some 2400 years in error (NILSSON 1964, 1968a). But the whole series has been reinvestigated in recent years, e.g. by JÄRNEFORS (1963) and NILSSON (1968b). The results supported DE GEER's count of late-glacial varves to within 19 years in one place and 31 years in another in central Sweden. The zero point of DE GEER's counting was not, however, in modern times but was an exceptionally thick varve, nearly 1 m thick, which he believed to represent the draining of a great ice-dammed lake at Ragunda, Jämtland, in northern central Sweden, supposedly when the dwindling Scandinavian ice sheet parted into two, which he took as the effective end of the ice age in northern Europe. The link from this event almost 9000 years ago to modern times was provided by a series of varves laid down in the delta of the Ångerman river near 63°N, which continually extended eastwards as the Baltic receded with the rapid rate of land upheaval ('isostatic recovery' from the former ice load) in that region (LIDÉN 1913, 1938). This series of river varves continues as a sequence right through the postglacial era to a date estimated by LIDÉN from the mean rate of land uplift as A.D. 920.[1] This estimate is the most tenuous part of the whole reasoning by which the Swedish varve chronology is fixed, and it now seems probable that the last of the recognizable Ångerman river varve sequences should be dated a little earlier.

FROMM (1970) neatly summarizes the position by saying that the sources of error in the Swedish varve chronology can be divided into three groups:

(1) Errors in the actual varve measurements or in the correlations between varve series in different places. These may amount to ±50 to ±100 years (or about ±1% at most) in the latest versions of the chronology.

(2) Missing varves and twin varves in one year. Errors due to these confusions could be up to about ±1%.

(3) Errors in bridging the varve sequence to the present time, estimated as between 0 and +300 years and most likely about 200 years.

---

1. The earliest 500 years of this varve sequence along the Ångerman river could be checked in the neighbouring valley of the Inndalsälv 50 to 70 km south-west.

Items (1) and (2) probably affect most other varve sequences in other parts of the world more or less similarly. Outer limits of the probable real ages of the Swedish varve datings in the works of LIDÉN, BORELL and OFFERBERG, JÄRNEFORS, and NILSSON suggested by FROMM are

> For 5000 B.C.: 5000 to 5560 B.C.
> For 6000 B.C.: 6000 to 6570 B.C.
> For 8000 B.C.: 7950 to 8625 B.C.
> For 10 000 B.C.: 9900 to 10 675 B.C.

TAUBER suggests that the overall range of error in the Swedish varve sequences be considered as 1 % for the postglacial part and 2–3 % for the late-glacial part. FROMM's article indicates where the Swedish varve measurements are to be found and how far they have been published.

By applying radiocarbon dating to materials dated on this and other varve chronologies, particularly materials dated at more than 5000 years ago, some insight has been gained into the time distribution of systematic errors in radiocarbon dating, which presumably stem from past variations in the natural occurrence of ^{14}C in the Earth's atmosphere (see especially TAUBER 1970, also STUIVER 1970 and pp. 60–8 herein).

DE GEER's picture of the manner of the latter stages of the decay of the Scandinavian ice can now be refined. It appears too complex to be described as bipartition of the ice sheet, of which there were doubtless various remnants for a time. The draining of the lake at Ragunda is now thought to have taken place in at least two stages, producing thick varves 80 years apart (BORELL and OFFERBERG 1955). The history of this and the various great lakes which preceded the opening up of the Baltic, the early flooding of the region with Atlantic water and formation of a series of salt water varves, can all be dated by the varve chronology. Few postglacial climatic events in the region have, however, produced marked, and therefore datable, variations in the varve sequences.

Other important varve chronologies include (a) the dating of corresponding events in Finland by SAURAMO (1923), (b) a lake in Taiwan (Formosa) studied by MIYADI (1939) and STUIVER (1970), and (c) the series of varves left by the retreat of the North American ice from Long Island, New York, to James Bay, northern Ontario, counted by ANTEVS (1947, 1953) but lacking a link to the present day, because assessment of the final retreat across Hudson's Bay to the present relics of the ice in Baffin Island involves vaguer methods of estimation and conjecture. Moreover, the cross-correlations of varve series between New England and Hudson's Bay are confused and uncertain, probably because (as TAUBER (1970) has suggested) the retreat of the long front of the great North American ice sheet would be interrupted here and there in an almost random manner by local halts and readvances, as the centre of greatest precipitation wandered from one longitude to another in runs of years affected by different meteorological situations.

A remarkable series of 9349 apparently yearly varves, presumably spanning the whole history of postglacial times, has lately been reported in a lake in Minnesota, U.S.A., called the Lake of the Clouds, 48°N, 91°7′W (W. A. WATTS at the April 1972 Cambridge Symposium on Quaternary Plant Ecology of the British Ecological Society). These varves consist of a seasonal alternation of pale-coloured iron carbonate and dark organic bands, but some are only a few hundredths of a millimetre thick and it may be impossible to guarantee completeness of the count.

A series of annual layers going back to 2294 B.C. in the mud deposits of a salty lake, Lake Saki, on the west coast of the Crimea (45°1′N 33°5′E) has also been measured (SCHOSTAKOWITSCH 1934) and appears to indicate rainfall variations in the area (see Appendix V, Tables 29(a–c), and fig. 16.24).

A sea bed core, described by OLAUSSON and OLSSON (1969) in the Gulf of Aden (11°9′N 44°3′E), shows yearly layering in millimetre-thick varves in the portion laid down during early postglacial times, when the rates of accumulation of wind-blown mineral grains and waterborne organic matter were apparently both greater than now; the absolute dating of the sequence depends, however, on radiocarbon measurements.

Stratigraphy of ice sheets

Recognition of yearly layers in an ice sheet is made possible by the similar sequences of grain size, hardness and density, normally culminating in an autumn layer which is hard and of high density overlying the coarser grained summer layer which may or may not show evidence of melt processes. The autumn layer is sometimes a single wind slab or it may be a succession of closely spaced wind crusts. It is formed as a result of the strong vertical gradient of temperature within the snow when the surface cools, causing upward transfer of vapour in the air trapped within the snow, leading to recrystallization at the surface; the upward transfer is hastened by the turbulence produced by winds blowing over the snow surface, which remains colder than the summer layer beneath (BENSON 1967). The amount of the annual accumulation is established by weighing. The strata remain recognizable, despite compression by the ice above, for many decades. Some year layers are difficult to identify, but identification may be helped by tracing them over some horizontal distance or in second and third pits. The layers could be counted back to the year 1760 at the South Pole in 1958 with an uncertainty of ± 13 years, at a depth of 26 metres. The year layers visually identified at the South Pole from 1760 to 1957 (GIOVINETTO and SCHWERDTFEGER 1966) and in northern Greenland from 1783 (LANGWAY 1967) establish histories of the accumulation of snow at these places and show secular changes.

Oxygen isotope measurements of the ice extracted from pits and borings indicate the regular seasonal variation of temperature and thereby make it possible to establish the yearly layers at much greater depths in the ice sheets, e.g. down to 411 m in northern Greenland indicating variations of the snow accumulation back to A.D. 934 (see pp. 76 and 93). There can be little confusion of the years within any sequence of a few decades identified by this means, but at this depth it is advisable to check the absolute age by some other method.

Radioactive isotope dating methods ('radiometric dating')

Physical methods of dating, which depend on measurements of the residual radioactivity in matter in which this activity has been decaying since some definable time of interest in the past, come next in precision after the methods we have so far discussed.

Radiocarbon dating

We have already paid some attention in Volume 1 to the significance of the slight variations in the amount of radioactive carbon in the atmosphere in the past, which are detected when radiocarbon dating is applied to objects of known age, as registering a history of variations of solar activity and/or of the intensity of circulation of the oceans. We shall here consider the subject as it affects the accuracy of radiocarbon dates.[1]

Atoms of the radioactive isotope of carbon ^{14}C are formed in the Earth's atmosphere as a result of the continual bombardment by cosmic rays, mostly protons from elsewhere in the Galaxy (those that constitute the corpuscles in the solar wind being of lower energy – see Volume 1, Chapter 10). The primary particles coming in vary widely in energy, but the peak frequency for particles from the Galaxy is in the range 10^8 to 10^{10} electron volts. Only those with energy exceeding 2×10^9 eV can pass through the whole atmosphere. Most are checked by collisions with atoms and molecules high in the atmosphere, splitting off free neutrons which are found most abundantly in the stratosphere. When these neutrons collide with atoms of nitrogen, the atmosphere's main constituent, the dominant reaction is one that produces an atom of radiocarbon: the atomic nucleus is changed by capturing the neutron and losing a proton (which appears as a free hydrogen atom):[2]

$$\tfrac{1}{0}n + {}^{14}_{7}\text{N} \rightarrow {}^{14}_{6}\text{C} + {}^{1}_{1}\text{H}.$$

About 1% of the collisions of neutrons with the atmospheric nitrogen atoms result in a different reaction, producing tritium (a radioactive isotope of hydrogen) instead and an atom of ordinary carbon:[3]

$$\tfrac{1}{0}n + {}^{14}_{7}\text{N} \rightarrow {}^{12}_{6}\text{C} + {}^{3}_{1}\text{H}.$$

The carbon atoms produced in these ways sooner or later become oxidized ('burnt') and from then on form part – a minute part – of the atmospheric carbon dioxide. Thus they become distributed throughout the atmosphere and oceans, entering into all the life

1. The possibilities of using radiocarbon to date the past were first worked out by LIBBY (1946).
2. In the chemical notation used in the formulae on this page the small numbers at the top give the weights of the atoms and those at the bottom the number of protons in the nucleus. (When atomic weights which are not whole numbers are quoted in the literature, this is due to the presence of different isotopes with differing atomic weights in most samples of it, their chemical properties being alike owing to their having the same number of protons in the nucleus.)

3. The whole tale of the bombardment of the atoms of atmospheric gases by cosmic rays includes the production of many radioisotopes: ^{3}H, ^{7}Be, ^{10}Be and ^{14}C by the action on atmospheric nitrogen and oxygen; ^{32}Si, ^{32}P, ^{33}P and ^{35}S from the argon in the atmosphere.

processes in which CO_2 is involved, and so are found throughout the biosphere and in the carbonates dissolved in the ocean and in river waters, etc., and deposited on their beds. The time required for the mixing through the different reservoirs of carbon represented by the atmosphere and the biosphere is evidently immaterial in relation to the average life of 8300 years before a ^{14}C atom disintegrates: for tests have repeatedly shown that the concentration of radiocarbon in living matter at any one time is all over the world closely the same.

A small proportion of all naturally occurring carbon consists of another isotope, ^{13}C, which is stable, i.e. not radioactive. The proportions of the different isotopes found in living wood today are $^{13}C/^{12}C \sim 1.52 \times 10^{-2}$, $^{14}C/^{12}C \sim 1.07 \times 10^{-2}$. Any fractionation (i.e. differential transmission of the isotopes) occurring when the carbon passes from one reservoir to another, or in the laboratory processes used in preparation of materials for dating measurements, must appear in a measurable anomaly of the ^{13}C concentration and the fractionation of the ^{14}C should be just double this. In practice, small corrections are found to be necessary and are applied on this basis.

The unstable isotope ^{14}C decays at a rate proportional to the number a of such atoms present at any time t, i.e.:

$$-\frac{da}{dt} = \lambda a$$

where λ is the constant of proportionality. Each ^{14}C atom breaks down into an atom of ^{14}N from which a β- particle and a neutrino split away. Hence, the number of disintegrations detected by a counter in the absence of other sources of α or β particles, or γ rays, is a measure of the amount of ^{14}C present.[1]

On average, cosmic rays produce two new radiocarbon atoms per cm^2 of the Earth's upper atmospheric surface per second, and there is increasing evidence (from the radioactivity found in meteorites and in moon rock) that the flux or galactic cosmic rays concerned has remained constant over periods much longer than the life of radiocarbon. Hence, the ^{14}C proportion of atmospheric CO_2 is continually replenished, as is that in all vegetable matter that breathes it in and all animal life that feeds on this. And so, throughout the biosphere, the same proportion of ^{14}C in living matter is maintained.

The basis of radiocarbon dating is that from the moment an organism dies communication with the atmosphere ceases and the decay of the ^{14}C in it is uncompensated. The decay follows an exponential curve, such that the number a_t of radiocarbon atoms still present in a sample at time t is related to the number a_0 originally present by the law

$$a_t = a_0 e^{-\lambda t}$$

1. α particles, streams of which are called α rays, are positively charged particles emitted from an atomic nucleus; they each consist of two protons and two neutrons, and on picking up two electrons from other matter to neutralize their charge they become ordinary atoms of helium. β particles, streams of which are called β rays, are free electrons emitted by disintegrating atoms.

whence the half-life (in German, Halbwertzeit) $T_{1/2}$, the time taken for a_t to fall to $\frac{1}{2}a_0$, i.e. $a_{T_{1/2}} = \frac{1}{2}a_0$, is given by $T_{1/2} = 1/\lambda \log_e 2$.

From measurements on ancient materials of known age the best estimate of the half-life of radiocarbon is now taken as 5730 ± 40 years (BERGER 1970, GODWIN 1962).[1] The half-life of tritium, being only 12·3 years, makes this product of the reactions with atmospheric nitrogen of no interest in connection with dating events in climatic history (though it is of interest in tracing atmospheric motions following the injections of artifically produced ^{14}C and ^3H resulting from nuclear bomb tests in recent years).

The experimental difficulties in obtaining meaningful and sufficiently sensitive measurements of ^{14}C for dating are formidable (LIBBY 1970, WILLIS 1961) because of the abundance of other radiation, some of it high energy particles, that has to be excluded from the counter. The radiation from the disintegrating ^{14}C atoms in the carbon dioxide gas carefully prepared from the sample of material to be dated consists of 'soft', low energy particles. Modern vegetation samples can be expected to give about 70 disintegrations per minute; samples of matter that lived 5730 years ago should therefore give 35 dpm, matter that is 11 500 years old about 17·5 dpm and so on. The measurements are made on the carbon dioxide in the counter, itself built of radiation-clean materials, inside a protective hut built of lead bricks and lined with iron plates designed to cut out the radioactivity in the walls and floor of the laboratory. Yet even this hut is penetrated by the highly energetic cosmic rays (cosmic ray muons) which are capable of penetrating many metres thickness of rock, and the background count due to these amounts to 800 per minute. These are prevented from interfering with the measurement by a ring of counters surrounding the central counter, containing the carbon dioxide from the sample, which they switch off every time they register a cosmic ray; it remains switched off only for a fraction of a millisecond and is operative for about 99 % of the time, and the count from the sample can be adjusted accordingly. A fairly constant rate of 13 counts per minute from background radiation of unknown origin gets through all these precautions. Counting is carried on for, say, 1000 minutes (16 hours, 40 minutes), sometimes longer when materials of great age are being studied.

Laboratories issue instructions to guide those gathering material to be dated, particularly as regards the avoidance of contamination by carbonaceous materials of quite different age before, during or after collection (clean polythene bags securely sealed are recommended) and the quantity required: 5 grammes of actual carbon are regarded as the minimum (corresponding to say 40 g of charcoal, 75 g of wood, 100–150 g of peat, 300 g of shell or 500–1000 g of bone).

Since the average life of a ^{14}C atom is 8300 years, a 1 % error in the count means an error

---

1. 5730 ± 40 years has since 1962 been accepted as the best estimate of the half-life of radiocarbon; but many dates have been explicitly calculated in terms of the figure earlier adopted, namely 5568 years, and about 3 % should be added to radiocarbon ages using the earlier figure to make them comparable. Further revisions may yet be proposed, however, since one carefully conducted more recent measurement has suggested a half-life of 5833 ± 127 years (quoted by BERGER 1970).

of 83 years. Because the decay is exponential, counts attributable to radiocarbon are detectable as an addition to the background with materials up to 50 000 to 70 000 years old at most; but the effects of contamination of the sample by older or younger matter introduce changes of count that give increasingly serious errors with increasing age: for a 0·1 % contamination the error is about 300 years at 30 000 years true age and 3000 years at 50 000 years true age. Material of infinite age with this contamination could appear to be only 57 000 years old.

The investigator collecting material for ^{14}C dating should develop a strategy for avoiding sites and samples liable to give false ages due to contamination *in situ* by older or younger carbon. Sediments may, according to their position, contain more or less allochthonous material – i.e. material derived from elsewhere – either of inorganic or organic nature, borne by water or wind. The effect of hard water, either on plants growing in it or on sediments contaminated by the carbonate deposited, is to introduce very old carbon (long out of reach of exchanges with the atmosphere) with a low $^{14}C/^{12}C$ ratio. Recent sediment in hard-water lakes may for this reason have an apparent age up to some thousands of years. On the other hand, the penetration of roots from above or the deposition of organic debris and humus may introduce much younger carbon, and old carbonate may be contaminated by contact with ground water or the carbon dioxide of the air. Shells, particularly their outer part, are commonly leached in this way and many datings performed on them are therefore unreliable. In the case of bones only the organic fraction should be used for dating, and pretreatment to remove humus is necessary. The errors incurred by neglect of these precautions may be large (OLSSON 1968). Usually the best ancient materials for giving consistent and representative dates are charcoal, wood and peat, the latter especially from raised bogs (GEYH and ROHDE 1972).

Radiocarbon datings of material from the Antarctic Ocean are commonly found to give excessive ages. Thus, the flesh of a recently dead seal found rotting on an Antarctic beach in 1971–2 gave a radiocarbon age of about 500 years, and bone from the same animal gave a radiocarbon age of about 150 years. In one extreme case of seaweed presumed to be recent an age over 2000 years was indicated. Errors of the order of +500 to +1000 years are common. These misreadings may be attributed to the admixture of very old carbon dioxide long out of contact with the atmosphere in water melted from portions of the Antarctic ice sheet and in deep ocean water brought up to near the surface in the circulation near the zone of the Antarctic Convergence (see Volume 1, pp. 326–7). LERMAN *et al.* (1970) have reported that radiocarbon ages of tree rings in the southern hemisphere appear somewhat too old and indicate that there is a deficiency of about 4·5 ± 1‰ in the ^{14}C content of the southern hemisphere atmosphere at 42°s as compared with the northern hemisphere at 42°n, which can be attributed to the more extensive oceans and resulting greater input of old carbon into the atmosphere. The difference between the hemispheres becomes greater at higher latitudes.

Radiocarbon ages are usually quoted with a ± one sigma error, which signifies the standard error of the count plus that of the background radiation count as well as of the

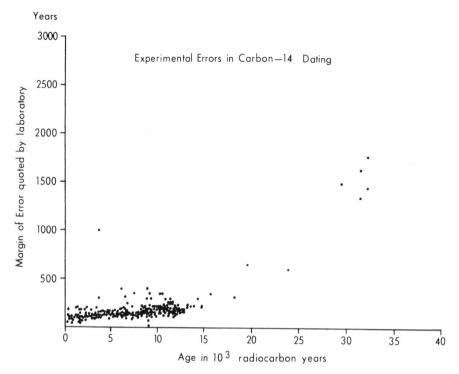

Fig. 13.10   Experimental errors (i.e. random error) in radiocarbon dating.
The plotted points are estimates of one sigma error published with the age results of measurements made in the Cambridge, Copenhagen and Washington University radiocarbon laboratories between 1959 and 1968.
(*From Radiocarbon, vols. 1–10.*)

modern value of ^{14}C concentration (the deviation from which interests us) – i.e. the combined sources of experimental error. Fig. 13.10 illustrates how this estimate of the margin of random error varies with the age of the material being dated. Even for material only a few hundred years old it averages about 100 years and ranges from 30 to 200 years. For ages about 10 000 years it is most commonly about 200 years and ranges from 100 to 400 years. With ages about 30 000 years this error margin has increased to $\pm$ 1500 years or rather more; and, owing to the shape of the radioactivity decay curve with time, one standard deviation of the count implies significantly different plus and minus errors in years. For materials of great age reliability of the measurement can be improved, when costs permit, by extending the count over longer periods and by laboratory techniques for enriching the proportion of the heavier isotopes; but the radiocarbon method appears to reach its ultimate limit somewhere in the age range 50 000 to 70 000 years, and at ages of over 35 000 years the error margin is often regarded as too wide and uncertain to be specified.

Various tests have provided satisfactory (indeed possibly surprising) evidence that carbon atoms, including the radioactive ones, do not migrate from year ring to year ring within a tree; though it is important to eliminate any mobile fluid (sap) from the sample to be dated. This is not in practice a source of significant error.

Besides the random errors of the measurements, however, allowance should be made for what is now known of the history of variations in the amount of ^{14}C in the atmosphere at different times past. Since the assumption of the radiocarbon method that the atmospheric concentration is constant, as constant as the flux of cosmic rays produced by the galaxy, is not strictly true, the deviation prevailing at the time of death of the organic matter studied will cause a false estimate of the age. These variations, first studied by DE VRIES (1958) and sometimes referred to as the 'DE VRIES effect', are tentatively ascribed to two natural causes:

(1) The variation of the intensity of the Earth's magnetic field, which had a maximum around 400–100 B.C. when it was 1·6 times the present strength and a minimum some 4000 years earlier when it was 0·6 times present strength. From irregular shorter-term changes it seems clear that when the field is strong it deflects some of the cosmic rays approaching the Earth, hence presumably decreasing the production of ^{14}C. Moreover, the long history of geomagnetic field strength shows a close negative correlation with the atmospheric ^{14}C amounts derived by radiocarbon dating of objects of known age. For many centuries between about 200 B.C. and A.D. 800 the atmospheric radiocarbon concentration appears to have averaged about 1% less than now, whereas around 4000–5000 B.C. it seems to have averaged 12% more than now. The long-term change of magnetic field strength appears to be cyclic, with a period of about 8900 years (BUCHA 1970).

(2) Variations in the frequency and intensity of solar disturbance, the magnetic fields associated with the solar wind being also capable of inhibiting the approach of cosmic rays from the galaxy (STUIVER 1961, DAMON 1968, SUESS 1968). It seems likely (BAXTER and WALTON 1971) that, in addition to the undoubted effect of this upon the production rate of ^{14}C, solar-induced changes of atmospheric circulation and of the exchange of air, including ^{14}C, between stratosphere and troposphere, and between the troposphere and the oceans, are important. A negative correlation between ^{14}C and sunspot number, with variations of ^{14}C concentration by about 3% within the 11-year sunspot cycles since 1890, has been demonstrated (from measurements on plants, such as cereals and flax, seeds, wool, wines and spirits identified as to their precise year of origin); and there seems to have been a long-term decrease of about 0·5% in the average atmospheric ^{14}C between 1890 and 1950, as the general level of sunspot number rose with the increasingly active cycles. There seem also to be fluctuations of this sort on time scales of about 100 to 400 years, perhaps oftenest 200 years (with positive deviations of atmospheric radiocarbon from the 1890 value amounting to about +2% around A.D. 1500 and A.D. 1700 – in each case during or just after some decades of very low solar activity).

The results, reported by SUESS (1970), of hundreds of measurements of the radiocarbon ages of the rings of bristlecone pine trees from the western U.S.A., independently dated by the tree ring chronology which is taken to approximate the true age, are shown in fig. 13.11 in the form of a calibration curve whereby the radiocarbon age of an object can be converted to probable true age (the margin of random error previously discussed still applies).[1] Notice on this curve that there are certain past dates at which the radiocarbon age is ambiguous: decay from different initial ^{14}C concentrations at several different dates could give the same residual radioactivity today. For example, a radiocarbon age of 4100 years B.P. (= 'before present') could result from any one of seven true ages between about 2550 and 2900 B.C.

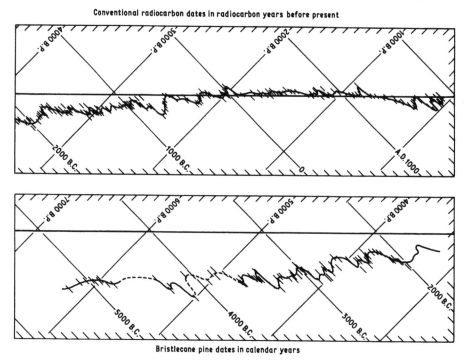

Conventional radiocarbon dates in radiocarbon years before present

Bristlecone pine dates in calendar years

Fig. 13.11   Relationship found between radiocarbon ages on the 5568-year half-life and true ages derived from bristlecone pine tree rings.
(*After* SUESS.)

The systematic error in radiocarbon dates attributable to these variations in the cosmic ray flux actually reaching the atmosphere is more difficult to establish beyond the time of the earliest independently dated tree rings about 7000 years ago. TAUBER'S (1970) comparisons of radiocarbon ages with the Scandinavian varve chronology suggest that beyond a period around 4000 B.C., when the error was greatest and radiocarbon ages come out about 700

1. Preliminary curves of the same kind as here illustrated were published by RALPH and MICHAEL in 1967 on the basis of measurements available up to that time.

years too young, perhaps because the minimum of the Earth's magnetic field around that time allowed an abnormally great incidence of cosmic rays and production of ^{14}C, the discrepancy becomes much less. Dates between 10 500 and 6000 B.C. suggested by the most refined varve chronology and radiocarbon dating measurements are in agreement with each other within the limits of random error of both methods. There is some likelihood that for greater ages than this the systematic error of radiocarbon dates varies cyclically in a manner similar to that observed within the last 8900-year cycle of the Earth's magnetic field strength.

Radiocarbon ages are conventionally quoted in years before present (B.P.), meaning before A.D. 1950, about which time this method of dating first came into use.

It is not sufficient to measure the deviation of the radioactivity in the sample of material to be dated from the level prevailing now or in 1950, because Man's activities in recent times have changed the concentration of ^{14}C in the atmosphere (and are still changing it) in addition to the continuing natural fluctuations on various time scales just discussed. The two chief causes of these Man-made changes are:

(1) The burning at increasing rates of ancient carbon, in which all ^{14}C activity has long been dead, in coal, oil, gas and all such 'fossil fuels', has been increasing the amount of carbon dioxide in the atmosphere (see Volume 1, Chapter 2) and diluting its radioactivity with 'dead carbon'. This industrial revolution effect, first pointed out by SUESS in 1953, and since known as the 'SUESS effect', should have reduced the atmospheric ^{14}C by about 10 % from the end of the nineteenth century to that time, were not some of this Man-made CO_2 taken up by the biosphere and in the ocean. The observed dilution of atmospheric ^{14}C attributable to this cause seems to have been between 2 % and 6 % by 1940 (HOUTERMANS, SUESS and MUNK 1967) i.e. perhaps half the Man-made output stayed in the atmosphere.

(2) Nuclear bomb explosions produce so much radiocarbon that from 1954 onwards the industrial dilution was substantially offset and in 1963 the amount of ^{14}C in the atmosphere, according to measurements made between 27° and 78°N, rose to twice (203 %) the natural level (NYDAL 1966). By 1965, with a cessation of most weapon testing, the excess had fallen to 70–80 % above normal. The 1963 ring in living trees is therefore clearly and easily identified by its ^{14}C activity, and any migration of ^{14}C atoms from it into the older wood should be easily traced.

To get round these difficulties the ^{14}C activity measured in materials being dated is expressed in terms of its departure from a nineteenth century level, that of wood grown in 1890, before Man-made disturbance of the atmosphere's ^{14}C reservoir became significant (it has been noted that wood grown in 1850 displays an activity 1·75 % below that of 1890, but this may be attributed to the natural variation of ^{14}C production probably associated with a group of vigorous solar cycles). To provide a datum for calibration, the National Bureau of Standards in Washington, D.C., keeps a sample of oxalic acid prepared in 1950 from West Indian sugar grown in that same year (when the Man-made effects were already appreciable). The 1890 activity used as the datum from which departures are measured has the inevitably odd description (repeated in many published radiocarbon dating reports) of '95 % of the N.B.S. oxalic acid standard', which represents the ^{14}C activity in 1950 of 1890 organic matter, i.e. after 60 years decay of its initial activity.[1]

Beyond the range of radiocarbon dating, there are no dating methods so far devised which give an accuracy sufficient to establish the climatic patterns that prevailed over particular periods less than some thousands of years in duration. The dating methods that are available for such eras will therefore be only briefly reviewed here. With each method described the problems about representativeness of the material on the site from which a sample is taken for investigation must be considered. And always complications due to fractionation when isotopes of different atomic weight pass from one reservoir to another may have caused an anomaly of the initial condition, the time from which radioactive decay is measured. These will not be gone into. Nevertheless, these radiometric methods of dating rocks and other materials of great age are of great importance to geology in providing a time scale to the great spans of the Earth's past which were hitherto only identified by name. Tree rings and varves may, of course, provide 'floating chronologies' which could give some indication of the year-by-year weather sequence, at the place where they were found, over some period beyond the range of precise dating.

Uranium series dating

The various naturally occurring isotopes of uranium (^{238}U, ^{235}U, etc.) decay, ultimately to lead, by passing through characteristic successions of atomic disintegrations forming elements of intermediate atomic structure and widely differing half-lives. In the disintegrations helium gas ($^{4}_{2}$He) is given off or trapped within the surrounding rock. Some of the elements produced have half-lives which make the quantity present of interest as a means of dating the substance in which they are found.

The principal successions and the half-lives of the elements concerned may be indicated as in the table at the top of p. 69 opposite. In this table d stands for days, h for hours, y for years and my for millions of years.

In theory the absolute amounts of any of these elements present, or the ratios of the quantities of different ones, may be used as an indicator of the time elapsed since the beginning of the process that produced them. Thus, assuming the accumulation of lead and helium, the two stable products, began at the moment of crystallization of a rock from volcanic magma, measurements of the amounts of either of these elements present will give an indication of the age of the rock. But in the case of helium measurements this must be regarded as a minimum age, since some of the helium produced may well have escaped. Ages obtained by measurements of the lead may be too old, falsified by pre-existing lead present in the molten volcanic material in some cases, e.g. if there was an influx of older rock material

---

1. I am indebted to Dr R. SWITSUR of the Cambridge University Radiocarbon Dating Laboratory for making clear these details about the oxalic acid standard used.

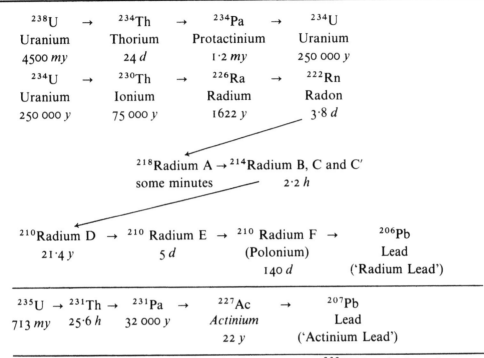

^{238}U	$\rightarrow$	^{234}Th	$\rightarrow$	^{234}Pa	$\rightarrow$	^{234}U
Uranium		Thorium		Protactinium		Uranium
4500 *my*		24 *d*		1·2 *my*		250 000 *y*

^{234}U → ^{230}Th → ^{226}Ra → ^{222}Rn
Uranium   Ionium   Radium   Radon
250 000 *y*   75 000 *y*   1622 *y*   3·8 *d*

^{218}Radium A → ^{214}Radium B, C and C′
some minutes   2·2 *h*

^{210}Radium D → ^{210} Radium E → ^{210} Radium F → ^{206}Pb
21·4 *y*   5 *d*   (Polonium)   Lead
            140 *d*   ('Radium Lead')

^{235}U → ^{231}Th → ^{231}Pa → ^{227}Ac → ^{207}Pb
713 *my* 25·6 *h* 32 000 *y*   *Actinium*   Lead
                     22 *y*   ('Actinium Lead')

Similarly, the heaviest member of the thorium sequence (^{232}Th) breaks down in stages to an isotope of lead (^{208}Pb), 'thorium lead'.

remelted in the volcanic episode that preceded final crystallization. Thus, the ratios of the quantities found of the various unstable (i.e. radioactive) elements produced in the uranium and thorium successions should give a more reliable indication of the length of time during which the production has been going on.

Of most interest for our studies is the *protactinium-ionium method* (ERICSON and WOLLIN 1966, RONA and EMILIANI 1969), which has been applied to dating the material in ocean sediments up to about 200 000 years old. This measures the ratio of the protactinium ^{231}Pa to the thorium isotope ^{230}Th which is known as ionium. At the short-range end of its applicability, where it is subject to a wider margin of uncertainty than the radiocarbon method, this method has given results in reasonable agreement with radiocarbon. The error margins indicated by two or three repeat measurements of the ^{231}Pa/^{230}Th ratio in Caribbean cores were:

For ages	*Error ranges*
7000–30 000 *y*	± 1000 *y*
About 70 000 *y*	± 2000 *y*
About 100 000 *y*	± 3000 *y*
About 150 000 *y*	± 6000 *y*

Land-derived clay particles while suspended in the ocean take up these isotopes in the same ratio as that existing in the ocean water, ultimately derived from any ^{238}U and ^{235}U present. Once buried in the bottom sediment, equilibrium with the sea water ceases and the ratios of the unstable isotopes present gradually change in accordance with their different decay rates. Thus, it is the date of burial in the sediment that starts the process whose duration can be measured.

Measurements of the ^{230}Th/^{232}Th ratio and of the quantity of ^{230}Th alone (known as the *Ionium method*) have also been tried but proved less reliable, giving dates consistently older than the protactinium-ionium method.

For ages beyond 200 000 years up to 2 to 4 million years, most reliance – all too much reliance – has had to be placed upon extending the results of the foregoing measurements by extrapolation of the sediment accumulation, assuming an effectively constant rate of sedimentation. ERICSON and WOLLIN, while admitting that this cannot be justified for sediments laid down in shallow water, e.g. on continental shelves, argue that it is justified for deep ocean sediments, where differences of sedimentation rate in different climatic regimes characterized by different rates of run-off from land should be small. The same assumption can hardly apply to the airborne dust deposited on the ocean (and ultimately on the ocean bed) in different climatic regimes. However, differences of sedimentation rate can be recognized in the texture of the deposit. Slow accumulation is marked by large numbers of foraminifera shells, rapid accumulation by large proportions of fine-grained material.

Other radioactive isotope methods

For rock materials of very great ages the following radioactive isotope methods in addition to those depending on the uranium–thorium–lead sequences are used.

Potassium-argon dating: A very small proportion of the radioactive isotope of potassium (^{40}K) exists alongside the common, stable isotope (^{39}K) in all potassium-bearing rocks and minerals. ^{40}K disintegrates, producing argon (^{40}A) and some calcium (^{40}Ca). The half-life of ^{40}K is 1300 million years. This very long time taken for half the ^{40}K to disintegrate makes the method applicable to dating some of the oldest rocks on the Earth, the decay time again being measured from the time when the rocks crystallized. The method depends on measuring the quantity of argon trapped in the rock, and thence the K/A ratio, and is therefore subject to error – liable to give a minimum age rather than the true age – if some of the argon has escaped since. It is thought to be subject to an error margin of about 100 000 years in the case of the youngest rocks to which it has been applied, at ages of about a million years.

Potassium-argon dating of volcanic flows covering glacial gravels and till in the upper Santa Cruz valley (50°s) in Argentina, supported by recognition of the sequence of normal and reversed polarity of the Earth's magnetism in the stratigraphy, indicates that at least one glaciation occurred in southern Argentina rather more than 3·6 million years ago. Further glaciations of the area could be identified between 1 and 3 million years ago (FLECK, MERCER *et al.* 1972).

Rubidium–strontium dating: This is another method applicable to rocks which do not contain the heavy metals of the uranium group. Rubidium is a metal in the same column of the periodic table as sodium and potassium. It occurs widely, but only in very small quantities, in two isotopes ^{87}Rb and ^{85}Rb. About 27% of it consists of the isotope ^{87}Rb, which decays very slowly to ^{87}Sr (strontium) by emitting a β-particle. The half-life of ^{87}Rb is 53 000 million years. This still longer half-life (than in the case of ^{40}K), as well as the very small quantities available to work with, means that the method is only suitable for dating the oldest rocks and to estimate the age of the solar system itself.

Thermoluminescent dating

Some minerals, e.g. quartz, have the property of acquiring thermoluminescence by exposure to radiation, whether this be of natural or artificial origin. This property is due to the trapping of electrons in the crystal lattice. When the mineral is heated, these electrons escape and are seen as the emission of a glow. Heating above 500°C gets rid of all this previously acquired accumulation of energy represented by the trapped electrons. Thus, measurements of the thermoluminescence of any substance when heated are a measure of the radiation dose it had accumulated since the last time it was heated above 500°C; the amount that had been accumulated depends on the nature of the mineral and the amount of radiation to which it had been exposed in the time elapsed. The rate at which the particular substance acquires thermoluminescence can be tested in the laboratory by exposure to radiation from an artificial radioisotope. This provides a method of dating the time at which volcanic rocks cooled or at which a given article of pottery was fired.

There are theoretical and practical difficulties, however, and so far the margin of error is $\pm 10\%$ and can probably never be reduced to less than $\pm 5\%$ (AITKEN 1970). The advantage of the method applied to pottery or other objects heated by Man is that it dates the actual process, and not some related matter such as the time at which a tree used for timber was felled, and that its error margin may be slightly less than that of the radiocarbon method for ages at which the latter has big systematic errors or gives ambiguous results.

A review and bibliography on thermoluminescent dating has been given by SEELEY (1975).

For additional information on methods of radiometric dating, see ALLIBONE (1970), ERICSON and WOLLIN (1966), MICHAEL and RALPH (1971), OLAUSSON (1971), RONA and EMILIANI (1969), ZEUNER (1958).

Other methods of dating

There are numerous other indications of the age of a deposit and of features or articles (some of which may be datable by independent evidence about them) observed with a deposit. The stratification, with the oldest layers at the bottom and the youngest at the top, provided it has not been disturbed, at least makes the *relative* ages of different levels clear. Some of the indications of age which we shall mention were used to obtain rough datings long before the invention of radioactive isotope measuring techniques. The latter may now be used to give some approach to *absolute ages* for sample levels within a deposit and so to indicate a time scale for the whole deposit. On this basis, with due care and suitable reservations, the older methods supported by a few fixed ages within a given deposit determined by radiometric dating may now be used to extend the time scale somewhat beyond the limit of the radiometric (e.g. ^{14}C) method used. They may also conveniently indicate, however roughly, ages within the times that, for instance, radiocarbon dating can cover without the need to resort to such cumbersome and costly measurements at every stage.

Archaeological dating

Man-made things – known as human artefacts – ranging from highly artistic pottery and metal ornaments to the most primitive bone tools from Palaeolithic times, which can be dated by radiocarbon tests or other independent means, have enabled archaeologists to build up a knowledge of the history of the discovery of new materials and methods of working, and of artistic styles, as well as of the places and times of origin of each. Wherever such objects are found buried in undisturbed soil or subsoil layers, identification of the culture to which the object or style belonged sets an approximate date or, more reliably, a *maximum* age for that point in the stratigraphy. This has been chiefly of use in increasing knowledge of the distribution of trade and communications in prehistoric times, and of the time lags involved in the spread of fashion and new techniques, rather than in dating the marks of climatic history; but it may supply useful corroboration of the latter in some cases. Moreover, even the trade routes used – for instance, when they crossed the Alps, the Sahara desert or a great sea – may have interesting climatic implications. In using archaeological knowledge for dating, however, one is limited more than in the case of the use of pollen zones by the 'time transgressive' element – i.e. by the lags in the arrival of new fashions and new techniques in regions remote from where they originated.

The earliest cultures of Man (*Homo sapiens* and the related types of Man that existed before the last ice age) are identified by the types of *flint, and later bone, tools and hunting weapons* that they made. The dates of similar developments in similar places may be subject to indeterminate, and possibly very long, lags (except where long-distance wanderings over treeless plains facilitated contacts). Animal life was abundant in relation to the numbers of human beings, especially in Europe south of the ice sheet during the glaciation and probably in North America and Asia as well (it may have been the primitive hunting tribes of men rather than the climatic change that was responsible for the extinction of the mammoth in postglacial Siberia). *Woodwork tools* of bone, antler and flint were already made by the Aurignacian people in the temperate interstadial period in Europe about 40 000 years ago.

The following list of dates of other important discoveries and advances in technique, compiled from CHILDE (1954, 1958) and CLARK and PIGGOTT (1965), and from information supplied by the Historisk Museum of Oslo University, is given partly as a background to later sections of this work dealing with archaeological evidence that may be thought to have a bearing upon climatic changes, but also because it illustrates cases where various lags in the spread of new methods are known to have occurred. In other cases, it may well be that similar inventions occurred independently in different parts of the world.

Boats for river transport, probably canoes of skin or birch bark, evidently go back to before 7000 B.C. and seem to have played a considerable part in the Maglemosian culture in the early postglacial Europe of that time, but somewhat similar craft were probably made and used long before that.[1]

Boats for sea crossings may even have been made by the Solutrean peoples between 20 000 and 15 000 years ago (i.e. during the last ice age), judging by the close similarity of articles used in Spain and North Africa, unless the Strait of Gibraltar was closed by a land bridge, which is not usually supposed. (See also p. 248.) Sauveterrian burials on islands off Brittany indicate that boats capable of a short sea crossing (sea level was still lower than now) were available about 8000 B.C. Crossings to Ireland were being made before 3000 B.C., at least over the narrow sea from Kintyre in Scotland to Ulster, and rock drawings in Norway suggest that skin boats were used for cod fishery and possibly for catching seals and porpoises in Maglemosian times, i.e. before 4000 B.C. and possibly as early as 5000 to 7000 B.C. Boats with sails on the Mediterranean and in Egypt around 3000 B.C. are depicted in records of the time. In the Bronze Age some Egyptian vessels were up to 60 m long and nearly 20 m in the beam; the Cretan vessels were much smaller. Bronze Age ships in Norway had high prows and sterns, like the vessels of Viking times, but were rowed: the pictures do not show sails.

Pottery: the earliest known evidence of pottery is from 6000 to 4500 B.C., but by about the latter date pots were used over a wide area from Fayum in Egypt to Iran and from Mesopotamia to Phoenicia and Turkmenia. The Starčevo pottery of southeastern and eastern Europe was already highly sophisticated before 4000 B.C. and sometimes showed painted designs. The oldest remains of pottery vessels in northern Europe are found in the kitchen middens of the Ertebølle culture in Denmark, left by people who were strand-loopers beside the Littorina Sea, between 5000 and 4000 B.C.

Earliest use of metals: Copper seems to have been worked with casts and cold-hammered in Egypt and Mesopotamia from about 3500 B.C. The earliest furnaces found, however, have been dated about 1500 B.C. in Egypt and 1000 B.C. in Europe. *Tin* was traded from Cornwall to Minoan Crete around 1500 B.C., but the *Bronze Age* had begun with the use of copper–arsenic and copper–tin alloys as early as 3000 B.C. in India, Mesopotamia, Asia Minor and

---

1. A conveniently accessible brief summary and survey of the most important culture names in Europe and their approximate dates is given by WOOD (1972).

Greece; it began in central Europe and Britain by, or soon after 2300 B.C. and by 1500 B.C. at the latest in Norway. The Beaker people brought the use of copper to Britain around 2400 B.C. By about 2000 to 1500 B.C. it may be said that almost all Europe was, at least for some centuries, culturally linked by an active exchange of wares.

The Iron Age began about 1200 B.C. in Hither Asia and spread rapidly, no doubt through the superiority of weapons made of iron but perhaps partly through climatically induced upheavals and struggles amounting to a dark age in human affairs. Soon after 1200 B.C. the use of iron had spread to Greece and thence to Etruria in northern Italy. By about 750 B.C. it appeared north of the Alps in the Hallstatt culture, and reached England and Norway about 600 B.C.; it did not appear in Scotland until around 250 B.C. and in Ireland later than that. Iron did not become common in Egypt till after 650 B.C.

Domestication of animals also seems to have progressed gradually. The people of the earliest towns and villages about Jericho and southern Turkey and Iran bred sheep and goats by around 7000–6000 B.C., but not yet cows. By 3000 B.C. the peasants all over the Near East, from Egypt to Mesopotamia, had yoked oxen to the plough. And by about 2500 B.C., if not earlier, asses and horses were harnessed to two-wheeled carts in Hither Asia. The first use of wheeled vehicles and horses was very early in the Neolithic in Poland, i.e. also by 2500 B.C. or earlier.

Sledges are known to have been used from quite early postglacial times, and by 2000 B.C. heavy sledges up to 4 m long, drawn by dogs, were in use from Finland to the Urals. Reindeer sledges had reached the Baltic by 1000 B.C.

Skis go back to Stone Age times in Norway and are clearly shown in Karelian rock drawings between 3000 and 2000 B.C.

The beginnings of agriculture go back to perhaps 9000 B.C. in those areas of the mountains in the Near East where the grasses from which modern cereals have been developed grow wild. The farmers of the Starčevo culture in southeast Europe had querns and ovens for processing cereals by 4000 B.C. Neolithic agriculture seems to have reached England and elsewhere in northern Europe before 4000 B.C. and possibly as early as 4200 B.C. It even appears that one of the early forms of wheat was preferred to barley and other hardier crops in England in the time of the Windmill Hill culture around 3500 B.C., and wheat was used in Norway also then or soon after. The earliest known cultivation of the olive, in Almeria in Spain, is dated around 3500 B.C. It appears that vine cultivation in the Near East may be of approximately equal antiquity. By 3000 B.C. intoxicants were widely used, and specialized pottery vessels had been developed for ceremonial drinking.

It seems from this survey that the lags in the spread of knowledge were great in early postglacial times. For instance, urban communities were established in the Near East, the earliest known being Jericho about 7500 B.C., about the same time as the earliest identified camps of hunters and fishers in northern Europe at Star Carr, Yorkshire. The first northern and central European villages appear in earliest Neolithic times, by about 4000 B.C. (WATERBOLK 1968). By 2000–1500 B.C. the greatest cultural lags became much less, but seem to have increased again after 1200 B.C. They may also have been much less among the peoples roaming the treeless plains in the ice age.

Flow method of dating the ice at any level in ice sheets

Striking successes seem to have been achieved in dating the ice below the levels where all usable traces of the year layers become obliterated by molecular diffusion in both the Greenland and Antarctic ice sheets by applying a simple concept of the flow of the ice (DANSGAARD *et al.* 1969a, 1969b, 1970, EPSTEIN, SHARP and GOW 1970). Fig. 13.12 represents the vertical cross-section through a domed ice sheet of typical profile (NYE 1959), resting on essentially flat bedrock, the long thin arrows showing the pattern of flow in the ice that is generally accepted. *I* represents the ice-shed, where the flow divides, and *C* is some point where a core drilled through the ice to the bottom is taken for investigation. The horizontal lines at *I* indicate how uniform increments to the ice, e.g. under constant yearly accumulation, *a*, become thinner and spread out horizontally as they go deeper into the ice sheet (see the discussion by NYE 1963). The horizontal arrows beside the column at *C* indicate the adopted horizontal velocity components (V_x) at different depths in the core, which appear to represent only a very slight simplification of the distribution of V_x with depth derived by more refined theory; the horizontal velocity component is taken as constant down to height ($y = h$) at which V_x starts to fall off linearly to O at the bottom ($y = O$), where the ice is frozen to the bedrock. The figure adopted for *H*, the total thickness of the ice at *C*, which was 1390 m in the Camp Century core (77°N 56°W 2000 m a.s.l. in north Greenland) and 2164 m in the Antarctic core taken at Byrd station (80°S 120°W 1530 m a.s.l.), was reduced by 20–30 metres to allow for the low densities of the snow and firn in the uppermost part of the core.

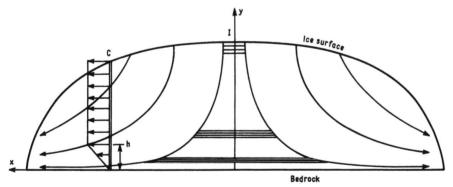

Fig. 13.12   Flow of ice in a typical domed ice sheet, seen in vertical cross-section with thinning and spreading by plastic deformation of the accumulation layers illustrated below *I*, the ice-shed, and horizontal components of motion in the ice below *C*, where a core was taken, represented by the length of the horizontal arrows. The ice is pictured resting on a horizontal bedrock surface, to which its base is frozen stationary.

Then throughout that part of the ice above h, where it can be taken that the horizontal velocity component does not change with depth, the equations of continuity and incompressibility lead to the results that the thickness a_y of a typical deformed yearly accumulation layer at height y above the bedrock is given by

$$a_y = a\frac{y}{H}.$$

The vertical velocity (V_y) at any level y is given by

$$V_y = -\frac{a}{h}\int_0^y dy.$$

This gives the age t of the ice at any level y as

$$t = \frac{2H - h}{2a}\log_e\frac{2H - h}{2y - h}$$

where $h \leqslant y \leqslant H$. And in the bottom layers of the ice

$$t = t_h + \frac{2H - h}{a}\left(\frac{h}{y} - 1\right)$$

where $o < y \leqslant h$.

The method has to assume constant yearly average accumulation on the surface of the ice cap – an assumption which may not be unreasonable, however, in the interior of Greenland and Antarctica where the cores were taken. H has to be estimated at the place where the snow was first deposited that formed the ice found at the level being studied.

Verification of the method as probably giving some approximation to real ages, and the justification thereby of the daring assumptions on which the theoretical derivation of the ages rests, may be regarded as indicated by the results. These have entailed a further implied discovery, which could not have been taken for granted in advance, namely that the sequence of changes of prevailing temperature given by oxygen isotope analyses (see pp. 97–8 later) of the ice substance point to about the same dates in Antarctica and Greenland for the onset (around 70 000 years ago), culmination (between 17 000 and 20 000 years ago) and ending (around 11 000 years ago) of the cold climates of the last ice age as are indicated by most evidence in Europe and elsewhere. If we may assume that the dates should really coincide, and that they correspond (as they seem to do) to global changes of climate and atmospheric circulation taking place more or less abruptly over the whole Earth, then the error margin attaching to dates of the ice derived from the flow model cannot be more than 5% and may be very much less.

DANSGAARD's dating by this method of the valuable north Greenland oxygen-18 curve has, however, been severely criticized by MÖRNER (1974) as unreliable, especially the part before 45 000 years ago, and a new chronology of it based on comparisons back to 35 000 years B.P. with better established chronologies for Europe and North America has been proposed. An interesting implication is the occurrence of important variations in the rate of ice accumulation, though these cannot yet be established in detail or with any great assurance.

Amino-acid dating

Organic chemistry may provide a means of dating shells and other organic detritus in the stratigraphy of a sediment back to about the mid Tertiary, perhaps 30 million years ago, through details of the amino-acids present, as first hinted by ABELSON (1956). These polymers, like proteins, are among the basic structures required by living organisms. Some amino-acids appear stable, and are present in similar amounts in Recent as in Tertiary shells of the same species, whereas others are relatively unstable. Thus alanine and glycine, which together make up 60 % of the amino-acids in the insoluble fraction of Recent *Mytilus* shells, and are each about equally abundant in them, were found to give a ratio of 15 alanine to 1 of glycine in similar shells radiocarbon-dated as over 30 000 years old. Since the decay of the different amino-acids depends only on temperature and time, and some have been studied in the laboratory at high temperatures (up to 225°c) where decay proceeds rapidly, it should be possible to deduce the decay rates applicable at the temperatures between 0° and 30°c prevailing on the Earth and apply the results to geochronology over millions of years. Very small quantities of the order of 100 milligrammes suffice for these tests. Experience of the method using different substances and reactions is needed to indicate error margins (see, for example, HARE 1969).

BADA *et al.* in more recent work (1973, 1975) have shown how the slow racemization[1] of amino-acids in human and animal bones can be used, subject to assumption of the overall average temperature experienced by such material at the site concerned, to date remains back to at least 120 000 years B.P. Alternatively, in cases where the dating can be established by radiocarbon – i.e. within about the last 40 000 years – the average temperature of the material over the time since deposition can be deduced within a few tenths of a degree from the observed progress of racemization. These possibilities arise because the change in the substance depends only on temperature and time. Dating beyond the limits of the radiocarbon method is possible because of the longer half-lives of these amino-acid reactions: about 15 000 years in the case of the racemization of aspartic acid and 100 000 years in the production of D-alloisoleucine from L-isoleucine, both at 20°c. Radiocarbon age determinations on younger materials at the same site can be used to calibrate the rate of racemization. Using the aspartic acid reaction, dates of bones of cave-dwelling men and

---

1. *Racemization* is the transformation of a substance – such as amino-acids produced in living organisms – with an asymmetric atomic structure of the molecules, and the property of turning the plane of polarized light, into a form with symmetrical arrangement of the atomic lattice or into a mixture with equal abundance of the mirror-image structure, in either case with loss of the optically active effect on polarized light. Substances so produced bear a similar relationship to the original substance to that which racemic acid bears to tartaric acid.

animals at the coast of Cape Province, South Africa, associated with a life that required high sea level (proximity of the sea to the cave), appeared accurate within 10 000 years at 120 000 B.P. and 5000 years around 65 000 B.P.

Pollen zone dating

It has long been recognized that a sequence of past changes in the vegetation – in its composition at any given place and in the geographical limits of genera and species – is registered in the pollen found in layered sediments of lake beds and at different depths in peat and undisturbed soil and subsoil. This is the case in all parts of the world, including the Arctic tundra, the lowlands of the temperate zone and the mountains in every continent. Moreover, a number of the changes seem to have taken place essentially suddenly, so that in any column of sediment or soil that may be examined distinct zones of different pollen assemblages – associated in some places with characteristic larger remains (e.g. tree stumps) and changes in the condition of the peat (e.g. dark, dense and full of humus where it was at some time exposed to a dry climate, much lighter and less humified where growth continued vigorously in a moist regime) – are more or less sharply marked off from each other. In Norway already a century ago the botanist BLYTT (1876a, 1876b, see also 1909) took the prehistoric changes of vegetation which he recognized in the stratigraphy of peat bogs and lakes to identify a sequence of climatic changes that evidently divided up the history of postglacial times. At first these changes were dated just by their position in the stratigraphy. The dating was later progressively refined as new methods became available. After the first steps in pollen analysis about the turn of the century, SERNANDER (1908, 1910) extended BLYTT's scheme to Sweden, where cross-reference to DE GEER's varve chronology then became possible. An obviously related sequence of vegetation, and presumably climate, changes was soon seen to apply in other parts of Europe, and the BLYTT–SERNANDER scheme began to be given a much wider currency, e.g. by GAMS and NORDHAGEN (1923) who also noted relationships to the succession of periods of dominance of different cultures in human archaeology and the possibilities of dating the later changes by reference to recorded events in human history in places as far away as the Near East and even China. Systematic pollen analysis and radiocarbon dating have since given much fuller knowledge of the distribution of the climate and vegetation changes and their timing.

The diagram in fig. 13.13 gives a survey of BLYTT and SERNANDER's named climatic periods, and of the dominant characteristics and designations of the pollen zones and related matters, now commonly recognized in the European lowlands from Ireland to Germany, together with present knowledge of the time scale as established by radiocarbon datings (uncorrected). The principal sources from which this diagram was compiled were FIRBAS (1949) and GODWIN (1956), also HAFSTEN (1960), IVERSEN (1954) and JESSEN (1949). More extensive tables showing the parallel course of events in other areas in Europe have been given by FLINT (1957, p. 397) and by GAMS and NORDHAGEN (1923).

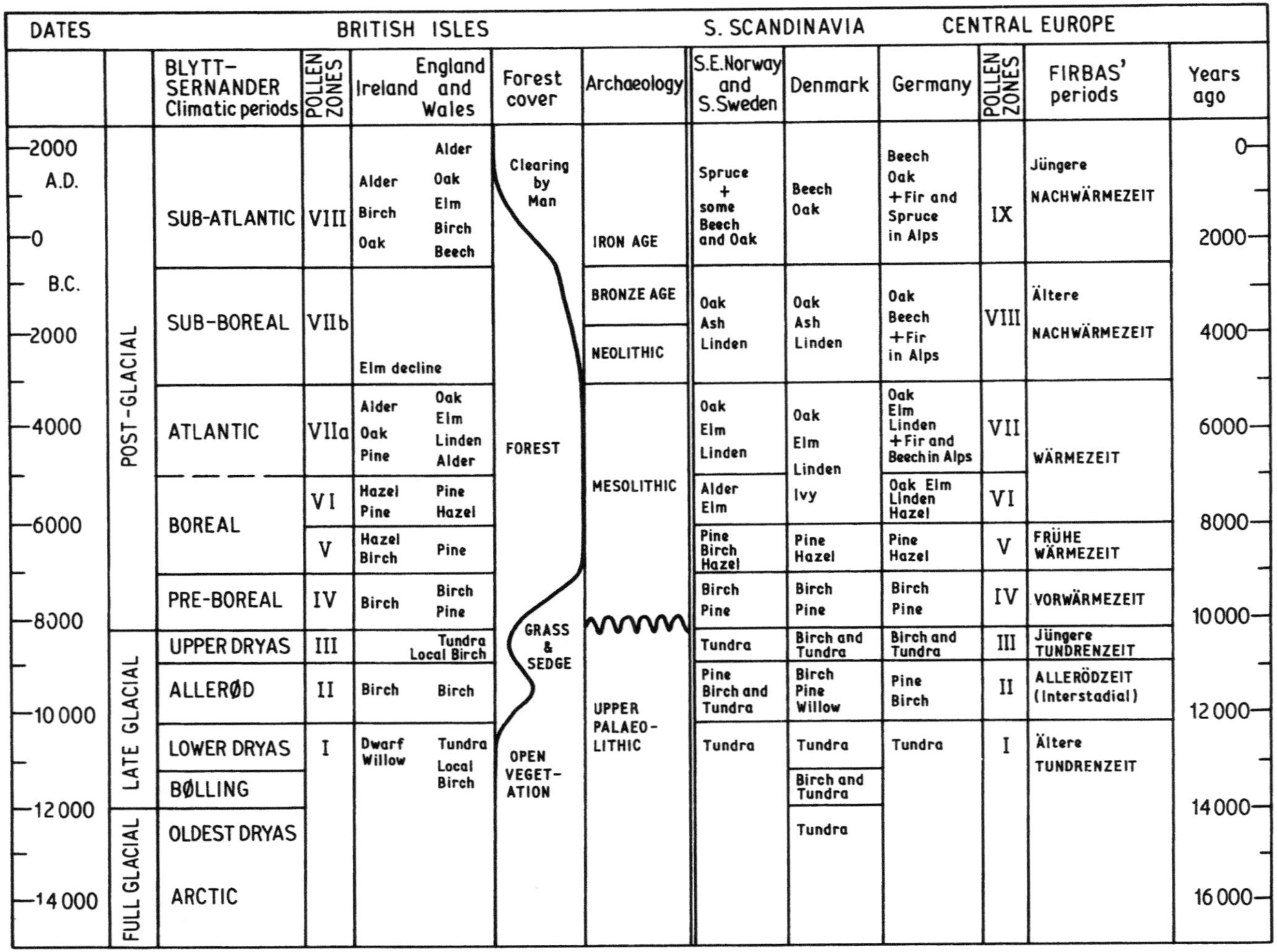

Fig. 13.13 Pollen zones (*after* GODWIN (*1956*) *for the British Isles and after* FIRBAS (*1949*) *for central Europe*) and the sequence of climatic regimes in Europe since the last ice age. Corresponding stages in the history of the vegetation, and in human archaeology in the British Isles, are indicated.

DATES
-2000	A.D.
0	B.C.
-2000	
-4000	
-6000	
-8000	
-10 000	
-12 000	
-14 000	

BRITISH ISLES

POST-GLACIAL / LATE GLACIAL / FULL GLACIAL

BLYTT-SERNANDER Climatic periods	POLLEN ZONES	Ireland	England and Wales	Forest cover	Archaeology
SUB-ATLANTIC	VIII	Alder Birch Oak	Alder Oak Elm Birch Beech	Clearing by Man	IRON AGE
SUB-BOREAL	VIIb	Elm decline	Oak Elm Linden Alder	FOREST	BRONZE AGE / NEOLITHIC
ATLANTIC	VIIa	Alder Oak Pine	Pine Hazel		MESOLITHIC
BOREAL	VI	Hazel Pine	Pine		
	V	Hazel Birch			
PRE-BOREAL	IV	Birch	Birch Pine		
UPPER DRYAS	III	Birch	Tundra Local Birch	GRASS & SEDGE	
ALLERØD	II	Birch	Birch		
LOWER DRYAS	I	Dwarf Willow	Tundra Local Birch	OPEN VEGETATION	UPPER PALAEO-LITHIC
BØLLING					
OLDEST DRYAS					
ARCTIC					

S. SCANDINAVIA

S.E. Norway and S. Sweden	Denmark
Spruce + some Beech and Oak	Beech Oak
Oak Ash Linden	Oak Ash Linden
Oak Elm Linden	Oak Elm Linden Ivy
Alder Elm	Pine Hazel
Pine Birch Hazel	Birch Pine
Birch Pine	Birch and Tundra
Tundra	Birch Pine Willow
Pine Birch and Tundra	Tundra
Tundra	Birch and Tundra
	Tundra

CENTRAL EUROPE

Germany	POLLEN ZONES	FIRBAS' periods	Years ago
Beech Oak +Fir and Spruce in Alps	IX	Jüngere NACHWÄRMEZEIT	0
Oak Beech +Fir in Alps	VIII	Ältere NACHWÄRMEZEIT	2000
Oak Elm Linden +Fir and Beech in Alps	VII	WÄRMEZEIT	4000
Oak Elm Linden Hazel	VI		6000
Pine Hazel	V	FRÜHE WÄRMEZEIT	8000
Birch Pine	IV	VORWÄRMEZEIT	10 000
Birch and Tundra	III	Jüngere TUNDRENZEIT	
Pine Birch	II	ALLERÖDZEIT (Interstadial)	12 000
Tundra	I	Ältere TUNDRENZEIT	14 000
			16 000

The pollen zone boundaries themselves have been used as markers for relative dating of the stratigraphy of deposits which show sufficiently clear similarity to the standard sequence of changes. This is the basis of some chronologies that were derived before radiocarbon datings were available. AUER (1958) even dared to use this method to date the Late-Glacial series of volcanic eruptions in southern South America by reference to climate and vegetation changes established in Europe, on the assumption that these were really world-wide times of change and of like thermal character in the northern and southern hemispheres' temperate zones. The same argument appears to be the basis of the dates given for pollen zones recognized in New Zealand (cited by MOAR 1971). This bold assumption has apparently been justified in retrospect by a number of independent researches, e.g. the literature pointing to abrupt, world-wide changes around 11 000 years ago (BROECKER et al. 1960, OGDEN 1967) and the application of a form of time-series analysis by BRYSON et al. (1970) to all the radiocarbon dates relating to postglacial environmental change published in the literature (i.e. in the ten volumes of *Radiocarbon* 1959–68). The latter study suggested that the dates were significantly clustered around the dates of the zone boundaries in fig. 13.13 as well as two more recent epochs about 750 and 1700 years ago (though the standard deviations of these dates are still somewhat uncomfortably large).

Closer study of the earliest and the latest of the epochs of change indicated by BRYSON's study, namely the climatic changes around 10 000 to 11 000 years ago suggested that at a site in Minnesota around 10 400 years ago spruce forest was replaced by pine in 170 years and in Ohio a largely spruce forest was replaced by oak in 1100 years. At another point, in Wisconsin, the transition from pine to oak appears to have taken 140 to 200 years (WEST 1961). And around A.D. 1200, when the plains in Iowa began to be affected by drought attributed to increased frequency of westerly winds (and the rain shadow of the Rocky Mountains), the oak largely disappeared in a century and grass pollen rose dramatically within 45 years. These are all pointers to a great rapidity of the climatic changes concerned, amounting to a sudden switch of prevailing circulation type. The response in terms of dying out of the old vegetation is also probably characteristically rapid, but the vegetation and pollen zone changes that depend on immigration of new species are likely to be 'time transgressive' or 'metachronous' – i.e. to show a lag that depends on distance from the former positions of the species concerned.[1] (This difficulty is likely to be less in America and China, where the climatic (i.e. thermal) gradient from north to south is always strong, than in Europe where the nearest stocks of the advancing species may have far to come.) NICHOLS (1967) has also found that the vegetation sequence defined by pollen analysis in northern Canada west of Hudson's Bay fits the timing and the character of the thermal changes implied by the European pollen zones. And a series of studies, also supported by

1. SMITH (1965) has turned attention to the sporadic and time-transgressive details of the impact of a climatic shift upon a previously established vegetation and to the inertia (stability) of an established vegetation over wide areas (see also WEST (1968) and the reference to 'subfossil' forest in this volume on p. 229).

radiocarbon dates, by HEUSSER (e.g. 1966) imply the same for northwestern North America and for southern Chile.

Another feature of peat bog stratigraphy which has been used as a marker for dating is the recurrence surfaces (GODWIN 1954, 1956, pp. 33–4). These register renewed growth of the bog in a moist climate after a dry phase, by a change in colour from dark to light. Five such recurrence surfaces (*rekurrensytor*) are recognized in Sweden: these have been defined by GRANLUND (1932) as RY I (dated about A.D. 1200), RY II (A.D. 400), RY III (600 B.C.), RY IV (1200 B.C.) and RY V (2300 B.C.). Similar features are certainly to be seen in the peat elsewhere in Europe and in Canada; some of them may relate to the same dates, but there are additional ones of more localized extent. The recurrence surface around 600 to 500 B.C. seems to be nearly everywhere the most marked; it is well known to the country people in Ireland and in northwest Germany (see p. 211) as dividing the lower dark 'turf' from the light-coloured and less dense upper turf. This colour boundary may perhaps be used as a date marker: but its date can hardly be fixed more narrowly than to within 100–200 years (OVERBECK *et al.* 1957) and may be open to confusion with features of quite different date in some peat deposits (see pp. 211–13).

Sedimentation rates on the ocean bed

It is now understood that the ocean bed is a dynamic environment. Slumping of sediments on slopes and drift and scouring of bottom deposits by water currents occur in many places. Moreover, in such cases the land-derived clay particles and the carbonates deposited as the shells and bones of marine organisms are not only of different origin but are affected differently as regards solution and transport so long as they are exposed to the action of the water. Too little account still seems to be taken of the fact that run-off from the land, and the supply of riverborne minerals and organic matter, and also the rate of circulation of the ocean water itself, must have changed significantly as ice ages succeeded warm eras and vice versa. The best sites for sediment examination are for these reasons probably on rises of the ocean floor in the deep regions, well away from the continental slopes. In such places, according to ERICSON and WOLLIN (1966), the rate of deposition of the sediment has remained reasonably constant through all the great climatic changes in the Pleistocene. Nevertheless, the rapid sedimentation of deposits on the continental shelves (if we confine attention to those places where they are undisturbed and provided that the variations in their rate of deposition can be established) must offer a more detailed record. They continue to attract interest accordingly.

Rates of sedimentation averaging as much as nearly 90 cm/1000 years over the last 10 000 years have been reported on the floor of the Gulf of Aden at a depth of about 800 m (OLAUSSON and OLSSON 1969), with evidence of 50% more rapid deposition than this and 70% greater aeolian component – suggesting a particularly windy climate – for about 700 years just in the so-called Pre-Boreal climatic era about 8000 years ago. There are not many cases where the changes of rate with changes of climate have been so clearly established. High rates of deposition have also been found near Java (average 54 cm/1000 years) and in the western Mediterranean near 5°w (30 cm/1000 years) as well as in the deep Indian Ocean near the African side (20 cm/1000 years) at depths of over 5000 m, near 53°E at $4\frac{1}{2}$° and 7°N (OLAUSSON *et al.* 1971).

ERICSON and WOLLIN (1966) state that a sedimentation rate of about 2·5 cm/1000 years is typical for the Atlantic; but this seems too broad a generalization, ignoring differences with latitude and climate. EMILIANI (1958) reports a mean rate of 8 cm/1000 years for a point near 35°N 44°w in the North Atlantic, and a rate of 11 cm/1000 years has been reported near the Mid-Atlantic Ridge about 40°N. An average of 3·3 cm/1000 years is reported by EMILIANI (1966) for the Caribbean and equatorial Atlantic, whereas 1 cm/1000 years seems to be typical for the equatorial Pacific. According to ERICSON and WOLLIN the mean figure for the latter area should be only about 0·3 cm/1000 years. About half the floor of the Pacific Ocean consists of red clay, a deposit found where the calcium carbonate has been dissolved: in this situation an average sediment thickness of as little as 1 mm may correspond to 1000 years. Samples of sediment from the bed of the Arctic basin indicate an accumulation rate of about 2–3 cm/1000 years and samples from the Antarctic Ocean around 0·7 cm/1000 years.

All these estimates are averages over the last 10 000 or several tens of thousands of years, usually based on ^{14}C datings or else on the ^{231}Pa/^{230}Th method. Some of them have been used as a means of dating the beginning of the Pleistocene over a million years ago and, not surprisingly, there have been differences of opinion about the age concerned by up to a factor of three (cf. especially EMILIANI 1955, ERICSON *et al.* 1964); though this may be partly a matter of how different authors have defined the Pliocene–Pleistocene boundary (ERICSON and WOLLIN 1968).

Before the days of radiometric dating attempts were made to estimate the rates of deposition of ocean bed sediments by two other means:

(1) By counting the magnetic spherules of iron and iron oxide present, which are presumed to be of extraterrestrial origin, and assuming a constant supply of this meteor dust.
(2) By assessing the proportion of coccoliths in the sediment and assuming a constant productivity of the minute organisms (*Coccolithophoridae*) that produce them.

Neither of the assumed constant rates on which these approaches were based, however, rests on firm knowledge.

Volcanic tephra layers (tephrochronology)

Within the fan-shaped area of deep deposit of debris (tephra) from a single great volcanic eruption, or the wider region covered deeply by layers of ash in a period of abnormal activity of the nearer volcanoes, the age of which has been established by radiometric dating or other means, this layer provides a 'fix' in the dating of the stratigraphy wherever it can be identified. In some cases, a whole sequence of tephra layers of known date provides a fairly recognizable chronology, though of no more than regional significance.

Most examples of the use of tephrochronology are to be found, not surprisingly, in and near regions of continual volcanic activity.

Probably the most extensive scale on which this method has been used is seen in the series of postglacial tephra layers found by AUER (1956, 1958, 1959, 1965) first in Tierra del Fuego and later over wide areas of Patagonia and used by him to date the stratigraphy of many peat bogs. AUER does not give details of the horizontal extent of the ash deposits from the different volcanoes, though it appears that the deposits in the Tierra del Fuego bogs are fairly uniform over the whole island. But although different volcanoes, and different types of ash, are involved farther north, in Patagonia, he states that 'with few exceptions eruptive volcanic activity . . . was rhythmic and practically simultaneous' (AUER 1956, p. 211) and he demonstrates the effective synchroneity of the four main eruptive waves that produced the layers catalogued below all the way from 54°s to 41°s by a diagram (AUER 1958, p. 83). The chronology and the series of waves of heightened activity of the south Andean volcanoes which it represents was specified by AUER as follows:

Layer I: A white rice-like tephra layer, commonly 1 to 3 cm thick, ^{14}C dated as about 9000 years old.

Layer II: A greenish, or yellow-grey, brown tephra 5 to 30 cm thick, the coarsest grained of all the postglacial tephra layers in Patagonia, chemically similar to earlier layers in Late-Glacial times. This layer was ^{14}C dated as 5500 to 5000 years old.

Layer III: A white tephra layer, commonly 3 to 10 cm thick, ^{14}C dated as 2500 to 2200 years old.

Layer IV: Light grey ash forming a stratum at or near the top of the pollen stratigraphy profiles of the Patagonian peat bogs. From the range of depths concerned, in relation to the depths of the ^{14}C dated layers I–III lower in the bogs, this layer IV must represent volcanic activity in the last 400 to 500 years.

A series of older layers, deeper in the bog stratigraphy and evidently related to similar bouts of extra volcanic activity, was identified by AUER as follows and dated much more tenuously by changes in the vegetation indicated by the pollens counted in the bog layers and appearing to indicate a history of sharp changes in the thermal climate resembling the established history in Europe. These earlier layers were specified as follows:

O_1: A stratum of pumice particles, apparently 17 000 to 16 000 years old.

O_2: Another stratum of pumice particles thought to be 14 000 to 13 500 years old.

O_3: An ash layer, apparently about 12 000 years old.

O_4: Another ash layer, thought to be 11 000 years old.

Other countries for which such chronologies of volcanic deposit layers can be established include Iceland, Japan and New Zealand. In Iceland up to 100 tephra layers may be found in a single postglacial soil section. Among the great tephra deposits whose extent and thickness over Iceland have been mapped are the Hekla eruption layers dated around 4000, 2000 and 750 B.C. (this last being the greatest Icelandic dust layer of postglacial times, covering about 80 000 km² of the land surface mainly north from the volcano and 260 000 km² total area when the deposit on the sea bed to the north and northeast of the island is included), also layers from the Hekla eruptions around the time of Christ and in A.D. 1104 and from the eruption of Öraefi in A.D. 1362 (which produced the greatest ash layer since the human settlement of the island). The great output of lava in Iceland also means that some lava layers cover considerable areas and provide a similar dated marker in the stratigraphy in the area of their occurrence, particularly that from the eruption of Laki and Skaptar Jökull in A.D. 1783; this, the greatest single lava flow on Earth in historical times, was emitted from a fissure 25 km long and covers nearly 600 km² (THORARINSSON 1944, 1958, THORARINSSON *et al.* 1959).

Japanese studies have established separate dated sequences of ash and pumice layers from the different volcanoes, e.g. the study of Mount Fuji and adjacent areas by MACHIDA (1964), that of the Ontake pumice by KOBAYASHI (1965) and others dealing with the history of Asama pumice and pumice layers deposited on Kyushu and Hokkaido islands about 33 000 years ago (*Japanese Committee on Tephrochronology* 1969).

In North Island, New Zealand, deposits of pumice and ash originating in the eruptive activity at Lake Taupo (38°s 176°E) around 20 000 years ago, as well as about 1800 and 700 to 1000 years ago, have been used as reference levels for judging other ages in soil sections (see e.g. SCHOFIELD 1965, LAMB 1970).

Similarly, the great eruption of Santorin (36°N 25°E) in the Aegean about 1500 and 1450 B.C. produced a thick ash layer over the regions about the eastern Mediterranean.

Other volcanic deposits useful in dating stratigraphic sections in which they are found include:

(*a*) The series of ash deposits on the bed of the northeast Pacific from Mount Mazama (near 49°N 120½°w), the oldest of which is dated about 6600 years ago (GRIGGS *et al.* 1970).

(*b*) The numerous ash deposits in the Antarctic ice sheet and on the sea bed from the eruptions of Mount Erebus (77½°s 167°E) and possibly other Antarctic volcanoes.

(*c*) Similar ash and pumice layers in the Arctic from the Iceland, Alaskan and Kamchatka volcanoes.

Dating of volcanic layers themselves has usually been determined from ¹⁴C dating tests performed on organic carbon – humus from peat or the remains of marine organisms on the sea bottom – immediately above and below the volcanic material. In the case of ash buried in an ice sheet the age may be determined either by the stratigraphy of annual ice layers (if that is readily established, e.g. in the uppermost part of the ice) or by the flow method described above (pp. 75–7).

In some parts of the world extensive deposits of thick strata of volcanic material, bearing witness to long periods in the past of much greater volcanic activity than now, can be roughly dated by their relationship to the magnetic reversals and their place in the stratigraphy. FUCHS (1947) found three such layers near the Rift Valley in East Africa apparently coinciding, at least roughly, with the onset of the main periods of pluvial climate there in the lower, middle and upper Pleistocene. KENNETT and WATKINS (1970) have indicated extensive ash layers in the southernmost Pacific near the Ross Sea, apparently associated with several times of frequent volcanic explosions in or near the Balleny Islands (66–68°s 162–165°E) in the last 700 000 years and some up to 3 million years ago. Dating of other points in the stratigraphy by interpolation between these volcanic layers would, however, have uncertainties about sedimentation rate added to the error margin of the dating of the ash layers themselves.

Geomagnetic reversals and archaeomagnetic dating

The Earth's magnetic field undergoes variations of strength, which are usually assumed to be cyclic with a period of the order of 10^4 years (cf. p. 65) between the minima (BUCHA (1970) gives some detail for the last 10 000 to 20 000 years).[1] At usually much longer intervals, the Earth's main (dipole) magnetic field reverses its direction (and is presumably very weak for some time during the reversal). A fossil record of the reversals is preserved in the remanent magnetization of the rocks, imprinted in them at the time of crystallization, as may be shown by heating rock or pottery and allowing it to cool in a magnetic field. The reversals are registered in sediments containing volcanic materials (ash or lava).[2] There is no apparent regularity or periodicity in the timing of the reversals, of which very many are recorded in the Earth's history and at least 27 in the last $4\frac{1}{2}$ million years (the total duration of either polarity seems to be the same, and the mean field strength in either case is the same). 'Epochs' of dominance of either 'normal' (as now) or reversed polarity seem most frequently to last about a million years; the longest 'epoch' recognized spanned 5×10^7 years, the shortest under 10^5 years. There are, however, also shorter episodes (known as 'events') of opposite polarity, typically lasting a few tens of thousands of years within the longer 'epochs'. The latest such major event so far known, the Laschamp event, gave reversed polarity between 20 000 and 30 000 years ago (see also p. 88).

1. This statement refers to the main dipole moment associated with rotation in the Earth's molten core. The non-dipole or irregular component, which is attributed to convective processes in the Earth's interior and amounts to 20% of the present field strength, has shown a variation on a scale of about 10^3 years, associated with the secular wandering of the magnetic poles (see for example, CHAPMAN 1951, AITKEN 1970).

2. Evidence of the reversals of the Earth's magnetic field preserved in ocean bed sediments was first demonstrated by HARRISON and FUNNELL (1964).

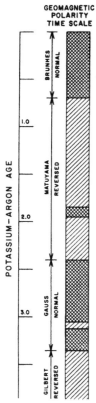

GEOMAGNETIC
POLARITY
TIME SCALE

Fig. 13.14   Geomagnetic polarity epochs of the last 4 million years. Cross hatching = 'normal' polarity, i.e. as now. Diagonal hatching = reversed polarity.
Two of the opposite polarity events are shown, one in the Matuyama reversed polarity epoch and one in the Gauss normal epoch.
(*Adapted and simplified from an early paper by* Cox.)

The magnetic reversals may also be marked by extinctions of species of fauna, since the Earth's surface is presumably subject to an increase in cosmic ray bombardment when the geomagnetic field is weak and this should cause an increase in the incidence of mutations: evidence of such effects in the minute marine fauna represented in ocean bed sediments has been reported (e.g. GLASS *et al.* 1967, KENNETT and WATKINS 1970).

It has been suggested that times of weakness of the Earth's magnetic field may be times of intense volcanic activity and that, through this, they may tend to induce ice ages. It may be possible to settle this, if dating techniques become precise enough to establish whether there is indeed a correlation between the reversals and these things.

A time scale of the geomagnetic reversals had been established by potassium-argon dating (Cox 1969) and is illustrated here in simplified form in fig. 13.12, showing little more than the main epochs recognized in the last few million years. In any probes of actual sediments from different places on the ocean bed the reversals will be found at somewhat different depths and vertical separations in the cores, but their identification is aided by the additional detail of reversal events which has become known since the stage of knowledge represented by fig. 13.14. This additional detail of opposite polarity events, and the names (mostly after places of discovery) by which they are known in the literature, may be listed with approximate ages as follows:

EPOCH	EVENT
Brunhes normal epoch	Stärnö event
0 to 690 000 *y* B.P.	*c*. 890 B.C.
	Laschamp event
	10 153 to 10 127 B.C.
Matuyama reversed epoch	Jaramillo event
690 000 to 2·43 *my* B.P.	890 000 to 950 000 *y* B.P.
	Olduvai events
	1·61 to 1·63,
	1·64 to 1·79,
	1·95 to 1·98 and
	2·11 to 2·13 *my* B.P.
Gauss normal epoch	Kaena event
2·43 to 3·32 *my* B.P.	2·74 to 2·86 *my* B.P.
	Mammoth event
	2·94 to 3·06 *my* B.P.
Gilbert reversed epoch	Cochiti event
3·32 to over 4·55 *my* B.P.	3·70 to 3·92 *my* B.P.
	Nunivak event
	4·05 to 4·15 *my* B.P.

This catalogue of magnetic epochs and events is useful in relating contemporaneous phenomena all over the Earth, because the reversals are a global phenomenon and the pattern of their history must be everywhere the same. Nevertheless, it has, of course, no greater accuracy than the radiometric (or other) method by which the time scale was established.

The shorter time-scale variations of the Earth's magnetic field may also be useful in dating, known as archaeomagnetic dating because of its application to the times with which archaeology is concerned. THOMPSON (1973) has shown, by comparing records obtained from Windermere and Lough Neagh, in northwest England and northern Ireland respectively, that the remanent magnetism in undisturbed lake sediments preserves a detailed record of the past changes of the Earth's magnetic field. A marked periodicity of about 2800 years was shown by the history of the declination, which was followed through six cycles in the Windermere sediment. No great variation and no such regularity was apparent in the inclination record. The intensity of the horizontal remanent magnetism showed just a single cycle, with its maximum differing from BUCHA'S and others' results (see p. 65) on baked archaeological material in being in the (warmest) mid postglacial times; this was associated by THOMPSON and by the late Dr MACKERETH with a direct relationship between this natural remanent magnetic intensity in unmetamorphosed sediments and the carbon content of the sediment, hence apparently affected by the intensity of biological activity in the lake at the time of deposition. Thus, it was just the strong swings shown by the history of magnetic declination which, dated in one lake by radiocarbon or other means, themselves provided a ready method of identifying approximate dates in the sediments of other lakes in the British Isles region.

The possibilities of this method are being enlarged and refined by comparisons with the Swedish varve chronology and other long vertical cores from other sediments, with results that tend to resolve the problems in either type of record, so improving the calibration of both (TARLING 1975) and offering an accuracy better than radiocarbon dating. It has already (NOEL and TARLING 1975) greatly sharpened dating of the short-lived Laschamp geomagnetic event (which is now seen as a quasi-reversal of the Earth's field, when the geomagnetic north pole made a brief excursion as far as the equator at 50–65°W, outside the usual range of secular variation, though the 60°W–120°E meridian seems to be a preferred path for the pole during polarity transitions), and has indicated the time around 860 B.C. as the date of a similar event detected in the remanent magnetism of Swedish varves.

Astronomical dating (Earth's orbital cycles)

The cyclic variation of the Earth's orbital elements described in Volume I of this work, as calculated by MILANKOVITCH (1930), must affect the amount of solar radiation available at different latitudes and seasons (and to a much less extent the total for the year), as recently recalculated by VERNEKAR (1968). There are good grounds for thinking that this must induce an alternation of warm and cold climatic periods on the Earth and that the timing of the last major cycle fits that of the last glaciation (LAMB *et al.* 1966, see also EMILIANI 1961 and SCHWARZBACH 1961, p. 218). If one accepts this as the cause, at least of the timing, of *all* the Quaternary glaciations, as first suggested by ADHÉMAR in 1842 and JAMES CROLL about 1860, this provides a time scale to which all fossil evidence of these ice ages should fit. This method of dating was advocated over many years by ZEUNER (e.g. 1958) and has been quite widely applied e.g. by BERNARD (1962), KUTZBACH *et al.* 1968).[1] The objection to it is the sweeping basic assumption that all ice ages are essentially controlled by the same principal cause and the extent to which it involves arguing in a circle, dating climatic evidence

---

1. A recent publication of the Geological Society of London included a proposal (EVANS 1972) to make more extensive and formal use of the MILANKOVITCH 40 000-year obliquity cycles as a means of dating than ever before, taking these cycles (counted back in time from the present cycle as No. 1) as the units of a time scale. The cold and warm phases of each cycle are to be labelled C and W respectively; on this scale the cool-temperate (Upton Warren) interstadial that divided the last glaciation is labelled 2W and the last (Eemian) interglacial becomes 3W. The earliest glaciation that is assigned to a particular cycle is labelled 24C and is thus about 960 000 years ago.

As radiometric dating methods lack the precision at the ages in question to test the association of glaciations earlier than the last one or two with the MILANKOVITCH cycles, this elaborate scheme must be regarded as fostering unprovable presumptions about the origins of many glacial episodes in the Quaternary.

by a climatic argument. Moreover, even if the assumption were entirely sound, an error of just one cycle of ellipticity, obliquity or precession in the identification of a given piece of climatic evidence would lead to an error of about 100 000, 40 000 or 20 000 years (respectively).

Weathering of the landscape

Besides the methods of dating depending on precise measurements, described in the previous sections, ZEUNER (1958, pp. 349 ff.), OLAUSSON (1971) and others have pointed out that estimation of the times required by certain weathering and geomorphological processes may supply a desirable check on the reasonableness of some of the results. Thus IVERSEN (1958) has described the normal cycle in the development of vegetation, and hence soils, in the course of a single glacial-interglacial cycle of typical duration in the Pleistocene, and OLAUSSON himself gives the cycle of development and subsequent silting up of pro-glacial lakes in regions such as the Baltic as between 5000 and 15 000 years.

13.3   Field evidence of past climates interpreted by physical science

In this section we shall survey types of evidence of climate in the past and methods of interpretation available by means of the physical sciences, including the Earth sciences.
    Increasing use is now being made in palaeoclimatology of the general principle that where more than one stable isotope of a commonly occurring element exists, as with oxygen, hydrogen and carbon, and the proportion of the heavier isotopes changes during changes of state (as from liquid to vapour, solution to gas, or during chemical and biological action) and depends on the temperature at which the change of state occurs, fossil evidence of that temperature is registered in the isotopic composition of the product. LIBBY (1972) has stressed the greater reliability of palaeotemperature determinations, and the likelihood of additional information, that could be gained by 'multiple overdetermination of temperature' using measurements of many temperature-dependent isotope ratios, including possibly $^{15}N/^{14}N$, $^{34}S/^{32}S$ and $^{41}K/^{39}K$ in addition to all those discussed in the following pages.

Oxygen isotopes and palaeotemperatures

    Of all the multifarious physical traces of former climate none offers the opportunity for more precise measurements or appears to yield more detailed information than the varying proportion of oxygen-18 in the ice at different depths in the polar ice sheets and in the carbonates deposited on the ocean bed. The variations are related to temperature.[1]

1. How oxygen isotope measurements might be used to
    determine past temperatures of the oceans was first
    worked out by UREY (1947).

There are three isotopes of oxygen, ^{16}O, ^{17}O and ^{18}O; they are present in water in all its phases, eight-ninths of the mass of water being oxygen. The proportion of oxygen-17 is so minute that it may be ignored in these discussions. Oxygen-18 accounts for about 2 parts per 1000. The most important isotope components of water are $^{1}H_2{}^{16}O$, $^{1}H^{2}H^{16}O$ and $^{1}H_2{}^{18}O$ and their average proportions in the world's oceans are respectively 997 680, 320, and 2000 parts per million. A fundamental treatment considering all these has been given by DANSGAARD (1964) (see also EMILIANI (1955), particularly as regards marine carbonates). Here we shall consider how the oxygen-18/oxygen-16 ratio is affected by atmospheric processes (the deuterium (^{2}H) and heavy oxygen proportions in fact undergo parallel variations). Oxygen-18 and deuterium are both stable isotopes, so that the proportion present does not alter with time. Measurements are made with a mass spectrometer.[1]

Fractionation occurring when oxygen is exchanged between different substances, or when water passes from one phase to another, is the process in which temperature registers an effect. This means that the transferences are selective: the proportions of the different isotopes change as the oxygen passes from one stock to the other, just as distillation separates substances of different volatility. This is indeed what happens to the oxygen isotopes in the evaporation at the ocean surface and in all subsequent condensation processes. An analogous selection takes place when biological organisms use ocean water, carbon dioxide and the soluble calcium hydrogen carbonate to build up the carbonate ($CaCO_3$) in their shells. The vapour pressure of the heavy isotope $H_2{}^{18}O$ is less than that of $H_2{}^{16}O$, but the difference gets less as the temperature rises (with the kinetic energy of the molecules increasing and their mass contributing a smaller proportion of it). The vapour pressure ratio $H_2{}^{18}O/H_2{}^{16}O$ is 0·99° at 0°C and 0·992 at 25°C. The corresponding $^{18}O/^{16}O$ fractionation factors for $CaCO_3$ and liquid water are 1·025 at 0° and 1·021 at 25° (these are the fractionation factors corresponding to equilibrium between the isotopes at the boundary surface between the phases; this equilibrium may take some time to be realized).

The result of the fractionation is that the $^{18}O/^{16}O$ ratio is always rather less in the atmosphere than it is either in the ocean or in the products of condensation, for during condensation of water droplets or ice from the vapour phase again the selective process operates towards the equilibrium value with more of the heavy isotope in the liquid or solid phase. Similarly, when carbonate is taken out of solution, there is a higher proportion of the heavy isotope ^{18}O in the carbonate than in the water. All these fractionations depend on the temperature at which the process occurs. Thus, the $^{18}O/^{16}O$ ratio in the water vapour in the atmosphere depends largely on the temperature of the ocean surface in the tropics, where most evaporation occurs; and the ratio found in the polar ice sheets depends on the ratio in the atmosphere and on the temperatures prevailing at the condensation level in the atmosphere at the time when the snow was formed over the ice sheet. The $^{18}O/^{16}O$ ratio in

1. The principle of the mass spectrometer is to measure the deflection of electrically accelerated atomic particles by a known magnetic field applied at right angles to their path, the particles being less deflected the greater their mass.

carbonates deposited on the ocean bed through the death of organisms in the sea depends on the temperatures prevailing in the water at the levels in which these organisms lived.[1] Hence, provided the $^{18}O/^{16}O$ ratios prevailing in the oceans and in the atmosphere are known, the measured departures

(1) in ice taken from an ice sheet indicate the temperature at which the snow formed in the air overhead;

(2) in ocean bed carbonate remains of planktonic species (i.e. plankton living near the ocean surface) indicate the water temperatures that prevailed in the surface layers when the organisms were alive.[2]

(3) in ocean bed carbonate remains of benthonic species (which live near the ocean floor) indicate the prevailing temperatures of the bottom water when these organisms were alive.

The departures are normally expressed as relative deviations, δ, in parts per 1000 (‰) from the $^{18}O/^{16}O$ ratio found in standard mean ocean water, 'SMOW':

$$\delta = 1000 \frac{^{18}O/^{16}O_{(sample)} - {}^{18}O/^{16}O_{(SMOW)}}{^{18}O/^{16}O_{(SMOW)}}$$

Nevertheless, it must be remembered that the fractionation factor is applied to whatever the ratio is in the particular stock from which the oxygen is passing. Hence, the $^{18}O/^{16}O$ ratio observed in any individual fall of rain or snow may be unrepresentative of the average conditions. Hence, also, interpretation of the ratio found in precipitation (and glaciers) in continental interiors is likely to be impossibly complicated because of recycling (repeated evaporation and precipitation) of the water on its way from the ocean. Complications also arise in precipitation in low latitudes, and in middle latitudes in summer, because of the range of heights and temperatures at which condensation occurs and the recycling which goes on within convective clouds.

Interpretation is likely to be most straightforward in the case of the snow deposited on ice sheets in high latitudes at points where these are nourished by moisture-bearing winds

1. SHACKLETON, WISEMAN and BUCKLEY report (1973), on the basis of the first measurements that have been possible on foraminifera collected live, that the oxygen isotope ratio in different species differs and in the case of *Globeriginoides ruber* differs so much (-0.50‰) from the sea water in which they live as to be equivalent to an error of $2.5°C$ (too warm). In another species, however, *P. obliquiloc-ulata*, the difference ($+0.06$‰) from the oxygen isotope ratio in the water was insignificant. If these fractionation differences are constant characteristics of each species, as they appear to be, it is important to identify the species and allow the appropriate correction in all palaeotemperature derivations from their remains in ocean bed deposits. Such corrections are sometimes referred to as 'calibration' of the palaeothermometer'.

2. The adjective *planktonic* is now commonly used to indicate species that live in and drift with the water near the surface. *Pelagic* was the adjective formerly applied to species (particularly seals and whales) dwelling near the surface of the deep ocean, whereas plankton properly describes the (small) organisms that float and drift with the water at any depth. *Benthic* or *benthonic* describes species (flora and fauna) which inhabit the depths of the sea.

direct from the ocean (anomalies observed near the coast of Antarctica are tentatively attributed partly to exchanges of H_2O vapour in the snow drifted by the continual strong katabatic winds from the interior and partly to the origins of the ice found near the edge of a great ice sheet).[1] DANSGAARD (1964) gives a plot of the present-day mean annual values of δ observed in precipitation against mean annual temperature, which show a nearly linear progression at places with oceanic climates from $\delta = -6$, $T = +11°$ at Valentia (Ireland) to $\delta = -18$, $T = -7°$ at Upernavik (west Greenland) and in the central parts of the ice caps to $\delta = -49$, $T = -49°$ at the South Pole.

The most useful, empirically derived equations for interpreting oxygen-18 measurements are:

(1) the mean annual temperature T_a of surface air today may be obtained from the mean annual δ_p value of the precipitation in the same area by the relation due to DANSGAARD (1964):

$$0.7T_a = \delta_p + 13.6.$$

We see from this that the $^{18}O/^{16}O$ ratio changes by 1 part per 1000 for a 0.7°C change of temperature.

(2) The mean temperature T_o at which the shells of foraminifera are formed in the ocean is given by the equation due to EPSTEIN et al. (1953):

$$T_o = 0.14(\delta_F - \delta_o)^2 - 4.3(\delta_F - \delta_o) + 16.5$$

where δ_F and δ_o are the deviations of the $^{18}O/^{16}O$ ratio in the foraminifera and in the water in which they live from the SMOW standard value. Because the observed values of $(\delta_F - \delta_o)$ range between about $+1$ and -1, the squared term is small and variations of T_o are mainly accommodated by changes in the value of the term with the negative sign: so negative values of oxygen isotope anomaly in the shells correspond to the highest ocean temperatures. A change of 1 part per 1000 corresponds in the case of shells to about 4.3°C.

Oxygen-18 measurements are subject to such small margins of error – standard errors of δ around ±0.2 per 1000 are attainable in work on ice, equivalent to about ±0.3°C – that they can be used to identify the seasonal variations of temperature of the snowfall which supplied the glacier ice in a sample. Hence, ^{18}O measurements make it possible to distinguish the year layers in the stratigraphy of ice sheets at depths far beyond where they can be

---

1. According to the accepted pattern of flow in an ice sheet (fig. 13.12), the farther inland the ice forms the greater the depths it goes to in the course of its subsequent flow. Hence, the ice near the surface which melts in the ablation season on the periphery of Greenland or Antarctica has formed at no great distance inland, and it is indeed found to have δ values corresponding to precipitation in the climate of the same area. Thus, the upper ice in a core near the Adélie Land coast of Antarctica had a δ value of -20 parts per 1000, whereas the bottom half of the 98 m core gave $\delta = -44$ corresponding to conditions much farther inland (case cited by DANSGAARD and TAUBER 1969).

identified by other means.[1] LANGWAY (1967) illustrates this by plots of δ against depth through 2 to 4 metre segments of an ice core in northern Greenland (77°N 56°W), each representing 5 to 10 years accumulation, which show the seasonal fluctuation of temperature clearly at over 400 metres depth where the snow is about 1000 years old. More recently, JOHNSEN et al. (1972) have shown that the seasonal fluctuation is still quite clear at 1000 m depth in the Greenland core, where the ice is 8300 years old and the year layers have been thinned, by plastic deformation and spreading, to a small fraction of their original thickness.

Fresh waters have lower values of $^{18}O/^{16}O$ ratio than ocean water, as is shown by present-day measurements, the negative value of δ increasing the colder the origin of the precipitation. Sea water near river mouths is rendered unrepresentative of the ocean by the fresh water admixture.

Interpretation of the ice age part of the $^{18}O/^{16}O$ record obtained either from an ice sheet core or from the carbonates deposited on the ocean bottom needs adjustment for the changed proportion of the isotopes then ruling in the oceans. At the present time we estimate (Volume 1, p. 482) that about 2·5% of the Earth's total water stock is outside the oceans as fresh water or ice, mainly (1·9%) ice. At times of maximum glaciation as much as 5 to 5·5% was apparently stored in the form of ice on land, and presumably in all 6% or rather more was outside the oceans as fresh water or ice which had been precipitated from the atmosphere at such temperatures that it was poor in ^{18}O. In consequence, the main ocean reservoir must have been significantly enriched in ^{18}O as compared with now.[2] SHACKLETON even suggested in 1967 that the δ changes in the ocean derived from measurements on the foraminifera on the ocean bed may be entirely attributable to the quantity of water removed to build the ice sheets and that the temperature of the oceans hardly changed at all. The

---

1. In principle, the isotope method therefore offers a method of indicating even short-term fluctuations of temperature and of precise dating of the ice to great depths, but since eight isotope determinations per year are required for this the cost of the many thousands of mass spectrometer measurements would probably be prohibitive. Moreover, good results are only possible in sites with a never-failing supply of winter and summer precipitation, a condition which makes the site of the northwest Greenland core peculiarly favourable. There may be a risk of counting an occasional year too many through a rare case of a long spell of very mild air in the winter half-year being mistaken for summer conditions or, in more continental situations, of missing a year through lack of precipitation in one season or the other; but it seems likely that these errors have not arisen in northwest Greenland. JOHNSEN et al. (1972) report that the amplitude of the seasonal fluctuation of δ values is 10‰ in freshly fallen snow in the northwest Greenland ice but sinks to 2‰ within a few hundred years, thus giving only a damped response to the seasonal temperature range.

The shorter time scale fluctuations, within the seasons, disappear in under 20 years, presumably due to the water vapour exchanges in the strong temperature gradients that exist in the surface snow as long as there are air channels trapped in it; once these are closed little more mass exchange can take place (i.e. only by slow molecular diffusion in the solid ice) and the annual fluctuations of δ still have an amplitude of 1·5‰ in ice that is many thousands of years old. The yearly layers are not clearly shown by the oxygen-18 measurements at most depths in the ice core at Byrd station (80°S 120°W) in the interior of Antarctica reported by JOHNSEN et al. (1972); they are discernible near the surface and, surprisingly, in a limited section of the core where the ice is believed to be about 15 500 years old. Snow accumulation in the area is barely half that at the Greenland site and certainly much more variable than in northwest Greenland.

2. This change – like all other changes in the watermass – would, of course, take many hundreds of years to reach the most stagnant parts of the deep ocean.

density and salinity of the water remaining in the oceans in the periods of maximum glaciation must certainly have been increased. The rapid changes of faunal species distribution and oxygen isotope ratio about 11 000 years ago registered in the sediments on the bed of the North Atlantic Ocean (BROECKER *et al.* 1960), however, demand both a quick rise of temperature by several degrees and rapid dilution of the ocean by water from the melting of the great northern ice sheets about that time (see also MERCER 1969).

Estimates of the $^{18}O/^{16}O$ ratio in the oceans at the times of continental glaciation, the required datum for calculation from the observed values of δ in $CaCO_3$ deposits and in ice of the temperature anomalies affecting the ocean itself and those affecting the precipitation over any part of the Earth at such times, are inevitably subject to some uncertainties and the figures used by EMILIANI, $\delta_{ice\ sheets} = -15‰$ and $\delta_{ocean} = +0\cdot5‰$ (the former taken as characterizing the great North American ice sheet and the Scandinavian ice, both in much more temperate latitudes than Antarctica or Greenland), have been widely criticized. DANS-GAARD and TAUBER (1969) report an average δ value of about $-30‰$ in the upper part (representative of present climate) of the core bored right through the inland ice at 70°N 56°W in northern Greenland and a return to similar values near the bottom of the ice, whereas the layers of ice between 10 000 and 70 000 years old gave lower δ values with an average as low as $-41‰$ for the ice formed during the last glacial climax 20 000 to 25 000 years ago. The latter figure must have applied to much of the northern part of the great North American ice sheet of that time also, but the same authors arrive at an average δ value of about $-30‰$ (considered as a conservative estimate) for the former North American ice sheet as a whole, on the basis that the precipitation from which it was formed was representative of temperatures about 12°C lower than now in the known position of the ice crest and at a height of 2 km above sea level. A mean δ value of $-30‰$ was also taken for the former Scandinavian ice sheet and for the excess ice present at the same time in other parts of the world, including the margins of Antarctica and Greenland. On the basis of these figures the change of mean δ value of the water that remained in the oceans should be $+1\cdot2‰$; this was considered a conservative estimate, though the true figure could be less if the temperatures typical of the condensation level where the precipitation that constituted the former ice sheets was formed were not as low as DANSGAARD and TAUBER assumed. Moreover, it is reasonable to assume, as EMILIANI did, that the composition of the ocean bottom water may have been affected by excess glacial melt water lowering its temperature and its δ value. Another approach to assessment of the mean isotopic constitution of the ice age oceans is by measuring the δ values in the bottom-living (benthonic) foraminifera deposited at such times and assuming that the temperature of the bottom water did not differ from that of today. But this also is to some extent arguing in a circle. On this basis, SHACKLETON (1967) concluded that the isotopic composition of the oceans showed an anomaly of $+1\cdot4$ to $1\cdot6‰$, which would imply that the temperatures prevailing in the surface waters were also effectively unchanged from today's, since the whole observed deviation of oxygen-18 in the pelagic foraminifera would be due to the changed composition of the water. DANSGAARD and TAUBER's figures would give an overall change of prevailing temperature of the surface waters of the equatorial Atlantic and the Caribbean between maximum glaciation and interglacial times of at most 2°C, compared with EMILIANI's derivation of 5° to 6°C.

It seems possible to obtain a credible independent estimate of the ice age temperature lowering in the oceans by a biological method (see pp. 237–43) using quantitative analysis of the relative abundances of different species of near-surface-dwelling foraminifera whose (present) temperature tolerances are known. The results (IMBRIE and KIPP 1971) are closest to EMILIANI's figures quoted above.

There is no doubt that the oxygen isotope ratios measured in the polar ice caps in ice that was laid down during the last major glaciation, between 70 000 and 15 000 years ago, imply a great lowering of temperature below present-day values. To estimate the actual temperatures and hence the amount of this lowering, however, the relevant δ value is the departure from the isotopic composition of the ocean in those times (*not* from today's SMOW). It must be dangerous to assume that the equation empirically derived for relating δ and temperature in present-day climates applies equally well in full glacial times, though it may hold good for calculations relating to other interglacials. For one thing, the position of the northwest Greenland core was many times farther from open ocean in the ice age than it is today and must have had a correspondingly continental climate. Despite these reservations, it seems to be implied that the annual mean temperatures prevailing over the part of Greenland where the ice in the core was formed were of the order of 20°C lower than today. However, recently careful measurements of the amount of air trapped in the ice (RAYNAUD and LORIUS 1974) have shown this to be much less in the ice from the times of widespread glaciation, suggesting that this ice was formed under lower atmospheric pressure, corresponding to a level of the ice surface 1300 m higher above sea level, than now. This extra elevation would account for half, or under some assumptions more than half, the lowering of prevailing temperature, so the temperature departure in northern Greenland from today's values, level for level, may have been between −7° and −10°C. At Byrd station in Antarctica (80°s 120°w) the temperature lowering indicated by the oxygen isotope measurements seems to have been of the order of 10° to 12°C, but no measurements indicative of the height at which the ice was then laid down are so far known to have been made.

The results that have been obtained from oxygen isotope work on carbonates could be illustrated by many examples from sea bed cores, of which many have already been examined. EMILIANI (1966a) cites the results of researches in a number of parts of the world which indicate that the temperature of the surface waters of the oceans in the middle latitudes of the northern and southern hemispheres has undergone a gradual decline (with superimposed fluctuations) over the last 150 million years, from an average of about 23° to the present 17°C. Similarly, the temperature of the surface waters in high latitudes, which supply the abyssal waters of all the world's oceans and have in general the same temperature as the abyssal waters, declined from about 14°C near the end of the Mesozoic (in the Cretaceous) era 75 million years ago to 11° in the mid Tertiary 30 million years ago and thereafter more steeply to about 3° in the last million years.

The many times repeated glacial-interglacial cycles of the Quaternary stand out from the oxygen isotope analysis of deep sea cores in the Caribbean and the generalized temperature curve for the last 420 000 years derived therefrom by EMILIANI (1955, 1966b), reproduced here in figs. 13.15(a) and (b). Results from the equatorial Atlantic and equatorial Pacific were very similar and have now been extended back through several more glacial-interglacial cycles (EMILIANI and SHACKLETON 1974). The dating of these cycles points to a recurrent cycle length close to 100 000 years, in keeping with the Earth's orbital variations.

The succession of δ values from oxygen isotope measurements in the Greenland ice core from Camp Century (77°N 56°W) by DANSGAARD et al. (1971), unconverted to temperatures but dated (according to the flow-model method), are shown in fig. 13.16. The corresponding record (EPSTEIN et al. 1970) from the ice cored at Byrd station, Antarctica (80°S 120°W), where precise dating by the flow method or in other ways is not yet possible because of the complicated flow over bedrock which is far below sea level at that point, may be seen in

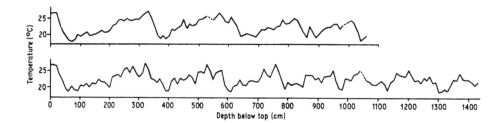

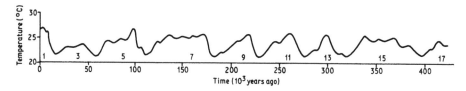

Fig. 13.15   Results of oxygen isotope analysis of deep sea cores in the Caribbean, δ $^{18}O/^{16}O$ as converted to temperatures of the surface waters by EMILIANI (1966b):
(a) Detail from two cores, both near 15°N 69°W, where the sea is about 4000 m deep, plotted against depth in the core.
(b) EMILIANI's generalized temperature curve for the surface waters of the central Caribbean, plotted against ages (determined by ^{14}C and $^{230}Th/^{231}Pa$ measurements) up to 420 000 years ago.
    (*Reproduced by kind permission of Dr* EMILIANI.)
    Note: Unconverted oxygen isotope variations in one of these cores are seen in fig. 13.21.

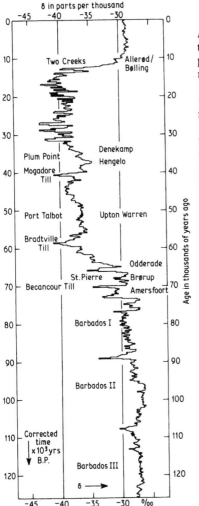

Fig. 13.16   Deviations (δ) per 1000 of $^{18}O/^{16}O$ ratio from the present-day standard mean ocean water (SMOW) value plotted against age at 200-year intervals in the ice in northwest Greenland through the last 125 000 years.

The measurements were made on ice from the Camp Century core at 77°N 56°W and ages determined by the flow method by DANSGAARD *et al.* (1971).

Tentative correlations with the Allerød and other warm interstadials recognized in Europe and with the Two Creeks ice re-advance in North America and other cold phases are indicated by their names at their known dates. The lower part of the diagram, before 70 000 years ago, registers the last interglacial.

(*Reproduced by kind permission of Dr* DANSGAARD.)

fig. 13.17; the estimates of date in this diagram are based upon assumed constancy of snow accumulation at about 12 grammes/cm²/year. The whole course of the last (Würm/Wisconsin) ice age is clearly represented, with the onset and end at about the same dates in both hemispheres. Many of the details appear to be in phase with the warm and cold stages (including among the former the long, cool temperate 'Upton Warren' interstadial) known in Europe and North America, the main ones being indicated by their names on fig. 13.16 (see also JOHNSEN *et al.* (1972) for more about the interhemispheric comparison). It is probably safe to treat this new evidence of far-reaching synchroneity of climatic vicissitudes as confirming broadly the dating of the ice cores. Fig. 13.18 shows the results from the Greenland core in more detail (DANSGAARD, private communication, 6 March 1972) for the

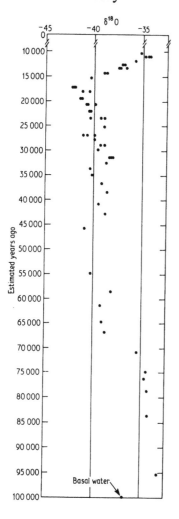

Fig. 13.17 Deviations (δ) per 1000 of $^{18}O/^{16}O$ ratio from SMOW values plotted against age (estimated by constant snow accumulation) in an ice core at Byrd station, Antarctica (80°s 120°w).

(*Reproduced by kind permission of Dr* A. J. Gow.)

last 1000 years. In two of the curves in this figure harmonic analysis has been applied to yield a forecast of implied temperatures for over 100 years into the future. In this diagram also there are some striking agreements with the course of climatic history recorded in manuscripts in Europe, particularly the cold conditions around the A.D. 1430s and even more sharply in the late 1600s as well as the outstanding warmth developing after 1850 to a climax in the first half of the present century. The climax of the medieval warmth in northern Greenland is, however, shown in the eleventh century A.D., or perhaps earlier, i.e. 100 to 200 years earlier than in Europe. This is not incompatible with the historical evidence from south Greenland. The suggestion that the same level of warmth recurred, though briefly, around A.D. 1380–1400 and 1470 finds an interesting echo in the (Lamb) summer index values in Europe (see pp. 31–4 and 562 in the Appendix), derived from manuscript reports of the

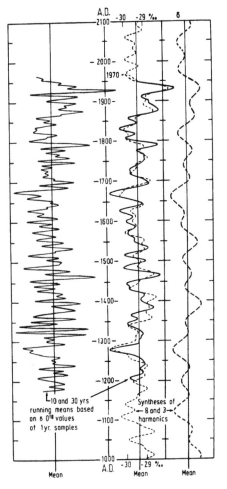

Fig. 13.18   Variations of the oxygen isotope ratio in the ice in northwest Greenland (77°N 56°W) since A.D. 1000.

Thin lines = 10-year means.

Thick lines = 30-year means.

Broken lines = The smoothed variations shown by the sums of the first three and the first eight harmonics.

The results of the harmonic analysis have been used to extrapolate the course of the same variations to the year A.D. 2100.

(*Reproduced by kind permission of Dr* DANSGAARD.)

seasons. There is some surprise again in the early date (*c.* 1930) of the peak of twentieth century warmth in north Greenland indicated by the oxygen-18 measurements. Clearly not every detail of the climatic sequence is identical in northern Greenland and Europe – summer sunshine in higher latitude anticyclones would account for some differences – but there is a broad parallel none the less.

Speleothems, i.e. stalactites and stalagmites in caves, composed of calcite deposited by water dripping from the roof of the cave and bringing the ^{18}O values representative of the atmospheric precipitation in the area, offer possibilities of palaeotemperature assessment in regions where there are no ice sheets. The oxygen isotope ratio found in the stalactite or stalagmite depends (1) on that in the water producing the deposit and (2) on the temperature at which the calcite crystallized. The speleothems examined by HENDY and WILSON (1968) in New Zealand caves show, in cross-section, a series of concentric growth rings, possibly annual. They grow much more rapidly than the deposits on the sea bed, and because of the rapid growth they can be used to derive high-resolution temperature profiles. The authors suggest that successive decade averages of temperature may be established to within 0·2°C. Their curve for the last 900 years in New Zealand (fig. 13.19) shows a remarkable extent of agreement with the course of average temperature prevailing in England over the same period, as derived by LAMB from the frequency of manuscript reports of extreme months (see pp. 31–4). The agreement seems about as close as that between the curve for northern Greenland (fig. 13.18) and England, though again there are differences of detail. The method, being applicable in regions where there are no ice sheets, offers prospects of contributing to knowledge of the detailed course of the more drastic climatic changes that have occurred in postglacial times.

Interesting deductions affecting the climatic history of the eastern Mediterranean and surrounding lands in postglacial times have been made by MARGARITZ and KAUFMAN (1973) from oxygen isotope measurements on the aragonite in shells on present and past shorelines along the coast of Israel. The present eastern Mediterranean has an anomalously high $\delta^{18}O$ value and is more saline than the western Mediterranean, both anomalies being attributed to the excess of evaporation over precipitation (estimated at 0·8 m/year at present). The anomaly of oxygen-18 seems to have come into being rapidly between 5000 and 4500 years ago and to have attained its present magnitude by some time in the first millennium B.C. The magnitude of the anomaly suggests that the temperature of the eastern Mediterranean has varied in parallel with that of the western Mediterranean, the eastern part being always about 4°C the warmer of the two basins and both having declined by about 2°C between 5000 and 2000–3000 years ago; also that the climate of the eastern basin was much more humid before the change.

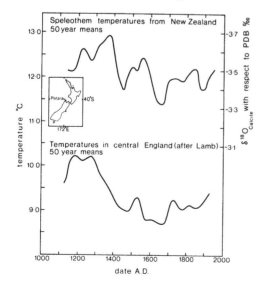

Fig. 13.19   Course of the 50-year mean temperatures in the Nelson area, South Island, New Zealand, since A.D. 1100, derived from oxygen isotope measurements on a stalagmite in a cave at Patarau.

Temperatures derived by LAMB for Central England, as shown in Volume I, p. 236, are repeated here for comparison.

(*Adapted from the original diagram kindly supplied by Professor* A. T. WILSON.)

WILSON and HENDY (1971) have taken the oxygen isotope record from polar ice sheets as essentially a measure of the temperature difference – in whatever climatic regime (glacial or interglacial) is for the time being prevailing – between the point concerned in high latitudes and the surface waters of the tropical oceans, where most water vapour is taken up into the atmosphere. Since it appears from this that the depression of temperature in the polar regions in the ice ages was greater than in the tropics, the ice sheet ^{18}O scale is inverted so that the curve rises with the tropics–pole temperature gradient, as in figs. 13.20(a) and (b). From this it appears that the temperature difference built up to a maximum in the final and culminating phase of the last glaciation, 25% above present values in the northern hemisphere and 20% above present values in the southern hemisphere. WILSON and HENDY suggest that this means that the general atmospheric circulation must have increased in strength correspondingly during the glaciation.[1] If this is true, and the strong thermal gradients must have produced very strong jet streams, it should presumably be interpreted as meaning a strong circulation concentrated in rather low latitudes: much more extensive areas in high latitudes than now were presumably more or less outside the strong circulation. Verification of the hypothesis is seen in the many evidences of a windy climate in glacial times in middle latitudes, in the wind-blown loess deposits in Europe and China and in the evidence of active, mobile dunes in central Australia despite a probably greater rainfall there than now.

Deuterium

The proportion of naturally occurring heavy hydrogen, ^{2}H, i.e. deuterium (a stable isotope), in atmospheric precipitation and assimilated in organic compounds in plants is found to be related to climate, decreasing with the long-term mean temperature. It may be possible therefore to use measurements of this in organic materials (e.g. wood and peat) for palaeotemperature determinations in regions where there is no ice. There are complications, however, associated with quite substantial changes in the proportion of the heavy isotopes

---

1. A word of caution seems advisable here, since the distribution by latitude of temperature changes in the middle and upper troposphere may differ from that near the Earth's surface. Figs. 7.25 and 7.26 (Volume 1, p. 305) showed that over the years between 1945 and 1965 when the general circulation was weakening, the cooling of the upper levels was greatest over the subtropical zone about 22° to 26°N; at the surface it was greatest in the Arctic, particularly near 80°N 60°E. Oxygen isotope measurements in ice from the Greenland and Antarctic ice caps are related to snow formed over those regions at various levels in the atmosphere, perhaps mostly a good deal nearer the surface than the upper air temperature measurements in recent decades, above referred to; the ^{18}O departures may be taken to represent the difference in temperature between the surface of the tropical oceans and the lower atmosphere over the polar ice sheets without giving any very reliable indication of the changes of temperature gradient in the upper air at the levels where the mainstreams of the atmospheric circulation are generated. The implication is that WILSON and HENDY's results are likely to be qualitatively true (i.e. as regards the *sense* of the past variations of strength of the main windstreams) when ice age and postglacial are compared, but may exaggerate the magnitudes of the changes; for the smaller climatic changes, as between 1945 and 1965, or even between medieval warmth and Little Ice Age, the change of wind strength need not even be in the right sense and apparently was not.

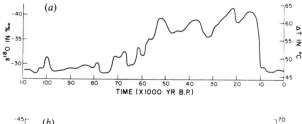

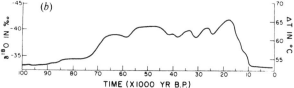

Fig. 13.20 δ values of the oxygen isotope ratio in the
polar ice sheets expressed as implied values of the
temperature difference between the polar regions and the
surface waters of the tropical oceans over the last 100 000
years.
(*a*) Northern hemisphere, based on the Greenland ice
   core at 77°N 56°W.
(*b*) Southern hemisphere, based on the Antarctic ice core
   at 80°S 120°W.
   (*After* WILSON *and* HENDY *1971; reproduced by
kind permission of the authors.*)

that accompany evaporation and condensation, so that the prevailing level of relative
humidity, the frequency of these phase changes, and the amount of precipitation and their
seasonal distribution may have to be considered.

Carbon-13

STUIVER (1970) reports that the $^{13}C/^{12}C$ ratio in the carbonates in at least some postglacial
lake sediments seems to vary linearly with the $^{18}O/^{16}O$ ratio. However, the temperature
dependence of the fractionation of this stable isotope of carbon is too small to account for
the variation observed; so fractionation during evaporation and photosynthetic changes
may also be involved. Hence, though they obviously register climatic changes, the variations
of carbon-13 considered alone seem unlikely to throw any clear light on these changes so far
as temperature is concerned. SHACKLETON (1970) has, however, shown how carbon-13 and
oxygen-18 measurements on shells, in particular of the common cockle (*Cerastoderma
edule*), may be used to interpret the nature of aquatic environments subject to seasonal and
secular changes controlled by climate. The shells used were found in a Neolithic mound on
the plain of Macedonia in northern Greece, 25 km from the present mouth of the river which
drains the area. The structure of the shells revealed seasonal changes in the ^{13}C and ^{18}O

fractions. Now, river water is low as compared with sea water on both these heavy isotopes. The bicarbonate in sea water has the proportion of ^{18}O found in the oceans, not the reduced values representative of atmospheric precipitation; and the bicarbonate in river water incorporates isotopically light carbon derived from the decomposition of plant matter. Therefore in estuaries the proportion of the heavy isotopes decreases with increasing distance from the sea, and in climates where the river flow varies greatly with the time of year the isotopic composition of the estuary water must also vary seasonally. The structure of the shells examined by SHACKLETON showed variations such as the Mediterranean climate must produce today but indicated that the point concerned was in Neolithic times on the estuary.

^{13}C and ^{18}O studies (MÖRNER and WALLIN in *Palaeogeogr., Palaeoclim., Palaeoecol.* 1976) on the bottom sediment of Lake Tingstäde Träsk on the island of Gotland in the Baltic have yielded a very detailed record of postglacial times from 10 700 to 1000 years B.P., which, interpreted as a mean summer temperature curve, indicates:

(1) A sharply reduced temperature, by about 8°C, lasting 500–600 years, presumably in the post-Allerød reversal, followed by an equally sharp recovery to the former level about 10 000 years ago.

(2) A slow and fluctuating rise to the postglacial warmest times about 1·5 to 2·0°C warmer than now, followed by a fluctuating decline.

(3) The decline was sharpest, and amounted to about 2°C, in the last millennium B.C.

(4) The subsequent fluctuations indicate a rise to temperature maxima about A.D. 400 and 1000.

Radium D (alias lead-210) in polar ice

Ra D, otherwise known as ^{210}Pb (lead-210), a radioactive element with a half-life of 21·4 years formed in the sequence by which the uranium in rocks breaks down to the stable isotope of lead, ^{206}Pb (see p. 69), can be detected in the year layers in polar ice sheets: the variations of the minute quantities of it present (when adjusted for radioactive decay) presumably indicate variations in the vigour or the pattern of the large-scale atmospheric circulation which effects the transport from other latitudes and/or the downput of snow which deposits the ^{210}Pb. LAMBERT et al. (1971) have studied these variations in the firn on the Antarctic ice sheet near the coast of Adélie Land and report that the concentration of ^{210}Pb rises and falls with the sunspot number, having been 30–50% higher over the 3 or 4 years closest to the 1957 and 1968–9 sunspot maxima than in the years 1962–3 just before sunspot minimum. The concentration of ^{210}Pb also rose sharply in every one of the 31 months during the years 1959 to 1969 when there was a shorter-term rise of sunspot number by more than 30%. The same studies showed a regular seasonal variation of the ^{210}Pb concentration, rising sharply in November to a maximum in the Antarctic summer, the

December and January averages being almost double those for all the months from March to October.

Because of its rather short half-life, studies of this particular unstable isotope are, of course, only applicable to throwing light on problems of the spread of pollution and of the circulation in modern times in areas where data are inadequate for other methods.

Underground temperature profiles

The vertical distribution of temperature with depth below the ground surface registers not only the flux of heat from the Earth's interior but also the slow downward propagation of temperature waves from the surface, as climatic changes at the surface vary, and have varied, the gradient from the Earth's interior. The velocity of downward passage of temperature changes from the surface varies with the thermal diffusivity of the rock, defined as $k/c\rho$ where k is the thermal conductivity, c the specific heat and ρ the density of the rock. All these items and the temperatures in the rock (after allowing at least ten times the duration of the boring operation for temperatures in the hole to recover from the disturbing effect) are susceptible of very precise measurements. So theoretical temperature profiles corresponding to various proposed past climatic histories can be calculated and compared with the observed temperature distribution with a view to selecting the history which gives the best fit.

The method can only be expected to give sensible results in areas of uniform rock, and without either large water bodies or concentrations of radioactive material near the hole. CERMAK (1971) has demonstrated the potentiality of the method with a study of boreholes up to 600 m deep in the Kapuskasing area (49–50°N 81–84°w) of northeastern Ontario bored directly into the pre-Cambrian gneissic and granitic rocks of the Canadian shield. Temperatures were read at 8 m intervals. Conductivity and density were measured at 4 m intervals. Temperature decreased with increasing depth in the first 100 m and below that increased with depth. The fine structure is shown best by plotting the vertical temperature gradient (which also simplifies interpretation because the heat flow is proportional to it) against depth. Key points of the model climatic chronologies for which theoretical temperature profiles were computed were: (1) a long duration of temperature of the bedrock surface just at the pressure melting point (= −2°C) of ice 3000 m thick in the last glaciation, (2) postglacial warmest time about 2°C warmer than the representative surface temperature (called the 'reference temperature') found by extrapolating the straight part of the temperature profile to the surface, (3) the surface temperature prevailing in the colder climates which set in from 1500 B.C. onwards, (4) the surface temperature in the warmest centuries of the early Middle Ages about A.D. 1000 to 1300, and (5) the cold Little Ice Age climate of about A.D. 1550 to 1800. If (1), (2) and (3) are regarded as firmly established, in view of their long duration, from botanical and other data, the values for (4) and (5) which give the best fitting curve provide independent evidence of the temperatures characterizing the two extreme phases of the climate within the last 1000 years. Moreover, the departures of the temperatures in the uppermost part of the column from the reference value should throw some light on the climatic changes (which have been strong in the region) within the last 100 years. The results suggested that the early medieval warm centuries produced mean surface temperatures 1·1° to 1·6°C above the reference temperature and the Little Ice Age cold period about 1·0°C below the reference value. The rather shorter-term changes within recent times appeared even more striking, temperatures having risen in two stages – the first step about 100 to 160 years ago and the second step just 40 years ago – to some 3°C above the reference temperature.

Unfortunately a substantial margin of uncertainty as to what these temperature deviations represent remains, since the mean annual temperature of the surface in the area studied (at present +5°C) is much higher than the average surface air temperature (at present about 0°C) thanks to the snow cover which shields the actual ground surface from the low temperatures prevailing throughout the cold season (the 'reference temperature' referred to in the analysis is at present +2·5°C). There is little doubt that CERMAK's study confirms the occurrence of a warm era in the early Middle Ages and of a colder climate than now from about A.D. 1550 to 1800. The method might be expected to produce results free of these ambiguities of interpretation in areas of uniform rock in climates where there is little or no winter snow cover.

The sand and dust fraction of deep sea deposits

Another quantitative measure of past climatic conditions is offered by studies of the different rates of deposition at different levels and dates within a column of sediment taken from the sea bed of biotic and abiotic minerals. The quantity of land-derived ('terrigenous') material in ocean bed deposits decreases with increasing distance from land, particularly the larger grain sizes. Far from land the fraction consisting of grains of more than about 60 μm diameter consists almost exclusively of shells and fragments of shells of foraminifera and coccoliths, etc. (ARRENIUS 1952), and the percentage of $CaCO_3$ can be used as a measure of how these bulk in the sediment. The smaller grains include windborne ('aeolian') and waterborne elements: wind-blown sand from deserts and sea shores, silts and soil particles (loess), pollen and particles derived from the vegetation, as well as waterborne silts and clays. Volcanic matter may also be found.

Fig. 13.21 shows plots of the percentage of small grain sizes and oxygen isotope ratio values found by EMILIANI (1966b) in a core taken from the central Caribbean. The scale of abundance of the smaller grains is inverted to show the degree of parallelism of the two curves, with its implication that the deposit of land-derived material increases when the climate gets colder. There is, in fact, an absolute increase in the amount of terrigenous material during the cold periods, not just in the ratio it bears to the marine biological element.

An increase of wind-blown material during the ice ages seems to have been rather widely found in ocean bed deposits in the middle and lower latitudes of the Atlantic and Pacific, as in middle latitudes on land. OLAUSSON and OLSSON (1969) used chemical analysis to isolate the phosphate (P_2O_5) content as a measure of the amount of sediment that was unequivocally of marine-biological origin and compared it with the amount of inorganic mineral of 16 to 62 μm grain sizes, identified under a binocular microscope, in cores from the Gulf of Aden (near 12°N 44°E) and the Mediterranean. In these areas there was a strong maximum of wind-blown material just at, or after, the end of the last glaciation, perhaps mainly in the Pre-Boreal. Wind-blown matter was deposited at much less than the present rate in the glacial period itself and all through the postglacial warmest times, particularly the late Boreal and Atlantic periods and in the later part of the Sub-Boreal. Probably the main reason was that moisture and vegetation were sufficiently present in the areas that are now desert to reduce the uptake of dust and sand.

PARKIN and others (e.g. 1972; see also Volume I, p. 436) have collected airborne dust in various latitudes over the Atlantic Ocean and studied its size distribution, which is related to the distance travelled with the airstream from the source of the dust. The colour of the dust was generally brown and clean-looking off the coast of the Sahara but grey due to admixture of black humus nearer the equator and on either side of it. Similar colour differences were found in the stratigraphy of cores of the sediment on the ocean bed. The sequence of changes of grain size in the history at points where cores were taken off the African coast clearly registered the interglacial/ice age/postglacial changes. In one core near 8°N 23°W the sequence of grain sizes was difficult to interpret, prevalence of smaller sizes during glacial times possibly being attributable to a better cover of vegetation and more moisture than now in the Sahara region; but another core near 21°N 23°W, which would presumably be within the Trade Wind system throughout the year, showed coarser grains and so probably stronger winds in glacial times than now (PARKIN 1974).

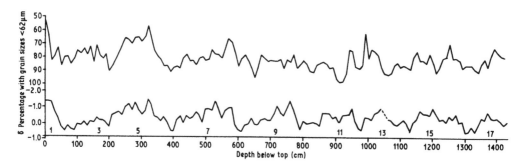

Fig. 13.21   Fraction with grain size less than 62 μm and oxygen isotope ratio anomaly plotted against depth in a core (P 6304-9) of sediment from the central Caribbean (Antilles Sea) near 15°N 69°W.

The curve of abundance of small grain size is inverted to show the parallelism with the oxygen isotope curve and implied temperatures (cf. fig. 13.15).

(*Reproduced by kind permission of Dr* EMILIANI.)

Studies of the grains of quartz sand in deposits from the bed of the Southern Ocean in the sub-Antarctic zone (MARGOLIS and KENNETT 1971), using an electron microscope, which makes it possible to distinguish ice-rafted grains with their still sharp outline, striations, etc., from grains subjected to mainly subaqueous turbulent abrasion, have indicated a long history of glaciation of the Antarctic continent. Ice-rafted grains point to glaciation of a substantial area in the Antarctic 50 million years ago, in the lower and middle Eocene, and in the Oligocene. No such evidence has been found for the early and middle Miocene, which were probably warmer (from faunal evidence); but from the late Miocene to the present day deposition of the ice-rafted type of grains suggests continuous glaciation of Antarctica.

Sediments and the chemistry of the waters

Measurements of the varying amounts of certain chemical elements with depth in a sediment (i.e. with date of deposition) undoubtedly reflect the influences of climatic history, and yield evidence independent of any we have so far discussed regarding the nature of the climatic shifts; but in few, if any, cases does it seem possible to derive climatic values from them.

(1) Inland waters

A report on chemical studies of the bottoms of the lakes in the English Lake District by MACKERETH (1965) gives an excellent insight into the possibilities of this approach to climatic history.

The lake sediments laid down during glacial times were purely mineral clay, often varved, the material being derived from the catchment basin. In the postglacial sediments 70–80 % is mineral and 20–30 % organic matter mainly from the soils of the drainage basin but including some matter produced within the waters of the lake.

The content of organic carbon in the sediments is the variable which bulks largest, rising rapidly in early postglacial times to reach its maximum towards the end of the warmest postglacial climate, somewhat before 3000 B.C. Decline then set in and a sharp fall of carbon content occurred about 2000 years ago. The course of this history was similar in the nutrient-rich ('eutrophic') and in the least productive ('oligotrophic') lakes investigated. More remarkably, the history appears closely similar in the English lakes and in an American pond, as seen in fig. 13.22.

The concentrations of the strongly alkaline metals, sodium (Na) and potassium (K), increased in direct proportion to the mineral content of the sediments. They were highest in early postglacial times, indicating great erosion of the loose glacial debris in the drainage basins and particularly rapid extraction of the soluble salts of the base-rich substances. Minimum concentrations of total mineral and of Na and K occurring in the mid postglacial warmest times clearly point to minimum erosion then, i.e. to soil stability associated with the

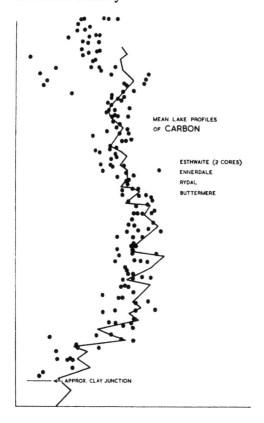

MEAN LAKE PROFILES
OF CARBON

ESTHWAITE (2 CORES)
ENNERDALE
RYDAL
BUTTERMERE

APPROX. CLAY JUNCTION

Fig. 13.22   Diagram by MACKERETH (1965) to show (by the dots) the variations of carbon content of the sediment in three English lakes through postglacial times (starting from the clay junction at the bottom of the sediment). The solid line shows the variations of carbon content in Linsley Pond, Connecticut (25 m above sea level, 6 km northeast of New Haven) with its depth scale adjusted to bring the clay junction to the same point.

(*From* VALLENTYNE *and* SWABEY *1955; reproduced by kind permission of Dr* MACKERETH *and the Royal Society.*)

densest vegetation.[1] The history of magnesium (Mg) concentration in the sediments presented the same picture. Calcium (Ca) is particularly readily dissolved from rock matter and is only deposited in high concentrations during times of great erosion; the main feature of the concentration of this element in the lake sediments was its great peak at the beginning of the postglacial and, unlike the other metals mentioned, little or no renewed increase of Ca in the most recent millennia is to be seen.

The best hope of deriving any strictly climatic values from the chemistry of lake sediments is associated with the halogen content. The amounts of the elements chlorine (Cl) and bromine (Br) – PENNINGTON and LISHMAN (1971) have used iodine (I) – in lake sediments may be thought to depend chiefly on the rate at which they are supplied by the rain falling over the catchment area, after being initially caught up into the atmosphere in salt spray from the ocean in windy weather (in the case of the English Lake District, therefore, the quantities of these elements might be an index of the prevalence of W'ly winds as well as of the amount of rain). Soil waters are in general richer in halogens than the rocks

---

1. The maximum of organic carbon reaching the lake sediments about this time must have been somewhat   depressed by this soil stability.

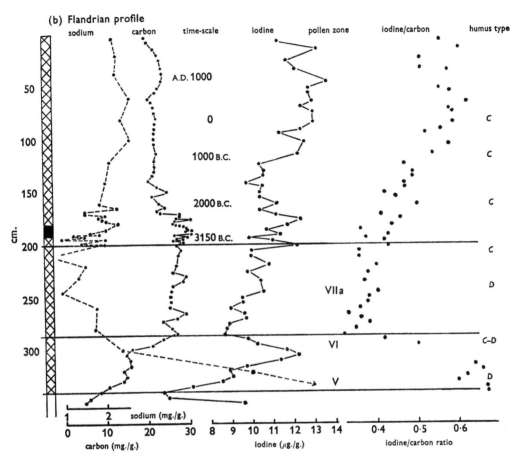

Fig. 13.23 Postglacial sediment chemistry profile from Blea Tarn, English Lake District.
(*From* PENNINGTON *and* LISHMAN *1971; reproduced by kind permission of the authors and the Cambridge Philosophical Society and Cambridge University Press.*)

from which the soils are derived. But the amounts of these elements held in the soils and in the lake sediments after deposition depend on such things as the amount of humus acids and soil colloids present and the relative proportions of halides and other anions which form soluble salts. These are presumably the reasons why, for instance, iodine values at different sites show no relation to the present-day rainfall differences between them (PENNINGTON and LISHMAN 1971), although the variations from one level to another in any one Lake District sediment seem to support the succession of rainfall changes deduced from vegetation history in England and Scandinavia. Fig. 13.23 shows the full postglacial profiles of sodium, carbon and iodine from the sediment in Blea Tarn, Langdale in the English Lake District, with the radiocarbon and pollen-zone time scales. It is clear, particularly from the iodine/carbon ratio, as well as from the sodium curve, that the iodine variations must be interpreted only partly in terms of rainfall variations and are also considerably affected by the history of erosion and the climate-induced progressive changes in the soils and swamps of the uplands. The prolonged upward creep of the I/C ratio, which this and other Lake District curves show, paralleled by the other halogens, is presumably derived from the progressive leaching of the soils of the wettest upland regions of the drainage basins. In the course of this process the upland soils were gradually left more and more acid and the forests there were replaced by blanket peat.

PENNINGTON and LISHMAN found that the history of iodine variations and of the I/C ratio in lochs in southwestern and in northwestern Scotland differed so much from that shown by the English Lakes as to leave a suspicion that the rainfall history of the western half of Scotland may differ in important respects from the succession which seems to apply to England and Sweden.

Differences of electrical conductivity of the waters constitute one of the most easily measured indices of different chemical composition in present-day lakes. H. H. BIRKS reports (Cambridge Symposium of British Ecol. Soc., April 1972) that measurements in a number of lakes in Minnesota have indicated that the flowering of different aquatic plant taxa appears to be correlated with the electrical conductivity of the water, highest conductivity going with the prairie region lakes and lowest conductivity with waters in the coniferous forest.

(2) Ocean water

The aspects of the chemistry of the oceans which most concern the evidence of past climatic changes and the changes in the ocean circulation that accompanied them are salinity differences and variations in the solubility of calcium carbonate.

Enough has been said in Volume I about the role of salinity. In some parts of the world salt deposits (on the surface or buried by later rock) remain as witness either to the effect in regions of warm climate of periods of lowering of sea level or of the drying up of inland seas.

The solubility of carbon dioxide and of carbonates increases the lower the temperature of the water. Being cold, the deepest waters of the oceans are rich in dissolved CO_2 and have a high capacity for dissolving carbonates on the ocean bed. The calcareous skeletons of marine fauna are partly dissolved as they sink to the bottom, and in the case of the microfauna a large portion is dissolved before reaching the bottom. Below about 4000 to 5000 m depth the water, being unsaturated in $CaCO_3$ derived in this way from recently living organisms, attacks the carbonate in the sediment. The critical depth below which solution exceeds supply may be called the 'calcium carbonate compensation depth' (OLAUSSON 1967). Carbonate sediments therefore tend to be limited to those parts of the oceans which are not exposed to the cold bottom water (the aragonite form dissolves faster than the calcite form, and the former therefore accumulates in the sediment only in rather shallow, warm seas). ARRHENIUS's (1952) study of the carbonate present in deep sea cores in the eastern equatorial Pacific indicated a lag of some thousands of years between the times of formation of the most aggressive bottom waters in high latitudes (i.e. the coldest waters in glacial ages with greatest dissolving action on carbonate deposits) and the times when solution in low latitudes was greatest. Most of the Pacific Ocean north of 8°N, predominantly deeper than 4000 m, lacks both Quaternary and upper Tertiary calcareous oozes. About half the South Pacific and Indian Oceans do have such deposits.

With the present geography of the Earth most of the products of weathering are drained by the rivers into the North Atlantic. This ocean receives about 70 % of the global total of Ca delivered by weathering (estimated by OLAUSSON (1971) as 122 × 10^{13} grammes of CaCO$_3$ yearly), the South Atlantic 8 %, the Indian Ocean 10 % and the Pacific about 12 %. Hence, more of the Atlantic than the Pacific is saturated with CaCO$_3$ and biogenic carbonate is quite widespread in the Atlantic oozes despite the production of cold bottom water in the North and South Atlantic. About half the present yearly input of CaCO$_3$ from the rivers is estimated as lost by solution and by mechanisms which transport it to the other oceans; and in the South Atlantic, the Indian Ocean and the South Pacific about twice as much carbonate is deposited as the rivers discharge into these oceans. This balance must clearly have changed with the major climatic changes from glacial to interglacial periods and vice versa. In glacial times by far the greatest cooling of the waters seems to have occurred in the Atlantic and the production of cold bottom water was probably much enhanced in that sector. This seems verified by the fact that the glacial phases are marked by minima of calcium carbonate in cores from the North Atlantic and by maxima of CaCO$_3$ in central Pacific cores (OLAUSSON 1967; see also carbonate and oxygen-18 curves published by EMILIANI 1955 and elsewhere).

Sea level

The mean level of the sea surface, averaged over its tidal variations, is taken as a fixed datum on topographical maps but is, in fact, subject to continual long-term variations of differing magnitude, some world-wide, some local, some due to 'tectonic' events in the Earth's crust and some due to climatic changes. Nowadays, the variations are monitored by numerous local tide gauges operated by institutions in each country and are the concern of an international body, The Permanent Service for Mean Sea Level of the International Union for Geodesy and Geophysics. Fig. 13.24 shows a record of more than a hundred years from Sheerness in the Thames estuary. Any sea level stand that is maintained with little alteration for a long time leaves a record of itself in the beaches formed. Hence, the history of sea level variations can be 'read' by studying the heights above or below present sea level of old beaches and river delta formations, provided that these can be dated and that the area has not been altered by local tectonic events.

Changes of sea level of limited regional extent associated with crustal movements that represent adjustments (for example) to changes of ice load on the land within, or near, the

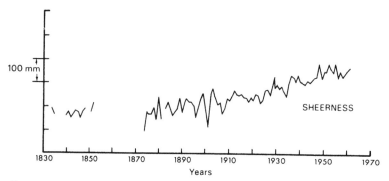

Fig. 13.24   Sea level changes at Sheerness (Thames Estuary) from the
1830s to the 1960s.
   (*From* ROSSITER *1972.*)

region concerned are called *isostatic*.[1] Changes of sea level of global extent, in which regional crustal adjustments are not involved, are called *eustatic*.

The total volume of water in the world's oceans varies:

(1) Because of variations in the amount of water stored on land, chiefly as ice and snow but also with changes in the levels of lakes and rivers and of the water table in the subsoil and porous rocks (see Volume 1, p. 346, regarding the durations of storage in these ways).

(2) With the expansion and contraction of the water in the ocean as its temperature changes (a warming of 1°C throughout the ocean should raise its surface by 60 cm).

Other changes of sea level relative to the land occur locally or regionally:

(3) Because of tectonic changes:

(*a*) Crustal warping associated with orogenesis (mountain building) and compensating movements over a wide area (e.g. the progressive downwarping of the North Sea basin that has been going on since the Carboniferous and continues today in the southern part in association with the Alpine orogenesis).

(*b*) Land depression and recovery, continuing over thousands of years, associated with the imposition and loss of the ice load of a glaciation.[2]

1. 'Isostasy' describes the equilibrium conditions whereby the continental blocks, being made of lighter rocks than the ocean floor, 'float' above the 5 km thick basalt material of the ocean floor that covers the Earth's mantle; but, because this makes a much greater total thickness of crust (about 32 km) in the continental blocks, the basalt layer is depressed by the overlying continental rocks (WILSON 1957). The equilibrium demands that this depression is greatest underneath mountains and loads of ice, which depress the continental rock layer as well.

2. Isostatic depression of the general ocean floor is now believed (BLOOM 1967) to have occurred with the added load of water when the great ice sheets of each glaciation melted. Depression should set in early and swiftly in each interglacial at coasts bounded by deep water near at hand, but should be a much more gradual process at coasts where the sea is shallow and should involve progressive submergence of such coasts, even when the eustatic rise of sea level has ceased. As the load is much more widely spread than that of the glacial ice sheets on land, the amount of the depression must be correspondingly less.

(c) In local collapses and upheavals in unstable parts of the crust, associated with volcanic eruptions and faulting (earthquakes).

(4) In short-lived storm surges due to extremes of atmospheric pressure and wind stress. A local reduction of the atmospheric pressure by 100 mb should raise the sea surface by about 1 m if equilibrium were attained. The combined effects of pressure-fall and wind, particularly when blowing over a long fetch into a shallow sea and at about the time of high spring tide, are liable to cause disastrous sea floods on coastal lowlands.[1]

Occasionally 'tidal waves' or 'tsunami', travel thousands of kilometres across the ocean from their origin in a volcanic upheaval, an earthquake or a storm at sea, and sweep over the coasts where they arrive in similar disastrous floods. In all these cases a storm beach or trail of sand and shingle is liable to be carried inland and leave a lasting mark of the event.

(5) MAKSIMOV's analysis (1971), illustrated by fig. 13.25, indicates that some changes of sea level represent a redistribution by latitude, a deformation of the Earth's ocean envelope similar to that of the equatorial bulge of the Earth itself, associated with the balance between gravitational and centrifugal forces and varying with a secular variation of the 'gravitational constant'.

Items (1), (2) and (3b) in the above list register the effects of climatic changes. Secular variations in the frequency of sea storm floods, particularly in an area like the North Sea not liable to tsunami, may provide evidence of times of prevailing high sea level.

There seems, judging from the observation and dating of old shoreline terraces which represent the highest sea levels reached during successive warm interglacial times, to have been a progressive lowering of world sea level during the last 0·8 million years by about 100 m. A fall of world sea level can in fact be traced much farther back, beginning apparently in the Miocene about 20 million years ago. The drop between then and the end of the Tertiary was already 70–100 m (TANNER 1968) when the large amplitude fluctuations caused by the glacial-interglacial cycles of the Quaternary set in. The history suggested is shown schematically in fig. 13.26. ZEUNER (1959) suggested with notable diffidence that the prolonged decline might be caused by deformation, a progressive lowering, of the ocean floors. It may be easier to justify the suggestion by J. W. HOLMES (1969) that the cause is a

1. Temperate latitude depressions have caused surges of sea level up to 3 m above the normal tidal level in the North Sea and up to 2 m in the South Australian Gulfs (EASTON 1970). The Bay of Bengal cyclone which caused the sea to inundate the Ganges delta area south of Dacca on 12–13 November 1970 to a depth up to 5 m, drowning an estimated 300 000 people (this estimate was subsequently amended by the Bangladesh authorities to 'about three-quarters of a million'), and a similar event with the same estimated number of lives lost in the Ganges delta area on 11–12 October 1737, were among the worst ever single disasters due to weather (cf. the lists given by LATTER 1969) in terms of loss of life. A calamity of similar dimensions occurred at Haiphong, North Vietnam, with a typhoon on 8 October 1881. A number of the great North Sea floods in the Middle Ages, before the modern sea defences in the Netherlands were built, are reported to have caused disasters of the same order (see pp. 121–5).

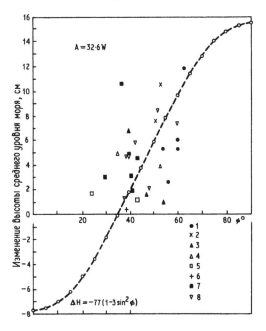

Fig. 13.25   Change of mean sea level from
1900 to 1950, plotted against latitude of the
observations.
    1 = Baltic Sea
    2 = North Sea
    3 = Atlantic Ocean (coasts of England,
        France and Portugal)
    4 = Atlantic Ocean (coasts of Canada and
        U.S.A.)
    5 = Atlantic Ocean (coasts of Gulf of
        Mexico and Argentina)
    6 = Azores
    7 = Mediterranean Sea
    8 = Weighted means
The curve was calculated theoretically and
adjusted to best fit by MAKSIMOV (1971).

long continued increase in the total volume of water stored as water and as ice on and in the
landmasses of the Earth in the soils and rock.[1] The outline view of the long history of world
sea level sketched in fig. 13.26 bears an obvious resemblance to the course of mean
temperature in middle latitudes as derived for Europe independently of sea level
considerations by WOLDSTEDT in fig. 14.9.

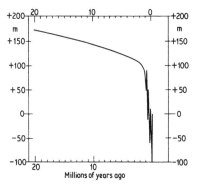

Fig. 13.26   World sea level changes since the late Tertiary
(Miocene).

(Ice-age-interglacial variations only schematically
indicated – in reduced numbers).

There was active mountain building in the Tertiary era and this brought to the surface
lava basalts which, especially in the colder climates since, have tended to weather down to

---

1. The decline of sea level appears to have proceeded so
steadily over the last million years, apart from the
superposed glacial-interglacial oscillations, as to
suggest the operation of a single main cause. This

was, in fact, ZEUNER's reason for thinking a slow
crustal process like ocean-floor spreading and continental drift must be at work.

fine-grained clays capable of absorbing up to 0·5 m depth of water for each metre depth of soil. The glacial climates of the Quaternary and their sequels in frost and thaw and water erosion have made it an era of rapid weathering. SCHELL (1970) has pointed out that the weak equator-to-pole and continent-to-ocean temperature gradients during the warm eras that continued well into the Tertiary would support only a sluggish atmospheric circulation, and hence the probably greater evaporation from the oceans in those times would have led to increased rainfall mainly over the oceans themselves. The land surfaces could have been characteristically drier than they have usually been since. In all the higher latitudes evaporation from the oceans may well have become greater in the Quaternary (just as it does in winter today over the oceans in middle and higher latitudes) owing to the stronger winds, and this must have favoured increased downput of rain (and snow) over land in the Quaternary.

Estimates of the glacial-interglacial eustatic changes of sea level, as reviewed by FAIRBRIDGE (1961), and incorporating latest estimates for the last glaciation, may be summarized in the following table. Antarctica was at no time free of ice since the Tertiary.

Table 13.1   Extreme sea level stands during the Quaternary

Glaciation	Interglacial	Sea level departure from present
Würm/Wisconsin		− 100 to − 130 m
	Eemian/Sangamon	+ 3 to + 18 m
Riss/Illinoian		− 55 m?
	Holstein/Yarmouth	+ 30 to + 55 m
Mindel/Kansan		− 5 m?
	Sicilian/Aftonian	+ 80 to + 100 m
Gunz/Nebraskan		+ 55 m
Donau/pre-Nebraskan		+ 130 to + 200 m

The estimates of lowest sea level in the earlier ice ages are the most uncertain figures in this table. Melting of all the ice still present on land today would raise the sea level to about + 60–80 m according to the latest estimates (see Volume 1, p. 484).

Much careful work has been done in recent years on identifying and dating interglacial high sea levels from beaches and coral terraces with a view to comparing the dates with the MILANKOVITCH orbital variables (e.g. MESOLELLA *et al.* 1969, SHACKLETON 1969, VEEH and CHAPPELL 1970, WARD *et al.* 1971). Marked interruptions in the development of coral reef terraces at Barbados between 10 000 and 80 000 years ago and between 130 000 and 170 000 years ago are attributed to (glacial) low sea levels, the high sea levels being about 82 000, 105 000 and 125 000 years ago. At Oahu, Hawaii, dating of fossil corals indicates that about 120 000 years ago sea level was about + 7·6 m and that that was the last time that the sea stood significantly higher than its present level (TEH-LUNG KU *et al.* 1974). The gradual uplift (averaging 30 cm per 1000 years) of the island of Barbados 13°2″N 59°5″W in the West Indies during the Pleistocene, with coral reefs forming just at the current sea surface level except during the cool phases that accompanied the glaciations in high latitudes, has produced a series of reef terraces. All the older terraces have been elevated above present sea level and can be dated either radiometrically or by their height. This makes possible a more precise derivation of the range of former sea level fluctuations and their dates than anywhere else in the world. The dating appears to support MILANKOVITCH's astronomical theory of the causes of the ice age-interglacial sequence (Volume 1, pp. 30–7). A moderately high sea level established at New Guinea and in the Ryukyu Islands around 40 000 to 50 000 years ago seems to agree with the long interstadial in the last ice age and with a feature of the MILANKOVITCH curves (see also p. 000).

To portray the course of world sea level changes during postglacial times has been the object of many analyses (notably by GODWIN *et. al.* 1958, FAIRBRIDGE 1961, SHEPARD 1963, SCHOFIELD and THOMPSON 1964, JELGERSMA 1966 and MÖRNER 1969). There are obvious difficulties in assembling enough reliable measurements from areas not affected by local or regional tectonic anomalies since the old beaches were formed, to establish the world average of sea level at different times past to within a metre or two. The quantity of such data that can be taken as contemporaneous, and the amount of detail of changes with time that can be reliably discerned, are also limited by the error margins of the dating methods used for each point of the world survey. The most cautious of these analyses, and therefore the most reliable, show only the main rise of sea level as the ice sheets of the last glaciation melted and the ocean became warmer: a very rapid rise until about 5000 years ago, then flattening off to the present day. This interpretation is illustrated here by the dots in fig. 13.27, due to GODWIN *et al.* (1958). SHEPARD's result is essentially the same.

SCHOFIELD took advantage of the steady, and still continuing, isostatic rise of the land formerly covered (to different depths) by the Scandinavian ice sheet to distinguish the uniform component attributable to the eustatic rise, or world sea level, from the local isostatic changes and thereby arrived at the rather more detailed curve shown in fig. 13.27.[1] This curve appears reliable (when the extreme range of results, all lying between the outer curves is considered) in indicating that world sea level reached its highest point, between 2

---

1. SCHOFIELD's analysis of series of postglacial beaches in six different areas of the coast of southwest and southeast Norway and southwest Sweden is as follows:

The heights above present sea level of the now uplifted strandlines formed anywhere in southern Scandinavia at any date since the last ice age represent the algebraic sum of the ocal isostatic recovery of the land from the former ice load and the general submergence resulting from the world-wide eustatic rise of sea level. An assumption, verified by ANDERSEN (1960), was that the isostatic uplift of

the land at different points in southern Scandinavia over any time interval has been at rates that are related by constant ratios between any two particular localities. Proceeding from this, for a date, j, thousands of years ago, let y_a; be the change of height of the strandline at locality a, f_a be the proportionality constant for locality a relative to the average height change for the area, x_j be the rise of world sea level. Then $y_a - x_j = f_a k_j + e_a$, where e_a represents the tandom error for that place and time and k_j is a constant for all the localities $a, b, c, \ldots$ for the time interval from j the present day.

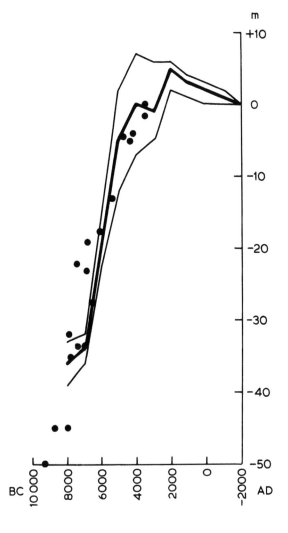

Fig. 13.27   Change of world mean sea level during the last 12 000 years.

The middle curve joins the average values calculated by SCHOFIELD and THOMPSON (1964) at 1000-year intervals from Scandinavian data, allowing for the isostatic variation at each locality. All the individual values calculated lay between the limits set by the outer curves.

The points superposed on this diagram are the ^{14}C-dated levels given by GODWIN, SUGGATE and WILLIS (1958) for positions near, or in, northwest Europe, the Gulf of Mexico, Persian Gulf, Australia and New Zealand.

Based on uncalibrated radiocarbon dates: the corrections applicable are seen in fig. 13.11.

and 7 m above present, around 4000 years ago. This means that, as should be expected, the highest stand of world sea level was attained about the end of the warmest millennia of postglacial times. There is consistency also in the slight temporary decline, or at least a hiatus in the rise, of sea level about 5000 years ago (indicated by the whole body of SCHOFIELD's data for that time), just when there is a good deal of evidence for the first important colder phase in Europe interrupting the warmest postglacial times (FRENZEL 1966).

The much more detailed history of fluctuations of world sea level indicated in the curves presented by FAIRBRIDGE (1961), though based on a particularly extensive world-wide network of points chosen as tectonically unaltered since the old beaches were formed, and

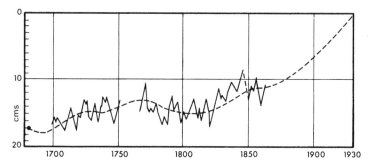

Fig. 13.28   The Amsterdam tide gauge record of mean sea level
from 1682 to 1930.
      (*After* FAIRBRIDGE *1961; data from* VAN VEEN *1954.*)

apparently agreeing with the expectations from much evidence of the fluctuations of
prevailing temperature on land, is obviously much less firmly substantiated by internal
consistency checks than the curves illustrated above and probably makes too great demands
of the dating techniques available.

The uneven distribution of actual sea level changes affected by regional crustal
anomalies may be illustrated by the following figures for recent years given by J. R.
ROSSITER (Institute of Coastal Oceanography and Tides, Birkenhead, unpublished
communication, 1970). The best estimate of the global average change of sea level since the
low stand about 1830 to 1890 (cf. fig. 13.24 which shows a long record for Sheerness in the
Thames Estuary) is thought to be that by 1970 it stood 8 cm higher, a rate which, if
continued, would be 10 cm/century.[1] The local rates over the same 80 years were equivalent
to +20 cm/century on the southern part of the North Sea coast of England (near the
Thames Estuary which fig. 13.24 represents) and in the Channel, +30 cm/century on part of
the east coast of the U.S.A., and −90 cm/century in the northernmost Baltic, where the
isostatic recovery is most rapid: the sea between northern Sweden and Finland would
become mostly dry within the next millennium at this rate.

The longest tide gauge record in the world is that made at Amsterdam from 1682,
illustrated here in fig. 13.28. Unfortunately, it is almost in the central part of the region
around the southern North Sea affected by crustal downwarping, the North Sea geosyncline
development. CHURCHILL (1965) investigated the vertical displacement of deposits in this
region that were formed at sea level 6500 years ago and found them 9 to 10 m below present
sea level, both on the Dutch and English sides of the southern North Sea, in the area between
the Thames and the Humber (north Lincolnshire) and in the Netherlands. Of this change, at
most about 6 m is attributable to the world sea level rise. So there must have been
downwarping of the land area of the order of 3 m or more in the 6500 years, a mean rate of 50

1. In fact the rise was fastest between about 1920 and   climatic period, and seems to have levelled off after
   1950, just during the early twentieth century warm   1950 (see fig. 13.24).

to, perhaps, 100 cm per 1000 years. In so far as the changes of sea level in the area in the last millennium agree with the probable world-wide (eustatic) changes, the local anomaly may have decreased in recent millennia and be small in modern times.[1] But there is a margin of uncertainty here and the land-sinking at Amsterdam since 1682 could be anything up to 12–15 cm. Hence the tide gauge curve (fig. 13.28) does not tell us with certainty whether world sea level was lower or higher than now at the beginning of the record, though it doubtless gives a sound indication of the year-to-year and decade-to-decade changes.[2] The two longest records of sea level measurements in the world, from Amsterdam and Sheerness, both come from the area of crustal downfolding associated with the North Sea basin development, and, in so far as this downfolding may be a smooth and continuous warping, it should be possible to distinguish analytically the steady long-term component of sea level change which represents this from the shorter-term fluctuations due to the eustatic changes associated with climate.

A. V. Šnitnikov (1969) has given a long history of sea level in the Pacific Ocean in the form of a graph for the average of three stations (unspecified), which yields the figures in the following table:

Table 13.2   Sea level changes in the Pacific Ocean
10-year average departures in cm from the level recorded in 1950–7
(After Šnitnikov 1969)

1807–16	− 11·8	1850–59	− 12·7	1900–09	− 9·8
1810–19	− 11·2	1860–69	− 12·2	1910–19	− 4·2
1820–29	− 9·6	1870–79	− 10·8	1920–29	− 4·7
1830–39	− 12·2	1880–89	− 11·1	1930–39	− 5·4
1840–49	− 12·9	1890–99	− 10·4	1940–49	− 0·6

Some coastal changes, particularly the incidence of blowing sand, suggest that sea level around the British coasts was generally lower by more than a few centimetres between A.D. 1600 or 1650 and 1700, and had been falling for some 300 years before that; other records suggest that it stood high, possibly a little higher than now, between A.D. 1000 and 1400 or later (i.e. during and after the medieval warm epoch) and also around A.D. 300 to 400 or a little later, which may be seen as marking another epoch of relative warmth.

These suggestions seem to be supported by the dates of known sea floods (which – other things being equal – must be more liable to occur when the general sea level stands higher

1. Sea level may be expected to fluctuate eustatically, up and down, with every fluctuation of climate or at least with those climatic fluctuations that are of world-wide extent. Crustal warping and the isostatic or regional changes of sea level must be much less liable to short-term variations up and down, and for this reason some interpretations of the history of coastal changes in East Anglia that postulate a sequence of such variations seem improbable.

2. The most recent assessment of the eustatic element in the Amsterdam tide gauge record (MORNER 1973) puts the implied sea level between 1680 and 1880 at 7–9 cm below the present level.

Table 13.3　Sea floods which caused losses of land and/or cost many human lives

North Sea coasts of England, Netherlands, Germany, etc.		Other coasts of Britain	
350–340 B.C.	Alternative range of dates given by ARENDS (1833) for the Cymbrian (Kymbrian) flood (next entry): perhaps this was an earlier flood also long remembered in legend as well as being originally reported by PYTHEAS within a few decades after the event.		
c. 120–114 B.C.	Jutland and NW Germany: the Cymbrian flood.		
		A.D. 68	Isle of Wight separated from England.
A.D. 245	Lincolnshire.		
333 or 334	Friesland: many villages lost.		
c. 400	Kent and Flanders (possibly the flood detailed in 419).	c. 400	Part of Goodwin Sands submerged (possibly 419).
		419	Solent, Southampton Water coasts devastated.
435	Friesland: hundreds died.		
516	Friesland.		
		520 (approx.)	W Wales: Cantref y Gwaelod lost. Perhaps also the province of Lyonesse off Cornwall. In both these areas cultivable land was lost to the sea.
533 and/or 570	Friesland: 3 days NW'ly storm, castle washed completely away.		
589	NE England (Hartlepool).		
800 and/or 806	Much of Heligoland lost. Hundreds died. Floods in Netherlands all winter.		
26.12.838	Friesland.		
Spring 860	Holland: Rhine-Lek channel to the sea opened.		
864	Holland: further devastation in the same area.		
1012	English and German coasts.		
28–29.9.1014	England, Walcheren and Flanders.		
1015–17	Great floods of North Sea coasts 'throughout these years'.		
1066	Oldenburg.		
1086 or 1087	England (Norfolk).	1087	Channel (Winchelsea).
1099	Kent, Thames Estuary and Holland: 100 000 estimated drowned.	1099	Further stage in formation of Goodwin Sands.

Table 13.3 – continued

North Sea coasts of England, Netherlands, Germany, etc.		Other coasts of Britain
1113	Flanders (emigration of population).	
1125	England: many villages destroyed.	
28.10.1135	Holland and NW Germany.	
9.1. and 16.2.1164	NW Germany: 100 000 estimated perished.	
3.11.1170	Holland; also much of Borkum lost.	
19.12.1181	Friesland: a town largely destroyed.	
1200	Friesland: 100 000 estimated drowned.	
1212–1215	'Extraordinarily many very severe North Sea storm floods.'	
1212	N Holland: enormous loss of life – estimated 306 000 drowned.	
1214	Brugge: sea flowed 7 km inland – many drowned.	
1216	Bremen: 20 000 died. Also much of Heligoland lost.	
17.11.1218	Holland and Germany: about 100 000 lives lost. Jadebusen basin formed (near Wilhelmshaven).	
16.1.1219	Friesland: estimated 36 000 perished. Land lost at Nordstrand (coast of Schleswig).	
Mid-Nov. 1236	English Fenland (Wisbech).	
(1230 or) 1238	S Jutland: an island (Bollertsand) and its villages lost, reduced to a sandbank off Rømø.	
1240	Schleswig: island of Nordstrand separated from mainland: 62 villages and many parishes lost.	
27.11.1242	Holland and Germany (Oldenburg); also sea flood in the Thames and all lower parts of London, up to 10 km from the river.	

Table 13.3 – continued

North Sea coasts of England, Netherlands, Germany, etc.		Other coasts of Britain	
18.10.1246	Belgium.		
November 1246	Friesland and Weser: thousands die.		
20.11.1248	Flanders, Holland and Friesland.		
28.12.1248	NW Germany, about Hamburg.		
1.10.1250	England about the Humber, also Holland and Friesland.	1250	Channel (Winchelsea).
1250–51	Netherlands: Zuyder Zee forming.		
21.3.1251	Lincolnshire and Kent: floods 2 m higher than 'ever seen before'.		
1253	Lincolnshire and Thames Estuary.		
28.1.1262	Friesland.		
16.1.1268	Friesland.		
1273	Friesland: 20 000 die.		
25.12.1277	Netherlands: the sea engulfed Reiderland; 50 villages lost.		
Many floods in 1287 (1 Jan., 25 Jan.– 2 Feb., 4 Feb., 14 and 17 Dec.)	Holland: Zuyder Zee enlarged, over 80 000 lives lost. Also Denmark, Friesland and Flanders, and England from the Humber to Kent. East Anglia: Norfolk Broads area flooded.	1287–8	Channel (Winchelsea destroyed – see entry describing the North Sea flood on 4.2.1288).
4.2.1288	Thanet, Romney and Winchelsea: all dykes demolished. Thames Estuary: villages far from river flooded. All of the Netherlands also flooded, except Walcheren, on 4–5 and 16.2.1288.		
16.1.1300	Half the former Heligoland lost and many former islands off the German coast.	1302 (approx.)	Solway.
25.11.1304	Netherlands, especially Walcheren.		
c. 2.2.1307	'All the English coast.'		
22.10.1313	Friesland: over 500 drowned.		

Table 13.3 – continued

North Sea coasts of England, Netherlands, Germany, etc.		Other coasts of Britain	
1321 (probably 11.11.1321)	Belgium.		
23.11.1334	England (Thames), also Belgium, Zealand, Holland and Friesland.		
1337	Netherlands: town of Rungholt and 14 smaller places lost.		
		1357 Summer	Sussex.
16.1.1362	Some of the present islands of the German Bight formed by coastal losses. Great losses also from Schleswig and the Danish coast; including many whole parishes. (In the next century the Council of Basle was informed that 60 parishes accounting for over half the agricultural income of the then Danish diocese of Slesvig (Schleswig) had been 'swallowed by the salt sea', presumably mainly in the floods of 1240 and 1362.)		
1364	England, Humber: former great port Ravenser (Ravensburgh or Ravenspur) mostly destroyed; several other towns in Holderness disappeared.		
9.10.1373	Friesland.		
16.11.1377	Holland, Zealand, Friesland and Flanders. 30 000 die. In NW Germany 32 villages lost.		
1400	Netherlands: Amsterdam and Enkhuisen opened up as ports. Humberside in England also affected.		
19.11.1404	Kent: many drowned. Also Flanders and Holland.		
1405 Autumn	Denmark: many thousands drown. Hamburg: 36 000 died.		
1412			

Table 13.3 – continued

North Sea coasts of England, Netherlands, Germany, etc.		Other coasts of Britain	
19.11.1421	Friesland and Holland: Zuyder Zee reached its greatest extent, 72 villages and 100 000 people lost (the dykes had been damaged by a storm in mid June 1421). England also affected.		
Sep.–Nov. 1422	Further devastation by storm floods in Friesland and Holland.		
31.10.1436	Schleswig-Holstein, NW Germany and N Friesland. Islands of Nordstrand and Pellworm, off Schleswig, parted.		
10.4.1446	Netherlands and N Germany: 16 parishes devastated near Dordrecht, over 100 000 lives lost.		
1.11.1470	All North Sea coasts: 10 000 perished.		
6.1.1471	Hamburg, Dithmarschen and Friesland: water an ell higher than in 1412.		
1475	England: towns about the Humber lost.		
16.10.1476	Schleswig-Holstein: 2 parishes destroyed.		
31.12.1476	Friesland.		
13.11.1477	Scheldt: 17 parishes destroyed. Old Ostende lost to the sea.		
14.1.1497	Holland: 28 parishes destroyed. Flanders and Friesland also affected.		
26–27.9.1509	Holland, Zealand and Friesland.		
16–17.1.1511	Friesland and NW Germany: Jadebusen enlarged.		
4–5.11.1530	Flanders, Zealand and Holland: 25 towns and many villages lost; also English coast (Essex and Kent/Thanet).		
2.11.1532	Netherlands: villages lost about the Scheldt. All coasts from Flanders to Hamburg affected.	1554	Lancashire (The Fylde).
4–5.3.1566	NW Germany: Dithmarschen, a village destroyed.		

Table 13.3 – continued

North Sea coasts of England, Netherlands, Germany, etc.	Other coasts of Britain
31.10– 2.11.1570 Netherlands: the great cities flooded, possibly as many as 400 000 drowned. Flood extended from N France to NW Germany	
5.10.1571 England: Humber, Lincolnshire and Fenland.	
19–21.8.1573 Friesland and NW Germany– Denmark: new bays formed in northern part. Many drowned.	
28.3.1578 Friesland–NW Germany: many lives lost.	
20.12.1593– 3.1.1594 Dutch and German coasts: several bays formed, many lives lost.	
	Early February 1607 Bristol Channel (Somerset-Gloucestershire): 4 villages flooded, 100 drowned.
31.10.1612 N Friesland: 4 parishes flooded and new bays formed.	
6.12 and 13.12.1612 Schleswig-Holstein (Nordstrand): bay formed – floods remain all winter.	
6.12.1630 Flanders and Holland: part of Flushing lost.	
21.10.1634 Netherlands, N Germany and Danish coast (Jutland): new bays formed, about 6000 people drowned. Former island of Nordstrand split into four parts.	
1–2.2.1643 N Friesland to Schleswig-Holstein.	
Nov. 1649 More of Heligoland lost.	
1 and 4.3.1651 Holland, Friesland, Zealand and Flanders.	
11 and 31.12.1651 N Friesland: hundreds drowned on both dates.	
14–15.1.1661 N Friesland and Hamburg: many buildings and people swept away.	
c. 30.9.1671 Flanders, Holland and Zealand: hundreds of people drowned.	
14–17.11.1675 17.11.1675 Coasts from Flanders to Schleswig-Holstein.	

Table 13.3 – continued

North Sea coasts of England, Netherlands, Germany, etc.	Other coasts of Britain
26.1.1682	Coasts from Flanders to NW Germany: many thousands drowned.	
22.10 and 22–23.11.1686	Friesland–NW Germany: over 1000 drowned in both these floods.	
1.1.1690	Zealand.	
1.10.1697	All NW German coast.	
Mar. 1714	Flanders, Holland, Zealand.	
25.12.1717	All coasts from Flanders to Jutland: tide 12 ft above normal, over 12 000 people drowned. England also affected.	
31.12.1720	Coasts from Zealand to NW Germany. Heligoland divided into 2 small islands.	*c.* 12.12.1720 (possibly Old Style) Lancashire: great tide with W'ly gales.
27.2.1736 1791, 1792	Lincolnshire and East Anglia.	* See footnote
3–4.2.1825	Danish coast and German Bight disaster.	
11.3.1883	Lincolnshire, the Fenland around Sutton (area not great).	
28.11.1897	N. Norfolk (area not great).	
6–7.1.1928	Sea flood in London: people drowned.	
30.11.– 1.12.1936	East Anglia.	
12–13.2.1938	East Anglia: new coast defences swept away.	
1.3.1949	N Norfolk: salt marsh created (area not great).	
31.1.1953	Netherlands; also Lincolnshire, East Anglia and the Thames Estuary: hundreds drowned in England and many more in the Netherlands, in all about 2000 perished.	
16–17.2.1962	NW Germany (Hamburg and Oldenburg): about 300 drowned.	
3–4.12.1973	NW Germany, around Hamburg: hundreds drowned.	

NOTE. Calendar dates in this table have been adjusted to the New Style calendar.

* There were very high flood tides in the Elbe at Hamburg on 22.3.1791 and several times during the exceptionally stormy month of December 1792, that on 11.12.1792 being the highest water level in the eighteenth century: but no report of loss of life is known to the author.

than at other times) in the following tables.[1] The situation revealed by these tables is complicated by the well authenticated increases of storminess in the North Sea in the thirteenth and late sixteenth and seventeenth centuries. The storms which destroyed the East Anglian port of Dunwich have been mostly omitted from the above list as they seem to have had more to do with erosion of the cliffs than with flooding – the dates generally do not coincide with widespread sea floods.

The nineteenth and twentieth century cases listed have generally involved much smaller losses of life and of land than those in earlier centuries, doubtless owing to more effective sea defence works.

Table 13.4(a) below summarizes the data for the North Sea and southern Baltic in numbers of severe sea floods each century. There seems to be some suggestion that the Baltic floods, which depend mostly on northeasterly storms, have undergone a long-term variation inverse to that of the frequency of North Sea disasters.

Table 13.4

(a) Numbers of severe sea floods reported each century

Centuries A.D.	English Channel coasts	North Sea coasts	Baltic southern coasts
0–99	1		
100s			
200s		1	
300s		1	
400s	1	1	
500s		2	
600s			
700s			
800s		1	
900s			
1000s	2	7	
1100s		3	
1200s	2	13	
1300s		4	1
1400s		4	
1500s	1	4	
1600s		3	4
1700s		1	6
1800s		1–3	2
1900–72		3	

1. As in the Bay of Bengal (p. 113), and on any other coast exposed to sea floods, the conditions for disaster involve coincidence of a storm surge of water driven into shallow seas by atmospheric pressure differences and wind stress, about the time of highest tides. The situation may be aggravated by a secular rise of world sea level (e.g. due to glaciers melting) and by local land sinking (as in the southern part of the North Sea area).

(b) Total numbers of sea floods around the coasts of the North Sea listed per century

A.D.	100–99	200–99	300–99	400–99
Number of floods	1	2	2	3

A.D.	500–99	600–99	700–99	800–99	900–99
Number of floods	7	1	1	5	2

A.D.	1000–99	1100–99	1200–99	1300–99	1400–99
Number of floods	12	15	54	35	50

A.D.	1500–99	1600–99	1700–99	1800–99	1900–74
Number of floods	116	128	68	(5)	(9)

Principal sources used:
ARENDS, F. (1833) *Physische Geschichte der Nordsee-Küste und deren Veränderungen durch Sturmfluthen seit den Cymbrischen Fluth bis Jetzt.* Emden (Woortman), 2 vols.
BRITTON, C. E. (1937) A meteorological chronology to A.D. 1450. *Geophys. Mem.*, **70.** London (H.M.S.O. for Met. Office).
BROOKS, C. E. P. (Unpublished) Tabulation from an unknown source, probably work done for Sir Napier Shaw c. 1920–25, found by U.K. Meteorological Office Library in 1964.
BROOKS, C. E. P. and GLASSPOOLE, J. (1928) *British Floods and Droughts.* London (Benn).
HENNIG, R. (1904) Katalog bemerkenswerter Witterungsereignisse von den ältesten Zeiten bis zum Jahre 1800. *Abhandl. der Preuss. Met. Inst.*, **II** (4). Berlin.
LOWE, E. J. (1870) *Natural Phenomena and Chronology of the Seasons.* London (Bell and Dalby).
VANDERLINDEN, E. (1924) *Chronique des événements météorologiques en Belgique jusqu'en 1834.* Brussels (*Mém. Acad. Roy. Belge*, 2nd ser., **5**).
WEIKINN, C. (1960–63) *Quellentexte zur Witterungsgeschichte Europas von der Zeitwende bis zum Jahre 1850*, Vols. I.1 to I.4. Berlin (Akad. Verlag).

Note:
The totals given in this table are minimum estimates, since all cases where floods have been reported in two or more areas in the same year have been treated as a single flood unless detail of the dates made it clear that there were two or more events. Any reports of storm floods which could refer only to rain flooding of a river have been omitted, although a few of these may have coincided with windstorms that produced North Sea surges. In the later centuries covered by the sources used separate floods reported within a few days of each other have been counted as separate events. Thus the list must be read with due allowance for the increase of reporting as time went on; no attempt has been made to develop a systematic adjustment for this because the increase of reporting was not a steady and continuous process, particularly over the first four or five centuries covered in the table. It seems possible that the real incidence of North Sea floods was similar during the 500s, 1200s and 1500s–1600s. It is probable that the decreases after each of those maxima and after the 800s were real. After the 1700s it is certain that some of the great reduction in the frequency of sea floods can be attributed to the increase of coast defence engineering.

A longer history of the North Sea coast of the Netherlands by JELGERSMA *et al.* (1970) indicates some further points which must be simply related to climatic changes. By about 3000 B.C. the North Sea, widening as the sea level rose with the world-wide melting of ice and warming of the seas, had reached a coastline about 10 km east of where it now is in the Netherlands. Sand dunes then began to form (the Older Dunes) but this activity had petered out by Roman times, presumably with the sand having become wet enough for vegetation to get a hold, though some activity was renewed later. The dune formation activity was cyclic, occurring either in phases of greater windiness or when the water table was lower: the active phases are tentatively dated 1700–1600, 1400–1100, 750–650, 500–350, 300–150 B.C. and (slighter) between A.D. 100 and 700 and 800–900. Periods of calmer seas with high sea level may be indicated by sedimentation off shore, particularly between about 2000 and 1700 B.C., 1100–750 B.C., in Roman times and between A.D. 900 and 1100. From about Roman times onwards the duneland became stabilized, presumably becoming moister, and gradually colonized by vegetation, ultimately with beech forest right to the coast (though this was felled by Man later). But from A.D. 1150–1200 onwards increasing storminess ultimately renewed the blowing sand, probably after sea level had fallen somewhat – the Younger Dunes formed particularly between 1600 and 1700 – and scouring of the sea bed off shore.

Taking a broader, summary view of the coasts of the British Isles and of the European continent between southwest Brittany and Denmark, there seems to be widespread evidence of blowing sand overrunning the coastlands in just three main periods:[1]

(1) Approximately 1700 to 500 B.C. or later.
(2) Between about A.D. 400 and 700.
(3) Between about A.D. 1300 and 1700.

Of these periods (3) was by far the most active except on the coasts of the Channel and of France affected by regional land sinking (and in those areas most of the sand is attributed to period (1)). Period (2) seems to have been the least important of the three, but there is evidence from the Channel Islands as well as the Netherlands. All three periods are believed to have been times of cooling climate, the main periods of cooling in the last 3000–4000 years; and of the three, period (2) is believed to have shown the least cooling. These episodes presumably indicate times of increased storminess in latitudes near 50–60°N on the Atlantic fringe of Europe and somewhat falling sea level, leading to the exposure of areas of beach sand to drying out. In origin the sands in most of the region were glacial outwash sands, submerged by the great rise of sea level between the end of the glaciation and about 2000 B.C.

Levels of lakes and inland seas

Variations in the level of the water in lakes and inland seas where the geography of the supply and outlet channels has not changed, and where tectonic changes do not enter in, register variations in the balance of downput of rain and snow versus evaporation over the catchment basin. Useful knowledge of such variations over past millennia and centuries has been built up by archaeological and geomorphological methods allied to radiometric or other methods of dating and for recent times by gauge measurements. Within about the last thirty years, however, human control of or human activities affecting the level of the larger lakes has become so general that there are few which clearly register natural changes any more.

1. Most of the evidence here referred to has only been published in a miscellany of local reports and much is unpublished. I am indebted to D. S. RANWELL of the Nature Conservancy, Norwich, for much of the information in this paragraph (personal communication, 2 May 1973).

The substantial changes, up and down, of the natural level of many such waters within recent centuries, and even in recent decades, however, speak eloquently for the continual play of climatic changes over a wide range of scales which are forever going on. These recent changes of level have also had serious practical consequences for human activities near the shores, particularly where the slope of the terrain is very gentle so that a change of water level by a metre or two may shift the shoreline by one or more kilometres.

In the western U.S.A., particularly in Utah, Nevada, the southeastern part of Oregon and the inland basins of California during the Quaternary ice ages there were innumerable lakes which have since dried up or of which only remnants are there today (FLINT 1957, pp. 226 ff.; WOLDSTEDT 1965, p. 208). The greatest of them was Lake Bonneville which stretched from about $37\frac{1}{2}°$ to $42\frac{1}{2}°$N and $112°$ to $114°$W in the Great Basin of Utah and attained an area of over 50 000 km^2, over ten times that of the present Great Salt Lake which is its remnant. Lake Bonneville was in existence at least three times, in different glaciations, and attained a maximum depth of over 300 m, marked by the highest of the old strandlines nearly that height above the present lake surface and showing that it overflowed to the Snake River basin and the Pacific. Erosion of this outlet later reduced the depth of the lake by a third, before its ultimate drying up in a different climate. The second biggest of these 'pluvial lakes' in U.S.A. was Lake Lahontan in Nevada, which attained about 25 000 km^2 and a depth of 170 m above Pyramid Lake, the main remnant of it today. The history of these lakes can be traced, and dated, from the fine series of terraces which mark their former strandlines. From radiocarbon datings by BROECKER and ORR (1958) it appears that the last main high stand of these lakes was from about 25 000 to 15 000 years ago, coinciding with the last, and greatest, climax of the Würm/Wisconsin glaciation, and that there had been a low stand going back to at least 34 000 years ago; subsequent high levels around 11 700 and, finally, 10 000 years ago presumably correspond to cold climate phases (Mankato and Valders readvances?) before and after the Two Creeks Interstadial in North America and the Allerød warm phase in Europe. A similar history has been established for Searles Lake, California, near $35\frac{1}{2}°$N $117\frac{1}{2}°$W (FLINT and GALE 1958).

Studies of Lake Ioannina, 469 m above sea level, in the northwestern mountains of Greece (39.6°N 20.9°E), have established that it stood at least 3 m higher than at present about the time of the last climax of glaciation in northern Europe some 20 000 years ago and spread along the valley floor to about three times its modern area before artificial drainage of the marshes (HIGGS et al. 1967).

The dated record of lake levels in eastern equatorial Africa between 5°N and 9°s over the last 20 000 to 30 000 years has been conveniently summarized by BUTZER et al. (1972) in the diagram reproduced here as fig. 13.29, together with the history of Lake Chad (13–14°N 13–15°E). All these lakes lack any outlet at the present time. All depend on the equatorial rains for their supply. Clearly their relationship to the glaciation is a complicated one, and the old idea of equating pluvial periods in Africa with glacial periods in high latitudes needs revising. Among the lakes here surveyed, all for which data exist so far back had high stands at or

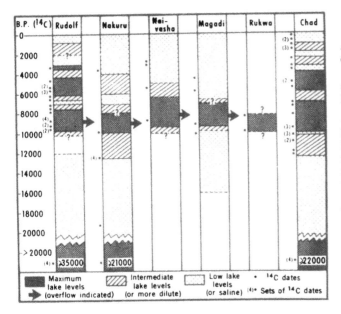

Fig. 13.29 Lake levels in equatorial and tropical Africa. Radiocarbon dated high stands and low stands.
Lake Rudolf   (4°30′–2°20′N 35°50′–36°40′E)
Lake Nakuru   (0°20′S–36°05′E)
Lake Naivasha (0°50′S 36°20′E)
Lake Magadi   (1°55′S 36°15′E)
Lake Rukwa   (8°S 32°20′E)
Lake Chad   (approx. 13°N 13°E)
   (*From* BUTZER *et al. 1972; reproduced by kind permission of the authors and of* Science.)

about the time of maximum glaciation 20 000 years ago and, more particularly, during the millennia immediately preceding the glacial maximum. They all seem to have sunk low between about 20 000 and 12 000 years ago when the ice sheets were beginning to recede, but had high stands once more between about 10 000 and 8000 years ago in early postglacial times, roughly corresponding to the so-called Pre-Boreal and Boreal phases in Europe, when reconstructions of the atmospheric circulation patterns prevailing (see figs. 16.10(*a*), (*b*), 16.11) suggest that the climatic zones in the European sector were displaced abnormally far north and the circulation had a prevalent meridional character. High stands again prevailed, at least in the northernmost lakes surveyed (Chad and Rudolf) between about 6000 and 4500 years ago, during the warmest postglacial times when the general circulation appears to have had a zonal character but with the subtropical anticyclone belt so far north over Europe as probably to allow a greater northward seasonal (monsoonal) excursion of the equatorial rains over Africa than now occurs.

From well-marked shorelines and other traces around Lake Chad today, it is known that during the last ice age climax around 22 000 years ago and for some time after (R. E. MOREAU, personal communication, December 1963) Mega-Chad was a great inland sea comparable with the present-day Caspian: it had arms extending to about 10°N 16°E, 13°N 11½°E, 18°N 14°E, and from there almost to 20°E (GROVE and WARREN 1968), the shoreline being about 330 m asl. or about 50 m higher than today. It seems to have been of almost the same size again 8500 years ago. Around 55 000 years ago Mega Chad is thought to have been still more extensive, with its shore line about the present 400 m above sea level contour. There was probably at those times a greater rainfall than now on the Ahaggar and Tibesti

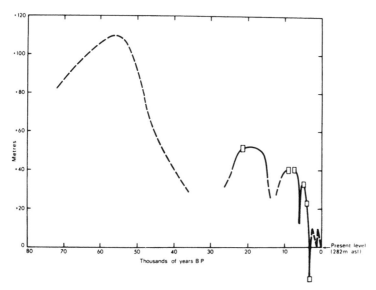

Fig. 13.30   History of the level of Lake Chad over the last 80 000 years. Scale in metres.

highlands in the central Sahara, supplying inflow to the Chad basin from the north. Inflow was, however, always *mainly* from the south; the quantity required has been thought to imply that the equatorial rain zone extended with little or no seasonal intermission to the north of the mountains of the Camerouns (which divide Chad from the Congo basin), to near 10°N. This seems to imply that in glacial times, or perhaps more particularly in the glacial advance phases, the equatorial belt of cloudiness and rains was broader than now. The history of the levels of Lake Chad throughout the last glaciation and since is surveyed in fig. 13.30. Within the last millennium the level was lowest (281–282 m) in the warmer phases of world climate, between about A.D. 900 and 1550, as well as in the present century, and rose to about 286 m in the climax of the Little Ice Age between 1550 and 1800 (MALEY 1973).

The Caspian Sea, which is at present the world's biggest inland sea, with an area of about 440 000 km² and stands about 28 m below world sea level, is surrounded by old shorelines around its basin up to 75–80 m above the present water surface, probably related to the early part of the last (Würm) ice age. The history of the level of the Caspian Sea over the last 80 000 years is indicated in fig. 13.31. During times of northern glaciation the extent of this sea was nearly doubled, therefore, extending north of its present limits to about 51°N and east till it may have joined with the, also expanded, Aral Sea. It was also nearly linked to the northwest with the Sea of Azov and the Black Sea, which was then a lake overflowing into the Mediterranean whose level was reduced with the lowering of water level in the world's oceans by the enhanced storage on land in ice sheets, etc. In the latter part of the Würm ice age the level of the Caspian is put at 30 m above today's. During the warmer phases of world

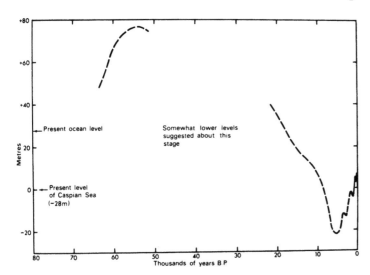

Fig. 13.31   History of the level of the Caspian Sea over the last
80 000 years. Scale in metres.

climate the level of the Caspian Sea has been relatively low. Submerged beaches show that the water surface was at different times up to 20–22 m below its present level, probably having been lowest 4000 to 6000 years ago (LEONTEV and FEDEROV 1953), indicating that the warmest postglacial times, in the so-called Atlantic epoch, though described as humid in northern and western Europe, must have been drier and more anticyclonic over the Caspian basin. Harbour works, fortifications and the quarries from which the stone came at the ancient city of Derbent on the western shore at the foot of the Caucasus, the building of which is recorded in documents (NIKOLAEVA *et al.* 1962), indicate that the level of the Caspian was 1–2 m lower than now during most of the first millennium A.D. and may have been as low as – 32 m above sea level (4 m below now) from the ninth to the twelfth century. After that, it rose sharply during the thirteenth and fourteenth centuries, reaching (according to GUMILYOV 1964, cited by CHAPPELL 1970) – 20 m above sea level. Later records show that, after a minimum about A.D. 1560 put at – 26 m above sea level, the level was rising again between 1606 and 1638–80 when it was – 23 m above sea level. By 1720–30 it had fallen to – 25·5 m above sea level, but soon rose sharply again and attained a new high stand of – 21 m above sea level between 1796 and 1810 (BERG 1934), from when it has fallen almost continuously to the present day.

The history of the levels of the Caspian Sea and of Lake Chad over the last 20 000 years may be compared in fig. 13.32. The levels of the Caspian Sea, which nowadays has no outlet, and of Lake Chad, with no outlet apart from underground seepage, may be taken as representative respectively of the variations of moisture available in middle latitudes (European Russian sector) and over northern and equatorial Africa. There may also be a

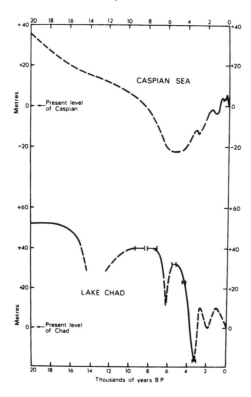

Fig. 13.32   Variations of the Caspian Sea and of Lake Chad over the last 20 000 years compared. Scale in metres.

meaningful parallel between the later rises of the level of Lake Chad and the spread of early settlements into the driest regions of southwestern Palestine around 6000 and 3000 B.C., reported by BLAKE (1969), and followed each time by withdrawal in the succeeding centuries, in the latter case withdrawal for good. There is, however, no general parallel between the moisture histories of middle and low latitudes, as represented by these two water bodies.

Fig. 13.33 gives a record of the recent variations of the Caspian Sea and of selected lakes in Europe, Asia, Africa and North and South America from various dates between 1600 and 1900 up to 1950. From this it is seen that the lakes in temperate latitudes – all those selected being east of an important watershed in the zone of prevailing W'ly winds – were all high in the nineteenth century and have been falling since.[1] It is recorded (FRASER 1970) also that

1. A year-by-year history of the level of Lake Titicaca, on the high plain (Altiplano) of Bolivia-Peru, from 1912 to 1970 (KESSLER 1974), shows that the mean water level, which underwent a net fall of about 1 m over that time (but was 3 m lower in the early and mid 1940s), decreases when the southern hemisphere belt of westerlies moves north and, as a result, the upper west wind components at that latitude (about 16°S) increase. There also seems to be an association with the 11-year sunspot cycle, which differs from that affecting Lake Victoria in east Africa (Volume 1, p. 441): Lake Titicaca has been highest about sunspot minimum in every cycle since 1912 except that between 1933 and 1954 it only underwent a single fluctuation, with its minimum about the sunspot minimum year 1944: this single fluctuation was of about twice the usual (1–2 m) range.

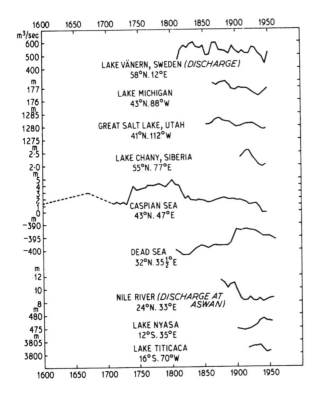

Fig. 13.33   Variations of lake levels in different parts of the world and of the Caspian Sea since A.D. 1600.

the lowlands of central and eastern Scotland in the sixteenth to the early nineteenth centuries were 'dissected to a far greater extent than at the present day by endless and immense watery tracts in the shape of countless lochs and lochans'. In Fife there were twenty lochs as big as Loch Leven is now and which have since been reduced, not all by being drained. Loch Leven itself has been reduced, partly by draining in the nineteenth century, to a quarter of its former size. The Dead Sea (KLEIN 1961), by contrast, was high just during the decades of strongest and largest-scale development of the middle latitude W'lies in the first decades of the twentieth century.

Mention has already been made in Volume I (Chapter 6, p. 216) of ŠNITNIKOV'S findings regarding fluctuations of lake levels in Asia with a period length of the order of 1800 years. In another work ŠNITNIKOV (1957), reviewing the available measurements made in the last two centuries and the reports by eighteenth century explorers of their observations and what they learnt from interrogating the tribesmen living on the Kirghiz steppes and farther north, has demonstrated a recurrent fluctuation in the levels of the water bodies throughout a vast region of Asia over periods ranging from about 20 to 45 years.[1] The range

1. The commonest length of the cycles in the last 250 years has been between 29 and 39 years, which led BRÜCKNER (1890) to the rainfall investigations from which he derived his suggested 35-year cycle in weather phenomena.

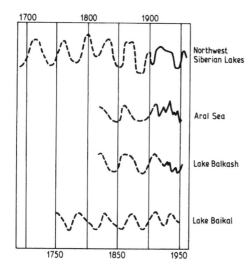

Fig. 13.34   Variations of the levels in the lakes in Siberia.
(*After* A. V. ŠNITNIKOV *1957*.)

of these fluctuations has been up to 6·5 m in the case of Lake Chany and is from 1 to 4 m in the case of many other lakes on the steppes between the Urals and the Ob and as far south as Kazakhstan. These ranges are so great that a number of the lake beds dry up and become cultivable or else turn into meadows and incipient forest, while others become so salty that the fish die and salt is extracted from them; then, in a few years, the lakes reappear, and fill with fresh water, and fish become abundant. ŠNITNIKOV states that the dry periods are in general periods of warmer climate in the region, with fine summers and little winter snow; the wet periods are both colder and snowier.

A diagram in SNITNIKOV's work demonstrates that the variations in level of the steppe lakes since 1838 have run parallel to the amounts of winter snowfall at Barnaul (53°N 84°E), with a 2- to 5-year lag. This history of summer rainfall is somewhat similar, but less closely parallel.

The great water bodies farther south, the Aral Sea, Lake Balkash, and Lake Baikal, are affected by the same rhythm, but the phase of the fluctuations in them is not identical with the lakes and marshes on the steppes.

The data are summarized in fig. 13.34 and in Table 13.5 below. A point of special interest is that, according to the verbal reports collected by travellers 50 to 80 years afterwards, the levels of the lakes on the steppes in western Siberia between the Urals and the Ob were exceptionally low in the 1690s (as they were again in the early 1880s). This implies a warm dry regime in western Siberia in the 1690s, coinciding with the coldest phase of the Little Ice Age climate and frequent northerly winds in western and northern Europe and Iceland. The moistest period in western Siberia was around 1790.

The variations of level of Lake Victoria ($\frac{1}{2}$°N–3°S) and Lake Malawi (9–15°S), formerly Lake Nyasa (data from PIKE 1965), in East Africa year by year in recent decades are shown in figs. 13.35(*a*) and (*b*) (see LAMB 1966 for more details). It is known that the level of Lake

Table 13.5   Years marking the phases of the fluctuating levels of the steppe lakes between the Urals and the Ob

				Dates A.D.				
Maximum levels			1710–20		1750–5	Early 1790s		
Minimum levels		1690s		1730–5		Mid 1770s		1805–10
Maximum levels	Mid 1820s		1866–8		c. 1895		1914	1946–51
Minimum levels		1853–4		1881–5		1901–2		1936–40

(*a*)

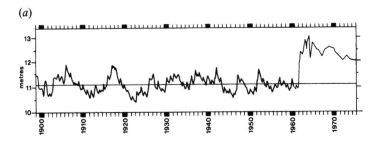

(*b*)

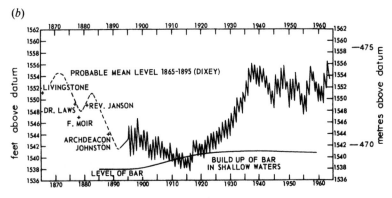

Fig. 13.35   Variations of the levels of (*a*) Lake Victoria and (*b*) Lake Malawi in recent decades.
   (*By courtesy of the Royal Geographical Society.*)
   The levels of Lake Victoria are monthly values from the beginning of the record to 1964, yearly values after 1964.
   (*The lower curve is adapted from* PIKE *1965.*)

Victoria, like that of Lake Nyasa, was high in the nineteenth century and that it fell 2·4 m between 1880 and 1898. Thus, both these lakes in equatorial Africa had a low stand which roughly coincided with the years of the strongly developed W'lies in middle latitudes between the late 1890s and the late 1930s, though the recovery of level of Lake Victoria was delayed until the second phase of the decline of the W'lies, from 1961 onwards, and was then abrupt.

The variations of a small lake with no outlet, El Junco, in impervious volcanic basalt in the crater of an old volcano on San Cristobal (Chatham Island) in the Galapagos group (near the equator at 90°w) with a continuous history of undisturbed accumulation of sediment longer than the range of radiocarbon dating, have been studied by COLINVAUX (1972). The lake seems to have had two periods of existence, before and since the last glaciation, and to have dried up during glacial times. It may reasonably be suggested (contrary to COLINVAUX) that this indicates that the intertropical convergence zone (ITCZ) and equatorial rains, which even nowadays are almost inactive over the cold water from the Humboldt current and Peru coast, were held south of the equator in that sector in glacial times by the greater vigour of the northern hemisphere atmospheric circulation relative to that over the southern (cf. LAMB 1959) and the prominence of the northern hemisphere's cold regime in that sector.

River levels

Records of the water level in rivers, and of their flooding and drying up, may register the history of the balance of rainfall and evaporation in their catchment basins so long as the river channel remains substantially unaltered and the moisture level in the soil either varies with the river or (e.g. in the case of the Nile in the desert zone) remains constant. The floods produced by rapid snow-melt, and occasionally by ice blockages, in the rivers in middle and higher northern latitudes are a special case which must be distinguished.

The very long records of the levels of the River Nile in Egypt are the example *par excellence* of the value of such records, being exceptionally informative regarding the variations of the equatorial rains in eastern central Africa and of the summer monsoon rains in the highlands of Ethiopia, where about three-quarters of the year's rain falls in the four months June to September. The lower Nile is fed by two main branches: the White Nile, which emerges from Lake Victoria and drains a large area of equatorial Africa, and the Blue Nile, which rises in the mountains of Ethiopia. Below the confluence of these two rivers at Khartum no further significant supply of water reaches the Nile as it flows 3000 km through the desert to the Mediterranean Sea. The seasonal variation of the level of the river in Egypt is very regular: each year the river is lowest in about April and May, rises rapidly in July and is high through late August and September. This pattern is due to the flood of water in the Blue Nile (and other tributaries which rise in Abyssinia) bringing down the water from the summer monsoon rains. The White Nile maintains a much more nearly constant flow through the year, corresponding to the much less marked seasonal variation of the rainfall near the equator, with its two maxima each year about April and November, though it does show secular variations. The yearly flood of the river has been the basis of Egyptian agriculture for at least the last 5000 years, and the height of the flood seems to have been measured and recorded in the Pharaoh's annals from the First Dynasty around 3100 B.C. (BELL 1970). The highest and lowest levels of the river measured each year at or near Cairo from the time of Muhammad, with certain gaps, have been published by Prince OMAR TOUSSOUN (1925), starting with the year A.D. 622.

The flood level of the Nile at Cairo can be taken to register the effect of the summer monsoon rains near 10°N. The level of the river at its low stage responds to the supply of the White Nile by the equatorial rains. The upper and lower graphs in fig. 13.36 show the course of flood level and low level stages of the Nile at Cairo by 50-year averages from A.D. 622, as worked out by BROOKS (1949), including his estimates which span the broken parts of the record. The progressive rise of both levels during the thirteen centuries of this record is attributable to the gradual silting up of the river bed and its flood plain, which has been going on throughout historical times and is generally put at 10 cm/century on average. This value is represented by the two straight lines superposed on the graphs in fig. 13.36, which should be taken as the zero lines from which the climatically induced variations of the river levels must be assessed.

From this record, it is the equatorial rains that show the bigger secular variations; though both curves vary up and down more or less in phase together from the 600s to the early 1400s, BROOKS reports that the broken record after 1370 needs to be treated judiciously, as it seems that it was only the highest maxima and lowest minima that were then recorded, a procedure which may be the cause of the apparent opposite course of the flood level and low level trends between about 1400 and the early 1800s, when the records once more become complete (though increasingly affected by the artificial influence of irrigation works). More details of the river levels are tabulated in Tables 30 and 31 in Appendix V. The flood levels of the Nile were high about 630, 850–930, 1100–1150 and 1400–1450 (especially the 1430s and 1440s) and again in the mid 1800s. There is obviously no consistent association with the variations of prevailing temperature in Europe, where the warmest phases were perhaps

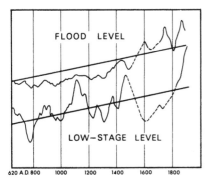

Fig. 13.36   Variations in the mean flood level and seasonal low level of the River Nile at Cairo since A.D. 622.

The straight lines represent the rise of both these seasonal stages of the river that would be expected from silting up of the bed at a uniform rate of 10 cm per century.

(*After* BROOKS *1949 from data given in* TOUSSOUN *1925.*)

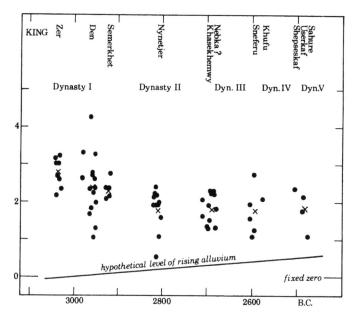

Fig. 13.37   Height of the Nile floods in Egypt in sample runs of years measured in ancient times.

(*Reproduced from* BELL (*1970*) *by kind permission.*)

sometimes due to a northern position of the subtropical high pressure belt and at other times associated with meridional patterns of the atmospheric circulation. The marked minimum in the levels of the Nile in the late 700s probably coincided with a cold phase in Europe, whereas the minimum around 950 to 1000 seems to have gone with a warm climate in Europe (apart from one exceptional winter in 1010–11, when there is reported to have been ice on the Nile at Cairo and on the Bosphorus). The low levels of the Nile prevailing about 1220 and 1300 probably coincided with two of the warmest phases of the early medieval warm climate in Europe, and the high levels of the Nile in the 1430s with the sharpest phase of the cooling which followed.

BROOKS notes that statements about the prevailing levels of the Nile flood at Cairo in earlier centuries are found in HERODOTUS' *History* and in at least two Roman writers, PLINY and AMMIANUS MARCELLINUS. From these, it is clear that the floods were generally high about 500 B.C. (a cool wet period in European history). There is much less certainty about the measurements indicating equally high levels around A.D. 100 and 300–400 (when Europe seems to have been enjoying a warmer and drier climate than before), because of confusion about the datum level used. It seems certain that the flood levels were not so high in the other centuries of the Roman era.

BELL (1970, 1971) has usefully analysed the earliest extant records of the Nile floods in stone inscriptions from the times of the first twelve dynasties. From this work it is clear (fig. 13.37) that, regardless of whatever assumption is made about the datum used for the measurements at Memphis, the average flood level of the Nile was always lower (by at least 0.7 m) and less variable from year to year from Dynasty II onwards, starting about 2900–

2800 B.C., than it had been in the First Dynasty. If a fixed datum was used, the decline in flood level was greater than this and increased as time went on. Sand dunes spreading from the western desert are known to have invaded the Nile valley in Middle Egypt, covering the alluvium with several metres of sand over a stretch 175 km in length (BUTZER 1959) at some time in the millennium that followed, but became fixed by a deposit of river mud in the high Nile phase about 500 B.C. or in the centuries immediately following. BELL'S (1970, 1971) analysis of texts reporting famines in Egypt indicates further periods of severity caused by low level of the Nile flood between 2180 and 2130 B.C., between 2000 and 1990 B.C., possibly again soon after 1786 B.C., and once more from 1200 B.C. onwards. A very wide area, possibly extending as far as Anatolia, Mesopotamia and the Indus valley, provides evidence of troubles attributable to aridity about the same times (some further details will be found in Appendix V, Table 31). The appearance of a 200-year interval between a number of these episodes is interesting.

The high floods of the nineteenth century A.D. were associated with a high level of the White Nile from equatorial Africa, keeping the river in Egypt fairly high even at its seasonal minimum, and they had the considerable year-to-year variability characteristic of the equatorial rains in that part of the world – part of which seems to be associated with the sunspot cycle (see Volume I, p. 441). It is likely that the high floods before 3000 B.C. owed more to the Blue Nile and the summer rains over Ethiopia, which were probably greater than now, since there is other evidence – e.g. in Lake Chad – that the rains reached farther north than they do now. The year-to-year variability of the river in Egypt appears (fig. 13.37) less than now for a long time after 2800 B.C., when the river level fell, and this may imply that the contribution from the equatorial rains and the White Nile became much smaller at that time.

In reality, the rate of silting of the river bed and flood-plain must vary somewhat with the variable behaviour of the river over the centuries, though this fact is glossed over by the lines shown as straight on fig. 13.36. The range of variations of the reported flood level and low level stages is, however, such that there can be no doubt that we discern real differences from decade to decade and century to century. It does not seem possible to establish the relative height of the floods in different millennia or between the beginning and end of the record, however, because of uncertainties about just where the measurements were made and whether portable gauges or a fixed mark were used.

There are no series of data for rivers in other climatic zones comparable in length or fullness with the unique Nile record.

Apparently, some records do exist for the great rivers of China.[1] A guide to Chinese work up to the 1940s on floods and droughts in that country, and on the data published up to that time, was given by SCHOVE (1949) with a useful commentary on the likely sources of error in making climatic deductions from the statistics (particularly due to the shifting

1. The floods of the Yangtse-kiang and the Yellow River (Huang-ho) in Honan province in June 1931, which drowned more than 1 million people, and the Yellow River flood in September–October 1887 when 900 000 were reported lost, are accounted the worst ever known human disasters due to weather.

centres of power, e.g. from the area about 32°N 119°E in the fourth century A.D. to around 40°N 117°E in the fifteenth century). The curves in fig. 13.38 were given by SCHOVE as indicators of moisture in China, central Asia and Syria respectively over more than the last 2000 years, that for Chinese raininess being due to YAO (1943, 1944) and that for the Syrian desert being taken from GRANT (1938). No great confidence can be felt regarding the general levels of these curves in different millennia (i.e. in their long-term trends), but the variations from one century to the next are probably reasonably well shown in most cases. CHU KO-CHEN (1926, 1961) states that the statistics of floods and droughts in China-within-the-Great-Wall (formerly known as China proper) indicate that the climate was comparatively dry from the fourth to the seventh century A.D. (from the time of the Three Kingdoms to the beginning of the Tang dynasty), it was wetter (i.e. more floods) in the twelfth and fourteenth centuries, and once more became drier in the fifteenth century and after (Ming dynasty).

Tabulations by YAMAMOTO (1972) of the numbers of floods and droughts recorded in the Iwate and Miyagi prefectures in northeastern Japan from A.D. 1300 onwards indicate a sequence at least partly inverse to that in China mentioned above and more like that in Europe. The fifteenth century and the period between 1750 and 1850 seem to have been both cold and wet in Japan, whereas the fourteenth and sixteenth centuries and the period 1600 to 1650 were markedly drier. Moreover, there are many records of the exhaustion of underground water and drying out of soil and subsoil in the period between A.D. 900 and 1200 in Japan.

BROOKS (1949, p. 306) attempted to derive a history of raininess in Britain, Belgium and Europe in general by half-centuries from 100 B.C. by compiling the numbers of reports of (a) 'storms and floods' and (b) droughts; but, although his index of raininess was aptly defined as

$$100 \times \frac{\text{number of reported storms and floods}}{\text{total number of reported storms, floods and droughts}}$$

to eliminate the effect of different frequency of reporting of such matters in different periods of history, his figures appear to be vitiated by the climatic inhomogeneity of the regions

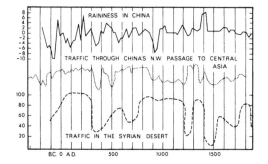

Fig. 13.38   History of moisture in different parts of Asia since 300 B.C.

Top curve: YAO's index of 'raininess' in China, $(R - D)/(R + D)$, where R = number of floods reported, D = number of droughts reported.

Middle curve: SCHOVE's index of the volume of traffic passing through the 'Jade Gate' from China to central Asia.

Bottom curve: GRANT's index of the amount of travel in the Syrian desert.

(*Reproduced from* SCHOVE *1949 by kind permission of the author and the Controller of Her Majesty's Stationery Office.*)

chosen. Through averaging together during overlapping 250-year periods the numbers of these events reported in regions as diverse as Britain and Italy, or even northwestern and southeastern Britain, and quoting the result as applying specifically to the middle 50 years of each time span, BROOKS's analysis must have muddled the data and blurred any real climatic differences which might be discernible (and which data from pollen analysis and rates of bog growth suggest are sometimes quite sudden in onset).

Some time ago the present writer tabulated (Table 13.6) the numbers of reports of floods and droughts in different parts of Europe separately, found in the collections of general weather data in just three sources (HENNIG 1904, BUCHINSKY 1957, TOPOR 1963), each believed to be the most comprehensive for the regions they cover. (Fig. 17.2 in Chapter 17 is based on a later and somewhat increased tabulation.) Though the numbers of reports for each region are rather small, it may be thought that an examination of the array of figures for tendencies which at one time or another appear to have been experienced in common by several regions could bring to light a realistic outline of the major features of the history of high and low levels of the rivers. The history which appears from this analysis is as follows:

Before 200 B.C. we have just one report (*c.* 340 B.C.) of floods in Spain (another in 181 B.C.) and one flood of the River Tiber in Italy, against two reports of serious drought years in Italy, the earliest report being for 428 B.C.

From 200 B.C. to A.D. 174 a shift to a moister climate in Italy seems indicated: there are 22 years with reported floods of the Tiber and sometimes of other Italian rivers, against only 2 reported drought years in Italy, in at least one of which (181 B.C.) Spain experienced floods.

After A.D. 174 and before 489 there are only two years with reports of floods in Italy.

Between 489 and 717 there are 18 years with reported floods in Italy, presumably marking a return to a moister climate. VITA-FINZI (1966) has noted that the onset of a long period of alluvial deposition by swollen streams and hill-wash which buried Classical sites in Greece, e.g. around Olympia, can be archaeologically dated (by buried coins, etc.) to the A.D. 500s. He believes that Italy, Sicily and North Africa were similarly affected (VITA-FINZI 1960, 1963, 1964), though this may have been a dry period in Asia Minor. In or about this same period there are the first reports from Europe north of the Alps: some notably rainy years with floods in Alsace and Germany just between 583 and 590 but otherwise a predominance of reports of droughts, in the 590s and mid 700s in Germany and Bohemia and in 678–81 in England. The 700s also seem to have become drier in Italy. *From about 800 to 900 or just after* there are much more frequent floods reported in Germany and again in Italy as well as a number reported in England, France and the Alps.

Between about 950 and 1150 there seem to have been remarkable swings between drought years and floods, both being more frequent than is typical of the longer record but particularly the cases of very low levels of the rivers north of the Alps; there were

Table 13.6 River floods (excluding snow-melt floods) and droughts in Europe by centuries

	England: Midlands, S and E		German plain		Bohemia		Russia		Spain		France		Alsace		Alps		Italy: N and central		Rumania	
	F	D	F	D	F	D	F	D	F	D	F	D	F	D	F	D	F	D	F	D
400s B.C.	1	—															—	1		
300s	—	—							1	—							—	—		
200s	—	—							—	—							2	1		
100s	—	—							1	—							3	1		
99–0 B.C.	—	—															6	—		
A.D. 0–99	1	—	1	1													6	1		
100s	—	—	—	—													6	—		
200s	—	—	—	—													2	—		
300s	—	—	—	—													—	—		
400s	—	—	—	1													2	—		
500s	—	—	2	2							4	—	8	—			7	—		
600s	—	4	1	—							1	1	—	—			6	—		
700s	—	—	1	1	—	1					—	—	—	—			3	—		
800s	2	—	8	1	—	—					4	—	—	—	4	—	6	1		
900s	—	—	2	2	—	3	2	1			—	—	—	—	—	—	—	—		
1000s	1	1	7	1	4	1	1	1					—	—	2	—	3	—		
1100s	3	1	16	5	2	—	5	1			2	—	1	1	8	1	4	7		
1200s	5	2	11	—	3	—	4	3	2	—	3	—	1	—	11	—	8	—		
1300s	3	—	26	2	6	—	8	11	2	—	1	—	3	—	11	1	3	—		
1400s	3	1	26	1	4	1	13	11	—	—	2	—	1	—	10	—	5	—		
1500s	3	3	22	6	11	4	10	6	—	2	3	—	1	1	14	—	3	—	4	3
1600s	1	3	15	1	6	1	9	4	3	2	5	—	—	—	14	—	11	2	6	10
1700s	—	1	7	—	3	1	21	5	3	—	2	—	1	—	9	—	5	3	7	11

F = floods, *D* = droughts.

runs of drought years between 988 and 1000 in England and on the German plain and again in the 1130s, when the drought spread to Italy.

From about 1150 (and especially after 1300) to 1500 or after most parts of Europe seem to have experienced the most obvious predominance of wetness and great floods in the whole record, except Russia where there were 22 great drought years in the 1300s and 1400s (11 in each century) against about an equal number of floods. Reports from Spain include 4 flood years between 1250 and 1333 and no exceptional droughts.

From about 1540 to 1750 or after reports of low river levels and droughts are predominant in the reports from England as well as in Rumania. Serious drought years were reported in Spain twice in the early 1500s and twice in the late 1600s, but otherwise flood years were predominant there. In the rest of Europe, including Russia, this seems to have been a time of high river levels and frequent floods – though droughts became about equally prominent in Italy after 1680. In Scotland wetness is believed to have continued, and indeed increased, over this period at least until 1700 or 1720. And in west Norway GROVE (1972) found strong peaks in the incidence of floods and landslides, as well as avalanches and rockfalls, in these years (see fig. 13.39).

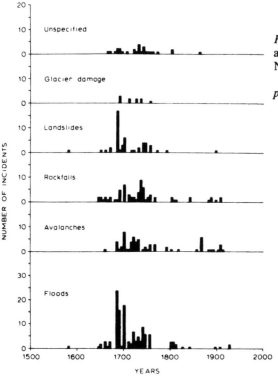

Fig. 13.39   Numbers of landslides, rockfalls, avalanches, floods and other disasters in west Norway by decades since A.D. 1500.
  (*After* GROVE *1972; reproduced by kind permission of the author.*)

The most general impressions of this history are that there was a very widespread incidence of maximum wetness about the end of, and after, the period of generally warm climates in the High Middle Ages, and the prevalence of high river levels continued in northern and central Europe more or less throughout the colder centuries that followed – the so-called Little Ice Age. There was a somewhat similar, but less pronounced, maximum of wetness in Italy and much of the Mediterranean (though apparently not in northwest Europe) after another period around A.D. 400 when there is evidence of a (probably less marked) maximum of warmth. River levels seem to have been lowest, and floods least frequent, in southern Europe in some of the colder climate phases, e.g. 500–100 B.C., A.D. 700s and after 1300, though the timing of these dry phases differed in different longitudes.

There seems to have been a generally moist period in the Mediterranean lasting most of the time from about A.D. 500 to the peak of the early medieval warm climate around 1100, though interrupted around A.D. 750–850 and in the 900s (see later, fig. 17.2). The Near East, and especially Asia Minor, seems to have undergone a long era of disastrous drought lasting perhaps from A.D. 300 to 800 or thereabouts (apart from a possible easing for a while in the sixth century). Farther south, reports from Baghdad of historic floods in A.D. 942, 1074, 1159 and 1174 indicate a maximum of moisture coinciding with the early medieval warm epoch of higher latitudes.

River ice, lake ice and ice on the Arctic seas

Records of the dates of freezing over and breaking up of the ice on various rivers and lakes, particularly in Sweden, Russia and the United States of America, over the last 100 to 200 years or more are printed in *World Weather Records* (U.S. Department of Commerce, Washington, D.C.) and in the early volumes of the Kungeliga Vetenskaps Akademiens *Handlingar* (from 1739), Stockholm. They yield evidence of the history of the winter climate in Europe and North America which appears consistent with other indications (see for instance the graph of length of the ice-free season on the River Neva in fig. 10.23, p. 431, in Volume 1). The same appears to be true of the records extending back to A.D. 1441 of the dates of freezing over of the small Lake Suwa in central Japan (ARAKAWA 1954).

Centuries-long records of the variations of the Arctic sea ice affecting the coasts of Iceland (see fig. 10.22, p. 430, in Volume 1; data given by THORODDSEN 1916, KOCH 1945), and of the Baltic ice affecting the port of Riga (BETIN 1957) and the Danish Sound (SPEERSCHNEIDER 1915), are similarly useful (e.g. later, pp. 515–20, 542–3, and see Tables 13–17 in the Appendix). Reports of ice on the Antarctic seas (LAMB and JOHNSON 1961, LAMB 1967) are more scattered, but also useful, though it seems that 'tidal waves' (tsunami) and volcanic eruptions may introduce a random element into the times of break-up of ice shelves around the limits of Antarctica and hence in the incidence of icebergs on the Southern Ocean. The effects of surges forward of parts of the Antarctic inland ice sheet and of glaciers ('ice streams') within it, which may occasionally take place for purely mechanical reasons, would also be largely unrelated to the pre-existing climate, though they could proceed to influence the climatic development for some time afterwards, according to A. T. WILSON even to the extent of inducing an ice age (see later pp. 319–23).

A selection of these data is given in tables in the appendices to this volume.

Ice sheet stratigraphy and snow accumulation

The amount of snow accumulating each year on an ice sheet may be deduced from the weight of ice per cm^2 in the year layers in the column of ice obtained by boring at any point, adjusted (increased) to allow for the thinning and spreading out of the buried layers under the pressure of the overlying newer ice in accordance with the expression on p. 78. It has to be realized that the quantity of snow accumulation indicated by this calculation relates not exactly to the place where the boring was made but to where this part of the ice column has flowed from since it was deposited on the surface of the ice sheet. For comparisons with other layers in the same column of ice to be meaningful, therefore, such studies must be made in extensive, fairly uniform ice sheets or in areas of weak flow, without faulting or crevassing within the ice, with little surface slope and hence in areas of rather uniform precipitation and not liable to much strong katabatic wind and drifting of the snow once fallen. The length of the snow accumulation record thus obtained could be as long as the series of unambiguously distinguished year layers in the ice (see p. 93) and some broad indications of the course of the temperature variation within the individual years may even be obtainable from the oxygen-18 measurements.

Fig. 13.40(a) shows the record of snow accumulation at the South Pole from 1760 to 1957 derived by one of these studies (GIOVINETTO and SCHWERTFEGER 1966). Fig. 13.40(b) shows an interesting parallel between this history at the South Pole and at another point, Southice (82°s 29½°w) on the Antarctic plateau, and at 77°N 56°w in northern Greenland, as regards the trends seen from the 1890s to the 1950s, with maximum accumulation at all these places in the years of strong general atmospheric circulation extending to high latitudes around the 1920s. It is known that the trends were different (and largely inverse) on the other (i.e. eastern) side of the ice cap crest in northern Greenland and at Byrd station (80°s 120°w) on the other side of the Transantarctic Mountains from the South Pole, in the Pacific sector of Antarctica (cf. fig. 7.23, p. 303, in Volume 1).

At these high latitude sites, high in the interior of Antarctica, the variations of snow accumulation register changes of precipitation. In a similar record of the last 200 years at Deception Island (63°s 60°7′w), however, off the coast of Antarctica, where the yearly ice layers were clearly defined by the dark bands of wind-blown dust which form each summer, the variations were found (ORHEIM 1972) to be correlated with the mean temperature of the summers ($r = -0.55$, significant at the 1% level). ORHEIM was therefore able to show the variations of ice accumulation since 1780 in the form of 10-year running means of the probable summer temperatures (reproduced here in fig. 13.41). The correlation on which the curve is based is not strong enough to give reliable temperature values, but the shape of the curve must be right and the temperature levels can be accepted as an approximation. The result indicates that a warmer climate has prevailed at Deception Island since 1880 apart from a colder period between the late 1930s and about 1950; this secular

(a)

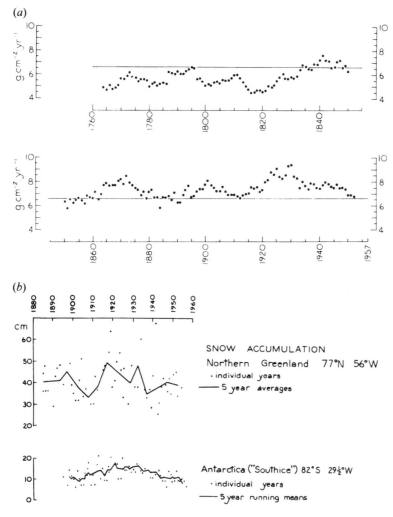

(b)

SNOW   ACCUMULATION

Northern   Greenland   77°N 56°W
· individual years
——— 5 year averages

Antarctica ("Southice") 82°S 29½°W
· individual years
——— 5 year running means

Fig. 13.40(*a*)   Snow accumulation at the South Pole, a record of the 10-year averages of annual accumulation from 1760 to 1957.
   (*After* GIOVINETTO *and* SCHWERDTFEGER *1966; reproduced by kind permission of the authors and of the* Archiv für Meteorologie, Geophysik und Bioklimatologie, *Vienna.*)
   (*b*) Snow accumulation on the ice caps in northern Greenland and Antarctica.

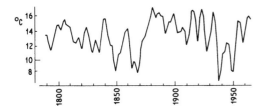

Fig. 13.41   Variations of the thickness of the yearly layers in the ice on Deception Island (63·0°s 60·7°w) in the form of 10-year running means of the estimated mean summer (December to February) temperatures.
(*After* ORHEIM *1972.*)

change, which roughly resembles that over most of the globe, is overlaid by cyclic variations of period about 11 years and 20 years.

Glacier variations

Glaciers advance and retreat in response to climatic changes. Old moraines which can be dated therefore provide clues to climatic history (see Plate III). Advances of glaciers in high latitudes (often referred to as 'polar glaciers' or 'cold glaciers') are likely to be related to increases of precipitation, commonly in periods of relatively high temperature (though still mainly below 0°C); rising temperature also increases the plasticity of the ice and hence the rate of flow. The increase of rate of flow becomes very rapid as the melting point is approached. Advances of glaciers in other latitudes (called 'temperate glaciers'), in which the temperature of the ice is always near the melting point except in winter near the surface, are mainly due to periods of colder climate shortening the melting season. In all latitudes the cause of most general advances and retreats of the glaciers is presumably a change in the balance between nourishment and wastage (ablation).

In all cases there is some lag in the response of the glacier to climatic change, a lag which differs at different points along the glacier's length and differs greatly from one glacier to another owing to the individuality of their terrains, with catchment areas at different altitudes and of differing extent, different slopes, differences of bedrock friction and different internal stresses within the ice. It is difficult, therefore, and sometimes impossible, to relate a particular advance of a particular glacier with the weather in particular years, though measurements of the varying mass balance in the upper part of the glacier year by year can usually elucidate the point.

Regular measurements of glaciers, chiefly their length, were instituted in 1894 by the *Deutsche und Oesterreichische Alpenverein* and were later extended under the auspices of the International Commission on Snow and Ice to glaciers in many other parts of the world. By the 1950s a good many measurements of the flow and also the mass balance of glaciers were being undertaken. The observational studies have been summarized by HOINKES in many writings (e.g. 1964), HOINKES and RUDOLPH (1962), and ŠNITNIKOV (1961) among others (see also the yearly reports over many years in the *Zeitschrift für Gletscherkunde* by KLEBELSBERT and KINZL, latterly also the summaries published each year in *Ice* by the International Glaciological Society, Cambridge).

It was first realized by DE MARCHI (1895) that waves travelling down glaciers at rates greater than the flow of ice were an important feature, and many cases have since been reported of the resulting surges forward of the glacier snouts even in years not generally favourable to glacier advance; thus in 1936–7 the Black Rapids Glacier in Alaska advanced 5 km in 5 months, and travelling waves have been observed in many glaciers in the present century, e.g. in the Mer de Glace and Glacier des Bossons at Chamounix. Ice velocities from 0·5 m/day up to 120 m/day have been reported in surges of valley glaciers. Theoretical understanding of the flow of glaciers and ice sheets has been advanced notably by GLEN (1952, 1953, 1956, see also 1967), ROBIN (1955) and by NYE in a series of papers (1953, 1960, 1961, 1963a, 1963b), resulting in a noteworthy increase in the importance accorded to kinematic waves and surges.

The ideas developed about the possible surges of ice sheets led WILSON (1964) to propose that the world-wide glacial–interglacial alternations of the last million years could be accounted for by periodic surges of the Antarctic ice sheet suddenly covering the Southern Ocean with a vast extent of ice and cold melt-water which would chill the whole Earth, via the oceans, before the previous heat balance could be restored and the Antarctic ice cap built up again. Efforts have been made to test the theory by looking for sudden rises of world sea level immediately before the onset of each glaciation, but the evidence remains unconvincing. An obvious objection to the theory is that there seems little likelihood that the Antarctic ice sheet would surge in all sectors, discharging from each of the main topographical basins defined by the underlying rock, around the same time; the evidence of variations of ice on the Antarctic Ocean in the nineteenth century suggests rather that lesser surges in this or that sector tend to relieve the stress of an ice build-up at fairly frequent intervals.

Surges forward, either when the ice at the base of the glacier reaches the pressure melting point (due to accumulated mass of ice above) or through some sudden release of internal friction, clearly play a great part in the advances of many, if not all, glaciers. It has been noticed that during the colder climate of recent centuries, often referred to as the Little Ice Age, between about 1550 and 1850, when in most parts of the world the greatest extent of ice since the last ice age seems to have been reached, the available accounts seem to imply that most of the advances of the glacier snouts took place in just a few catastrophic periods of about 10 to 15 years duration. Thus, for example, BRAY (1971) finds that in the northwestern United States, British Columbia and Alberta over half the glacier advances of the period 1580–1900 took place either between 1711 and 1724 or between 1835 and 1849.

It seems likely that (even apart from periods of general warming of climate like the first half of the twentieth century when most glacier measurements so far were made) temperate glaciers characteristically spend most of their history with their snouts thinning and retreating or else moving but little. The snouts of these glaciers are at altitudes where ablation dominates, and when the ice is thin enough retreat of the snout in any normal or warm summer may be inevitable. Ultimately the recession of the snout ceases because of replenishment of the ice in the upper part of the catchment basin and a wave of thicker ice

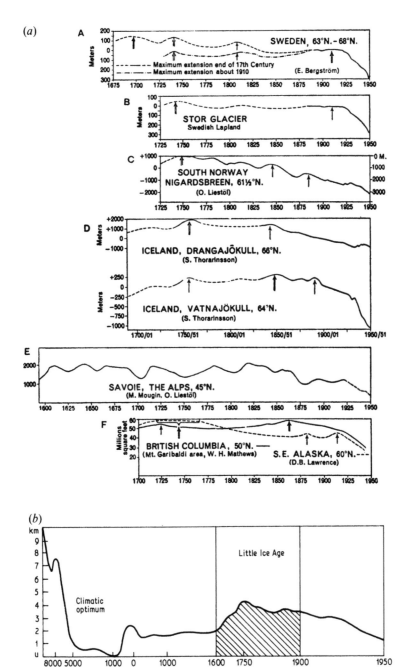

Fig. 13.42 (*a*) Variations of glacier termini since A.D. 1600.

(*b*) Variations in the length in km of the glaciers in Iceland and Norway since 10 000 B.C.

(*After* H. W. AHLMANN *1953*.)

passing down the glacier, sometimes ending in a catastrophic surge of the snout. Hence, the response lag of the glacier snout to colder years differs from that for warmer years, being characteristically 3 to 30 years according to NYE but in some cases even longer. These points are illustrated by an investigation by BRADLEY and MILLER (1972) in Baffin Island. The 1960s in that area were a decade of colder summers and milder, snowier winters; during this decade the snouts of existing glaciers generally continued to retreat, but in the upland plateau region, 400 to 600 m above sea level, net accumulation of snow and ice set in, the quasipermanent snowbeds increased in size and two small glaciers formed.

In sum, all these variations mean that individual glaciers generally give no precise or unambiguous indication of climatic events. However, the statistics of advance and retreat among a 'population' of many glaciers in one region of similar climate and exposure to the prevailing winds may be expected to correspond well with the history of climatic changes, albeit with some lag, characteristically of a few decades for advances and less for glacier retreats. Fig. 13.42(a) summarizes the general variations of length of the glaciers in different regions of the northern hemisphere during the last 400 years, which AHLMANN (1953) deduced from descriptive manuscripts of the times and dated moraines. Advanced stands of the glaciers are seen to have been quite general between 1610 and about 1700 and between 1810 and 1860; glacier maxima also occur in the eighteenth century, but the timing differs more from region to region, mainly between 1720 and 1750. The longer history of the glaciers in Norway and Iceland over the last 10 000 years is summarized in Fig. 13.42(b), also from AHLMANN (1953). Some recent investigators of glacier histories in North America, Scandinavia and elsewhere have thought to link the great glacier advances that culminated between A.D. 1650 and 1850 in the so-called Little Ice Age, and the earlier advances (which

Table 13.7 Percentage of glaciers in the Italian Alps observed retreating in different years

Year						1925	1926	1927	1928	1929
%						94	65	81	94	96
Year	1930	1931	1932	1933	1934	1935	1936	1937	1938	1939
%	71	89	69	93	91	92	64	83	89	72
Year	1940	1941	1942	1943	1944	1945	1946	1947	1948	1949
%	67	67	89	75	78	78	84	92	80	95
Year	1950	1951	1952	1953	1954	1955	1956	1957	1958	1959
%	95	86	87	79	82	81	74	78	87	80
Year	1960	1961	1962	1963	1964	1965	1966	1967	1968	1969
%	50	60	68	45	79	44	49	54	43	43
Year	1970	1971	1972	1973						
%	51	56	20	53						

were in some areas quite comparable with the Little Ice Age positions) between 1300 and 400 B.C., with the first ('Piora oscillation') advances of the Nachwärmezeit around 3500–3000 B.C. and even with the Cochrane readvance at the end of the last glaciation in North America (taken as 6500–5500 B.C.) as common evidence of regularly recurring climatic fluctuations at about 2500 year intervals. At the present time, however, it would be premature to accept this as a firm conclusion owing to possible failure to recognize the proper fit of other glacier advances at present thought to be of generally lesser magnitude (e.g. about A.D. 500–700) into the sequence.

Recent Soviet work on dating organic material buried in glacial deposits on Spitsbergen is reported to have established that the glaciers there had dwindled to their present size in the period between A.D. 200 and 1200 and that after that they grew to cover the whole archipelago. Recession set in again in the last hundred years.

Table 13.7 registers the (rather quick) response of the glacier snouts in the Italian Alps to the warm years of the twentieth century, which in central Europe culminated between 1933 and 1952, and to the significantly colder years which set in about 1960 (the number of glaciers investigated ranged from 25 in the war years to between 100 and 150 at most other times).[1] The retreat had been considerable in the thirty years before 1925 and the glacier tongues had presumably been thinning. By 1975 the glaciers in other parts of the Alps had gone over to advance, and in early 1976 it was reported that 'all the glaciers in west and north Norway had been increasing for about the past ten years' whereas the glaciers in eastern Norway and in Spitsbergen continued to shrink.

The decade averages in Table 13.8 show the nature of the changes in more detail.

Table 13.8   Changes in position of the snouts of the glaciers in the Italian Alps
Decade averages

Decade	% Advancing	% Stationary, hidden by snow or otherwise uncertain	% Retreating
1925–9	7·6	6·4	86·0
1930–9	9·6	9·1	81·3
1940–9	10·6	8·9	80·5
1950–9	6·4	10·7	82·9
1960–9	10·9	35·6	53·5

1. The data in this table, believed to be the longest series of the kind available anywhere in the world, were extracted from *Ice*, the news bulletin of the International Glaciological Society, Cambridge, England, the original source being the *Bolletino* of the Comitato Glaciologico Italiano. Unfortunately, according to a recent note by C. CESCA (*Bolletino*, II serie, No. 20, 1972), there is some doubt about the quality of the figures given prior to that for 1972, owing, it is alleged, to overconfident counting of glaciers whose behaviour was uncertain or which could not be clearly observed because they were covered by snow. Revised figures proposed for the percentage retreating in recent years are: 1965 74, 1966 65, 1967 70, 1968 51, 1969 55, 1970 62, 1971 63, 1972 20. The percentage of the glaciers in the Italian Alps which were advancing is given in this latest note as 1965 13, 1966 23, 1967 16, 1968 17, 1969 27, 1970 14, 1971 19, 1972 54.

There may be particular interest in the study of the history of glacier variations on the three great mountain masses high enough to bear snow and ice which are nearly evenly spaced around the equatorial zone, the Andes, the great mountains in East Africa, and in New Guinea. The glaciers on the east side of Ruwenzori in East Africa reached down to 2000 m above sea level in the last ice age. Ice accumulation occurs at present on the Carstensz Mountains in West Irian, on the island of New Guinea, down to 4400 m; but in the last ice age glaciers descended the deep, shady valleys at one point reaching 1700 m above sea level 10 000 years ago (PETERSON and HOPE 1972). Such a great advance presumably owed much to very great precipitation in the East Indies region with the warmest ocean surfaces in the world stretching far to east and west; and it is suggested that the greatest extent of the glaciers there occurred at the very end of the glaciation, when deglaciation elsewhere and rising sea level caused the sea to return to the area between New Guinea and Australia. Suggestions of greatest growth of the ice on the south side of New Guinea seem to indicate a change in the prevailing winds in that sector of the equatorial zone.

The glaciers on the mountains of West Irian-New Guinea have been retreating in the present century, as have most other ice masses in the tropics, and the ice cap which existed on Mount Wilhelmina in 1938–9 had disappeared by about 1960. Retreat of the remaining glaciers on New Guinea from 1960 to 1972, however, has in most cases not been very marked (from a report by G. S. HOPE and J. A. PETERSON to the Australian and New Zealand Association for the Advancement of Science, 1972).

The variations of the glaciers of the Southern Alps in South Island, New Zealand, have been reviewed by WARDLE (1973). The Franz Josef Glacier and the Fox Glacier termini (near $43\frac{1}{2}°$s) have been surveyed at intervals since 1893, and WARDLE'S study included the variations of 18 glaciers during postglacial times, using radiocarbon and tree ring dating. Glacier advance episodes were dated (1) before 4730 years B.P., (2) between 2570 and 2160 B.P., (3) between 1510 and 1095 years B.P., (4) in recent centuries, when the main maxima seemed to be before A.D. 1620 and about 1780 and 1830. There is an obvious and strong correspondence between this whole series of dates and those generally obtained in the northern hemisphere. The prevailing retreat of the glaciers since 1830 was interrupted by minor advances between 1890 and 1920, and in the 1930s, early 1950s and mid 1960s.

Correspondence of the variations year by year over the period 1946–71 of the mass balance of the ice on Deception Island (63°s 60°40′w) with the glaciers in the northern hemisphere appears to be inverse (i.e. negative correlation coefficients), as seen in Table 13.9 (from ORHEIM 1972). This is in marked contrast to the longer record of the New Zealand glaciers. ORHEIM also found some indications of negative correlation coefficients (statistical significance level only about 10–20%) between the summer temperatures at certain places south of 50°s on opposite sides of the southern hemisphere. However, there are many instances of variations of temperature and temperature-dependent climatic phenomena in widely separated parts of the globe which are linked by negative correlation coefficients when yearly data are compared but by positive correlation coefficients between the averages for successive half centuries or longer periods.

Table 13.9   Correlations between annual mass balances of glaciers from 1946 to 1970 (after ORHEIM 1972)

	Sarennes	Hintereis-ferner	Stor-breen	Stor-glaciären	Polar Urals (IGAN glacier)	South Cascade	Deception 63°s 61°w
	1949–70	1952–70	1949–70	1946–70	1958–67	1953–71	1946–71
Aletsch 46°N, 8°E	0·89 22 <0·001	0·91 17 <0·001	−0·09 22	0·15 25	0·32 13	−0·07 17	−0·41 25 <0·05
Sarennes 45°N, 6°E		0·73 18 <0·001	−0·31 22 <0·20	−0·31 22 <0·20	0·34 13	−0·27 17	−0·54 22 <0·01
Hintereis-ferner 47°N, 11°E			0·18 18	0·12 18	0·36 13	0·07 17	−0·34 17 <0·20
Storbreen 62°N, 8°E				0·69 22 <0·001	0·26 13	0·16 17	−0·04 22
Storglaciären 68°N, 19°E					0·04 13	0·39 17 <0·15	−0·01 25
Polar Urals (IGAN glacier) 67°N, 65°E						0·02 12	0·07 13
S. Cascade 48°N, 124°W							−0·06 18

Reading vertically, the numbers show: the correlation coefficient, the number of years correlated, and the probability of the correlation occurring by chance. Only probabilities <0·20 are shown. The Aletsch record is from 1946 to 1970. Data are missing for 1954 for South Cascade Glacier.

Permafrost

Development of perennially frozen ground depends on the thermal climate of the ground surface as determined by (1) the radiation balance there, (2) the heat transported by the winds and summer rain as well as winter snow, and (3) the blanketing effects of a snow layer on the surface, vegetation cover and so on. More specifically, its development is advanced by the length and severity of the frost season and is counteracted by the length and warmth of the thaw season and the amount of rain that falls during it; development of the layer of frozen soil is also hindered by the thickness of any snow that covers the ground when the frosts are severest. The term is commonly taken to include any ground at temperatures below 0°c but without ice because of its dryness (or even because of high salt content in the water it contains). The frozen layer may be found at some depth below the surface in the warm season or after a secular warming of climate. The layer near the surface that is subject to seasonal thaw is known as the *active layer*, because geomorphological effects are liable to be considerable in it. Under surfaces bare of vegetation the seasonal thaw may spread down a metre or two, especially in porous, well-drained materials; but in poorly drained peat and under a vegetation mat it may be extremely shallow.[1]

Perennially frozen ground is generally to be found wherever the overall average surface air temperature is below −2° to −3°c (FRENZEL 1967, p. 33) and is liable to start forming wherever temperatures have been as low as this for more than a year. Already formed permafrost (Swedish *tjäle*) seems generally to persist undiminished while the mean air temperature of the years remains below −1°c. Fossil permafrost remaining from the cold climates of the ice age and early postglacial times is often now deeply buried but still affects the drainage of the overlying layers.

GRIGORIEV'S (1930) frost coefficient G expresses the influence of the factors other than vegetation and soil by the formula

$$G = \frac{\sum T_{\text{cold season}}}{S.R. \sum T_{\text{warm season}}}$$

where $\sum T_{\text{cold season}}$ stands for the sum of the negative monthly mean temperatures in the course of the year,

$\sum T_{\text{warm season}}$ stands for the sum of the positive monthly mean temperatures in the course of the year,

S is the average snow depth in cm in the month when the snow is thickest,

R is the total rainfall in mm in the months that have positive mean temperatures.

When G is $\leqslant -9 \cdot 5 \times 10^{-5}$ the ground becomes perennially frozen. With values of G between $-8 \cdot 5$ and $-9 \cdot 5 \times 10^{-5}$ frozen ground persists from year to year in patches, where the local conditions favour it.

The thickness of the permafrost layer in the ground in Alaska, Canada and Siberia is 200–400 m over a wide zone, and a thickness as great as 600 m has been reported at Nordvik in Siberia.[2] Permafrost is generally continuous today (fig. 13.43) north of the Arctic circle in

---

1. Thermal conductivity is least in well-aerated soils, and is several times greater in wet soils, ice and solid rock (see Volume 1, pp. 14–15).
2. A frozen layer just 1 m thick in places where the ground was poorly covered by snow was reported in

parts of southeast England in February 1963 at the end of the most prolonged and severe frost period in modern times; in the snowier places the depth of the frozen ground was 30–50 cm.

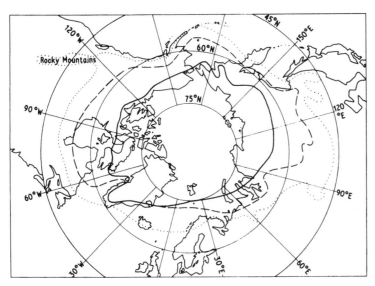

Fig. 13.43   Present distribution of permafrost.
Continuous line = Limit of continuous permafrost.
Broken line = Limit of discontinuous permafrost.
Dotted line = Limit of sporadic permafrost.
(*Adapted from* R. F. BLACK *in* Bulletin of the Geological Society
of America, *1954.*)

Canada and northern Asia; continuous permafrost extends south to 62°N in Greenland and
in Siberia between 140 and 150°E. Outside this region there is a wide zone, approximately
limited over the continents by the 540-dekametre isopleth of mean yearly 500 mb height in
the overlying atmosphere (cf. fig. 3.2, p. 76 in Volume 1), where there is much perennially
frozen ground but where also islands of completely frost-free soil exist in summer. Sporadic
occurrences of permafrost are found as far south as 46°N in eastern Siberia, near 50°N across
most of Canada from the Rockies eastwards and 66°N in Scandinavia (FLINT 1957, after
BLACK 1954).[1]

Above the frozen layer, the soil of varying depth which thaws out in summer supports
forest inland in northern and eastern Siberia; but, because of the impervious layer below, the
surface of many of the flatter, low-lying areas tends to remain waterlogged, and therefore
chilly, all summer and supports only tundra vegetation.

Since ice wedges (see p. 164) are only active today in climates where the yearly mean

1. It is reported (MONGAIT 1961) that in the early Iron
Age burial mounds at Pazyryk about 1650 m above
sea level in the Altai Mountains near 50°N 90°E,
some 600 km southeast of Novosibirsk, and at Bash-
Adar nearer Novosibirsk in a river valley, in central
Asia, although both places are outside the area where
the ground is generally frozen, ice developed at the

foot of the burial shaft and preserved the bodies of
men and horses and much fine clothing to the present
time. At a number of places in northern Siberia the
bodies of the now extinct woolly mammoth (*Mam-
muthus primigenius*) were preserved in the permafrost
in good condition (possibly from the postglacial
climatic optimum) to modern times.

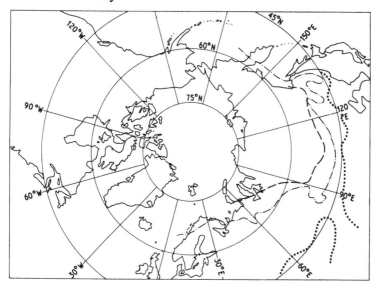

Fig. 13.44   Limit of occurrence of permafrost today
(dashed/dotted line) and in the warmest postglacial times (thick
broken line).
(*After* B. FRENZEL *1959.*)

temperature is $-6°$ to $8°$C, or below (PÉWÉ 1966), and the extreme limit of *pingos* and *palsar*
which is associated with annual temperatures around $0°$C is close to that of permafrost, the
more extensive region that was affected by permafrost in glacial times can be mapped from
the occurrence of fossil relics of these features. Fig. 13.44 shows for comparison the limits of
continuous permafrost deduced by FRENZEL (1959) for the last ice age, for the warmest
postglacial times and for the present day. From this it is apparent that the extent of
permafrost has increased materially in the cooler climates of the last 3000 to 5000 years,
though there has also been a recession of the permafrost in the warmer periods of the present
century and there is continual change towards equilibrium with the current climate.[1]

4. A vivid personal account by the commander,
BENCKENDORF, of a small Russian steamship
engaged in survey work in 1846 near the mouths of
the Lena and Indigirks rivers in northeast Siberia
(given in English by H. N. HUTCHINSON in *Extinct
Monsters* (London, Chapman & Hall, 1893), pp.
205–9) recorded the appearance of the landscape in
an exceptionally warm summer and of a mammoth,
released from the grip of the permafrost, being swept
down the rivers: 'We steamed ... up the Indigirka;
but there were no thoughts of land; we saw around
us only a sea of dirty brown water, and knew the river
only by the rishing and roaring of the stream. The
river rolled against us trees, moss, and large masses
of peat. ...' After eight days sailing up the river the
vessel was stopped, but the landscape remained
unrecognizable because of the floods and the current
continued to tear up the soft ground. At this point a
mammoth with its elephant's head reared at times
out of the water was washed against the ship and
secured by ropes for long enough for some examination of it to be made before the soft ground sank
under the water and the animal disappeared. In the
same work (p. 201) the gradual emergence from the
ice of another mammoth, still frozen, on the shore of
a lake near the Lena river between 1799 and 1803,
observed by a Tungus fisherman, is described. By
1806 this animal had been eaten, presumably by
wolves, bears and foxes so that only the skeleton
remained. These conditions surely suggest that the

Geomorphological evidence

Numerous relics of the action of different climatic regimes in the past may be recognized in the forms of the landscape. Most of this evidence is purely qualitative, except where it indicates (dated) former positions of the mean 0°c isotherm or of the extreme limit of occurrence of forsts, the former depth (thickness) of ice sheets and glaciers, and so on.

Former ice sheets

The greatest depth reached by ice sheets (and glaciers) in the past is indicated by the height above sea level of the upper limit of ice-smoothed rock faces and, in some places (including Antarctica), by former side-moraines high up on the mountains. The mountain tops which projected above the ice sheet, as 'nunataks', show evidence of continuous weathering. Thus, WOLDSTEDT (1958, p. 7) is able to show a cross-section through Scandinavia with a profile of the mountains and of the upper surface of the former inland ice. By considering this evidence together with the evidence from raised beaches and former local sea level of the magnitude of the depression of the land by the ice load, the distribution of former ice thickness and the contour of the upper surface of the ice sheet over the flat lands as well as over the mountains can be obtained. To do this, the total depression of the land must be calculated from the present height of the highest raised beaches and the further uplift still to be expected (by extrapolating the history of isostatic recovery to the point at which the continuing uplift will apparently cease).

By these means, JAKOVLEV (1956) has been able to derive the topography of the north European ice sheet, with domes up to 3500 m over the northern Baltic and 4000 m over Novaja Zemlya, at its maximum development in the great Saale (Riss) and Elster (Mindel) glaciations, here reproduced in fig. 13.45; and WOLDSTEDT (1958) quotes 3000 m, again over the northern Baltic, for the summit of the Scandinavian ice in the last (Weichselian/Würm) ice age. Similarly, SCHYTT *et al.* (1968), by studying the present height above sea level of the strand lines of the Barents Sea 6500 years ago, dated by radiocarbon tests on the driftwood left on the beaches, find unmistakable evidence of the occurrence in the last glaciation of a great ice dome grounded on the bed of the Barents Sea.

The directions of flow of the former ice sheets (which in some cases are seen to have changed during the course of their development) are registered by the grooves ('*striae*') which still mark the surface of rock which was smoothed and scored by stones and rocks embedded in the ice. These movements provide further evidence of the contres where the greatest thickness of ice was built up (cf. fig. 13.45).

---

mammoths found intact in the permafrost must have been preserved ever since their deaths, by the re-advance of the permafrost in the last few millennia since the postglacial warmest climates.

Several other finds of mammoth bones in earlier centures, recorded in the same book, e.g. in central Europe, had given rise to legends about giants who had stalked the mountains in days long past.

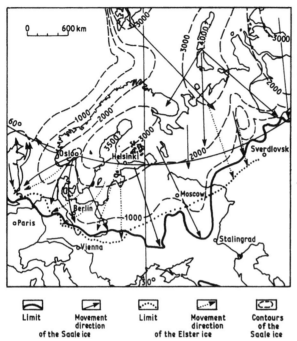

Fig. 13.45 Height contours for the upper surface of the North European ice sheet at the maximum development of the Riss (Saale) glaciation (last-but-one ice age) and movements registered by the ice over the bedrock.

The dotted lines refer to the limit and ice movements derived for the previous, i.e. Mindel (Elster), glaciation. (*After* JAKOVLEV *1956, as presented by* WOLDSTEDT *1958.*)

The characteristically shaped *roches moutonnées* also reveal the direction of movement of the ice. These are rounded knolls of the solid bedrock which formerly underlay the glacier or ice sheet, smoothed and shaped to a gentle reverse slope on their up-glacier side and scored by the stones embedded in the base of the ice; on the down-glacier side of the knoll, where cracked and loose pieces of the rock were dragged away by the ice and there was less smoothing action (or none), the rock face is steep.

Former glaciers and greater extensions of existing glaciers

The common traces of these are their former *moraines* – ridges of debris, consisting of an unsorted mixture of sand, gravel, stones and larger boulders, which are of sharp-edged, angular shapes unlike river gravels. The former *end-moraine* and *side-moraines* of the glacier sometimes remain as continuous ridges over long distances far down the valley and back along the mountainsides above any existing remnant of the glacier. They are usually colonized within a few decades after the ice retreat by vegetation, which preserves them against stream erosion. They can often be dated by radiocarbon measurements on relics of the previous vegetation such as trees pushed over and buried by the advancing ice and debris. The same kind of ill-sorted rubble which formed the ground-moraine, or underlying surface over which a glacier or ice sheet once moved, also provides characteristic evidence. It is known as *till* (or *boulder clay* in the older texts).

Other signs of former glaciers and ice sheets are:

(1) *Drumlins:* mounds and ridges of debris, elongated in the direction of flow of the ice, largely formed of *till* or 'ground-moraine' material, underlying the glacier or ice sheet, which was dragged along with the ice and got held back at the point where it now is by some obstacle such as a larger block (which may still be present).

(2) *Eskers:* long, narrow, and somewhat winding ridges up to 30 m high of gravel and sand which were laid down by streams flowing underneath the glacier when it melted.

(3) *Kames:* steep-sided, isolated mounds or short ridges formed of the sediment from streams and stagnant water in the crevasses and hollows in a melting glacier or ice sheet. Kame terraces sometimes occur where there was running water along the side of the ice, bounded on the other side by a hill or mountain slope. Such terraces are usually in shorter stretches than the gravel terraces of ordinary rivers.

(4) *Sandur:* outwash plains consisting of moraine material carried away by the rivers from the melting glacier and redeposited within a few kilometres or tens of kilometres, before the grains have become rounded or size-sorted to the extent normal in river gravels.

(5) *Striae* and *smoothed rock faces:* mostly on the former glacier bed but also the side walls of U-shaped, formerly glacier-filled, valleys.

(6) *'Hanging valleys':* side valleys which join a U-shaped main valley at levels often far above its floor, are evidence of the deepening of the main valley by the glacier eroding it.

Fuller descriptions and pictures for identification of these land forms may be found in FLINT (1971) or SPARKS and WEST (1972).

Former snowline

Corries (alias *cirques* or, in Wales, *cwm*), each one a steep, commonly precipitous, sided indentation in a mountainside, shaped like a half bowl and with a slight hollow in the base, are commonly formed near the lower limit of perennial snow on the shady and lee sides of mountains, where snow swept off the tops by the wind collects. The process that produces these features is the melting by day and freezing by night all through the summer ablation season of melt-water that gets into the cracks in the rock from the firn and snow-beds collecting in the hollow, summer being the season when melting as well as refreezing is most active and when the freezing limit varies least from day to day and week to week. A pre-existing slight dip in the ground may be needed to start the process. Corries tend to be formed on the northeastern sides of mountains in the northern hemisphere belt of westerlies and on the southeast sides in the corresponding zone in the southern hemisphere. Where the permanent snow has since disappeared, the hollow in the corrie floor is commonly occupied by a tarn or small lake. Their widespread occurrence in the British Isles, in New England and in Tasmania bears witness to the former occurrence of permanent snow-beds and glacier heads at lower levels and in lower latitudes than today: from them the level at the relevant time of the snow-line and the direction of the winds which used to scour the snow off the tops (and collect it in beds that were most persistent where the cirques formed) can be deduced.

A study by DERBYSHIRE (1971) of the distribution of cirques evidently formed in glacial times in the hills of Tasmania, which rise to about 1600 m, by snow-beds (or glaciers) which owed their existence to snow blown off the tops and collecting in sheltered places on the lee slopes and especially liable to survive on the side sloping away from the midday sun, indicated that the prevailing winds there were broadly westerly in the ice ages, just as they are now.

HASTENRATH (1971), taking advantage of the great north-south extent of the Rockies-Andes range which offers a transect through all the latitude zones of the Earth's wind circulation, has demonstrated the information to be gained from a survey of cirque floor heights between about 37°N and 35°s (see fig. 13.46). The tectonic instability of these ranges does not seem to have vitiated the results, if we dare judge so from the agreement of the findings with meteorological expectation; this may be partly due to the large number of points sampled (perhaps the likeliest kind of error is that a general subsequent uplift of the whole of the western ranges may have decreased the apparent lowering of the snow-line there in the ice ages). Taking the cirque bottom as an approximation to the snow-line during the Quaternary ice ages and comparing this with the present snow-line in the western ranges (cordillera), the lowering of the snow-line (where the mountains are high enough for permanent snow today) was found to be about 700 m at 37°N and rather less than 900 m

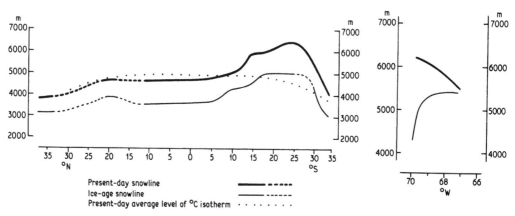

Fig. 13.46   Snow-line at the present day and in the ice ages along the western ranges of the Rockies and Andes from 37°N to 35°s.

The right-hand diagram shows the variation in the height of the snow-line across the Andes now and in ice age times at latitude 29°s.

(*After* HASTENRATH *1971*.)

about 19°N, but was consistently greater than 1000 m between 10°N and 7°S. The lowering decreased to about 700 m at 12°S, but then increased all through the present desert zone and was more than 1500 m at 30°S. The snow-line at the present day is highest in the desert zone, particularly about 25°S, where it is above 6000 m and much above the level of the mean 0°C yearly temperature isotherm. The highest snow-line during the glaciations seems to have been about 5000 m, between about 27°S and 17°S. The ice age snow-line indicated by the cirque floors descended to about 3500–3700 m throughout the equatorial section, between about 5°S and 10°N, and then rose to a maximum of 4000 m around 20°N. This distribution seems to demand temperatures lower by some 5°C at most points and more W'ly winds than now near the equator (W'ly components prevail today on the west coast of the Americas between about 5°S and 10°N). It also appears that the desert zone was displaced, or extended, somewhat nearer the equator than now in South America. That the prevailing temperate zone W'lies of the southern hemisphere reached a few degrees of latitude nearer the equator than now is apparently confirmed by HASTENRATH's surveys of the snow-line and cirque heights on transects from west to east across the mountains. The slope of the snow-line from west to east across the mountains (constancy of latitude being taken as an approximation to constancy of mean temperature of the year) reveals the direction from which the prevailing winds (really the greatest wind transport of moisture) comes, the snow-line being lower by some hundreds of metres (at 32°S today by more than 1000 m) on the windward side. The windward side is clearly shown as the western side south of about 30°S today; but at 29°S already the snow-line is about 600 m lower on the eastern side of the Andes and continues so at least to 9°S where the difference is about 100 m. However, the glacial cirques show that during the ice ages the snow-line was about 100 m lower on the western than on the eastern side at 29°S, a sure indication that the belt of middle latitudes W'lies at those times reached a few degrees nearer the equator than now (though not as far as 25°S where the next survey was made).

Frost action

The repeated freezing and thawing of water in the cracks in rock is the chief agent of mechanical weathering. It splits off material, ranging in size up to immense boulders, which falls or slides down the steeper slopes. *Frost weathering* is most rapid in moist climates where the mean temperature over much of the year is near the freezing point. The smaller debris is subsequently washed away by stream erosion, but the large boulders and chunks of the cliff stay *in situ* at the foot of the hill or perched on the slope unless carried away by a glacier.

Frost cracks, striking nearly vertically down into uniform rock, clay or silt, and commonly appearing as a *polygonal pattern* when the ground is seen from above, are formed by the shrinkage of already frozen ground when the frost becomes intense. Hoar frost collecting in the crack widens it at the top into a wedge-shaped section (*frost wedge*), and alternating frost and thaw in the warmer seasons help to deepen the crack. Finally, when the ice wedge melts in a warmer climate, foreign debris and sediment is washed or falls into the crack, forming a *fossil ice wedge* (or '*ice wedge pseudomorph*'). Ice wedges appear to be active today – e.g. in Alaska – where the annual mean temperature is $-6°$ to $-8°$ c or below.

Other forms developed in frost climates and often still traceable[1] after thousands of years are.

(1) *Patterned ground*, commonly 5- or 6-sided polygons, outlined either by the cracks described above or by the walls of collapsed pingos and *palsar* (see below). The size is commonly 20–30 m across, but in some areas up to 100 m or more and in others only 1–2 m. If the ground slopes, parallel stripes, up and down slope, occur instead of polygons.

(2) *Pingos*, ice-cored mounds, commonly a few metres across, but in parts of Canada and Greenland today up to 100–200 m across and 40 m high, formed by the freezing of water issuing from a spring, or water collecting above an impervious substratum, already almost sealed over by ice into a lens-shaped mass of ice trapped between impervious layers of rock or ice and frozen soil above and below. When the ice melts, the mound collapses but parts of its walls remain as a visible pattern in the landscape.

Palsar are a special, smaller-scale type of pingo, characteristically a few metres across, formed in peat moors in frost climates, e.g. in Swedish and Finnish Lapland (where the name comes from). As the mound grows, the wind gradually becomes more and more effective in blowing the snow off the windward side till the plants there die and the soil thins, ultimately exposing the ice core to the summer sun so that it melts and the *palse* falls in (after a life cycle of some decades).

These ice mounds (*palsar* and pingos) are a permafrost type phenomenon, and are seen to form near the southern limit of such phenomena, in areas where the yearly mean temperature is about 0°c or somewhat below, but where there is some movement of water beneath the frozen layer. An essential condition, provided by continental–interior Arctic climates, is that the earth becomes deeply frozen each autumn before it is blanketed by much snow.

(3) *Solifluction*, or the creeping flow of water-saturated earth, which goes on in the surface thawed layer even on slopes as gentle as 2° (FLINT 1957, p. 198) where the subsoil is frozen for most of the year and impermeable to water, results in deposits, sometimes called '*head*', of material of jumbled sizes all derived from the nearby slope and often showing traces of stratification which has been disturbed.

(4) *Cryoturbation* and *involutions* are the names applied to the contortions and occasional overturning of the layering visible in sections through deposits that have suffered frost heave, through differential freezing and thawing and percolation of water.

1. As, for instance, polygons and remains of pingos or *palsar* from glacial times, still visible in Wales (Cardiganshire) and in Norfolk (Thetford Heath), England, today.

(5) *Scree,* or *talus,* consisting of stones, rocks and boulders of all sizes, resting on a mountain slope beneath a frost-weathered rock wall. This material collects on slopes up to 25° to 40° if dry, but readily slides downhill if disturbed. If the supply fails through less frost action, vegetation may get a hold on the scree.

(6) Appearances of *stone streams* and so-called *rock glaciers* are thought to be respectively the result of washing out of the smaller-grained element of a solifluction deposit and the melting away completely of the glacier which once carried the rocks. Both are therefore evidence of former frost climates.

(7) *Dry valleys* in the chalk hill landscapes in southern England and parts of Germany, where the rock is so porous that no surface streams run, are thought to have been formed by stream and river erosion (especially of frost wedge debris) in times when there was still a permanently frozen, and therefore impermeable layer below.

(8) *Cold water in subterranean aquifers:* BAKKER (1969) cites reports of low temperature of old artesian water dated 25 000–35 000 years B.P. from water-bearing strata in the northwestern Sahara and in western Egypt.

Relics of postglacial warming

(1) *Kettle holes:* round depressions, from 3 m to 30 m deep and 5 m to 500 m across, where glacial gravels, etc., overlay an isolated mass of ice left behind by a retreating glacier or ice sheet and collapsed when the 'dead ice' melted.

(2) *Pro-glacial lakes:* dammed by the old moraines as the ice melted and retreated. Often their former beds are now dried up or filled with peat by the accumulation from thousands of years of growth of bog plants.

(3) *Rivers now shrunken and meandering about a wide valley bottom,* which today are far too small to have formed the valley they occupy. (*Sandur* outwash plains are a special case of this.)

(4) *Landscape changes due to isostatic uplift in postglacial times:* AARIO (1969) has examined the abandoned channel of the Hinkua river which used to drain Finland's lake plateau (Päijänne) northwards into the Gulf of Bothnia, before that area had risen so much as to turn the drainage southward. From careful measurements he concludes that the discharge in the warmest postglacial (Atlantic) times was only about 70% of what it is now. This appears reasonable even with a slightly greater rainfall than now, if the evaporation appropriate to a 1·5°C higher yearly mean temperature is calculated.

River terraces and downcutting

Gravel terraces formed higher up the valley sides than the present course of the river are attributed to periods of slackened flow and aggradation of the river bed, particularly in the lower part of the course of the river, at times of high sea level.

Renewed erosion, deepening the course of the river and downcutting the valley, occurs when a lowering of sea level increases the overall fall of the river.

Formerly wetter climates in the deserts and arid zone margin

Well developed river systems now dry, raised beach systems surrounding shrunken lakes and dry salt deposits marking the former lake floor bear witness to a formerly greater excess of rainfall over evaporation in their catchment areas.

Formerly drier climates

Witnessed by intrusions of blown sand beyond the present limits of desert and sand dunes now fixed by vegetation.

Warm climates

The most widespread physical mark left by warm climates is a red or reddish colour in soils and exposed sediments. Red earths, loams and the laterites which harden on exposure to the atmosphere seem to develop where the yearly mean temperature is over 16°c and rainfall over 600 mm in areas of carbonate rocks, or over 1000 mm with silicate rocks, but not where the rainfall is much greater. They are therefore typically a product of the savanna zone. The iron and aluminium oxides become concentrated in the surface layers, as the silicates and carbonates are gradually leached out, mostly downwards; at depth the colouring may become blotched with yellow and grey. With greater rainfall and thicker vegetation in the humid tropics humic acids remove the iron and the red colouring is lost.

Windy climates

The traces of windly climates are exposed bare rocks in places where this could not be due to water or ice, loess dunes formed of blown dry earth and sand, and storm beaches where the sea has overrun a windy coast.

Prevalent wind directions may be deduced from the orientation of dunes, subject to the recognition of characteristic lengthwise and cross-wind shapes, the latter being crescent-shaped, 'barkhan' dunes with the horns of the crescent pointing down wind.

On the northern part of the west coast of Alaska, at Cape Krusenstern near Kotzebue, J. L. GIDDINGS (*Aspen Conference* 1962) studied a plain, 3 km wide, formed over the last 6000 years by about 110 beach ridges, which now enclose a lagoon in front of the original mountainous coast. The gravel comes from river mouths farther north. Modern observation shows that the beach ridges are formed by wave action in the autumn storms: northwest and south of the cape when the wind is NW and near the cape when the wind is SE. For about the last 500 to 900 years the deposition has been on the northwest and south sides; for 1000 years before that, it was mainly near the cape, as was also the case during the second millennium B.C. - these data hint at a 2000 years long oscillation of the wind circulation.

Indications of changes of prevailing direction of the surface winds on the east coast of Scotland (Firth of Forth) have been reported by SMITH (1971) from the orientations of sandbanks, now existing as spits of land just above sea level, which can be approximately dated by the known sequence of changes of sea level and radiocarbon tests on associated peat and beach fauna. It is deduced that W'ly winds were dominant prior to 4000 years ago, but that the climatic change between that time and about 2500 years ago was associated with a lowering of sea level (at least in that area) and by periods of N'ly storms.

For further information on geomorphological evidence, the reader should consult the specialist works in this field: e.g. SCHWARZBACH (1961), FRENZEL (1967), WEST (1968), FLINT (1957), MACHATSCHEK (1969), OLLIER (1969), STODDART (1969) and YOUNG (1972). For identification pictures, CORNWALL (1970) and THORÉN (1969) are particularly recommended.

Soils and their origins and implications

Soils are the product of weathering of the local rocks and the generations of vegetation that have grown on and thrust their roots down into the fine-grained material during the centuries or millennia that the particular layer has been forming, together with bacterial action breaking down the organic matter and the chemical action of the organic acids and mineral alkalis and salts in solution. They register the effects of past climates in so far as these determined the rate of supply of mineral by weathering and the abundance of organic matter as well as the nature of the plant cover, fauna and bacterial communities. The water present in the soil, its rate of flow and the drainage and hollows in the underlying rock surface, also determines the removal of minerals in solution and by mechanical removal of the smaller grain sizes and of the organic material, thus changing the character of the soil and making it inhospitable to some elements of the vegetation. These matters are not uniquely determined by climate but by many local differences of water supply from rivers and streams, and of the slope and permeability or non-permeability of the underlying rock. Thus, although the soil integrates the total history of climate and vegetation in the area since the soil formed, the vegetation has itself been to an increasing extent determined by the soil development and the climatic sequence may be hard to read aright from the outcome.

Soil cover on the heights is generally subjected to greater rainfall and less evaporation than on the lowlands. In the wettest places, especially in pockets of poor drainage in the bedrock on the heights, but also some extensive lower moors in oceanic climates, the soluble calcium carbonate is gradually leached out and lost (together with some other basic salts) and the remaining soil becomes more and more acid (due to organic, humic acids) as time goes on. This increasingly selects the vegetation that can grow in such places, until only bog plants thrive and an acid peat gradually accumulates – though in periods of drier climate the bog surface may be partly recolonized for a time by some species of trees (e.g. birch and pine). The mineral left behind, however, when the carbonates have been dissolved out, is largely silica (quartz), SiO_2, which has no value for the nutrition of plants. Soils on the slopes drain and dry out quickest. In the valley bottoms the richest elements collect. On the whole, the fen peats of the valley bottoms and lowlands are kept alkaline by the supply of calcium continually brought in by the water which has leached it from the higher places in the drainage basin.

A high ratio of organic to mineral content is characteristic of wet soils and climates warm enough to favour vegetation growth. A high ratio of mineral to organic content marks the soils of cold climates.

Flowing water carries away the fine-grained materials, leaving a coarser sandy matrix and stones behind. Where the flow slackens, or is weak, the fine silt settles to form clay with only minute pore spaces and poor drainage.

Classification of soils usually stresses the following types:

(1) *Podsols:* the soils of the coniferous forest and sub-Arctic heaths, produced in cool, or cold, moist climates. Through leaching, the upper layers are pale grey; the lower layers are richer in humic acids and the sesquioxides of iron and aluminium (Fe_2O_3 and Al_2O_3).

(2) *Peats:* extremely rich in organic matter (humus), the soils of bogs and marshes formed in poorly drained places especially on impermeable (and largely insoluble) rock – hence the low mineral content – and on top of podsols in wet upland, where the action of leaching and humic acids makes the peat acid. Fen peats in low-lying places may not become acid, thanks to a continued supply of base minerals with the water.

(3) *Brown earths:* the soils of the temperate deciduous forest. Thanks to the warmer and drier climate, the top layers are not so leached, the organic matter and iron are retained and oxidized.

(4) *Black earths (chernozems):* black to dark grey soils, rich in humus, formed in continental climates with strong evaporation in the warm summers, so that leaching is minimal and indeed calcium carbonate may be carried upwards by capillary action in summer. These are soils of the steppes and prairies.

(5) *Chestnut soils:* resembling the chernozems but brown rather than black, with less humus but again rich in calcium carbonate, formed in the drier parts of the Russian steppes and North American Great Plains.

(6) *Mediterranean soils* and *laterites* of similar climates in the southern United States and elsewhere (named after the Latin word *later*, a brick, because of their habit of setting hard on exposure to the sun and air): resemble the brown earths, but develop a red colour because of the drier summers and consequently less hydration

of the ferric oxide. They appear, however, to form only in climates where there is a wet season in part of the year, allowing regular capillary action.

(7) *Desert soils:* carbonate-rich or alkaline through the upward transport, when wet, of dissolved minerals in the capillaries associated with the continual strong evaporation near the surface, but poor in organic matter.

The acidity which characterizes the fossil soils immediately below the coal measures of the Carboniferous era in Europe is evidence of the swampy, leached condition of the land on which the forests of that era grew.

Something is known of the rate of formation of soils from a few places where the beginning of the process can be dated precisely, as where lakes were drained in early modern times. DIMBLEBY (1965) quotes 2000 to 4000 years as the period needed for the production of a fully matured soil characteristic of one or other of the recognized types. Nevertheless, the development is most rapid in the first few hundred years and 'becomes logarithmically slower as the soil matures'. The onset of the cooler, wetter conditions of the Sub-Atlantic period in the last millennium B.C. caused a widespread, massive and rapid degeneration of soils in the British Isles, producing blanket peat on many upland areas of high rainfall and soils of reduced fertility on land which had been a source of subsistence for centuries.

Buried soils which have been insulated from exposure to air and surface water by an impervious layer above, and so preserved, e.g. from a previous interglacial or warm interstadial period, may indicate broadly the type of climate which prevailed. Soil profiles even seem to be preserved when covered by a subsequent development of peat (such fossil soil layers in permafrost areas may indicate periods when the earth thawed out). But much more precise indications of the sequence of vegetation and implied climatic developments can be had from pollen analysis (see pp. 193–207).

Soil sequences, where preserved, may provide a continuous record of responses to different climates, in some cases back over a million years, though complicated as regards dating by changes of accumulation rate.

For further treatment of soils and their climatic significance, the reader may consult RUSSELL (1957), FRENZEL (1967), SCHWARZBACH (1961) and WEST (1968).

Volcanic debris as a tracer of past climatic regimes

A number of authors in recent years have traced the history of volcanic activity in terms of the frequency of volcanic ash and larger forms of tephra in ice-sheets and on the ocean bed. KENNETT and WATKINS (1970) report that many of the greatest maxima of volcanic material in the cores they examined from the southwest Pacific between New Zealand and Antarctica seemed to coincide with the times of reversal of the Earth's magnetic field in the last 4 million years. The cause of this may be related to upper mantle activity, which could produce a wave of volcanism, particularly in oceanic areas. They suggest that the scale of the volcanic activity must have produced sharp climatic changes (see Volume 1, Chapter 10) and that these sharp climatic episodes are a likelier cause of the faunal extinctions observed to be associated with the times of geomagnetic reversal than biological mutations brought about by the temporary increase of cosmic ray bombardment of the Earth. However that may be, it is interesting to note the agreement of their data with those of A UER from Tierra del Fuego and Patagonia (see Volume 1, p. 432) and R UDDIMAN (1973) from the North Atlantic in finding an important maximum of volcanic activity between 9000 and 11 000 years ago, about the end of the last glaciation in much of the northern hemisphere and possibly triggered off by the changed disposition of stresses on the Earth's crust due to rapid melting of the ice sheets. G OW (1970) and H AMILTON and S ELIGA (1972) reported a sustained maximum of dust concentration in the ice sheets in Antarctica and Greenland more or less coinciding with the last glaciation, but particularly with the build-up to the last (in most areas the *main*) climax of the last glaciation between 30 000 and 20 000 years ago and with the end of the glaciation about 10 000 years ago (H AMILTON and S ELIGA demonstrated that calcium (Ca) concentration in the Greenland core varied with volcanic activity in recent times and so could be used as an indicator of volcanism in the more remote past).

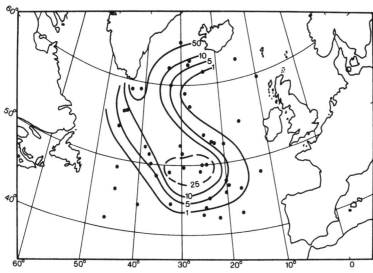

Fig. 13.47   Distribution of sand-sized volcanic dust grains of late glacial age in the North Atlantic ocean bed sediment.
    The concentrations referred to by the numbered isopleths are thousands of grains in a sediment column of 1 cm² cross section in the latest great volcanic dust layer, dated about 9300 years B.P. (the quantities have been adjusted for differing sedimentation rates at different points, which affect the volcanic material and other sediment about equally).
    The dots show the points where the sediment was sampled.
    (*After* W. F. R UDDIMAN *and* L. K. G LOVER *1975, by kind permission.*)

RUDDIMAN's (1973) analysis of the distribution on the bed of the North Atlantic of volcanic tephra, apparently originating in Iceland and Jan Mayen, with peak activity about 9300 years ago, is particularly interesting. By concentrating on sand-sized material (0·2 to 2 mm diameter by definition) too large to be carried far by the winds, but found as far from its source as the mid Atlantic near 50°N 'see fig. 13.47), RUDDIMAN deduced the course of an ice-bearing ocean current prevailing at that time; the tephra is presumed to have been carried by the ice and deposited when and where the ice melted. The same current could be verified by the distribution of polar foraminifera, and by studying the history of the latter in the ocean bed stratigraphy the advances and weakening of the current with time were indicated.

YOSHINO and TABUCHI (1975) have shown how the direction and shape of volcanic ash deposits from the Japanese volcanoes yield information about the direction and strength of the winds prevailing in the middle troposphere (generally at 5 to 8 km height) at the time of the eruption. Least complications from seasonal and short-term wind variations arise in the northern island, Hokkaido. Analysis of the deposits from the eruptions of the last 60 000 to 70 000 years indicated prevalence of W'ly jetstreams throughout, but with variations of strength. The W'lies appeared generally strong in recent centuries (which were warm in Japan), but weaker in the colder period in Japan between 500 and 1000 years ago, strong in the warmest postglacial times and weak at times during the last ice age about 13 000 and between 40 000 and 50 000 years ago (actually these were times of decreasing glaciation); the W'lies appeared strong during the times of increasing glaciation 60 000 and 20 000 years ago.

13.4   Field evidence of past climates derived from the biological sciences

Reconstruction of the facts of past climatic regimes draws on evidence from the most diverse fields of science and learning. The biological sciences have made, and continue to make, a very great contribution. It becomes clear, however, that all the numerical values for the elements of climate which we can derive from these or other sources, including the physical sciences, and which we may use either directly to understand past environments or indirectly in mathematical models of the atmosphere and oceans, to obtain fuller physical understanding of their heat economies and circulation regimes, are no better than estimates.

The estimates derived may, however, be good enough to establish the main features of the climatic sequence, provided that:

(1) The observed distributions of the biological species used were truly a response to climate.

(2) The values derived therefrom for the climatic elements are soundly based on (a) an understanding of the whole complex of factors, not all directly climatic, that govern the biological occurrences, and (b) known climatic thresholds which set limits to

those occurrences, and that these thresholds have remained constant – i.e. that the species have not developed so as to change the limits of their climatic tolerance.

(3) The reconstructed circulation patterns and climatic distributions are derived in a straightforward manner from the physical and dynamical laws governing circulation and climate.

Ultimately, corroboration should be sought from other, independent types of evidence. The circulation patterns derived by constructions based on the data regarding one element of the climate (so far usually temperature) are open to the test that they must be capable of explaining the distributions in the same climatic epoch of other elements (precipitation, storminess, prevailing cloud cover, etc.).

Because there are so many complex interactions in the circulation of the atmosphere and oceans and in the energy and moisture exchanges that produce climate, the nearest approach to certainty about past regimes is likely to be attained by those forms of analysis, and in respect of those items of the resulting diagnosis, which demand the fewest assumptions and the least input of elaborate theory. It is the short chains of cause and effect that can be established most securely.

Biologists' studies, particularly species analysis of the pollens found preserved in peat and lake bed sediments, and similar analysis of the remains of the microscopic organisms which lived in the surface waters of the sea or in the depths of the ocean and which largely constitute the organic fraction of the sediments on the ocean bed, have (with the aid of various methods of dating) by now contributed an enormous number of time cross-sections registering the effects of climatic history in most parts of the world. What these probes actually provide is of course a register of the changing populations and distributions of the various species, the changes being only partly (though often largely) a direct response to climate and partly determined by such things as biological competition and the effects of previous climatic and other conditions in modifying the soil and so on. A further stage of analysis is therefore needed to clarify the manner in which the climate has operated. The sorting, retrieval and various stages of analysis, including even the large-scale spatial analysis, of this ever growing mass of information urgently calls for the setting up of centralized data banks and computer methods, the use of principal component analysis techniques and so on. Numerical methods offer the further advantages of consistency and comparability through the use of standardized definitions of what constitutes a similar assemblage of species and the boundaries of such assemblages, which mark the times and places of rapid change. The introduction of these methods, which is now proceeding, has been discussed and demonstrated in papers on various aspects by BONNY (1972), GORDON and BIRKS (1972), WEBB and BRYSON (1972), and in relation to marine biological data by IMBRIE and KIPP (1971).

The presentation in this chapter attempts to survey briefly the different types of biological evidence of past environments, the development and potentialities of their use, and what they can be expected to tell of past climates.

Climatic responses of biological species

Limits: thresholds and tolerances

The simplest climatic responses for our purpose are where a species meets its extreme limit in respect of one element of the climate. This may be a severe frost, a prolonged drought, a long hot summer, too high a maximum temperature, or such a circumstance as exposure to parching winds (low relative humidity) when the subsoil is dry or frozen. Too short a growing season may also be the determining factor, though this seldom produces a sharp limit since seedlings may become established in the occasional warmer summers. For broad surveys, it is commonly assumed that the probability of these critical conditions occurring is usually (though it certainly is not always) sufficiently indicated by monthly or seasonal mean temperatures and precipitation values averaged over many years. Plants and animals, however, for a variety of reasons, seldom occur in abundance near the limits of the regions where they could survive: they are commonly limited by competition with other species that are more successful in the marginal areas. And where the climate is changing in a direction which brings new areas within the range of tolerance of a given species, some long time may be required before the new areas are all colonized. The response to improved conditions is bound to be quickest with those species that are most mobile, particularly the minute things such as pollen and spores that are borne by the winds and ocean currents. The response to deteriorations will be in some areas immediate and drastic, a patchwork of catastrophes.

The diagnostic advantage of micro-organisms and analysis of large populations

Generally speaking, human beings and the larger animals are too adaptable to different climates, and move about during their individual lives over too great distances, for evidence of their presence to supply firm data on the past climate of any place. There are exceptions, however – e.g. the implications of travel across what are now waterless deserts or stormy oceans and over high mountain passes that have at times been blocked for long periods by ice and snow. On the other hand, the microfauna, especially insects and the micro-organisms in the sea, but also certain fish, nesting birds and those elements of the flora, etc., which are spread most readily by windborne pollen and spores, may tell a great deal, provided that:

(1) The limits of their observable ranges at the times concerned were controlled by climate.
(2) The climatic conditions that are limiting for the species today are known.
(3) These limiting climatic conditions have not changed in the course of time through adaptation (acclimatization) or evolution of the species.

There is another advantage of dealing with microfauna and the microscopic pollen and spores of plants, namely that they commonly produce very large populations and variations in their abundance may be firmly established by statistically significant changes in the density of population shown by large counts. Examination of the changes of frequency from layer to layer in a sediment gives a finer resolution in time, the quicker the rate of sedimentation at the sites chosen.

In general, the fauna, from the minutest forms to the larger vertebrates, are found to accompany the vegetation which provides their preferred habitats.

Evolutionary changes of tolerance

Biologists wishing to use their data to establish the nature of past environments have had to concern themselves with what can be observed regarding the rates of evolution of species as well as the present climatic tolerances of species now living. Valuable surveys of these topics are given by F RENZEL (1967, pp. 2–7, 36–45) for both flora and land fauna and, more briefly, by W EST (1968, pp. 123, 346); COOPE's (1970) discussion of the interpretation of insect evidence is particularly wide ranging; for foraminifera and other marine organisms the reader is referred to ERICSON *et al.* (1964), ERICSON and WOLLIN (1966) and WISEMAN (1966). It emerges that over the last 300 000 years or so, through several ice ages and warm interglacial times, the remains found of botanical species, land vertebrate fauna and marine life are chiefly such as can be confidently identified with species still living somewhere in the world. For the insect remains examined from the last 30 000 years (and probably much longer) this is overwhelmingly the case, though there have been (*a*) losses of varieties through extinctions and (*b*) development of a smaller number of new varieties through crossing. There has been more evolution among some land animals than among other genera. FRENZEL (1967) cites bears and the elephants and mammoths in Eurasia as examples. ZEUNER (1959, pp. 309 ff., 333–7) gives further details and estimates 500 000 years as required for the evolution of an essentially new species; the greatest range of mammals adapted to cold conditions is found in the last glaciation, though other evidence suggests that previous glaciations were more or less equally cold.

It is usual to assume that species morphologically identical had the same climatic tolerance in the past as today. There is at least one case, however, where a species seems to have significantly extended its climatic range: the species of foraminifera *Globorotalia truncatulinoides* has spread from the subtropics to the sub-Antarctic within the last 300 000 years (KENNETT 1968a, 1970), though this has not been matched by a spread of the same species to cold seas in the northern hemisphere and may well turn out to depend on biological competition from other creatures or the lack of it. Another case, mentioned on p. 192, is the apparent adaptation of a number of species of radiolaria about 700 000 years ago to lower temperatures than before, therewith greatly extending their latitude range. This event coincided with the reversal of the Earth's magnetic field from the Matuyama to the Brunhes epoch, so some effect from mutations under abnormal cosmic ray bombardment of the Earth may be suspected.

The assumption that the species used as indicators have not changed their tolerance is most likely to be true where the same assemblages of fauna and flora are found in the past and at the present day.[1] It is recognized as less safe to assume that no change over tens of thousands of years in the tolerance limits of a single species has occurred than in the case of large assemblages such as those characterized by the names 'mixed oak forest' or Arctic tundra.

It is also much less safe to argue from the apparent absence of a species, evidence of which may yet be found, than from the positive evidence of those species known to have been present. Furthermore, the absence of a species does not in itself prove that the climate was unsuitable for it.

Extinctions and their causes

With regard to extinctions, a curious feature of the late Pleistocene extinctions of megafauna noted by Professor P. S. MARTIN of the University of Arizona, Tucson (unpublished lecture, given in Cambridge University on 18 October 1965), is that it was typically the latest evolved (therefore dominant and best adapted) member of a genus that disappeared, perhaps because its precise adaptation made it particularly vulnerable to climatic change. This tended to mean that the whole genus was lost. Many extinctions of species have, of course, been brought about by Man: this applies particularly to the hunting down of the larger fauna in postglacial and recent times in North America (HAYNES 1966), Africa (MARTIN 1966) and Eurasia (see p. 72), but there are examples also in Australia and New Zealand. It is also due to Man, not climate, that the lions, tigers and other large mammals which formerly ranged widely in the temperate zone are now confined to limited areas in Africa and southern Asia – not the regions which they like best but those which Man likes least and is least prepared to conquer (SPARKS and WEST 1972).

Another, non-climatic reason for extinctions and for new species appearing seems to be the exposure to increased bombardment of the Earth by cosmic rays, causing a higher incidence of mutations, during reversals of the geomagnetic field (in the weak phase of the

---

1. Some may be inclined to speculate whether animals in the wild have sometimes been responsible for changing the vegetation cover – e.g. deer eliminating the seedling trees of the forest – that would otherwise be in equilibrium with the prevailing climate. It seems, however, that, apart from the effects of temporary and rather localized population explosions of this or that herbivorous species, we may recognize climax assemblages of fauna and flora which represent the ultimate balance of Nature in response to a given climate and terrain. SIMMS (1971) has pointed out an element in Nature's balance which is widely missing today: in the landscape, before Man attained his present ascendancy, large carnivores (such as bears and wolves in Britain and much of northern Europe) kept a check upon the numbers of deer, boar and wild cattle that grazed and damaged the seedling trees. It is when Man disturbs that balance, either directly or through introducing certain animals and suppressing others, that the typical adaptation to climate is obscured. The effect of this may be observed in the Scottish Highlands where 'the wooded islands of many of the lochs stand as living monuments to the old forest which has completely vanished from the surrounding moorlands. The islands though accessible to deer have escaped the ravages of fire and sheep ...' (FRASER DARLING and BOYD 1964, p. 69).

Earth's field). Evidence for this hypothesis of UFFEN's (1963) was first spotted by HARRISON and FUNNELL (1964) in two equatorial Pacific ocean bed cores, in which the extinction of one species of radiolaria (*Pterocanium prismatium*) coincided more or less exactly with the geomagnetic reversal just under 1 million years ago. It has since been found that the disappearance of discoasters and a number of species of diatoms, foraminifera and radiolaria coincide with the magnetic reversals of the last 3 million years registered in ocean bed cores (GLASS *et al.* 1967, HAYS *et al.* 1969).

Other times of extinction of many animals seem to coincide with bouts of exceptional volcanic activity and may therefore have been caused by the accompanying sharp disturbance of climate (KENNETT and WATKINS 1970), as is possible in the case of the three principal periods of extinctions of reptiles and amphibians: at the ends of the Permian, the Triassic and the Cretaceous. Another suggestion is a supernova explosion sufficiently near our solar system and the resulting upset to climate. The latter possibility should be considered, according to RUSSELL and TUCKER (1971), in connection with the many extinctions (various species of phytoplankton, foraminifera, mollusks, marine turtles and the disappearance of the dinosaurs) at the end of the Cretaceous and beginning of the Tertiary, where there is thought to have been a geologically brief interval of conditions colder than had been experienced for more than 100 million years. It has been pointed out by BAKKER (1972) that the weight of evidence now is that the dinosaurs were warm-blooded (endothermic) and therefore had a body heat economy adapted to warm or equable climates but vulnerable to sudden, prolonged cold, as the most nearly similar, naked animals (rhinos, hippos, elephants, armadillos) of the tropics today would be. Hence, there is ground for thinking that this extinction was due to a climatic change, even though we cannot put figures to the limits of temperature tolerance (or other climatic limitations) of the extinct species.

Observable climatic limits of species at the present day

To give a more precise picture of the climatic limits for existing species, we may illustrate the distribution of a species which is affected by two climatic variables, m and n, by isopleths such as those drawn in fig. 13.48. In most real cases, however, the abundance is at least

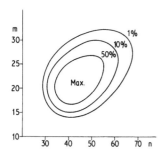

Fig. 13.48   Representation of the abundance under different climatic conditions of a biological species affected by two variables, m and n, such as absolute maximum temperature and average yearly rainfall.

The isopleths are drawn at 50%, 10% and 1% of the greatest frequency of the species which occurs with m values about 20–22 and n values 40–50. The species practically does not occur with m values below 13 or above 35, or with n values below 29 or above 67. But even within these ranges the limits and the conditions for the higher population densities are strongly affected by the value of the other climatic variable.

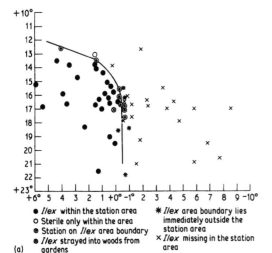

(a)

● *Ilex* within the station area
○ Sterile only within the area
◉ Station on *Ilex* area boundary
⊗ *Ilex* strayed into woods from gardens
✳ *Ilex* area boundary lies immediately outside the station area
✕ *Ilex* missing in the station area

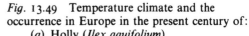

Fig. 13.49   Temperature climate and the occurrence in Europe in the present century of:
(*a*) Holly (*Ilex aquifolium*)
(*b*) Ivy (*Hedera helix*)
(*c*) Mistletoe (*Viscum album*)
Places 'within the station area' were within 20 km and at about the same height.
(*After* IVERSEN *1944. Reproduced by courtesy of Danmarks Geologiske Undersøgelse.*)

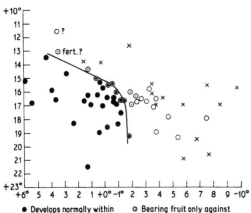

(b)

● Develops normally within the station area
◉ Mostly sterile, only exceptionally bearing fruit in the tree-top
⊙ Bearing fruit only against favourable exposed cliffs
○ Never bears ripe fruit
✕ *Hedera* missing within the station area

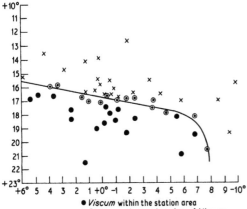

(c)

● *Viscum* within the station area
◉ Station on the area boundary of *Viscum*
✕ *Viscum* absent from the station area

somewhat affected by many more variables besides, and the population densities associated should be represented by a multidimensional figure. Nevertheless, it is clear that there must be ascertainable absolute limiting values of the more important climatic variables.

There is not yet nearly as much knowledge of the climatic limits for species living at the present day as would be desirable for past climatic reconstructions. Three good examples are, however, given by the data shown in fig. 13.49, from a detailed analysis by IVERSEN (1944) of the records of meteorological stations where holly, ivy and mistletoe grow, or do not grow, nearby. The curve for holly indicates that it is rare in places where the normal temperature of the coldest month is more than very slightly below the freezing point; evidently it will not stand much winter frost, as was verified by the killing off of holly in Denmark in the cold winters of 1939–42 except at sites near the sea or protected by an overgrowth of woodland. Ivy is also sharply limited by low winter temperatures, though it extends a little farther in this direction than holly; both plants seem to require moderate summer temperatures, the warmest month averaging between 13° and 19°c in most places where they are found, though up to 22°c in the most continental habitats occupied. Mistletoe demands more summer warmth, the warmest month mostly between 16° and 19°c, though up to 22°c in some areas; but the winter temperatures can be lower than with holly or ivy.

It seems probable that studies like those of IVERSEN cited above, if carried out for the main forest trees, would yield much useful information. In some cases, perhaps, the climatic thresholds would be blurred by the effects of competition between species and of widely differing local terrains. In practice, however, the limits of the main assemblages are sufficiently indicated by their associations with certain monthly mean isotherms and isohyets. Indeed, some of the climatic values which appear to limit the assemblages in question have already entered into the most widely used classification of climates, that of KÖPPEN (see Volume 1, pp. 509–14).

Apparent limiting conditions for certain species and assemblages are listed in Table 13.10 and in the following paragraphs, as samples of what may be learnt in this way.

Table 13.10   Apparent climatic limits of the ranges of vegetation species living today

Holly (*Ilex aquifolium*)	Mean temperature of the coldest month not below −0·5°c, warmest month not below 12·5°c.
Ivy (*Hedera helix*)	Mean temperature of the coldest month not below −2°c, warmest month not below 13°c.
Mistletoe (*Viscum album*)	Mean temperature of the coldest month not below −8°c, warmest month not below 15·8°c.
Bracken (*Pteridium aquilinum*)	Wide temperature tolerance, but demands forest-free, open land to give light.
Ling (*Calluna vulgaris*)	Thrives in drier, colder, more continental conditions than the bell heather (*Erica tetralix*) – no figures available.

Table 13.10—continued

Mediterranean heather (*Erica mediterranea*) which also occurs in Ireland	Normal lowest temperatures not below 3°c.
Mountain avens (*Dryas octopetala*)	Normal highest temperatures not above 23°c in British Isles habitats, not above 27°c in Scandinavia.
Parsley fern (*Cryptogamma crispa*)	Normal highest temperatures not above 24°c.
Sedge grass (*Cladium mariscus*)	Demands warm summers; will not tolerate long severe frosts – no figures available.
Dwarf birch (*Betula nana*)	Normal highest temperatures not above 22°c in British habitats, 27°c in Scandinavia.
Dwarf willow (*Salix reticulata*)	Normal highest temperatures not above 21°c in Scottish habitats, 23°c in Ireland and northern England, 26°c in Scandinavia.
North European or least willow (*Salix herbacea*)	Limited by normal highest temperatures 28°c.
Birch (*Betula*) and pine (*Pinus*) forest	Mean temperature of warmest month not below 10°c (somewhat higher in oceanic habitats).
Spruce (*Picea abies*)	Demands more soil moisture than birch and pine and is less able therefore to tolerate warm, dry summers; mean temperature of the warmest month should not be above 18° to 19°c but must be above 12·5°c.

Birch, pine and spruce all have the capacity to survive long severe winter frosts: spruce seed can actually spread with the blowing snow in severe, windy winters.

Other plants of the heath and forest floor in the sub-Arctic forest zone which have commonly been used as indicators of climatic character include the crowberry (*Empetrum nigrum*) which thrives in more oceanic types of cold climate than *Empetrum hermaphroditum*, the mosses which oust the berry plants and dominate the damper, shadier parts of the forest floor, and the pale grey lichens, particularly *Cladonia* spp. (the 'reindeer lichen'), which becomes dominant in the drier places.

Alder (*Alnus*)	Demands more summer warmth than birch or pine but reaches southern Lapland today, normal July temperatures over 14°c; also demands more moisture than the birch or pine.
Hazel (*Corylus*)	Thrives on a rather cool oceanic climate, normal July temperatures over 12° to 13°c (reaches Orkney today), does not tolerate very warm, dry summers.

The boundary between mixed oak forest and pine forest may be determined, other things being equal, by the number of freezing days – a normal expectation of 60 to 90 days yearly with no thaw has been suggested as the criterion in central Europe, but it is certain that the figure is lower near the oak–pine boundary in Scotland: it seems that the true nature of the decisive factors has not been satisfactorily determined.

Table 13.10—continued

Mixed oak forest	Mean temperature of the warmest month not below 12° to 14°C in oceanic regions, not below 16° to 18°C in continental interiors. The eastern limit in Europe appears to be controlled by the severer winter frosts.
Mongolian oak (*Quercus mongolica*)	Warmest month not below 20°C.
Elm (*Ulmus glabra*)	Similar requirements to oak, but sensitive to spring frosts after flowering; fruiting fails in years with late frosts in East Anglia today.
Linden (or lime) trees (*Tilia*)	Earlier in postglacial times an important element in the mixed oak forest zone; appears to demand a little more summer warmth than the European oaks.
Ash (*Fraxinus*)	Demands more light* than most trees of the temperate forest zone and is less tolerant of waterlogging than the oaks; ash trees have a competitive advantage therefore on dry, calcareous soils.
Beech (*Fagus sylvatica*)	Late spring frosts and low summer temperatures (July temperature averages below 12°C) appear to control its present western and northern limits in Europe. Evidence that between A.D. 1000 and 1500 beeches were abundant in the Harz, the Bohemian mountains, Schwarzwald and the Vosges up to heights where the present July temperature is 11°C is taken by FIRBAS (1949) as an indicator of some climatic cooling in the later centuries. The eastern limit in Europe appears to be controlled by spring–summer evaporation excess and by winter cold.
Hornbeam (*Carpinus*)	A deciduous tree of rather continental range, seems to require normal July mean temperature not below 16° and coldest month not below −6°C.
Southern (New Zealand) silver beech (*Nothofagus menziesii*)	An evergreen beech occupying the drier valley bottoms and lower slopes below the areas dominated by conifers. There is also a colder-living mountain beech (*Nothofagus cliffortioides*) which tends to occupy the ridge crests and slopes of dry aspect in South Island, New Zealand (no figures for climatic tolerances known).
The New Zealand conifers – rimu (*Dacrydium cupressinum*), the podocarps, e.g. matai (*Podocarpus spicatus*), and kaikawaka (*Libocedrus bidwillii*)	Characteristic in wetter habitats than the beeches, the matai in fertile lowlands of moderate rainfall and the other two species in areas of high rainfall (no figures available), kaikawaka more particularly in the uplands.
Eucalyptus species in Australia	CHURCHILL (1968) studies the temperature and rainfall requirements of three species which occur widely in southwestern Australia. Diagrams similar to those in fig.

Table 13.10—continued

	13.49 showed that the occurrence of karri (*Eucalyptus diversicolor*) was limited to places with normally over 750 mm rainfall and was sparse at places with under 1000 mm annually; over 175 mm seemed to be required as the normal contribution of the wettest month and over 12 mm in the driest month.
	Marri (*E. calophylla*) tolerated normal annual rainfalls down to 450 mm, with wettest month over 60 mm and driest month over 5 mm.
	Jarrah (*E. marginata*) tolerated normal yearly rainfalls down to 375 mm, with wettest month giving over 80 mm and driest month over 5 mm.
Saguaro, the giant tree cactus of Arizona (*Carnegiea gigantea*)	Northern limit defined by the occurrence of a single day without thaw during the life of the plant. Overnight frosts do not matter, but total kill results from 30–36 hours without the temperature rising above 0°c (HASTINGS and TURNER 1965).
Northern limit of general vine cultivation in Europe	Approximately defined by 18°c July isotherm, but dates of last frosts in spring and October temperatures may be more critical.
Northern limits of citrus fruits in Europe	Approximately defined by 22°c July isotherm.
Tropical grasses of the savanna zone	Given sufficient moisture, maximum growth occurs at 25° to 30°c.

It is believed that the natural rubber, *Hevea Brasiliensis*, is another tree that will only flourish within narrow climatic limits, though these have not been determined; it too could perhaps be an indicator of past climatic vicissitudes, and its sensitivity may be economically important.

---

* In seeming contradiction of this, a record of the widths of the yearly growth rings in ash trees in the South Downs in southern England (H. G. CHAPPELL, personal communication 28 March 1968) from 1887 to 1965 rather closely parallels the variations of cloudiness. Growth was about twice as great in the cloudier summers before 1895 and in the 1960s, and was also increased in the 1920s, as compared with the rest of the record. The explanation may be that the chalky soils become too dry in the sunnier years.

GODWIN (1935) has uttered a useful warning against taking such limits too strictly, especially against applying them to every individual tree. The presence or absence of competition from other species is very important, and there is a wide range of climatic conditions between those that are optimal for a given tree's growth and those that may bring about death or prevent successful seeding and regeneration. The conditions mentioned in Tables 13.10 and 13.11 may, however, reasonably be adopted as first working assumptions for climatic reconstructions.

The main sources used in compiling Table 13.10 were CONOLLY and DAHL (1970), FIRBAS (1949) and GODWIN (1956), in all of which much further information will be found. We may add the following references to the vegetation of China, given by CHU KO-CHEN (1973):

(1) The northern limit of bamboo groves (*Bambuseae*) naturally growing in China approximately coincides with the January mean monthly temperature isotherm of 0°c.

(2) Lychee (*Litchi chinensis*) cannot stand minimum temperatures below about −4°c.

(3) Orange (*Citrus sinensis*) and mandarin (*C. reticulata*) trees cannot stand minimum temperatures below −8°c (American and European cultivators quote figures of −4° to −7°c, and even −2°c will damage the fruit which has not been picked).

(4) Plum trees (*Prunus*), native to the oak forest zone of China, are limited by absolute minimum temperatures below about −14°c.

All the species listed have been used to derive information about climatic history from the changes in the geographical range of their occurrence detectable in the past record. Nevertheless, further details about the climatic events which actually limit the species and control its abundance are desirable in nearly every case.

The different temperature tolerances of some species in the British Isles and in Scandinavia, mentioned in the table, point to the effect of some other unidentified variable influencing the situation.

By scanning the cases listed in Table 13.10, it will be appreciated that there is less risk of erroneous climatic conclusions when the limits of like assemblages at different epochs are compared than when the limits of a single species are relied on. In dealing with a well recognized assemblage of vegetation, and the animal and insect life that goes with it, it may not be unreasonable to suppose that the thermal limit can be defined to about 1°c. Some valid deductions may be made from observed changes in the occurrence of a single species in cases where the limits of its climatic tolerance seem universally the same or where the deductions are supported by another, independent line of evidence.

A few examples of the miscellany of diagnostically useful climatic thresholds affecting the animal kingdom are given in Table 13.11. This short list includes samples of genera, e.g. the beetles and foraminifera, whose remains commonly remain identifiable after many thousands of years, and of others such as the fish and barnacles, manuscript records of which may indicate the variations of sea temperature over recent centuries or decades. More of the background and the possibilities of this branch of the subject, as well as some more examples, are given in the following paragraphs.

Table 13.11   Apparent climatic limits of the ranges of sample faunal species living today

Beetles	
Ground beetle (*Oodes gracilis*)	Mean air temperature in July normally not below about 18°C.
House (timber) longhorn beetle (*Hylotrupes bajulus*)	Regions with mean July temperature normally above 16·5°C.
Water beetle (*Dytiscus semisculatus*)	Regions with normal July air temperature averaging 16° to 16·5°C or above, the mean water temperature probably not below 18°C in the warmest month.
Freshwater tortoise	
European pond tortoise (*Emys orbicularis*)	Mean air temperature in July normally not below 18·5°C.
Marine life	
Foraminifera (*Globerigina pachyderma*)	Predominance of left-coiling tests of this widely distributed species where prevailing sea temperatures are low, in April below 7·2°C. In warmer waters right-coiling forms predominate.
Cod (*Gadidae*)	Abundant in seas where the surface water temperatures are above +2°C and below about 15°C.
Herring (*Clupea harengus*)	Rarely in water below +3°C or above about 13°C.
Barnacles (*Balanus balanoides*)	A northern variety, scarce where waters near the shore are normally warmer than about 18°C in the warmest month or above 8° to 9°C in the coldest month.
Chthamalus stellatus	A southern barnacle which reaches its present northern limit in Scotland, where the temperature in the coldest month is about 5°C.
Corals (*Madreporia*)	Active reef-building varieties demand water temperatures always above 18°C.

Sources used in the compilation of Table 13.11 include BEVERTON and LEE (1965), BRIDEN and IRVING (1964), COOPE (1970), DEGERBØL and KROG (1951), ERICSON and WOLLIN (1966), JAKOBSSON (1969), LINDROTH (1943), SOUTHWARD and CRISP (1954), TEICHERT (1964) and WHITE (1954).

Secular changes of range and their climatic implications

A rare case where the limits of tolerance of Man himself may yield reliable information on climate in the past comes from the old Icelandic *Landnámabók*, describing the Viking settlement of southwest Greenland in the years between A.D. 985 and 1000. Of one of the leading original settlers, a cousin of Erik the Red, named Thorkel Farserk, who settled at Hvalseyjarfjord, it is related that: 'He was extremely strong. Erik the Red once visited him. As he wished to entertain his cousin well, but had no serviceable boat at hand, he had to swim out to Hvalsey to fetch a full-grown sheep, and carry it home . . . it was a distance of well over 2 miles' (quoted by KROGH 1967). I am informed by Dr L. G. C. E. PUGH of the Medical Research Laboratories, Holly Hill, Hampstead, from his studies of the endurance of Channel swimmers and others who have attempted to swim the Firth of Forth, the Bosphorus and the Nile in cold conditions, that 10°C would be a fair estimate of the lowest temperature at which a person not specifically trained in long-distance swimming could swim 2 miles, and even then he would have to be fat. It is to be noted that Thorkel did not just manage to do it, but brought home a sheep as well. As the average temperatures in the fjords of that coast in August in the warm years around 1950 were +3° to +6°C, it seems that the water must have been at least 4°C warmer in the year concerned in Viking times.

Most secure climatic deductions depend on occurrences of considerable numbers of this or that species. Even then, it must be borne in mind that climatic changes which open up favourable opportunities commonly evoke a slower response than adverse conditions.

Critical factors for the ranges of birds may be severity of the winter frosts and depth and duration of snow cover, date of the spring thaw affecting hospitability of the nesting sites, warmth and length of the summer, and whether there is too little or too much water in the habitat. But no precise climatic figures for the limits of range can be given. Often the changes presumably operate through climate affecting the abundance of the birds' insect food. By some long succession of extensions of range, similar to the recent ones discussed below, the now habitual global pattern of seasonal migrations of birds must have been established (or re-established) in the 15 000 to 20 000 years since the maximum extent of the last glaciation. Remains found indicate that at that time ptarmigan (*Lagopus mutus*) as well as thrushes (*Turdus viscivorus*), which are still familiar in Britain, though the former is now confined to the Scottish Highlands, were resident in central England. Extensions of the ranges of European birds, apparently associated with the nearly world-wide warming of climates in the present century, include the establishment of breeding pairs of blackheaded gull (*Larus*) in Iceland from 1911, swallows (though of only one species, *Hirundo rustica*) in the Faeroe Islands and Iceland in the 1930s, starlings (*Sturnus vulgaris*) in Iceland from 1941 and the fieldfare (*Turdus pilaris*) in Greenland and on Jan Mayen Island (71°N) after 1937. The turtle dove (*Streptopelia turtur*) is known to have spread from southern and eastern England to Wales, eastern Ireland and southern Scotland since 1800, apparently registering recovery from the colder centuries of the Little Ice Age. Some species were still extending their range northward in the 1950s after the peak of the warm period was over: among these, it is believed, was the lapwing (*Vanellus*), which was breeding in Iceland by 1960, and the more modest northward expansion of the black redstart (*Phoenicurus ochruros*) in England. However, factors other than climate also change the habits of birds: following the increasing habit of British and American seagulls in this century of urban scavenging and following the plough, it is reported that some species now only go to sea in their mating season (HARRIS 1964). The cause of the rapid spread of the collared dove (*Streptopelia decaocto*) from southeast Europe to the British Isles during the present century has not been satisfactorily explained but may well be partly climatic (e.g. increasing aridity in parts of the former range).

Although many cases can be cited where changes in the ranges of bird species are more or less clearly associated with a climatic change, in other cases the causes are more obscure. A number of species, e.g. both British and Scandinavian nesting blackbirds (*Turdus merula*), seem to have been extending their range in every direction in recent decades; it may be suspected that this dynamism of the species indicates that the birds are profiting in some new way from the age of Man's affluent society, e.g. the spread of suburban dwellings and the increased productivity of the land under modern agriculture.

The increased frequency of westerly wind situations over the Atlantic and the intensity of the general wind circulation, which accompanied the climatic warming in the first half of the present century, seems to have resulted in an increased frequency of the arrival in the British Isles of errant species (known as 'accidentals') from North America.[1] While there are dangers in arguing from isolated occurrences of rarities (NELDER 1962, NICHOLSON *et al.* 1962), there are obvious cases where anomalies can be put down to abnormal winds (LANDSBOROUGH THOMSON 1964, also in HARRISON 1968),[2] as must also apply to the recorded instances, scattered over many centuries, of the arrival of locusts in various parts of Europe near 50°N.

The time of greatest predominance of W'ly winds in Britain, between about 1900 and 1950, associated with intense and large-scale development of the North Atlantic low pressure systems with their centres passing near Iceland, seems to have had other interesting consequences affecting bird migration. In just those decades there were particularly few additions to the list of known regular (European) passage migrants and/or winter visitors in the British Isles, whereas in Greenland the reverse was the case. From 1820 to 1890 the 'first records' of bird species in Greenland were predominantly American varieties and the decade average for first reports of European birds was less than two; the latter averaged six per decade from the 1890s to 1940, but then dropped to two again. Among vagrants in Greenland, European varieties were 60 % of the total in the 1920s, having increased from the 1890s to that peak share of the total; by the 1940s their share was under 50 % and continued falling into the 1960s (WILLIAMSON 1973).

The spread in recent decades into northern and western Europe of species of ducks and grebes and other breeding birds of shallow food-rich lakes is thought to be connected with the increasing desiccation of the region of Soviet Asia about the Caspian and Aral Seas. The occasional irruptions into western Europe and the British Isles, notably in 1859, 1863 and 1888, as well as into China, of Pallas's sandgrouse (*Syrrhaptes paradoxus*) from its normal habitat in the dry regions of central Asia may be associated with excessively dry years there.

1. American robins and cuckoos are among the species reported.
2. The winds concerned need not have lasted for more than a few days. Examples are supplied by the large arrivals on the east coast of Britain of birds migrating from Scandinavia in some Septembers with notable northeasterly winds and by the huge numbers of birds often blown off course by tropical cyclones.

Changes in the ranges of numerous resident and migratory bird species in Europe during the first half of the present century, and their probable associations with changes of prevailing temperature and moisture, and with Man's alterations of the landscape, were studied in some detail by KALELA (1949).

A rather general southward movement since 1960, and the reappearance in Scotland after an absence of half a century or more, of north European birds, such as the snowy owl (*Nyctea scandiaca*), which nested in Shetland in 1968 after a gap of 60 years, and the great northern diver (*Gavia immer*), which began nesting in Scotland again in 1970, may be attributable to the cooling of the Arctic in the last two decades.[1] Variations reported in the frequency of the snow bunting in Scotland over many years probably have a similar origin. The arrival in most northern and eastern parts of the British Isles in 1968 of large numbers of nutcrackers (*Nucifraga caryocatactes*), normally resident in northeast European Russia and Siberia with some extension to southern Sweden and Norway, may be associated with the same cause and with the frequent NE'ly winds of 1968. Another case may be the return of the golden eagle (*Aquila chrysaetos*) to nest in the Lake District in northern England from 1970 onwards. Similarly, birds for which England is near the northern end of their range, such as the wryneck (*Jynx torquilla*) and the woodlark (*Lullula arborea*) have been less frequently seen there in recent years than earlier in this century. Yearly figures for the numbers of herons (*Ardea cinerea*), counted since 1933 in the Thames valley, and in the Cheshire and Lancashire lowlands in northwest England, show sharp declines after severe winters, amounting to about 30% in 1940–2, over 50% in 1947 and over 70% in 1963; the numbers gradually recovered over the intervening years and were highest in 1950–4 and 1960–1.

In Africa south of the Sahara, in the zone affected by increasing drought in recent years, a similar southward tendency of the ranges of bird species has been noticed (FRY 1973). At Zaria (11°10′N 7°40′W) in Nigeria documentation of the bird fauna goes back to 1915. Additions to the list from that date until 1962 do not indicate any trends; but from 1962 onwards seven species, all of them northern varieties and adapted to the arid zone, have been added and, as the trend has continued up to 1973, have built up their populations. The newcomers include a lapwing (*Vanellus tectus*), the European turtle dove (*Streptopelia turtur*), a starling (*Spreo pulcher*), and (so far only as a vagrant) the stork (*Ciconia ciconia*). No southern species has extended its range to that area in these years.[2]

For insects, particularly beetles, some more precise figures for their thermal limits are available (e.g. in LINDROTH 1943, 1945). Thus, *Calosoma inquisitor*, a beetle which lives on oak trees and preys on caterpillars, seems limited by mean July temperatures below 15°C. For a water beetle, *Dytiscus semisculatus*, normal July temperatures about 16° to 16·5°C represent the lower limit. The carabid ground beetle, *Oodes gracilis*, is limited to regions where the mean July temperature is above about 18°C. COOPE (1969a, 1969b, 1970) has used these, and many other beetles whose present known range is only in northernmost Europe and Siberia,

---

1. The mean temperature for the whole year at Franz Josefs Land (82°N 53°E) 1960–9 was 4°C lower than for the three preceding decades.

2. A further observation on changes in the distribution of bird species (added at the final proof-reading stage) is placed at the end of this chapter on p. 279.

to indicate the probable July temperatures in England in the mid Weichselian (last ice age) interstadial and the course of postglacial summer temperature history.

A house longhorn beetle, *Hylotrupes bajulus*, which seems to be limited to areas where the mean temperature of the warmest month is above 16·5°C, did great damage to the woodwork in houses in late Georgian times in London, and then died out, but became a serious pest again between 1934 and 1953, in the warmest group of summers of the present century. It is probably continually reintroduced with shipments of timber from the Baltic, and the decline of its activity in England after about 1800 and the 1950s can probably be attributed to cooler summers.

Similar deductions may be possible from the occurrences of moths and butterflies. Recognitions of species occurring in Britain were proceeding apace during much of the nineteenth century, as the science became organized; but between 1880 and 1930 only one new species of moth was added, whereas between 1930 and 1960 fourteen species new to this country were recognized, probably due to warmer summers (MERE 1961). After 1960 certain species, which in Britain are on the fringe of their main, continental (warmer summers) range, were disappearing from the country.[1]

Winter temperatures also limit the ranges of insect life. Mr J. E. SATCHELL of the Nature Conservancy reports (personal communication, 24 February 1965) that a species of wood ant which was formerly widely distributed in southern Ireland declined abruptly in the mild winters about the 1920s, probably owing to disruption of its breeding habits, and survived only in the area of central Ireland where frosts are most frequent. The elm bark beetles that carry the Dutch elm disease thrive on mild winters, and the disease spread in Europe and the United States in the runs of mild winters in the 1930s and 1971–5. HEPTING (1971) mentions other forest diseases, many associated with fungi, which spread under similar conditions.

Many of the rusts, mildews and other pathogen-caused diseases which infect trees, grasses and grain crops are encouraged by a variety of weather factors, notably by wet summers. Potato blight in the British Isles is triggered by multiplication of the organisms in periods of over 48 hours in which the air temperature remains continuously above 10°C with relative humidity above 90% (SMITH 1956, see also SMITH and WALKER 1966).

Locusts (*Schistocerca gregaria*) breed after rains in the desert zone and fringes of the deserts. Once airborne they are carried by the winds, which generally convey them towards the intertropical convergence zone (RAINEY 1973), where the equatorial rains are generated in the up-currents fed by the convergence. There the locust swarms commonly multiply. Temperatures above 50°C appear to be fatal to locusts. They are most active in air temperatures between 25° and 35°C. They become lethargic at air temperatures below 15°C, though capable of short flights at temperatures down to 9°C; the effect of sunshine may give the insect a body temperature 10°C above that of the ambient air. Thus, they are occasionally

---

1. The same is reported to be true of certain lizards and
   toads.

carried alive to Britain and central Europe by S'ly winds. (The long records of occasional locust invasions – and other insects – may be informative as regards unusual wind flow.) Frost does not kill locusts, but they die after sinking into snow and becoming frozen in.

Certain human and animal diseases are limited by the effect of climatic conditions upon the insect which is host to the pathogen (MARTINI 1952). The malarial mosquito (*Anopheles*) demands temperatures over 16°C as well as a terrain with stagnant waters for its breeding, and the July isotherm for 16° seems to mark the extreme northern limit of malarial districts (which prior to about A.D. 1680, and again at times in the nineteenth century, included the English fenland and north German marshes); but the parasite largely fails to develop within the lifetime of the mosquito at temperatures below 19° to 20°C. The bubonic plague flea (*Xenopsylla cheops*) is sensitive to temperature and humidity: its activity is markedly speeded up at temperatures between 20° and 32°C, but it lives four times as long with relative humidities about 90% as it does with relative humidities below 30%.

The occurrence of mollusks – the soil mollusca (land snails) of a great range of sizes, and the freshwater and marine mollusca, whose shells are preserved almost indefinitely (save for the effects of solution of the carbonate, particularly in acid environments, or of shattering) – can be used for a variety of climatic deductions, including water temperatures, salinity and the limit reached by sea water in estuaries. European soil mollusk species have also been classified as indicators of various types of woodlands, grasslands and marshes (LOZEK 1965). This is a big subject which cannot be opened up here. The reader may refer to DRAKE (1960–2), SPARKS and WEST (1972) and WEST (1968). The variations over very long periods of time in the percentages of different species in the fossil molluscan fauna in a deposit are commonly presented in the form of diagrams resembling pollen diagrams. Among the more important conclusions derived from these studies is that the water of the River Nile in Nubia was colder than at present during glacial times (quoted by VAN ZINDEREN BAKKER 1969).

A reptile, the European pond tortoise (*Emys orbicularis*), has been used as an indicator of mean July temperatures of 18·5° to 20°C or more, defined by its present limit in Germany and France respectively (DEGERBØL and KROG 1951). Its occurrence in the warmest postglacial times in Denmark and East Anglia seems to establish that summer temperatures were about 2°C above present levels. The larger mammalian fauna mostly cannot give such unmistakable indications. Camel remains found in Alaska and tiger in the New Siberian Islands about latitude 75°N are attributed to the warmest postglacial times (ALISSOW *et al.* 1956, p. 327) when there was certainly less ice than now on the Arctic Ocean, but probably indicate summer wanderings in temperate, not tropical conditions. Remains of hippopotamus, lions, elephants and probably hyaenas, have been found in interglacial gravels of the river Thames under Trafalgar Square in London: these animals are known to be capable of living in the open in the temperate climate of today, and the present restriction of their range must be attributed rather to the depredations of Man.

Just a few examples must indicate the climatic deductions that can be made from life in the sea.

The abundance of the cod (*Gadus morhua*) and other *Gadidae* is limited to regions of the ocean where the surface waters are not below 2°C, and it seems likely that the near-failure of the cod fishery about the Faeroe Islands in the seventeenth and eighteenth centuries marked an extreme advance of the polar water (indeed an airborne radiation thermometer showed that waters with temperature only 3°C reached to within 100 km of the Faeroe Islands from the north in April 1968 (PICKETT and ATHEY 1968)). In northern waters the abundance of cod fluctuates with the sea temperature, and near the southern limit of its range cod becomes gradually scarcer as the temperature rises: fig. 13.50 shows how the cod increased in the North Sea in the late 1960s when water temperatures fell.

Available records of the longer history of the cod fisheries in northern waters evidently indicate long-term fluctuations of water temperature and the changes in the ocean currents that have been at least partly responsible for them. Cod were very plentiful off the west coast of Greenland around 1820 as far north as Disko Bugt. Again in the 1840s cod were caught in abundance south of 61°N off that coast and spread northwards, ultimately reaching Disko Island. The cod then decreased in 1849, and by 1852 there were no cod at all. There was little ice near Iceland from 1840 to 1854. The cod fishery returned to those waters only in 1945.

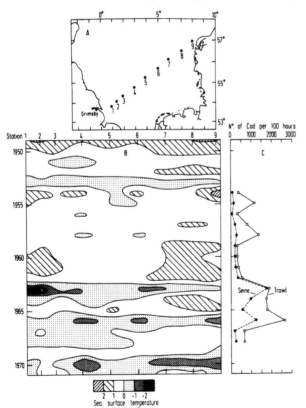

Fig. 13.50   Cod catches and water temperatures in the North Sea.

(*a*) Positions of temperature measurement stations along a line across the North Sea.

(*b*) Mean sea surface temperature anomalies in winter along this line from 1949 to 1971.

(*c*) Year-class strength of North Sea cod each year from 1954 to 1968, measured by the subsequent catch of two-year-olds per 100 hours effort by the Grimsby trawl and seine fishery.

(*From* DICKSON *and* LEE *1972; reproduced by kind permission.*)

Fluctuations in the catches of the salmon fisheries on the coasts of Norway and Canada have been reported since 1870, with suggested regular periodicities in the range 8 to 11 years.

Marked variations in the distributions of the Atlantic herring (*Clupea harengus*) have been known to occur since their sudden disappearance from the Baltic in the sixteenth century. It has been suggested that the Baltic variations have most to do with the variations of tidal force (Volume I, pp. 221–2) and the winds which control the amount of saline, and well oxygenated, Atlantic water which enters the Baltic from time to time through the Danish Sound. BEVERTON and LEE (1965) have noted the substantial twentieth century shifts of the border zone between the herring and the pilchard (*Sardina pilchardus*) near the British Isles, in phase with the changes of prevailing water temperature, and have recorded the history from A.D. 1500 to the present time of the Norwegian and Swedish herring stocks which apparently moved north whenever the oceanic polar front moved north and warm Atlantic water increased in the Baltic. Different herring populations in the Atlantic have different ranges, some present stocks spawning off south and southwest Iceland, some in March and some in July, the largest stock spawning off west Norway in March, and others farther south (see papers by CRAIG 1960, JAKOBSSON 1969, PARRISH and CRAIG 1963, PETTERSSON 1914). All move north after spawning to feed. The coldest-living stocks feed in the mixed Atlantic and Arctic watermasses near the oceanic polar front between Iceland and Spitsbergen between June and August. An interesting study by JAKOBSSON (1969) of how the herring stocks adapted themselves to the advances of the polar water near Iceland in the 1960s established that the fish occasionally moved across a tongue of this water in temperatures down to −1°C but that they never stayed long in waters below +3°C; apparently the food they depend on was absent in the colder and less saline water. The main concentrations of feeding herring were always in the uppermost 50 m (mainly at about 20 m) in temperatures between +4 and +6°C. Another interesting aspect was that the herring moved rapidly away from their habitual feeding grounds when the water became too cold, often entering new areas in the Atlantic water far to the northeast; but, in spite of this, they returned to their usual wintering areas off east Iceland and west Norway in September. Plainly the fish are able to adapt themselves to new migration patterns and register a very quick response to changes of water temperature regime.

SOUTHWARD and CRISP (1954) reported how the relative abundances of a southern and a northern barnacle in waters around the British Isles showed a delicate and quick response to recent changes of prevailing sea water temperatures (see also CRISP 1965, KITCHING 1950). The southern species *Chthamalus stellatus* finds its best feeding conditions in waters between 5° and 30°C, and its northern limit essentially coincides with the February isotherm of 5°C. The northern species *Balanus balanoides* ranges in waters between 0° and 18°C; after receding for some years, particularly between 1948 and 1951, it proceeded to advance again at the expense of *Chthamalus*.

The mussel *Hiatella arctica* has been present in all the world's oceans since early Tertiary times. STRAUCH (1968) has shown that in the present seas the shell size of adult specimens increases as the water temperatures fall. The increase of size becomes rapid when the winter temperatures fall below about $3°c$, where the mean is 15 mm, to a mean size of 40 to 50 mm at $-1°c$. Further analysis suggests that the mean shell size of this marine mollusk could be used to indicate the prevailing water temperatures in both the warmest and coldest months.

R. S. GLOVER of the Scottish Marine Biological Association has reported (personal communication, 2 December 1969) that a survey of the plankton sampled by vessels between North America, Europe, Iceland and Greenland between 1950 and 1968 showed that the spring flush of the phytoplankton was becoming progressively later, the change amounting to perhaps 20 days over the period, and that this was apparently adversely affecting the abundance of zooplankton. The change was tentatively associated with a decline in the amount of solar radiation: there was an increase in cloudiness in the spring months in Shetland, but it was suspected that the decline in measured strength of the solar beam itself (see Volume 1, fig. 2.3(a), p. 23) may also be playing a part.

For palaeo-oceanography, studies of the abundances of different species of for-aminifera and radiolaria (which are both single-celled animals) as well as their calcareous secretions (coccoliths), and of diatoms (which are single-celled plants among the group known as algae) have yielded much information. The foraminifera build 'tests' or shells up to 2 to 5 mm long, and occasionally 20 to 50 mm, some species from the calcium carbonate dissolved in the sea water and others from sand grains and the discarded calcareous shells and remains of other species. Coccoliths and the long extinct discoasters are found as microscopic plates and stars respectively of calcium carbonate which were secreted and deposited on the ocean bed by extremely minute organisms (the *Coccolithophoridae* and *Discoasteridae*, respectively). There is only one species of foraminifera in Arctic waters, *Globerigina pachyderma*, predominantly coiling to the left ('sinistral coiling'), whereas the same species is found but predominantly right coiling ('dextral coiling') in warmer waters: the dividing line today coincides approximately with the $7·2°c$ April water surface isotherm in the North Atlantic (ERICSON and WOLLIN 1966, p. 89). It has been found (STEUERWALD and CLARK 1972) that increases in the percentage of larger forms of this species are also a characteristic of cold water epochs. Other species (e.g. *Globorotalia truncatulinoides* in the sub-Antarctic zone) also show temperature-dependent predominance of left coiling or right coiling varieties. The indications of this work were that the Arctic Ocean had never in the past 3 million years been much warmer than its present, postglacial condition. The microscopic plates and stars respectively of calcium carbonate which were secreted and soluble in sea water, particularly the cold water of the ocean deeps, than are the calcareous tests of foraminifera. Radiolaria and diatoms are therefore far more widely found on the bed of the Southern Ocean and much of the Pacific than foraminifera. They are represented by many species living in all the oceans, including the Antarctic, today. Studies of the percentages, or ratios, of different species can readily be used therefore to give indications of past temperature. J. D. HAYS reports (verbal communication, 18 May 1973) that plots of the percentages of different recognized assemblages of radiolarian species against surface water temperatures in the southern hemisphere oceans at the present day indicate that an error margin of no more than $\pm 1.5°c$ should apply to temperatures derived in this way, but KEANY and KENNETT (1972) report that the temperature tolerance of some radiolaria seems to have changed about 700 000 years ago. MCINTYRE (1967) has used the boundaries of 'warm' and 'cold' species of coccoliths to map the displacement of sea surface temperature values in the North Atlantic in the last glaciation (see fig. 13.69, p. 241). Calibration of the ratios of the different coccolith species against present-day surface water temperatures suggests a probable error of rather less than $\pm 1.2°c$.

Corals (*Madreporiae*) can be used as indicators of palaeotemperatures of the warmer seas. Though there are species of coral living as far north as 35°N in Japan, where the water temperature in February is only 12°C, and 35°S on the east coast of Australia, and still colder-living species at a few points on the west coasts of Europe between latitudes 51° and 69°N, the main reef-building corals demand temperatures of normally 21°C and over (SCHWARZBACH 1961, p. 24) and cannot tolerate extreme minima below 18°C. Most active reefs are therefore between latitudes 30°N and 30°S. They further need good, clear sea water of normal salinity. Modern, and all Mesozoic and Cainozoic, corals build their reefs of calcium carbonate in the form of aragonite (TEICHERT 1964). The reefs are not subject to much solution in the warm seas, and many fossil reefs survive on land from past geological eras as evidence of the former existence of warm seas and the effects of continental drift and pole wandering (BRIDEN and IRVING 1964, SCHWARZBACH 1949). The dated periods of coral growth and interruptions to growth during the Pleistocene at the Bahamas, near latitude 25°N, mark out the interglacial and glacial periods (MESOLELLA *et al.* 1969); during glacial times the waters of the western Atlantic at that latitude were evidently too cool. Today there are about 35 species of corals active there.

Bermuda (32°3′N 64°9′W) is an ancient submarine volcanic mount with active coral building now and in earlier interglacial periods over the shallow submerged platform, which the summit of the mount presents. It is perhaps the northernmost example today of such extensive activity, the number of species found being only 10 as against 60 to 180 in the tropical areas of reef building. The hills on the island appear to be dunes formed of coral sand blown by strong winds at times of lowered sea level when much of the coral was exposed. The dunes show a layered structure, each layer 2–20 mm thick, possibly the result of the fresh calcium carbonate sand being cemented by sea spray and rains during and after each gale. This is evidence of a windy climate, but not necessarily of tropical hurricanes, during the glacial periods farther north.

Approaches to the climatic record of the past

Pollen analysis, spores and macrofossils

No branch of field observation has contributed more to our perception of the past climatic record than palynology, the study of the pollens of seed-bearing plants and the analysis of fossil pollen assemblages, together with the spores of lower plants, and consideration of such 'macro' remains as are to be found. Macroscopic fossils, or macrofossils, are by definition those plant remains that are big enough to be seen by the naked eye and studied with, at most, the use of low magnifications. They range from seeds, fruits and leaves up to tree stumps and even the trunks of fallen trees which have been preserved. Preservation, either of macro- or of microfossils, is likeliest in waterlogged environments, such as peat bogs and lake sediments, where lack of oxygen excludes the usual destructive organisms. The bigger objects are, of course, relics of vegetation that grew on the actual site of the find or close at hand; the pollen and spores deposited include some that may have come great distances.

The beginnings of pollen analysis (FÆGRI and IVERSEN 1964) go back to the recognition of fossil pollen grains of pre-Quaternary age by GÖPPERT in 1836. Systematic methods, involving counting the relative frequencies of different pollens at various levels in a deposit were introduced by VON POST (1918), who first saw their possibilities for establishing the past composition of the vegetation and, in particular, the history of its spread after the last glaciation, and presented the first pollen percentage diagrams. ERDTMAN (1921), writing in German, first demonstrated the possibilities of the method to a wider readership outside Scandinavia and extended the method, including some study of the modern pollen rain and its long-distance transport (1937). Among the leaders in applying pollen analysis to other parts of the world, extensive contributions to knowledge and interpretation of the vegetation sequence are particularly found in the work of FIRBAS (1949) covering central Europe, GODWIN (1956) in Britain, JESSEN (1949) in Ireland, NEJSTADT (1957) covering the territories of the Soviet Union and AUER (1958, 1965) for southernmost South America.

The method has been extended to other parts of the world by CRANWELL and VON POST (1936), who introduced it in New Zealand, and under the coordinating effort of VAN ZINDEREN BAKKER (1969, and with COETZEE 1970) in eastern tropical Africa and southern Africa and elsewhere, by VAN DER HAMMEN (1963, 1968) in northern parts of South America, by SEARS (1937) in the United States and later by DAVIS (1961) and by WRIGHT (1966), whose work also includes parts of the Middle East, by TERASMAE (1958) and later NICHOLS (1967a) in Canada and FREDSKILD (e.g. 1973) in Greenland, by BEUG (1967) in Turkey, SINGH (1963) in India and by TSUKADA (1967) in Japan. WALKER'S (1966) review of the position in New Guinea, Australia and New Zealand, and HEUSSER'S (1966) for western North and South America, are still of value. In the case of New Zealand, MOAR (1973) has surveyed the results back to the early stages of the last glaciation and reappraised the postglacial evidence.

Pollen counts are initially plotted, using conventional symbols (FÆGRI and IVERSEN

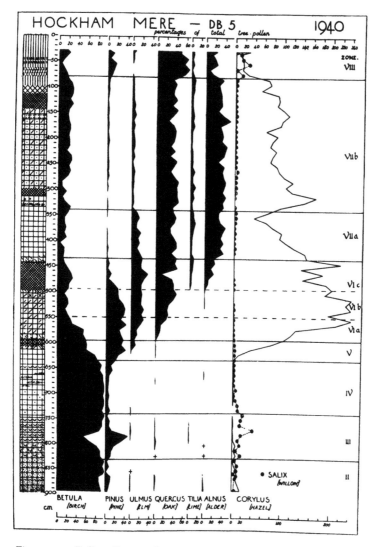

Fig. 13.51   Pollen diagram from a postglacial lake deposit at Hockham Mere, Norfolk, England (52·5°N 0·9°E).

Percentages of birch, pine, etc., in the total tree pollen (*TP*), excluding hazel, read from left to right. The abundance of hazel pollen, on the same scale, is shown by the open curve (note the large amplitude variations of this plant). The symbols in the left-hand column refer to various types of lake mud up to 100 cm depth and fen peat in the top 100 cm.

(*Originally published in* The History of the British Flora *by* H. GODWIN *and reproduced here by courtesy of the Cambridge University Press.*)

1964, p. 87; FIRBAS 1949, p. 9) for the various species, usually to show along horizontal lines the percentage of each species in some total count at each level in the deposit. Each plotted line is called the pollen spectrum at the level concerned. Counts of the species identified in some 100 to 300 grains are usually made at about 10 cm intervals down through the deposit. This may give a time resolution of the order of 100 years in the peat of raised bogs and of a few centuries in many other locations. When the points for any one species are joined up, they form curves representing the varying proportions of that species in the pollen throughout the depth of the deposit. For convenience in comparing different pollen diagrams, the different species curves should be presented in a standard order and spaced out, usually with the tree pollens at the left and the non-tree pollens towards the right of the diagram. Sample pollen diagrams to illustrate the postglacial history of the principal trees in East Anglia (Norfolk) and northern Scotland are shown in figs. 13.51 and 13.52. The East Anglian diagram shows how GODWIN (1956) has been able to define the sequence of pollen zones within each of which the pollen spectra remain similar and which are bounded by significant changes in the abundance of this or that major species; these zones have been found to be widely applicable throughout England and are now known to be very similar in their timing and nature to the sequence elsewhere in the present broad-leafed deciduous forest regions of Europe (see fig. 13.13). Distribution maps for the different tree pollens in each of the successive stages of postglacial vegetation history defined by the pollen zones not only give clues to the climatic history (BIRKS et al. 1975, GODWIN 1975), but also show the routes by which the different tree species arrived in Britain from the continent of Europe or, in some cases, from ice age refugia in the north and west.[1] Different zones have to be used for the present pine forest areas, as in fig. 13.52 and elsewhere in northern Europe. Fig. 13.53 shows a pollen diagram for the now essentially treeless Orkney Islands; long-distance transport of tree pollen is registered in this diagram which also details the principal non-tree species. Here also special pollen zones have been defined, but the correspondence with GODWIN's standard postglacial zones for England is shown in brackets at the right of the diagram; in mid postglacial times, when tree pollen was abundant, evidence of 'macro' remains has been found (TRAILL 1868), showing that some tree species (probably birch, hazel and pine did grow in Orkney.

Perhaps the most general impression from pollen analysis and pollen zonation schemes is of the comparatively long stretches of time during which the composition of the vegetation remained substantially unaltered and the rapidity of the changes whenever they occurred. Chi-squared tests, etc., can be used to establish where statistically significant changes occur, but may often be considered unnecessary.

Practice varies as regards the total pollen sum to which the percentages relate. Often they are expressed as a percentage of the total tree pollen (TP) – also known as arboreal pollen (AP) – and the herbs as percentages of the non-tree pollen (NTP or NAP), though

1. Pollen maps of North America for the different main tree species at different epochs during postglacial times have also been produced (BERNABO et al. 1977, WEBB et al. 1976).

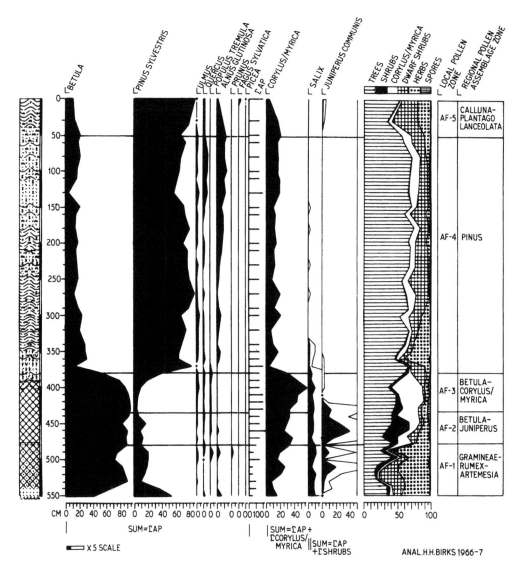

Fig. 13.52 Pollen diagram from a peat bog in Abernethy Forest, northern central Scotland (56·2°N 3·6°W).

Percentages of the different tree species in the total tree pollen (*TP*) and at the extreme right the percentages of pollens of trees, shrubs, herbs etc. in the total pollen sum (*TP + NTP*).

Appropriate local age zones have been adopted; the bottom zone is considered to be Late Glacial, all the others are postglacial.

(*After* H. H. BIRKS (*1970*); *by kind permission.*)

sometimes their numbers are also converted to percentages (liable to exceed 100%) of the tree pollen. In some cases, as at Hockham (fig. 13.51), where hazel (*Corylus*) pollen is very abundant, this is not included in the tree pollen total and is expressed separately as a percentage (at times exceeding 100%) of the total pollen (TP) of the other trees.

Fig. 13.54 shows a diagram, quite similar to that for Hockham Mere in the postglacial (fig. 13.51), but representing the counts at another place in eastern England of pollens from the Hoxnian (last but one) interglacial, of the order of 200 000 years ago. WEST (1968, p. 303) has put forward a simple and useful standard scheme by which all interglacial periods in middle latitudes can be subdivided similarly into four main zones: I pre-temperate (immediately following the previous glaciation), II early temperate, III late temperate, IV post-temperate. The corresponding, apparently regular, sequences of climate, soil development, vegetation and fauna in a country like Britain are illustrated in fig. 13.55, presented in

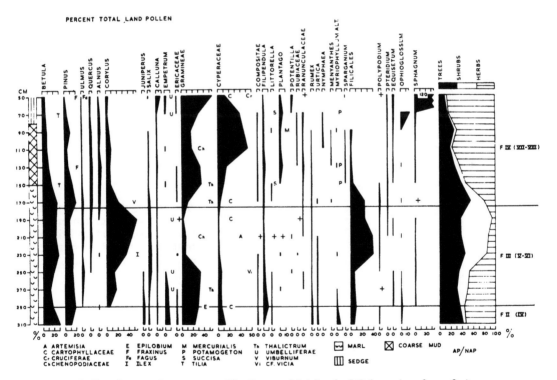

Fig. 13.53   Pollen diagram from peat at The Loons, Mainland of Orkney (59·1°N 3·3°w).

Percentages of the different species of trees and other vegetation represented are percentages of the total pollen sum (*TP + NTP*).

Appropriate local pollen zones are indicated by the horizontal lines dividing the diagram: the equivalent standard pollen zones, starting with IV, the Pre-Boreal or early postglacial, are indicated by the Roman numbers in brackets (cf. fig. 13.13).

(*After* N. T. MOAR (*1969b*); *by kind permission.*)

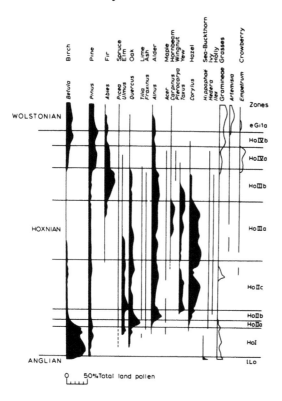

Fig. 13.54   Pollen diagram from a
Hoxnian interglacial lake deposit at Marks
Tey, Essex, England (51·9°N 0·8°E).

Horizontal lines across the diagram
indicate the zones I–IV according to
WEST's scheme (see text) and some
subdivisions.

(*After* C. TURNER (*1970*); *reproduced
here by kind permission of the author and
of the Royal Society.*)

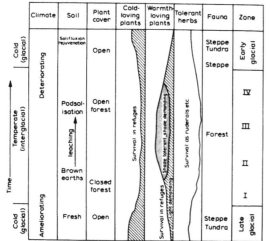

Fig. 13.55   The regular features of the
climate, soil and vegetation sequence
through interglacial periods in Britain, as
schematically drawn by R. G. WEST.

(*Reproduced with the author's kind
permission.*)

the order in which their characteristics are registered in the depths of a deposit. The
postglacial sequence also appears to fit this pattern, with the present situation somewhere in
zone III.

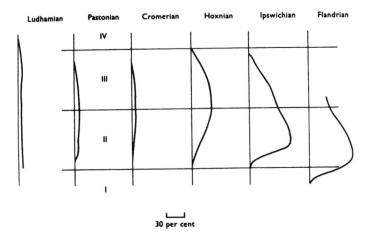

Fig. 13.56   Characteristic curves representing the different history of abundance of the pollen of the oceanic climate indicator species *Corylus* (hazel) in East Anglia in the last five temperate/interglacial times, as well as in the early Pleistocene ('Ludhamian') temperate stage and the postglacial ('Flandrian') to date.
   (*From* WEST (*1970*); *reproduced by kind permission of the author and the Cambridge University Press.*)

Fig. 13.56 picks out interesting differences, indicated by pollen diagrams, between successive interglacials as they have affected East Anglia. The differences seem to be largely in the degree of oceanic influence in the climate.

Absolute pollen frequency diagrams (DAVIS 1963), presenting counts of the total number of pollen grains of each species per cm^3 of the deposit, would probably always be preferred to the usual relative (percentage) frequency diagrams, were it not for the great labour involved in bigger counts and the laboratory difficulties of securing completeness. The advantage is most obvious when one considers a climatic change in which *all* pollens become more (or less) abundant. No change will be registered in the relative frequency diagrams, so that a major change of climate could be overlooked.

The choice of sampling sites must greatly affect the representativeness of the results obtained. Lake beds and peat bogs are favoured because they are places where the pollen deposited is likely to be preserved and not to have been disturbed by Man. Places where great depths of deposit accumulate are likely to give the finest time resolution for samples taken at the closest manageable interval. Dates and sedimentation rates should, of course, be fixed by radiocarbon dating; sedimentation cannot be assumed constant, though this had to be done in the past and may often give a rough first approximation to the age-depth scale. Lakes in which there is a strong stream flow may receive a considerable waterborne contribution, which complicates the interpretation of the area of origin of the pollens. On the other hand, in some cases the regular seasonal variation of the water transport brings an annual layer of largely mineral content to the sediment which makes it possible to date the samples taken by counting the year layers (varves). Too rapid bottom currents can, however, erode the sediment and overturn its time stratification:[1] on these grounds OLDFIELD (1970) considers that ombrogenous bog sites should be preferred to lake sites. NICHOLS (1967b) warns that confusion or disturbance of the pollen record is particularly liable to have occurred in lakes with rich bottom faunas, lakes with a history of floating islands of organic matter, lakes subject to freezing to the bottom and those that have gullies in their bottoms or are subject to earthquakes and, of course, any shallow water-bodies that have been used by cattle.

The sampling technique involves cutting a 'monolith' at an open section or taking a core from the deposit with tools specially designed to achieve the least possible disturbance at the walls of the sample. In the laboratory the samples taken from the monolith or core may be briefly boiled in dilute potassium hydroxide (KOH) to dissolve some of the humus and clarify the sample. Other chemical processes may be used to remove lignin, cellulose and mineral matter. Filtering and centrifuge techniques are also employed before the pollen content of what was initially perhaps 0.5 cm^3 of the sediment is presented on a slide under the microscope and examined under magnifications of 400 to 1000.

The different composition of the 'pollen rain' at different epochs in any area tells us most directly something about the history of the vegetation. Implications about the climate are a deduction from that.[2] And there are many other things that may be learnt, some of which affect what can (or cannot) safely be deduced about the history of the climate. The botanist's aims may also be to learn about the different migration rates of different plants moving into regions which have become more hospitable than before and about the competition which then develops between the genera and species which were first established and the later arrivals, until a climax community is arrived at which is in equilibrium with the climate prevailing and with the community of animals and microfauna for which it provides the habitat.

Once the climax state is reached, the species composition of the vegetation undergoes little change until the climate changes again, until the soil deteriorates through leaching or becomes enriched, or until the balance is upset in some other way, e.g. by coastal or volcanic disasters or by the intervention of Man. The vegetation history revealed by pollen analysis provides evidence of these things too, and has been used as an aid to archaeology in unravelling the record of human settlement, forest clearance and the development of agricultural practices.

Spores are known from as far back as the Silurian geological period, over 400 million years ago, pollen grains also from sediments dating well back into the Palaeozoic. The great

1. Sediment that has been moved by water from where it was originally deposited is known as 'derived' material. Alteration of the stratigraphic sequence by water movement or biological activity is described as 'reworking'.
2. The riddle of how far the sudden decline in the abundance of elm in the mid postglacial about 3000 B.C. may be attributable to a climatic change, and how far to the activities of primitive Man, will be dealt with in Chapter 16, p. 414. A possibly similar elm decline can be traced in the middle of other inter-glacial periods.

possibilities of pollen analysis are due to the tremendous resistance to decay (except through oxidation) of the complex carbohydrate substance which constitutes the exine or outer wall of pollen grains and spores, and also to the great diversity of forms which aids recognition. From the Tertiary and earlier geological periods, however, it is difficult to be sure of that identity of the finds with modern vegetation classes, or plant taxa, whose climatic associations are known.[1] Only in cases of identity with modern species is it reasonable to assume that their climatic requirements can be known with any approach to precision. Nevertheless, broadly tropical and temperate forest classes may perhaps be distinguished from earlier geological times.

Palynology may therefore throw light on the evolution of plant species and on the antiquity of those species now living. It also indicates in broad terms some of the gross climatic changes that accompanied continental drift and pole wandering in the course of geological time. The floras of northwest Europe in the early Tertiary (Oligocene and earlier) are recognized as largely (over 50%) of tropical and subtropical affinity (WEST 1968, ZAGWIJN 1959); by the late Tertiary (Pliocene) temperate forms were heavily predominant. But the Pliocene forests of this part of the world were rich in genera, such as *Sequoia* and *Tsuga* (the hemlock spruce), which after the first cold stages of the Pleistocene disappeared from Europe and are now found native only in eastern Asia and North America – i.e. in sectors of the northern hemisphere where their southward migration before the advancing cold climates was not barred by mountains, deserts and seas (SPARKS and WEST 1972, p. 191). Pollen diagrams tracing the history of plant species in England through the various interglacial times in the Pleistocene, discussed by WEST (1970) and SPARKS and WEST (1972, p. 177), are largely made up of species now native. They show broadly parallel histories in each case but with some interesting differences, e.g. in the abundance of the oceanic species *Corylus* (hazel) and the more continental types *Tilia* (lime), *Picea* (spruce) and *Carpinus* (hornbeam) in different interglacials and the stage-by-stage elimination of the Tertiary flora.

Representation of different pollen taxa in the pollen deposited on a site is not in the same proportions as the abundance of the various plant genera and species in the neighbourhood. This is because of great differences in the pollen productivity of different species, in the size, shape and weight of the pollens and in the manner in which they are spread. Some aquatic and other specialized plants release no pollen at all into the atmosphere; others (known as zoogamous species) rely on birds, bats and insects to carry their pollen, and some of these are so effectively pollinated in this way that their pollen production does not need to be large and little of it enters the atmosphere or is found in deposits. An enormous range of plants, however, including most trees, rely on the wind to carry their pollen – they are called anemogamous – and produce vast quantities of pollen. One shoot of hemp produces over 500 million pollen grains, one 10-year-old branch of birch,

---

1. The word 'taxon' means any recognized group,
   genus or species.

spruce or oak about 100 million, and pine about 350 million. Certain of the zoogamous species, including the lime or linden tree (*Tilia*) and ling (*Calluna*), produce pollen in comparable amounts and are well represented in deposits.

Many studies have been made of the dispersal of pollen. Rates of fall in calm air range from about 40 cm/sec for heavy pollens to 2 cm/sec for the smallest, most buoyant types. Even the heaviest pollen grains are therefore liable to be carried up to great heights in the atmosphere by the most vigorous up-currents in cumulus clouds. On the other hand, 1 mm of rainfall is enough to wash out practically all pollen grains from the air it falls through. Given no rain, the limiting distance reached by just 1 % of the grains should range from 1 to 500 km for different species according to theoretical calculations. Pollen brought down by rain may, however, become airborne again later, and some travels very great distances before final deposition. FÆGRI and IVERSEN (1964) quote the following figures for pollen caught on Ocean Weather Ship M at 66°N 2°E in the northernmost Atlantic, 450 km from the coast of Norway, in the course of one summer season:

Grasses (*Gramineae*)	5·5 per cm²
Pine (*Pinus*)	5·5 ” ”
Birch (*Betula*)	4·7 ” ”
Oak (*Quercus*)	0·8 ” ”
Spruce (*Picea*)	0

TYLDESLEY (1973) has studied in some detail the meteorological situations which bring tree pollen to the treeless Shetland Islands (60°N 1°W), 250 km from Scotland and 380 km from Norway, pine pollen on isolated days in concentrations up to 30 grains/m³, birch up to 20 grains/m³, in the surface air on tracks from the British Isles, Scandinavia or the continent across the North Sea: the deposition rates in Shetland appeared to range up to 3 grains/cm² on the days concerned.

MOAR (1969a) has reported the deposition of Australian pollens (*Acacia, Eucalyptus, Pinus* and *Casuarina*) and a red dust of Australian mineral type in the snow at 2100 m above sea level on a glacier near Mount Cook in New Zealand, a distance of over 1600 km from source. The spores of wheat rust fungus have also been borne across the Tasman Sea by the wind. Discussion of the initial colonization of the remotest islands in the Southern Ocean by vegetation, probably of South American origin, has sometimes favoured the carriage of the seed by the waters of the ocean itself.

A fine review study of the probable sources and representativeness of the pollen deposited in lakes and bogs in forested areas has been made by TAUBER (1965). Starting from the work of SUTTON (1932, 1947, 1953), GREGORY (1945) and PASQUILL (1961) on atmospheric turbulence and the dispersion of gases, spores and other minute particles, TAUBER considers the following three components which make up the pollen supply:

(1) The pollens in the air in the trunk space of the forest; this trunk space fraction is called f_t.

(2) The pollens carried above the canopy of trees (f_c).

(3) The pollens higher up in the atmosphere which are brought down by rain, the 'rainout' component (f_r).

By definition, $f_t + f_c + f_r = 1$.

These three fractions roughly correspond to those often more loosely described as the local, regional and long-distance components respectively of the 'pollen rain'.

TAUBER criticizes the name 'pollen rain', which is much used to describe the total pollen fall over whatever area is sampled, because it may lead to a false picture of the transport of pollen in the atmosphere. The net motion is nearly horizontal, because the prevailing horizontal winds are many times greater than the fall speeds and pollens are as readily carried up to great heights by convection as are cloud droplets and the smaller sizes of drizzle. Convective up-currents are, however, usually less active (due to evaporative cooling of the air) over a forest than over most types of open, dry ground, especially bare or grassy hillsides facing the sun. Species giving out their pollen in the warmest months (e.g. *Pinus* and *Tilia*) must, however, tend to have a bigger proportion of their pollen than other trees carried aloft by convection and so contributing to the long-distance transport. Wind velocity characteristically drops at the windward edge of a wood to 40–70% of the free wind velocity; and inside the wood, in the trunk space, it usually falls to 10–20% of the free wind velocity unless there is little undergrowth. Towards the lee edge of the wood the wind velocity characteristically rises again to about double its lowest value in the wood and increases further immediately beyond the trees. Wind velocities equal to those above the canopy are, however, usually not attained to leeward in distances less than 20–30 times the height of the trees. Before the trees are in full leaf wind velocities inside a forest may be up to double the values indicated above. Eddies and down-mixing of the air from above the canopy are important in the lee of a forest at distances up to 20–30 times the height of the trees, though not in the case of very open woodland with much flow through the trunk space. In general, natural forest, perhaps especially where the trees are big (as, presumably, in the deciduous forests on the plains of the northern hemisphere continents in the warmest postglacial times), has more openings produced by fallen trees, and perhaps by fires, and may usually release more pollen than plantations of cultivated timber. Alder, aspen, elm, hazel and many willows release their pollen before the trees are in leaf, and of the big deciduous forest trees only lime (*Tilia*) flowers when the trees are in full leaf and winds inside the forest are at their minimum.

At heights below 2 m in a forest the wind motion rapidly decreases to zero, so pollens emitted by the undergrowth will not be transported far – characteristically less than 100 m. Within the forest area the numbers of pollen grains in the air during emission are greatest at, and just above and below, the level of the tree crowns, about double the numbers at the 4 m level in the wood or 10 m above the tree tops. Estimates of the canopy pollen just above the tree crowns are that of the light pollens 50% will have originated within 250 m and 75%

within 5 km, and of the heavy pollens (e.g. beech) 75 % will have originated within 100 m and 90 % within 1 km.

There are, thus, many differences, and the interpretation of a pollen diagram must even depend on the frequency of rainfall over the sampling site. Undergrowth in a wood and bushes growing around the edge of a lake filter out the pollen which has come more than 10 to 100 m in the lowest air and add a great contribution of their own to the pollen fall in the immediate vicinity. TAUBER's calculations suggest as a general guide the following approximate values for the three fractions of the pollen deposition:

(1) In the middle of small lakes or bogs of 100–200 m diameter:

$$f_t = 0\cdot8$$
$$f_c = 0\cdot1$$
$$f_r = 0\cdot1$$

(2) In the middle of big lakes or bogs of several km diameter:

$$f_t = 0\cdot1$$
$$f_c = 0\cdot7$$
$$f_r = 0\cdot2$$

These figures appear to be supported by careful examination of many botanists' pollen diagrams and by experiments with pollen traps in actual lakes. The main difference between the two cases is in the size of the area represented by the pollen catch: with small lakes the effective source area (from which 80 % of the pollen grains collected come) should be thought of as extending up to 300–1000 m from the site (within the forest itself, however, ANDERSEN (1970, 1973) finds that a major portion of the tree pollen is deposited within 30 m). For the pollen that falls on big lakes the effective source area may extend to 30–100 km around. Because of the presence of a small long-distance element, the evidence of macrofossils is often needed to establish for sure whether a particular plant grew in the neighbourhood.

Various suggestions have been made about how the percentage of a given pollen in a deposit might be corrected for the distinctive species differences of pollen production and transport to give a better representation of the percentage of the surrounding area occupied by that species. DAVIS (1963) called the ratio (pollen percentage):(area percentage) the R value. Unfortunately, these R values are far from constant; and, if there is some long-distance transport of the species, R becomes very large as the percentage of the local area occupied by the species approaches zero. Better results are obtained by taking the ratio, or relative value, of R for the species under consideration to the R value for a standard reference species, such as beech (*Fagus sylvatica*) which flowers well on a great variety of soils and habitats in middle latitudes. ANDERSEN (1973) uses an improved version of the same solution, defining the pollen deposition from a given tree species at any sample point as

$$p = aP + p_o$$

where a is the area covered by tree crowns of the given species in the locality, P is the pollen productivity factor for the species (high values for abundant pollen producers) and p_o stands for the quantity of pollen of the same species arriving from outside the locality. P values for each species are obtained in practice from measurements of p and a at various points within a forest, a technique which yields estimates of p_o as well. The figures from various forests become comparable if the P value at any point for the given species is divided by the P value at that point for a constant reference species, such as *Fagus sylvatica*. Nearly constant values of this ratio P_{rel}, the 'relative pollen productivity', were obtained for each of the main tree species in various stands of deciduous forest in Denmark. This led to correction factors to be applied to the percentages of different pollens in a deposit as follows:

Quercus, Betula, Alnus, Pinus	$\frac{1}{4}$
(oak, birch, alder, pine)	
Carpinus	$\frac{1}{3}$
(hornbeam)	
Ulmus, Picea	$\frac{1}{2}$
(elm, spruce)	
Fagus, Abies	1
(beech, fir)	
Tilia, Fraxinus	2
(lime, ash)	

These values are quite similar to the earliest 'free-hand' estimates suggested by IVERSEN (1947).

In the case of shrubs which grow under bigger trees as well as in the open, wide differences in their pollen productivity must occur. ANDERSEN's suggestions for correction factors to be applied to their pollen deposits are in line with IVERSEN's for

Corylus	Growing under bigger trees	1
(hazel)	Growing in the open	$\frac{1}{4}$

Use of these correction factors changes the picture of the constitution of the forests through postglacial time. It is found that not oak but lime (*Tilia*) was the most important tree in Denmark in Atlantic and Sub-Boreal times. Catastrophic destruction of the *Tilia* forest thereafter may be attributable to burning and grazing by Neolithic farmers, which seems to have allowed some increase in the frequency of birch, hazel and oak. The subsequent spread of beech to its present dominance in the natural woods of Denmark seems to have occurred at a time of diminished human activity. In England the effect of these corrections seems rather to be to make elm of equal importance with oak in the forests of the Atlantic period, limes, alders and birches being the next most important trees.

Similar quantitative corrections have not yet been developed for the representation of pollens from plants other than trees in the deposits. But the improved estimates of the total area covered by trees give a better idea also of the total extent of open areas.

The most far-reaching – indeed ambitious – method of derivation of past climatic conditions, yielding estimates of detail, a quantitative palaeoclimatology, from pollen analysis has been developed by WEBB and BRYSON (1972). A scheme of canonical correlation analysis, due to HOTELLING (1936) and developed by GLAHN (1968) and MORRISON (1968), is used to relate a matrix of pollen data from just eight key species that were present throughout, from the periods to be investigated to the present day, to various matrices of climatic data for the same place. The choice of pollen types (species) is limited by the requirement that the standard deviation, and range, of each pollen type in the modern record must not be exceeded in the past record. (It was found that improbable results were obtained if this restriction was waived. Nevertheless, there may be some reservations about estimates derived without including, say elm pollen, for periods in which elm seems to have thrived and been much more abundant than at the present day.) Transfer functions, or canonical regression coefficients, are derived from the correlations of those pairs of variates that are found to be significantly correlated. The equations so derived yield estimates of the climatic variables directly from the pollen data. The climatic variables calculated in this way by WEBB and BRYSON for points in Minnesota and Wisconsin at roughly 2000-year intervals since the last glacial maximum include estimates of July temperature, precipitation and evaporation during the growing season, and the average yearly duration of the dominance of each of the principal airmasses, Arctic, Pacific and tropical (Gulf of Mexico) air. In general, the results tend to confirm the climatic succession already pictured from classical pollen analysis and must be held to give some deeper insight into the nature of the climatic regimes involved, particularly as regards the atmospheric circulation. Rather similar quantitative methods have been successfully developed by IMBRIE and KIPP (1971) for ocean palaeotemperatures (see pp. 241–2).

The longest pollen-analytical records at present available from which climatic history may be derived are the 130 000-year core from a lake in Macedonia (van der HAMMEN et al. 1971) and one (Grande Pile, 47·6°N 6·5°E) in eastern France near the foot of the Vosges, of probably still greater age range, which seems to extend down to cold climates before the Eemian interglacial (WOILLARD 1975).

Necessarily such an automated calculation procedure ignores complications and variations in the vegetation's responses to climatic shifts at particular times. The argumentation can indeed only be valid after a given climatic regime has prevailed long enough for the vegetation development to attain equilibrium with it, i.e. for a climax vegetation to develop. The rapidity of the major climatic changes of postglacial times in the central parts of North America has been studied by BRYSON et al. (1970) in terms of the standard deviation of radiocarbon dates related to each transition (the argument is that radiocarbon dating tests are equally likely to have used vegetation material from immediately below or above the horizon that marks the change). The standard deviations found ranged from 230 to about 500 years. The major climatic shifts may be supposed, therefore, to have been completed within about this span of time. In the same paper there are reported a number of cases where the rates of the vegetation's response to the climatic change have been estimated from the sedimentation rate and the height interval that spans the change in the pollen distribution in a peat, or subsoil, column. Many of the vegetation transitions indicated by changes in the pollen spectra, when closely timed in this way, appear to follow an exponential curve. This means that full attainment of the new climax vegetation takes a very long, and strictly undefinable, time; but a 'half-life' for the transition can be defined as the time when the percentage of a given pollen type has changed half-way from its former value to the level ultimately reached. Estimates of half-life ranging from 77 to 200 years are reported by BRYSON *et al.* (1970) for the vegetation changes following the major changes of climatic regime in postglacial times, including the arrival of the earliest postglacial regime. Immigration of some tree species into Britain and northern Europe from the distant refugia occupied in glacial times took some thousands of years, partly because the ground was first occupied by species whose pollen is abundant and carried farthest. The rapidity of the changeover in the pollen record in the most abrupt cases may well imply that natural forest fires removed the remnants of the former forest after some event which had killed many of the trees.[1]

Mosses

Moss spores have contributed less than macroscopic remains to our knowledge of past assemblages in Britain (DICKSON 1967). Though there is an extensive moss flora, records of the spores other than the *Sphagnum* species are few.

About 115 species of moss remains from the last glaciation have been found in over 50 deposits in the British Isles. Most are from the last stages of the glaciation, but there is evidence that a fair proportion (not less than 30%) survived the glaciation in refuges in the south and, perhaps, the west of the British Isles. Indications of the climatic shift are gained by comparing the present and past positions of the limits of the various species. Thus, remains of *Polytrichum norvegicum* from the time of the last glaciation are found in southern England, whereas its present occurrence in Britain is restricted to heights above 900 m in the Highlands of Scotland. *Polytrichum alpinum*, which is more widespread in the mountains of the north and west of the British Isles, occurs in late Weichselian (Würm) glacial deposits from the lowlands as far south as 53°N. Of the more southern mosses in the present British flora, *Antitrichia curtipendula* has been found in many Late-Glacial deposits in western parts of the British Isles from Cornwall to southwest Scotland, the Isle of Man and near the coast of Ireland.

The abundance of *Sphagnum* found in the Chelford interstadial deposit on the Cheshire plain from early Weichselian (Würm) times, dated about 57 000 years ago, is taken as a sign that the climate was more temperate than in the millennia before and after. Six other British

---

1. Forest fires are a recurrent feature of the natural vegetation development. A study (SWAIN 1973) based on charcoal layers in a lake sediment in Minnesota indicated an average frequency over the last thousand years of fires in the surrounding forest every 60–70 years. Over 90% of the coniferous forests in one area investigated in northeastern Minnesota had been burnt at least once in the last 300 years. Generally reduced quantities of charcoal in the individual fire layers since about A.D. 1550 were tentatively attributed to the cooler, damper climate of the Little Ice Age period.

mosses which DICKSON (1967) believes to be the likeliest species to have survived the entire Weichselian in Britain are missing from the Chelford deposit, which shows a generally acid-soil (calcifuge) flora, presumably indicating leached conditions and a wet climate in the area in the preceding millennia.

Bog growth

The stratigraphy of peat bogs may provide a valuable record of the sequence of wetter and drier climates, particularly when the succession of layers of somewhat different character is dated by sufficient radiocarbon tests to determine the changes in the rate of growth of the peat.

Bogs are essentially areas of wet, spongy ground, consisting largely of the decaying remains of earlier generations of the moisture-loving *Sphagnum* and/or other mosses, heathers, berry plants and myrtles that formerly grew on its surface together with a relatively small proportion of originally wind- and waterborne mineral matter. Their growth is favoured whenever and wherever a constantly high water table is maintained by poorly drained hollows and pockets in impervious rock beneath, by stagnant shallow lakes and slow-moving streams, or by a continually wet, cloudy climate. They are classified as

(a) *Topogenous mires*, where the topography of water-holding basins is the main factor.

(b) *Soligenous mires* due to concentration of the surface drainage.

(c) *Ombrogenous mires*, where the bog plants are sustained by the high precipitation/evaporation ratio and ground water supply is not the main factor. *Sphagnum* is usually prominent in these.

Some topogenous mires in low-lying areas are rich in mineral nutrients (i.e. they are 'eutrophic') received with the water from higher up the valleys: these mires are called fens. Most other mires have become leached and are poor in nutrients (i.e. they are 'oligotrophic'); the peat becomes soured by the humic acids developing in them, which are not neutralized by base minerals – these mires are the ones properly called bogs. A characteristic calcifuge flora, however, thrives on them, particularly *Sphagna* but also other mosses and liverworts, heathers, etc.

In very oceanic climates with high rainfall frequency, the soil water level may be generally high enough – not only where drainage water accumulates – to maintain the growth of peat-forming plants even on sloping ground, except where the slopes are steep. GODWIN (1956, p. 30) reports that in the west of Ireland, beyond approximately the annual rainfall isohyet of 1000 mm, this is so even on slopes of as much as 15°. The bog which then covers much of the terrain is called *blanket bog*.[1]

With rather less oceanic climate, as in the lowland plains of central Ireland east of the 1000 mm isohyet (GODWIN 1956, pp. 30–1), the climax type of mire development is the *raised bog* (German *Hochmoor*, Swedish *Högmosse*). On nearly flat land, often where

---

1. In Canada also, particularly in the lowlands around Hudson's Bay, there is a widespread shallow ombro-genous peat resembling the blanket bogs of northern and northwestern Europe and known as *muskeg*.

topogenous mires have first formed, the active growth of the bog plants in the wettest places gradually raises the level of the bog surface there, while at the sides *Sphagnum* growth is hindered by more base-rich water in the soil. Large raised bogs may be several kilometres across and present a low domed surface in which the centre may be from one to several metres higher than the edges.

After runs of particularly wet and/or cold years, with increased precipitation/evaporation ratio, as when a shift towards colder climate takes place, raised bogs and bogs where there is a little slope down-valley are liable to become so supercharged with water that the strength of the plants' root fibres gives way and the bog 'bursts'. *Bog bursts* caused disasters near Charleville, County Cork, in south Ireland, in June 1697, and in the Solway Moss (55°N 3°W), near the Scottish border, in November 1771 and December 1772 when over 200 ha of farmland were covered with black, peaty mud up to 4 m deep which engulfed many farm dwellings; slight movement threatened another burst close to the same place about 1850, but seemed to be prevented by the areas of drainage and tree planting undertaken after the previous incident. It is reported (*Gentleman's Magazine*, June 1773, pp. 265–6) that before the 'irruption' of the Solway Moss, the place was impassable on foot even in summer and that poles could be driven 4 to 6 m or more into the watery mass. Another case was reported at Haworth bog on the Pennine moors near Bradford, Yorkshire, on 2 September 1824. It may be significant that these, the only reported cases known to the writer of the peat bogs menacing the local human communities in the British Isles, came after great rains towards the end of, or just after, notoriously cold decades, the 1690s, 1760s, 1810s and 1840s.[1]

Radiocarbon dating may be vitiated in the wettest parts of peat bogs by water erosion and transference of humus. OVERBECK *et al.* (1957) doubted whether it was possible to determine the age of bog material to within 100 to 200 years. That is doubtless still true if real ages are meant, but since that time many more radiocarbon age tests have been done, and where these are sufficiently numerous in a sample column of peat (with the margins of experimental error known) it may be possible to establish ages closely enough to reveal variations in the rate of peat growth.

Most existing peat bogs seem to have started to grow at times ranging from the latter end of the warmest postglacial times around 3000 B.C. to the onset of the substantially cooler climates of the 'Sub-Atlantic' period, in the last millennium B.C. In some areas destruction of surviving woodland on uplands in western Britain by Neolithic Man (specifically in the areas of upland Wales where megalithic tombs are numerous) seems to have hastened the onset of blanket bog growth (MOORE 1973). The dates of initiation of peat growth are determined by radiocarbon dates in the lowest peat, where it rests on layers with more mineral content. In

---

1. The sample is far too small to prove a case, but the association suggested is physically reasonable and seems to tie in to some extent with the incidence of avalanches and landslides in Norway (fig. 13.38). Although no statistics are available, and despite all modern improvements in artificial drainage, there seems to have been a possibly similar increase in the incidence of landslides in England in the late 1960s following a run of colder summers.

some cases the peat rests on glacial materials and peat growth can be seen to have taken place in the earliest postglacial times. In some other cases the start of peat growth occurred in a warm, moist climate in the Atlantic period itself, i.e. in the warmest postglacial times.

In Canada (NICHOLS 1969) radiocarbon dates for the basal peat are generally between 4500 and 1500 years ago, i.e. in the latter part of the warmest postglacial times or since, though mostly grouped around the dates associated with evidence of episodes of climatic cooling widespread in the northern hemisphere 3500 and 2500 years ago. There is another, smaller group of onset dates 700–300 years ago. But in the Arctic areas of northern Keewatin and Quebec onset of peat growth seems rather to have occurred in the warmer periods around A.D. 400 and 1000, when the wind circulation may have brought warmer, moister air to an area which at other times was too cold for peat growth.

Rates of peat growth vary widely from site to site and region to region. Average rates over the past 5000 years up to 8 cm/century have been quoted (MOORE 1972) for the raised bogs in west Wales at Borth (52·5°N 4·0°W, 3 m asl.) and Tregaron (52·2°N 3·9°W, 160 m asl.) and between 1·2 and 3·5 cm/century for the blanket bogs on the uplands in the same area. D. WALKER has given 5 cm/century as typical for the peat remaining from around 2000–0 B.C. on the Pennine uplands in northern England. MITCHELL (1956) estimated, on the few radiocarbon dates then available, that raised bogs in Ireland grew at an average rate of 8·5 cm/century between about 2500 and 1500 B.C. and 10 cm/century between A.D. 500 and 1700; in the 2000 years between 1500 B.C. and A.D. 500 about 145 cm were formed, but there may have been considerable variations within that time in its rate of growth. TURNER (1965a, 1965b) has traced a history of variations in the rate of peat growth at Tregaron with the aid of many radiocarbon dates. Applying the bristlecone pine calibration (p. 66, fig. 13.11) to these dates, it appears that there was a period between about 1300 and 900 B.C. when the surface of the bog was generally dry, not growing, and possibly being eroded; this was followed by an abrupt change to extremely wet conditions, when 90 cm of peat was added within 400 years, a growth rate which may at times have reached 25 cm/century. The rate then slowed down and seems to have averaged as little as 1 cm/century between late Roman times (about A.D. 400) and A.D. 1200. Radiocarbon datings from some other bogs in western parts of England, Wales and Scotland suggest that peat growth was at a standstill during at least part of the Roman occupation of Britain (A.D. 43 to 410) and again between A.D. 900 and 1200 but that there was quite rapid growth, perhaps 10 cm/century, at some time between about A.D. 450 and 650. There was another renewal of rapid growth of the peat about A.D. 1200 or 1250, since when the top 50 cm of the peat at Tregaron has formed.

As a result of these variations with time in the rate of growth of peat, most bogs show a stratigraphy of *recurrence surfaces*, where regrowth overlies peat layers which had been partially decomposed to form humus ('humified peat'), becoming darker in colour and of somewhat denser structure in the process. Over wide areas of northwestern and central Europe from Ireland to Germany, Denmark and Sweden, as well as east into Russia and Siberia, cuttings in the peat bogs reveal one very prominent boundary horizon, now generally known among students of peat stratigraphy as the *Grenzhorizont* (after WEBER 1900), which marks off the older, deeper, dark turf (or peat), the *Schwarztorf* (which makes the best fuel), from the younger, lighter, overlying layers, the *Weisstorf*. This was ascribed already by WEBER to a climatic change, dated tentatively in the middle of the first millennium B.C. GODWIN (1956) suggests that some of the difference of colour may be due to a difference of the bog plant community, with much more ling (*Calluna*) and cotton grass (*Eriophorum*) growing in the drier, more continental climatic conditions when the older turf was formed. Certainly, the upper *Sphagnum* peat is of the same kind as that which has continued to form to the present day in the cool, oceanic climate of Ireland. Doubts have since arisen as to whether the most prominent *Grenzhorizont* is everywhere of the same age. GRANLUND (1932) established the existence of a whole series of rather similar boundary layers (Swedish *rekurrens-ytor*) in the raised bogs of southern Sweden, which he labelled and approximately dated archaeologically and by other means as RY I around A.D. 1200, RY II A.D. 400, RY III 600 B.C., RY IV 1200 B.C. and RY V 2300 B.C. These dates have stood up well to comparison with subsequent radiocarbon tests; though A.D. 1250 to 1300 may be a more widely representative timing for RY I, the latest one of the series.

GODWIN (1954) believes it possible to rule out GRANLUND's idea that the recurrent regrowths of the peat were due to a natural oscillation within the bogs depending on a limiting steepness of the surface leading to natural drainage: in a large raised bog over 2 km in diameter an increased height in the centre of a metre or so 'must be too slight to be held responsible for the entire difference between continued growth and total cessation. . . . By contrast . . . how powerful must be the effects upon the central bog surface of those fluctuations of rainfall and evaporation which we know to occur year by year, decade by decade, and which we may by inference suppose also to have operated over longer cycles.' Moreover, the main recurrence surface of around 500 B.C. is equally pronounced in bogs of all sizes. Typical recurrence surfaces suggest a catastrophic increase of wetness broadly about that time. The formerly dry, and sometimes cracked, bog surfaces, over which trees – particularly birches and pines – had sometimes begun to grow, became flooded and large permanent pools appeared with aquatic *Sphagna* among the decaying hummocks which still bore different species. In the Somerset levels, in the west of England, the population built wooden trackways to keep open the ancient ways across the flat land and later introduced boats. The remains of these wooden objects, in Somerset, commonly about 700–500 B.C. and 350–100 B.C., have made it possible to date this climatic change (GODWIN and WILLIS 1959) and others in other areas by radiocarbon tests.[1]

The evidence of drying out of north and west European bog surfaces for 200 to 400 years in late Sub-Boreal time extends to Ireland and includes stumps of birch and pine in the peat (JESSEN 1949, p. 259). But there, as to some extent in many parts of Europe, additional recurrence surfaces, some of them quite localized, confuse the picture. A good perspective on

1. Another group of wooden trackways in the Somerset
   Levels around Glastonbury is dated 2900–2800 B.C.

the subject has been given by MITCHELL (1956), whose key diagram is reproduced here as fig. 13.57. Differences of bedrock topography and drainage underneath the peat vary the sensitivity of different bogs to changes of the precipitation/evaporation ratio and so introduce a spread of dates into the responses of different areas to any change of climate. In Ireland, particularly, there are additional recurrence surfaces, perhaps because of that country's immediate exposure to moisture-bearing winds from the Atlantic Ocean. This is also the likeliest reason why the timing of the main groups of recurrence surfaces there – prominent about 1500 and 800 B.C. and A.D. 500, according to MITCHELL (1956) – differs from the rest of Europe.

Another type of regional peculiarity affects the sequence of recurrence surfaces near the Baltic, e.g. in Finland and Russia, where the water table level varied with the changes of sea level, affected by the balance between the postglacial eustatic rise of world sea level and the strong isostatic uplift of the land surface which had underlain the European ice sheet.

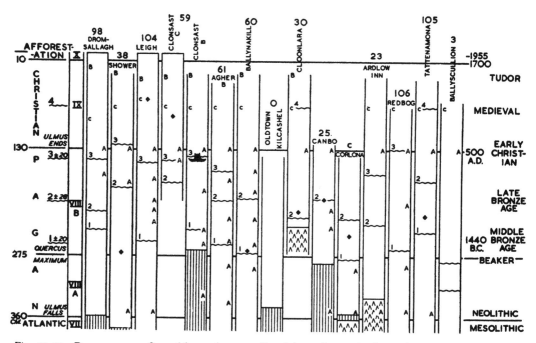

Fig. 13.57   Recurrence surfaces (shown by wavy lines) in various raised peat bogs (named at the tops of the columns) in different parts of Ireland.

These renewals of growth are seen to differ in their timing from bog to bog, but are commonest around 1500 B.C., 800 B.C. and A.D. 500.

The vertical shading indicates fen peat, the peaked symbols wood peat; where the columns are blank, the peat is ombrogenous.

A layer with tree stumps which grew around A.D. 350–400 is indicated in one bog in central Ireland. The other symbols indicate levels with evidence of human activity.

(*Diagram from* MITCHELL (*1956*) *reproduced here with the author's kind permission.*)

Nevertheless KHOTINSKY (1971) reports the existence of some probably consistent recurrence surfaces in peat bogs throughout the forest zone of Russia and in western Siberia, as well as similar appearances in the peat bogs in the Far East and in Alaska.

Table 13.12 summarizes the data on fluctuations of climate over several millennia from the peat bogs in Europe and sets beside them some archaeological information relating to climate in Greece and the Near East, where the sequence is clearly not in phase with northwestern Europe though the changes in the two regions tend to be contemporaneous. Correction of the radiocarbon dates by the bristlecone pine calibration (fig. 13.11) has brought many of the dates into straightforward agreement. There seems to have been a sequence of well-marked wet phases in Europe north of the Alps back almost to 3000 B.C. and possibly earlier, those which set in about 2200, 1200 and 500 B.C., as well as in A.D. 1250 to 1300, bringing the sharpest changes and being the most widely registered. There seem to have been several centuries of drier conditions in northern and western Europe at least between about 1800 and 1400 B.C., possibly from 2000 B.C. until about 1200 B.C., and again, at a more modest level, from about 80 B.C. till A.D. 530. There may have been preferred (i.e. commonest) intervals of close to 200 or 400 years between successive wet phases and between the corresponding drought maxima in Egypt.

Dendroclimatology

Interest in what may be learnt from tree rings goes back at least to LEONARDO DA VINCI, but the systematic application of their study in the present century to decipher a record of past climate was begun by A. E. DOUGLASS who started about 1901 as an astronomer searching for traces of associations between sunspots, tree growth and weather. It soon turned out that their main interest was as a potential record of climate and its effects upon the prehistoric peoples of the southwestern United States. DOUGLASS described his method in an article in the *Monthly Weather Review* in 1909. Many years later, in 1937, the Laboratory of Tree Ring Research at the University of Arizona, which is still the leading centre for this study, was founded under his direction. The work was early taken up by HUNTINGTON (1913), who used it to explore a possible parallel between the record of droughts and rainfall variations over the last 3000 years in California and in central and southwest Asia in about the same latitude, and later by ANTEVS (1925). Both these authors attempted refinements and corrections by working with large numbers of trees – a development which has now been carried a great deal further with modern computerized techniques (see, for example, FRITTS 1971, 1976, FRITTS *et al.* 1971).

The growth rings of trees, which are predominantly year rings (see pp. 52–5 and Plate I), show the tree's response to variations of its environment, particularly those due to weather but also the effects of exposure and shelter provided by surrounding trees, the attacks of insects and disease organisms, etc., and any changes in the soil whether due to

Table 13.12 Climatic fluctuations registered in the stratigraphy of peat bogs and some parallel data

Note: Dates derived from radiocarbon measurements are rounded approximations, after correction by comparison with Egyptian and Scandinavian varve chronologies, using the calibration curves published by RALPH and MICHAEL (1967) and SUESS (1970). Those radiocarbon dates the most probable value of which required correction by more than a few decades are marked.*

Ireland (data from JESSEN 1949, MITCHELL 1956, SMITH et al. 1971)	Britain (data from GODWIN 1956, GODWIN and WILLIS 1959)	Scandinavia (data from GRANLUND 1932, TAUBER 1965)	Germany and Central Europe (data from FRENZEL 1966, OVERBECK et al. 1957)	Greece (after CARPENTER 1966)	Nile (data on annual floods in Egypt from BELL 1970, 1971)
			3400–3000 B.C. cold phase		
c. 2800* B.C. wetter, peat growth setting in		2800 B.C. wet phase beginning			
		c. 2500 B.C. probably dry			
c. 2400 B.C. becoming wetter, pines decline, bog growth increasing					
c. 2200 B.C. further spread of bog growth		2200 B.C. wet phase beginning			
					2180–2130 B.C. low floods
c. 2000* B.C. further spread of bog growth					
					2000–1990 B.C. low floods 1786 B.C. ff. low floods
c. 1500 B.C. renewal of bog growth			1400 B.C. ff. adequate moisture		

Table 13.12—continued

Ireland (*data from* JESSEN 1949, MITCHELL 1956, SMITH *et al.* 1971)	Britain (*data from* GODWIN 1956, GODWIN and WILLIS 1959)	Scandinavia (*data from* GRANLUND 1932, TAUBER 1965)	Germany and Central Europe (*data from* FRENZEL 1966, OVERBECK *et al.* 1957)	Greece (*after* CARPENTER 1966)	Nile (*data on annual floods in Egypt from* BELL 1970, 1971)
	c. 1200 B.C. onset of wetter conditions	*c.* 1200* B.C. wet phase beginning	1250–1200* B.C. wet phase	1230–1100 B.C. cultural decline and de-population, possibly in-dicating dro-ught worsening after 1100 B.C.	*c.* 1200 B.C. low floods phase sets in again
c. 800 B.C. re-newal of bog growth			*c.* 900* B.C. wet phase beginning		
	c. 700–500 B.C. conditions becoming wet-ter	*c.* 500 B.C. wet phase beginning	*c.* 600 B.C. wet phase beginning		
			c. 400 B.C. wet phase beginning		
	c. 350–100 B.C. wet conditions		*c.* 150–100 B.C. wet phase begin-ning		
B.C. —— A.D.	60 B.C.–A.D. 50 becoming drier				
		A.D. 400–500 wet phase begin-ning	several in-dications of fur-ther recurrence surfaces (re-growth of bogs) in N. Germany A.D. 565–95 and A.D. 690–770		
c. A.D. 500 re-newal of bog growth					
		A.D. 1200–1300 wet phase begin-ning			

climate or accidents. If the aim is to single out a climatic record, there is a problem of interpretation.

Interpretation of the differences of ring width from one year to another in terms of climate is easiest in the following cases:

(1) Trees growing near the poleward limit of forest, where the variations are mainly due to differences of summer temperature and duration of conditions warm enough for growth.

(2) Trees growing on high ground near the upper tree limit, where the variations are also predominantly due to differences of summer temperature and duration of the season for growth.

(3) Trees growing near the arid margin (i.e. near the lower tree limit on the mountains in Arizona and southwest Asia), where the variations are mainly caused by lack of moisture in some years.

In the places such as the lowlands of England and Germany, which are in the middle of the present habitat for the deciduous broad-leafed trees of the mixed oak forest, diagnosis of the weather conditions which caused a specially narrow (or broad) ring is much more difficult. In some years a narrow ring may be due to drought, in another year to lack of sunshine and insufficient warmth. Even in these countries, however, some sites may be found where there is always enough moisture in the ground or, on the other hand, where the growing season is always warm enough but the soil moisture varies greatly.

HUNTINGTON (1913) has described how as early as 1911 and 1912 he measured the rings of 451 trees, all *Sequoias*, in the forests of the Sequoia National Park on the lower slopes of the Sierra Nevada in California (36–37°N near 119°W). Many of the trees were over 2000 years old and some were over 3000. In all 105 000 rings were measured. In a bad year exposed trees may fail to put on any wood on one side, so measurements and counting have to be done along more than one radius. A graduated correction also has to be made for the broader rings put on by a young tree (which has a narrow stem and small circumference); these broad rings are found near the centre of the trunk of the full-grown tree. HUNTINGTON found that the average *Sequoia washingtoniana* acquired a radius of about 2·5 cm in its first ten years; at an age of 200 years the average ring width was about 90% of that in the young tree, at 500 years about 60% and at 1700 years about 30%. Adjustment of the measured ring widths in each tree by an appropriate factor for its age, derived from these considerations, produced a ring width index for each year which was independent of the age of the tree and yielded comparable values that could reasonably be averaged among all the trees in the region investigated. From this work HUNTINGTON obtained the curve which is here reproduced in fig. 13.58, indicating the probable history of rainfall variations in the Californian Sierra Nevada since 1300 B.C., though he pointed out that the appearance of greater variability before 200 B.C. may be due to the small number of the very old trees available in his sample (his curve for the supposedly parallel variations of rainfall in Asia refers to various regions between Palestine and central Asia and, having been put together from miscellaneous data on famines and lake levels, is open to the criticism of subjectivity; nevertheless, it bears a good deal of resemblance to the results of later work on the changing level of the Caspian Sea and to the lowest curve in fig. 13.37 in this book).

We may list the following principles and methods of tree ring work aimed at dendroclimatology, largely from FRITTS (1971):

(1) Choose trees and sites in which ring width is limited, directly or indirectly, by the stress of weather. The aim here is to find maximum ring-width variability, to maximize also the variation that is common to many trees, and at the same time to minimize the variability arising from non-climatic causes.

(2) Each ring-width series must be adjusted for the age of the tree. This adjustment is by considering the individual ring widths in relation to an exponential curve of mean growth plotted against age. The expected ring width, y, is approximated by the curve of an expression $y = a\,e^{-bx}$, where x is the number of years after the tree's period of maximum growth in its youth and a and b are constants. The measured width of the individual ring is then divided by the value (y) corresponding to the tree's age read off the smooth curve, giving a ratio which is called the ring-width index (the process has been computerized and is described by FRITTS et al. 1969). Ring-width index values are mostly in the range 0 to 2.

(3) The sensitivity of a tree ring record is defined by the variability from ring to ring, V, where

$$V = \frac{1}{n-1} \sum_{i=1}^{i=n-1} \left| \frac{2(x_{i+1} - x_i)}{x_{i+1} + x_i} \right|$$

n is the total number of rings and x is either the ring width or the ring-width index in the ith year. This may be used to test not only the sensitivity of different trees but the amplitude of the variations from one year to the next in different parts of a long series of tree rings.

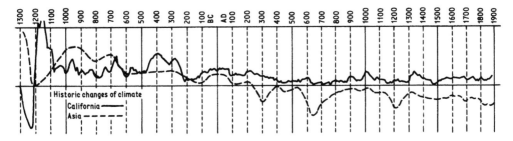

Fig. 13.58   Rainfall variations in California and southwest-central Asia from 1300 B.C. to A.D. 1900, as presented by HUNTINGTON (1913).

(4) Longer-term variations and quasiperiodicities may be explored by calculating the autocorrelations, or lag correlations, between the ring-width index x_i and the value L years later (x_{i+L}) within any given series.

There is some evidence (FRITTS 1971, p. 428) that long-term changes of climate are less faithfully recorded in the ring series of trees in arid sites – perhaps because of competition among the trees on the site gradually changing the density of the forest and hence the availability of moisture – than in the ring series from trees near their thermal limit on the heights or in high latitudes.

In arid regions it is best to use trees that are in the driest sites. In the polar regions the best trees for the purpose are on high ground within 100 m of the upper tree limit.

(5) Cross-dating (see p. 54) with the ring series found in fallen logs and dead trees (e.g. used in buildings) of greater age may be used to extend tree ring series farther back in time.

(6) Master tree ring series, built up by averaging the ring-width index values from many individual trees (and from more than one radius in each tree) and cross-dating, give much more reliable dating and better representation of the trees' common response to the external limiting factor that we are interested in. Repetition of the tree ring measurements in these ways is known as replication. FRITTS reports that 'literally hundreds of ... tree ring chronologies throughout western North America show ... synchrony ... within a 300-mile (480 km) radius and in some cases up to 1100 miles (1770 km)'.

Variability of width from ring to ring is found to increase with the age of the tree and as the mean ring width decreases. It also increases the higher in the stem of the tree the samples are taken.

The dendroclimatologist thinks of a tree as a filter – FRITTS calls it a 'window' – through which a certain complex of climatic behaviour is converted into a ring width. Different tree species each have their own characteristic responses, or filtering properties, though at forest-border sites in the arid southwestern United States Douglas fir (*Pseudotsuga menziesii*) and Ponderosa pine (*Pinus ponderosa*) behave much alike and can be used together. HUSTICH in 1945 presented a series of correlation coefficients (generally between 0·50 and 0·65) between various combinations of the observed climatic elements and tree growth in northern Finland (Lapland). SIRÉN (1961) gives a series of equations for calculating a mean ring-width index for trees in the same region, which includes besides the observations of the particular year also the mean temperature of the last ten growing seasons before it. Early work in this field yielded much stronger correlations between 10-year climatic and tree ring values than between those of the particular year which were sometimes quite small. The actual nature of the associations has been made clearer by more recent studies at the Laboratory of Tree Ring Research at Tucson. FRITTS (1966) has published a set of

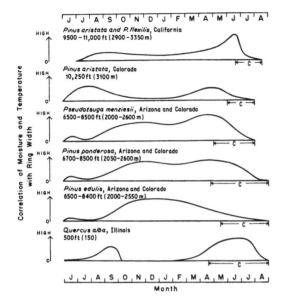

Fig. 13.59   Correlation of ring widths in trees in the United States with weather during the growing season and in the preceding 15 months:

Top two curves: correlation of ring widths in trees at high-level sites in the southwestern U.S.A. with temperature.

Remaining curves: correlation of ring widths in trees in semi-arid sites with rainfall.

(From FRITTS *(1966); reproduced by kind permission.)*

curves (reproduced here in fig. 13.59) showing the variations of the correlation coefficients between ring width and precipitation, and also temperature, over 15 months prior to and including the season of growth of the ring for several different species of tree at various sites, including some near the arid margin and some near the upper tree line. Stepwise multiple correlation techniques applied to studying the relationships between weather items over about 15 months and ring width in each of a number of different tree species have improved the position further, until regression equations have been developed which in some cases account for over 80 % of the variance (e.g. FRITTS 1962).

Douglas fir and bristlecone pine (*Pinus aristata*) have the advantage for climatic research that there is little autocorrelation in the ring widths from one year to the next: the response is largely to the weather of the given year (cf. fig. 13.59).

In the case of Douglas fir, on the gently southward-sloping Mesa Verde (Colorado plateau, $37\frac{1}{2}°$N 108–109°W), FRITTS *et al.* (1965) found that the variation of ring width is largely correlated (positive correlation) with precipitation over the 13 months June to June and to a limited extent inversely correlated with the temperatures over the same period. In this light, the overlapping records of ring widths in the area from A.D. 442 to 1962, reproduced in fig. 13.60, may be interpreted as an index of moisture variations on the Mesa Verde.[1] BROOKS (1949) remarked that, apart from the generally low values in the A.D. 1200s

1. An important objective in this particular research was to throw light on the drought in the late A.D. 1200s which is thought to have been a factor in the disappearance of the prehistoric Indian peoples from the Mesa Verde about that time. The tree-ring series does indeed indicate a drought from about 1273 to 1289, particularly severe at its beginning. This drought was greater and more persistent than any since 1673 but seems to have been surpassed by at least ten other cases seen in fig. 13.60.

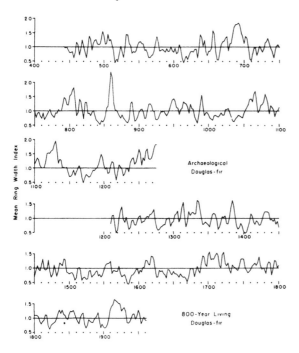

Fig. 13.60 Ring width index series from Douglas fir trees (*Pseudotsuga menziesii*) on the southwestern Colorado plateau (Mesa Verde), 2000 to 2500 m above sea level, from A.D. 440 to 1964: 5-year means plotted at the middle year.

Ring widths are directly related to precipitation and inversely related to temperature over about 13 months from June to June.

(*From* FRITTS *et al. 1965; reproduced by kind permission.*)

and high values in the 1300s, the variations on the Colorado plateau seem to have been mostly opposite to those found by HUNTINGTON in California. The tree ring record of Ponderosa pine in eastern Oregon from A.D. 1268 to 1935, in HEPTING (1971), evidently represents a region with yet another rainfall history: a common feature between the Oregon record and that of white pines in Idaho, farther east but also in the Rocky Mountains region near 45°N, is the increasing dryness indicated after about 1860, particularly from about 1910 to the dry years about 1930.

V. C. LaMARCHE (personal communication, 11 May 1973) has produced a ring width series from bristlecone pine (*Pinus aristata*) at the upper tree line in the White Mountains, California, from A.D. 800 to 1960 (fig. 13.61(*a*)) that shows a remarkable parallel (correlation coefficient connecting the successive 100-year averages of the two series $r = +0.58 \pm 0.21$) with the mean temperature in central England derived by the present author (Volume 1, p. 236, and pp. 31–4, 561 of this volume). By plotting together the sequence of tree ring widths at the upper and lower tree lines in the White Mountains (fig. 13.61(*b*)), LaMARCHE has presented a history of the changes of prevailing temperature and moisture of the climate there over the last 1200 years. Another curve (fig. 13.62), kindly supplied by LaMARCHE, presents successive 100-year averages of tree ring width in the bristlecone pines at the upper tree line (effective control: temperature) in the White Mountains from 3500 B.C. to A.D. 1950. Note the general warmth indicated by high growth from the beginning of the record to about 1300 B.C. and the briefer periods of rapid growth 200 B.C.–A.D. 300 and around A.D. 1200 and 1850–1950

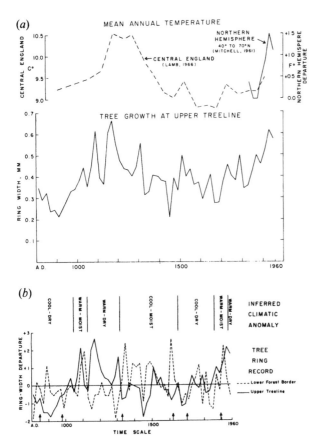

Fig. 13.61   Tree growth in the White Mountains, California since A.D. 800.

(*a*) Ring widths (20-year means) at the upper tree line compared with the temperature values (50-year means) for central England derived by LAMB (1965).

(*b*) Ring widths (20-year means) compared at the upper tree line (where summer temperature is the main control) and at the lower tree line (where rainfall is the main control).

The ring width values were obtained from bristlecone pines (*Pinus longaeva* and *P. aristata*).

(*Diagrams kindly supplied for this book by* V. C. LA MARCHE *personal communication,* 21 January 1974.)

(the full sequence of these ring widths from 3431 B.C. to A.D. 1969 is given, in the form of 20-year averages, in Table 19 in Appendix V). Work by STOCKTON and FRITTS (1971) indicates that stream flow, watershed run-off and the incidence of flooding in the western and mid-western United States can be calibrated with the records of tree growth in tree ring series from the western States: thereby very long proxy records of the probable history of these items may be reconstructed.

The longest and most informative tree ring series from near the northern forest limit, where the basic control is temperature of the summers, is that given by SIRÉN (1961) for pines (*Pinus sylvatica*) at sites distributed over northern Finland (68–70°N 22–29°E) from A.D. 1181 to 1960, reproduced here in fig. 13.63 (the yearly data are tabulated in Appendix V, Table 20). Analysis of this series led SIRÉN to diagnose periodicities of about 70, 90 and 200 years as its main constituents, a result which he has suggested may be used for prediction. Fig. 13.64 indicates a general parallel between the long records of ring widths in Scandinavia and those of the Siberian larch in the northern Urals. HUSTICH (1956) has given a comparative table of tree-ring index values for each year from 1880 to about 1940 in Alaska,

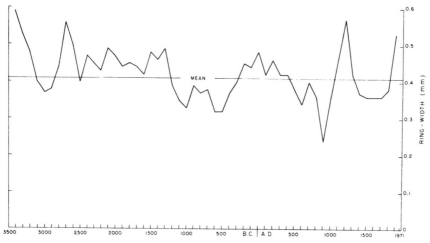

Fig. 13.62   Ring widths in bristlecone pine (*Pinus aristata*) near the upper tree line (temperature controlled) from 3500 B.C. to the present, 100-year means.

(*Diagram kindly supplied for this book by* V. C. LA MARCHE, *personal* communication, 21 January 1974.)

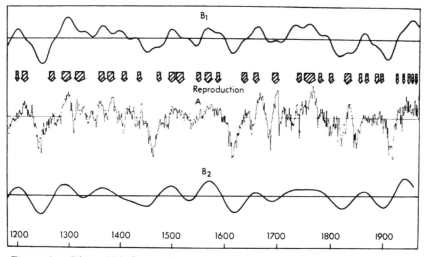

Fig. 13.63   Ring width fluctuations (curve *A*) in trees near the northern forest limit in Lapland (north Finland), from A.D. 1181 to 1960.

Arrows mark the times when natural regeneration of the forest stands was successfully occurring. Note how these were also times of accelerated growth of the existing trees, as shown by the ring width index curve.

The smooth curve B_1 was synthesized from the six most prominent quasiperiodicities found in the ring width series (cycles of 33, 72, 92, 111, 204 and 342 years); B_2 was synthesized by using just the three most prominent quasiperiodicities (72, 92 and 204 years).

(*From* SIRÉN (*1961*); *reproduced by kind permission.*)

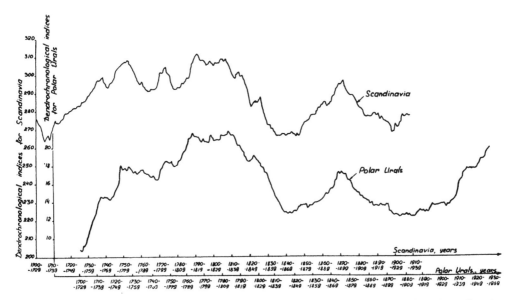

Fig. 13.64   Comparison of running 30-year means of tree ring widths in northern Scandinavia and in the northern Urals (*Larix sibirica*).
(*From* ADAMENKO *1963*.)

Labrador, Norway and Finland (no simple long-distance correlation between these areas could be noticed).

Farther south in Europe there are some potentially valuable tree ring series, none of which have yet been so fully analysed for climatic information as the series from the United States or Finland. The German oak 1000-year chronology, tabulated in Tables 21 and 22 in Appendix V in this volume, certainly contains some points of climatic interest. There can be little doubt that the remarkable sequences of low values (narrow rings) in the periods A.D. 910–947 and 980–1012 indicate an extraordinary incidence of droughts. Indeed, it seems possible that the main response in this series is to moisture variations. A second point of interest is that J. M. FLETCHER (personal communication, 10 April 1973; see also FLETCHER 1974) reports that a simple index[1] constructed from ring widths of dated English oaks used in chests in Westminster Abbey showed 70–75% agreement with the master German oak chronology in the early Middle Ages for some centuries prior to A.D. 1250, and averaged 63% between 1250 and 1400, but showed much poorer agreement (50–55%) in the period 1400 to at least 1550.[1] This is what might be expected if the atmospheric circulation had a higher frequency of zonal patterns (presumably westerly winds and zonally extended anticyclones) before 1400 than in the centuries which followed. HUBER and JAZEWITSCH

1. A simple + or − index was used to indicate whether a ring
   was wider or narrower than the one which preceded it.

(1958) found that the year-by-year variability of ring width shown by an 80-year sample of German oak from Neolithic (Sub-Boreal) times was significantly greater than the variability over the years A.D. 1870–1950; this enhanced variability may also be taken as an indicator of a period with reduced frequency of westerly winds (see also p. 412).

HUBER (1964) shows a diagram giving the average radial growth obtained from tree rings in larches near the upper tree line in the Berchtesgaden area (47·6°N 13·0°E), indicating variable growth rates between A.D. 1330 and 1490, then generally rapid growth which tailed off after 1560; since about 1590 growth has been generally at only half the rate that prevailed between 1490 and 1560. About half the sample showed more rapid growth again between 1770 and 1810 and between 1850 and 1950.

Archaeological records of tree rings from Novgorod (58·5°N 31·3°E), reproduced by THOMPSON (1967), seem to indicate generally narrow rings (frequent droughts?) around A.D. 930 and 970 to some time about 1000–1030, 1120 to 1170 or rather later, and again in the mid 1300s.

Two tree-ring records for over 300 years from different latitudes in the Far East are illustrated in fig. 13.65. They show a striking negative correlation with each other, and the author (OUTI 1964) reports periodicities of about 5, 10, 20 and 100 years in the series. The two longest-term periodic components appeared to be correlated with the south–north difference of atmospheric pressure (i.e. the gradient for W'ly winds) over Japan.[1]

Tree rings in Java from A.D. 1514 to 1929 in the species *Tectona grandis* (Appendix V, Table 23) were analysed by H. P. BERLAGE and DE BOER and published by the latter (1951). The cooler, wetter years were found to correspond to the broader rings, drier warm years to narrow rings. The analysis showed a prominent cycle of mean length 3·4 years over the whole data series, which the authors equate with the Southern Oscillation and thereby regard it as having been in operation throughout the four centuries. The best marked other periodicities identified had period lengths of 51 and 89 years.

Plans to extend this kind of work to the southern hemisphere have been contemplated for some time at the Tree Ring Research Laboratory at Tucson, Arizona. Some exploratory work by E. SCHULMAN in 1950 in Chile, where *Auracaria* and *Austrocedrus* live to 500–700 years and *Fitzroya* with ages of 2000–4000 years have been reported, showed promising results. New Zealand also has trees which grow to great ages.

Applications of tree ring work in the western United States to dendroclimatology, under the leadership of Professor H. C. FRITTS at Tucson, include tentative extensions of local stream flow and lake level records back in time (STOCKTON 1971) and the construction of tree growth and implied rainfall maps of western North America for each decade from A.D. 1000–9 to 1190–9 (ROBINSON and DEAN 1969) and for overlapping 10-year periods from 1501–10, 1506–15, etc., to 1931–40 (FRITTS 1965). ROBINSON and DEAN produced

---

1. OUTI (1961, 1962) has also published yearly ring width values for cypress trees at two sites in Formosa (Taiwan), near 24°N 121°E, from A.D. 860 to the present century, and finds evidence therefrom of variations of the North Pacific anticyclone and of associations with the variations of solar activity.

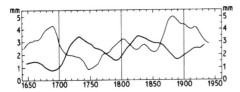

Fig. 13.65   History of tree growth from
A.D. 1640 to 1940 (thick line) at Yamagata,
Japan (38°N 140°E) and (thin line) on the
Taipingshan mountains in Taiwan
(Formosa) (near 23°N 121°E): 25-year
running means.
    The trees used were respectively *keyaki*
(*Zelkova serrata*) and Formosan cypress
(*Chamaecyparis*).
    (*After* OUTI (*1964*); *by kind
permission.*)

mean maps (unpublished) for every decade from A.D. 680 to 1279 and for each individual
year from 1100 to 1279.

    The most ambitious aim of tree ring dendroclimatology is no less than to reconstruct the
atmospheric circulation patterns of individual years and decades in the prehistoric past.
Collaboration between Professor FRITTS and Dr J. E. KUTZBACH, Director of the Center
for Climatic Research, University of Wisconsin, Madison, has set the lead in this
development. FRITTS *et al.* (1971) and LaMARCHE and FRITTS (1971) have applied
multivariate statistical analysis to the ring widths of trees in western North America, and
proceeding to identify eigenvectors of the variation of (*a*) tree growth over that area and (*b*)
the sea level atmospheric pressure distribution, have derived atmospheric circulation
patterns over half the northern hemisphere from the relationships between the two sets of
eigenvectors. Pressure and tree ring data for 1900–62 were used to establish the most
important eigenvectors of each and the relationships between them. The next step is to apply
these relationships to the tree growth patterns of years before pressure observations were
made, in order to reveal the pressure and wind circulation distributions which they indicate.
There is, of course, an underlying assumption that the most important eigenvectors of the
period 1900–62 were also the ones that mattered in the periods for which the reconstructions
are attempted. This assumption can be tested in the case of tree growth, and LaMARCHE and
FRITTS's study (1971) has shown that for the period since 1700 it holds good for the first
three eigenvectors of the tree data but not for the fourth eigenvector. On this basis, pressure
distribution maps for four 3-month seasons of the year ('winter' taken as January to March,
'spring' as April to June, 'summer' as July to September, and 'autumn' as October to
December) have been estimated for each year 1700–1899 – though the individual years' maps
are recognized as liable to sizable errors in areas where the relationships used are not very
strong – and averages for each pentade and decade.

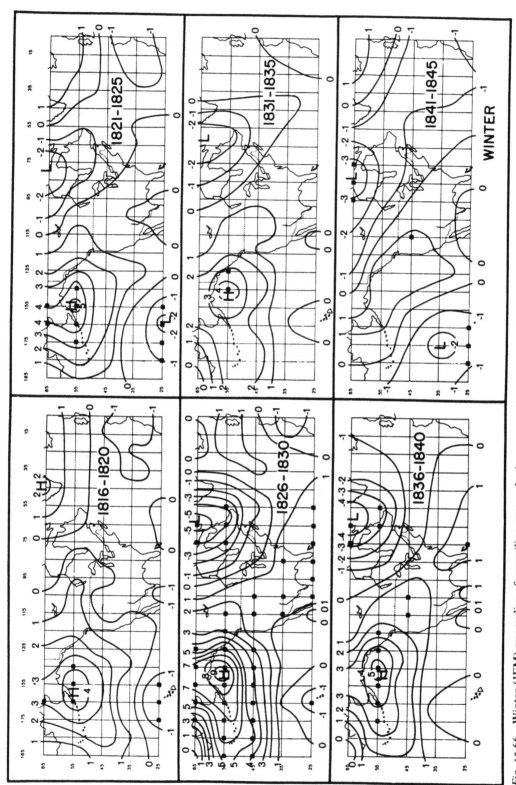

Fig. 13.66 Winter (JFM) anomalies of prevailing atmospheric pressure at mean sea level, by 5-year periods from 1816 to 1845, over much of the northern hemisphere, as indicated by correlation with tree ring widths in the western United States.
The anomalies are expressed as departures in millibars from the average pressure for the same months for the period 1899–1939, 1945–62.
The square dots mark departures that are more than twice the standard error.
(From FRITTS *(1971): reproduced by kind permission.)*

Samples of the maps derived covering the winters of the periods 1816–45 are reproduced here in fig. 13.66 (see also fig. 18.25 later in this volume). The maps indicate that the period was characterized by more meridional circulation than 1900–62 with westward displacement of the Aleutian and Iceland low pressure areas and probably some weakening of both. Verification is difficult except near the European fringe, but the general conclusions mentioned are in keeping with what else is known of the northern hemisphere climate of those times.

The percentage of the variance of pressure accounted for in 1900–62 by the relationships with the tree growth data from western North America ranged from about 10% to 50% in different parts of the half hemisphere. It was strongest in all four seasons over the subtropical belt of the oceans and over much of North America from the western U.S.A. to central northern Canada. It was also strong near Kamchatka and the Bering Sea in January to June.

Apart from ring widths, studies of wood density variations (e.g. by POLGE 1970) and isotope measurements (LIBBY and PANDOLFI 1974) to establish past temperatures are likely to be increasingly pursued and seem capable of yielding information about rainfall and temperature variations that occurred within the growing season as well as from year to year (see also LERMAN 1974).

Professor V. C. LaMARCHE of the Tree Ring Laboratory at Tucson, Arizona, informs me (personal communication, 9 June 1976) that a small minority of years in the long record of bristlecone pine near the upper tree line in California and in Swiss pines in the Alps show moderate to severe frost damage due to unseasonable frosts late in the growing season (temperatures no lower than $-2°$ to $-5°C$ are required to produce the damage, the date of which within the season can be approximately determined by its position within the ring). These rare cases of August–September frosts seem to occur specifically in years when important volcanic dust veils are known to have been present in the stratosphere. The one possible exception in A.D. 1941 may perhaps be taken to indicate that the eruption of Santorin in the Aegean in 1939–41 was more serious than indicated in the LAMB (1970) Dust Veil Index assessments, possibly due to the difficulties of observation and reporting during the war. Thus the frost-damaged rings could serve as a marker to confirm datings and also register useful indications of the weather and (perhaps) volcanic eruptions. Years mentioned by LaMARCHE as showing prominent frost damage in many trees were: 1626 B.C., A.D. 1453, 1601, and in more recent times in 1835, 1837, 1841, 1884, 1902, 1906, 1912, 1941 and 1965. The first year in this list was mentioned because it is the only case within several centuries in the second millennium B.C. and may therefore indicate the true date of the great Santorin eruption which has been usually attributed by archaeologists to the fifteenth century B.C. and associated with the fall of Minoan Crete.

The use of tree stumps and fallen logs above or beyond the present tree line, dating them by cross-correlation of the ring series or by radiocarbon, to discover the history of past changes of the tree limit represents quite another branch of dendroclimatology. Few, if any, of these vestiges of former forest that have by chance remained in position undestroyed to our own times are likely to represent the extreme edge of the forest at the time when they were established. What can be deduced from their occurrence therefore is a minimum estimate of the displacement of the tree line. Most references to vegetation history in this book will be found in the chapters dealing with the successive climatic regimes, but a few examples cited here will illustrate the relevance of these studies.

In the Alps GAMS (1937) records that the upper tree line in the mid 1300s was about 200 m higher than in this century. Similarly, in their studies of the central European hill country from the Erzgebirge, on the border of Czechoslovakia, and the Deutsche Mittelgebirge to the Vosges, FIRBAS and LOSERT (1949) report that the upper tree line was lower by 100 to 200 m with the onset of colder climate in the late Middle Ages. LA MARCHE (1972, and personal communication 9 May 1973) has concluded from extensive studies of remains of dead trees up to 150 m above the present upper timber line in the White Mountains of California and in the Snake Range in Nevada (see LA MARCHE and MOONEY 1967) that a gradual decline of the tree line began about 500 B.C., but that the sharpest drop set in in the fourteenth century A.D. and culminated between A.D. 1600 and 1850 in a tree line lower than today's.

The Tuktoyaktuk Peninsula (69–70°N 130–134°W), just east of the Mackenzie River in northwest Canada, has been examined by RITCHIE and HARE (1971). At present the peninsula is dwarf birch tundra and is dominated by the Arctic airstream even in July. But tree remains and pollen analysis results combine to show that in the warmest postglacial times 5500 to 8500 years ago it was under closed spruce forest, which is taken to imply dominance by the warm, moist Pacific airmass in the Julys of those times. The authors concluded that the Arctic front was generally over the Beaufort Sea in the Julys of the warmest postglacial times, about 350 km north of its position nowadays. Plate IV illustrates this by a sizable stump of spruce *Picea glauca* still there today.

The association between the seasonal mean positions of the main airstream boundaries, defined by streamline analysis, over North America and the boundaries of the natural vegetation – particularly the association between the July position of the Arctic front and the northern forest limit – has been clearly demonstrated by BRYSON (1966) – see pp. 393–6. By a tentative analysis of frontal positions in modern Julys over Eurasia, to arrive at a median position of the Arctic front over eastern longitudes, KREBS and BARRY (1970) found near coincidence between this and the forest–tundra boundary in that sector also. These limits run in general from west to east across northern Siberia, whereas over much of Canada the alignment is from northwest to southeast; that in both cases their association with each other is maintained strengthens the case for believing that the association is of necessity. It is probably a two-way relationship, with the vegetation distribution a natural response to the temperature and radiation conditions associated with the airmasses on either side of the Arctic front, but the differences of surface roughness, albedo and moisture helping to maintain the general position of the front. It is reasonable to take the shifts of the vegetation limits under other climatic regimes in the past as indicating the positions of the airstream boundaries which prevailed in the times concerned.

Important climatic deductions have been made from the record of tree-line displacements in many other parts of the world, but the record has had to be established from pollen analysis since larger remnants have not survived. The results will be referred to at appropriate places in later chapters. In the lands about the Mediterranean, and doubtless in other countries where a considerable human population existed at an early date, Man's disturbance of the forests seems to have hastened their disappearance as the climate became drier and subsoil moisture sources were used up. The consequences included loss of soil cover (washed away in the occasional heavy rainstorms) and alluvial deposition in the lower parts of the river valleys, where the silt-laden rivers extended their deltas seawards (e.g. BRICE 1955). These forests were in a 'subfossil' condition – a term coined by Professor AMIRAN of the Hebrew University of Jerusalem – before they were destroyed; they had till then survived some hundreds of years or longer after the climate to which they were adapted had ceased (see also footnote p. 80).

Insects

Insect remains are the most prolific source of information on past climates from among all the land fauna. Their small size, the great diversity of species, the good preservation of certain hard parts (e.g. the abdomen and male genitalia) make it possible for large numbers to be identified in a deposit. Thus, we have the prerequisite for similar treatments to those used with fossil pollen and another, independent pathway to precise interpretation of the ancient environment. Of the nearly 2 million faunal species known to science about half are insects, and many of these are present in vast numbers. Among the fossil insects, most of the identifiable fragments belong to the beetle order (*Coleoptera*): these are the most useful in palaeoecological work, because their skeletons are robust and, as they are among the best documented orders of insect, identification is relatively straightforward. In lake deposits *Diptera* (two-winged flies) remains are more abundant but usually represented by larval fragments. Some work has been done also on Ants (*Hymenoptera*).

Beetle faunas have the further advantage that they show that there has clearly been no evolution among *Coleoptera* species during the latter half of the Quaternary. Species morphologically identical with today's are found from at least the last 350 000 years. COOPE (e.g. 1970; see also, especially, COOPE *et al.* 1971) has vigorously pursued, and expounded, the use of fossil *Coleoptera* as palaeoclimatic indicators in a series of papers. The distinctive contributions to knowledge of past climatic conditions that have come from this approach will be illustrated here largely from his results for the British Isles. A Netherlands beetle fauna from the middle of the last ice age, about 33 000 years ago, much resembling the British glacial faunas, including species such as *Helophorus jacutus* which now lives no nearer than northeast Siberia, has also been described by COOPE (1969b) and its implications regarding climate and the rates of spread of winged and flightless beetles discussed.

This is a rather new tool in palaeoclimatology, because formerly entomologists too readily assumed that insect species would not be identifiable from isolated fragments and, moreoever, that there must have been too much evolution of species to make such identification likely.

COOPE argues that it is safe to assume that the ecological and climatic requirements of beetle species have not changed, because not only the species themselves have been constant but the assemblages of species have also remained consistent. The climatic interpretations derived from fossil *Coleoptera* have been based almost entirely on comparing the geographical distributions of species today and in the past.

Competition between different species is believed to play little part in determining the limits of the geographical distributions (unlike the situation among plants), at least in middle and high latitudes, because seasonal changes of the climatic environment, particularly temperature, generally limit the beetle species before the density of different populations becomes important. It has been thought strange that even beetles which sometimes live in microhabitats with their own microclimates can be used as indicators of the broad-scale distribution of world climates, but these microclimates presumably only occur at all widely within certain macroclimates. In fact, the geographical range limits for beetle species (other than those – misleading ones for our purpose – which are dependent on decaying matter, and the heat generated therein, for their habitat) do largely parallel the isopleths of summer temperature.

In order to get an indicator which is somewhat independent of the climatic indications from the vegetation, most interest is attached to carnivorous and omnivorous beetles. Woody plants provide some of the special habitats used, particularly by those beetles which require shade, but the plant species is often a matter of indifference.[1]

Proceeding from these arguments, COOPE and his co-workers can report a striking series of findings which throw light upon the course of the last glaciation in Britain and even upon the patterns of the atmospheric circulation.

A rich insect fauna has been studied from a deposit at Chelford (Cheshire, England), radiocarbon-dated about 60 000 years ago, which marks a temperate stage in the early part of the last ice age. The beetle fauna is consistent with a Boreal forest environment like that of south-central Finland today.

The *Coleoptera* found in deposits in various parts of central England from the short temperate stage in the middle of the last (Weichselian) ice age raised an interesting problem. Whereas beds with radiocarbon ages of over 43 500 years ago yielded characteristic Arctic assemblages, those dated approximately 43 000 to 40 000 years B.P. had beetle faunas which today are associated with mean July temperatures of 15°C and over. This 'thermophilous' fauna was difficult to reconcile with the pollen evidence of the vegetation, which seems to have been a treeless open grassland: no forest returned in spite of the evidence from the beetles of a climate with warm summers, about as warm as today's in north Germany. One possible explanation, suggested by F. G. BELL (1970), is that the winter climate of the time

---

1. COOPE mentions in this connection species that live under the bark of dead trees and those that inhabit floating mats of dead reeds on the surface of lakes.

was decisive: the effects of very severe winters on trees, especially through cryoturbation and solifluction would be much more adverse than upon beetles, which hibernate. The combined effects of soil disturbance, low winter temperatures and parching winds drying the shoots and reproductive parts may have prevented the growth of any tree species, though permitting the beetles to survive. It may be, however, that this climate did not last long enough for the trees to spread from their glacial refugia (see fig. 15.16, p. 343).

Beetle species observed in England during the colder stages of the last two glaciations are found today either in northern Scandinavia, sometimes at high levels, or in Siberia. Some of the species of *Helophorus* are today restricted to eastern Siberia beyond the Lena river and Lake Baikal.

A beetle fauna found in exposed silts and sands near the foot of Dimlington Cliff in Yorkshire, England, radiocarbon-dated about 18 000–18 500 B.P., i.e. in the Full Glacial about the beginning of the last glacial climax, consists of few and generally Arctic species, these being typical for a pond with a little aquatic vegetation and some mosses, but little else, round about (PENNY *et al.* 1969). Thus, at that time the indicated vegetation and insect fauna are obviously compatible.

Great difficulties in reconciling the evidence of the flora and of the beetle fauna arise in the climatic stages which followed the glacial climax. Towards the end of the coldest stages, a fauna dated 13 490 radiocarbon years B.P. was a characteristically Arctic assemblage, indicating average July temperatures probably not quite up to 10°C, though it differed from the earlier cold faunas in the disappearance of those that are eastern (Siberian) species today. Yet before 12 000 years ago, when the landscape was still treeless, the beetle faunas from England, Wales and the Isle of Man belong today in regions with mean July temperatures somewhat above 17° – i.e. already warmer than now in the places where the faunas have been found. It is a thoroughly modern, west European beetle fauna like that of southeast England today. COOPE *et al.* (1971) believe that the beetle fauna are in this case registering a quicker, and therefore more accurate, response to the climatic change than the vegetation and that the broad-leafed deciduous trees could not migrate to England from their glacial refugia beyond the mountains in southern Europe within the time that this first warm climate lasted.

The discrepancy between the temperature curves indicated by the beetle fauna and by the flora, covering the times that have been associated with the Bølling and Allerød warm stages, is seen in fig. 13.67. There is agreement between the two lines of evidence about the declining temperature trend which followed and which culminated in pollen zone III, the roughly 500 years of cold climate sometimes known as the post-Allerød stage. COOPE suggests that summer temperatures in England were about 10°C at that time. In the rapid warming which followed, the northern *Coleoptera* types disappeared from England and July temperatures near 17°C are soon indicated.

A beetle fauna examined by OSBORNE (1969) from the deposit at the bottom of a deep shaft of late Bronze Age time near Stonehenge in Wiltshire, England, radiocarbon-dated 3330 ± 90 years B.P., was found to be nearly the same as the modern fauna of the area apart

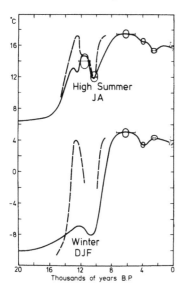

Fig. 13.67   Probable course of prevailing temperatures in central England over the last 20 000 years.

Full lines = mainly from botanical evidence.

Broken lines = amendment suggested by analysis of the beetle fauna.

Ovals = margins of uncertainty of the dating and temperature values derived.

Horizontal bars = apparent duration of the conditions from which the temperature values were derived.

Dots = mean temperature values derived for individual centuries of the last millennium from reported weather or actual temperature measurements (see fig. 6.6, Volume 1, p. 236).

from one more southern species and one or two other hints of a slightly warmer environment.

Some past climatic regimes indicated in these ways may be unmatched anywhere on the Earth today. That is not improbable, however. The prominence of extreme continental, and far northeastern, beetle species in England at various stages in the last glaciation surely indicates continental-type climatic regimes, with winds from the east and north predominating over westerly winds from the ocean. By way of example, figs. 13.68(*a*)–(*d*) show the present European distributions of four beetle species found in a silt deposit 6 km southeast of Leeds (Yorkshire, England), tentatively dated around 32 000–38 000 years B.P. (GAUNT *et al.* 1970). Other species found have east Siberian distributions today. As regards the temperate stages within the last glaciation, there may also have been a considerable frequency of southerly wind components; infrequency of westerly winds as compared with today seems again to be implied.

The variations in the occurrence in England of beetles, butterflies and moths from different European continental and sometimes African sources have a distinct interest as a register of abnormal winds and temperature trends (DE WORMS 1958, HURST 1964, MERE 1961). In some cases the variations are of practical importance, as with the insect pests of agriculture and the longhorn beetles (see p. 187) which damage roofs and woodwork (WHITE 1954). Other references to insects will be found on pp. 186–7 and 343[1].

1. ADDYMAN *et al.* (1976) give examples of beetles which succumbed to the Little Ice Age of recent centuries in Britain, and regard the abundance of the beetle *Aglenus brunneus* (Gyll.) and of the bug *Heterogaster urticae* (F.), which is a ready flyer, in medieval York as evidence of prevailing temperatures above those of the present day in the early Middle Ages and in Roman times, when it was also abundant. These writers believe it also possible to derive from the beetle faunas evidence of the extent of an urban heat island in medieval cities. The death-watch beetle *Xestobium rufivillosum* (Deg.) is among those which seem to have been severely curtailed in the sixteenth and seventeenth centuries in England.

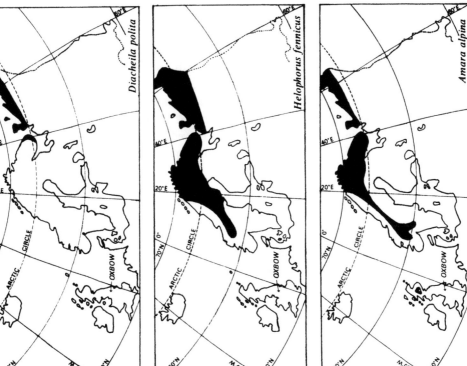

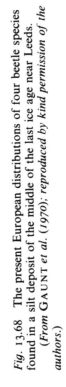

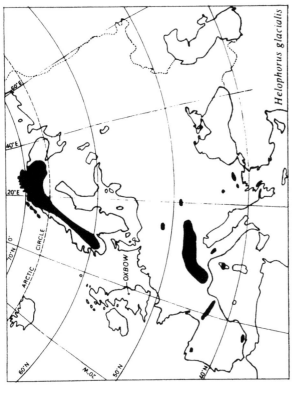

Fig. 13.68   The present European distributions of four beetle species found in a silt deposit of the middle of the last ice age near Leeds. *(From* GAUNT *et al. (1970); reproduced by kind permission of the authors.)*

The records of the occurrences of locusts (*Locusta migratoria*) in various parts of Europe, sometimes in sufficient numbers to damage the harvests, afford an example of an insect history over the last 1200 years that may be informative about the incidence of S'ly and E'ly winds (which are mostly quite rare in summer). The years in which locust invasions known to have occurred in central Europe from the ninth century to the nineteenth are given in Table 13.13, from MÜLLER (1953), quoting from a survey by RUDY (1925).

Table 13.13   Locust years in central Europe

Century	Years with locust invasions reported
9th	803, 844, 864, 872, 873, 874, 875, 887
10th	None
11th	1090, 1091
12th	1101
13th	None
14th	1333, 1335, 1336, 1337–1341, 1353, 1354, 1359, 1363, 1364, 1366, 1371, 1373, 1376, 1388
15th	None
16th	1527, 1528, 1531, 1541, 1542, 1547
17th	1693
18th	1719, 1747–1749, 1752, 1759–1761
19th	1851, 1858, 1874, 1875

From our other sources of evidence about past temperatures it is probably not right to equate the groups of locust years with greatest warmth, as MÜLLER tends to do. The safer conclusion is that the warmth of some other periods, notably the early Middle Ages between A.D. 960 and 1310, the 1730s and the first half of the present century was largely produced by situations – probably of anticyclonic and cyclonic W'ly types over the British Isles and Germany – in which winds from the SE quadrant played little part. In the cooler periods the total absence of locusts could be attributed either to frequent N'ly wind components or to the lower temperatures.

Birds

The ranges of bird species, and of the nesting sites successfully occupied, certainly change when the climate changes, not only with the gross differences between glacial and interglacial climates, but also with the smaller shifts and fluctuations within this and other recent centuries (see pp. 184–6 and review articles by CRISP 1959, HARRIS 1964). Records of differences of range that were regularly maintained for some time in the past are therefore of interest to climatology, though they seem unlikely to supply more than confirmatory evidence of climatic differences already known by other means.

One may speculate that it should be possible to establish in outline the radical shifts of whole species populations of birds, as of insects, which were maintained in glacial times, provided enough identifiable field evidence could be collected; but it seems unlikely that any systematic method for deriving past distributions of global climate from bird evidence alone could be worked out. The numbers of specimens available for analysis could in no way compare with the myriads of pollen grains, spores and insect remains to be found in deposits.

MOREAU (1963, 1969) notes the disjunct distributions of birds and animal species in present-day Africa. Particularly in the case of forest types whose identical populations are now found in isolated habitats on the mountains. Thus the montane plants of the Cameroon highlands are largely the same as those of the East African highlands. It would apparently take a temperature lowering of 5°c to spread the forest and link these habitats into one – perhaps an indication of a minimum value for the ice age temperature-fall in equatorial Africa.

Generally, the critical features of past climatic regimes are ascertainable by such a variety of lines of evidence as to check and confirm each other and that climatology should rather be expected to throw light on the probable distributions of birds.

Land animals

With mammals the position is much the same as with birds. Their ranges are affected by climatic variations, but the numbers of specimens from the past available as evidence do not lend themselves to systematic methods of deriving climatic data. Moreover, the size of most animal remains means that they must rest in a deposit of some depth and that their bulk may have disturbed the age stratification of the silts or boggy material in which they lie. SHOTTON (1962) gives a number of examples of the use of vertebrates as climatic indicators, particularly those that show the climatic shifts in Britain and northern Europe during, and since, the last glaciation, though remarking that many of the species which lived in the periglacial environment must have migrated great distances seasonally.

All we can expect from mammal evidence therefore is that the past distributions of animals and climate must be consistent with each other. Climatology is as likely to throw light on zoogeography as vice versa.

The most obvious cases in which the influence of climatic changes upon animal distributions can be traced, however broadly and indirectly, include:

(1) The immigrations of animals, and of Man, into the Americas by way of the Bering land bridge from north Asia when world sea level was low in glacial times. (There must have been similar freedom of movement between Australia and New Guinea at times of low sea level during the Pleistocene ice ages, though deeper straits still divided that enlarged continent from Asia. The last land connections between Australia and the rest of the world are believed, partly on evidence of the state of faunal evolution, to have been severed in the Tertiary.)

(2) The recolonization of Britain and Ireland from the east, as the climate became warmer and the ice melted away after the last glaciation. The immigration of mammals ceased when the rising world sea level first cut off Ireland and later Britain by submerging the Northsealand.

(3) The movements of animals and men across the Sahara in the warmest postglacial times, when we know that the levels of Lake Chad and of the Nile river were high. By implication, the water table must have been generally higher than now and the oases more extensive. BUTZER (1958) has studied the evidence, much of it from prehistoric rock drawings, of the presence of elephants, antelopes, giraffes, ostriches, hippopotami, crocodiles and other varieties inhabiting the oases, the Tibesti highlands and, particularly, the Nile valley in upper Egypt – areas where they have apparently not been known since some time between 3000 and 2500 B.C. (see also LAMB 1968). MOREAU (1963) reports fossil evidence of northern mammals, including bear (*Ursus arctos*) and elk (*Cervus elephas*) in the region Morocco–Tunisia, in the late Pleistocene, presumably during the European glaciations.

Further studies of the present and past distributions of animals in Africa, and reasoning about past climate and vegetation distributions, will be found in MONOD (1964) and MOREAU (1969). The latter devotes sections to forest birds, mammals, fishes and frogs, and gives a large bibliography.

Some examples, with maps, of the responses of animals and birds to the climatic fluctuations of the present century will be found in CRISP (1959) and SCHULTZ (1972).

Mollusca

With mollusks one comes once again to an order where there are many species and great numbers of individuals, so that statistical analysis is possible. The preservation of their calcareous shells is almost unlimited in basic environments, provided that the grain size and movement of the matrix in which they are incorporated does not shatter them beyond recognition. They degenerate and disappear quite rapidly in acid soils. The order includes snails and slugs (which have no shells), clams, oysters, mussels, squids and octopuses. The shell sizes range from grains of sand upwards. All the main classes have been recognized from as early as the Cambrian era. ZEUNER (1958) notes that the rate of evolution of marine mollusks has been notably slower than among mammals.

The shallow water molluscan faunas of the Pacific coast of North America in the Tertiary, studied by ADDICOTT (1970), consist largely of warm water taxa whose descendants now live farther south. A tropical marine climate with a wealth of molluscan species is indicated for middle latitudes of the eastern Pacific in the Eocene; a change to less diverse faunas in the early Oligocene suggests a cooler phase, followed by a return to warm water faunas in the late Oligocene to early (or middle) Miocene, after which a progressive shift towards colder aspects sets in. Moreover, the south to north gradient of the warm water forms and of the counts of genus diversity increases steadily after the warm peak of the Miocene.

The earlier studies of mollusks were almost entirely in terms of species lists and their geographical distributions. Modern studies exploit the possibilities of frequency analysis, commonly plotted as percentage frequency histograms for each species counted at 10 cm intervals down a deposit. Examples are given in SPARKS and WEST (1972) and WEST (1968). Sometimes, as with pollen analysis, absolute frequency diagrams are used.

Mollusks may be classified into marine and terrestrial groups. Recognition of the forms which are characteristic for dry land, woodland, marshes, clear running water, etc., is important to our purpose. KERNEY *et al.* (1964) made much use of molluscan frequency diagrams in deriving a history of the environment on the chalklands in eastern Kent in late glacial and postglacial times: this revealed some drying out in Allerød times and a marked reversion to a wet regime with pools and flooding on the chalk (only possible with permafrost below) in the Post-Allerød (pollen zone III). The temperature tolerances of the different mollusks were considered well enough known to suggest the prevailing air temperatures in southern England in those times: mean July temperatures in zone III perhaps as high as $+12°$ or $+13°C$ and average January values no lower than $-6°C$ (some molluscan species are too widely tolerant to be used in this way, others appear to give clear indications derived from their modern ranges).

Another example of use of the association of particular mollusks with different temperature ranges is DE HEINZELIN's (1967) conclusion, as cited by VAN ZINDEREN BAKKER (1969), that 'during the later part of the upper Pleistocene [i.e., presumably, in glacial times] the water of the [Nile] river in Nubia was definitely colder than at present'.

The value of studies of molluscan land faunas is that they give independent evidence of the nature of the immediate environment of a site at each epoch in the past, i.e. its nature as regards warmth and wetness, the vegetation and, particularly, the soil type. In general, the molluscan evidence is more local in its significance than the pollen and is free of any component related to the long-distance pollen transport. It may therefore help in interpretation of the pollen spectra from samples collected at the same place. Some mollusca may be used as evidence of the nature of the water in a river or stream, as regards temperature, basicity and clearness or stagnation. Much use has been made of the distinction between fresh water and salt water groups of molluscan species to determine past variations of the tidal limit in estuaries and of brackishness in the waters of the English fenland.

Life in the sea

With marine life forms, the fauna and the flora of the sea, as with those that inhabit the world's land areas, the micro-organisms (which include the simplest, one-celled forms of life, protozoa and algae) are at the mercy of the climate of the water, its temperature, salinity, pollution, etc., and the amount of sunlight penetrating it. There are optimum conditions for each species in which its productivity is greatest, and there are limiting conditions beyond which the species dies. The larger and more highly developed forms of life are also limited in their ranges by the temperature and other conditions prevailing – particularly as these affect the abundance of the organisms they feed on – and, of course, by the oxygen content of the water; but they can generally migrate at will in search of their accustomed foods and the water climate they thrive in.

Simple quantitative methods can readily be developed for deriving climatic values implied by the abundances of different species among the micro-organisms whose remains largely constitute the organic carbonate fraction of the sediment on the ocean bed (there is, of course, the usual basic assumption that identity of species means that its climatic requirements in the past were the same as now). The *Meteor* expedition to the tropical Atlantic in 1925–7 was the first to raise cores which penetrated more than the top 30 cm of the sediment, and the analysis of these (SCHOTT 1935) was the pioneering work which used the foraminifera species present to deduce past climate. The suggestion that the remains of these protozoa could be used in this way, however, goes back to PHILLIPI (1910, p. 568). The Swedish Deep Sea Expedition of 1947–8 in the equatorial Pacific brought back the first long cores; the general variations of calcium carbonate content down these cores were analysed (ARRHENIUS 1952). EMILIANI's (1955, 1966) application of oxygen isotope analysis to derive the water temperature history of the tropical Atlantic through the ice ages and interglacials of the last 425 000 years (see fig. 13.15) can now be compared with the numerous diagrams of percentage abundances of different species of foraminifera, and of left- and right-coiling directions (see p. 191 in this book), published by ERICSON and his co-workers (e.g. 1954, 1966) and many others since, and with the advanced methods developed by IMBRIE and KIPP (1971) for deriving temperature from the species analysis. Such comparisons throw light on how much of the oxygen-18 anomaly in the ice age part of the ocean bed sediment is attributable to temperature change and how much is due to the loss of isotopically light water deposited in the ice caps (see pp. 93–5).

As early as 1950 WISEMAN and OVEY, by studying the relationship between the present geographical distribution of near-surface living foraminifera and annual mean sea surface temperature in the Atlantic Ocean from 69°N to 67°s, and taking account of the distribution of dredged erratic stones from the sea floor, sketched a preliminary map of the general limit of drift ice in the North Atlantic at times of maximum glaciation. Their suggested course of the limit from near 33°N in the western Atlantic to 42°N on the coast of Europe seems not to need much amendment in the light of all the additional cores surveyed since.

Histograms and frequency curves, showing the changes in relative abundance of warmer and colder water groups of species of foraminifera and other micro-organisms with depth in the ocean bed sediment, have by now been published for very many points in the world's oceans from the Arctic to the Antarctic. Separate curves for several different species of forams in cores at 26 different sites in the Atlantic between 30°N and 30°s, published by

ERICSON *et al.* (1964, 1966), give a clear record of several glacial and interglacial regimes and the notably quick changes between them. Arctic Ocean sediment cores representing the last 700 000 years have been analysed in terms of depth (time) graphs of the varying abundance and coiling direction of *Globerigina pachyderma*, published by HUNKINS *et al.* (1972); STEUERWALD and CLARK (1972) have given curves also for the benthic foraminifera. These studies suggest that during the last million years or so the Arctic Ocean, at least in the region about 75–85°N 140–170°W, has never been much warmer than it is at the present day; during the last 80 000 years no great change in the pack-ice cover of the central Arctic is indicated, but before that there had been six cycles in which conditions like the present alternated with colder conditions when the pack-ice cover was extremely tight, allowing no open water even in summer. The cycle length was close to 100 000 years.

There is a restricted choice of species and of their calcareous secretions (coccoliths) available for analysis from the bed of the cold oceans in high latitudes. Only one species of foram, *Globerigina pachyderma*, is present in sufficient abundance in Arctic or Antarctic cores. Coccoliths are destroyed by cold waters of salinity less than 20 parts per 1000 and so are not even available in the Newfoundland Banks area of the Atlantic. In the Southern Ocean, even in the sub-Antarctic zone, however, there are many species of radiolaria. The remains of the largely silicious diatoms are also useful in cold oceans. Studies of cores from the sub-Antarctic will be found in the works of HAYS (e.g. 1967, 1969), KENNETT (1968a, 1968b, 1970, 1972), KEANY and KENNETT (1972) and others: some of their probes go far down into Tertiary sediment.

Indian Ocean cores, particularly from the Arabian Sea, have been submitted to chemical, mineralogical and foraminiferal analysis in a number of works by OLAUSSON and his associates (e.g. 1971).

Cores from the South Pacific west of Chile in the region 27–33°S 102–109°W, where the sedimentation rate averaged only 1 cm/3000 years but which penetrated the last 1 200 000 years, have been analysed by ERICSON and WOLLIN (1970) in terms of the ratio of the abundance of the *Globorotalia menardii* species to the total foraminifera population. The sequence of counts at 10 cm (approximately 30 000 years) intervals indicated temperature fluctuations in that region largely in antiphase with the Atlantic and perhaps, therefore, with the rest of the world. The authors supposed that the southeast Pacific region concerned was predominantly influenced by the output of the Antarctic continental ice via the Humboldt Current. This would imply that the Antarctic ice was generally more extensive at times when the Atlantic, north and south of the equator, was warm. Another recent discussion of Pacific Ocean cores is included in WOLLIN *et al.* (1970).

In the North Pacific Cascadia Basin, cores between 43° and 47°N at 126–128°W, in nearly 3000 m depth off the coast of Oregon, have been analysed by GRIGGS *et al.* (1970). Judging from the fact that foraminifera are more abundant relative to radiolaria today in the colder northern water, and radiolaria are more abundant in the warmer southern water (as has also been found in the western North Atlantic between 11° and 41°N), graphs of the foraminifera/radiolaria ratio in species counts at different levels in the top 3 to 8 m of the Cascadia Basin sediment indicate two cold phases in the postglacial record, one 4000–5000 years ago and the other about 2000–2500 years ago. The earlier of these two cold phases is also apparent in plots of *G. pachyderma* frequency off the coast of California. It is possible that these cold phases should be interpreted in terms of increased upwelling, bringing the colder, nutrient-rich bottom water to the surface. MOORE (1974) found that the radiolarian species assemblages in the Cascadia Basin at the climax of the last glaciation affecting that region about 24 000 years B.P. were like that now found off southern Alaska, indicating surface water temperature in summer about 11°C, i.e. 4°C below the present normal level off Oregon; the oscillations in the area in postglacial times probably had a range of about 1°C.

A history over the last 225 000 years of southward penetrations in the eastern Atlantic, between Iceland and Madeira, of the cold polar water has been derived from analyses of the foraminifera and coccoliths present by MCINTYRE *et al.* (1972). This showed two strongest glacial advances, around 30 000–15 000 years ago and 160 000–135 000 years ago, and four other, shorter and less severe, cold periods. Warmest postglacial conditions were about 6000–4000 years ago.

The cores discussed in the preceding paragraphs differed greatly in the length of time they surveyed, depending largely on the differences of sedimentation rate. Now that first samples have been surveyed from most parts of the globe, increasing attention is being given to coring the ocean bed sediment in places where rapid sedimentation offers opportunities of finer time resolution from species counts at any given spacing down the core. KENNETT and HUDDLESTUN (1972a) have analysed the foraminifera in 28 cores in the western part of the Gulf of Mexico (19–27°N 91–97°W) where the sedimentation rates were 10–30 cm/1000 years and sampling was carried out at 10 cm intervals. This revealed two main glacial times and three major volcanic ash layers, which came (1) just at the end of the earlier glaciation about 135 000 years ago, (2) towards the end of the interglacial about 90 000 years ago, and (3) in the early part of the last glaciation about 75 000 years ago. In another publication, the same authors (1972b) found that the warm water foraminifera disappeared from all the cores within 350 years about 90 000 years ago, with the onset of a cold phase which culminated around 84 000 years ago. The date appears to coincide with a sharp drop around 90 000 B.P. of the oxygen-18 values in the Greenland ice sheet, which took place within about 100 years, according to DANSGAARD's dating.

THIEDE (1971) has made numerous probes along a line out to 12°W from the coast of Portugal near 38°N, where the sedimentation rate is over 20 cm/1000 years, and reports that there was no gradient of temperature westwards in glacial times, whereas today upwelling produces lower temperatures near the coast than farther out. Off the coast of Morocco to near 10°W, at 33–34°N (where the sedimentation rate of only 3 cm/1000 years makes the time resolution coarser), the gradient of sea temperature in glacial times seemed much the same as now and, indeed, little or no change of temperature level was indicated. Farther out from that coast, however, MCINTYRE (1967) found evidence of one of the biggest anomalies of

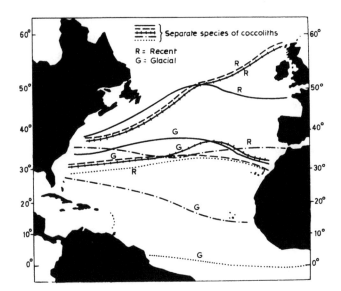

Fig. 13.69 Boundaries of five species of coccoliths in the ocean bed sediment in the North Atlantic, from recent 'R', and glacial (last ice age), 'G', sediments, the latter up to 60 000 years old.

(*From* McIntyre (*1967*); *reproduced by kind permission.*)

sea temperature marking glacial times. Fig. 13.69 illustrates how simply the different distribution of coccolith species in glacial and recent times indicates the ice age displacement of the sea surface temperature isotherms.

The most powerful systematic methods of deriving past climatic values from species counts in deep sea sediments have been worked out by Imbrie and Kipp (1971).

First, it is necessary to understand the limitations affecting the data and their time resolution. The deepest parts of the ocean are unsuitable for sampling because of the amount of carbonate dissolved by the cold waters there. High ratios of benthic to surface-dwelling foraminifera in the sediment are evidence of solution, which is selective in its effect. Interpretation may also be confused where rapid currents in the deep ocean transport material from afar. Similarly, steep slopes where slumping occurs must be avoided. The best sites are at depths no greater than about 3000 m, on the tops and flanks of gentle rises in the ocean bed. In an ideal core-top sample 1 cm thick, taken where sedimentation is at a rate typical for the Atlantic of about 2 cm/1000 years, it might be supposed that the sample would represent the last 500 years. In fact, however, the activity of burrowing animals mixes in older material from deeper in the sediment. The magnitude of this effect can be tested near an unconformity in a sediment. Imbrie and Kipp conclude that intense burrowing is limited to 2 to 3 cm, though some extends to 5 cm or beyond. Typical Atlantic core-top samples therefore represent the last 2500 years, or at most the last 4000 years (Ruddiman *et al.* 1970), though with most material from the last 1000 years.

Imbrie and Kipp's method is based on establishing multiple correlations between the core-top microfaunas and the present average temperatures and salinities in the surface waters. The method is akin to that applied by Webb and Bryson (1972) to pollen spectra sampled on land (see p. 206). Basic assumptions are that the relationships between average temperature, salinity, etc., and the ecologically significant aspects of the climate of the surface waters are essentially linear, and that the ecosystem (the plankton as a whole) of the ocean has not changed significantly during the period studied, however long that may be. The raw data, first from the core-tops and then from the rest of the cores, were sorted and collated into five assemblages recognized as typical for different regions of the Atlantic Ocean, north and south of the equator:

(1) *Tropical assemblage*, at present dominant between 30°N and 30°S in the western Atlantic and between 10°N and 10°S near the coast of Africa.
Dominant species: *Globigerinoides ruber* and *G. sacculifer*. *Globorotalia menardii* and *Pulleniatina obliquiloculata* are important secondary elements in the assemblage.

(2) *Subtropical assemblage*, at present dominant between the range of (1) and latitudes 40–45° north and south of the equator.
Dominant species: *Globorotalia truncatulinoides, G. falconensis, G. inflata* and *G. calida*.

(3) *Subpolar assemblage*, at present dominant approximately in the zone of the North Atlantic Drift to 66°N and as far south as the Antarctic Convergence.
Dominant species: *G. pachyderma, G. bulloides* and *Globigerinita glutinata*.

(4) *Polar assemblage*.
Dominant species: *G. pachyderma*.

(5) *A gyre-margin assemblage*, which is at present prominent all around the periphery of the subtropical anticyclones in the North and South Atlantic.
Most characteristic species: *Globorotalia menardii* and *Pulleniatina obliquiloculata*.

Mapping the areas of dominance of these different assemblages suggests that temperature is the main control, except in the case of the gyre-margin assemblage. The last named is most abundant with certain combinations of salinity and winter temperature (both elements increasing or decreasing together). The assemblages are recognized in terms of the abundance of each species expressed as a percentage of its own total observed abundance range. Species which never rise above low abundances are not used. It is correlations between the percentage abundances of the assemblages (not the individual species) and the physical parameters of the ocean that form the basis of the derivations of palaeo-oceanographic conditions.

The final step is to write what are in essence regression equations, which transform assemblage abundance groupings into climatic values. These are then applied to the data of observation at each level in the core. Fig. 13.70 illustrates the results in terms of the last 25 000 years history of prevailing sea surface temperatures (*a*) in August derived from a core taken at 54°15′N 16°50′W (in a depth of 2393 m) and (*b*) in February from a core from 24°48′N 75°55′W (in a depth of 3110 m).

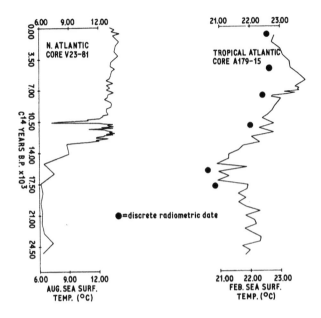

Fig. 13.70   Sea surface temperatures
over the last 25 000 years, as
calculated by IMBRIE and KIPP
(1971) from foraminiferal species
populations.
Core V23-81
   at 54°15′N 16°50′W
   August temperatures.
Core A179-15
   at 24°48′N 75°55′W
   February temperatures.

The encompassing surveys of implied past conditions in the ocean which can be derived
by these means are seen in the maps of the Atlantic Ocean results in Chapter 15, p. 350.

A further stage in the analysis of palaeotemperatures from either the species
composition or oxygen isotope measurements on deep sea cores, and which is applicable also
to the variations of carbonate and of organic carbon within them, has been demonstrated by
PISIAS *et al.* (1973). Variations of all these items must be attributed to climatic changes or to
the ocean current changes that go with them. Spectral analysis of sample cores from the
North Pacific (43°N 126½°W) and North Atlantic (53°N 22°W and 54°N 17°W) show just three
significant peaks indicating quasi-periodicities at around 400 (actually 370–380) years, 1300
and 2600 years, present alike in the glacial and postglacial parts of the core. The authors note
that these period lengths closely resemble periods found in the ice of the Greenland ice-cap
near 77°N 56°W by DANSGAARD *et al.* (1971), and in glacier advances elsewhere by DENTON
(1972), as well as in the fluctuations of atmospheric radiocarbon, which are known to be
associated with fluctuations in solar activity.

13.5   Evidence from human history and archaeology

Climatic fluctuations and changes touch Man and affect the environment of human activity
in so many ways that the evidence of them is abundant. But, because Man is so adaptable and
so apt to personalize the blame for his misfortunes by finding scapegoats and making victims
of his fellows, the evidence of climatic causes is commonly obscured and may be recorded

only incidentally, if at all, in personal quarrels, riots, wars, invasions and economic pressures and distress.[1] In most cases, it is likely that better knowledge of the climatic record will lead to fresh understanding of human history rather than that the latter can contribute firm data on past climate, though further documentary research is much needed.

There is limited scope for the development of systematic methods for deriving climatic information from human history, though some attempts have been made, e.g. by the analysis of wheat prices (BEVERIDGE 1921, 1922, 1929) and wine harvest records (LAHR 1950, MÜLLER 1953, LADURIE 1971). The analysis of prices, of course, entails some device for allowing for progressive monetary inflation, which runs through the record as an almost continuous trend but at a very variable rate. The record must also be considered in relation to scarcities produced by other than directly climatic causes: e.g. by wars or depopulation of the farmlands due to plague and other human epidemics, or by the influence of calamities in neighbouring regions. If these misleading cases can be allowed for, grain price records may provide a useful indication of year-by-year weather variations and even of runs of a few consecutive years of favourable or unfavourable weather for growth. BEVERIDGE'S (1921, 1922, 1929) famous study of wheat prices in England (at Exeter) from A.D. 1316 to 1820 led ultimately – by means of periodogram analysis – to the indication (for the span of some 300 years over which the data were considered most reliable) of a statistically significant tendency for fluctuations with periods around 15·3 and 5·1 years; a number of other period lengths also looked interesting (particularly around 12, 17, 20, and 36, 55 and 68 years), but their significance was not proven (YULE and KENDALL 1950, pp. 644–5). Autocorrelation is liable to arise in the data not only from persistence or trends in the weather but also from the abundance of seed following a good harvest and the lack of it after bad years.

HOSKINS (1964, 1968) has also carefully reviewed the history of grain harvests and prices in England from A.D. 1480 onwards. HOSKINS's method is to classify the harvests according to the departure of the average price of wheat over the following 12 months from the 'norm' represented by the 31-year moving mean. Harvests are then classified as follows:

Abundant	price more than 30% below the norm
Good	price 10–30% below the norm
Average	price 10% below to 10% above the norm
Deficient	price 10–25% above the norm
Bad	price 25–50% above the norm
Dearth	prices over 50% above the norm

From this analysis HOSKINS was able to identify many individual good and bad years and

1. A few historians even claim to have 'proved' that no change of climate was involved in the cultural decline and other disturbances of the late Middle Ages in Europe, because the events were plainly brought about by economic and social pressures. There may be some analogy here to the preference which leads some meteorologists to insist that the explanation of climatic changes lies within the ocean–atmosphere system and is unlikely to involve solar or cosmic influences outside the Earth. There is great need for open minds and a careful integration of knowledge.

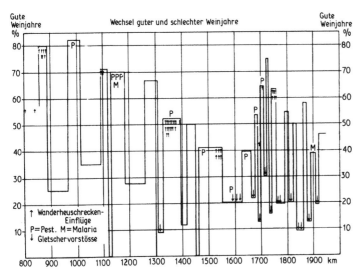

Fig. 13.71 Graph given by MÜLLER (1953) to represent the percentage of wine harvests in Baden (southwest Germany) classified as good in different epochs over the last twelve centuries.

Superposed symbols indicate
↑ Locusts reported
↓ Glacier advances
P Plague
M Malaria
(*Reproduced by kind permission.*)

quite a number of runs of good and bad years, which must have had very severe effects on the welfare of the country.[1] It is, however, impossible to reveal more prolonged climatic shifts and trends by this method of analysis, since they are automatically masked – i.e. eliminated – by the changing level of the moving mean or 'norm'. Accordingly, it is not surprising that HOSKINS found no traces of climatic change, except that the abnormal variability from 1680 to 1730 stood out.

Other possibilities of identifying the effects of climatic fluctuations and changes exist in the records of payments in taxes and tithes and of emergencies which called for the revision or relief of taxes.

It seems likely that the long records of wine harvests in France, Germany and Luxemburg, classified according to quality and abundance, are related to a much more

1. Runs of bad harvests in southern England were noted in 1481–3, 1500–3, 1519–21, 1527–9, 1549–51, 1692–3, 1727–8; and very bad harvests in 1594–7, 1646–50, 1657–61, 1695–8, 1708–11. Runs of good harvests were noted in 1492–5, 1537–42, 1566–71, 1582–4, 1591–3, 1601–6, 1618–20, 1626–8, 1643–5, 1665–7, 1669–72, 1685–90, 1700–2, 1704–7, 1730–3, 1741–4, 1758–60; very good harvests in 1509–10, 1537–9, 1546–8, 1652–5, 1687–8, 1704–7.

nearly constant datum than any price index.[1] Analyses of these, e.g. by MÜLLER (1953) here illustrated in fig. 13.71, and of the available descriptive accounts of seasonal weather (e.g. by EASTON 1928, LAMB 1965, TITOW 1960), as numerical series, have indicated a climatic cooling in the late Middle Ages paralleling the (mostly more recently evaluated) evidence of the lowering of the tree line on the mountains of Europe, advances of the glaciers and of the sea ice and in oxygen isotope measurements, etc. Fig. 13.70 reproduces MÜLLER's (1953) assessment of the variations of frequency of good wine years in Baden over the last twelve centuries. EASTON's winter index also indicated a periodicity of about 89 years.[2]

Many events in the past story of mankind, which we know of from historical documents or from archaeology, register the effects of the climatic changes and fluctuations which have by now been established on other evidence. The nature of the evidence in human history and archaeology is of course very diverse, and must be carefully reviewed in each case for what light it may shed on the impact of climatic events on the environment and on life, but usually it will provide no more than corroboration of the climatic data which already rest on firmer physical or biological evidence.

The most useful pioneer works in the history of climate are those by HUNTINGTON (1907), BROOKS (1926, second edition 1949), UTTERSTRÖM (1955), THORARINSSON (1956) and LADURIE (1971). Many specialist works in geology, geomorphology, glaciology, palaeo-oceanography, palaeobotany and forest history (e.g. FIRBAS 1949, GODWIN 1956, KOCH 1945, PETTERSSON 1914, THORODDSEN 1916–17), cited in the bibliographies to other chapters of this work, could be added to the list.

It will be our policy in the later chapters of this book to include accounts of what can be discerned of the effects of each climatic regime and each major shift of climate on the vegetation and the animal world and on Man. In this chapter we shall proceed to survey a number of particular cases in which the archaeological and historical evidence of the impact of climatic changes seems clear, though limited, and in which, therefore, this source of evidence and its evaluation can be illustrated.

The list of matters in which responses to climatic shifts can be seen includes:

(1) The poleward limits of human settlement and of the cultivation of various crops.
(2) The height of the upper limits of human settlement and of cultivation, seasonal pastures, etc.
(3) The limits of human occupation, cultivation and pastures in arid lands, in desert oases, and even in areas which are now full desert.

---

1. The records of the French wine harvest dates are unfortunately not suitable for use as a climatic index, at least in the eighteenth century, when the grape picking was delayed by royal decree in order to get a stronger wine (thus giving a spurious indication of cool summers). However, MANLEY (1958) found a reasonable correlation before 1710 between the tem-peratures in central England and the wine harvest dates in northern France; so the record may be acceptable for earlier times.

2. EASTON allocated a numerical temperature index to each winter according to its description in the manuscript records of western Europe; see Table 5 in Appendix V.

(4) Changes of land use, the abandoning of areas formerly grazed or cultivated and of settlements, migrations and the exodus of populations from areas formerly occupied.

(5) The routes of migration, travel and trade, sometimes over passes and desert regions which at other times could not be traversed.

(6) The building of wooden trackways across areas which were becoming increasingly marshy and which, in some cases, ultimately had to be crossed by boat.

(7) Changes in the fortunes of lake dwellings, which in prehistoric Europe (as at Federsee, near Biberach, in the German Alpenvorland in Schwaben, and near Glastonbury, Somerset, in England) were sometimes built on piles in the lake or right at the lake edge. Similarly, the levels of old harbour works which were afterwards submerged by the Caspian Sea or the Mediterranean.

(8) Ancient voyages over seas that were at other times likely to be too stormy or beset by ice.

(9) The record of famines, and of human (and animal) health and diseases, variations in the normal length of life, normal stature and overall numbers of the population.

(10) Specific features registered in art, architecture and industry.

It has sometimes been maintained that the whole rise and fall of civilizations may be linked to climatic changes which either undermined the organized pattern of supply of foodstuffs and other raw materials or set in motion barbarian invasions. We shall mention cases where these troubles seem to have been at work, but one must also concede (cf. TOYNBEE 1946) that human societies may go through a cyclical evolution – through the discovery and development of a given way of life, and the increase of ease, comfort and tolerance until loyalties and energy weaken and they are unequal to further challenges from invaders or climate – a cycle which may have its own time scale.

Evidence of the impact of climate in the ways listed above will be mentioned at the appropriate places in the chapters which follow. Here we shall only enlarge on a few items about which more must be said at this stage.

Migrations and settled civilizations (see fig. 1 in Appendix V)

The earliest radiocarbon dates for the presence of Man in North America, dating a hunting culture in eastern Oregon and on the Great Plains around 13 000 years ago resembling a culture developed 2000 years earlier in Siberia (FLINT 1971, p. 785), suggest that the immigration was from Asia and that it took place when the last glaciation was near its maximum development. This seems to imply that entry was by way of the Bering Strait land bridge which existed while world sea level was lowered by the masses of water 'locked up' on land in the great ice sheets; the strait reopened about 12 000 years B.P. Beringia, as it is sometimes called, when sea level was 100 m below its present stand, had been a country nearly as broad as present-day Alaska, occupied by tundra and grassland and patches of Boreal forest. It was not glaciated, and its climate was probably not unlike that of western Alaska today, owing to the circulation of water from the warm North Pacific Drift. However, at some stages, around 25 000 to 20 000 years ago, and possibly as late as 14 000 years ago, the way south from Alaska may well have been blocked by the joining of the Cordilleran (Rocky Mountains) and Laurentide ice sheets and by the former reaching the coast; a coastal strip, freed by the early stages of retreat of the Cordilleran glaciers and since submerged by the eustatic rise of world sea level, seems the likeliest route. It has to be presumed that in some way the ice age helped the arrival of Man in North America; but the dating is not precise enough to determine whether the immigrants could have walked in over the dry land of Beringia or whether part of the route necessitated crossing shallow water with the aid of logs, rafts or boats or by walking on floating ice.

The first arrival of aboriginal men in Australia, somewhat before 35 000 years ago, also apparently occurred in the times of greatly lowered sea level during the last ice age but would have necessitated crossing water in some of the straits between Asia and Australia.

The Bering Strait land bridge had existed (HOPKINS 1967) in early and mid Tertiary time, allowing the passage of land fauna in both directions between Asia and America but separating the marine faunas of the Pacific and Arctic Oceans. A sea connection opened through the Bering Strait temporarily in the later part of the Miocene, but during most of the Pliocene the land bridge existed again and mammals passed across it. Between 4 and 3·5 million years ago the strait was again open. Since then its fluctuations seem to have been determined by the glacial and interglacial changes of sea level, with at least six ages when the strait existed.

In postglacial times variations in the range of activities of the Eskimo cultures, and their contacts with their southern neighbours, seem to respond to the prevailing temperature level and the sea ice conditions. GIDDINGS (1960) has given a chronological table of the cultures about the Bering Sea, in which the earliest dated finds are from around 3000 B.C. Later, the Olvik cultures in the last five centuries before Christ seem to have been in contact with the peoples farther south in Asia and to have introduced the use of iron from them (LEVIN and SERGEYEV 1964). The earliest iron object found so far in the Soviet Arctic territories, on the Chuckchi peninsula, is from the first millennium A.D. (ARUTYUNOV et al. 1964).

The first great spread of Eskimo activity over all the far northern regions seems to have coincided with the early medieval warm epoch. Ellesmere Land was first occupied about A.D. 900, and about the same period Eskimo culture was spreading from the Bering Sea westwards in the Siberian Arctic and out to the New Siberian Islands. During the eleventh century the Thule culture spread along the coastal areas from Alaska to north Greenland: whaling was an important activity in the economy of the Thule Eskimos. After A.D. 1200 all this was progressively given up, and the equipment was lost. The Eskimos moved south in Alaska and in Greenland, and by the sixteenth century Ellesmere Land was uninhabited (CHARD and GRIFFIN 1962). In the colder centuries that followed one or two cases were recorded when a single Eskimo in his kayak, doubtless driven to flee for refuge on an unknown course by unexpected ice and storms, turned up in Scotland (on the last occasion, in 1843, in the River Dee near Aberdeen), though none of these survived the ordeal.

Farther south, in the heart of the great northern hemisphere continents, movements of peoples can be traced which seem to go hand in hand with variations in the moister climate. GRIFFIN (1961) has traced the spread of people with an agricultural economy north and northwest up the Illinois, Mississippi and Missouri valleys from A.D. 700 or 800 until at least 1000 into southern and western Wisconsin and eastern Minnesota. There was some similar movement into parts of Iowa and South Dakota and in the Rocky Mountains region also to occupy much of Utah and eastern Colorado and New Mexico. Just at the time of climatic cooling in higher latitudes, from about A.D. 1200 onwards, withdrawal southwards and to moister regions from all the areas here mentioned set in, and pollen and soil studies indicate that the climate became drier. By the time the Spaniards penetrated the region that is now in the southwestern United States, the Pueblo Indians and their agriculture were concentrated along the Rio Grande river, and a parallel history of expansion and retraction of the Hohokam in southern Arizona was all over. Moisture changes farther east, in the lower Great Lakes region and New York State, were doubtless much less significant (and may even have followed a different course), and there was no parallel decline of the Iroquoian culture of Ontario and New York; indeed the climax there seems to have been between A.D. 1400 and 1600. In the northern Midwest, GRIFFIN reports evidence of a similar oscillation about 1000 years earlier. The sedentary Hopewell people and their culture were spreading north and west up the river valleys into northern Indiana, Michigan, Wisconsin and Minnesota between about 300 B.C. and A.D. 300, followed by withdrawal and cultural decline in the centuries after that, until A.D. 700 or later. Some of the Hopewellian village sites on the western part of the Great Plains are covered by a dry soil type separating this level from the humus soil of A.D. 800–1200. Also in the arid southwestern United States there was a development of settled agricultural villages in the 500 years or so after 300 B.C. Sparser data on soil profiles in the southwest and on the high, western plains indicate yet another, earlier oscillation on about the same time scale, with a dry climax about 800–300 B.C., preceded by alluvial deposition between about 1200 and 800 B.C. Other oscillations may be detected even before that, for which cultural associations can be suggested. More data and more precise dating of both climatic and cultural indicators are needed, however, before these suggested associations in very ancient times can be convincing.

In the case of Eurasia it becomes quite clear that knowledge of the sequence of changes of moisture and temperature level, which can be established from the past variations of lake levels and the history of the vegetation, can throw light on the history of Man rather than vice versa. The timing of the major population movements makes it reasonably certain that they were connected with the major events of the climatic history. But whereas the *Völkerwanderungen* which culminated in the barbarian invasions of the Roman empire (and parts of China) seem to have been triggered by increasing dryness, and presumed failure of

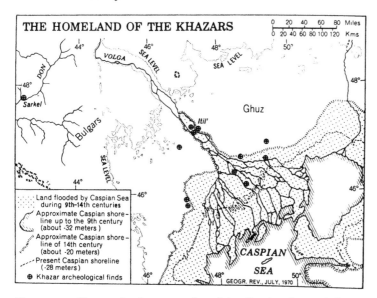

Fig. 13.72 Changes in the geography of the Caspian Sea and the lower Volga basin from the ninth century to the present day, indicating the well watered basin in the thirteenth and fourteenth centuries A.D., about the time of the conquests of Genghis Khan.

The 'Sea level' line marks the contour that corresponds to the level of the oceans. Heights mentioned are relative to ocean level.

(*Reproduced from* CHAPPELL (*1970*) *by kind permission of the author. Printed here by courtesy of the* Geographical Review (*American Geographical Society*), *New York.*)

the pastures, in central Asia towards A.D. 300 (cf. HUNTINGTON 1907), the Mongol hordes of Genghis Khan, whose conquests between A.D. 1205 and 1225 spread to south Russia, the Indus and the gates of Peking, seem to have flourished in a time of moist climate and plenty (GUMILYOV 1968, cited by CHAPPELL 1970, p. 370), which probably produced a population explosion. One aspect of this medieval moisture maximum in inner Asia is shown in fig. 13.72.

Much farther back in time we may notice some aspects of human history where the influence of the temperature and moisture regime of the 'postglacial climatic optimum', or *Wärmezeit*, is seen. One of these is that there was apparently no insuperable desert barrier to migration, or other weather difficulties, sufficient to prevent the development, or perhaps the spread in earliest Neolithic times – from some previously more localized homeland, possibly in the Caucasus region – of the Indo-European community of peoples, with some approach to unity of language, all over Europe and southwest Asia and parts of India before about 3000 or 2500 B.C. After that separate development – evidently in more isolation from each other than before – of the Indian, Persian, Armenian, Slavonic, Baltic, Greek, Latin,

Celtic and Germanic language groups proceeded. Most, perhaps all, of the later waves of migration which brought first one and then another of these linguistic groups to dominance in this or that part of Europe and southwest Asia seem from their dates to be related to climatic changes. More refined dating of both would probably throw light on the connection in each case. Archaeological evidence from China, in particular a Neolithic village of the Yangshao culture of 5000–6000 years ago, apparently supported by many inscriptions on oracle bones from as late as the period 1400–1100 B.C., indicates (CHU KO-CHEN 1973) that through the warmest postglacial times the winter climate there was generally about 5°C warmer than now, (see footnote, Chapter 16 p. 389), the whole year 2°C warmer than now, rain was commoner than snow in winter, the various species of the natural fauna and flora ranged farther north than now, and crops like rice could be sown about a month earlier. The evidence that the limit of bamboo was about 3° of latitude farther north than now may be the most important point, for it means that bamboo groves were then vastly more extensive than now in the Chinese lowlands as far north as the Yellow River (Hoang-Ho) basin instead of stopping at the mountains near 32°N. Since bamboo was the most important raw material, providing food in its young shoots, while its full grown stems were used for building and for writing on, not to mention hats and other garments, furniture and musical instruments, its abundance must have greatly facilitated the development of civilization in China.

Other features of the postglacial warmest times which we shall notice in their proper context in later chapters include the undeniably greater availability of water in northwest India down to the time of the Indus Valley (Harappan) culture, in 2500 or earlier to 1750 B.C. (SETH 1963, SINGH 1971), and in Egypt and the Sahara before 2500 B.C. (BUTZER 1958). Radiometric dating also shows that the earliest Alpine lake dwellings in Switzerland (at Thayngen and Burgäschisee), like those at Federsee in the Alpenvorland, were constructed between 4000 and 3500 B.C. Other construction phases followed at Federsee around 3200, 2800, 2400 B.C. and after, and again about 800 B.C., presumably all these being stages when the lakes were well developed, though the areas have since become largely peat bog. GAMS (1937) writes of these sites in the Alpine region that a period of flooding about 2000 B.C. divided the older from the younger series of lake settlements (*Pfahlbausiedlungen*). The wet period after 900 B.C. ultimately brought the whole of this life style to an end with increasing flooding, rapid peat growth and bog bursts.

Ancient ocean voyages and discoveries (see fig. 1 in Appendix V)
Voyages of discovery in any age are likely to have been of two kinds:

(1) Accidental discoveries of islands and coasts beyond the sea, resulting from vessels being blown or carried off course by winds and currents, or by boat parties desperately seeking escape from storms, ice or warlike invaders of their homelands.

(2) Expeditions deliberately organized for exploration and exploits of fishermen, whalers and traders deliberately extending their range.

The accounts of discoveries of the former class are sometimes informative regarding the winds, weather and ice conditions met with in a particular year (there is an example shown in fig. 13.2). The range, and success or failure, of the expeditions in class 2 may tell us more of the climatic conditions prevailing in the times when they were undertaken.

The Norse discoveries of eastern Iceland about A.D. 860, of Greenland (Gunnbjørn's Skerries on the east coast near 65°N) about the year 930 and of America (Vinland) in 986, and the Portuguese discovery of the Azores in 1432, were all due to ships being blown off course.[1] It is known that reports exist of Japanese fishermen from time to time having similar experiences in the Pacific. But many Viking expeditions, like those of the Irish monks before them to the Faeroe Islands soon after 700 and to Iceland in the 790s (and of St Brendan about A.D. 540 ranging widely across the Atlantic to places which cannot certainly be identified but apparently reaching the Greenland ice and possibly the Azores and Sargasso Sea), must be classed as expeditions with exploration as an aim.[2] It seems clear that conditions for sailing in northern waters were more commonly favourable in Viking times, especially after about A.D. 900, than they had been for some while before that or than they subsequently became; storms as well as ice made communications between Norway and Iceland and Greenland more difficult from about 1200 onwards.

The most frequent Viking sailings between Greenland and Vinland seem to have been between about A.D. 1000 and 1020, but the decline after that early stage seems more reasonably attributed to difficulties with the native Indian inhabitants – a potentially overwhelming commitment – than to climate. Visits, presumably to collect timber, and perhaps furs, etc., continued, and a bishop sailed there about 1120. The last report of such a voyage is one of a party of Greenlanders in 1347, who were storm-driven to Iceland on their return. By that time the recommended sailing route between Iceland and Greenland had been changed to a track farther south than before (see p. 6). During the heyday of the medieval Norse colony in southwest Greenland cod was fished and pollen analysis indicates that in some localities corn was grown; the situation for the latter was clearly that of the extreme margin of possibility. The *King's Mirror* records that some among the wealthiest

1. The discoveries here listed are the first sightings reported by NADDOD (in the *Landnámabók*) in the case of Iceland, and in the case of Greenland by GUNNBJØRN a Norse sailor who was driven south and then too far west to within sight of Greenland on his crossing to Iceland, and by BJARNI HERJOLFSSON who in 986 was also carried at first south and then too far west, to the coast of 'Vinland', on a voyage from Iceland to southwest Greenland. The first landings and the expeditions which established settlements in each case came some years later. The source mainly used for the information in the next paragraphs is JONES (1968).

2. Medieval reports are also known (TAYLOR 1956) of an expedition fleet sent by the British King Arthur in A.D. 531 exploring northwards from the northern isles of Scotland. The reports allege that the expedition carried over 2000 people, including 400 women, and established groups in both Greenland and Iceland, though nearly half perished when five of the ships were driven on to rocks in a storm. The expedition is alleged to have met inhabitants already established in Iceland, who warned of dangerous seas and currents farther north. If there is any substance in these reports, this earlier population in Iceland must have died out between that time and 790–860; the epoch is thought to have included a marked cold period. A party of Greenlanders, however, who reached Norway in the deteriorating conditions of the late fourteenth century, in A.D. 1364, knew of this legend and claimed to be descended from these first immigrants.

and most prominent people in Viking Greenland sowed grain, 'but the great majority in that country do not know what bread is, having never seen it' (cited by K ROGH 1967). (Modern Eskimo sheep farmers there grow barley, oats and rye as fodder: they ripen only in exceptional years.) The Old Norse Greenlanders pressed their explorations far to the north and northwest into the Canadian Arctic.[1] Cairns were built on Washington Irving Island (in 79°N) and in Jones Sound (76°35′N) which leads through to the remoter parts of the Canadian archipelago. There is even evidence (R. L. I VES 1953) to suggest that one Viking ship got right through the Northwest Passage and ultimately reached the Gulf of California, where its fair and red-haired occupants astonished the local Indians who preserved the legend. The northernmost penetrations of the Viking Greenlanders were probably mostly at the same time when the Eskimos were operating farthest north, and had even reached northeast Greenland, since few encounters between them are reported.

East of Greenland, and in the European sector of the Arctic, voyages seem also to have been pressed farther north about the same time. There had been sailings back and forth along the Norwegian coast for many centuries, and OTTAR (OTHERE), whose home is believed to have been either in Malangen or Kvaløy (69½°N) in north Norway, told King Alfred in England how about 870–80 he had sailed north beyond the limit of the whalers of those days, evidently to the White Sea, which he explored till he found the land all cultivated by Lapps on the far side of a big river mouth. HARALD HARDRÅDE, who was king of Norway and England – he was killed at Stamford Bridge in 1066 – and whose exploits ranged from the far north to Byzantium, is reported by A DAM OF BREMEN (Book IV) to have explored, evidently some time between 1040 and 1065, 'the expanse of the Northern Ocean' with a fleet of ships to a point where to the north and northeast, beyond the limits of land (Spitsbergen or Novaya Zemlya?), he reached ice up to 3 m thick and 'there lay before their eyes at length the darksome bounds of a failing world, and by retracing his steps he barely escaped in safety the vast pit of the abyss'. In 1194 Icelandic annals record the discovery of a land called Svalbardi, a name meaning 'cold side (or coast)', 4 days' sailing northeast of Iceland and presumably therefore Spitsbergen. Seals, walrus and whales were hunted in that direction, and there are good grounds for supposing that this hunting ranged far east towards Novaya Zemlya.

The contrast between the successes of these expeditions in the early medieval days of expansion and the disasters which dogged the European attempts to find and force the Northwest and Northeast Passages to the Pacific from A.D. 1500 until the present century (HAMPDEN 1958, KIRWAN 1959) parallels the many other indications of the climatic change which had occurred between those times.

Voyages traversing great stretches of the Atlantic Ocean are known to have been made in every century since much more ancient times. Many were regular trading voyages; others, like the Phoenician circumnavigation of Africa under Pharaoh N ECHO between 610 and 595

1. An expedition reached Melville Bay (76°N on the west Greenland coast) in 1267, finding traces of Eskimo occupation, and returned safely. A runic inscription on a stone near Upernivik (73°N) recorded a visit there as late as 1333.

B.C., and HANNO's voyage to the west coast of Africa near the equator about 425 B.C., were deliberate voyages of exploration. Doubtless there were also accidental landfalls on previously unknown coasts and islands, when ships were blown off course. A hoard of Carthaginian coins found in 1749 in a vault on the island of Corvo in the Azores points to knowledge of those islands between 600 and 140 B.C., and it has been suggested that fugitives from the Roman destruction of Carthage in 146 B.C. may have reached not only various islands in the west but also Brazil and central America – a possible source of apparent links between the pre-Columbian art of those areas and ancient Egyptian forms (though the resemblances perhaps need indicate no more than a parallel development on both sides of the ocean maturing later in the Americas) (ASHE 1962, CARY and WARMINGTON 1929). The Phoenicians had first reached the Atlantic soon after 1200 B.C., and founded a colony at Gades (Cadiz) in 1110 B.C.

There may well be implications about warm climate and less stormy winds than now in the seafaring enterprises of earlier peoples in northern waters in late Neolithic and Bronze Age times. Amber from Denmark (Jutland) was traded by sea to the Mediterranean. There was shipping from Cornwall, and perhaps from Brittany, to Scandinavia and probably also to the Mediterranean, as early as about 2000 B.C.; and the megalithic monuments were established about that date, or in some cases a few centuries earlier, in western and northern Britain and in Denmark, south Sweden and Norway as well as at Stonehenge. The apparent uses of these monuments as astronomical observatories (THOM 1964, HAWKINS 1966, LAMB 1974) also suggest a greater frequency of cloudless skies. It is clear that the western channels and sea routes near the Hebrides and Orkney Islands were most used (FOX 1932, pp. 74–5); possibly the Straits of Dover were still so much narrower and shallower than now that they were a dangerous passage at that time, but the traffic suggests that the Pentland Firth was less dangerous than now. A little of the trade also spread far up the coast of Norway,[1] where in the early Bronze Age most imported wares seem to have come in by the west coast. CLARK and PIGGOTT (1965) remark that the small semi-subterranean stone-built houses of the Skara Brae period in Orkney reproduce in startling detail recent Eskimo types. These and other features of the culture suggest contacts with other peoples in the tradition of the Circumpolar zone. It is not impossible therefore that Atlantic crossings by the northern route may have taken place at that time; isolated finds of megalithic structures in eastern North America (ASHE 1962), if authenticated, may be memorials of such a link.

By the late Bronze Age the main routes for trade between northwest Europe and the Aegean were overland, and presumably in connection with this pattern of trade southeastern and eastern England and southeast Norway were becoming the most inhabited areas in those countries (FOX 1932, HAGEN 1961). It was about 1800 B.C. that traffic across the Brenner Pass became important; but BROOKS's (1949, p. 301) suggestion that it was only then that it first became traversable after the ice age seems to be wrong.[2] What is certain is that Caithness

1. As far as Troms, i.e. 68–70°N (HOLMSEN 1961, p. 49; NORDHAGEN 1933).

2. It cannot be reconciled with GAMS's (1937) summary of knowledge of the height of the tree line and

and Orkney became important chiefly in two epochs, around 2000 B.C. and A.D. 1000, both coinciding with periods of warm climate.

Memory of the northern sea routes evidently persisted long after the peak of activity on them in the early Bronze Age. PYTHEAS explored the western coasts of Europe between 325 and 285 B.C., most likely between 310 and 306 B.C. (CARY and WARMINGTON 1929). From Cornwall he went north by the old western route through the Irish Sea and Hebridean Sea to northern Scotland, where he observed high seas running '80 cubits high', presumably at the north coast, and was told of an island, 'Thule', 6 days' sail farther north. PYTHEAS's book *On the Ocean or a Description of the Earth* has been lost, but his writing about Thule made a great stir among ancient geographers and is therefore known at second hand through a number of sources, some of them coloured by their authors' prejudices. It is not quite clear that PYTHEAS went to Thule himself, but his description gives various details and an observation of the latitude ($66\frac{1}{2}°$N) which is accepted as reliable, since he was a skilled astronomer. From the latitude given, and from the fact that one day's voyage beyond Thule the sea was frozen, the island seems clearly identified as Iceland (though Norway has been suggested). He speaks of inhabitants who had no crops or cattle and subsisted on wild berries and honey and something translated as 'millet', which they threshed in covered barns because of the continual rains. He also claims to have observed a strange substance 'in which earth and air are suspended', presumably volcanic pumice, a 'sea lung', on the sea's surface around Thule.

The dates now indicated by radiocarbon tests for the Polynesian migrations which settled the islands in the Pacific spread over a long period of time during which climate changed in the higher northern latitudes, but nothing is known of relevant disturbances of regime between 20°N and 20°s in the wide Pacific sector. Only if the jetstream system and the subtropical anticyclones were displaced nearer the equator in the coldest climate phases, tropical cyclones may have been less frequent at such times.[1] Quite small changes of sea level could also presumably be important to the population on many a coral atoll, especially if a rise of general sea level coincided with an increase of storminess or in the incidence of tsunami of volcanic origin. Fiji was inhabited by 1200 B.C., islands in the Tongan group in western Polynesia by about 400 B.C. and the Samoan group by about 200 B.C. The earliest radiocarbon dates in eastern Polynesia are from the Marquesas between 124 B.C. and A.D. 200, though there are some indications there and in Tahiti of settlers before 500 B.C. Easter Island came later, about A.D. 600. Hawaii and New Zealand were reached towards A.D. 750,

recession of the Alpine glaciers during the warmest postglacial millennia. The extent of the glaciers in the Glockner area during the Bronze Age is estimated to have been only one fifth of that during the seventeenth and nineteenth centuries A.D.

1. A computer modelling study (LEVISON *et al.* 1972) has been applied to the problems of possible drift migrations of the Pacific islanders and the chances of making safe landfall. Climatic change was examined in terms of a shift of the prevailing wind zones north or south by 5° of latitude. The results suggested that the migrations could not have been made by pure drift with wind and current, but demanded some skill in navigation. Climatic change was thought to make no difference; but the specification of climatic change in the model ignored the possibilities arising from changes in the frequency of meridional wind components.

and all the widely scattered groups of eastern Polynesia by A.D. 1000.[1] Events which may well have depended in some way on climatic variation can, however, be recognized (1) in a Rarotongan tradition of voyages into the Antarctic about A.D. 650 (RAESIDE 1948), which now seems to have been a time when the Polynesians were also ranging more widely than before into the eastern Pacific, also (2) in an era of fires, remembered in Maori legend, which destroyed the forests in eastern South Island, New Zealand, and in some eastern districts of the North Island, probably between A.D. 1000 and 1450. One forest blaze is alleged to have been started by a Maori chief, Tamatea; but the warm dry climate of the eleventh to fourteenth centuries is believed to have been unfavourable to the podocarp forest and stopped its regeneration (HOLLOWAY 1954).

Salt

Salt, which today is among the cheapest of foods and, like water, is served free of charge in the restaurants in all civilized countries, was formerly so highly prized as to be the object of massacres and the slave trade. Like water, salt (NaCl) is vital to the human body and to the bodies of animals, and its concentration in the blood is regulated to within narrow limits by the kidneys and sweat glands and by the body's alternations of thirst and craving for salt.[2] Dr M. R. BLOCH of the Negev Institute for Arid Zone Research has drawn attention (1970) to the archaeology of salt, and his study shows incidentally how, with due caution, this archaeology may throw light on climatic history and vice versa.

   An important thesis is developed by BLOCH, starting from the fact that the earliest towns, notably Jericho, and agricultural settlements which we know of, from about 8000 B.C., in the neighbourhood of the Dead Sea with its accessible salt deposits, seem already to have eaten bread and made leather, which is hardly possible without the use of salt. These settlements must be presumed to have had forerunners which have not yet been found. Since the date of establishment of this advanced culture in the Jordan valley comes in the time of quickest rise of sea level (up to 10 cm/year) at the end of the ice age – and perhaps just after a pause associated with the post-Allerød (Younger Dryas) cold centuries – it is not unreasonable to suppose that the same way of life had previously been more easily developed

1. The Maoris arrived about A.D. 1350. An earlier Polynesian people is known to have once been numerous in southern Otago.

2. When forced to live with too little salt, the body takes on a little less fluid than is released. So the total blood content falls as a result of the mechanism which strives to regulate salt concentration. The flesh gradually becomes lean and then wastes away. The diseases which follow include apathy, cramp, sleeplessness and, ultimately, coma and death. BLOCH (1970) is inclined to attribute to salt deficiency the backwardness which developed in African populations and in other lands in the tropical rain forest zone by comparison with Europe and Asia.

Man becomes addicted to a high level of salt intake; and, if the supply fails, his energy and sexual activity are the first things to be affected. It seems clear that maintenance of the numbers of the population may be affected. BLOCH points to the innumerable signs of Man's desperation for salt, demonstrated, for instance, in the word 'salary' (and 'salty' wit) so ancient that PLINY found it necessary to explain that soldiers were originally paid in salt. And in recent centuries the price of a slave in Africa south of the Sahara was from one to four blocks of salt.

along the coastal plain of Palestine and Syria, which till after 9000 B.C. must have been many kilometres broader than now, where salt could readily be produced by the evaporation of sea water on the sea shore (this was done in very early times by the primitive inhabitants of the coasts of Africa and Europe, even in latitudes where great half-baked clay pans had to be invented to evaporate the water with the aid of artificial heat (RIEHM 1960)).

The postglacial rise of the sea must have driven the inhabitants of low-lying coastlands in all parts of the world to settle in new places, where they could find food and reorganize their salt supply. After a few centuries, their former settlements lay metres deep under the sea and are presumably now up to 50 m below sea level. The danger at times of storms must have made it seem that nowhere beside the sea was safe and no salt pans could be maintained. The Jordan valley must have been among the few good sites in the world that could be found so near the former coastal plain. It offered cultivable land, fresh water as well as salt, and even lakes with fish. In the world in general, and in tropical lands in particular where no such sites existed, one may suppose that there was a great decline of population.

At an early stage, the organized trade in salt from the limited inland sites where it was available involved transport by ass and camel and caravan trains with armed protection along routes provided with fortresses. On this account, and for simpler reasons as well, whenever there was a pause in the inroads of the sea, as when a cooler phase of world climate halted the recession of glaciers and stabilized or temporarily reversed the trend of sea level, one may suppose that men would gladly revert to coastwise settlement.

BLOCH, in fact, traces a long history of alternations of cultural dominance between inland and coastal cultures, which seems related to climate controlled fluctuations of sea level. Thus, the Jordan valley was apparently abandoned for some time around 5000 B.C. and again about 3500 B.C., between 2200 and 1800 B.C. and about 1500–1200 B.C.[1] In at least the last three of these intervals it is known that coastal civilizations flourished in the Mediterranean and elsewhere. For the rest of the time in those millennia the more continuous developments of civilization were inland, as was again the case after 1200 B.C., when sea level was still high and possibly higher than it had been for a few centuries previously. The mining of salt at Hallstatt in the Salzburg area of Austria, i.e. in the heart of the European continent, goes back to about 800 B.C. or earlier.

The great period of Greek and Phoenician sea trading and establishment of maritime colonies coincided with the lowered general sea level associated with the cooler climate between about 600 and 100 B.C. Harbour works from about 500 B.C. at Naples and in the Adriatic indicate a sea level more than 1 m below that of the present. There is similar evidence from various points on the north African coast, the coasts of the eastern

1. It would, at first sight, be hard to account for a drift of population back to the sea coasts around 2000 B.C., which on most evidence seems to have been the time of highest postglacial sea level or, at least, the time when world sea level reached its highest postglacial stand. This implies, however, that sea level ceased to rise and became stabilized. The intervals when the sea coasts were abandoned after 2000 B.C. may have been as much due to increased storminess and storm floods as to any fluctuations of sea level lasting for many years.

Mediterranean and the Aegean and in the Crimea – from ancient harbour works and salt pans – though a minor rise of sea level may have occurred about 400 B.C., when the harbour at the mouth of the Tiber was moved to Ostia and the older Roman coastal salt pans were lost (there was also a shortage of salt for some time in Greece about the time of the Peloponnesian war in 431 B.C.). But through these centuries, the coastal civilizations managed to maintain themselves against the threats of inland empires. The Graeco-Roman era saw a great development of the sea fisheries, which would not have been possible without salt to preserve the catch. Salt fish, and salted and smoked fish, pork and sausages, were transported over great distances (e.g. fish from the Crimea and northwest Europe, and sausages from Athens) to Rome where they were considered great delicacies.

Salt began to be in shorter supply in the first centuries A.D., perhaps because of the growth of population. Sea level was also rising again, however, and by around A.D. 400 may have been more than 1 m above the present level. The harbours of Rome (Trajan's harbour) and Ravenna (Classis) were moved to places which today are 1 km or more inland. One can point (BLOCH 1965) to radiocarbon dated beach and estuary lines in Brazil, Ceylon, Crete, England, the Netherlands and Palestine which indicate a world-wide (eustatic) sea level high stand at about A.D. 400.

Another epoch of lowered sea level (perhaps almost to the level of 500 B.C.) can be established around A.D. 700. Salt became more widely available again at that time and was used once more for the long-distance shipping of fish. Many peat-burning centres of salt production were active (HALLAM 1959–60) in Lincolnshire and elsewhere on the North Sea coasts from then until after A.D. 1000, including some which had existed earlier and been out of use for 1000 years. The medieval warm epoch, however, was apparently accompanied by a gradual rise of sea level, which together with the increasing storminess from about A.D. 1200 onwards wrought great losses of land around the North Sea and flooded the peat cuttings in East Anglia, which became the Norfolk Broads.

Population, health and disease

It has to be expected that better knowledge of climatic history has far more to contribute to understanding of the history of human well-being, health and disease than vice versa. Only where, as often happens, the weather of the seasons which caused crop failures and famine is described in accounts of these disasters, and similarly with the weather that preceded or accompanied some epidemics, does this field of inquiry produce specific information about weather and climate.[1] The temperature and weather controls of the bacilli and insects concerned in the outbreak of certain diseases have been mentioned (p. 188), and doubtless

---

1. The frequency of famines in different epochs may at first sight seem likely to indicate climatic fluctuations, but famines depend as much on the number of mouths to feed as on the weather, and transport facilities or the lack of them may also play a part. Mapping all the information available for each individual year may, however, help towards diagnosis of the weather pattern.

there is much more to discover in this important field of biometeorology. Nevertheless this knowledge can only yield hints of why a particular disease reached epidemic proportions in a given region at some particular time. It may well be also that weather, and its effects upon vegetation and crops, has controlled the population bursts among rats and insects which led to outbreaks of plague or of widespread diseases of cattle and crops. But, in general, we shall get only corroboration or qualitative pointers to events in climatic history from study of these matters. Yet some interesting relationships appear, which might reward further investigation.

If we view first the longest time scales, the development of the different races of Man, notably in the matters of body size, shape and pigmentation, seems itself to have been a process of selection and adaptation to climate and the environment that goes with it. The white or pink-skinned peoples descended from the earlier inhabitants of northwest Europe fare best in – and are perhaps the race best adapted to – cool, cloudy climates without very much sunshine. Their skin is of a type that burns easily when exposed to strong ultraviolet radiation from the sun and sky. The brown and black races have a skin which is not vulnerable to damage in this way. Their physical type was presumably developed by peoples living between the deserts and the equatorial rain-forest in Africa and Asia and was equally well suited to life in Australia. The third skin type, which may be olive white, sallow, yellow or brown, readily tans, or darkens, when exposed to the sun and becomes paler again in the cloudier seasons or when living indoors. With Man, as with the other warm-blooded animals, body size tends to be bigger among those living in the colder climates of their range than in warmer regions.[1] The average weight of the peoples of northern Europe today is from 70 to 80 kg, that of the native peoples in Arab lands and southern Asia is only 50 to 60 kg. The native inhabitants of cold climates tend to be either big or stocky (thick set), those native to hot climates tend to be lean and to have long fingers. Negroes have been found to be more vulnerable to frostbite than white people, but tolerate humid heat better than the latter because their body temperature regulates itself with less sweating and evaporation from the skin.[2]

1. Since the surface area of an object varies with the square of its radius, whereas the volume increases with the cube of the radius, larger bodies have a smaller ratio of cooling surface to bulk.
2. Man's power of acclimatization, however, can confuse these issues. Among the most remarkable cases of acclimatization are the Arctic fishermen who manipulate wet lines and nets in temperatures about the freezing point and the ability (reported in the last century) of the native inhabitants of Tierra del Fuego and the Australian desert to sit and even to sleep unclothed in the frost. Acclimatization to high mountain climates, with reduced oxygen pressure, like other acclimatization processes, takes time, and mountain sickness is the reward of many who ignore this. Acclimatization to centrally heated environ-

ments appears to increase the vulnerability of the white races to exposure to cold.

Man's comfort in the open depends not only on temperature but on the wind speed and relative humidity. The simplest among a number of formulae developed to express the cooling power of the air, or windchill, is HILL's 'cooling factor'. This is given by H, in thousands of a calorie lost per cm^2 of skin per second, as

$$H = (0.14 + 0.47\sqrt{v})\,(36.5 - T)$$

where v is the wind speed in m/sec and T is the air temperature in °C. If the skin is perspiring, the figure 0.14 in the first bracket must be replaced by 0.21 and T is replaced by T', the wet bulb temperature. Values of H between 5 and 8 are comfortable; $H \leqslant 3$ is hot, $H \geqslant 15$ is described as cold, $\geqslant 20$ extremely cold.

There is a valuable literature on climate and human comfort and the energy of nations. The reader should refer particularly to HUNTINGTON 1924, MARKHAM 1942, BROOKS 1951 and LANDSBERG 1962; the last-named gives further bibliographies: on 'comfort indices' on p. 246 and on various applications in planning and design of cities and housing, in engineering, and in industry and commerce, etc., on pp. 405–6.

Some of the diseases of mankind are also known (from the study of bones, skeletons, mummies, bodies preserved in peat bogs, cave paintings, etc.) to have a long history[1] (WELLS 1964). In some cases connections with climatic history are clear, in other cases obscure but perhaps still present.

The evolution of the diseases classed as treponematoses, related to syphilis, in Man has been traced back by HACKETT (1963) to the apparently world-wide distribution of pinta established in very early times, presumably with the migrations of peoples into Australia and the Americas during the last ice age. Pinta, which is primarily a skin infection passed by children in countries where few clothes are worn, is now confined to the Indians of those parts of the humid tropics of Central and South America where there is least contact with other races: it had been general there when the Europeans first arrived about A.D. 1500. Yaws, which is also chiefly a skin infection passed by children where little clothing is worn, but which affects the deeper tissues and the bones more than pinta, seems to have developed as a mutation within the Old World at some time when the postglacial rise of sea level had cut off connection with Australia and the Americas; it is therefore a disease of the humid tropics in the Old World, where pinta has not existed since perhaps 10 000 B.C. Endemic syphilis, a mucous infection also passed mainly by children, affects more parts of the body than pinta or yaws and seems to have developed around 7000–5000 B.C. as a mutation from the latter which established itself in the arid tropics from northwest Africa to central Asia (Mongolia) and independently in corresponding climates south of the equator, in Bechuanaland and perhaps in central Australia. It may have been an adaptation to the cold desert nights in climates too dry for the yaws virus but in temperatures where human resistance was lowered. A further mutation may have produced venereal syphilis in the Old World as long ago as 3000 B.C., though it was otherwise diagnosed, on some evidence usually as leprosy (see especially WELLS 1964, p. 102). The emergence of this variety of syphilis seems to coincide with the beginning of the wearing of more clothes: presumably the earlier virus forms would have died out when clothes hindered their passage from host to host. Whatever the part, if any, played by the Columban discovery of America, there is no doubt that yet another mutation produced a graver form of syphilis in Europe between A.D. 1490 and 1510, which had not previously been present either in Europe or the Americas.

In the diseases discussed in the previous paragraph we see the effect of climate, and climate-induced changes of sea level, chiefly as providing barriers which isolated within very

1. Osteo-arthritis can, in fact, be recognized in the bones of reptiles of the Jurassic age, 150 million years ago.

broad regions each new form produced by successive mutations. With many other diseases variations of their prevalence, or the outbreak of epidemics, may be related to the weather prevailing at a particular epoch or in a given year. WELLS mentions that the Anglo-Saxons in England were often severely crippled with arthritis, whereas the Bronze Age burials in Dorset show little sign of it. This may be a measure of the cooler, damper climate of the seventh to ninth centuries A.D. compared to the second millennium B.C., though the author supposed it had to do with Saxon engagement in heavier farm work (the latter is probably a questionable comparison).

Malaria, known in medieval England as the agues and in Germany as *Wechselfieber*, was established in the marshy districts of central and northern Europe as early as the tenth century and reached a peak in the medieval warm epoch, particularly in the 1150s (MARTINI 1954, pp. 220–1). It may have been imported by armies returning from ravaging Italy. It was at its worst in the warmest years and in the spring of the following year, but dry sites and dry houses and rooms were always relatively healthy places. There is little or no mention of the disease in Europe north of the Alps in the fifteenth century, but it was present from about A.D. 1500 onwards, reaching even Scotland and Norway, and there were epidemics in individual warm years even in the seventeenth century (Oliver Cromwell died of the agues, probably contracted in the English fenland near his home, in 1658). There is some suggestion that in Italy the incidence of malaria got worse in the cooler climatic phases, probably because of increasing extent of marshes; gradual depopulation of the Campagna, Latium and southern Etruria between 500 and the 130s B.C., and again between A.D. 600 and 800, has been tentatively ascribed to this. Malaria seems again to have increased in Italy with the increasing wetness in the thirteenth and fourteenth centuries A.D. The chilly summers after A.D. 1690 seem to have eliminated the disease in northern Europe for a time, but it was present again in the warmer years in the eighteenth and nineteenth centuries: peak frequencies in Sweden in 1847–9 and 1856 reached almost two cases per 1000 of the population.

Some sort of loose parallel between the waves of malaria and of the plague (or pest) in central and northern Europe has been alleged (MARTINI 1954, p. 222). There were, however, at least two tremendous epidemics of the plague which devastated the population of Europe, beside which in retrospect other outbreaks seem of rather little account, though in both cases a series of recurrences at intervals of from 4 to 12 years followed in the half century after the great epidemics (mortality was by far greatest in the initial outbreaks in each case).[1] The first known great epidemic, the plague of Justinian's reign as emperor, which has sometimes been ascribed to the pneumonic (lung) form of the pest rather than bubonic plague, spread across

---

1. The other reported plagues of antiquity are now generally diagnosed as due to other diseases. That which afflicted Athens in 430 B.C., in the second year of the Peloponnesian war, was 'almost certainly' typhus. The pestilence which caused up to 2000 deaths a day in Rome, and affected much of the Roman empire, in A.D. 144–6, and again in 171–4, is put down largely to smallpox (WELLS 1964). In parts of the empire, e.g. in Egypt, a third of the populace is estimated to have died. Another great pestilence which began about the year A.D. 250 seems not to have been identified (BONSER 1963).

Europe from Egypt and the Near East in A.D. 543–7; mortality was on a scale comparable with the Black Death of just eight centuries later (the disease long failed to reach the British Isles, presumably because the black rat (the preferred host of the fleas which carry the bacillus) was not then present in these islands and there were virtually no rat-infested ships such as played a role in spreading the Black Death in the fourteenth century; the Justinian plague finally reached Ireland in 664 and spread to England, where outbreaks continued at least until 688). The visitation of the late Middle Ages, the Black Death, which was the bubonic (glandular) plague, spread over Europe from the Crimea in 1348–50, having perhaps started in central Asia in 1338–9, reaching China and India by 1346. It reached Iceland about 1400. Probably between one fifth and one third of the population of Europe died in the initial outbreaks. Both the great plagues, in the sixth and fourteenth centuries A.D., occurred in times of increasing climatic disturbance, i.e. of intermittent great storminess and wetness alternating with periods of drought and great heat, at the end of an epoch of several centuries of warm climate in Europe. Both have also been attributed to the chaos prevailing in periods of cultural breakdown. The two diagnoses may not really be in conflict. In both cases the recurrent outbreaks finally ceased in the colder centuries which followed: severe winters helped to quell the individual epidemics.

Infantile paralysis (*poliomyelitis*) is known from ancient Egypt, though people in subtropical countries where it is endemic seem to have some immunity to its severer forms. A 'lameness' mentioned in the Annals of Ulster in the year A.D. 708 is thought to mean an early epidemic in the British Isles. Its outbreaks in Europe and the United States of America became notably more frequent and severe in the warm epoch of the first half of the present century and were particularly severe in some of the outstanding fine dry summers, as in 1947 in Europe, though doubtless these epidemics owed a good deal to infection picked up in areas in North Africa and India where the disease is endemic and carried to Europe and North America by the movement of armies and other travellers. The peak of the occurrence of this disease in both Europe and North America, before its control by inoculation, was in the years of generally warm summers in the period 1942–53, particularly in the driest of those summers.

The incidence of the deadly sleeping sickness, spread by the bites of the tsetse fly, which gave West Africa the name of 'the White Man's Grave' before the modern control measures of inoculation, clearing the bush and spraying DDT, evidently underwent important natural variations from time to time, which probably depended on climatic variations. The fly requires a certain level of humidity and the shade given by the forest and bush vegetation near the river banks. It is found all the way from Sierra Leone to Lake Victoria, but it spreads only in the rainy seasons. A memorial in Sierra Leone indicates that the death rate among Europeans was far greater in the 1860s and 1870s than either before or since.[1] The

---

1. The average expectation of life of a European at that time in West Africa was six months. Yet the Jesuit and Methodist Christian missionary societies never had less than twelve men waiting in London to go out to replace those who died.

continuing record of rainfall measurements at stations on the West African coast begins only in the 1880s. It is known that rainfall decreased from then until the 1920s and after that had a tendency to increase. The incidence of sleeping sickness was also thought to be increasing again in the 1930s, but events were soon overtaken by more effective artificial control of the disease. Farther north in Nigeria and the other territories between the Atlantic and the Sudan the wettest years are those when the summer monsoon rains spread farthest north and may not be the same as the years of greatest rainfall on the coast of the Gulf of Guinea: recent evidence suggests that between 10° and 20°N the years of most rain coincide with strong zonal index of the circulation over middle latitudes and high frequency of westerly weather type in the British Isles (WINSTANLEY 1973). This would indicate that the 1860s and early 1870s were probably unusually wet years with the rivers and other water bodies relatively well filled, a condition repeated around the 1920s. Perhaps this sequence, rather than rainfall at the coast, is the clue to the history of the White Man's Grave before control of the mosquito.

Another disease of which we have some knowledge back to about 3000 B.C., and the varying incidence of which, at least in middle and higher latitudes, probably goes with climatic variations, affecting sunshine and the amount of ultraviolet light, is tuberculosis. It was virulent in ancient Greece about 400 B.C., according to a description by HIPPOCRATES; galloping epidemics of tuberculosis occurred after dry summers with more than usual frequency of N'ly wind and after rainy autumns with much S'ly wind. RUSSELL (1958) thinks that these accounts of its virulence mean that it was either new to Greece or had got worse about the time of HIPPOCRATES. It seems to have become less fatal in the Mediterranean world by the first century A.D., a time when the climate seems to have been becoming warmer and sunnier. Much earlier cases have been recognized in the Neolithic population of northern Europe, and in a mummy from ancient Egypt, but nothing is known of its frequency in those times. Indirect indications may perhaps be gained from statistics of the average length of life, and particularly from the frequency of deaths between the ages of 15 and 35 years, among the peoples of former times.

The past history of total numbers of the population of any country or region is hard to establish before the days of censuses and registers of births. Estimates may be made from the numbers paying tax and from the number of households, but these are wide open to errors in the assumed average numbers living together in one family or in one house at any given time.[1] One of the best modern studies of the subject is that by RUSSELL (1958).

Many studies have made it clear that European societies before the present century generally ensured that the numbers of their population were at least roughly regulated to their circumstances by a stern insistence that no one should marry until he could support a family. RUSSELL reports that in the late Roman empire (fourth century A.D.) not more than

---

1. An average of 5 per household has been assumed in many historical studies. This figure is almost certainly too high, since the man-wife-three-children unit lasts only a few years in the career of any family and some allowance should be made for old, widowed and unmarried people living alone or in small households. RUSSELL's figure of 3·5 is probably a better general estimate.

39 to 44% of the population of marriageable age were in fact married. Amongst the serfs in the ninth century A.D. the proportion was 32% in southern France and 29% in Germany. There was an upward surge of marriages after the Black Death, when jobs were plentiful and waiting to be filled. But the figures in France and central Europe between 1427 and 1454 gradually fell from 43–49% to 32–39%. WRIGLEY (1966) reported that the average age at first marriage in Colyton, Devon (England), from 1560 to about 1645 was 27 years. Thereafter the average age of women at first marriage rose to 30 years about 1700 (though the age of the bridegrooms apparently did not change). After 1720 the age at marriage began to fall again, until by 1880 it was 26 and in the 1820s and 1830s it was only 23. Whatever the detailed reasons for these variations, they all go in the direction that would be expected if they were in fact controlled by the mean annual temperature curve.

A similar association is suggested by such figures as are available for length of life. The following average ages at death are given by COMFORT (1969):

Mesolithic people in Europe 31·5
Neolithic, Anatolia   38·2
Bronze Age, Austria   38
Classical Greece   35
Classical Rome   32
England A.D. 1276   48
England A.D. 1376–1400   38

It must be granted that other factors besides climate, particularly the proportion of the people studied who lived in towns, could have influenced the variations shown by the figures in this list. Nevertheless they show an average length of life in the two samples from the warmest postglacial times which exceeded by about 10% (or more) the figures for the later civilized times in Classical Greece and Rome. Similarly the average length of life in the warmest part of the early Middle Ages in England was impressively long and exceeded by a substantial percentage that found a century later.[1]

We find the same tendency in the figures for average stature of young adult men in Iceland, given by BERGTHORSSON (1962):[2]

Period (A.D.)	Mean height (cm)	Number of specimens sampled
874–1000	173·2	39
1000–1100	171·8	55

1. If the figure given for England in 1276 is truly representative for Europe in the medieval warm epoch, this would readily explain the evidence of a great increase population from the ninth to the thirteenth centuries.
2. It seems likely that the same climatic changes also affected general health and the average age of attaining puberty in particular. Only Norway has statistics on this that go back more than 100 years, however, so our knowledge is limited to the years of generally improving (warming) climate and is confused by the possible effects of modern comfort and child care. TANNER (1962) quotes the following figures for the average age of menarche in Norway: 1847 17 years, 1871 16½, 1880 16, 1888 15½, 1908 15, 1920 14½, 1932 14, 1948 13½, 1959 13·06 years.

Period (A.D.)	Mean height (cm)	Number of specimens sampled
1100–1563	172·0	17
1650–1796	168·6*	23
1700–1800	166·8	7
1952–1954	177·4	1463

* This figure may be unrepresentatively high, because it came from a rich people's graveyard.

In Greenland the average stature of the people buried in the graveyard of the Old Norse colony at Herjolfsnes, initially similar to the Iceland population, had fallen by the fifteenth century to 164 cm.

Estimates of the total population of Iceland tell much the same story. THORARINSSON (1961) gives the figures in Table 13.14.

Only in communities living in areas or circumstances particularly vulnerable to climatic change do we find evidence like this of the population dwindling to a small fraction of its former size. We have mentioned in Chapter 12 (p. 13) that in 1784 the total evacuation of Iceland was seriously considered.[1] Long before that the old Norse colony in Greenland had completely died out. And in the upland parishes in Scotland in the successive harvest failures in the 1690s it has been estimated that about one third of the population died.

Table 13.14   Population of Iceland at different epochs (BULL 1916, from THORARINSSON 1956, 1961)

Date (A.D.)	Population	
1095	about 77 520	(from tax records)
1311	about 72 420	(,, ,, ,,)
1703	50 358	
1784	about 38 000	
1801	47 240	
1901	78 470	
1960	177 292	

We shall see in a later chapter how the history of the withdrawal of cultivation from the heights and from the northern limits which it had attained in Europe in the warmest period at the height of the Middle Ages, and of the villages and lonely habitations that were abandoned in England, Scandinavia and Germany, parallels the decline in the level of prevailing temperatures from the thirteenth century to about A.D. 1700. In Iceland, too, the

1. The revival of Iceland's population by 1800, seen in Table 13.14, can safely be put down to the effect of the warmer years in the 1790s. The great increase between 1900 and 1960 came in a more prolonged and still warmer time, but it doubtless owes something to the modern technology of comfort and health.

same withdrawal is seen; there the inhabited districts never exceeded a quarter of the area of the country, mainly the coastal fringe and the broader lowlands in the southwest and the valleys in the north and east of the island, but at one or two points in the early days settlement did extend to the inland plateau. The latter places were later abandoned.

It is a pity that many historians in western Europe (e.g. SALTMARSH 1941, SLICHER VAN BATH 1963), well-informed in their own field, have denied that climatic change was at work in the general decline of agriculture in the late Middle Ages and the accompanying economic and cultural decline. This has left them with a needlessly confused and complicated story, in which most weight is usually given to the Black Death, although it is admitted that the decline began at least 40 years earlier with the famine years around 1315 and even the later phases of the recession, especially in the fifteenth century, do not notably fit the dates of the plague. The trouble seems to be that their accounts of the phenomena were written before the problem of deriving the temperature record had been solved. The climatic references in their works are mainly to BROOKS, whose pioneering studies in the field of climatic history had been directed largely at rainfall variations without much attention to temperature. Subsequent research has derived the course of prevailing temperatures, first from statistical analysis of the documentary reports of prevailing weather, and later from oxygen isotope measurements on the Greenland ice and the trees in Europe, and from changes in the height of the upper tree line, etc. (e.g. figs. 6.6 in Volume 1 and 13.18 in this volume). The sustained warmth of the twelfth and thirteenth centuries probably meant that the soil was generally drier, because of the longer periods of evaporation, despite greater rainfall than in the colder centuries which followed. This diagnosis is supported by the evidence of the peat bogs, their growth rates and recurrence surfaces, indicating a marked increase of wetness after about A.D. 1300. STEENSBERG (1951) showed that the archaeological dating of this change can be considered precise. The climatic history so revealed seems to 'fit' in considerable detail the events reported by the historians.

It seems to be universally agreed that climatic changes account for many of the changes of fortune which affected agriculture and health in the northern countries. The general parallel with the course of the temperature record is evident in Sweden, where the population statistics have long been good and estimates have been extended back to the sixteenth century by careful local studies.[1] There may be some climatic implication in the point, noted by UTTERSTRÖM (1955), that the population of Sweden and Denmark, especially the latter, recovered within two or three generations after the Black Death, whereas the numbers in Norway – especially but not only in the colder mountain districts – were not made good for over 200 years. The warm decades from 1500 to 1540, lasting on to some extent to the 1550s, which are recognizable over most of Europe in the summer wetness/dryness index (Appendix V, Table 4), and in the winter index also in the early decades of that century, evidently had

---

1. The important sources are G. SUNDBÄRG's *Bevöl-kerungsstatistik Schwedens* (1907) and local studies in the provinces of Dalarna and Ångermanland. A brief summary of the results is given by UTTERSTRÖM (1955). See also MANLEY's commentary (1958).

their effect in the prosperity that marked the reign of GUSTAVUS I (1523–60), when crop failures were rare and not many severe winters were mentioned. Economic activity and the numbers of the population both increased considerably. Calamitous years followed at the end of the century, particularly between 1595 and 1603, with a high death rate. A parson's account of this period from Värmland, quoted by UTTERSTRÖM (1955), describes the distress of those years in central Sweden in details that are very similar to the accounts of the 1690s both in Scotland and Sweden (cf. Chapter 12, p. 11). Despite the increase of marriages and children born among the survivors, and the fact that there were good harvests and mostly milder winters in Sweden from 1604 to about 1620, the population of the country seems to have been less in 1620 than in 1590. An increase of population occurred through the middle of the seventeenth century, though it may have been interrupted in the years of economic catastrophe just around 1630, 1649–52, 1675–7, and certainly in the 1690s. The increase was reckoned at 25 % between 1630 and about 1695, when several years of harvest failure brought the rise of population in Sweden to a halt. The number of births was much larger between 1660 and 1690 than in 1690–1720. Sweden's population began to increase again, with a notable spurt in the 1720s and 1730s, which were warm decades; over the eighteenth century as a whole the increase was much greater than in the seventeenth century.

The population changes in northeast Scotland since about 1700, described by WALTON (1961), also parallel the fluctuations affecting the warmth or coldness of the climate. After the high death rate in the 1690s there was a general increase of population, which, apart from individual famine years in 1709 and 1740–1 (both caused by weather), continued until the 1770s. This was checked by years of scarcity in the early 1780s, particularly in the poorer and harsher Highland glens, and emigration contributed to the decline of population between then and 1800. Population increased again over the first half of the nineteenth century; but, though the situation was beginning to be affected by industrialization, the population of the region began to decline again with the difficult years for agriculture in the second half of the nineteenth century.

The prosperity, and hence the numbers of the population, in southern Europe, and in England also, during the sixteenth century seems to have gained by the difficulties and partial eclipse of the countries farther north. There was a great increase in the population of the Mediterranean countries over the period 1450 to 1600; the increase was still going on between 1550 and 1600, though more slowly, but it then ceased and in some areas reversed (BRAUDEL 1949, cited by UTTERSTRÖM 1955). LADURIE's (1971) attitude to the effects of the colder climate of the period after 1550 in southern Europe is curiously ambivalent, though he finally (1971, pp. 233–7) accepts the facts. His own researches on the agricultural history of the upland districts of the Languedoc in southern France had started his interest in the subject, because again and again the records of poor harvests, famines and shortages there in the sixteenth and seventeenth centuries were found to be accompanied by reports of severe winters and wet summers (between 1599 and 1603 cold weather destroyed much of the olive plantations in Provence). Yet LADURIE insists that these were only passing incidents and there was no trend; the behaviour of the Alpine glaciers he regards as a separate matter, and he criticizes UTTERSTRÖM and other historians for introducing climate to account for troubles which could be as well explained by economic developments. Nevertheless, LADURIE's own survey of the reports in local documents of the winters in southern France produces the figures in Table 13.15.

Table 13.15   Winters in southern France in the sixteenth century

	Very severe	*Severe*	*Cold*	*Mild*
	Numbers of episodes described as			
1500–49	1	3	8	2
1550–99	7	12	16	2*

* Both the mild episodes reported in the second half of the century were before 1570.

There was a devastating famine in France in 1587 due to torrential rains, and in Sicily in 1591, and there were disturbances in the countries ruled by the Turks in the 1590s associated with general distress in southern Europe. Between 1596 and 1601 there were crop failures in the districts of the Moldau and Wallachia, due to summer rains, which caused famines in Constantinople and shortages in Italy. The population growth in Spain was first checked by the plague in 1599–1600, but in 1694 was about 25% less than it had been a century earlier, according to BRAUDEL (1949).

Another source of historical information illustrated by LADURIE (1971, pp. 273–4, 378–80) is GIRALT's (1962) count of the number of days each year in the sixteenth century when prayers for rain are reported in the council records of Barcelona. From this it seems that droughts caused most anxiety in the summers and autumns between 1520 and 1550, and autumn droughts were again an anxious matter between 1561 and 1566 and 1575–84; the occurrences in winter and spring were much more evenly distributed over the century. Spain may well have shared in the summer and autumn wetness of other parts of Europe after 1550, and especially after 1590.

The vicissitudes of climate in Europe in the Middle Ages and since can also be traced in the reports of plant diseases, locusts and caterpillars, etc., which affected crops and the natural vegetation, and in the murrains of cattle and sheep. There was a great decrease in the numbers of sheep reared in Spain after 1560, and especially after 1600, associated with a general decline of agriculture and of the agricultural population in Castile; and though these can be associated with economic and international affairs (perhaps even including disturbance of the economy by imports of treasure from the New World), the coincidence with the dates of climatic decline in northern Europe and glacier advances in the Alps and elsewhere is remarkable.

Engineering works, irrigation, architecture, etc.

Differences of prevailing climate in the past are sometimes registered in the structures which men built to meet the needs of the time or in what happened to those structures afterwards.

Proto-Suez canals

The first canal dug at Suez to link the Mediterranean and the Red Sea may be a case in which climate played a part, in that what we now know of climatic history indicates that this was done approximately when world sea level reached its highest postglacial stand, and the project may therefore have suggested itself just because it then for the first time looked feasible. It is thought to have been early in the twentieth century B.C., under the pharaoh SESOSTRIS I, that a canal – albeit a fresh water canal – was dug from the Nile delta to the Red Sea near where Suez now stands. Another canal was made (or perhaps the earlier one was put in order again) under RAMESES II (1304–1237 B.C.).[1] This canal fell into decay and was restored at intervals which seem to have more to do with the changing ascendancy of empires (e.g. restoration begun by the pharaoh NECHO II around 600 B.C. and completed a century later by the Persian conqueror DARIUS) than with climate. It was restored and reopened for the last time in the seventh century A.D. and finally abandoned a century later after being blocked for military reasons.

Coastal habitations in northwest Europe

I. A. CRAWFORD (personal communication, 21 March 1973) has drawn my attention to the information on storms, and bearing upon changes of sea level and climate, established from archaeology and historical documents relating to a formerly inhabited site at Udal on the island of North Uist in the Outer Hebrides ($57°32'$N $7°30'$W), close to a northwestern shore open to the Atlantic (it is an area where there is little isostatic effect on sea level; any changes are likely to represent world-wide sea level changes). The site was occupied, and shows the architecture and household artefacts of almost 4000 years of continuous occupation, until it and a number of other townships in the islands were deserted after a great gale in A.D. 1697.[2] Trouble with drifting sand – great bands of the nearly white, loose sand from the shore burying the earlier cultural levels – seems to have started around A.D. 1400, evidently marking an important increase of storminess and presumably a lowering of the sea level widening the strand (though it is known that the tidal range was greater around A.D. 1400 than at any time since). There was a return to more stable conditions for a time, estimated as

1. The account of the exodus of the Israelites from ancient Egypt (II Moses 13:17–14:31), supposedly about 1230 B.C., surely indicates that the isthmus of Suez was narrowed at that time by an area penetrated by a tongue of the Red Sea, in which the sand of the desert was awash, sometimes as tidal shallows but sometimes rather more deeply.

2. Reported from the Harris estate papers by A. MORRISON in the *Transactions of the Gaelic Society of Inverness*, **45**, 47–8 (1967–8).

starting around 1500. But by 1542 the storms were increasing again, and the site became uninhabitable with the depth of sand that covered it in the gale of 1697. It was finally buried by 6 m of sand.

The better-known disaster in northeast Scotland in the winter of 1694–5, when in a violent NW'ly storm the Culbin sands on the north-facing coast of the Moray Firth, near Findhorn (57·6°N 3·7°W), overwhelmed farms and farmland over an area of 15 km² to depths of up to 30 m in moving sand, is also believed to have been the culmination of a series of storms of blowing sand in the area. It seems that the incidence of serious sand drift probably being some time early in the fifteenth century (see also Chapter 17, p. 460).

Bridges, irrigation works and other building structures, soil moisture, etc.

Valley cutting by streams in Hadramaut (16°N 50–52°E) in southern Arabia, which can be linked to beach terraces of a much lowered sea level on the coast of the Arabian Sea, indicates greater rainfall there in ice age times. LEIDLMAIR (1962), who has examined the area, reports that the filling of the valley and ravines suggests that the moist epoch was not characterized by prolonged rains but by occasional heavy downpours which were simply commoner than the similar events today. There were no perennial rivers. Undated irrigation channels, clearly built long before Islamic times and possibly therefore in the postglacial warmest millennia, drew their water supply at points 2 m above the present stream bed where it would not be possible to run water into them today. This must indicate one or more postglacial epochs when rainfall was sufficient to supply these irrigation works and for the stream to cut its bed down further (see footnote in Chapter 17, p. 428).

There are many indications of an era of rising lake levels and greater stream activity throughout the Mediterranean region setting in in post-Roman times (VITA FINZI 1964),[1] with greater erosion and deposition of alluvium along the lower stretches of the water courses, which had ceased by Turkish times (fifteenth century A.D. and after). The wetter phase seems to have begun about the sixth century A.D. and may have been separated in time from another wet phase in the Middle Ages. It is easy to suggest that the collapse and overwhelming of the structures built in the Roman empire for irrigation and control of the streams in the wadis of North Africa can be attributed to cultural breakdown, neglect and barbarian destruction; but the ubiquity of the evidence – including even remote sites, e.g. in the Epirus mountains – suggests a climatic cause. There seem to have been perennial streams in many places where there are none today. VITA FINZI hints that one aspect of the wetter phase (or phases) may have been a more even distribution of the rainfall over the year than now obtains. Confirmatory evidence of other kinds can also be found, e.g. a Roman villa believed to have been built about A.D. 300, the remains of which are now submerged in the

1. Evidence of this in Spain, Morocco, Algeria, Tunisia, Tripolitania, Egypt, Italy north of Rome, Sicily, Greece and Jordan is mentioned by VITA FINZI (1964, 1966). See also HIGGS and VITA FINZI (1966, p. 9).

lake at Castel Gandolfo in the Alban Hills south of Rome. The villa had been built to face the lake, and remains of a small stone landing stage have been found. Old writings exist which report a rise in the level of the lake in the post-Roman period.[1] Evidence of the moist period in the sixth century A.D. may be seen in a phase of flourishing agriculture in the Negev desert near Gaza. WILLIAMS (1973) reports a Byzantine farmstead with extensive buildings, water tanks and fields from that time near Tell Fara. There seems no doubt, however, that its demise was brought about by the warlike events of the following century, after which the untended land was reduced to desert. Though overgrazing is blamed for that, a renewed dryness of the climate may have played a part, since the Bedouin with their sheep and goats had always been present in the area.

A moister regime than now in the Middle Ages seems to be attested by the bridge with twelve arches which was built in A.D. 1113 over the river Oreto at Palermo (Plate V), a structure far too big for today's needs. And the great Arab geographer IDRIS (A.D. 1100–66) described the rivers San Leonardo and Erminio in Sicily also as navigable, which would now be impossible even for the vessels of those times (quoted from BUTZER 1958, p. 12).

There is also archaeological evidence bearing upon moisture changes in northern Europe, some of which will be referred to in appropriate places in the chapters on the history of climate. The history of the medieval water mills in east Kent where the streams are now too small, which was studied by MEYER (1927), has been cited by BROOKS (1949) as evidence that the rainfall was heavy in southeast England in the eleventh and twelfth centuries and became much less by the beginning of the fourteenth century; but, from most other evidence, this seems to be almost the reverse of the truth. It has been suggested that the water level in the chalk was still high in Roman times and has been falling generally since, perhaps by as much as 15 to 20 m. More recent studies of the rainfall in medieval England suggest that, though there were undoubtedly some heavy rains and floods, and probably a somewhat higher average rainfall than now, this was offset by the evaporation in the many warm and dry seasons, particularly in the tenth century and between about 1100 and 1310. Indeed, subsidence which has affected the Norman architecture of some old English churches and cathedrals (Carlisle is an example), and not the later parts of the structures, has been thought to point to excessive drying out and compaction of the soil (R. M. HIGGINS personal communication, 20 November 1967).[2] And archaeological digging at York Minster

---

1. From a report in *The Daily Telegraph*, London, 19 December 1972.
2. Professor J. HEYMAN, in a letter to *The Times* (28 December 1974), refers to 'the large number of collapses which occurred in the twelfth and thirteenth centuries' and suggests that no more may have been involved than the normal compaction of the soil under a heavy building, so that there is a critical period for any building within 10 or 20 years of its erection. He cites the west tower of Ely cathedral as another case where the settling of the soil prior to about 600 years ago has produced great distortion of the abutting arcades in the building. This leaves open the question whether the repeated coincidence of the date when settling ceased is to be attributed to effective cessation of the era of building large buildings or to a change of the prevailing physical conditions of the soil associated with the indisputable change of climate.

has suggested a number of considerable fluctuations in the water table since Roman times (J. A. LENG personal communication, 15 November 1973).

Remains of old banks and ditches, presumably enclosing and draining fields for cultivation on Whitmoor Common, an area which is now wet heath and sphagnum bog, near Guildford, Surrey, tentatively dated by coins found to the third to fifth century A.D., may indicate that that was a rather drier time; and enclosures are also known to have been made there in the Norman period (eleventh to thirteenth century) (G. J. THOMAS, personal communication, 25 April 1972).

A sharp change to wetter soil conditions in England is suggested by archaeological work by D. G. and J. G. HURST (personal communication, 29 September 1966) on various large moats which were dug in the fourteenth century: e.g. at Ashwell (Hertfordshire), Northolt (near London) and Milton (near Lymington, Hampshire). In some cases much smaller drainage ditches of thirteenth century date were found under the moats. At Upton, in the Cotswolds near Blockley, Gloucestershire, where the houses built in the twelfth and thirteenth century were in the valley bottom, building in the next two centuries moved up the valley side; and at Moreton, near Bristol, in Somerset, the whole village was moved away from the valley bottom about that time.

A deduction of quite another kind may be possible from the bridge with many stone piers (Plate VI) which APOLLODORUS of Damascus built between A.D. 101 and 106 for the Emperor TRAJAN across the river Danube at the Iron Gates. It stood for almost 170 years at a point where in most recent centuries such a structure might be expected to be carried away by ice in one or other of the severer winters. It was reputedly destroyed in the end by the Dacian tribes when the Romans withdrew from the area.

Mining

Mining for flints in the chalk in England, France, Belgium, Sweden, Egypt and elsewhere goes far back into prehistory, to the early Neolithic (CHILDE 1942). Mining for metals (e.g. gold in Ireland) goes back to about 2000 B.C.; tin and copper were won from the mountains of Europe and the Near East around, or soon after, that date. In the climatic deterioration between about 1000 and 500 B.C., accompanied by advances of the glaciers and lowering of the upper forest limit, many of the previously busy mines in the Alps, as well as traffic routes over the high passes, were abandoned (GAMS 1937, FLOHN 1950). In the warm centuries during the Middle Ages the glaciers receded and the forests spread up the mountains to above their present limits, and the long abandoned mine workings – e.g. the gold mines in the Hohe Tauern in Austria (47°N 12–13°E) – were started again. This was when the water supply ducts from high up beside the Aletsch glacier in Switzerland (Chapter 12, p. 8) were laid. These, and other water-lead installations (e.g. in Saasthal and in the Dolomites) of that time in central Europe, bear witness to anxiety about water shortage and droughts in the Middle Ages.

Again there is evidence of a turn of the climate of middle and northern Europe towards greater wetness after A.D. 1300. Underground water began to cause great difficulties in the silver mines in central Europe: at Goslar in 1360 it was reported that water had been increasing in the mines in the Harz Mountains for more than 50 years, and all attempts to get rid of the water had failed. In Bohemia similar trouble seems to have begun in 1315, and led to some mines being abandoned in 1321. Attempts to reopen mines at Rammelsberg in the early 1400s, which had been abandoned in the previous century, also failed (STEENSBERG 1951).

Art

Cave-dwellers' paintings from the last ice age in central France (the best-known at Lascaux in the Dordogne are thought to date from around 16 000 to more than 20 000 years ago, spanning the time of the climax of the last glaciation) and from other districts in central France and northern Spain (e.g. at the Altamira caves near Santander) convey a record of the large fauna with which the artists were then familiar in those regions. These included bison, horses, an animal like a Shetland pony, red cattle, elephants or mammoth, and reindeer, as well as rhinoceros and steinbok (ibex). Some of the pictures also show fish (salmon), birds and some grasses.

Similarly, rock drawings in the Sahara (Plate VII) and the Nile valley and the present desert areas between the Nile and the Red Sea tell of the fauna that must have been seen there between 5000 and 3000 B.C. or somewhat after. BUTZER (1958) has reported the faunas revealed by these drawings. In the Egyptian Nile valley elephants, giraffes and ostriches are depicted, all of which are missing from the art of dynastic times; lions and wild sheep, which were also frequent in the rock drawings become much rarer in the pictorial record of dynastic Egypt. Drawings and paintings from the fourth and fifth millennia B.C. in Tibesti and the Fezzan, as well as elsewhere in the Sahara, show elephants, giraffes, ostriches, rhinoceros, antelopes, etc., and in some of the oases a kind of buffalo, hippopotamus and crocodiles. Most of this fauna has since disappeared, presumably because of increasing aridity of the regions with falling water table and decline of the oases. BUTZER (1958) has attempted to define the modern limits of the ranges of elephant, giraffe, hippopotamus and rhinoceros in terms of annual rainfall and thereby to estimate the position of the mean rainfall isopleths 5000–7000 years ago. Fig. 13.73 reproduces his map derived in this way.[1] If, however, the extent of the oases in early Neolithic times owed much to fossil water from aquifers which had been charged in much earlier millennia, during Europe's glacial times, this map may be

1. WICKENS (1975), following work by WARREN (1970), uses the orientation of the sand dune systems to establish past shifts of the climatic belts over the Sudan and so deduces shifts of the isohyets 250 km north of their present positions between about 6000 and 3000 B.P. (cf. BUTZER's map in fig. 13.73) and up to 4° of latitude north of today's positions between 20 000 and 15 000, and again between 12 000 and 7000 years ago, whereas they possibly moved to a little south of today's positions between 7000 and 6000 B.P.

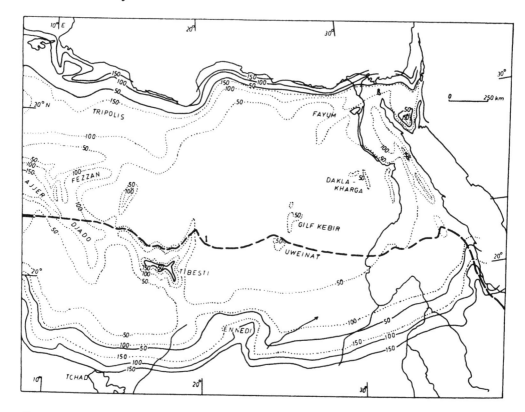

Fig. 13.73   Average rainfall in the eastern Sahara, as derived by BUTZER.
Full lines = today's regime.
Dotted lines = 5000–3000 B.C.
The isohyets give the per annum averages in millimetres from many years' data.
The heavy broken line marks the present northern limit of summer rain.
(*From* BUTZER (*1958*); *reproduced by kind permission of the author and of the Akademie der Wissenschaften und Literatur, Mainz.*)

in error, suggesting more rainfall than did occur in the Saharan region during the warmest postglacial times.

Stone Age rock engravings and paintings from postglacial times in Scandinavia show wild animals: bear and elk and whale, and backgrounds of forest and water. Hunting scenes and travel on skis are depicted. This Stone Age art is represented in most parts of Norway as far as 69°N and in Sweden and south Finland, but did not spread into Russia beyond Karelia and the approaches to the White Sea (CLARK and PIGGOTT 1965). The drawings from the Bronze Age are distinguished by domestic objects, including boats with high stem and stern, with both ends ornamented like the Viking ships of a later age, and apparently rowed by large crews. They are found graven on rocks among meadows and fields suitable for cultivation (HAGEN 1961).

From the foregoing paragraphs we learn just a few climatic implications in outline:

(1) That the fauna of the European plains and other open country south of the ice caps during the ice age included not only many present inhabitants of the temperate and sub-Arctic zones but also a few species that are now confined to the tropics (evidently the present restriction of range of the latter types is not due to climate).

(2) That there must have been more extensive oases and ground water in the Sahara desert in the warmest postglacial times, and probably more frequent rainfall, though the rains still came as occasional heavy storms.

(3) That the human inhabitants of Scandinavia in the warmest postglacial times were used to travelling over snow in winter and were used to seafaring in summer, though their boats were best adapted to the fjords and sheltered waters among the islands of the Norwegian coast and Baltic.

Some correspondence with the variations of prevailing climate within the last 400 to 500 years seems unmistakable in the history of European painting. In particular, the onset of the period of colder climate in the sixteenth and seventeenth centuries made an obvious mark on the imagination of PIETER BRUEGEL THE ELDER in ways that had far-reaching consequences; and it looks as if changes from one half century or so to another in the prevailing cloudiness or otherwise of the summers from that time on in northern and western Europe produced an exaggerated (or 'amplified') response in the minds of the artists and the prevailing fashions in landscape painting. The modern developments of abstract art, and earlier symptoms of that trend, of course, produce a situation in which any influence of the changing face of nature is suppressed or obscured.

PIETER BRUEGEL THE ELDER's famous picture 'Hunters in the Snow' (Plate VIII) seems to have been the first picture ever painted in which the landscape (albeit it an imaginative landscape) was the centre of interest instead of a mere background to some allegorical story or to a personal portrait.[1] This picture started the fashion of the Dutch and Flemish winter landscape pictures. It was painted in February 1565 and undoubtedly records the impression made on the artist by the winter of 1564–5, the longest and severest winter that anyone then living in the Netherlands had experienced. It was, as it turned out, a symptom – which was later to be repeated a good many times – of the deterioration of the general winter climate in Europe which was then beginning. BRUEGEL was so impressed by it that he set about recasting many religious themes in the setting of the bitter northern winter and so dramatizing the exposure and poverty in the circumstances of Christ's birth. An example is given in his 1563 and 1567 versions of the visit of the three kings to the Holy Family in an open shed. The earlier version of this theme is without any indication of weather

1. M. J. FRIEDLANDER, in *From van Eyck to Bruegel* (London 1956), wrote (pp. 135, 139) that PIETER BRUEGEL THE ELDER 'was at bottom an illustrator.... In the art of the fifteenth century ... the human beings [were] large, the landscape ... small. The human being ... the centre of interest ... everything else is accessory. ... BRUEGEL reverses the relationship. ...'

either in the scene itself or in the clothes worn by the people; the later version clearly places the shed on the outskirts of a Dutch village beside a frozen river and with snow falling. Many more Dutch winter landscape pictures were painted in the years that followed; sometimes different artists even repeated the same landscape in different winters. The elder BRUEGEL himself also used the remaining four years of his life in painting a series of landscapes to dramatize each season of the year. And so, it seems, a tradition of landscape painting was born.

A study by the author (LAMB 1967) suggests that the Dutch and English landscape painters have left us an interesting record of the variations in the prevailing character of the summers, though tending to exaggerate the variations (the two countries have a similar summer climate). The meteorological observations at Oxford show differences in the average cloudiness on summer afternoons in different years within the period from 1880 to 1960, varying from 5-tenths to 8-tenths of the sky covered with clouds, when the averages for the best and worst individual summers are compared, and from 6-tenths to 7-tenths as between the best and worst groups of summers. Average figures for the sky cover in a survey of 200 pictures by representative Dutch and English landscape artists of different periods in the leading London collections were as follows:[1]

Period	Average cloud cover (in tenths of the visible sky)
1550–68: paintings by BRUEGEL (who may have been still influenced somewhat by memories of his years in Italy	$4\frac{1}{2}$
1590–1700: paintings by RUYSDAEL, HOBBEMA and others	7 to $7\frac{1}{2}$
1730–88: paintings by RICHARD WILSON	5 to 6
1790–1840: paintings by CONSTABLE and TURNER (omitting TURNER'S less representational works)	7 to $7\frac{1}{2}$
1930s: British artists, representational works only	$5\frac{1}{2}$

The English Medieval Vineyards

It is appropriate to include in this section a note on the cultivation of the vine and on wine-

1. NEUBERGER (1970) repeated the exercise with a much larger sample, over 6500 paintings, but perhaps unfortunately included pictures from all over Europe, including the Mediterranean countries and northern Europe in the same survey. The results showed surprisingly low cloudiness (average 33%) in the fifteenth century, though this may be explained by a high proportion of Italian works at that time. The figure obtained for all Europe rose from 50% for 1500–49 to an average of 80% for 1550 to 1699. The figures obtained for the period after 1700 showed smaller variations, averaging about 75% in the eighteenth century and around 70% since. Both NEUBERGER (1970) and BRIMBLECOMBE and OGDEN (in *Weather* 1977) have found impressive evidence in paintings from the fifteenth century to the twentieth of the increasing pollution of the atmosphere of London and other European cities. The backgrounds become more and more frequently hazy and the prevailing rendering of the sky colour changes from blue to yellowish and pinkish greys. On this and other types of historical evidence BRIMBLECOMBE and OGDEN conclude that pollution of London's air rose sharply with the introduction of the burning of coals from Newcastle in Tudor and early Stuart times and was at a fairly constant high level from about 1690 to 1900.

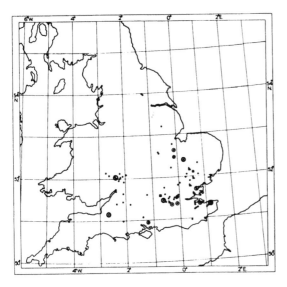

Fig. 13.74   Positions of vineyards cultivated in England between A.D. 1000 and 1300.

   Sources: W. CAMDEN, *Britannia, Or a Chronological Description of Great Britain and Ireland* (first edn 1586), translated into English, revised and digested by E. GIBSON (London 1722); H. C. DARBY *et al.*, *The Domesday Geography of England* (in several vols), (Cambridge 1952 ff.); H. K. ELLIS, *A General Introduction to Domesday Book* (London, Commissioners of Public Records, 1833); E. HYAMS, *Vineyards in England* (London 1953); G. ORDISH, *Wine Growing in England* (London 1953); A. L. SIMON, *English Wines and Cordials* (London 1946); *Ordnance Survey* large-scale maps; and various local works.

• Vineyard, usually 1-2 acres or size not known.
▲ Vineyard, 5-10 acres.
■ Vineyard, over 10 acres.
○ Denotes evidence of continuous operation for 30-100 years.
Ⓞ Denotes evidence of continuous operation for over 100 years.

making in England, particularly in the Middle Ages when many vineyards were maintained in central, southern and eastern England south of about 53°N, since consideration of them has been used (LAMB 1965) in deriving information about the temperatures then prevailing. The distribution of the former vineyards is indicated in fig. 13.74. Their northern limit parallels but lies about 500 km farther north than the modern limit of commercial vineyards in France and Germany. The success of the English vineyards in the period from about A.D. 1100 to 1310 was such that the French trade sought to have them abolished under an early treaty of peace with England (ELLIS 1833). Some of these vineyards were, in fact, operated for several hundred years, though decline set in and very few new ones were started after 1310. From WILLIAM OF MALMESBURY's description about the year 1150 of the vineyards in the Vale of Gloucester and at Thorney near Peterborough, the vines were cultivated in the open, trained up upon poles and not protected by walls.[1]

   For a good grape harvest the vine makes the following climatic demands: (1) freedom

1. WILLIAM OF MALMESBURY described the vine-yards in many districts of England in his book *De Pontificibus*, written about A.D. 1150. Of the vale of Gloucester he wrote 'here you may behold highways and publick roads full of Fruit-trees, not planted, but growing naturally. . . . No county in England has so many or so good vineyards as this, either for fertility or for sweetness of the grape. The wine has no unpleasant tartness or eagerness; and is little inferior to the French in sweetness' (from the translation in CAMDEN's *Britannia* in the English edition by E. GIBSON, London 1722). The description is thought to have applied particularly to the hilly country near the borders of Gloucestershire, Herefordshire (Plate IX) and Worcestershire where the soils are similar to the red grits and marls of the Rhine and Mosel valleys, which the English wine of this district probably resembled most in flavour. Ledbury area had some of the vineyards that were longest main-tained.

from late spring frosts, especially at and after flowering, (2) sufficient sunshine and warmth in summer, (3) not too much rain, (4) sufficient autumn sunshine and warmth, in regions where the summer is only just warm enough, to raise the sugar content, and (5) a dormant winter season, in which frosts are seldom serious.

Severe frosts have affected the northern limit of vineyards in France and Germany in a few of the most extreme winters – cultivation was never resumed in some parts of France after 1709, and the winters of the 1430–40 decade may have had a similar outcome both in France and Germany. Probably a prolonged period of frost with extreme temperatures below −20°C to −25°C is required for this to happen. Such temperatures have occurred in some of the former English vine districts, but it seems likely that (with the possible exception of Ely in the 1430–40 decade) such frosts only occurred after cultivation had already been given up. Severe frosts are less likely to have been the limiting factor at any stage of the

Table 13.16   Temperatures (°C) representing the modern northern limit of commercial vineyards

	Average lowest in May	Extreme lowest in May	Average temperature of July	Average temperature of October
Reims (1891–1956)	+0·5	−2·2	18·9	10·3
Luxemburg (1854–1956)	+1·7	−3·3	18·1	9·7
Köln (1851–1930)	+3·3	−0·5	18·6	10·5
Frankfurt/Main (1851–1936)	+2·8	−1·0	18·6	9·7
Apparent lowest acceptable values (approx.)	+1·0	−2·0	18·0	10·0

Table 13.17   Shortfall of prevailing temperatures (°C) in 1921–50, below the values in the bottom line of Table 13.16, in what were in 1150–1300 by report the best English wine districts

District and station used	Average lowest in May	Extreme lowest in May	Average for July	Average for October
Fenland (Cambridge)	1·5	2·5	1·0	0
Hereford–Gloucester border (Ross)	0·5	0·8	1·3	0
Middle-lower Severn and Thames valleys (Oxford)	0·5	0	1·0	0
London (Greenwich)*	0	0·2	0·2	0
Mid Essex valleys (Halstead)	1·5	1·9	0·7	0
Kent (Canterbury)†	0·1	1·9	0·6	0

* The Greenwich temperatures are doubtless raised by the artificial climate of the built-up area of London.
† Canterbury data 1921–44 only.

climatic decline in England than on the continent. The figures in Tables 13.16 and 13.17 suggest that it was cool springs, May frosts and to a less degree lack of summer warmth that in the end told against the vine in England (see also p. 460).

Table 13.16 gives relevant climatic values near the modern northern limit of commercial vineyards, and Table 13.17 indicates how far twentieth century temperatures in the best of the former English vine districts appear to fall short of the lowest values required (it is true that individual enthusiasm has succeeded in operating, often (but not always) in walled gardens, isolated vineyards in specially favourable sites in the south of England at one time or another in most centuries since the Middle Ages; these have never continued long after the retirement or death of the founder.) It is reasonable to deduce that in the medieval warm period, at least between A.D. 1150 and 1300, and probably between 950 and 1050 also, the temperatures prevailing in the English vineyard districts were higher than twentieth century values by about these amounts. Frosts in May were surely much rarer.

In central Europe also vineyards were found considerably farther north than now. Between 1128 and 1437 wine was produced in East Prussia, even at Tilsit (55°N), also in Brandenburg and in south Norway. In the Black Forest (Schwarzwald) there were vineyards up to 780 m above sea level, whereas the highest vineyards in Germany today reach 560 m near the Boden See (Lake Constance). The English medieval vineyards were nearly all less than 100 m above sea level, though one exceptionally favourable site is suggested at 200 m in Herefordshire. These figures for the height difference between the upper limit of vineyards then and now, taken in conjunction with the usual average lapse rate of temperature with height in the summertime of 0·6 to 0·7°C/100 m, suggest that in central Europe the average temperatures over the summer half-year in the medieval warm epoch were 1·0 to 1·4°C above those in this century and that in England the difference was fractionally less, probably in the range 0·7 to 1·0°C.

"FURTHER OBSERVATION ON BIRDS (see p. 186):
    The Hermit Ibis (Geronticus eremita), also known as the Bald Ibis, may provide an interesting example of an impact of climate on bird distribution that has been observed over a long period of time, although it is not quite clear that climatic change alone is responsible for the changes observed. Today the species is confined to North Africa, Arabia and Asia Minor, where it is a bird of dry country feeding on beetles and nesting on cliffs, and is described as a declining species. We know from Konrad von GESNER (Historia animalium, Zurich 1551–58) that this bird formerly bred in central Europe, coming as a regular yearly migrant to Germany with the storks and returning to the south again in July. It must have done so over a long time, presumably from the earlier Middle Ages, since it had acquired a familiar German name, Waldrapp, indicating that it was known in the woods and also for its croak or bark-like call (Rapp). It is reported to have become extinct in central Europe in the seventeenth century, its disappearance thus coinciding with the climax of the Little Ice Age.

    This and the similar case of another bird, the Steinhuhn, which in the sixteenth century still nested on the cliffs of the Rhine valley but today is confined to southern Europe, Asia Minor and Arabia, are mentioned by F. HAMM in Beiträge zur Naturkunde Niedersachsens, 4 (1), 1–14, 1951. (See also A new dictionary of birds, edited A. LANDSBOROUGH THOMSON, p. 390: London and Edinburgh (Nelson).)"

Chapter 14

Climate and the long history of the Earth

One can approach diagnosis of the climatic regimes of past epochs either by theoretical constructions based on physical and dynamical laws, using little more than an outline knowledge of the dispositions of land and sea and mountain ranges and any necessary allowance for differences of heat flux at earlier stages of the history of Earth and sun, or from observation of the traces left by former climates in the fossil record. These may, rather loosely, be called the theoretical and the empirical approach respectively. Certainty and understanding only begin to appear when, and in so far as, these two approaches can be seen to verify each other and help sort out the evidence.

Through long stretches of geological time, in periods before the Quaternary, it seems that the Earth had only warm climates, with no great ice sheets; though it is now thought that there may have been many periods when any islands or bigger landmasses which were then near the poles bore a cover of 'permanent' ice. Indeed, this may have been true in most eras.

It is essentially more difficult – because more different from our own experience and the observations at our disposal – to construe the meteorology and the global circulation patterns of the atmosphere and oceans which prevailed in warm eras with little or no ice at the poles than to reconstruct the patterns of the times of greatest glaciation in the ice ages.[1] The latter only involve adjusting the amount of permanent ice, and its effects, from the situation of the present day. To attempt to reconstruct the patterns of warm ice-free eras,

1. The word era is used here and elsewhere in this work in its *Oxford English Dictionary* sense to mean a (long) period during which a particular state of things prevailed; and is not restricted to the con- ventional usage of geology. The latter is, however, indicated in the Table of Geological Eras in Appendix V.

without the constraint of a strong thermal gradient related to the position of an ice margin, we have to rely much more on our theoretical understanding of the laws of atmospheric flow. Some insight may be gained by observation of the atmospheres of other planets (see HIDE 1971 and literature listed therein) and from laboratory experiments with fluids in a rotating dish under controlled conditions of heating and cooling (FULTZ 1961, HIDE 1970). Some possibilities of verification do exist, however, particularly through the discovery of geological evidence of the positions of desert belts (sand and salt deposits/evaporites) in the periods concerned, and of the prevailing wind directions in them (extent of the Trade Winds), registered by fossil sand dunes of various (longitudinal, transverse, barchan and bi-directional) types (MCKEE 1964). Sometimes there are even surviving ripples in the sand surface, though these must be carefully distinguished (by their grain structure) from water ripples before being attributed to some prevailing wind behaviour. Permian sandstones in Europe and North America show examples of these features. SCHWARZBACH (1961, pp. 58–9, 73) also describes fossil evidence of drops of heavy rain and evidence of lightning strikes, which are perhaps more likely to have been preserved from the rare convection shower and thunderstorm activity on ancient dried-out surfaces in arid regions than from any precipitation in wetter climates. Isotope work also offers increasing possibilities of diagnosing the temperatures which prevailed.

During the warm ice-free eras there would still be a general thermal gradient from the equator to the polar regions, though the overall temperature range at the surface must on a yearly average have been only about a half to three-fifths of that now existing;[1] in mid winter it was probably a little under half of the present range. In the upper air, owing to the presumably much weaker surface temperature inversions over the polar regions, the ratios near the 10 km level should have been as much as 80–85 % of the present equator–pole range on a yearly average and 65–70 % in mid winter.[2] If at any epoch the solar radiation reaching the Earth's surface and lower atmosphere was stronger than now, that would strengthen the equator–pole temperature difference, producing somewhat higher percentages of the present range than those here suggested; but the intensity of the present range depends greatly on the high albedo of the persistent snow and ice in, and near, the polar caps.

Fig. 3.20 (Volume 1, p. 120) and the theoretical considerations underlying it suggest that the warm era conditions estimated in the previous paragraph would shift the prevailing position of the axis of the subtropical anticyclone belt on average 5° to 10° poleward of its present latitude. The tendency of a weaker flow of the upper W'lies around the circumpolar vortex to produce shorter wavelengths and an increased number of troughs and ridges

---

1. Estimates by FLOHN of the distribution of mean temperature by latitude in warm ice-free eras and in glacial times, as well as today, appear in fig. 3.21 (Volume 1, p. 121, of the present work). Other estimates are worked out later in this chapter and presented in fig. 14.5. The reader may also be interested to refer to BROOKS (1949, Chapter 1, pp.

31–45) and to the commentary on BROOKS theory of the effect of a polar ice cap on the temperature distribution by latitude given in Volume 1 of the present work, on pp. 256–7.
2. For comparison, the mean temperatures prevailing in the present century, latitude by latitude, are given in a table in the Appendix.

around each hemisphere would be largely offset by the effect of the higher latitude of the mainstream of the W'lies (see fig. 3.5, Volume 1, p. 89). In the absence of persistent polar ice, however, there may have been more frequent passages across the highest latitudes of cyclonic storms, even if they were mostly rather weaker than those which occur now in those regions. And in such situations, considerations of scale suggest there would be another depression sequence (and a second jetstream aloft) in, or near, subtropical latitudes. At such times, very short wave lengths might be expected to prevail in the lower latitude system. Conditions generally prevailing between the subtropical high-pressure belts and the equator, with warmer oceans and (both local and poleward) temperature differences weaker than now, would probably be characterized by rather smaller-scale systems and more chaotic circulation patterns than in the present era. Convection rains in those latitudes might therefore be more localized and sporadic, but violent. The total amount of moisture in the atmosphere would be greater than in the present era; precipitation should presumably have been greater also, especially in all the higher latitudes and in those latitudes near the tropical circles which were reached by greater seasonal advances of the 'equatorial' or monsoon rains than now occur.

Many points of theoretical interest depending on detail in the climatic regimes of the remote past may be impossible to resolve. Among these are the questions of how much of the required meridional transports of heat, kinetic energy and momentum in the warm eras was achieved by eddy activity and how much by meridional components in the mean circulation of the atmosphere. A question of more specific interest is what part, if any, tropical cyclones played in these exchanges and, particularly, how their frequency compared with today. The writer has argued elsewhere that tropical hurricanes were probably less frequent than now in warm ice-free eras, despite the increased amount of water vapour in the atmosphere, because insufficient vertical gradients of temperature would be developed in airstreams which underwent less heating from the surface in the course of transfer from high to low latitudes. And since, in the ice ages, the intertropical convergence zone probably never passed far enough from the geographical equator – except perhaps in the Pacific – for the Coriolis force to start the initial rotation, tropical storms may well have been few then also. The implication is that such storms may be most frequent in intermediate climatic regimes like the present.

Mathematical models of the global atmosphere can hardly throw much light on such questions while their mesh is too coarse to show the development of tropical hurricanes.

Causes and time scales of climatic changes

It is beyond the scope of this book, and would in any case be premature at the present time, to attempt specific reconstructions of the world climatic regimes of many different epochs of the geological past. What can be attempted is to indicate what the different regimes depended on, and how we may come to a better understanding of them.

The farther one goes back into the geological past, the more fragmentary the field evidence becomes and the wider the margins of uncertainty of dating, so that it becomes increasingly difficult to establish the synchroneity of evidence from different parts of the Earth. Hence, in the remotest geological eras, there must be great uncertainty that any true global pattern can be recognized.

Changes and fluctuations of climate are most conveniently considered in different classes according to the time scales involved, which correspond to the action of different processes and different underlying causes. We may define the following classes:

(1)   Changes due to the development of the sun (luminosity waxing) and fluctuations of solar output induced by rotations of the galaxy: time scales 10^9–10^8 years.

(2a) Changes in the major geography, i.e. the Earth's crustal drift (continental drift and pole wandering) and development, changing the positions and heights of mountain ranges, and changes in the Earth's rate of rotation: over periods in the range 10^9–10^7 years.

(2b) Changes of composition and radiative characteristics of the Earth's developing atmosphere over the same time ranges.

(3)   Climatic fluctuations (e.g. between glacial and interglacial regimes) on time scales of the same order as the astronomical variations shown in the Earth's orbital arrangements: i.e. 10^5–10^4 years.

(4)   Recurrent changes, many of which appear to be cyclic, and possibly associated with fluctuations of (a) solar output, or solar disturbance, (b) the energy channelled towards particular parts of the Earth (especially around the magnetic poles) due to the effect of interactions between the Earth's, sun's and galactic magnetic fields on the travel of energetic particles, or (c) the range of the combined tidal force of sun and moon: the periods range from several times 10^3 years down to less than 1 year.

(5)   Evolutions over 1 to 10 years in the patterns and intensity of the global atmospheric circulation, and in the temperature regime and ice extent at the surface of the Earth, which follow those great volcanic explosions that establish persistent dust veils of hemispheric or global extent in the stratosphere. It is not yet established whether any regularities exist in the occurrences of much longer-term waves of enhanced frequency of great volcanic activity of wide, and possibly global, extent: time scales apparently in the range 10^2–10^4 years.

(6)   Changes generated by variations – however caused – in the circulation and internal heat economy of the oceans, including both surface ice and the deep ocean, on time scales ranging from a few weeks up to 10^3 years.

(7)   Fluctuations and changes on periods up to some weeks duration, associated with variations in the circulation and internal heat economy of the atmosphere (or atmosphere and oceans). Some of these seem to be more or less regular cyclic variations, e.g. on time scales of 7, 15 or about 30 days. Some may also be related to

an external 'trigger' (or exciting agency), e.g. a solar flare or a particle invasion of solar or cosmic origin.

(8) Variations of the atmosphere's transparency due to cloudiness, smoke from forest fires, etc., or from heavily populated and industrial areas, changes in the amount of carbon dioxide and other gases, or aerosols, water vapour, etc.; time scales very various.

(9) At the end of this list come the regular effects of the Earth's yearly passage round its orbit, including all the stages of seasonal variation of the radiation budget and the responses produced in the atmospheric circulation, the finer seasonal structure of the year, and lastly the regular effects of night and day as the Earth rotates on its axis.

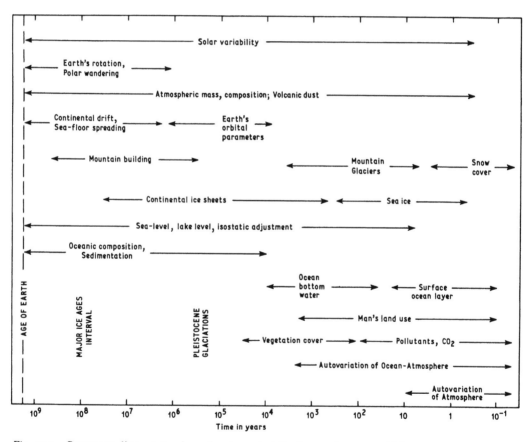

Fig. 14.1   Summary diagram to show the characteristic time-scales over which changes in various influences upon the global climate operate.

(*Adapted from a diagram by* W. L. GATES.)

The time scales characterizing the working of the various influences and processes are summarized in the diagram in fig. 14.1. All the fluctuations on the shorter time scales in the list above are liable to give rise to longer-lasting effects through the storage of heat in the ocean and smaller water bodies and through the albedo changes produced by snow and ice.

This chapter is mainly concerned with the changes of climatic regime in classes 1, 2(a) and 2 (b). But it must be supposed that fluctuations in the other classes – i.e. on all the shorter time scales – were always going on as well.

Changes in the sun and beyond the solar system

There can be little doubt that as the composition of a star like the sun changes in the course of its life history its luminosity changes. There must be much less certainty about the rate of change of its output, perhaps even about the direction of change at a given time. According to Cox (1968) the sun's luminosity must have been increasing over the ranges of time with which we are concerned, though the rate of increase is not known. Öpik (1965) has suggested that during the pre-Quaternary ice ages over 250 million years ago the sun's output may have been less than 88% of its present value. Glaciation should in that case have extended to the equator, which does not seem to be borne out by the evidence of such glaciations as occurred in Ordovician and Permian times, 400 to 250 million years ago; but the suggestion may provide an explanation of the evidence of possibly much more extensive and severe Pre-Cambrian ice ages (Harland 1964), two or more times, between 500 and 2000 million years ago. Because of the difficulty of dating the evidence closely enough to establish firmly the synchroneity of the evidences of glaciation in diverse parts of the globe in those remote times, it may not be possible to distinguish whether this evidence should be interpreted as pointing to occurrences of world-wide glaciation or to more numerous ice ages occurring in different areas in association with the drift of the continents over hundreds of millions of years. Moreover, it is not clear how the Earth could recover from complete ice cover (or even from a situation with ice covering all the land areas) because of the enormous loss of solar heating due to the high reflectivity (of the order of 80%) of the ice and snow, except in the course of the slow build-up of the sun's luminosity in the course of its life cycle.

The most elementary consideration of the seething motions which must go on inside the sun suggests that the sun must also undergo variations of output on many shorter time scales. Öpik (1958) regards it as certain that the sun 'flickers' in this way but difficult, or perhaps impossible, to suggest what the time scales should be.[1] He regards the evidence of past ice ages on the Earth at fairly uniform intervals of 250 to 300 million years as evidence of one of these flickering processes. Umbgrove (1947) was the first to suggest that solar

1. It is presumably no less difficult to suggest the range of the fluctuations of solar output, except that they must have been small enough to allow the continuance of life on Earth over some hundreds of millions of years past.

flickering over periods of this order could be due to gravitational disturbance associated with the regular motions of the other systems in the galaxy rotating about its axis.[1]

Another suggestion, by DILKE and GOUGH (1972), is that overstability in the sun causes its core to mix every few hundred million years (probably now about every $2 \cdot 5 \times 10^8$ years), starting a new burning cycle each time after the mixing event. Before the mixing occurs luminosity may have fallen by about 5%.

Composition of the atmosphere

During the history of the Earth the composition of the atmosphere and, therefore, its radioactive properties have changed. The earliest part of the story is bound up with the development of the Earth itself, and need not greatly concern us here. Hydrogen and helium constituted 99% of the whole, but were able largely to escape from the exosphere while the dimensions were hundreds of times those of the present planet and its envelope and there was no such concentration of mass as now exists (see, for instance, KUIPER 1957). If it is presumed that the protoplanets were of about the same composition as the sun, it is hydrogen and helium, and the other inert gases, that are most deficient in the present Earth. These considerations indicate that almost all the original atmosphere must have been lost to space. Arguments based on the atomic weights and relative abundances of the different inert gases still present suggest that the escape of the original atmosphere, including most of those gases, took place from a very rapidly rotating Earth. The present atmosphere seems to have developed from volcanic exhalations from the Earth's crust, which in the early part of the Earth's story were probably occurring at an enormously greater rate than at present.

Ammonia, water vapour and methane, as well as carbon monoxide and carbon dioxide, must always have been prominent among the gases emerging; sulphurous gases would also be present. There is reason to suppose that no great amount of molecular oxygen may have existed in the atmosphere until plant life was present to release oxygen in the process of photosynthesis. Before there was much oxygen the solar ultraviolet rays would penetrate deep into the atmosphere and decompose water vapour to free atomic oxygen, which would thereupon make a proportion of ozone, until the latter became sufficient to shield the lower layers from further penetration of this radiation which would inhibit the development of life.

It used to be thought that the amount of carbon dioxide in the atmosphere had fluctuated widely during the Earth's history, warming the global climate through its

---

1. This theory has since been supported and further developed by STEINER (1967). DAMON (1968), however, relates the occurrence of ice ages at about 250 million year intervals to crustal warping (epeirogeny), aggravated at 40 million year intervals by mountain building associated with a cycle of that length in the mantle; 'during periods of maximum regression of the epicontinental seas, climates are cooler and more differentiated.' He estimates that at the extremes of continental upwarping and mountain building the total area of the continents probably exceeded that of today, whereas at the time of maximum levelling of the relief transgression of the seas may have reduced the total continental area by as much as one third.

'greenhouse effect' (see Volume 1, p. 49) as it increased and thereby provoking a luxuriant plant life, which ultimately reduced the CO_2 to the recent level (i.e. before the industrial era) by fixation of the carbon in the plant tissues. It was realized that the slow production of limestone on the ocean bed also contributes to the removal of carbon dioxide from the atmosphere. These ideas were produced about the turn of the century (ARRHENIUS 1896, CHAMBERLIN 1899) to explain the apparent global warmth of the Carboniferous period – the coal measures of that period being mostly the product apparently of a type of tropical rain-forest, tree ferns, often with giant leaf fans and no growth rings. But the distribution of Carboniferous climates is now understood in terms of continental drift. It was also thought that glacial climates might be explained by periods when the Earth's atmosphere had become poor in carbon dioxide. There is no doubt that carbon fixation by plants, exchanges between the atmosphere and the ocean (CO_2 being released to the atmosphere when the ocean becomes warmer) and fixation at the ocean bed in carbonates, regulate the amount of carbon dioxide in the atmosphere; but, for this very reason, it is probable that the fluctuations of the global total (before Man's intervention in modern times) were within much narrower limits than was suggested by those who would press this as causing great variations of climate in the geological past.

Before the existence of algae in abundance, or the more advanced forms of plant life, there must have been a long phase in the earlier history of the Earth when carbon dioxide was accumulating in the atmosphere, ultimately perhaps reaching a greater proportion than now. One might expect therefore an early stage far back in the Pre-Cambrian when, with very little CO_2, the Earth was liable to more severe glacial climates than at any time since. And, less certainly, a stage may have been reached in the later Pre-Cambrian when more abundant CO_2 substantially moderated any fluctuation that tended to produce an ice age.

The Earth's crustal drift (continental drift and polar wandering): evidence from palaeomagnetism

As early as 1686 ROBERT HOOKE suggested polar wandering as a possible explanation of the evidence, from finds of fossil ammonites and turtle shells, that Europe had had a tropical climate at some time in the geological past. It has sometimes been alleged that in 1620 FRANCIS BACON noting the correspondence in shape between South America and southern Africa – just then seen on reliable maps for the first time – had suggested that they might have once been close together; but no such suggestion in fact appeared in his *Novum Organum* according to HALLAM (1973), who attributes the first clear assertion of this, supported by a map, to SNIDER PELLEGRINI in 1858, perhaps influenced by A. VON HUMBOLDT. This idea of continents drifting apart was revived by ALFRED WEGENER (1912), and expounded in terms of the geological evidence of similar rock formations on either side of the South Atlantic and with the concept of the lighter, largely siliceous rock of the continent blocks floating on the denser, iron- and magnesium-rich materials underlying both continents and oceans. Evidence of a certain plasticity of the mantle beneath the crust was seen in the slow isostatic rebound of the landmasses which had borne a great ice load in the last glaciation. Palaeoclimatic evidence for a drift of the continents, which included changes of latitude, was mustered by KÖPPEN and WEGENER (1924). Despite the fact that this hypothesis offered a much simpler explanation than any other put forward of the (in some cases impressively) matching coastlines and geology of various continents, particularly on either side of the North and South Atlantic, and got rid of the absurdity of a proposed regime producing a giant equatorial ice cap and (it was supposed) no glaciation in the Antarctic during the Permian and upper Carboniferous eras (elaborated in all its complexity by BROOKS 1949, Chapter 15, pp. 247–262), continental drift and polar wandering were not generally accepted until the evidence of palaeomagnetism, accumulating rapidly from the 1950s onwards, was seen to support it.

The development of the subject of palaeomagnetism, and the unfolding of Earth history that has come from it, probably owes most to NAGATA (1953), BLACKETT (1956), RUNCORN (see, for example, 1962, 1964), NAIRN (1961), BRIDEN and IRVING (1964).

It became known in 1925, from the work of CHEVALLIER on the dated lava flows of Mount Etna in Sicily, that when molten lava cools, the iron oxide compounds (ferrites) in it take on a magnetization in line with the ambient magnetic field and remain from that time on as a fossil record of the Earth's magnetic field direction at the time.[1] Both the declination and the dip angles (i.e. the departures from true north and from the horizontal) agreed with historical observation in the years concerned. The iron oxides in sedimentary rocks are also found to be magnetized in line with the Earth's field at the time when they were deposited as fine particles.[2] Study of the weak remanent magnetism in rocks of any age may therefore be taken as a fossil compass, indicating the direction of the Earth's field at the time of their formation, so long as metamorphic and other tectonic processes such as tilting have not disturbed the evidence since. Carrying the observations of rock magnetism back before the first measurements of declination by Man (around A.D. 1550) showed a continuation of the sort of changes observed since in the course of what is known as the geomagnetic secular variation, as the magnetic pole wanders around high geographical latitudes. And extending the survey over rocks of the last 20–25 million years showed the magnetic record averaging out as an axial dipole field, the equivalent of a magnet in the centre of the Earth aligned along the rotation axis (i.e. in line with the geographical poles).

The palaeomagnetism of rocks has thus produced a body of evidence about the behaviour of the Earth's magnetic field leading to the theory of the origins of this field in the motions in the Earth's fluid core. These motions must be governed to some extent by the

1. CHEVALLIER, R. (1925) *Annales de Physique*, 4, 5. Paris.
2. When molten lava cools, at temperatures around 600°C the atoms begin to develop alignment with the Earth's magnetic field. This is not necessarily the alignment of the whole rock crystals. The magneti- zation of sedimentary rock comes about through the minute iron compound particles rotating towards alignment with the Earth's field while the substance is still a wet slurry. Hence, both types of rock register the Earth's magnetic field at the time when they are laid down.

rotation of the whole system, so that on a long-term average the geomagnetic axis lies along the Earth's rotational axis, though fluctuating about it with a standard deviation of some 11° of latitude.

A complication was discovered in that 50% of the measurements of remanent magnetism in the older rocks were 180° out, indicating that the Earth's magnetic field has reversed many times, flicking over at intervals of about a million years (see Chapter 13, pp. 85–8). But, as these are precise reversals, they are no hindrance to use of the measurements as pointing in the direction of the pole at the time when the rocks were laid down. The angle of dip is preserved and, regardless of the polarity changes, makes clear which hemisphere the place of observation was in at the time the rocks were formed.

When one plots on a map the pole positions indicated by the palaeomagnetism in the rocks of any era, the results obtained from the rocks within any one continent block are consistent within the margin of error of the measurements and the range of positions produced by the secular variation. But the positions indicated by the rocks in different continents differ, as illustrated by the paths of the North Pole in fig. 14.2 given by the measurements on rocks in North America and Europe. To explain this, one must suppose that the length of the base line from America to Europe, at the ends of which the pole directions were registered, has changed; i.e. the continents have drifted relative to one another. The map also indicates that, regardless of the discrepancy so produced, the North Pole lay in the great Pacific Ocean from the late Pre-Cambrian, and possibly earlier, until some time in the Permian, a span of probably more than 300 million years.

Triassic rocks in Britain, red sandstones, about 200 million years old, are magnetized with a dip appropriate to latitude 20°N. Deeper down in the rock structure of Europe lie the coal beds, formed apparently from tropical rain-forest plants or trees with no annual growth-rings. The remanent magnetization in the rocks of this Carboniferous climate in Europe, as in North America, from around 300 million years ago, shows zero dip, in comformity with a position near the equator at that time. The dip observed in the Old Red Sandstone rocks of the Devonian and late Silurian eras, around 400 million years ago, places Britain in the southern hemisphere desert zone near 20°s. Thus, over all those millions of years Europe seems to have drifted more or less steadily northwards to its present position.

By contrast, in the lands with evidence of Permo-Carboniferous glaciation (India, Australia, South Africa, South America, . . .) the remanent rock magnetism indicates much more complex paths and generally quicker drift. India, now centred about 20°N, seems to have moved fastest, coming from 50°s only 130 million years ago; the Himalayas are now regarded as the product of its collision with the continental block of Asia. Rock magnetism in central Australia, now about 20°s, indicates a course which crossed the equator between 500 and 400 million years ago, reached 60–70°s between 200 and 100 million years ago and has returned from there. Its climate underwent rapid cooling between 300 and 200 million years ago.

If the Earth's rotation axis remains fixed in attitude, as its motion around the sun

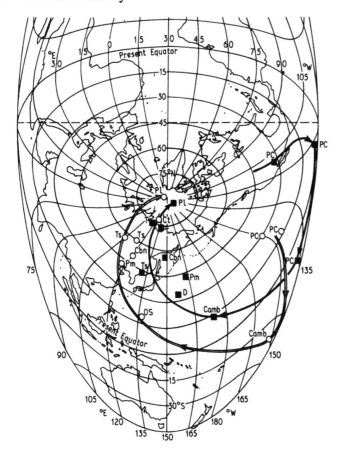

Fig. 14.2 Pole-wandering paths indicated for the North Pole from the palaeomagnetism still measurable in rocks in North America (bold line) and northwest Europe (broken line).

PC = Pre-Cambrian
Camb = Cambrian
  OS = Ordovician or Silurian
  D = Devonian
Cbn = Carboniferous
Pm = Permian
Ts = Triassic
Ct = Cretaceous
  T = Tertiary (Eocene to Oligocene)
  Pl = Upper Tertiary and Pleistocene

(*Adapted from a map compiled by* A. G. KOMAROV *in Priroda, Moscow, 1960.*)

proceeds through the ages, while the crust breaks into continental blocks (and smaller fragments) which part from each other and glide over the mantle and collide, we have both continental drift and the appearance of polar wandering. This is what seems to have happened, though some conflicts of evidence (particularly in the earlier eras) and difficulties of interpretation continue to crop up (see, e.g., *Nature*, **227**, 776, and **241**, 497–8, London, 22 August 1970 and 23 February 1973). The evidence of palaeomagnetism is generally supported by, and helps to explain, the evidence of climates in each era from the distribution of corals, deserts, salt deposits and glaciation. A difficulty, however, lies in the fact that the climatic zones each cover too broad a band of latitudes to corroborate the palaeomagnetic latitudes with adequate precision.[1]

The whole evidence for continental drift continues to be challenged by some, notably A. A. MEYERHOFF in the United States. One of the difficulties may be that at some stages of the

---

1. SCHWARZBACH (1961, pp. 25, 186, 187) reproduces world maps of the salt deposits and coral reefs formed in (respectively 5 and 10) different periods during the last 500 million years.

Earth's history, e.g. in the Permian, the magnetic field was not axial (*Nature* **227**, 776, London, 22 August 1970). It seems to be true that not only the palaeoclimatic, but also the palaeontological and geological evidence of past eras is best explained by the crustal drift positions indicated by palaeomagnetic data, but that each of these other lines of evidence individually yields too imprecise an interpretation to clinch the matter. Proof may ultimately elude us, but the continental drift data provides the most attractive hypothesis in that it is the simplest and is at least generally supported by the widest range of independent types of evidence.

The thesis which makes sense of the movements of the continental blocks is a system of convection in the upper mantle. Material rises in the mid-ocean ridges and from there spreads to either side: the concept known as sea-floor spreading (HESS 1962). The new oceanic crust so formed becomes magnetized in the direction of the Earth's field as soon as it cools, and the direction of this magnetization persists in the rock concerned. Because of the variation with time of the Earth's field this is evidence of the time when the rock solidified and, hence, the rate at which it has spread sideways. The rates are from 1 to 9 cm/year. The eastern and western parts of Iceland, which has been formed – probably in the last 13 million years – by volcanic lavas outpoured from the mid-Atlantic ridge on which it lies, are drifting farther apart at a rate of a few centimetres a year. And the rate of widening of the North Atlantic Ocean since it began about 200 million years ago has averaged 2–3 cm/year. The continental blocks are carried along with this movement, and the great orogenic belts are formed where these continent blocks, or the ocean plates with which they are associated, collide. Mountain-building in such belts in general marks the location of the downward limb of the convection cells in the mantle, where oceanic crust with or without the continental blocks that float upon it converges from either side and ultimately descends ('subduction') towards the Earth's interior. These are the concepts of plate tectonics, which vindicated the theory of continental drift with the measurements of VINE and MATTHEWS (1963); see also VINE (1966).

Most of the present zones of formation of new oceanic crust have been in existence for the last 100 million years or more; but tracing the drift back much farther indicates that throughout most of the Permian and Triassic periods all the great continents were clustered together in one supercontinent, Pangaea, and that the Atlantic and Indian Oceans did not exist. The Pacific was much vaster than now.

It has been suggested that the individual convection cells in the Earth's mantle may have a characteristic life span of about 200 million years. Before the Permian, 300 million years and more ago, the pattern of drift was certainly different and the large areas of the present continents which have not been deformed since Pre-Cambrian time were moving towards each other to form Pangaea. The older mountain belts remain as witness of where the continental collisions of earlier geological eras took place. These were the great mountain chains which in their time must have exerted the chief barrier effects upon the general circulation of the atmosphere.

Various attempts have been made at mapping the continent positions in different geological periods reassembled by using the palaeomagnetic evidence. Computer methods have been used to discover the arrangement which would give the best fit of the shelf edges of the continental blocks, reassembled in the cluster around the South Pole in Carboniferous times, which was first suggested by WEGENER as the reality of the supercontinent already guessed at earlier by E. SUESS and called Gondwanaland.[1] The results of these best-fit experiments have not always agreed with the geological structure and have therefore proved controversial. Those reconstructions which make sense of the geology have gained most acceptance and have needed least revision. Uncertainty applies even more to the positions in the eras of more than 400 million years ago. Probably the best, and most widely accepted, reconstructions are those of SMITH et al. (1973), based on rotations of the continental blocks on principles laid down by BULLARD et al. (1965). Their maps of the Cambrian, Carboniferous, Permian and Eocene geographies respectively about 500, 340, 250 and 50 million years ago, are reproduced in figs. 14.3.[2]

Summary of the sequence of geographies

Major features of the sequence of geographies, which developed through the drift of the continent blocks, and about which there is general agreement, may be listed as follows. There must be less certainty about the detailed statements in the second section of each item below.

(1) About 500 million years ago, and for long after that, the northern hemisphere consisted largely of ocean, with the pole in the midst of that ocean. Antarctica straddled the equator. The continents were mostly in two clusters: one cluster in process of forming Euramerica in which Europe and North America became joined by 400 million years ago and, by 300 million years ago, northern Asia as well, and the other cluster being Gondwanaland, already existing and composing most of the rest. The two clusters were nearest together at about 30°S and 45°S, where they were separated by broad meridional ocean channels.

Most of the mountain belts of that time of which evidence remains seem to have been in low latitudes, between 30°N and 30°S. The Appalachian-Caledonian orogenic belt seems to have been zonally orientated between 20° and 30°S.

(2) The two continent clusters gradually moved towards each other, mostly southward, except that northern/eastern Asia (Angaraland) long remained separated from the rest (Gondwanaland) by what was in Devonian times an ocean ring (Tethys Sea) between the equator and 30°N. Much of the southern hemisphere was still land,

1. The name Gondwanaland seems to have been coined by the Austrian geologist EDUARD SUESS in his *Antlitz der Erde* in the early 1900s for a great continent embracing South America, Africa, India and Antarctica in their present positions, the intervening parts having supposedly foundered.

2. SMITH et al. (1973) give both Mercator and polar stereographic maps of both hemispheres showing the estimated positions of the present continent blocks about 510 (Cambrian/lower Ordovician), 380 (lower Devonian), 340 (lower Carboniferous), 250 (Permian), 220 (Triassic), 170 (Jurassic), 100 (Cretaceous), and 50 million years ago (Eocene).

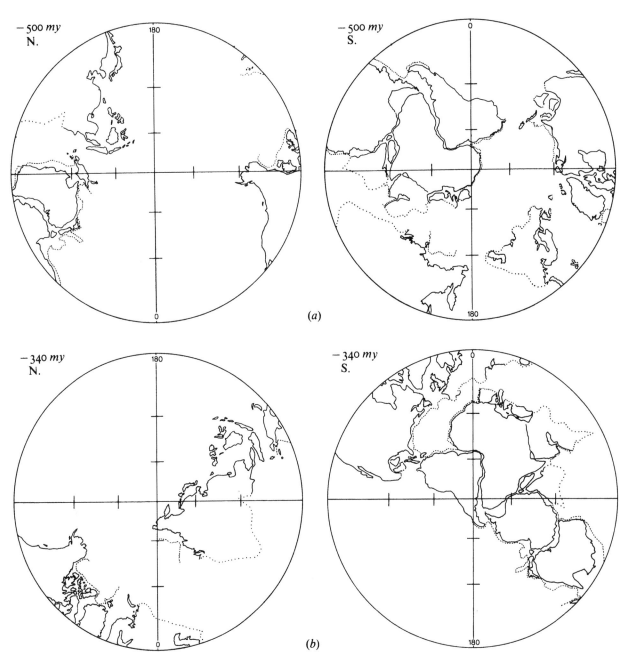

Fig. 14.3 Geographical maps showing positions derived from palaeomagnetic data for the continent blocks
(*a*) About 500 million years ago, in late Cambrian-early Ordovician time.
(*b*) About 340 million years ago, early in the Carboniferous era.
(*c*) About 250 million years ago, in the Permian era.
(*d*) About 50 million years ago, in the early Tertiary (Eocene).

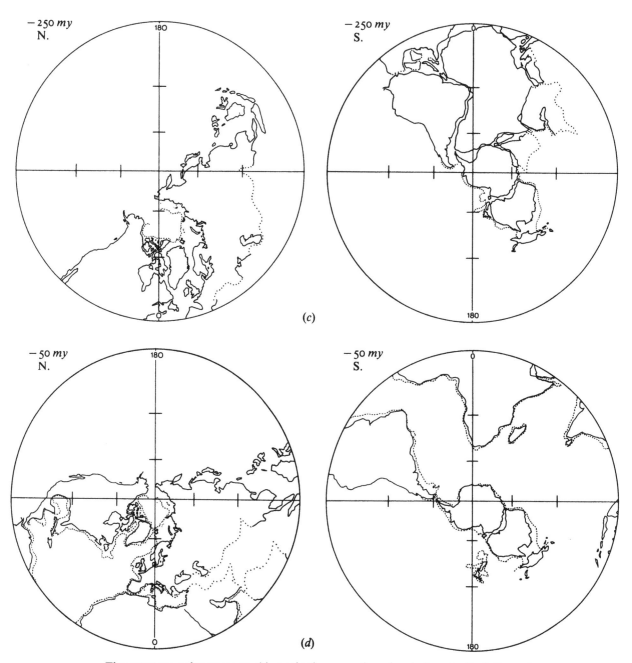

The maps are polar stereographic projections: northern hemisphere to the left, southern hemisphere to the right.

The circumference of each map is the equator. Latitudes 30° and 60° are also marked. The longitudes are arbitrary.

(*Reproduced from* SMITH, BRIDEN *and* DREWRY (*1973*) *by kind permission.*)

The labels beside the top left segment of each map indicate the approximate date in millions of years (*my*) before present and the hemisphere to which the map refers.

mostly south of 30°s, and the Gondwanaland cluster gradually moved across the South Pole, which between 400 and 350 million years ago (in the Devonian) lay in the region where southernmost South Africa and South America joined.

There seems to have been a meridional sea channel in Devonian times flanked by parallel mountain ranges, extending from 30°s to high southern latitudes, between eastern North America and western South America. This connected two great extensions of the northern hemisphere ocean on either side of the Earth.

(3) Between 400 and 250 million years ago Angaraland (joined with North America and Europe as 'Laurasia') and Gondwanaland drifted in such a way that, by the Permian era, they had probably joined together in a single vast landmass which has been called Pangaea. At the same time what is now the northeastern tip of the former remained near the North Pole and part of Gondwanaland (near the junction of South America, Africa and Antarctica) remained close to the South Pole. Hence, the land distribution of the Permian consisted of one supercontinent spanning all latitudes on one side of the planet. This supercontinent was narrowest at the equator, where its extent may have been just 30° to 60° longitude. A huge gulf, the 'Tethyan Gulf', broadening eastwards, stretched east from this land neck, between Africa-Arabia-India-Australia on the one hand and the main block of Asia (Angaraland) on the other.

The mountain ranges were probably more or less meridionally oriented and in the western sectors of the land regions north and south of the equator.

(4) Through the Mesozoic the broad features of the above land cluster, spanning all latitudes on one side of the Earth, persisted while North America and Europe drifted northwards, part of the former (Alaska) approaching the North Pole. Though part of Antarctica was by the Mesozoic always near the South Pole, through most of the era both poles seem to have been (just) within the one vast ocean.

By the late Mesozoic (Cretaceous), there may have been once more a narrow, but now continuous sea channel (Tethys Sea) in a nearly east–west orientation near 20°N, passing through the present Mediterranean. This situation lasted well into the Tertiary era (HOPKINS 1972) and must have produced an anomalous circulation of the ocean surface currents: the North Atlantic and North Pacific were cul-de-sacs on one side, and the South Pacific on the other side, of a current which could encircle the globe in low latitudes. Ultimately this current came to be blocked by the creation of central America and the isthmus of Panama.

There were apparently meridional mountain chains near the western margins of the great land regions and stretching through most latitudes in both hemispheres. This alignment must have acted as a barrier (more similar to the mountains of eastern Asia than to the Rockies and Andes chain today) to the zonal circulation of the atmosphere and prevented the formation of any circumpolar ocean current.

Meridional heat transfer in the ocean was probably a greater proportion of the whole than now, and for much of the duration of those eras it reached freely to both poles.

(5) It was only during the Tertiary that the following important features of the present geography were attained:

1 South Pole centrally placed in Antarctica.

2 North Pole within the nearly land-girt Arctic Ocean.

3 Separation of Australia and South America from Antarctica, thus producing the circumpolar Antarctic Ocean ring, with the establishment of the circumpolar ocean current, the great West Wind Drift, for the first time, about 30 million years ago.

Glaciations

The principal ice ages before the Quaternary show evidence of dependence on the continental drift and pole position results described above, i.e. the traces observable today are all in places that were in high latitudes in the eras concerned (except, apparently, a good deal of the Pre-Cambrian evidence). Glaciation probably did not, however, reach equally extended limits in all the eras when the continent positions were suitable. There is room therefore for supposing that changes of solar output (e.g. due to the evolution of the sun itself and to variations associated with forces in the Galaxy), or of the solar radiation reaching the Earth's surface (e.g. due to changes in the Earth's atmosphere), played a significant part.

The following list of the principal ice age times has been compiled from summaries given by W. B. HARLAND (at the Twenty-First Interuniversity Geological Congress, held at Birmingham, England, in January 1974) and W. L. GATES (in a report to the United States National Research Council in 1974):

Circa 2250 million years ago ⎫
„   950   „   „   „ ⎬
„   750   „   „   „ ⎫ The Late Pre-Cambrian or Varangian ice ages
„   650   „   „   „ ⎭
„   450   „   „   „   Ordovician ice age
„   250–350 „   „   „   Permo-Carboniferous ice ages
„   0–1   „   „   „   Pleistocene ice ages

The very first symptom of the climatic decline towards the Pleistocene ice ages can perhaps be recognized in the cool phase around 50 to 70 million years ago, when many ancient monsters became extinct, at the end of the Mesozoic and beginning of the Tertiary (see Table 1 in Appendix V). There were still some periods of great warmth to come, however, after that and the continuing decline of temperature levels did not set in until the middle to late Tertiary.

It is difficult to make deductions from the evidence of the various Pre-Cambrian ice ages in the above list because of uncertainty of the dating of evidence as contemporaneous, lack of firm knowledge of the continental positions (and of their extent), difficulties of firm identification of some of the evidence, etc. Nevertheless, W. B. HARLAND insists that much of the evidence of the late Pre-Cambrian ice ages was certainly in low latitudes at the time. He believes that there was widespread deposition of marine tillites on the then equator, that the evidence is too widespread to be accounted for by just occasional very big icebergs reaching the equator and that some of the boulders deposited had not come very far from their point of origin. Hence, the Pre-Cambrian ice ages appear to have been much more severe, and probably more prolonged, than the Pleistocene or Permo-Carboniferous ones. On the other hand, S. K. RUNCORN has suggested that the Earth's magnetism in early (e.g. Pre-Cambrian) eras may have been multipolar, far different from a dipole field; if so, the evidence of palaeomagnetism would not be capable of fixing the positions of continents in those times, and it might be necessary to use the palaeoclimatic evidence to correct the results of palaeomagnetism.

Evidence of the Ordovician ice age is more copious and better preserved in the Sahara than any from Permian or Pleistocene times there. The region was about the South Pole at the time according to the palaeomagnetic evidence. Glacial formations of Ordovician age around the great Hoggar massif in the central Sahara include U-shaped valleys, floored with tillites overlying grooved and smoothed rock surfaces. There is evidence that the sea penetrated from the northwest and that some of the grooves were made by grounded ice floes and bergs rotating. The biggest part of the ice sheet seems to have extended south of the massif.

Mountain-building (orogenesis)

It used to be thought that the Earth had undergone a sequence of world-wide mountain-building events, followed by many millions of years of degradation of the relief by weather erosion, and that the implied times of high relief would be conductive to glaciation. Mountain-building and volcanism are now interpreted in terms of plate tectonics.[1] This suggests that the timing and location of orogenesis may be random, depending on the accidents of where continents or ocean plates happen to collide. However, if there be preferred sizes for the convection cells in the mantle and preferred rates of drift, these could result in certain elements of regularity in the occurrence of mountain-building. Nor is this necessarily the whole story, for various investigators (e.g. KENNETT and THUNELL 1975) have reported signs of world-wide, or nearly world-wide, rhythms in volcanic activity. KENNETT has deduced from analysis of over 80 cores with volcanic ash layers drilled from deep sea sediments all over the world that volcanic activity had a peak in the Miocene, and a

---

1. Volcanism is especially related to (a) activity of the mid-ocean ridges and (b) mountain-building in the zones of subduction where the ocean plates and continent blocks collide.

greater peak in the mid Pliocene, and has increased in the Pleistocene and Recent (i.e. postglacial) times to 13 times the general average frequency and twice the Pliocene peak.

Climatic implications

Pole wandering and drift of the continents produce great changes from one era to another in the land fraction in different latitude zones (fig. 14.4).

Fig. 14.4 Percentage of land by 10°-latitude zones.
(*a*) Present Earth.
(*b*) With the North Pole in mid Pacific but with the continents in their present relationships.
(*c*) With the supposed geography of about 400 million years ago, the South Pole in Gondwanaland.
(*Adapted from* A. COX *1968*.)

A first crude approach to evaluating the effect of such different land and sea distributions upon climate, developed for the purpose by FORBES (1859) and SPITALER (1886) in the last century, is still as illuminating as it is simple. From the mean temperatures prevailing in the nineteenth century at observing stations between 60°N and 50°S, SPITALER derived the following formula for the overall mean surface air temperature of each latitude zone, averaged around the globe:

$$T_\phi = -2 \cdot 43 + 17 \cdot 6 \cos \phi + 7 \cdot 1 \cos 2\phi + 19 \cdot 3n \cos 2\phi$$

where T_ϕ is the temperature in °C, ϕ is the latitude and n is the proportion of land in the given latitude zone. This expression gives a reasonably good fit with the twentieth century mean temperatures of the latitude zones (see Table V in Appendix V) over the range of latitudes for which it was devised, and consideration of the discrepancies in high latitudes is informative.[1] It seems that the area of the pack-ice in the central Arctic should be treated as if it were about 60 % land and that the great inland ice-sheet in Antarctica intensifies the land effect by about 20–25 %.[2]

SPITALER himself ventured to apply the formula to calculate the temperature of an entirely land hemisphere, which resulted as 11·1°C, and that of an entirely water hemisphere, which came out as 22·8°C; his figure for a land hemisphere should doubtless be lowered by 1° to 2°C by applying the corrections needed for the effect of the ice and its albedo in high latitudes. The curves in fig. 14.5 show the results from calculating the mean air temperatures at sea level for each latitude zone in the situations described by diagrams (b) and (c) in fig. 14.4 and compared with the observed temperatures of the present epoch. No corrections have been applied in case (b), since it may be that there would only be seasonal thin ice anywhere in high latitudes, its formation being limited by deep convection as well as by horizontal currents, when the poles lay in the midst of a great ocean with unhindered heat transport from other latitudes in ocean and air alike. It may well be that a positive correction should have been applied in this case; the surface temperatures in the polar region are likely to have been raised rather more than is indicated by the formula based on present land and sea conditions owing to the much reduced opportunity for strong surface inversions of

1. The errors in deriving the mean temperatures
   for a series of 10° latitude zones in the present epoch
   by using SPITALER's formula are as follows:

	Latitude north								
Central latitude of zone	85°	75°	65°	55°	45°	35°	25°	15°	5°
Temperature error (°C)	+11·4	+5·7	−2·4	−0·8	+0·2	0	−0·9	−1·2	+0·1

	Latitude south								
Central latitude of zone	5°	15°	25°	35°	45°	55°	65°	75°	85°
Temperature error (°C)	+0·9	0	0	−0·1	+1·2	+3·9	+7·5	+4·0	+4·5

Both HANN and VOEIKOV rightly warned against the incautious use of such formulae as those developed by FORBES and SPITALER; but, used with due consideration of the discrepancies observed and of the sources of error to be expected, as attempted in what follows in the present text, they may be as informative as, and produce results closer to certainty than, many more elaborate models which can now be constructed. The latter must also, of course, be pursued for the distinctive insights which they may yield. One must be aware, however,

that these, by their nature, attempt to explore the effects of more complex interactions and feedbacks, e.g. between ocean and atmosphere and between different depths in each, which it may be impossible to specify with the precision implied by the elegance of the computations based upon the assumed values.

2. The temperature at 85°N given by the formula becomes right if 0·57 be added to n. Similarly the temperature for the Southern Ocean ice zone comes right if 0·6 be added to n. An addition of 0·23 is needed at 85°S.

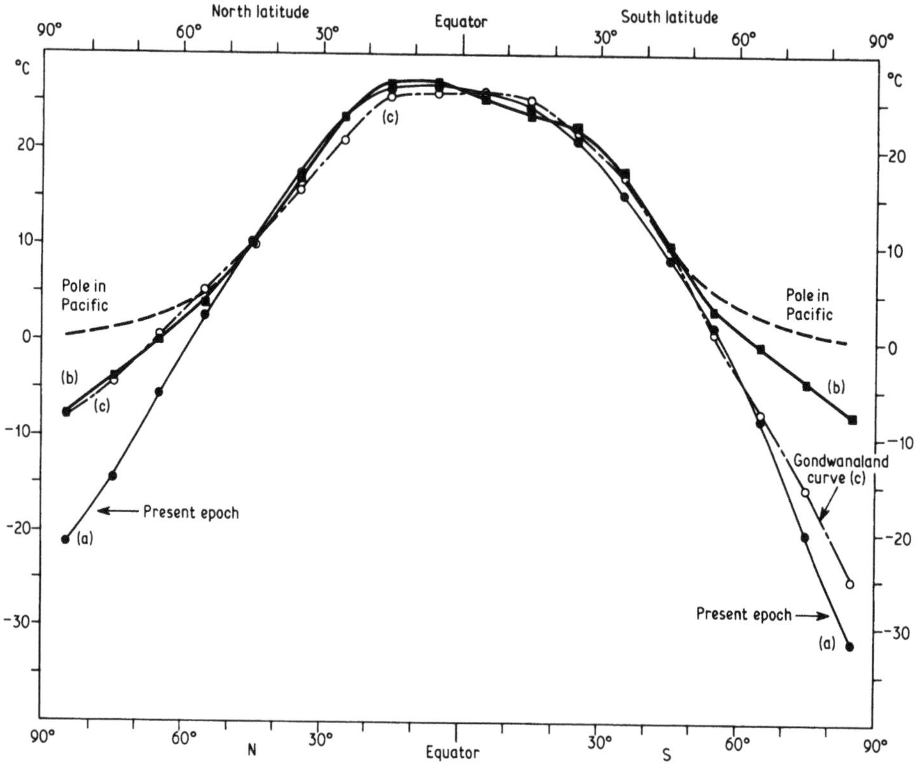

Fig. 14.5   Mean surface air temperature by latitude zones:

(*a*) Present epoch (observed).

(*b*) With both poles in a great ocean spanning most latitudes and land distribution as in fig. 14.4(*b*). Calculated by SPITALER's formula with no corrections for ice in high latitudes.

(*c*) With the Gondwanaland geography specified in fig. 14.4(*c*). Calculated by SPITALER's formula (adding 0·20 to *n* at 85° and 75°s, 0·15 at 65°s and 0·10 at 55°s for the probable effect of ice).

temperature to form over a great ocean accessible to warm water. An overall mean temperature close to 0°C at the poles seems likely on account of this. Because of the somewhat reduced land fraction near the South Pole in fig. 14.4(c), as compared with the present, the Gondwanaland situation gives temperatures there a little less cold than today. This would be expected with those geographies of the geological past which allowed the circulation of the great ocean to reach rather near the South Pole, while most of the land lay in the opposite sector of the hemisphere. But there may have been some eras (e.g. the Ordovician, with the pole in Africa) in which the pole lay deep inside the continent and, at the same time, the land fraction was big in all latitudes in that hemisphere; at such times the mean temperature of that polar region, remote from the ocean heat transport, should have been as low as, or lower than, in Antarctica today.

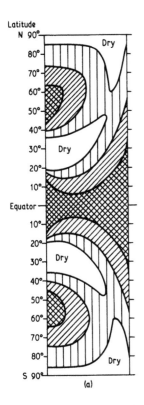

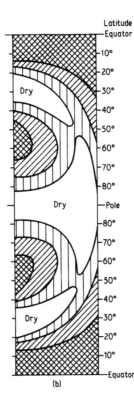

Fig. 14.6 Precipitation zones over an ideal 'rectangular' continent, 45° of longitude in width, extending (*a*) from pole to pole, (*b*) across one pole to the equator on either side. Shading in four zones from light (dry) to dark (wet).

(*Adapted from a diagram by Dr* PAMELA ROBINSON *1972.*)

All the temperatures calculated by the above procedure must of necessity correspond to the radiation regime of intermediate stages of the Earth's orbital (MILANKOVITCH) cycles resembling the present. In a hemisphere with mostly ocean in latitudes greater than about 50° it is likely that the temperatures prevailing would vary only within about ±1°C of this average condition in the course of the orbital variations. But for a hemisphere with a large land fraction in the higher latitudes, as in the Gondwanaland situation, the range of temperatures occurring in the Pleistocene between glacial and interglacial periods may be a reasonable guide. And, as in the Pleistocene, the low temperatures of the cold phases would depart further from the temperatures of the regime represented by our calculations than would those of the warmest interglacial times. Hence the overall average conditions for the hemisphere concerned, in a Gondwanaland-type geography, taken over one or more glacial–interglacial cycles, might be between 2° and 3°C colder than indicated by our calculations.

Another very simple model which offers an instructive insight into the distribution of climates in geological eras with other geographies than now is shown in fig. 14.6. An idealized distribution of precipitation over a 'rectangular' continent, actually a continent bounded by two meridians some 45° longitude apart, which stretches from pole to pole, constitutes fig. 14.6a. The distribution here shown is adapted from one which was outlined in an unpublished paper given at the British Association for the Advancement of Science 1972 meeting by Dr Pamela Robinson, from study of the rainfall distribution over the existing continents, particularly Africa which is at present almost symmetrically disposed across the equator (the first such model was presented long ago by Köppen as a pear-shaped continent, modelled upon Africa). In Tertiary times Africa was 12° latitude farther south than now. To elucidate the changes of rainfall pattern to be expected from the drift of the continent to other latitudes, fig. 14.6(b) indicates the effect of moving an idealized, truly rectangular continent until it is centred on the pole. The same rainfall zones are recognizable within much the same latitude bands as in fig. 14.6(a), but the breadth of the wet and dry zones is changed by nearness to, or remoteness from, the ocean. The most important element in these constructions is the observation that the spread of latitudes affected by either (1) the seasonal motion of the equatorial rain-belt or (2) tropical storms and other heavy rainfall systems is greatest towards the eastern sides of the continents, whereas the Trade Winds and the near-lying oceanic subtropical anticyclones induce dry zones, oriented as shown in the figure, towards the subtropical range of latitudes at the west coasts. Another wet region appears towards the western sides of the continents in middle latitudes, where westerly winds prevail (North and South America, southern Africa and Asia conform most nearly to this pattern today). Dr Robinson goes on to apply these ideas to the geography of the Upper Triassic, as

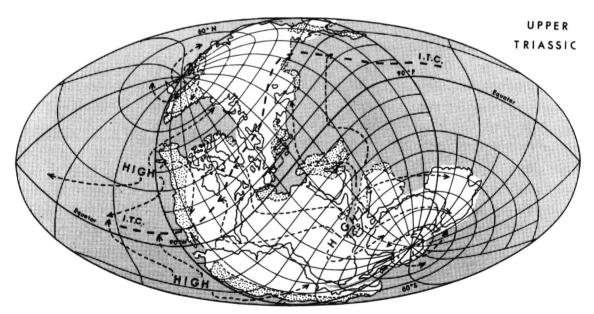

UPPER

TRIASSIC

Fig. 14.7   Geography and climate in the Upper Triassic.
Winds and intertropical convergence position in July.
The stippled areas are known to have been mountainous.
(*From* Robinson (*1972*); *reproduced by kind permission of the author.*)

shown here in fig. 14.7, in which the prevailing winds derived by the above considerations and the intertropical convergence zone position in the northern summer (July) have been sketched. Comparing this map with the evidence of Triassic deposits, we find that evaporites and/or fossil sand dunes indicate dry climates all along the long land tracks of the wind arrows, including the great continental-interior area of the intertropical zone. Coal measures of Triassic age occur in the probable monsoon-rain region between latitude 20° and 50°N (in present-day Russia) towards the eastern limit of the Tethyan Gulf but not on the southern margin of the same gulf, where the relief may have been too high for suitable forests. Triassic coals, which originated either as peat deposits or as trees with growth rings, occur (a) around the shores of the great gulf between 45° and 70°N (which has become the Arctic Ocean today), where the middle latitudes W'lies would blow in from the great ocean, and (b) in the zone 60–80°s in Australia and Antarctica, where the polar E'lies would blow in from the same vast ocean.

ROBINSON's map (fig. 14.8) of the upper Permian also leads to the suggestion of a dry climate in the huge continental-interior section of the intertropical zone and monsoon rains between 10° and 40°N north of the Tethyan Gulf, but this time shows monsoon rains also

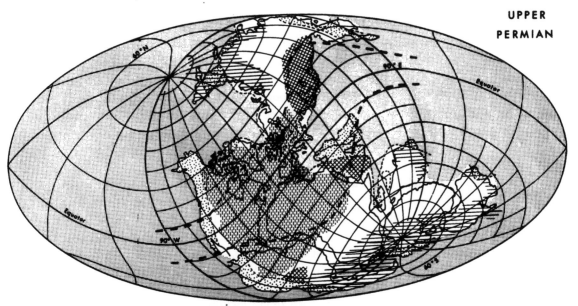

UPPER

PERMIAN

Fig. 14.8   Geography and climate in the Upper Permian.
Positions reached by the intertropical convergence zone in the northern and southern summers indicated by the broken lines.
Relief shown as in fig. 14.7.
Vertical dashed shading = all seasons dry.
Cross-hatched = summer monsoon rainfall.
Horizontal ruling = all seasons had precipitation.
(*From* ROBINSON (*1972*); *reproduced by kind permission of the author.*)

south of the Gulf in similar latitudes (in what is now Arabia and the Yemen). Persistently wet temperate climates are indicated in the zones of middle latitudes westerlies (in what are now northern Siberia and northern India–Australia) as well as right across the South Pole of those times, from eastern Australia to South America and southern Africa, a 1000-km-broad belt along the fringe of the vast ocean.

Changes of climate during the Tertiary era

By the beginning of the Tertiary important features of the present global geography had come into being (fig. 14.3d). The poles were in Antarctica and within an Arctic Ocean with a shape resembling that which it has now, though they were not quite in the positions they occupy today. However, the Atlantic Ocean was still much narrower than the ocean we know and the Pacific much wider than today. Also the Tethyan Gulf still extended eastwards as a continuation of the Mediterranean basin, between India and Arabia on the one hand and the main block of Asia on the other. Moreover, Europe was still some 20° latitude south of its present position. Africa and eastern North America were also somewhat farther south, while South America and Australia had not effectively separated from Antarctica.

Diagnosis of the temperatures prevailing during the Tertiary is proceeding, with the increasing use of oxygen isotope measurements as well as biological species abundances and characteristics. As yet many of the details are the subject of unresolved controversy, and even the main course of global temperature has hardly been firmly sorted out from the regional differences. Sea level changes may play an important part in settling this fundamental item, once enough measurements have been established for regions that have not been uplifted or downwarped since. It is clear, however, that the persistent cooling which led to the Pleistocene (Quaternary) glaciations did not set in until much later in the Tertiary, from about the middle and late Miocene onwards – i.e. it belongs to the last 20 million years approximately, though there had been some previous cooling episodes, e.g. about the Eocene-Oligocene boundary 40 million years ago. A selection of curves indicating the probable course of prevailing temperatures is reproduced in fig. 14.9. Tectonically stable areas have been sought in choosing sea level evidence. The steady northward drift of Europe has contributed to the longer history of cooling there.

A very striking feature of this geological period is the evidence of warm floras in middle and high latitudes. Most Tertiary coals are in such latitudes. The places in which evidence of a rich Tertiary flora, with forest, has been found include northern Alaska, Greenland, Spitsbergen and Antarctic islands in 62–64°s. In Europe the extinction of the last elements of the diverse flora of Tertiary times, which have now been lost, did not take place until some more of the first Quaternary glaciations had come and gone and come again. The evidence of warmth in the early to middle Tertiary in many cases seems to go beyond what can be explained by the palaeo-latitudes of the places in question. Even in the late Tertiary (Pliocene) HOPKINS et al. (1971) report a rich coniferous forest on the Seaward Peninsula of

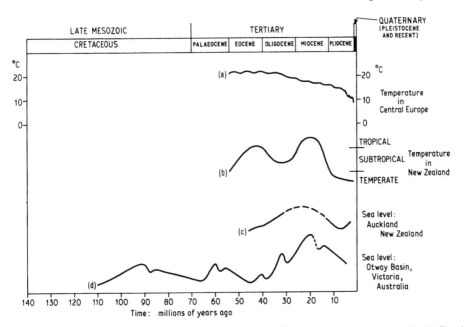

Fig. 14.9   Various indicators of the course of prevailing temperatures through the Tertiary era:
(a) Europe (*adapted from* WOLDSTEDT *1954*; (b) New Zealand (*after* HORNIBROOK *1968*).
(c) Sea level changes in the relatively stable area of New Zealand, near Auckland (North Island).
(d) Sea level changes in the Otway Basin area of southwestern Victoria, Australia.
        (*Curves (c) and (d) are from* GLENIE *et al. 1968*.)
        Note: The ages shown for the boundaries of the various epochs into which the Tertiary is divided
have here been adjusted to give the best possible agreement between the various authors' works used.
Many later works place the beginning of the Pliocene at only 5 or 7 million years ago, when the global
cooling became sharper than before.

Alaska, near the Bering Strait, apparently before any extensive tundra had developed. In the
northwestern United States (Oregon and Washington) WOLFE (1971) reports that early to
middle Tertiary floral assemblages show evidence of a tropical climate, judged by the present
temperature associations of the physiognomic characteristics of the plants (leaf types, etc.).
A major and rapid cooling of the climate there seems to have taken place in the Oligocene,
and there had probably been a previous major fluctuation in the late Eocene.
        A study of the marine molluscan faunas of the Pacific margin off the coast of the United
States (ADDICOTT 1970) indicates a tropical regime in the Eocene, with the great diversity of
taxa that is typical of a warm climate, succeeded by a cooler-water species and less diversity
in the early Oligocene; at latitudes 34–37°N the cold water indications become so marked as
to indicate a marked occurrence of upwelling on that section of the coast. During the later
Oligocene and in the Miocene these are replaced by marked dominance of unusually warm-
water species on the same section of the coast, with evidence of a strong gradient of sea
temperature just north of there. Also off east Asia WOLFE (1971) reports evidence of

warming in the middle Tertiary south of about 43°N and cooling north of there and suggests a similar division of the trends at 43°s. This may mean no more than that the cooling which led to the Pleistocene glaciations began first in the higher latitudes of both hemispheres; it had been preceded by a warming in all latitudes, which may have continued for some time equatorward of latitude 43°, or, more probably, there was a rearrangement and strengthening of the ocean currents which changed the locations of upwelling. Some estimates (FRAKES and KEMP 1972), based partly on oxygen isotope determinations, indicate average temperatures as high as 33° to 37°C in the western Pacific and Indian Ocean region between latitudes 20°N and 20°s in parts of the Eocene and Oligocene,[1] though these seem several degrees too high to agree with the earlier estimates of global mean air temperature at about 22°C, and absence of ice at the poles. Oxygen isotope measurements on belemnites from Alaska and Siberia indicate that surface water temperatures averaged 14 to 16°C in the ocean in high northern latitudes in the late Cretaceous. The temperature of the bottom water in all the main oceans presumably therefore was about 15°C at that time. Measurements on benthonic foraminifera (*Globerigina* spp.) shells in the equatorial Pacific trace the falling temperature of the bottom water: prevailing temperatures indicated were 10·4°C in the Oligocene, 7°C in the Miocene, 2·2°C in the late Pliocene (EMILIANI 1961).

In the early Tertiary the configuration of the continents was still markedly meridional, especially the western boundary of the great cluster which has been called Pangaea. It has been so for nearly 200 million years. By about 30 million years ago Australia seems to have moved far enough away from Antarctica for the great circumpolar ocean current, the West Wind Drift, to become established, isolating the Antarctic continent from the reach of the warm ocean currents which for so long had transported heat south from low latitudes.[2] This should probably be regarded as constituting the last important piece of the present global arrangement of land and sea which has brought about, or permitted the Quaternary ice ages.

There is evidence that glaciation proceeded far enough in Antarctica to bring ice to the sea's edge in the Ross Sea region about the beginning of the Miocene, some 25 million years ago and perhaps earlier in the Eocene (see p. 107), but this situation seems not to have been maintained during the warm period in the middle Miocene. The first evidence of ice-rafted debris – i.e. debris deposited from icebergs – on the bed of the Southern Ocean is dated only 4 to 5 million years ago (KENNETT and BRUNNER 1973). The core which shows this, on the Macquarie Ridge south of New Zealand, records at the same time the appearance of abraded quartz grains on the ocean bed, indicating active ocean bottom currents; this presumably marks a considerable increase – perhaps effectively the beginning – of the formation of the

1. Their study is of interest also for their attempt to sketch the prevailing ocean surface currents in the geography of the mid Tertiary.
2. Rather more detail of Australia's separation from Antarctica, beginning about 55 million years ago, has been indicated by KENNETT *et al.* (1972) with maps of the probable course of the ocean currents in that sector at two stages: (*a*) the eastward water drift still diverted north around Australia and New Guinea in the early Eocene, 53 million years ago, and (*b*) the West Wind Drift passing directly through a channel, some 5 to 10 degrees of latitude wide, between Tasmania and Antarctica in the late Oligocene, 30 million years ago.

cold Antarctic Bottom Water. The substantial increase in occurrence of both these items did not begin, however, until a little over 2 million years ago. Perhaps, the earliest substantial results of the great decline of world temperatures which led into the Pleistocene were a great growth of the ice-sheet in Antarctica and of mountain glaciers in the northern hemisphere around 10 million years ago (DENTON *et al.* 1971).

The fact that so many millions of years passed between the attainment of a global arrangement of landmasses and oceans favourable to glaciation and the first passage of whatever threshold triggered large-scale glaciation in both northern and southern hemispheres in many of the Earth's orbital cycles argues for a very slow process of global cooling as an essential element in the change. It may well be that this process was the necessarily gradual cooling of the world's oceans through their whole depth at a time when there was not yet any surface freshwater layer on the Arctic Ocean in which sea ice could readily form (see Volume 1, pp. 340–2).[1] Until this layer of low salinity was produced, cooling at the surface of the Arctic, as of the Antarctic, Ocean would immediately set up convection throughout the ocean's depth. The first appearance of the layer of low salinity on the surface of the Arctic Ocean, providing a surface where very low temperatures are developed at the ice surface and much incoming radiation is rejected by the reflectivity of the snow, may not have come until the first substantial glacier-melting event on the northern landmasses occurred.

It remains to be demonstrated whether the sequence of global warming in the early and mid Tertiary, followed by prolonged cooling (and falling world sea level) which continued at least until the last glaciation, demands some effect from a long-time-scale flicker of the sun in addition to the slowly maturing effects of a changing geography and reorientation of the currents in the world's oceans.

---

1. Since the total mass of the world's oceans is about 270 times the mass of the atmosphere, and the specific heat of the ocean water is about 4 times that of dry air at constant pressure, more than 1000 times as much heat is lost when the oceans cool by 1°C throughout their depth as compared with the same temperature change in the atmosphere. If, in a long-lasting change of climatic regime, the change affecting the dry land were equivalent to a fall of temperature by 1°C throughout the top 100 m of earth and rock, this would involve a gain or loss of the order of one-fiftieth of the heat required to change the temperature of the oceans by that amount.

Chapter 15

The Quaternary ice ages and interglacial periods

Progressive cooling of the Earth's climates from the middle of the Tertiary era onwards led in time to symptoms of glaciation.[1] Evidence so far known suggests that this happened first in Antarctica (DENTON et al. 1971). Ice-rafted quartz grains found in deep sea cores from the Southern Ocean (MARGOLIS and KENNETT 1971) indicate the occurrence of icebergs in the early and middle Eocene and again in Oligocene times, probably formed by calving at the coast of Antarctica from glaciers originating in the various mountain ranges: e.g. on the islands in what now (thanks to its ice cover) constitutes West Antarctica, on the Transantarctic Mountains (if they had been uplifted by that time), on the Antarctic peninsula (Grahamland), on various sub-Antarctic islands, and in southern Patagonia. Rocks of Eocene, Oligocene and in some places Miocene age, however, contain fossil leaves and wood fragments, as well as pollen, indicating a rich flora with deciduous and coniferous trees in both the Antarctic peninsula (ADIE 1964) and Ross Sea regions (CRANWELL et al. 1960, MCINTYRE and WILSON 1966). This biological evidence, and the available evidence of world sea level, suggests that no ice sheet of continental extent was present in Antarctica earlier than 40 million years ago and possibly not until long after that. Hence, the position of Antarctica over the South Pole – though doubtless a necessary condition – was not sufficient to cause extensive glaciation until some tens of millions of years had passed. The additional factors that were needed may have included the migration of the South Pole to near the

1. Average temperatures prevailing in different latitudes of the Earth at the present time are given in
   Appendix V, Table 2.

centre of the continent, the entry of the North Pole into the nearly land-ringed Arctic Ocean, uplift of the Transarctic Mountains and a wave of volcanic activity creating great dust veils in the atmosphere, or some extraterrestrial event such as a long-term 'flicker' of the sun (discussed in Chapter 14).

At some stage during the cooling after the mid Miocene great ice sheets did come into existence in Antarctica, certainly by 7 million years ago.[1] It seems probable (LAMB 1959, p. 19) that the first continent-wide ice sheet formed on the central plateau of East Antarctica, which was probably 2500 to 3000 m above sea level before the ice load reduced its level by isostatic adjustment to the present height of the subglacial bedrock (which still exceeds 2000 m above sea level in the heart of East Antarctica). It was probably only after this great ice sheet was there that the climate could become severe enough for the glaciers in the high mountains of West Antartica to produce thickening and coalescing ice-shelves on the deep ocean among the islands of what was then an archipelago, and for these ice shelves to become grounded and ultimately fill the ocean deep in that area (MERCER 1968, DENTON *et al.* 1971, pp. 278–81). The West Antarctic ice sheet also existed by 7 million years ago. Among the early signs of waxing glaciation are changes during the later Miocene in the ocean-bed sediment in the equatorial Pacific, such as increased grain size and a more unsorted sediment, indicative of increasingly active bottom currents perhaps due to the first production of Antarctic Bottom Water related to an Antarctic ice sheet as it is today.

Several glaciations of varying extent in southernmost Argentina prior to 1 million years ago, with ice sheets extending into the upper Santa Cruz valley (near 50°s), have been traced by FLECK *et al.* (1972) and MERCER *et al.* (1973). The earliest of them, dated by K-Ar method on overlying volcanic extrusions (basaltic lava), was rather more than 3·6 million years ago and was presumably made possible by the cooling of the oceans that had by that time taken place. There was apparently no glaciation in Argentina between then and 2·6 million years ago, and probably none at any time during the entire Gauss normal polarity epoch of the Earth's magnetic field (i.e. until 2·43 million years ago or later). There were more glaciations between 2·1 and 1 million years ago, the last of them probably more extensive than any that have occurred in that region since. This was when the till of the Strait of Magellan glacier was laid down.

Earliest known evidence of northern hemisphere ice-fields, coming into being as the cooling progressed through the latter part of the Tertiary era, is from the mountains and coast of what is now southern Alaska. The sequence of tillites in some of the mountain areas there goes far back into the Pliocene and possibly into the Miocene, more than 10 million years ago (FLINT 1971, p. 535). This means that ice sheets occurred there while the mountain ranges were being uplifted and attaining great heights. Glacial marine sediments, also of Pliocene age, occur along the present coast of the Gulf of Alaska (DENTON and ARMSTRONG 1969).

1. N. J. SHACKLETON, on the basis of ocean-bed sediment studies, places the growth of the great ice sheet in East Antarctica a little after the mid Miocene, about 13 million years ago.

The Pliocene–Pleistocene boundary

It is clear that, superposed on the prolonged global cooling which led to the Quaternary era, there were always fluctuations in the extent of glaciation; and these fluctuations could be described as glacial–interglacial cycles. Not surprisingly, therefore, there has been much debate about how to define the Pliocene–Pleistocene boundary and at what point to place it in the time scale.

The most obvious difference in the Pleistocene is that the glaciations were more severe and extensive, particularly on the northern hemisphere continents but probably also in Antarctica, than those in the late Tertiary. Until recently, however, this fact could not be clearly known, and the beginning of the Pleistocene was defined by evolutionary changes in biological organisms, e.g. the extinction of the Discoasters, or by the appearance or disappearance of certain species among the microfauna of the sea. A classic definition was the first appearance of the deep-water foraminifera *Anomalina baltica* and the mollusk *Arctica (Cyprina) islandica* near the base of a 1600 m thick (marine) sedimentary rock sequence at Le Castella in Calabria, southern Italy.[1] The Pliocene–Pleistocene change has similar characteristics in sedimentary rock sequences in central and northern Italy also, marking the immigration of a cold water fauna into the Mediterranean. Other definitions were based on land fauna, e.g. the appearance of *Equus* (horse) in North America. But these definitions gave different ages for the boundary, even though they doubtless each corresponded to a major climatic event, which was accompanied by sweeping changes in the environment that provoked migrations.

A deeper insight is offered by SHACKLETON and OPDYKE's (1976) analysis of the oxygen isotopes and carbonate in a section (fig. 15.1) from the bed of the equatorial Pacific Ocean, core V28–239, taken at $3°15'$N $159°11'$E, which for the first time penetrates beyond, indeed far beyond, the boundary. The whole length of the core goes back to about 2 million years ago, dated by the recognized magnetic reversals present. Accepting that the oxygen isotope curve gives an indication of the changes of ice volume on land, and of the oceans having consequently been rather over 1‰ richer in oxygen-18 at the last ice age maximum than today, the lower curve in fig. 15.1(a) indicates some ten glacial maxima in the last million years as great as, or greater than, the last one. Before that, the cold periods, or maxima of glaciation (the troughs in the curve), were less intense and the curve is dominated by shorter-period oscillations, among which a 40 000-year periodicity seems prominent. The level of the interglacial warm peaks of the curve is nearly constant throughout.

A power spectrum analysis of the last 500 000 years of the curve by HAYS, IMBRIE and SHACKLETON (1976) indicated that it was largely made up of fluctuations with period lengths close to 100 000 years (the dominant one), 40 000 years and 20 000 years. The same clear result has since been arrived at by other workers using data concentrated in the

---

1. The Eighteenth International Geological Congress in 1948 recommended that the base of the Calabrian stratum should be taken as defining the Pliocene–Pleistocene boundary.

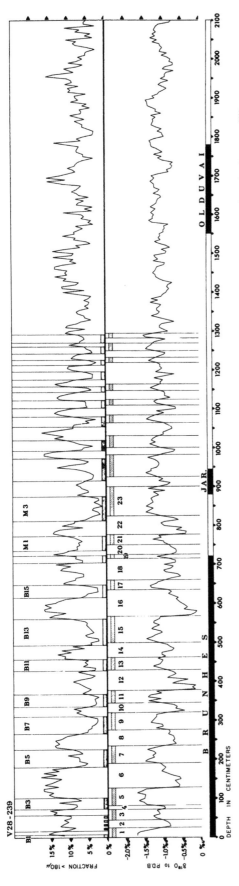

Fig. 15.1   Analysis of a deep sea core (V28-239) from the equatorial Pacific at 3°15′N 159°11′E at a water depth of 3490 m covering the last 2 million years. The geomagnetic stratigraphy indicates a time scale (shown in units of thousands of years at the bottom of the diagram) which implies that the sediment here examined was laid down over the last 2 million years.

The upper curve registers the percentage of the sediment made up of coarse particles (>180 μm cross-measurement). The coarse fraction of the sediment consists largely of the shells of foraminifera and therefore increases with the productivity of these creatures, increasing with increasing water temperature.

The lower curve plots the variations of the $^{18}O/^{16}O$ ratio in the species *Globeriginoides sacculifera*. The variations of oxygen isotope ratio provide another indicator of the swings between glacial and interglacial regimes, the ratio increasing as the Earth goes into a glacial regime. The numbers above the lower curve identify the stages of the Quaternary climatic sequence according to EMILIANI's system.

These curves should be compared with the shorter record of another equatorial Pacific core in fig. 15.4, from 1°1′N 160°29′E, where sedimentation was much more rapid (1200 cm in the last 700 000 years as against 735 cm in core V28-239) so that any time interval is represented by a greater length of core in fig. 15.4 and more detail can be discerned.

(*Diagrams kindly supplied by Dr N. J.* SHACKLETON *who performed the analyses.*)

southern hemisphere and covering other spans of years. These figures are so close to the periods of the Earth's orbital variables as to constitute a remarkable vindication of MILANKOVITCH'S (1930) hypothesis regarding the origin of the Quaternary glacial-interglacial cycles (see also CHAPPELL 1973).[1] It is only curious that the 100 000-year time scale is always more prominent than the theory suggests. The abruptness of some of the changes in the course of the glacial–interglacial–glacial sequences cannot, however, be simply explained in this way. Also it will be noticed that all curves such as those presented in fig. 15.1 and the later time sequence diagrams in this chapter show numerous superposed shorter-term fluctuations.

In view of the strong indication of an association between the alternation of ice age and interglacial climates and the Earth's orbital variables (explained in Volume 1, on pp. 30–7, 492–3) there is no need to invoke in this connection LORENZ's suggestion that the Earth's climatic regime may vary in this way because of an indifferent stability, described earlier (on p. 46) as 'almost intransitivity', and that the variations could be independent of any external exciting agency.

Causation of the glacial–interglacial cycles in the Quaternary

There have been many theories about the cause, or causes, of the alternations between glacial and warm interglacial ages in the Quaternary. Several of those that are still held (and vigorously pursued by attempts to prove them) in certain quarters, and which thereby give rise to continuing controversy, doubtless correspond to real processes which do take place but on a scale which falls far short of explaining the magnitude of the phenomenon. These cycles, regularly occurring as regards their grossest features, but which show many superposed irregularities, are – like most lesser climatic changes – certainly polycausal. With modern techniques of more accurate dating of the phenomena, and estimating the magnitudes of the temperature and other effects involved, it is beginning to become clearer which mechanisms contribute most. It is now generally agreed that the first requirement was a suitable geography with enough land in high latitudes (see Chapter 14).

The variations in the Earth's orbital arrangements (see Volume 1, pp. 30–7, 492–3)

The suggestion of an explanation along these lines is almost as old as the recognition of the occurrence of past ice ages, going back to JOHN HERSCHEL in 1830 and having been worked out, first in terms of precession alone, by the French mathematician J. F. ADHÉMAR in 1842, and later with the inclusion of ellipticity variations by JAMES CROLL in 1860. CROLL (1875) believed that periods with long cold winters and short hot summers in high latitudes would

---

1. There may be a point of some interest to the question of causation of the glacial-interglacial cycles in the change of relative importance of the obliquity and ellipticity cycles about one million years ago (fig. 15.1). Was there, for example, some change in the tilt of the Earth's rotation axis about that time?

favour the growth of glaciation; the precession of the seasons, he supposed, would cause ice ages to occur in the northern and southern hemispheres alternately. Both these latter suppositions are now seen to be wrong; it is mild, snowy winters and cold summers in high latitudes that are, in fact, most conducive to glaciation, and the ice ages involved cooling of the whole Earth so that the timing is largely the same in both hemispheres. The first clear assessment of these factors and comprehensive working out of the astronomical variables and their implications in terms of the radiation available at different seasons and latitudes over the last 600 000 years (done before the age of computers) was that by MILANKOVITCH (1930). Subsequent refinements and recalculations have now extended knowledge of these variables back to 1 million years before present and one million years into the future (VERNEKAR 1968, 1972).[1] An excerpt from VERNEKAR's results, covering periods within 105 000 years before and after the present is seen in fig. 15.2. The timing of the last glaciation, most of the stages of which can be assessed by radiocarbon dating, seems to fit the MILANKOVITCH hypothesis that the phenomenon is mainly determined by the summer radiation available in latitudes 60–70°N, where the surface is largely land, some of it mountainous; though allowance has to be made for a lag of 5000 years or more for the melting of the great ice sheets while the summer insolation was at its strongest. CALDER (1974), surveying the oxygen isotope record of the last 860 000 years from the ocean bed sediment in the equatorial Pacific, has suggested that the best fit is with the variations of insolation available at 50°N (see also Chapter 20, pp. 687–8, 697). The more general vindication of this theory by spectral analysis has been referred to above (pp. 310–12).

The magnitude of the temperature changes at the surface of the Earth to be expected from these variations calculated by MILANKOVITCH does not fit well (see Volume 1, p. 34), but no allowance was made for the albedo change produced by the snow and ice. When such allowance is made, variations of global mean temperature of the order of 5–6°C appear likely to accompany the extreme range of the radiation changes associated with variations of the orbital parameters.

The theory does not go into the resulting changes in the circulation of atmosphere and oceans and cannot therefore explain the geographical distribution of temperature anomalies during the Quaternary ice ages.

The SIMPSON theory

A very different hypothesis, attributing the glacial-interglacial cycles to periodic oscillations in the intensity of solar radiation, presumed to originate in the output of the sun itself, was put forward by Sir GEORGE SIMPSON (1934) and further elaborated in a later paper (1957).[2]

1. Similar calculations by A. BERGER of Louvain include further details.
2. C. G. ABBOT's calculations at the Smithsonian Institution, Washington, in the 1920s had indicated some variations in the solar constant, including a 23-year periodicity with an amplitude of somewhat under 0·5%. SIMPSON supposed that long-term fluctuations of the sun's output could also occur and could have a bigger amplitude.

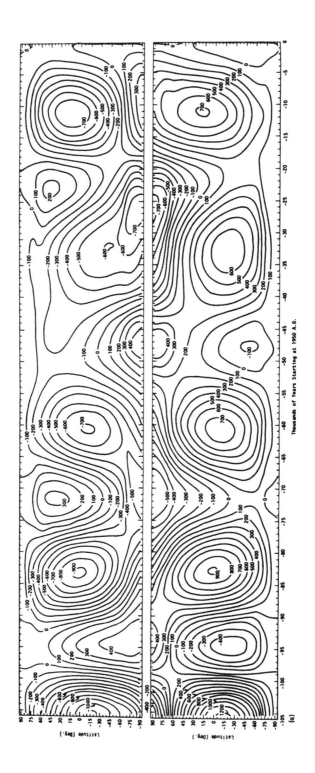

Thousands of Years Starting at 1950 A.D.

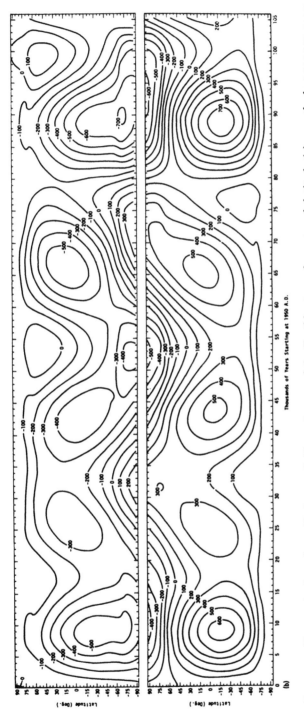

Fig. 15.2   Changes in the solar radiation available at the top of the Earth's atmosphere at each latitude (*a*) over the last 105 000 years before A.D. 1950, and (*b*) over 105 000 years into the future, as calculated by VERNEKAR (1968).

Upper diagrams: northern caloric winter/southern caloric summer
Lower diagrams: northern caloric summer/southern caloric winter.

Note: Following MILANKOVITCH, the *caloric summer* is defined as that half of the year in which the daily amount of solar radiation available exceeds that on any day in the *caloric winter* half of the year.

The radiation departures from present conditions are given in *canonical units*, as used by MILANKOVITCH, who took the solar constant as his unit of radiation and calculated the quantity available per 100 000th part of the year.
1 canonical unit = 0·0576 langleys/day. (1 langley = 1 gramme calory/cm²).

(*Curves reproduced by kind permission of Dr* VERNEKAR.)

SIMPSON suggested that the mechanism was an increase in the solar radiation which would first raise the surface temperature everywhere over the Earth, most of all in low latitudes. This would result in increased strength of the global wind circulation and increased evaporation from all water surfaces. Hence, he supposed, cloudiness and precipitation should increase and with the cloudiness global albedo should also increase. The first results would be a small net increase of temperature, averaged over the whole Earth, and a great increase of cloudiness and of the downput of rain and snow. Glaciation would come about because of the increase in the amount of snow on high ground in high latitudes. When the solar radiation increased further, there would come a point at which the proportion of rain as against snow increased even in high latitudes sufficiently to stop the growth of the ice sheets which had formed and then to cause them to melt. A warm and wet interglacial would then occur.

With the declining phase of the solar oscillation the reverse sequence of changes would occur. Towards the minimum of solar output global mean temperature would fall until the evaporation, and hence the atmosphere's water vapour content, was insufficient to nourish and maintain the ice sheets. The Earth would then pass into a cold dry interglacial with conditions like the present.

This ingenious hypothesis seems not to accord with the observational evidence now at our disposal, particularly since it has become clear that all, or nearly all, parts of the Earth were colder than now during the ice ages. Nevertheless, the paradox on which it is based may well play a part – though rather as a complicating detail – in climatic fluctuations; warming of ocean surfaces probably does increase cloudiness and lower the temperatures prevailing over some regions, but this should be mostly in low latitudes and especially where mountains cause increased cloudiness by forced ascent of the air.

On another aspect of the theory, M. C. MCCRACKEN (personal communication, 7 February 1972) points out that cloudiness is related to relative, rather than to absolute, humidity, and relative humidity is related to circulation patterns, adding that 'intuitively one can go further and see that the type of cloudiness would also change [as solar radiation increased], with less stratus and more cumulus, thus probably decreasing the cloud fraction, contrary to SIMPSON's hypothesis'.

Volcanic activity

BENJAMIN FRANKLIN suggested as long ago as 1784 that a great veil of volcanic 'smoke', spread by the winds far and wide over the Earth after a major volcanic eruption, might by weakening the penetration of the sun's rays to the surface of the Earth, reduce and shorten the summer's warmth and lead to a severe winter following. The possibility that much volcanic activity, maintaining frequent veils of dust in the high atmosphere, might cause an ice age has from time to time been widely discussed (e.g. ABBOT and FOWLE 1913,

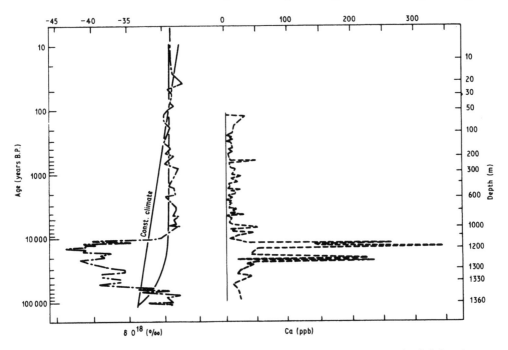

Fig. 15.3   Calcium concentration (right-hand curve) and oxygen isotope ratio (left-hand curve) at different depths in a core from the Greenland ice sheet at 77°N 56°W.

The oxygen isotope curve, registering the variations of prevailing temperature over the ice during the last 100 000 years, shows the course of the last ice age. As the ice sampled at the test point has flowed during that time from higher up the ice dome, allowance must be made for the lower temperatures prevailing higher up: this is done by relating the oxygen isotope ratios observed to the line which represents the effect of the progressive difference in height of origin of the ice under conditions of constant climate.

Comparison of the oxygen isotope and calcium (dust) curves make it plain that there was much more dust deposited from the atmosphere during the ice age. Presumably dust reaching northwest Greenland was mainly produced by volcanic activity, which must therefore have been enhanced during the ice age. Notice, however, that the enhancement set in *after* the climatic cooling had begun and that the most abundant dust occurred in the middle and later parts of the period of widespread northern hemisphere glaciation.

A similar relationship between the incidence of volcanic dust and the course of the last glaciation was found in an ice core taken from the Antarctic ice sheet at 80°S 120°W.

(*Diagram from* HAMILTON *and* SELIGA (*1972*); *reproduced by kind permission.*)

HUMPHREYS 1913, ARCTOWSKI 1915, WEXLER 1956) and has lately been pressed again by BUDYKO (1969). The present author has argued elsewhere (LAMB 1970), however, that, although volcanic activity does have effects of the right kind, its incidence is likely to be too occasional to cause an ice age (see also Volume 1, pp. 430–3.)

There is some evidence that the glacial–interglacial changes of climate, and some more limited variations in the ice load on land, may cause stresses in the Earth's crust which unleash a period of abnormally enhanced volcanic activity. It is likely that these intensify the cold and cause some prolongation of periods of cold climate, and they may cause some readvance phases which interrupt the general trend towards deglaciation at the end of an ice age. Actual evidence of this effect in the course of the last ice age is seen in fig. 15.3 (see also Chapter 13, p. 171); and it seems that a world-wide wave of frequent explosive volcanic activity produced some notably cold decades (e.g. A.D. 1690–9, 1810–19) in the latter part of the so-called Little Ice Age in recent centuries and the recovery period thereafter. Similarly KENNETT and HUDDLESTUN (1972) remark in connection with the abrupt, but rather short-lived, glacial episode around 90 000 years ago (see p. 335), and the widespread volcanic ash layer associated with it, that the volcanic ash maximum falls about 1000 years later than the catastrophic events that mark the effect of the cold climate in the biosphere and add: 'it would not seem that increased volcanism caused the catastrophic climatic change but that factors related with this climatic change caused the volcanism'.

There is no known reason to suppose that fluctuations of volcanic activity over the Earth should have a regular periodicity such as to cause the regular recurrences of major cold climate events observed in the Pleistocene record, as seen in fig. 15.1.

Theories of origin within the balance of the ocean-ice-atmosphere system

(1) *BROOKS' theory:* C. E. P. BROOKS (1925, 1949) elaborated a theory of the unstable growth and decay of polar ice caps based on the cooling power (associated with the albedo change and radiation properties) of floating ice formed on a polar ocean due to some initial slight fall of surface temperature below the freezing point. Once an area 10°-latitude in radius was covered, the ice would quickly spread over an area nearly 25°-latitude in radius; but again, once summer melting reduced the area to less than the critical 10°-latitude radius the remaining ice would quickly decay (see Volume 1, pp. 256–7). BROOKS' calculations showed that the theory should apply quite similarly to the growth and decay of ice sheets in the highest latitudes on land.

The temperature changes which triggered these developments might arise within the heat economy and circulation of the oceans and atmosphere or they might be imposed by changes in the radiation supply reaching the surface of the Earth. It seems likely that the importance of this mechanism lies in its tendency to amplify the climatic fluctuations produced by the effects of the orbital variations, by solar output variations or by volcanic dust, upon the radiation supply. Among these, the orbital variations, and perhaps fluctuations of the sun, may introduce the observed quasi-periodicities.

The effect of the albedo of ice on the equilibrium state (extent of ice cover) on the Earth to correspond to any particular percentage reduction of solar radiation has since been considered in several papers by BUDYKO (e.g. 1969), who calculated that a 1·5% reduction in solar radiation would be sufficient to start a new ice age and that a 5% reduction should institute a self-amplifying ice increase leading to glaciation of the whole Earth. More detailed calculations by GORDON and DAVIES (1975) suggest that, with any reduction of solar radiation, fluctuations of the ice margin as well as some general increase would ensue, and that the situation might become unstable once the solar intensity decrease approached 3%.

(2) *WEYL's theory:* P. K. WEYL (1968) pointed out that the fact that the salinity of the Atlantic Ocean at the present epoch exceeds that of the Pacific may be the reason why the northernmost Atlantic does not readily freeze over in winter south of 75°N, whereas the ice limit reaches nearly 60°N in the Pacific. The greater salinity of the Atlantic may be associated with the steady moisture transport by the Trade Winds westwards across the isthmus of Panama, while the return flow of moisture eastwards from the Pacific in the zone of westerlies is checked by the Rocky Mountains. A general weakening of the atmospheric circulation should therefore tend to reduce the loss of water substance via the atmosphere from the Atlantic and lower the salinity of that ocean, enabling the Arctic ice to spread farther south in the Atlantic sector.

The theory is expounded in more detail in Volume 1 (pp. 340–2). The magnitude of the water vapour flux is found to be in reasonable accord with the significance attached to the mechanism proposed, so far as available upper air observations in the Trade Wind zone permit such an assessment. The mechanism doubtless plays some part in all climatic fluctuations involving significant changes in the water vapour exchanges between ocean and ocean, either amplifying or reducing the variations in the extent of sea ice according to whether the wind circulation is strengthening or weakening, or shifting so as to encounter less or more of the mountain barriers, as the climate cools and warms.

No less important (see Volume 1, pp. 311, 324, 340) is the extent of the discrete layer of low salinity and low density water in which the pack-ice forms on the surface of the Arctic Ocean. This layer, also studied by WEYL (1968), is largely supplied by the great rivers of northern Eurasia and Canada which debouch into the Arctic Ocean.

(3) *EWING and DONN's theory:* In 1956 EWING and DONN proposed that from the time when the North Pole came within the largely land-girt Arctic Ocean, while that ocean was still open water, ice sheets would form and grow on the surrounding lands. These ice sheets would continue to build up until the lowering of sea level, which they caused, cut the Arctic Ocean off almost completely from the circulation of warm ocean currents from the Atlantic owing to the shallowness of the Iceland-Faeroes-Scotland sill (the Wyville-Thomson ridge). The Arctic Ocean would then freeze over. As a result the ice sheets, being starved of their moisture supply, would melt away, as ablation under the summer sun exceeded the moisture supply. The warmth of the surrounding continents in the interglacial period which thus ensued, and the warm winds in summer from those lands, would eventually melt the floating ice on the polar ocean, and the whole cycle would begin again and must repeat itself as long as the present geography of the land and ocean continues.

This theory failed to account for the following facts (which were not fully established at the time when it was put forward):

(a) Owing to the small moisture content of saturated air over an ocean with surface temperatures of the order of 0° to +5°C, ice sheets of the magnitude concerned could hardly be formed by moisture from an Arctic source.

(b) All northern Alaska remained unglaciated. The Alaskan ice formed on the southern side of the great mountains.

(c) Evidence from the sediments of the Arctic Ocean bed indicates that that ocean has been largely ice-covered probably throughout the last 700 000 years and certainly through the last 70 000 years.[1] Evidence has recently been reported (HERMAN et al. 1975) that there was a rather milder climate in the Arctic, with the ocean seasonally ice-free, in the early Pleistocene more than 700 000 years ago.

The theory has since been revised (DONN and EWING 1968); it is now suggested that a (partly?) ice-free condition of the Arctic Ocean is important only in initiating ice sheet growth on the northernmost parts of the continents and cooling the surrounding region to the point where further growth of the ice occurs on the lands where winds from other latitudes bring sufficient moisture. The Arctic Ocean would soon freeze over as the ice cover on surrounding lands became extensive, and would play no direct part thereafter in nourishing the continental ice masses. The end of the process would be reached when the Earth had cooled so far that the air no longer supplied enough moisture to sustain the ice sheets against summer melting.

The theory has been explored (SHAW and DONN 1971) in terms of a theoretical general circulation model – a thermodynamic model due to J. ADEM – which indicated a realistic distribution of temperatures over the hemisphere with the Arctic Ocean ice-covered as at present and suggested that, once that ice cover was removed, it would not return unless and until the surrounding continents were ice-covered once more.

The main interest of the theory may be in details such as this. There is no satisfactory reason to suppose that the cyclic changes would repeat themselves to the time scale of the observed fluctuations. Nor can either the mechanism of initiation of the ice sheets or their geographical distribution be so well explained as by southerly winds and moisture taken up from the oceans in lower latitudes.

The thinking underlying the theory, especially in its original form, seems to have been based on the idea, widely held a few years ago, that the great continental ice sheets could only have attained their ultimate extent by slowly building up into a great dome, which became unstable and slipped, or surged, forward over vast areas which would otherwise have continued to enjoy a temperate climate. WEERTMAN (1964) calculated that 15 000–30 000 years must have been required for the build-up in this way of the great northern hemisphere continental ice sheets, before they could reach their maximum height.

(4) *A. T. WILSON's theory:* There is much evidence that the great continental ice sheet

---

1. KELLOGG (1976) reports that examination of the foraminifera in cores from the Norwegian Sea indicates that the sea also was ice-covered through- out the last 150 000 years excepting only the period 127 000–110 000 B.P., in the Eemian interglacial, and postglacial times.

in Antarctica, which is estimated at present to contain about 24×10^6 km³ of ice (or 89 % of the present world total of glacier ice), has undergone variations of thickness and was at some time (or times) much thicker than today. Old side-moraines, and evidence of glacier action, on the rocky sides of mountains (nunataks) are found 300 to 800 m above the ice sheet in East Antarctica and up to 30 m above the present ice surface far inland in West Antarctica, up to 600 m above it near the coast (MERCER 1962). The record of successive glacial advances and retreats can be traced (DENTON et al. 1971) in the now ice-free Taylor and Wright valleys which cut through the mountains at the west side of the Ross Sea. The ice sheet in East Antarctica is dammed up against the west side of the mountains, and during its thickest phases (at least five different times) advanced over the passes into those valleys. These fluctuations seem to have been out of phase with the northern hemisphere glaciations, despite evidence that the variations of prevailing temperature in the Antarctic were in phase with those in the northern hemisphere and the rest of the world.[1] The ice sheet appears now to stand at its greatest height since more than 50 000 years ago, probably since some time before the last northern hemisphere glaciation.

WILSON's theory (1964) starts with the premise, advanced by ROBIN (1955, 1962), that even large ice sheets are liable to surges and that the base of the Antarctic ice sheet, at least in its central, thickest areas in East and West Antarctica, must be at its pressure melting point today – i.e. it supposedly rests on a layer of water. WILSON suggests that the Antarctic inland ice is inherently unstable. His suggestion that it builds up during the warm interglacial periods, as at present, agrees with the evidence that snow accumulation at the South Pole and elsewhere on the ice sheet in East Antarctica – as in northern Greenland and on some other glaciers in high northern latitudes – rose to a maximum (refer to fig. 13.39(a), (b)) in the decades of greatest global warmth and strongest wind circulation in the first half of the present century. The bottom of the ice sheet at some stage reaches its pressure melting point and is also heated by the geothermal heat flux from the Earth's interior. A surge begins, and once the ice starts to flow its base gains further heat by friction and the movement gathers momentum. So it flows out over the surrounding sea as a great ice shelf, which may extend as far as the Antarctic Convergence in the ocean near latitude 50°S. According to WILSON's reckoning, this vast ice shelf, increasing the albedo of that part of the world ocean from an average value of 8 to 80 %, would decrease the Earth's total heat intake by 4 %. This must lead to a cooling of the whole Earth and permit the snow to accumulate and form the great ice sheets on the northern continents. The size of the Antarctic ice sheet is critical; it must lead to enough cooling of the Earth to introduce the northern hemisphere ice sheets, which then make a further contribution to cooling the Earth. It is because the flow of ice from Antarctica cannot be maintained, however, that the supposed great ice shelf on the Southern Ocean ultimately melts and allows that ocean to warm up; gradually extra heat is spread to all the other oceans, and with the aid of wind transport this brings the ice age to an end.

---

1. A contrast is observed in the Ross Ice Shelf which expanded at least four separate times into a grounded ice sheet or ice dome. Its advances, the 'Ross Sea glaciations' (DENTON et al. 1971), seem to have been in phase with the northern hemisphere glaciations, all within the last 1·2 million years, seem to have been in phase with the northern hemisphere glaciations.

WILSON himself (1964, p. 149) and others (e.g. VAN DEN HEUVEL 1966, HUGHES 1970) have suggested various ways in which the timing of this proposed cycle of great Antarctic ice surges may be linked to radiation changes due to the MILANKOVITCH orbital cycles.

An obvious test of the theory is whether or not there is evidence of the sudden, great rises of world sea level, of the order of 20–30 m, which should mark such a surge of the whole Antarctic ice sheet just at the very end of each interglacial age. J. T. HOLLIN (e.g. 1965) has devoted much effort to exploring this aspect and has added some proposed refinement to the theory. It seems more probable that a surge might start from just one of the least mountainous sectors of the coast of Antarctica, say from Wilkes Land or the Filchner Ice Shelf in the Weddell Sea, and capture the drainage of half the ice sheet, reducing the thickness of that half by 50% and spreading an ice shelf of the typical 200 m thickness out over the ocean to latitude 55°S. This would dispose of 70 000 years accumulation of snow, built up at a constant rate of 3 grammes/cm²/year, as is typical of central Antarctica at present; and the whole cycle would fit reasonably well with the recurring glaciations of the last million years. Ignoring isostatic effects, such surges should raise sea level by 20 m. HOLLIN reports evidence from the end of each of the last two (Hoxnian and Ipswichian) interglacials in southern England of marine transgressions reaching various levels between 6 and 25 m above present sea level, accompanied by deposits whose special nature – some of them are famous for the quantity of remains of whole animals contained – suggests a very rapid, disastrous event. More tentatively, because the areas concerned have been subject to more crustal warping, HOLLIN adduces corresponding evidence from many parts of northwest Europe (e.g. Holland, Germany and the Baltic) and eastern North America. The case seems, however, to involve some special pleading, and it can hardly yet be regarded as proved that such abrupt rises of sea level have indeed characterized the end of even the last interglacials.

The severity of the different glacial cycles of the last 700 000 years or so appears more nearly constant than the variations of insolation due to the orbital cycles would suggest, even though the timing seems to be controlled by the latter. This recurrence of similar severity may point to the operation of some such mechanism as that proposed by WILSON, triggered by the changes of radiation due to the orbital cycles.

WILSON's theory has been substantially adopted by FLOHN (1969) as one important aspect of the glacial–interglacial fluctuations, which are nevertheless seen as polycausal. This much, in principle, is probably right; but there seems little doubt that the basic control of the timing of the Quaternary glaciations is in the (MILANKOVITCH) Earth's orbital variations and it is an open question as to how much of the Antarctic ice sheet surges at any one time. Evidence of the great masses of ice encountered by shipping on the Southern Ocean at various times during the nineteenth century suggests rather that surges occur from individual drainage basins in Antarctica, one at a time, and produce great extensions of the ice shelves, which break up within a few years in the Southern Ocean storms. Similarly, DENTON et al. (1971) while accepting the reality of Antarctic ice sheet surges in individual sectors report that 'evidence from deep sea cores from the Southern Ocean suggests . . . that possible surges did not form large ice shelves and thus did not trigger Quaternary ice ages'. Likewise, R. A. BRYSON reports (personal communication, 22 April 1974) that surges of the Laurentide ice sheet in North America can be seen to have affected only limited areas of ice, characteristically measuring only 250 by 400 km.

There is little doubt that for some years around 1960 a widespread preoccupation with the profile of glaciers and ice sheets and their flow under gravity led to an ungrounded supposition that slow build-up of a dome followed by catastrophic surging was the only process by which ice sheets could have come to occupy the great areas which they had at various times during the Quaternary. One of the most probable and instructive, derivations from this consideration of the profile of ice domes is one that has little to do with surges. This is HOLLIN's (1962) account of the glacial history of Antarctica itself, in which it is argued that the area of the ice sheet in Antarctica is controlled by the 'grounding line', where the ice shelves around the perimeter begin to float. The world-wide lowering of sea level during the northern hemisphere glaciations must have shifted the grounding line north. This displacement would provide the base from which a larger Antarctic ice dome could begin to build up.

(5) *Various other suggestions* about the causes of onset of glaciation or its end point to processes which may or may not go on, either amplifying the effects of radiation variations or complicating their course.[1] Thus, NEWELL (1974) suggests that during glacial periods poleward heat transport by the oceans was reduced while heat transport by the atmosphere may have been intensified. The latter depends, however, on the patterns of the global atmospheric circulation as well as its intensity. There is some reason to suppose that the atmospheric circulation was intensified during glacial periods in the lower middle latitudes, but it may well have been weakened in high latitudes (paralleling the probable changes in the ocean currents).

Hypotheses of this type say nothing, of course, about the timing of ice ages and interglacials.

The whole Pleistocene sequence: naming and numbering of warm and cold stages

The various ice ages and interglacial periods of the last 1 million years, and the warmer and colder interstadials or stages within them, have long been known by the names of certain type sites, usually the places where the first definitive evidence was found. This has led to a proliferation of names, and of different naming schemes in many different countries, which complicates the subject and constitutes a formidable hindrance – not to say a psychological barrier – to the newcomer and to the physical scientist interested in learning the facts as an essential preliminary to attempting to understand and explain the climatic record. The justification for the parallel naming schemes for like events in neighbouring regions is that one must not presume identity or simultaneity of the phenomena in different places before it

---

1. See, for example, the mention in Volume 1, on p. 433, of a suggestion by BLOCH that great volcanic activity may bring about the end of a glaciation by blackening the ice caps.

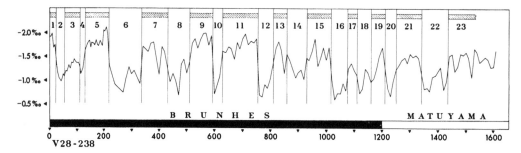

Fig. 15.4 Oxygen isotope record of the last 870 000 years, from measurements on *G. sacculifera*, in an equatorial Pacific sediment core (V28-238), taken at 1°1′N 160°29′E (water depth 3120 m), identifying the stages of the Quaternary climatic sequence by numbers according to the system developed by EMILIANI (1955) and originally based on tropical Atlantic and Caribbean cores. Odd numbers mark the main warm stages, even numbers the main cold stages.

(*Curve kindly supplied by* Dr N. J. SHACKLETON; *from* SHACKLETON *and* OPDYKE *1973.*)

can be proved. The names in many of these local schemes really have the status of preliminary labels for the internal working purposes of the laboratories most concerned.

The solution to the problem must be sought in reliable datings of the stages, which some day in the future should make it possible to adopt an international standard naming convention of just a few schemes, applicable in widely separated parts of the world. It is only now for the first time that dating methods are bringing the possibility of such a far-reaching simplification within sight. In the meantime, the best available standard for comparison of the named climatic stages in different parts of the world is the system of numbered stages developed by EMILIANI (1955) and illustrated here in a Pacific ocean bed core in fig. 15.4.

An early classification of glaciations traceable in the Alps, developed by PENCK and BRÜCKNER (1901–9), identified just four glaciations there, which were named Günz, Mindel, Riss and Würm. Insistence on one interglacial, the Mindel-Riss interglacial, being much longer than the others was a feature of the PENCK-BRÜCKNER scheme.[1]

This famous classification undoubtedly led many over the years since to attempt to align new evidence of glaciations, wherever it was found, to fit the supposed history of four such events. To sort out the resulting confusion, it became necessary to adopt *ad interim* local names for the sequences of events which could be identified, and ultimately dated, in other places. And the lists of names grew longer as more and more climatic stages were recognized. Table 15.1 gives a tentative identification of the correspondences between the naming systems used to identify Quaternary deposits in several principal regions. The correspondences between the stages named can be regarded as certain in the Upper Pleistocene, but become progressively more doubtful farther down the table.

1. The lengths of the last few glacial-interglacial cycles have been carefully re-examined by SHACKLETON and TURNER (1967).

Table 15.1   Names of the principal warm and cold climate stages recognized in Quaternary deposits in different regions of the northern hemisphere and their probable correspondences

Notes: Youngest stages at top of table.
Interglacial periods in italics.
Times of widespread glaciation in normal print.
Most of the place names appearing in the table are often used in their adjectival form in the English-language literature.

		North America central sector	Britain	North European plain	European Russia	Alps
HOLOCENE (or RECENT)	Postglacial		*Flandrian*	*Flandrian*		
PLEISTOCENE	Upper	Wisconsin *Sangamon* Illinois	Devensian *Ipswich* Wolstonian (or Gipping)	Weichsel *Eem* Saale	Valdai *Mikulino* Moskva *Odintsovo* Dniepr	Würm *Riss-Würm* Riss
	Middle	*Yarmouth* Kansan *Afton*	*Hoxne* Anglian (Lowestoft) *Cromer* Beeston *Paston*	*Holstein* Elster *Cromer*	*Likhvin* *Morozov*	*Mindel-Riss* Mindel *Günz-Mindel*
	Lower	Nebraskan	Baventian *Antian* Thurne *Ludham* Walton	Menap *Waal* Eburon *Tegel (Tiglian)* Brüggen *Pre-Tiglian* Amstel	Odessa	Günz *Donau-Gunz* Donau *Biber-Donau* Biber

The full climatic sequence demands in addition recognition of many substages, e.g. the various climaxes of the last two ice ages which have been most briefly named Riss I, Riss II, Würm I (about 60 000–70 000 years ago) and Würm II and III (respectively 20 000–25 000 and 15 000–20 000 years ago). Many of these have, however, also been given local names: Riss I = Dniepr in Russia, Drenthe in north Germany and the Netherlands, Riss II = Moskva and Warthe in these two regions respectively. Würm I = Kalinin in European Russia and is sometimes known as the Stettin stage in north Germany and as Iowan in North America; the late Würm climaxes correspond to the Ostahkov stage in Russia and the Pommeranian and Brandenburg moraines stages on the north German plain. There are further naming systems in use in Ireland, Siberia and the Far East. In New Zealand the following names are in use:

> *Aranui interglacial* = Postglacial times
> Otira glaciation,   = Würm II and III
>   with two climaxes
>   respectively about 14 000–17 000 years ago
>   and some time before 22 300 years ago
> *Oturi interglacial, probably contemporary*
>   *with the mid Würm warmer interstadials*

Waimea glaciation

Terangi interglacial

Waimaunga glaciation

Authorities used for the naming systems include EMILIANI and GEISS (1957), FLINT (1971) where many further details will be found, MITCHELL *et al.* (1973), MOAR and SUGGATE (1973), SCHWARZBACH (1961), SPARKS and WEST (1972).

Table 15.2   Identification of the northern hemisphere glaciations on the northern continents with EMILIANI's numbered stages found in deep sea sediments

Stage number	Alpine glaciation
2	Würm II and III
4	Würm I
6	Riss or Riss II
8	?
10	Mindel
12	?
14	Günz

The probable alignment of the Alpine glaciations with EMILIANI's numbering of the principal cold stages recognized in deep sea cores (fig. 15.1) during the Quaternary is indicated in Table 15.2, based on EMILIANI and GEISS (1957).[1] There is clearly increasing difficulty as one goes back through the Quaternary in fitting the names to all the cold stages in which glaciation must have been extensive, at least in the heartlands where ice sheets have repeatedly come and gone. Even the longest lists, in the British and European plain columns

1. Much detailed knowledge of the corresponding sequence in Japan and its comparisons with other parts of the world may be expected to come from the comprehensive series of studies of Lake Biwa (35°N 136°E) launched by the University of Kyoto (HORIE 1974).

of Table 15.1, do not supply names to cover all the principal cold stages recognized in the ocean record.

In the oxygen isotope records from the ocean bed many of the cold stages register about equal severity, yet so far only the last three cold stages named in Table 15.1, the Mindel/Elster/Anglian, Riss/Saale/Wolstonian, and Würm/Weichsel/Devensian, have been clearly recognized as glacial in Britain – i.e. ice sheets having spread over the lowlands of Britain. The earliest occurrence of permafrost in the Rhine basin in the lowlands of central Europe is dated about 700 000 years B.P., according to BUTZER (1974). According to the pollen evidence, there were also some differences in the climatic character of successive interglacials, at least on the western margin of Europe (cf. Chapter 13, pp. 197–9) despite much basic similarity of the time sequence through each (ref to figs. 13.54 and 13.55). These differences and the survival in Europe of some elements of the warm Tertiary flora (Chapter 13, p. 201) until the beginning of the middle Pleistocene – e.g. *Tsuga* (hemlock spruce) seems to have been present for the last time in the forests of northwest Europe in the Pastonian interglacial, though *Sequoia* and some other species were wiped out much earlier in the Pleistocene – suggest greater severity of the cold stages towards the late Pleistocene. This increasing severity, if real, parallels – and may have been partly caused by – the progressive lowering of sea level (cf. Chapter 13, pp. 113–15) which presumably laid bare more of the continental shelf in the later glaciations and thereby made it easier for the continental ice sheets to spread over some of the land areas (e.g. Northsealand) exposed during the ice age climates.

An important component of the sequence shown by the oxygen isotope records of the last 250 000 years from the ocean bed, and the timing of epochs of highest sea level, seems to have been diagnosed by EMILIANI (1969) – see also EMILIANI and RONA (1969) – who noticed that several separate high sea level stands tend to occur in each interglacial and that the spacing of them at about 20 000-year intervals coincides with the precession cycles (see fig. 15.5). Sea level reached a maximum within a few thousand years after each time that the

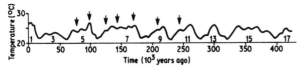

Fig. 15.5   The relationship of times of high sea level (marked by the arrows) in the last 250 000 years and EMILIANI's generalized palaeotemperature curve derived from oxygen isotope work on ocean bed sediments in the Caribbean and tropical Atlantic.

Dating by EMILIANI, using uranium series measurements and [14]C. Numbered stages as in fig. 15.4.

(*Diagram from* EMILIANI *and* RONA (*1969*); *reproduced by kind permission.*)

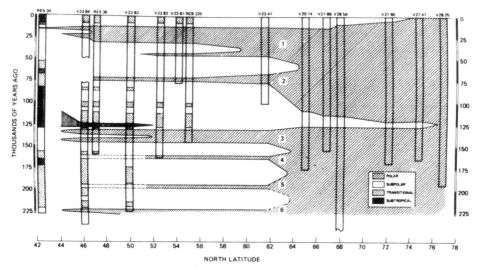

Fig. 15.6   North–south migration of polar water in the eastern Atlantic at 20°w during
the past 225 000 years, as shown by analysis of fourteen deep sea cores.

Dominance of the polar water is indicated by diagonal hatching and the other, less
cold, watermasses are indicated by the key.

(*Reproduced by kind permission of Dr* A. MCINTYRE.)

northern summer solstice took place at perihelion. The lag after the date when the warmest
summer month was at perihelion was usually 1000–3000 years. This suggests that the
Greenland ice undergoes substantial melting at such times once an interglacial is fully
established, possibly raising world sea level nearly 10 m above its present stand, and that the
Greenland ice sheet waxes significantly during those parts of an interglacial (as within
± 5000 years of the present time) when the Earth is near aphelion in the northern summer
months. The Greenland ice may be more significant than the great Antarctic ice sheet in this
regard because of its extension into relatively low latitudes.

An analysis by MCINTYRE *et al.* (1972) of the changes of prevailing watermass in the
eastern North Atlantic between 15° and 28°w over the last 225 000 years, from the remains of
the surface-living foraminifera and the coccoliths deposited on the ocean bed, despite
difficulties in reconciling the datings before 50 000 years ago with other studies, is of interest
in indicating that seven or eight times the polar water advanced from the Iceland region to
south of 51°N and in at least five of these instances reached 45° to 42°N in the eastern Atlantic
(see fig. 15.6).

There is much interest also in attempts to trace the variations of precipitation, wind
patterns and windiness during the Pleistocene. Early suggestions that pluvial periods – i.e.
periods with greatly increased rainfall – affected most of Africa during northern Europe's
glacial periods, and could be equated with them as regards timing, are now known to have
represented a gross oversimplification of the distribution of raininess both in space and time.

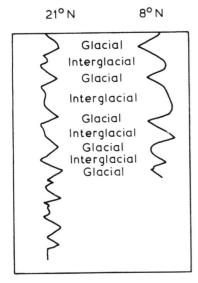

Fig. 15.7   Variations of abundance of the coarse fraction of windblown dust in the Atlantic ocean sediment from cores taken at 21°18′N 22°41′W (V23-100) and 8°18′N 22°45′W (V19-301), off the African coast.

Note the antiphase relationship between the windiness variations apparently implied at these two points in the course of the glacial-interglacial cycles.

The incidence of glacial and interglacial climatic regimes was determined by species abundance of the foraminifera *Globorotalia menardii* and by oxygen isotope measurements.

(*Schematic diagram after* D. W. PARKIN, *personal communication, 9 October 1974.*)

Probably the early stages of each glacial period brought a great increase of rainfall in the lower middle latitudes (e.g. Mediterranean), while the seas were still warm but cyclonic activity was increasing in those latitudes. (We shall gain further insight into this when we come to analyse the last ice age later in the chapter – see pp. 352–3.) In the later stages of each ice age much less water vapour could be taken up into the atmosphere from the colder seas, and in most latitudes the amount precipitated must have decreased. An analysis reproduced by KUKLA (1970) of the sequence of soil types in central Europe (Czechoslovakia) through the last million years shows dry, wind-blown loess soils coinciding with each of the coldest phases. Persistent dryness in the Mediterranean region and in Australia between about 27 000 and 23 000 B.P. seems to be indicated by the absence of radiocarbon dates (implying minimal vegetation) for both regions, noted in a statistical examination performed by GEYH and ROHDE (1972) (see also Chapter 16, p. 387). There was also a minimum of radiocarbon dates in Europe, but in the United States and the Caribbean the trend was opposite, perhaps because of cyclonic activity forced south by the advance of the ice.

Lakes in Africa and elsewhere could nevertheless be maintained at much higher levels than now because of reduced evaporation (except in very windy regions) in the cooler air and under cloudy skies. The same factors no doubt helped maintain the greatly expanded Caspian Sea and the deep ice age lakes (Lake Bonneville and Lahonton and others) in the now arid regions in the western United States.

The succession of variations of windiness indicated (PARKIN 1974 and personal communication, 9 October 1974) by the coarseness and composition of the wind-blown dust in ocean-bed sediment cores from the Atlantic at two different latitudes, 8°N and 21°N (both

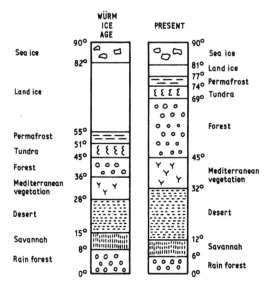

Fig. 15.8   Climate and vegetation zones now and in the last ice age in the Europe/Africa sector, 0–15°E.

(*Redrawn, after* BÜDEL *1951.*)

near 23°W), off the African coast show opposite variations with time through the glacial–interglacial cycles (fig. 15.7: see also Chapter 13, p. 106). It seems clear that the glacial regimes were windier at 21°N, and less windy at 8°N, than interglacial times. PARKIN supports his diagnosis by observations of the colour of the dust in the deposits, which was browner and cleaner looking in the windy regimes, greyer due to an admixture of humus and charcoal from vegetation (bush fires, etc.) in the regimes diagnosed as less windy. The same differences are found in the dusts collected over the tropical Atlantic in windstreams from desert and savanna or forest regions today (PARKIN *et al.* 1972). These observations of glacial–interglacial variations may be compared with BÜDEL's (1951) analysis of the latitude-spread of the vegetation zones in the Europe-Africa sector in the Würm ice age and now (see fig. 15.8). During the glacial regime the equatorial rain-forest and savanna zones were both expanded, so that their limits lay farther north than now; the desert zone was narrowed and shifted slightly south, so that its axis was very close to 21°N where the ocean core suggests increased windiness and full desert conditions on the neighbouring land. It seems possible that the glacial regime at 8°N was not less windy than today, but that the reduced dust sizes there agree with the evidence of more vegetation attributable to a moister regime in North Africa south of about 15°N. This moister regime could be explained by reduced evaporation and a significant frequency of frontal cloud-sheets penetrating from the north owing to greater mobility and frequent replacements of the subtropical anticyclone cells.[1] The evidence of this paragraph might be summed up as indicating stronger prevailing

1. This pattern is suggested by comparison between developments in the subtropical anticyclone belts in the northern hemisphere and in the windier southern hemisphere (e.g. over Australia) today (see also Volume 1, p. 337, section on the Doldrums).

winds in ice age times in most of the lower latitudes. A parallel change seems also to have applied in middle latitudes as far north as the periglacial zone in central Europe. But it may well be that, in general, the higher northern latitudes were less windy than now, except where strong katabatic winds descended from the great ice domes.

QUINN (1971) reports evidence from ocean-bed core studies that the narrow zone of cold water along the equator in the Pacific Ocean was still more strongly developed (with stronger gradients of temperature to north and south) in glacial times than today. The accompanying zone of dry climate near the equator probably spread across the whole Pacific. One must conclude that the upwelling in low latitudes near the west coast of South America was stronger in glacial times than today and, hence, that the wind circulation in low latitudes was also stronger. NEWELL also reports a southward displacement of the climatic zones in the North Pacific in glacial times, with the intertropical convergence generally south of the Galapagos Islands, i.e. just south of the geographical equator. Intensification of the Equatorial (westward drifting) Current is verified in a lowering by 5–6°C of the ocean surface temperatures off Peru, indicated by the foraminifera in glacial times, also intensification of the dry zone along the equator in the Pacific so that, for example, aridity spread even to the upper levels of the Galapagos Islands (SIMPSON 1975). As much of the Pacific seems to have experienced a smaller temperature change between glacial and interglacial times, not exceeding 2°C, the temperature contrast between the cold water spreading along the equator from the upwelling off Peru and the warm water surfaces of the North and South Pacific must have been greater in glacial times than now.

The last 125 000 years

We can now gain some acquaintance with the sequences by which the global climatic regime passes into and out of ice ages by closer analysis of the last glaciation, for which most evidence is available and for which dating errors are likely to be least. Plotted graphs of the variations of temperature indicators, such as oxygen isotope ratios and the abundance of various foraminifera species or of coccolith carbonate, with depth or time show many abrupt shifts of direction, many of the minor ones being probably ascribable to experimental error. But there are also some great shifts, implying changes of prevailing temperature by several degrees, which appear to have been very sharp, and are present about the same date in so many curves, that it cannot be doubted that the descent into the next glaciation involved several drastic events. We may also hope to gain some insight into the causation of these changes by study of their timing and by analysis of the circulation mechanisms in the atmosphere and oceans which seem to have prevailed at each stage.

A convenient presentation of the history, essentially a temperature record, of the last 125 000 years is given in fig. 15.9 in which DANSGAARD's oxygen isotope curve from the Greenland ice sheet at 77°N 56°W is placed alongside the less securely dated interpretations of similar measurements at Byrd station in West Antarctica (80°S 120°W). As in fig. 13.16

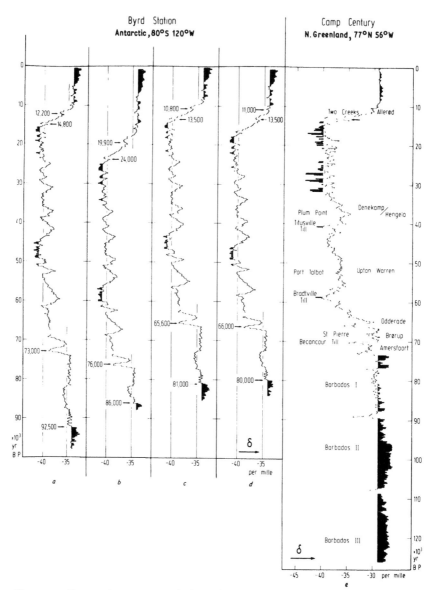

Fig. 15.9   Oxygen isotope records from Antarctica and Greenland for the last 120 000 years approximately.

The four curves to the left of the diagram represent the results of measurements of segments of an ice core from Byrd station (80°s 120°w) each covering about 200 years. The different estimates of the dating result from different assumptions about the rate of accumulation, total thickness of the ice sheet and divergence of the ice flow.

The right hand curve represents measurements at Camp Century in North Greenland (77°N 56°w) and indicates proposed correlations with recognized warmer stages in Europe (to the right) and cold stages in North America (to the left).

(*Reproduced from* JOHNSEN *et al. 1972; by kind permission.*)

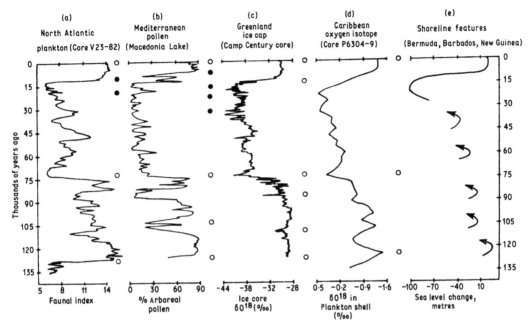

Fig. 15.10   Comparison of climatic records of the last 135 000 years, plotted so as to indicate warmth to the right and cold to the left.

(*a*) North Atlantic, west of Ireland, based on foraminifera identification (SANCETTA *et al. 1973*).

(*b*) Macedonia, abundance of tree pollen (VAN DER HAMMEN *et al. 1971*).

(*c*) Greenland ice sheet, oxygen isotope measurements (as in fig. 15.9).

(*d*) Caribbean Sea, oxygen isotope measurements on surface-dwelling foraminifera found in the sediment (EMILIANI *1968*).

(*e*) Generalized sea level curve (BLOOM *1971*). Solid round dots = dates controlled by carbon 14.

·   Open circles = dates fixed by stratigraphic horizons of known date.

(*Diagram kindly supplied by Drs* W. L. GATES *and* J. S. PERRY *with the permission of the U.S. National Academy of Sciences.*)

(p. 97), the conventional names of some of the more widely recognized warmer stages in Europe, times of high sea level at Barbados and coldest stages in North America, are entered on the diagram. Detailed sequences for other widely separated places in the northern hemisphere are given in fig. 15.10, and fig. 15.11 provides a perspective of the different time scales discussed in this and later chapters (see also discussion in NATIONAL ACADEMY OF SCIENCES 1975).

The warmest period of the last interglacial (the Eemian interglacial) – and, as is commonly the case in interglacials, this was the earliest warm period – was at its height around 120 000 years ago. This warmest phase, which corresponded to the postglacial period up till now, lasted altogether some 10 000 years. It was ended, according to most datings, by an abrupt cold episode, in which the ocean-bed oxygen isotope work suggests that there was

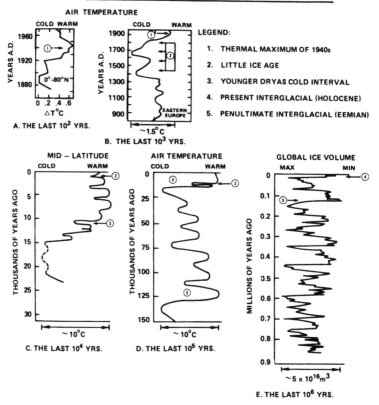

Fig. 15.11   Indicators of global temperature change: different time spans compared.

(*a*) Successive 5-year mean observed surface air temperatures, last 100 years (MITCHELL *1963*).

(*b*) Winter severity index for Europe near 50°N 35°E, last 1000 years (LAMB *1969*).

(*c*) Generalized temperature changes in middle latitudes of the northern hemisphere, from tree line studies, last 10 000 years (LAMARCHE *1974*).

(*d*) Middle latitudes northern hemisphere air temperature variations, from diagnosis of sea temperatures and from pollen records on land, last 100 000 years (compiled by W. L. GATES from the curves in fig. 15.10).

(*e*) Global ice volume changes from isotope measurements on plankton in a Pacific Ocean sediment core, last million years (as in fig. 15.4).

(*Diagram kindly supplied by Drs* W. L. GATES *and* J. S. PERRY *with the permission of the U.S. National Academy of Sciences.*)

extremely rapid growth of ice on land, about 115 000 years ago.[1] The Eemian interglacial is believed to have been the first time that England was separated from the European continent (DESTOMBES *et al.* 1975). During the warmest time in this Eemian interglacial, with sea level up to 18 m higher than at present (see Chapter 13, p. 115), an enlarged Baltic offered a through passage for the sea from the Atlantic, over much of Danish Sjælland and the Gulf of Finland, to the White Sea (GERASIMOV 1969). Scandinavia was an island, and there may have been further sea passages south of Jutland, across south Sweden (Skåne), and across middle Finland. All northern Europe presumably had a much more oceanic climate than in postglacial times.

Another warm period, according to most records not quite equalling the former, followed and was ended by another abrupt cold episode about 90 000 years ago.[2] This episode is by now well documented (e.g. DANSGAARD *et al.* 1971, KENNETT and HUDDLESTUN 1972, SHACKLETON 1969) and is seen as a sharp feature in fig. 15.9. The Greenland ice core record suggests that conditions changed from full interglacial half way to full ice age temperatures within 100 years. KENNETT's study suggests a corresponding change about the same time to a cold climate in the waters of the Gulf of Mexico which was completed in less than 350 years (see also p. 318). Within about that time a great deal of northern America, and probably of Europe also, must have become permanently snow-covered. Another study (DUPLESSY *et al.* 1970) suggests that the temperatures prevailing in southern France dropped 2·5°C at about this stage of the interglacial. The recovery which followed took in all these cases some 2000 years or so (cf. fig. 15.10). Despite the severity of the temperature drop, little more ice can have been formed than could be melted in 2000 years.[3] Once again the warm period which occupied the succeeding 10 000–20 000 years was somewhat less warm than its predecessor, and around 70 000 to 75 000 years ago it was ended by events which produced the first climax of the Würm/Weichselian/Wisconsin ice age.

The total length of the Eemian interglacial could thus be put at about 50 000 years, though the unbroken stretches were only somewhat over 10 000 years. SHACKLETON and TURNER (1967), carefully assessing and comparing the sequences derived from deposits on the ocean bed and in England and Germany, found similar durations for the Hoxnian and Cromerian interglacials (refer to Table 15.1, p. 325).

Onset of a full glacial regime

The array of curves in fig. 15.12 gives a closer view of the main stages of the last glaciation and, together with data for other places named below, allows us to deduce the manner of its

1. 108 000 B.P. in the Greenland record in fig. 15.11.
2. Conditions as warm as the postglacial are found to have been rare in the Quaternary, occupying no more than 8% of the record, namely only the warmest parts of the interglacials.
3. If the figure for France, with no great glaciers near, be taken as approximating to the global average, the

drop of the prevailing temperatures around 90 000 B.P. was about half that corresponding to a full ice age development. The date reported by DUPLESSY *et al.* for the event in France was about 97 000 B.P., but the similarity of the sequence suggests that it was in fact the same event reported elsewhere between 91 000 and 89 500 B.P.

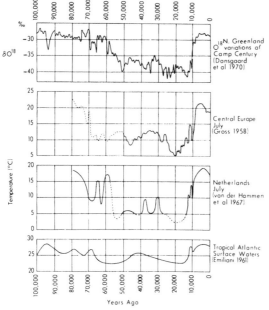

Fig. 15.12   Average temperatures over the last 100 000 years, derived from various indicators.

(*a*) Northern Greenland: oxygen-18 variations at Camp Century (77°N 56°W) (*after* DANSGAARD *et al. 1971*).

(*b*) Central Europe: mean air temperature in July, from evidence of the flora, fauna and soil mollusca (*after* GROSS *1958*).

(*c*) Netherlands: mean air temperature in July, from evidence of the flora and soil formations (*after* VAN DER HAMMEN *et al. 1967*).

(*d*) Mean sea surface temperature in the tropical Atlantic (all year) from $^{18}O/^{16}O$ ratio in the remains of planktonic foraminifera in the ocean bed deposits (*after* EMILIANI *1961*).

onset. These curves indicate substantial agreement about the dating of the last, and indeed the main, climax of the glaciation 15 000 to 20 000 years ago; though the Greenland curve indicates two main climaxes during the Late Würm there, the earlier one 23 000 to 25 000 years ago, the last one about 14 000 years ago (it is reported that the main climax of this glaciation near the Pacific coast of North America was about 24 000 B.P., and it is clear that in Europe there was a substantial cooling about that time). The temperature curves also show a good correspondence over the whole time since the last glacial maximum; though EMILIANI's diagnosis of the course of events in the tropical Atlantic shows only one of the two Late Glacial warm stages recognized in Europe and in the Greenland ice.

The shape of the temperature curve through the main cooling stage in the Early Glacial seems broadly agreed, especially as to its abruptness, but the dates suggested range from about 57 000 to 70 000 years ago and the sharp cooling is now thought to have set in 70 000 to 75 000 years ago (cf. 15.10).

It seems to be agreed that traces of moss assemblages, tree remains, and other flora and fauna found in England from about 57 000 to 59 000 years ago (according to COOPE, SHOTTON and STRACHAN (1961), DICKSON (1967), SIMPSON and WEST (1958)), organic layers in sediments in New York State estimated to be 52 000–64 000 years old (MULLER 1964) and in Denmark from 55 000 to 59 000 years ago (ANDERSEN 1961), and similar deposits found in the Netherlands (ZAGWIJN 1961), represent one or more early interstadials (i.e. reversions to somewhat warmer climate).[1] They seem – from the establishment of

---

1. Two warm 'interstadials' have been named, successively, Amersfoort and Brørup, after the places at which evidence was found in the early Würm deposits. WEST *et al.* (1974) recognize these interstadials also across the English Midlands as the Wretton and Chelford interstadials respectively.

forest in England (dominance of pine with some birch and a little spruce at the Chelford, Cheshire, site investigated by SIMPSON and WEST), in New York State (pine and spruce) and in southeastern Quebec (dominance of spruce) – to have brought summer temperatures nearly up to present levels (July average probably $+12°C$ or even a little higher) in Cheshire, England. But the winter temperatures evidently remained low: those typical of regions where the same tree species flourish today are low and a mean January temperature in the range $-10°$ to $-15°C$ might be indicated for Cheshire (WEST et al. 1974). The interstadial climate therefore seems to have been highly continental in these continental margin areas, probably indicating the continued presence of an already formed, extensive ice sheet not far away and remarkably little wind flow from the Atlantic over Europe.

By about 50 000 years ago an early maximum of the 'Irish Sea ice', and of ice domes over the northern and western parts of the British Isles, has been suggested. COOPE, SHOTTON STRACHAN (1961) regarded this as the time of greatest spread of the Irish and north British ice sheets in the last glaciation, but this view has been much modified by later work (SHOTTON 1967). It now appears that in the Severn Valley, Worcestershire (English Midlands), and on the east Yorkshire coast, as on the continent, the ice extent was greatest between about 25 000 and 17 000 years ago. Moreover, at least between about 45 000 and 38 000 years ago, and again about 30 000 years ago, evidence of mosses, pollen, and insect fauna in deposits examined at Upton Warren in the lower Severn Valley, at Fladbury near Evesham, at places in the Lea Valley near London, and at Hengelo and Denekamp in the Netherlands, indicate much more temperate climates. The assemblage of Arctic or near-Arctic and more southern species occurring together in these deposits (see, for instance, WEST 1968, p. 331) appears to represent conditions that occur nowhere at the present time and may indicate a climate that does not occur on the globe today, with an ice sheet persisting farther east in Europe and over the Scandinavian mountains while western Europe experienced some kind of temperate climate. More evidence of the distributions occurring in this interesting middle Würm period is awaited.

The similar shapes of the temperature curves, despite some uncertainty about the precise dates, plainly suggest that at some time between about 70 000 and 60 000 years ago ice sheets were quickly established over northern and northwestern Europe, probably including parts of the British Isles, and over some extensive hinterland in North America (presumably enveloping the Hudson's Bay region), all within 1000–5000 years. It further appears that these ice sheets were then big enough to survive a warm interstadial, which lasted long enough (presumably centuries or longer) and brought summer temperatures high enough for the reestablishment of forest in southern Britain and southeastern Quebec, though the ice sheets continued to exert a strong control on the winter temperatures.

Such a rapid constitution of the ice sheets at a time when the seas were initially almost as warm as now (EMILIANI's curve (fig. 15.12 bottom section) suggests a deficit of about 2°C) seems to demand a great prevalence of northerly surface winds and a meridionally extended upper cold trough, in most months of the year, over the sector Norwegian Sea-British Isles-

Scandinavia. On meteorological grounds, in terms of the wave pattern and wave length in the upper W'ly circumpolar vortex, this probably implies a similar, or even more marked, meridional trough and surface N'ly regime dominant over the Hudon's Bay and western plains sector of North America. These southward-extended troughs in the upper W'lies persistently occupying positions west of their smaller amplitude counterparts of the present day would presumably require either a much weakened or a much expanded circumpolar vortex, or both, with the main thermal gradients in the continental sectors at lower latitudes than now.

The very rapidity of establishment of the ice sheets (though they need not have attained more than a fraction of their final thickness in order to survive a 500- to 1 500-year-long warm interstadial of the kind indicated just after they were formed) limits the argument about types of meteorological situations that could produce the ice in the regions concerned. Indeed, it suggests a solution, because only one type of circulation pattern seems to meet the requirements.

Our picture of the large-scale atmospheric circulation prevailing at the onset of the main glaciation is one that should be called into being if there were a sharp reduction of the radiation supply (by say 5 %) reaching the Earth's surface at all seasons of the year at a time when the seas were still warm. See, for instance, WEXLER's (1956) calculation of the thermal pattern and upper winds over North America and the western Atlantic, reproduced here as

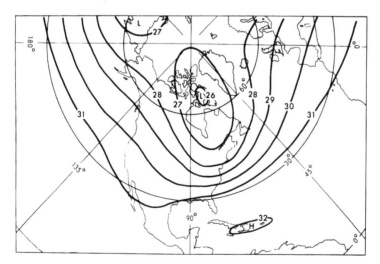

Fig. 15.13   Mean height of the 700 mb pressure level in January estimated for 20 % weakened solar radiation.

Heights marked in hundreds of metres.

The trough in the upper W'ly flow over eastern North America is very much sharpened and extended south in comparison with the situation in normal radiation.

(*After* WEXLER *1956, redrawn.*)

fig. 15.13, that should accompany a hypothetical 20 % reduction of insolation, whether due to a veil of volcanic dust in the stratosphere or a change in the output of the sun. LAMB (1963) noted that the actual anomalies of the circulation observed in the so-called Little Ice Age climate around A.D. 1800 appeared to be similar to WEXLER's calculation, though weaker. The surface water temperature of the North and South Atlantic at that time appear to have been, on overall average, about 0·5°C lower than now, an anomaly that could be produced by a less than 1 % deficiency of insolation operating over a long time.

Despite the occurrence of waves of enhanced volcanic activity during all the cold climate periods which we have been discussing – during the Würm/Weichselian glaciation and the Little Ice Age in recent centuries and in the cold episode 90 000 years ago as well – the timing indicates that this was probably not connected with the onset of the glaciation (cf. p. 318). The decreased strength of the solar beam at the times of rapid onset of a colder regime, implied by the atmospheric circulation here worked out, was therefore more likely to have been due to a fluctuation of the sun.

EWING and DONN (1956) suggested that an ice-free Arctic Ocean was a necessary condition for the supply of moisture to the ice sheets which formed over North America and northern Europe, but an open Arctic Ocean with surface water temperature around 0° to + 5°C would exert a vapour pressure only one-fifth of that of the tropical seas, or about a third of that of the Atlantic Ocean near 45°–50°N, and probably only 20–40 % greater than over the Arctic Ocean in summer today. Moreover, recent research (HUNKINS and KUTSCHALE 1965, KU and BROECKER 1965) indicates that, contrary to EWING and DONN's hypothesis, the Arctic Ocean has been largely ice-covered throughout (at least) the last 70 000 years. So it seems clear that the main moisture supply did not come from the oceans in high latitudes.

The picture here developed (fig. 15.14) of a highly meridional circulation pattern during the main onset of the Würm ice age, with great frequency – possibly actual prevalence – of N'ly surface winds in sectors near 90–100°W and 10°W–10°E, differs only in degree from the anomaly that apparently produced the recent cold centuries.[1] It was evidently more extreme as well as more persistent than the latter. The upper cold troughs presumably extended farther south and were more narrowly concentrated (higher frequency) over the sectors mentioned than in the Little Ice Age. The proposed circulation pattern would imply a bigger reduction in the average value of the zonal index (which measures the strength of the zonal component in the upper W'lies or circumpolar vortex) than was found around A.D. 1800, and might suggest a corresponding reduction of solar radiation receipt.

With this proposed prevailing wind pattern the beginnings of the Laurentide ice sheet in northern Canada would find a straightforward explanation, since the prevailing surface N'ly winds would constitute one side of a belt of extraordinarily frequent cyclonic activity steered north and northeastward from the Gulf of Mexico and western Atlantic toward Labrador

1. This circulation pattern is quite similar to that independently put forward by FLOHN (1969, p. 225).

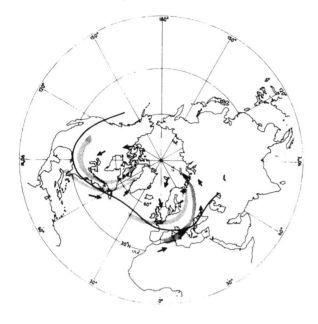

Fig. 15.14 Suggested pattern of northern hemisphere circulation during the onset of the Würm/Wisconsin glaciation, during formation of the Laurentide (northeast America) and Scandinavian ice caps.

Thin continuous line, suggested average course of 5300 m isopleth of 1000–500 mb thickness in January; short arrows = prevailing surface wind directions; broad, shaded arrows = most frequent tracks of surface cyclones.

and recurring northwestward over the Canadian Arctic. This would represent diversion of the cyclone sequences familiar at the present day (which cross the Atlantic toward Iceland and northern Europe) to the northward tracks described, over and near eastern North America. The main moisture-bearing winds would be diverted from their present trans-Atlantic paths and steered north towards Arctic Canada, and they would doubtless deposit snow from September to May or June over the region about 55–70°N with a frequency unknown today. Deposition might be mainly as snow even in summer, given a reduced radiation supply. Thus, the moisture source for the Canadian ice sheet would appear to be the Gulf of Mexico and the tropical Atlantic, a conclusion which may be necessary anyway to meet the indicated rapidity of ice sheet growth.

Examination of the rainfall, and the water equivalent of the snow, in each month of the year on the lands about the Hudson's Bay region between 55° and 70°N, taking only the months with the most cyclonic and southerly surface wind flow patterns, between 1955 and 1968 (i.e. those cases in recent years which came nearest to the suggested patterns of glacial onset), these showed that the total downput was typically 150–300% of normal over substantial areas east, north and northwest of Hudson's Bay, and sometimes 400–500%. This regime therefore seems capable of raising the annual total downput from the present 20 cm to 60 cm. This would presumably cut WEERTMAN's (1964) estimate of 15 000–30 000 years for the growth time required by a major ice sheet to 5000–10 000 years. In the case of the former Laurentide ice sheet it is easy to envisage the still more rapid covering of a large area by ice, initially in an ice sheet of rather flat profile, because of the vast, nearly uniform terrain which seems likely to have provided not just one but many starting points, where snow accumulation would first survive a few summers and then grow: the growth rate would not necessarily depend on the building up and unstable flow of an ice dome, as assumed by WEERTMAN.

Initially, the eastern fringe of North America would retain the warmth of a temperate climate with this circulation arrangement, and might even become warmer for a time owing to the increased frequency of southerly winds on the eastern side of the track of frequent cyclonic storms, and these conditions could return in the warmer interstadials. Gradually, however, growth of the continental ice sheet would strengthen the thermal gradients near its border, tending to increase the zonal index and wave length in the upper W'lies. These tendencies, and the growth of the ice sheet itself, would ultimately displace the northbound cyclone tracks and the belt of southerly surface winds farther east, out over the Atlantic.

The other proposed upper cold trough, also meridionally extended, in the European sector, should be understood as a dynamically induced accompaniment (downstream wave in the upper W'lies) of the American trough. It would suggest a similar prevalence of cyclonic developments steered north and northeast over central and eastern Europe, depositing snow at most times of the year over the region between northern Scandinavia and Novaya Zemlya. Parts of Russia or western Siberia would initially remain warm, with prevalent S'ly winds.

It seems to be implied that there would be a dearth of cyclonic activity, and probably mainly anticyclonic conditions, over the central and northern Atlantic, Most of the extra feeding of the Greenland and Iceland ice sheets would probably come from occasional advances of the American meridional upper trough to an eastern position, steering some depressions from mid Atlantic towards Greenland and Iceland.

Once the great continental ice sheets had been brought into being in the Early Glacial period here discussed, they apparently continued to increase in thickness till a maximum around 45 000–50 000 years ago; though the rate of accumulation, particularly in the central regions which had become remote from much cyclonic activity, would be much slower than in the first 1000 years. The later maximum of this glaciation around 20 000 years ago seems to have been quite similar, though in most respects it was the main climax. This we shall examine in more detail, which the greater abundance of field data makes possible.

Mid Weichselian climatic variations

Between the climaxes of low temperature and ice-cover in the Early and Late Glacial (cf. figs. 15.9–15.12) there were many thousands of years of variable, rather more temperate conditions, although these were generally colder than now. This is a peculiar situation, much more difficult to interpret by reference to present-day conditions than are the full interglacials. More data are needed before any reliable global reconstruction of such a regime can be attempted. For much of this time the seas continued cold; and the great ice sheets, though somewhat dwindled, presumably remained in existence. The latter point is verified by the oxygen isotope records from the sea bed (e.g. curves (a) and (d) in fig. 15.10) if

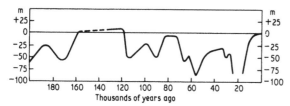

Fig. 15.15   History of world sea level over the last 200 000 years, according to VEEH and CHAPPELL (1970).

these are read as a measure of the amount of water substance abstracted from the oceans and 'locked away' as inland ice. Only for a time around 45 000–40 000 years B.P. (Upton Warren interstadial) the glacial ice seems to have been reduced to 30–50 % of the maximum excess over today's amount. This means that the total volume of ice on land, which today is about $33 \times 10^6 \text{ km}^3$, and at the Early and Late Glacial maxima probably amounted to 70–73 $\times 10^6 \text{ km}^3$, was perhaps reduced in the mid Weichselian warmest time to 45–53 $\times 10^6 \text{ km}^3$. It is to be noticed that this occurred some 5000 years after a maximum of the summer radiation at 65°N comparable with that around 80 000 years ago and only slightly less than the maxima 125 000 and 10 000 years ago, which probably bore a similar time relationship to the warmest phases of the Eemian interglacial and of the postglacial era respectively.

Reconstructed histories of world sea level, e.g. from dated sequences of coral reef terraces around the steep shores of New Guinea and in the Ryuku islands, volcanic islands where there has been long-continued gradual uplift, support this diagnosis (VEEH and CHAPPELL 1970). Sea level apparently stood 2 to 10 m above present 120 000 years ago before falling abruptly to much lower levels, followed by relative high stands around 80 000 and 40 000 years ago (fig. 15.15).

The most detailed evidence so far of the mid Weichselian climatic sequence is that from England in the studies of COOPE and others (see pp. 230–1). There is qualitative support for the sequence of warmer and colder phases from pollen analysis in both England and the Netherlands, but our knowledge of the extremity of the warming that occurred around 42 000 years ago relies on the evidence of the insect fauna, which was evidently able to adjust to the new climatic pattern to an extent that the vegetation character was not. The sequence may be summarized as follows (from COOPE *et al.* 1972 and MORGAN 1973) and is illustrated in fig. 15.16. After the early warm (Brørup/Chelford) interstadial about 60 000 years ago, when pine, spruce and birch grew in the English Midlands and prevailing July temperatures must have been 14–16°C, there was no forest flora in England for about 50 000 years, until the return of a predominantly birch forest at the time of the Allerød oscillation. July temperatures in England dropped to 10° or below and an Arctic insect fauna came in during the glacial climax around 50 000 B.P. Rapid warming occurred about 43 000 years ago, the Arctic insect fauna was more or less eliminated and replaced by such a warmth-loving assemblage that for about 1000 years average July temperatures in England as high as

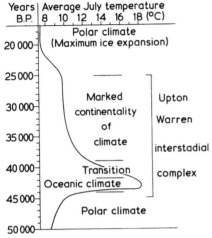

Fig. 15.16   July temperatures prevailing in England and the climate of western Europe between 50 000 and 17 000 years ago, as derived from insect evidence.
(*Diagram kindly supplied by* G. R. COOPE.)

17–18°C – i.e. somewhat warmer than today – may be deduced (COOPE and ANGUS 1975). But despite this favourable climate the landscape remained open and treeless. COOPE'S suggestion that the warm conditions did not last long enough for forest to spread so far from the remote refuges to which it had been confined beyond the Alps and the mountains of central France seems reasonable. The reversal of the climatic tendency which followed was fairly rapid, and by 38 000–36 000 years ago the insect communities of the Arctic tundra had been re-established in central England. The lack of organic material giving radiocarbon dates between about 30 000 and 15 000 B.P. suggests the development then of more extreme cold conditions, resembling Arctic desert.

Another indication of the warmth of the mid Weichselian interstadial (or interstadials)[1] comes from the discovery of date-palm pollen, of age about 40 000 years, from a site at Shanidar in the Zagros Mountains of Kurdistan, well north of and higher above sea level than dates mature at the present day (BRAIDWOOD 1963).

Understanding of the long strange sequence of climates in the northern hemisphere temperate zone mainly colder than now, which spanned the time between about 50 000 and 25 000 years ago may be one of the most difficult problems to solve. It is an important one, however, in that there is increasing evidence that Earth history shows far longer cold-climate periods without (or almost without) major continental ice sheets than with them.[2]

The last climax of glaciation

The distribution of field evidence from pollen analysis and geomorphological studies on land and in the shallow marginal seas, from the microfaunal assemblages represented in deposits

1. Doubtless there were more fluctuations of shorter period and smaller amplitude than those shown in fig. 15.16. Two warm interstadials named Hengelo and Denekamp, after the sites where they were found, have been recognized in the Netherlands. The dates suggested, between 35 000 and 40 000 B.P. and 30 000–35 000 B.P. respectively, are subject to the

rather wide uncertainty of the radiocarbon method at that range. The whole warm interval is widely known as the Upton Warren interstadial, after the site in the Severn valley in the English Midlands from which much of the first evidence came.

2. Since this chapter was written, world summaries of the conditions prevailing during the mid Weichselian

Fig. 15.17   Northern hemisphere geography at the time of the last climax of glaciation about 18 000 years ago.

on the ocean bed, and from oxygen isotope studies of many environments, now makes it possible to attempt reconstruction of the patterns of the prevailing temperature and atmospheric and ocean circulation at the final climax of the last ice age and in the most distinctive regimes that have occurred since.

The ice age geography of the northern hemisphere, recognition of which is basic to these studies, is shown in figs. 15.17–19 for the time of the climax 17 000–22 000 years ago as well

---

climatic fluctuations have been surveyed in publications by (1) A. DREIMANIS and A. RAUKAS in an article entitled 'Did middle Wisconsin, middle Weichselian and their equivalents represent an interglacial or an interstadial complex in the northern hemisphere?', pp. 109–20 in *Quaternary Studies* (*Select Papers IX INQUA Congress*), edited by R. P. SUGGATE and M. M. CRESSWELL (Wellington, *Royal Society of New Zealand Bulletin* No. 13, 1975),

and (2) H. H. LAMB in Climatic Analysis) opening the general discussion in the last pages of the Proceedings of the Joint Discussion Meeting of the Royal Society and Royal Irish Academy on The Changing Environmental Conditions in Great Britain and Ireland during the Devensian (last) Cold Stage, edited by R. G. WEST and G. F. MITCHELL (London, Royal Society, 1977).

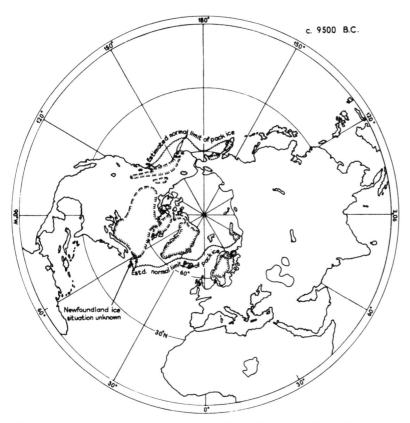

Fig. 15.18   Northern hemisphere geography at the time of the Allerød
(warm) interstadial 11 000–12 000 years ago.

as at the height of the Late Glacial (Allerød) warm interstadial in the tenth millennium B.C.
and in the colder, post-Allerød regime which lasted about 600 years (approximately 8800–
8200 B.C.) in the ninth millennium.[1]

The most noteworthy features of the geographies of these epochs are:

(1) The two great ice domes, amounting to high ice plateaux, over much of North
America and northwest Europe:

At the last ice age maximum the crests of these ice plateaux are believed to have been

1. Sources used for these geographies are:
*American sector, including Bering Strait, as well as
Greenland and Iceland:* FALCONER et al. (1965),
FLINT (1957), IVES and ANDREWS (1963),
WOLDSTEDT (1965).
Eurasian sector: FRENZEL (1960, 1967),
GERASIMOV (1965), WOLDSTEDT (1958),
WRIGHT (1961).
Coastlines, sea level, sea channels, and landbridges:

FRENZEL (1960), GERASIMOV (1965),
WOLDSTEDT (1958, 1965), ZEUNER (1958); and by
deduction from the sea levels indicated by
FAIRBRIDGE (1961), JELGERSMA (1966),
SCHOFIELD (1964), SHEPARD (1963).
*Caspian Sea, Lake Mega-Chad, North Africa, former
lakes in North America, etc.:* FLINT (1957),
FRENZEL (1960), LEONTEV and FEDEROV (1953),
SCHNEIDER (1967), WOLDSTEDT (1958, 1965).

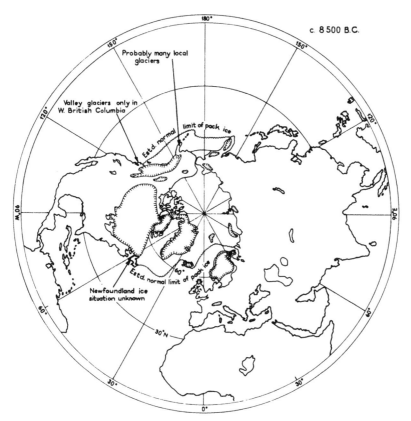

Fig. 15.19 Northern hemisphere geography during the post-Allerød cold
time between about 10 200 and 10 800 years ago.

2000–2500 m above sea level over Hudson's Bay, as well as over the northern Rockies, and
about 2500–3000 m over Sweden and the northern Baltic. Whenever these great ice barriers
lay in the path of the mainstream of the meandering upper W'ly winds, they must have
induced great lee troughs and tended to produce a train of large waves downstream around
the hemisphere, as the Rockies and the high mountain plateaux of Asia do today. Since,
however, the thermal pattern over the northern hemisphere in the presence of these great
areas of ice must have displaced the main flow of the upper W'lies most of the time into a
course south of the ice, the barrier effect upon the upper winds should only occasionally
come into play. Presumably, the main flow of the upper W'lies would generally attain its
lowest latitude just as it passed the ice dome; the axis of a cold trough should then generally
be near, or not far east of, the axis of the ice sheet. The nearest parallel in the world today
must be the effect of East Antarctica on the southern hemisphere W'lies (cf. LAMB 1959).

The surface winds would no doubt be continually affected by the great barrier, and a
katabatic drainage of the cold air down the slopes would go on whenever the winds in the

neighbouring free air were not too strong to allow it. There must also have been occasional föhn (warming) effects in winds that crossed the barrier and topographical convergence producing continual strong-wind effects near steep portions of the edge of the ice (see Volume I, p. 283: frontal funnel gales). And probably throughout the glaciation there was either a great dale in the ice surface or an ice-free gap between the glaciers and ice sheets which covered most of the Rocky Mountains (mainly on the Pacific side) and the great Laurentide ice sheet spreading from northern and eastern Canada to within 100 km of the Cordillera. This dale, or gap, in the ice right through from latitude 50° to 65°N, along a line paralleling the mountains near the Mackenzie River valley, presumably channelled the surface winds off the main ice sheet into a nearly permanent SE'ly wind sometimes of great strength – a wind regime resembling that of the coasts of Antarctica at the present day.[1]

(2) The great inland lakes and seas:

(a) Lake Mega-Chad, which is today within the Sahara.

(b) The enormously enlarged Caspian Sea occupying the lowland basin southeast of the Scandinavian ice massif.

(c) Other lakes not now present, or much larger than now, in Central and East Asia.

(d) Ice-dammed lakes, some of considerable size, particularly the one in the area of the present southern North Sea.

(e) Similar ice-dammed lakes in central North America. The largest of these, Lake Champlain, lasted for about 2000 years from 9500 to 7500 B.C. with gradually changing shape and position in the lower St Lawrence Valley.

(f) Lakes Bonneville and Lahontan, and others in the great basins west of the main watershed of the U.S. Rockies, especially between 38°N and 42°N.

(3) Enlargement of the continents brought about by the lowering of sea level estimated at 100–130 m below present at the ice age maximum and about 40 m during the Allerød oscillation.

This much reduced the opening for Atlantic water into the Mediterranean and closed the north and northwest European seas. It also closed the channels between the Asian landmass and Novaya Zemlya and the Siberian Arctic islands, also those between Canada, the Canadian Archipelago and Greenland as well as the Bering Strait.

Hence, the Arctic Ocean lost its present shallow link with the Pacific. The volume of water over the submarine sills between Greenland, Iceland and Scotland that stand between the Arctic Ocean and the Atlantic must have been reduced by a quarter to a third.

(4) At the culmination of this, as of earlier ice ages, the Scandinavian ice probably blocked the northern exit of the North Sea, so that the rivers Rhine and Thames probably drained into a great pro-glacial lake in the southern North Sea area which overflowed by a

---

1. The regime in this area was apparently not so hostile as to prevent the gap between the ice-masses being used as a route by which human wanderers entered the Americas (see Chapter 13, pp. 235, 248).

course in or near where the Channel is now, to reach the Atlantic.[1] In the Late Glacial stages, when the ice had retreated and left a great plain (Northsealand), these rivers are believed to have joined and drained north into the bay near the Dogger Bank moraine. It was not until about 6000–5000 B.C. that the Northsealand largely disappeared and Dover Strait came into being (SHOTTON 1962) and later still that the sea reached its present level.

(5)   Much more important, on the scale of the hemispheric thermal pattern features that affect the course of the general wind circulation, is that the geographies of the Allerød warm stage and of the Post-Allerød halt and readvance stage are so similar that it seems necessary to invoke some external cause for the difference of prevailing temperature level between them.

Fig. 2.4(c) in Volume 1 (p. 28) showed that in the case of the 500–600 years of colder climate, between about 1300 and 1900 A.D., an episode similar in duration to the Post-Allerød, the course of the estimated temperatures in England was nearly paralleled by an inverse variation in the amount of radioactive carbon (^{14}C) present in the atmosphere. A correlation coefficient of -0.80 was obtained, when a 150-year lag was allowed in the temperature variation. Since increased production of ^{14}C, due to cosmic ray bombardment of the Earth, is apparently associated with quiescence of the sun, it may reasonably be inferred that the temperature variation over those centuries was paralleled, and partly preceded, by a solar variation.

Southern hemisphere geography in the last ice age differed less from the present geography, despite the growth of glaciers on all the high mountains and valley glaciers reaching the plains in some areas. The expansion of the Antarctic continental ice sheet was insignificant in relation to wind circulation changes over the hemisphere. There were doubtless stages at which the sea ice on the Southern Ocean spread significantly farther north than at present, but W'ly winds still prevailed in Tasmania (see Chapter 13, p. 162) and, at

---

1. KELLAWAY et al. (1975) have marshalled evidence which suggests that in the Saale glaciation (see Table 15.1, p. 325) the English Channel was at some stage occupied by a tongue of ice advancing east from an ice mass seated on the continental shelf in the Western Approaches, while the ice advancing south through the Irish Sea and St George's Channel may have been similarly related to an ice massif seated off western Scotland and another occupying the Rockall Bank. The severe lowering of temperature at the climax of the last ice age in these same areas (see fig. 15.21), which would be exposed to a much greater supply of atmospheric moisture than points farther north and east, suggests the possibility of development of some of the same geography of ice again in the climax of the last glaciation, as recently as 20 000 years ago (see also map and discussion in JOHN 1965).

These suggestions have not yet, however, won general acceptance. The position seems to be that

KELLAWAY's thesis based on evidence from sea-floor geomorphology in the Channel and in the Strait of Dover, as well as interpretation of features found in southern England and northern France, appears reasonable in the light of known meteorological and (especially) ocean current variations. The ice age advance of polar water from the East Greenland and East Iceland Currents towards the eastern North Atlantic produced much the strongest negative temperature anomaly in the Western Approaches than appeared anywhere in the world's oceans (cf. also the Little Ice Age situation around the Faeroe Islands, mentioned in Chapter 18, pp. 507–13), and this may well have induced the building of a grounded ice sheet or dome from the heavy precipitation in that region. But the KELLAWAY thesis is not as yet obviously supported by geological evidence from the southern North Sea or the Southwest Approaches to the Channel.

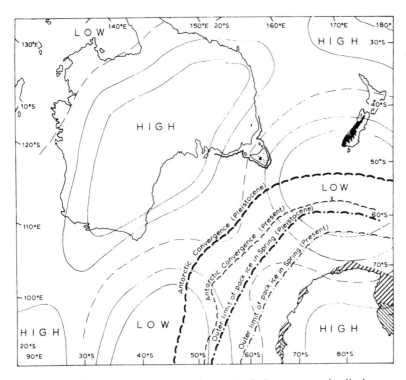

Fig. 15.20   Reconstruction of the winter circulation pattern, ice limits
and geography of the Australian sector of the Southern Ocean during the
last glacial maximum.
   Black areas = ice cover in New Zealand and Australia.
   Heavy hatching = possible extent of shelf ice at glacial maximum.
   Light hatching = shelf ice now.
   Heavy single line = coast of southeast Australia at glacial maximum.
   (*From* DERBYSHIRE (*1971*); *reproduced by kind permission.*)

least at some stages in Late Glacial times, did so in southernmost South America. VAN
ANDEL *et al.* (1967) consider that evidence in the Timor Sea region, off northern Australia,
requires the equatorial rainbelt to be 'compressed' in glacial times – i.e. held close to the
equator, with little seasonal migration – and that therefore the W wind belts of both
hemispheres were probably 5° to 10° of latitude nearer the equator than now. The indications
farther south are, however, that despite prevailing temperatures in Australia up to 6°C lower
than today, and perhaps 8° lower in Chile near 41°S (HEUSSER 1974), changes in the
prevailing latitudes of the ice limit and of the depression tracks on the Southern Ocean were
rather minor (cf. fig. 15.20). Indeed, it seems possible that both were a little farther south
than now in Late Glacial times when the northern ice was retreating more slowly than that in
the southern hemisphere, and that the S'ly position of the storm belt kept breaking up and

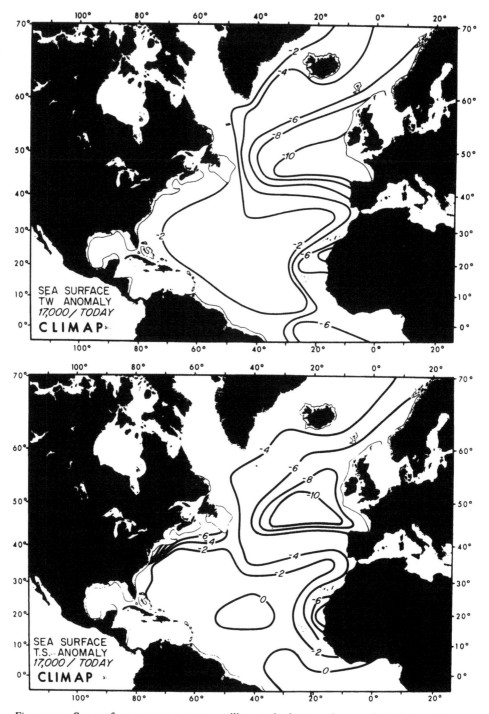

Fig. 15.21   Sea surface temperatures prevailing at the last maximum of glaciation about 15 000–17 000 years ago, as departures (°C) from today's values.

(*a*) In winter.

(*b*) In summer.

(*After* MCINTYRE (*1974*); *reproduced by kind permission.*)

dispersing any northward protrusions of the ice on the Antarctic Ocean. (An analogous situation seems to have developed during the Little Ice Age, judged by observations between A.D. 1770 and 1830.)

The greater extent of the land at the time of maximum lowering of sea level, due to the accumulation of ice in the great ice sheets, is for most purposes adequately indicated by drawing the coasts at the present 50 fathom (or 100 m) depth line. The most significant anomalies in the southern hemisphere so produced were:

(1) Broad land links between Australia and New Guinea and between other land areas north of Java, almost completing a broad link with southeast Asia.

(2) A great broadening of South America eastwards between about 40°s and 55°s, in places to almost twice its present width.

The land-bridge in the Timor Sea region north of Australia seems to have been a dry grassy plain, a steppe or savanna (WEBSTER and STRETEN 1972, p. 22). Pollen studies (J. J. GROOT in Quaternary Discussion, Cambridge, 15 November 1965) of the distribution of forest and grassland, etc., on the now largely submerged Argentine plain indicate that the ice ages had a drier climate than the interglacial periods and hence that the W'ly wind direction was at least as prevalent in glacial times as now, though the reverse observation applies at 52–55°s in the region of Tierra del Fuego.

The departures of prevailing temperature at the height of the last glaciation from present conditions are known in most detail for the surface waters of the North Atlantic Ocean (figs. 15.21(a) and (b)), thanks to foraminiferal studies of the ocean-bed sediment cores by McINTYRE(1974).[1] In many, perhaps most, parts of the world a surface temperature lowering of 4–5°C from present values seems to be indicated, including what may be deduced from the lower height above sea level of the vegetation zones in many parts of Africa and similarly in New Zealand and the southern hemisphere continents.[2] A much greater lowering

1. Since this chapter was written, a complete world map of the ocean surface temperatures prevailing at the time of maximum glaciation about 18 000 years ago, derived from studies of the planktonic biota (foraminifera, radiolaria and coccoliths) represented in cores from the ocean bed, has been published by the CLIMAP Project members in an article on 'The surface of the ice age Earth' (*Science*, 191 (4232), 1131–7; Washington, 19 March 1976). The most prominent features of the map are:

(1) Marked equatorward displacement of the polar front in the waters of the Southern Ocean, as in the North Atlantic but not in the North Pacific.

(2) Expansion of the surface area of the polar watermass in all oceans.

(3) Minimal change in the subtropical gyres in all oceans within which limited areas with temperatures slightly warmer than today's are suspected.

(4) Steepened thermal gradient across the polar fronts, presumably marking the axis of the W'lies in ice age times in both hemispheres.

(5) Increased areas of sea ice – at all seasons in the North Atlantic but especially during the southern summers in the Southern Ocean.

(6) The average departures of sea surface temperatures slightly warmer than today's are suspected. the North Atlantic, $-1.7°C$ South Atlantic, $-0.8°C$ northern Indian Ocean, $-1.3°C$ southern Indian Ocean, $-2.3°C$ North Pacific, $-2.9°C$ South Pacific. The overall average departure for all the world's oceans from today's values was just $-2.3°C$.

2. In more detail (largely from VAN ZINDEREN BAKKER 1969) temperature lowering in Africa at the climax phases of the last glaciation seems to have been about 5°C in the Atlas Mountains (in northwest

of the snow-line, actually by 1800 m, in the Zagros Mountains in southwest Asia and by amounts from 1400 to 2100 m in the Sierra Nevada in southern Spain, the Atlas Mountains of northwest Africa and in Ethiopia and parts of the Himalayas (WRIGHT 1961) has been attributed partly to greatly increased precipitation in those areas associated with the prevailing pattern of cyclonic activity in the ice age.[1] By contrast, in the Alps the upper tree-line dropped more than the snow-line (FIRBAS 1950), an indication that the climate there was drier than now.

In northern Australia, according to BOWLER et al. (1975), the Full Glacial climate seems to have been not only colder but also probably drier than today's. The dry plain then joining Australia to New Guinea may have been the reason why the ice age snow-line was higher in western New Guinea than in the east. At the same time all southern Australia was more under the influence of the Southern Ocean W'lies, so that rainfall was sufficient and evaporation low enough for many lakes that are now dry to exist between 40 000 and 25 000 years ago. By 16 000 B.P. the lakes were already drying up fast and sand dunes were forming, though there was a wetter phase that affected both northern and southern Australia about 10 000 years ago and lake levels continued generally high until 5000 B.P.

BONATTI (1966) reports pollen evidence of a treeless dry steppe in north Mediterranean lands, including northwest Spain, central and northern Italy (e.g. Po Valley) and Macedonia (Greece) around 15 000 B.P., in some places accompanied by tropical steppe fauna, e.g. horse, ibex and *Elephus primigenius*. This manifestly contrasts with the suggested increase of precipitation farther south, which contributed to the great lowering of the snow-line on the Sierra Nevada and the Atlas Mountains. Other evidence of a much wetter regime just south of the Mediterranean[2] comes from radiocarbon dating tests on artesian water, supplying the oases and in the aquifers, below the Saharan desert in Egypt (KNETSCH et al. 1962) and under Sinai and the Negev (VALÉRY 1972), which indicate that this water came from the atmosphere mainly between about 20 000 and 25 000 years ago, though some of the dates appear to go back to the more variable mid Weichselian climatic regimes 25 000–35 000 years ago.[3]

It is clearly not right to equate the ice ages with pluvial periods affecting Africa as a whole. Probably there have never been such periods. There was a much more interesting geography of the rain (and snow) distribution. Displacement equatorwards of the strongest

---

Africa) and 5·5°C in the mountains of Lesotho and Natal (in South Africa), judged by evidence of periglacial phenomena (e.g. frost-shattering of rocks). Similar evidence (at only 900–1000 m above sea level) points to a lowering of 14°C below today's temperatures in Cyrenaica. On Mount Kenya, in eastern equatorial Africa, the snow-line depression was 760–900 m and the lowering of the tree-line amounted to 1000–1100 m, from which a temperature lowering of between 5° and 9° may be deduced. The larger drop of the tree-line probably indicates a

drier climate than now. Lowering of the vegetation zones in New Guinea in the Late Glacial, about 12 000 years ago, seems to have amounted to 500 m.

1. These figures may be compared with the figures more widely typical of other parts of the world: snow line lowering 1200 m in the Alps and Pyrenees (e.g. FLINT 1971, p. 607).

2. Conditions were evidently very dry, however, at 21°N (see pp. 329–30).

3. Dating tests actually depend on the dissolved carbon dioxide in the water.

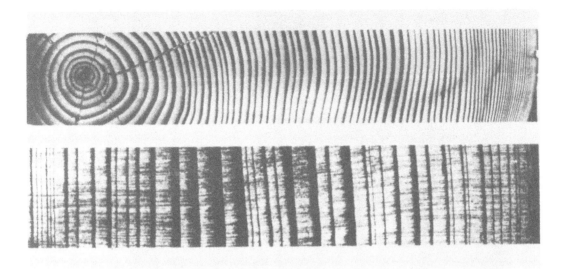

PLATE I Yearly growth ring sequences in (*above*) a complacent tree and (*below*) a sensitive tree. (*Reproduced by courtesy of the* N.C.A.R. Bulletin, *Boulder, Colorado, and the Laboratory of Tree Ring Research, University of Arizona, Tucson.*)

(*a*)

(*b*)

PLATE II (*a*) (*left*) Varves, formed in the bed of a former lake, seen in the clay used in a brickyard at Vaksala, near Uppsala in Sweden (59°52′N 17°41′E about 15 m above sea level). These varves, which date very approximately from 7500 to 8000 B.C., are now at a quite shallow depth below the surface, ranging down to 4 m below it.

(*b*) (*right*) Varves 2–3 m below the surface in a gravel pit in central Sweden (62°26′N 16°28′E 75 m above sea level), where the clay rests on top of the gravel of an esker along the river Ljungan. These varves date from about 7000 B.C.

(*Both these photographs are by Dr* JAN LUNDQVIST *of Sveriges Geologiska Undersökning, Stockholm, and are reproduced by his kind permission.*)

PLATE III End-moraines – seen in this air photograph in several arcs – before the snout of the Donjek Glacier in the Yukon (Canadian Northwest Territories) register various advances of the glacier in the course of climatic history.

The furthest forward moraine marks the greatest postglacial advance of the glacier, some 200 to 300 years ago, and is quite dark in colour with the vegetation that now clothes it. The later moraines and small ridges, resulting from minor advances since, are yet bare and appear light in colour. Melt-water streams have cut through the moraine systems to flow into the outwash plain at the bottom. Pro-glacial lakes are also seen.

(Air photograph taken in 1961 by AUSTIN S. POST *of the U.S. Geological Survey. Kindly supplied by Drs* G. H. DENTON *and* A. S. POST *and reproduced with their permission.)*

PLATE IV   Tree stump (*Picea glauca*) in the north Canadian tundra.

The stump, radiocarbon dated about 4940 years ($\pm$ 140) B.P., is seen still standing on a steep bank on the Tuktoyaktuk Peninsula (69°7′N 133°16′W) which borders the Arctic Ocean (Beaufort Sea) east of the delta of the Mackenzie River in extreme northwest Canada. This tree in what is now tundra shows wider growth rings than the nearest present-day spruce forest 80–100 km farther south, near Inuvik in the lowest part of the Mackenzie River valley.

(*Photograph kindly supplied by Professor* J. C. RITCHIE *of Scarborough College, Toronto University, and reproduced with his kind permission.*)

PLATE V   Medieval bridge (Ponte dell'Ammiraglio) at Palermo, Sicily, built in A.D. 1113 by George of Antioch, Admiral to Count Roger, to span a much larger river than now exists in the area. The River Oreto, which has since been diverted, is believed to have been navigable to this point. (*Photograph kindly supplied by General* F E A, *Director, Servizio Meterologico, Aeronautica Militare Italiano, Rome.*)

PLATE VI (*a*) (*above*) The Roman bridge, a wooden structure on stone piers, built for the Emperor Trajan, across the River Danube at the Iron Gates, as pictured on Trajan's Column in Rome. (*Reproduced by courtesy of the British Museum.*)

(*b*) (*below*) The River Danube at the Iron Gates today.
(*Reproduced by courtesy of the Yugoslav National Tourist Office, London.*)

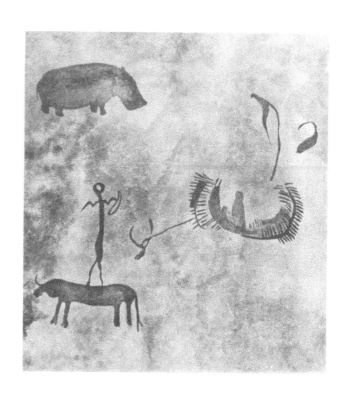

PLATE IX Site of an English medieval vineyard near Ledbury, Herefordshire, showing the terraces still visible.
(*Photograph by* H. H. L A M B.)

PLATE VII (*opposite, above*)·Cave painting, dated 3500 B.C., at Aounret in the Sahara, indicating a much wetter climate than exists today.

It shows a hippopotamus hunt, with canoes.
(*From* The Search for the Tassili Frescoes: the Story of the Prehistoric Rock-Paintings of the Sahara *by* H. L H O T E, (*Hutchinson, London, 1959). Originally published as A La Decouverte des Fresques du Tassili by* B. A R T H A N D – *Paris, Grenoble.*)

PLATE VIII (*opposite, below*) Pieter B R U E G E L the Elder's famous picture *Hunters in the Snow*, painted in February 1565 during the first of the rather frequent severe winters of the next 200 years.

An imaginary landscape, evidently suggested by the dramatic weather of that winter, it seems to have been the first true landscape painting.
(*Original in the Kunsthistorisches Museum, Vienna.*)

(a)

(b)

PLATE XI Submerged forest seen at low tide at Borth, Cardiganshire, on the west coast of Wales.
 Evidence both of the lower sea level and of forest that reached right to the western coasts about 6000 years ago, when the trees grew.
 The species represented in the stumps include alder, birch, oak and pine.
(*Photograph kindly supplied by Dr* J. A. TAYLOR, *University College of Wales, Aberystwyth.*)

PLATE X (*a*) (*opposite, above*) Driftwood (*Picea* sp.) lying on a raised beach at South Cape Fjord, Ellesmere Land. Driftwood is abundant on this beach and has given (uncorrected) radiocarbon dates of about 4700 ± 140 years B.P.
 (*b*) (*opposite, below*) Raised beaches on the east side of a fjord (Zordragerfjord) in Nordaustlandet, Svalbard. Maximum amounts of driftwood here occur on the beaches at 12·5 to 13 m above present sea level, giving dates about 6000 years B.P.
(*Photographs kindly supplied by Dr* W. BLAKE, *Terrain Sciences Division, Geological Survey of Canada, Ottawa.*)

PLATE XVI   Frost fair on the Thames in London in the winter of 1683–4.

There are many pictures recording the freezing of the Thames in various winters between this time and 1814. In some cases many prints of the same picture, which were sold on the ice, still exist. An original of this one is in the possession of the U.K. Meteorological Office.

PLATE XIV   (*opposite, above*) Ridge and furrow, marks of thirteenth century strip cultivation up to a height of 320 m above sea level on the north side of Redesdale, Northumberland (in northern England near the Scottish border).
(*Photograph by* H. H. LAMB.)

PLATE XV   (*opposite, below*) Svinafell in Öraefi, southern Iceland. One of the grandest farms established on a favourable south slope in the earliest times, now hard by a great glacier (Svinafellsjökull) seen in the middle background of the picture.
(*Photograph by Professor* SIGURDUR THORARINSSON, *reproduced by his kind permission.*)

PLATE XVII   All that remains of Daintoun, or Upper Davidstown, Dumfriesshire, in the Southern Uplands of Scotland (55°18′N 3°13′W) a village on a north-facing slope about 275 m above sea level, which was abandoned between 1690 and 1710.

The rectangular shapes on the hillside opposite, seen looking south from Eskdalemuir Observatory, are shown up by the low-angle light of the setting sun: they are the wall-footings of former dwellings.

(*Photograph by* I. J. W. POTHECARY, *reproduced by his kind permission.*)

PLATE XVIII   (*opposite*) The Rhone glacier, Switzerland. Two views from near the same point.
    (*a*) About 1857, just before the great retreat,
    (*b*) About 1960.
(*Pictures kindly supplied by the Swiss National Tourist Office, London.*)

PLATE XIX   The frozen Bodensee in February–March 1963.
Picture commemorating the great winter of 1962–3, which in various parts of central and western Europe was the coldest since 1830 or 1740.

PLATE XX   The frozen Baltic off Malmö in February 1963.
Swedish icebreaker clearing a passage for two fishing vessels.
The Baltic was similarly frozen in 1966.
(*Photograph copyright Planet News, reproduced by permission.*)

thermal gradients, and therewith of the zones of the general wind circulation and climate, would cause the extreme north and south of Africa to experience more cyclonic activity and more frequent rainfall at times when the polar ice caps were expanded. But zones of frequent anticyclonic influence and (at least, relative) dryness would lie between these belts and the equator (cf. fig. 15.8) even though the desert development may mostly have been less intense than today. Fossil dunes in West Africa 500 km south of the present southern edge of the Sahara, believed on latest evidence to have been formed between about 21 000 and 12 000 years ago (A. T. GROVE on 'The Late Quaternary in Africa', personal communication, May 1974), point to a long period of dry climate in Late Glacial times when the Sahara extended far south of its present limits or was displaced 'bodily' south. Other fossil dunes, dated 10 000 to 12 000 years B.P., mark a northward and eastward extension of the Kalahari desert in southern Africa about the same time (R. E. MOREAU, personal communication, 21 December 1963) – in other words, the deserts in both hemispheres reached nearer to the equator than now. There were also variations with time during the course of an ice age, some of which we have discussed on pp. 328–9. NEWELL et al. (1975) have recently concluded, from mathematical modelling of the mean zonal distributions of radiative processes, circulation and heat exchanges in the atmosphere, in the known (temperature and albedo, etc.) conditions of 20 000 years ago, that the total global rainfall was probably about 10% less than now.

The anomaly of prevailing surface temperatures (compared to now) was probably greatest near the edge of the ice sheet in central Europe (FLOHN 1952, KLEIN 1953), where it may have amounted to $-15°$ or $16°C$, and has been estimated at $-10°$ to $12°C$ elsewhere on the European plain and in England; $-9°C$ has been suggested for Greece, Italy and Spain. On the high mountains, above the surface temperature inversions which must have been frequent over the low ground, a temperature anomaly closer to the global average (probably about $-4°C$) may be assumed. In central North America, where the forest limit was generally within about 100–300 km of the edge of the ice sheet, an even stronger south-to-north thermal gradient must have existed than occurs there today (and still more so in comparison with anywhere else in the world) today. Temperature lowering over southern parts of North America may – as in the Pacific and, probably, in parts of central and eastern Asia – have been less than the global average, perhaps only about 2°C. In Wisconsin and Minnesota, within 300–500 km of the ice edge, WEBB and BRYSON derived, by the method described on p. 206, summer temperatures about the last glacial maximum 3°C lower than today's and about 4·5°C lower than in the warmest postglacial times (the same method yielded summer precipitation values at maximum glaciation in that area of North America near the ice limit about 10% greater than today, but values lower than today's in the warmest postglacial times). MORAN (unpublished thesis 1972) finds evidence that at the time of maximum glaciation winter temperatures in the United States Middle West may have been actually somewhat milder than today's, a situation perhaps attributable to the windy ice age climate and the frequency of dynamic warming (föhn effect) in air descending over the

Rockies and over the great ice-massif itself. In far nothern North America the temperature lowering, east and north of the Rockies, was greater than it was south of the ice. RAMPTON (1971) derives the following figures for the amount that prevailing July temperatures fell below today's from pollen analysis of pond sediments in the southwestern Yukon: 31 000–27 000 B.P. 5–9°C; 27 000–13 500 B.P. 7–7·5°C; 13 500–10 000 B.P. 6·7°C; 10 000–8700 B.P. 4·5°C.

An indication not only of the lowering of temperature in the last glacial maximum but of a greater lapse rate of temperature with height is reported by WRIGHT *et al.* (1973) from the Chuska Mountains in northwestern New Mexico: the lowering of the upper tree line was 900 m, that of the lower limit of the ponderosa pine was less than 500 m, both as compared with today. The check upon the lowering of the lower tree line was not due to any greater aridity than now. That the climate was, in fact, moister in that region than today is indicated by pollen evidence (OLDFIELD 1965) that coniferous forest covered the now arid high grassy plains, Llano Estacado, in eastern New Mexico and Texas from about 20 000 to 14 000 years ago.

Temperatures prevailing over wide areas of the Pacific, particularly in the lower latitudes, seem to have been no more than 1–2°C below present levels.[1] A feature of some interest in figs. 15.21(*a*) and (*b*) is the region in the central Atlantic near latitude 20°N where temperatures prevailing at the time of maximum glaciation of the northern continents differed little from today's (and in one area were possibly marginally higher than today's). A somewhat similar feature appeared also in the Little Ice Age period in recent centuries, when the Atlantic Ocean surface between latitudes 15°N and 40°N was mostly warmer than now, while both higher and lower latitudes showed negative anomalies (see Volume I, fig. 8.9, p. 334).

The upper atmospheric temperature and thickness patterns over the northern hemisphere at the glacial climax and in the late glacial phases between 10 000 and 8000 B.C. have been derived by LAMB and WOODROFFE (1970), and from them the prevailing patterns of sea level barometric pressure and winds may be deduced as described in Chapter 13 (pp. 41–4). The results are presented here in figs. 15.22–24. The main features are:

(1) Splitting and displacement of the main centre of the circumpolar vortex aloft to a position over the northern part of the Canadian ice sheet and a corresponding shift of the main polar anticyclone area at the surface to cover Greenland and northeast Canada.

(2) Great meridionality of the mean surface wind and pressure distributions, particularly in the Atlantic sector, with the polar anticyclone extending far south towards mid Atlantic and blocking most of the west to east, or southwest to northeast,

1. Nevertheless, PORTER *et al.* (1977) report old moraines and other marks of former glaciations on the peak of Mauna Kea (19·9°N 155·4°W 4206 m), Hawaii down to the 3500 m level, culminating about 20 000 years ago, and in earlier similar episodes about 55 000, 135 000 and 250 000 years ago. The latest of these ice-caps had disappeared by 9080 ± 200 years ago. The ice reached a similar extent on all sides of the volcano (which had continued active and produced subglacial eruptions).

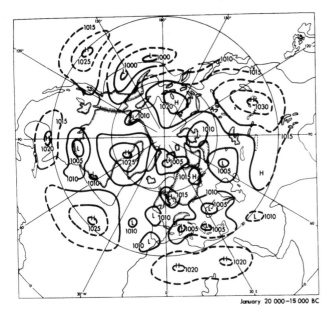

January 20 000–15 000 BC

Fig. 15.22 Probable distribution
of barometric pressure (and
implied prevailing surface winds)
at the last glacial maximum
(*a*) January.
(*b*) July.
    (*From* LAMB *and*
WOODROFFE *1970.*)

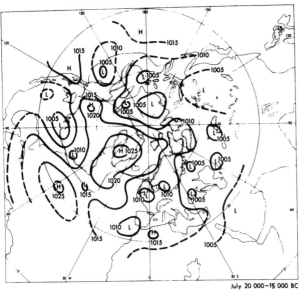

July 20 000–15 000 BC

progression of cyclonic systems which is so prevalent at the present day. These aspects appear very marked at the glacial maximum and successively less marked in the later epochs illustrated: in this sense, as well as chronologically, but not as regards indicated temperatures in Europe and eastern North America, the Allerød warm epoch appears truly intermediate between the glacial maximum and the final glacial stage.

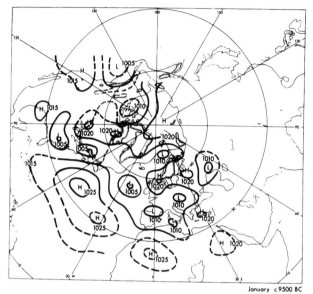

January c.9500 BC

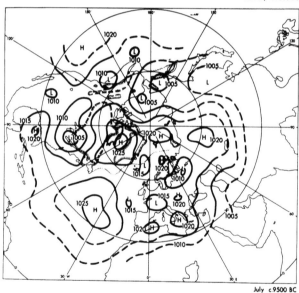

July c.9500 BC

Fig. 15.23 Probable distribution of barometric pressure (and implied prevailing surface winds) during the Allerød warm epoch
(*a*) January.
(*b*) July.
(*From* LAMB *and* WOODROFFE *1970.*)

(3) Little seasonal change of vigour, or of latitude, of the main circulation features, rather as in the southern hemisphere temperate and polar zones today. This applies especially to North America and the western Atlantic; some weakening and a more northern position of the cyclonicity in summer over Europe do appear.

(4) A cyclonic regime over western Siberia, both in summer and winter, particularly at the glacial maximum.

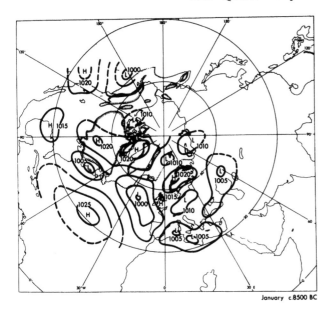

January c.8500 BC

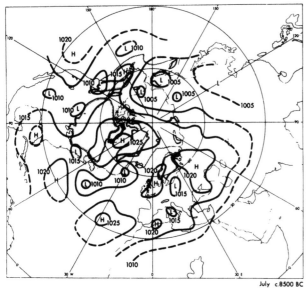

July c.8500 BC

Fig. 15.24 Probable distribution of barometric pressure (and implied prevailing surface winds) during the post-Allerød cold stage (Zone III of the European pollen diagrams).
(*a*) January.
(*b*) July.
  (*From* LAMB *and* WOODROFFE *1970*.)

(5) Reversal of the circulation over the inner Arctic, probably in the ocean as well as in the atmosphere, by comparison with the present day. Cyclonic (counterclockwise) rotation seems likely to have predominated over the polar region between north Greenland and the coast of Alaska and Beringia (Beringland). The main heat transport and locus of southerly surface winds entering the Arctic seems to have been from the Pacific over Alaska, not from the Atlantic and Europe as today.

(6) In the two Late Glacial phases the circulation was apparently still strong over the American and west Atlantic sector, but was weakening farther east (though it must have been becoming stronger over the higher latitudes of the Atlantic sector). As it weakened, there seems to have been a shortening wave length in the upper W'lies implied by the spacing between the cold sector over America and the position of a region of surface northerly wind components over the east Atlantic and Europe. This shortening of wave length appears to have brought S'ly surface wind components farther in over the North American Atlantic seaboard in the Allerød warm stage than previously and a partly similar development of rather more areas of net S'ly wind over central and western Europe. In the cold climate stage that followed, a further retrogression (westward shift) of the pattern over Europe and the eastern fringe of the Atlantic is indicated, especially in winter, and a turn to more E'ly wind components; over America and the western Atlantic the suggestion is rather of increased south–north thermal gradient (and pressure gradients), leading to eastward development of the circulation features and attainment of a longer wave length in the upper W'lies, so opening up a depression track into the Arctic by way of Iceland and the East Greenland Sea.

(7) The small difference in the general geography of ice and snow, open land and water, between the Allerød warm epoch and the colder epoch that followed can hardly be invoked as requiring such a sharp lowering of temperatures as has been derived from the field evidence in certain regions. It seems more probable that a significant change in the radiation budget must have occurred.

A recomputation (WRIGHT and LAMB 1974), using the later knowledge of the surface temperature field represented by fig. 15.21, produced only minor differences in the derived circulation: viz., in summer strong W'ly wind flow near 40°N across the Atlantic and more cyclonic activity in the Mediterranean than here shown and in winter a strong ridge from the Greenland anticyclone towards northwest Scotland, producing a more N'ly and E'ly wind situation over the German plain, between the north European ice sheet with its own smaller anticyclone and the Alps.

These reconstructions appear to find some support in the fossil evidence of an ice-bearing ocean current in Late Glacial times directed far south from the Denmark Strait, between Greenland and Iceland, to mid Atlantic (fig. 13.47).[1]

---

1. Attempts with advanced mathematical models to derive the probable large-scale atmospheric circulation pattern prevailing at the times of maximum glaciation have so far produced at least partly discordant results. One such exercise, by J. WILLIAMS et al. (1974), indicated the mainstream of the upper W'lies crossing over the southern part of the North American (Laurentide) ice dome but in other respects a hemispheric pattern largely resembling the maps here presented. Another model placed the prevailing position of the upper W'ly jetstream south of the edge of the Laurentide ice (as here) but indicated an unbroken sweep of E'ly surface winds between 30–40°N and 60–70°N across Asia and Europe and between 35°N and 60°N across America; this same model, however, greatly overemphasizes

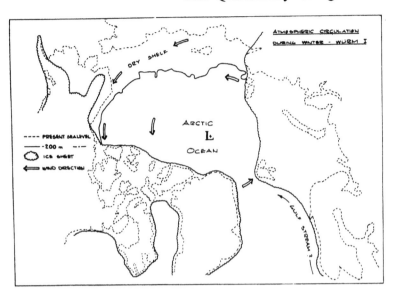

Fig. 15.25 Distribution of ice, land and open water and prevailing winds in winter over the central Arctic during the first climax of the last glaciation, suggested by OLAUSSON (1971).
(*Reproduced by kind permission.*)

The proposed reversal of the prevailing circulation of winds and ocean currents (ice drift) over the central Arctic during the height of the glaciation of the northern hemisphere continents may have interesting implications regarding the occurrence of open water in the Arctic Ocean. We have noted (p. 320) that the Arctic Ocean was essentially ice-covered throughout the Würm/Weichselian/Wisconsin glaciation and that moisture derived from open water surfaces in high latitudes can hardly have played any significant part in building the great ice sheets on land. It does seem, however, that divergence of the surface ice and water drift in a cyclonic gyre over the central Arctic would demand upwelling and may well have maintained an area of open water near the Pole, in the same way as the open North Water is maintained in Baffin Bay-Melville Bay near northwest Greenland today. Thus, OLAUSSON's map, shown in fig. 15.24, seems right in principle and applicable also to the final climax of the last glaciation, but should not be taken to imply open water over the whole Arctic Ocean: the extent of the sources of cold air over the surrounding ice fields, which were probably mostly rather low lying, as in northernmost Greenland today, would guarantee a climate that produced heavy pack-ice on most of the enclosed ocean.

An observation by Dr G. S. BOULTON (personal communication, 21 June 1973)

the winter continental anticyclones in today's climate and produces a spurious prevalence of E'ly winds in winter over the same latitudes.

apparently not only verifies the reversal of the ocean drift during the Full Glacial millennia but approximately dates the change of the present regime. The coastal spits of West Spitsbergen, which all trend northward with the current along the coast today, all trended southward before about 11 000 years B.P. according to datings of the raised beaches concerned.

In addition to the accumulating evidence (see footnote p. 348) of ice sheets thick enough to be grounded on the continental shelf west and southwest of the British Isles, possibly as ice domes fed by the abundant moisture in the winds from the Atlantic, there seems to be evidence from the pattern of postglacial isostatic uplift (fig. 15.26) that an ice dome of substantial mass was grounded in the central area of the present Barents Sea.[1]

After the warmth of the Late Glacial Allerød period, in the tenth millennium B.C., when summer temperatures in England may for a time have been as warm as, or warmer than, today's (the estimates range from +14 to +17°C for July), and mean July temperatures as high as 14°C have been suggested in Denmark (IVERSEN 1973, p. 27), there was a sharp reversion between 8800 and 8200 B.C. approximately in these parts of Europe. The retreat of the Scandinavian ice sheet was halted, as was that of the remaining ice in the western Highlands of Scotland – indeed, the period is known in Scotland as the Loch Lomond Readvance stage (SISSONS 1974). Small glaciers reappeared in the Lake District of northwest England (MANLEY 1959), and there seem to have been localized plateau ice caps in eastern Scotland (SISSONS 1974). MANLEY suggests that the summer temperatures in northern England may have dropped as much as 4–5°C from the values ruling in the previous millennium. KERNEY et al. (1964) derived July temperatures prevailing in Kent (extreme southeast England) around 12–13°C in this Post-Allerød time and January temperatures averaging perhaps −6°C, giving about +3°C for the annual mean in that area.[2]

MERCER (1969) has suggested that this fluctuation, though drastic, was largely confined to Europe and may therefore be attributable to cooling of the Atlantic by rapid break-up and removal of the ice shelves and larger massifs which had formed in the Arctic Ocean area in Full Glacial times. Other evidence, however, suggests that this climatic fluctuation was similar in nature, and very similar in time scale, to the course of the Bølling Oscillation just 2000 years earlier (cf. fig. 15.12) and to the oscillations that have occurred since the warmest postglacial times and which produced sharp cooling of the climates in Europe between 1000 and 500 B.C. and between A.D. 1300 and 1700. The amplitude of the

1. Iceland was almost completely covered by a single ice sheet, as were the smaller islands (e.g. Grimsey) and island groups such as Shetland (HOPPE 1968, 1971).
2. An interesting map of the proposed geography of coastline, forests and ice in northern Europe during the Allerød interstadial is given by GERASIMOV (1969), though it is probably not correct in omitting any surviving ice in Scotland. On this map a strip of sparse forest some 200 km wide envelops Wales and crosses northern England, continues (narrower) along the northern shore of Northsealand to Jutland, occupies the south of Norway, crosses Sweden just south of the ice edge to near Stockholm and continues up the eastern edge of the ice to the land still at that time above sea level (some of it incorporating areas of dead ice) around the present mouth of the White Sea. South and east of this, a 1000 km wide belt of forest, mostly pine and birch, stretched across all Europe north of the Alps to cover all Russia north of Moscow.

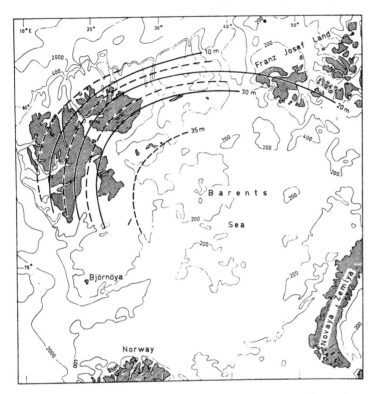

Fig. 15.26   Isostatic recovery during the last 6500 years shown by
raised beaches around the Barents Sea. The isopleths show the
present height in metres of a raised beach identified by a common
volcanic pumice layer dated 6500 years B.P. The maximum uplift
indicated in the central part of the Barents Sea suggests that an ice
dome was seated there at the time of maximum glaciation.
   (*From* SCHYTT *et al.* (*1968*); *reproduced by kind permission.*)

Allerød Oscillation in Europe was doubtless increased by the continued presence of the ice
sheet in Scandinavia at a time when central and southern Europe, and most of the Atlantic
Ocean, had already become warm. The repetition and the element of regularity in these
climatic fluctuations on a time scale of the order of 2000 years points to an external cause,
such as solar fluctuation or variations of the tidal force, though break-up of the Barents Sea
ice-mass may have acted, as suggested by MERCER, to increase the severity of the Late Glacial
throwback of the climate in these areas in the ninth millennium B.C. The phenomenon seems
not to have been so localized as MERCER suggests. SAARNISTO (1974) reports evidence of a
cold regime around Lake Superior in North America between 11 000 and 10 000 years ago,
named the Algonquin Stadial, which halted or reversed the ice retreat and produced new
end-moraines in that region; further evidence from that region and from the ocean is noted by
BROECKER *et al.* (1960); VAN GEEL and VAN DER HAMMEN (1973) report evidence of a

sharp cooling and lowering of the upper tree line by some 800 m on the eastern Cordillera of Colombia (near 5°N 74°W) at the same date;[1] and farther south, in the lake district of Chile, near 41°S 72–73°W, HEUSSER (1974, p. 311) reports, from palynological evidence, that the last 2000 years of the last glaciation, which is known there as the Llanquihue glaciation, saw a great oscillation with a 4°C rise of the summer temperatures, and spread of warmth-loving plants (beech and hydrangea) until about 9000 B.C., which was checked by conditions about 1°C cooler in the following millennium. Though some of these observations might be explained by sharp cooling of the whole North Atlantic Ocean, the occurrence of enhanced volcanic activity affecting the Iceland region, Europe north of the Alps (Eifel) and the southern Andes and Tierra del Fuego suggests that this, too, may have played a part and that the climatic fluctuation was world-wide.

Some evidence (MOORE 1974) suggests that there was a sequence of no less than three great oscillations, each on a roughly 2000-year time scale, which affected the course of deglaciation in Europe and perhaps elsewhere, between 15 000 and 10 000 B.P. These oscillations may also be detectable in the course of the rise of world sea level during the early stages of deglaciation (see graph in VAN ANDEL et al. 1967, p. 752); and in the history of Lake Biwa, Japan (35°N 136°E), as in northwest European Russia, there are signs of five oscillations between 20 000 and 10 000 years ago (HORIE 1974, p. 233). Evidence from the western half of the British Isles, summarized by PENNINGTON (1973), indicates that there was no climatic cooling there between the warmth of the Bølling and Allerød periods sufficient to check the advance of the vegetation. It seems that the climatic records derived from the ocean, and from these most oceanic areas of Europe, show fewer oscillations than the more continental places farther east.[2]

Evidence on the wind circulation in the Late Glacial Allerød and Post-Allerød periods is not confined to the reconstructions here presented in figs. 15.22 and 23. In Europe fossil wind data are present in the shape of dune formations recognized in the loess deposits of the European plain (POSER 1954) and wind-eroded and furrowed rocks and strand terraces in southern Sweden (Å. HILLEFORS, personal communication, 24 January 1969). POSER'S results suggest that westerly surface winds prevailed in the summers over the north German plain, turning to NW'ly over Poland and N'ly over Austria. The lack of radiometric dating means, however, that the dating is uncertain and this wind regime could refer to the earliest postglacial period, which followed the end of the Post-Allerød cold episode, before vegetation fixed the light soil. In southern Scandinavia the Late Glacial winds were clearly E'ly or NE'ly.

1. From a very thorough analysis of the moraines and the vegetation history of the last 18 000 years shown by radiocarbon-dated pollen diagrams from lake sediments and peats in a high valley (Valle de Lagunillas in the Sierra Nevada del Cocuy) at 3800 to 4400 m up in the Colombian Andes, GONZALEZ et al. (1966) report that the major climatic events in high altitude tropical South America during at least the last 12 000 years (and probably throughout the sequence) seem to have been synchronous with those in mid- and high-latitude North America and Europe.

2. This may go some way to explain the apparent discrepancy between the course of the temperatures derived from beetle evidence in the west Midlands of England and the botanical evidence discussed on p. 231 and illustrated in fig. 13.67.

Chapter 16

Postglacial times

The rise of world sea level over the last 10 000 years, seen already in fig. 13.27, gives an overall view of the course of deglaciation.[1] It is, however, a trend which must have been lagged on the trend of world temperature in such a way that the highest sea level[2] – which probably occurred around 4000 years ago – should coincide with the *end* of the period of highest temperature, which reduced the glaciers and ice sheets to their postglacial minimum. It was after 2000–1500 B.C. that most of the present glaciers in the Rocky Mountains south of 57°N were formed (MATTHES 1939) and that major readvance of those in the Alaskan Rockies first took place. And at their subsequent advanced positions – probably around 500 B.C.[3] as well as between A.D. 1650 and 1850 – the glaciers in the Alps regained an extent estimated in the Glockner region at about five times their Bronze Age minimum, when all the smaller ones had disappeared (GAMS 1937, p. 168).

Our knowledge of the course of prevailing temperatures was derived first from palaeobotany – from the shifts of the vegetation zones indicated by pollen analysis and the distribution of macro remains – and has since been amplified by foraminiferal analysis of cores from the ocean bed, as well as by oxygen isotope work on those cores and on ice from

1. An attempted portrayal of a longer sea level history, over the last 20 000 years, based upon observations in the Gulf of Mexico, was given by CURRAY (1961) and redrawn by VAN ANDEL *et al.* (1967) with knowledge of observations from other parts of the world. It indicates a rise starting from about 110 m below the present level at 20 000 B.P., with several large amplitude (20–40 m) oscillations in about the first 10 000 years. A minimum level around 14 000 B.P. is placed at −90 m.

2. About 3·5–5 m above present sea level according to latest assessment by LABEYRIE (in MITCHELL and STEPHENS 1974, p. 116).

3. Also around A.D. 450–700, according to latest evidence.

the remaining ice sheets, and other studies. There is broad agreement between the results, illustrated, for example, in figs. 13.4(c), 13.16 and 15.10; see also fig. 16.6.

We do not know how or where the postglacial warming began. But the fact that postglacial times can be counted from as early as 14 000 B.P. in New Zealand (SUGGATE 1965), and that the oxygen isotope curves from Antarctica (figs. 13.17, 15.9) – despite all difficulties of dating there – clearly show the change beginning even earlier than that, surely means that the oceans were already warmer than in the immediately preceding millennia and were transporting more heat.[1] Importance of the oceans in bringing the first warming to high latitudes may also be implied by the fact that in northwestern North America the maximum extent of glaciation was rather more than 20 000 years ago. Since the response of the surface temperatures of oceans and ice sheets to increased insolation is characteristically slower than that of dry land surfaces, it is probable that the earliest postglacial warming took effect on those continental land surfaces – mainly in low latitudes and in the southern hemisphere – which were farthest from the extensive ice surfaces.

The course of deglaciation

It was in the lands on either side of the North Atlantic (and to some extent, no doubt, in that ocean itself) that deglaciation was longest delayed, most of all in North America because of the huge mass of the Laurentide ice sheet.[2] Thus, the cold regime persisted longest (and had been extended farthest south) in, and near, the very same ocean sector of the northern hemisphere where the postglacial warming ultimately spread farthest north and, from 7000 B.C., or earlier, had penetrated farthest into the Arctic.[3] This is the greatest example that is fully demonstrated of the amplification of climatic shifts in the Atlantic sector, explained in Volume 1, pp. 336–8. We shall later see it at work also in the Little Ice Age and in the twentieth century fluctuation which produced the warming in the 1920s and the cooling in the 1960s.

The progress of deglaciation in Europe is illustrated here in fig. 16.1 by isochrones marking positions of the edge of the retreating Scandinavian ice sheet. There have been numerous more detailed presentations for different parts of Europe of the history of the ice limit and the end-moraines which it produced.[4] Details from two regions may be quoted to illustrate the scene. In Scotland, after the Post-Allerød stage – when an ice sheet still covered

1. At Marion Island (47°S 38°E) the coldest time was before 16 000 B.P. (VAN ZINDEREN BAKKER 1973).
2. Estimates of the volume of the Laurentide ice sheet at its last maximum (Volume 1, p. 484) range from about 50% to 80% of the ice now present in Antarctica.
3. Indeed, the evidence of the shoreline in West Spitsbergen (Chapter 15, p. 360) indicates that the North Atlantic Drift water was heading north along that coast from about 9000 B.C.

4. The most comprehensive survey of these sources and illustration of their results has been given by WOLDSTEDT (1958). Many maps are also given by FLINT (1971). A summary for Scotland is given by CASSELDINE and MITCHELL (1974). Examples from Scandinavia are GJESSING (1960, 1966) and HILLEFORS (1969).

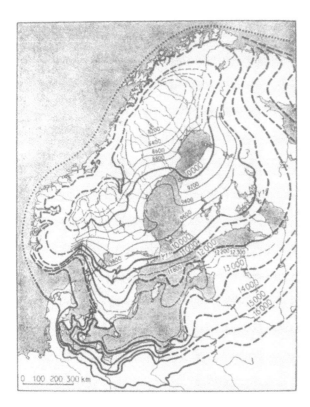

Fig. 16.1   Positions of the retreating edge of the main north European ice sheet from about 17 000 to 8200 years ago. Dates based on Swedish varve chronology.

(*After* DE GEER *1954*.)

much of the western Highlands between 56° and 57½°N and ice tongues filling the major valleys (e.g. the Great Glen and the upper part of Strath Spey) dammed back lakes in the side-valleys (e.g. Glen Roy and Glen Feshie), and there were still numerous local ice caps in the far northwest, in the Cheviots and on the eastern Grampian plateau behind Glen Clova – final disappearance of the ice was probably rapid. Small glaciers, which had reformed in the Lake District in northwestern England about 8800 B.C., probably disappeared around 8200 B.C. (see MANLEY reference in Chapter 15). In Norway and southwest Sweden also a great moraine, the Ra moraine, marks where the ice halted in the ninth millennium B.C. within 10–20 km inland from the south coast and near the head of Oslofjord. The ice sheet's crest still overlay upper Gudbrandsdal and Østerdal, well south of the present watershed, and so the drainage of the melting phase (both above and below the ice) ran up the present valleys in central Norway to spill over to the Atlantic. The ice seems not to have retreated from the fairly low-lying Kuusamo region in northeast Finland until about 6500 B.C. (HICKS 1975). The Scandinavian ice was reduced to about its present remnants by around 6000 B.C.

The retreat of the ice was followed by a complex sequence of developments in the Baltic and North European region due to the competing effects of the (eustatically) rising world sea

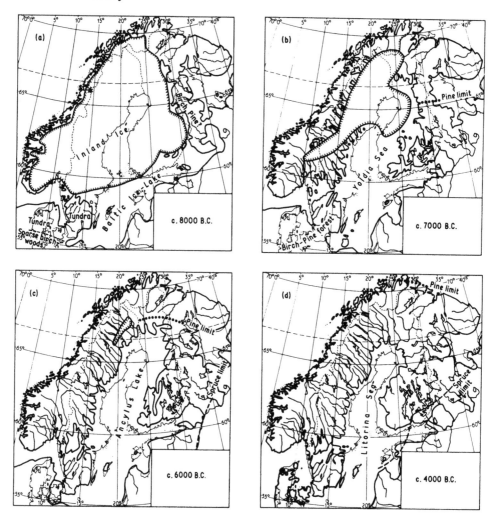

Fig. 16.2 The postglacial development of the Baltic. (*a*) 8000 B.C. (*b*) 7000 B.C.
(*c*) 6000 B.C. (*d*) 4000 B.C.
   (*After* HULTÉN *1971*.)

level and the (isostatic) recovery of the land from the former ice load. Five stages may be distinguished.[1]

1. Maps due to FROMM (1953) and SAURAMO are reproduced by FLINT (1971), WOLDSTEDT (1958), and ZEUNER (1958) A series of maps of northern Europe for *c.* 16 000, 12 000, 8000, 7000, 6000, 4000, 2000 and 300 B.C., with vegetation limits, is given in HULTÉN (1971, pp. 36–7). A selection is shown here in figs. 16.2(*a–d*). Maps of the Baltic entrance at several stages are given in IVERSEN (1973, p. 58).

(1) Until around 8000 B.C. a great pro-glacial lake, the *Baltic Ice Lake*, stretched from 15°E 200 km wide along the ice front to Lake Onega and a narrow extension to 36°E nearly joining the White Sea. There was a narrow extension north along the ice sheet's eastern limit in Finland to about 68°N. The western end of this lake was blocked by the tundra-covered peninsula which joined all Denmark to southeast Sweden.

(2) Next the rising sea level opened a channel across southern Sweden near 58½°N by which Atlantic water flooded in to form a sea area, the *Yoldia Sea*, with only slightly greater extent and similar limits to those of the previous great ice lake. This lasted from about 7700 B.C. to around 7000 B.C.

About 7000 B.C. England was still joined to the continent by a broad plain, Northsealand, the northern coast of which ran from Yorkshire near 54°N (Flamborough Head) to Jutland near 56°N and extended north around the Dogger Bank (1–4°E) (GODWIN 1956, p. 324) – see fig. 16.3.[1]

(3) By 6000 B.C. the rapid melting of the ice had enlarged the water area to a width of about 400 km all through the Gulf of Bothnia basin to near 67°N; but the land-rise had once more barred any connection with the Atlantic, so that this was a fresh-water lake, the *Ancylus Lake*.

(4) By 5000 B.C. the rise of world sea level consequent upon the melting of the great North American ice sheet had once more allowed Atlantic salt water to enter the Baltic region, this time by way of the present Danish Sounds, forming a more extensive, and more saline, sea than now, the *Litorina Sea*.

It was about this time, or perhaps as early as 6000 B.C., also that the complete separation of England from the continent provided another passage for Atlantic water to reach the North Sea.[2]

(5) The Litorina Sea underwent various fluctuations of extent. Soon the link with Lake Ladoga was lost, and gradually the continuing land-rise narrowed the Bothnia Gulf and made the channels to the Atlantic shallower. Declining salinity had changed the fauna and flora of the sea by 2000 B.C. (the *Limnaea Sea* stage) and the water surface was nowhere more than 200–220 km wide in the northern part.

1. A series of maps showing the coasts of the Channel between Essex and Holland between 9600 and 8000 years ago, when it was developing as a link with the southern North Sea, has been published by D'OLIER (1972). According to D'OLIER, around 7600–7300 B.C. there was a channel under half the present width, with its axis near the present continental side. There was still a wide plain farther north joining Norfolk and Lincolnshire to Holland. Over the next 1000 years the water area was enlarged by the formation of broadening estuaries on the English side, while some flat islands were left. This process continued until the link joining England to the continent was finally severed. The Thames estuary at 6000 B.C. had barely half its present width and the Essex and Kent coasts still stretched 20 km (and in places farther) east of their present positions.

2. Ireland had clearly been separated from Britain earlier because of the greater depth of the channels than of the southern North Sea. Nevertheless, persistence of an area that was mostly land between southern Ireland, Cornwall and France for some time after the Post-Allerød cold episode seems to be indicated by the immigration of the 'Lusitanian' element in Ireland's flora, e.g. the Arbutus tree (*Arbutus unedo*) which is limited by the mean January isotherm of 4·5°C and is killed by temperatures below −10°C.

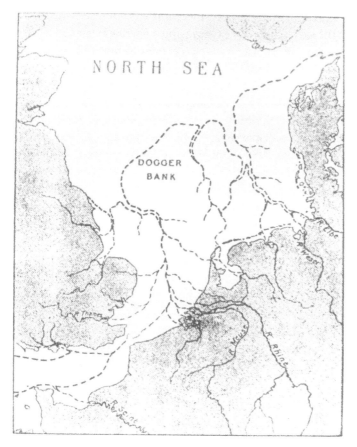

Fig. 16.3   The geography of Northsealand about 7000 B.C.
(*First published by* CLEMENT REID *in 1902; also in*
GODWIN *1956*.)

By 300 B.C. the geography had become quite similar to that of the modern Baltic, differing from it chiefly by a somewhat greater inlet near Stockholm and a slightly wider sea area than now farther north.

It can now be seen as a mistake to attribute to these regional changes of geography in Europe the greatest differences between the successive climatic regimes with which this chapter is concerned. But they must have caused differences that were important all over Europe north of the Alps, particularly because there can have been little winter ice on the salty Litorina Sea.

The first spill of (salty) Mediterranean water into the Black Sea had come about 7300 B.C. Several others occurred about 6500 B.C. and later, each time producing for some while a chemical stratification of the Black Sea waters leading to anoxic conditions which stifled life

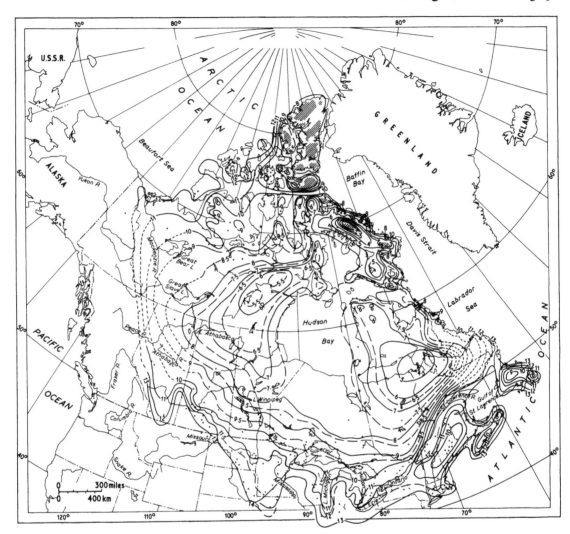

Fig. 16.4   Isochromes of the withdrawal of the North American (Laurentide) ice sheet.
All dates B.P. (before present).
Dots show where radiocarbon dates have been established.
Moraines, coastlines and all other available field evidence used.
(*Reproduced from* BRYSON *et al.* (*1969*); *by kind permission.*)

at the bottom (similar deoxygenation disasters have occurred in other seas, notably the Baltic, on many occasions when climatic conditions introduced a stable stratification of the waters and no currents or convection renewed the oxygen in the deep layers). It was not until the high sea levels in the warmest postglacial times, from about 5000–4500 B.C. that the Black Sea was permanently linked with the ocean (DEGENS and HECKY 1974).

Fig. 16.5   The Cockburn and Cochrane moraines formed at the edges of the North American ice sheet during its readvance phase about 8000 years ago.

(*After* FALCONER *et al. 1965.*)

The course of the melting away of the great North American continental (Laurentide) ice sheet is traced in fig. 16.4.[1] We shall demonstrate (see pp. 381–4 and figs. 16.8, 16.10(*a*) and (*b*)) that the much later survival of the Laurentide ice sheet than of its European counterpart – so that in the so-called Boreal period (fig. 13.13) North America was still essentially in an ice age regime while Europe was warm – had a major influence in causing the climatic pattern of that stage. About 6500–6000 B.C. there was a major halt, or readvance, of the Laurentide ice which produced the moraine systems (now equated by radiocarbon dating: FALCONER *et al.* 1965) long known as the Cochrane moraines in southern Canada and the Cockburn moraine system around the then northern edge of the ice sheet, clearly delimiting the ice, as seen in fig. 16.5. Very soon after this readvance the sea got into Hudson's Bay (ANDREWS and IVES 1972) by way of Hudson Strait, north of Labrador, and the ice sheet was quickly reduced to three segments, the last of which disappeared from Keewatin, North West Territories, and from central Labrador-Quebec around 3000 B.C., while vestiges remain in Baffin Island today.[2]

1. Numerous studies of the detailed course of the retreat and readvances of the North American ice sheet since the last maximum of the Wisconsin glaciation have appeared (e.g. those by WRIGHT 1971a, IVES 1960, MANLEY 1955 and ANDREWS and BARRY 1972).

2. Another interpretation is that the entry of the sea to Hudson's Bay immediately preceded the Cochrane Readvance which had the nature of a surge caused by the penetration of water beneath the ice.

Outline of the climate and vegetation sequence in Europe

The sequence of climatic regimes through the period of deglaciation and since, as recognized in Europe, first from botanical studies, is summarized in the following paragraphs. The scheme follows that outlined in fig. 13.13. The rapidity of the changes from one regime to the next, and of the vegetation changes provoked thereby, has been discussed in various places (e.g. Chapter 13, pp. 80, 207).

The Oldest Dryas (or tundra) time

Early stages of the ice retreat. The open landscapes of northern and central Europe were dominated by plants of the Arctic tundra: grasses on the higher ground, sedges in the wetter places, dwarf willow, saxifrages, club mosses and lichens and, among the flowers, *Dryas octopetala*.

The Bølling oscillation (or interstadial)

By 12 000 B.C. temperatures were rising rapidly (see fig. 13.67). Pollen data from Bølling lake indicate that birch and rowan became established in favourable sites in Denmark. Pines advanced from the southeast as far as Poland and middle Germany. Tundra persisted in the less favourable sites. In western parts of the British Isles there seems to have been no important setback after this until around 9000 B.C. But elsewhere in Europe the tundra became dominant once more towards, or around, 10 000 B.C. (in the so-called *Older Dryas period*).

The Allerød oscillation (or interstadial)

General warmth between 10 000 and 9000 B.C. Birchwoods became prominent in England and Denmark, as well as in Germany where pines also became general for a time. Oaks, followed soon afterwards by elms and hazel, became established in southeast France (Rhone valley).

The Younger Dryas (or tundra/Post-Allerød) time

A sudden fall of the prevailing temperatures from about 9000 B.C. ushered in a colder climate that lasted six or seven centuries all over Europe. Tundra once more dominated the landscape of much of Europe from northern England to the German plain. Trees did not entirely disappear from the most favourable sites even in Denmark and a good deal of birch and pine probably persisted in south Germany.

The Pre-Boreal period

A renewed rise of temperature setting in about 8300 B.C. led to the sustained warm climates of postglacial times. Juniper scrub seems briefly to have become the dominant vegetation in much of Denmark, as it had been as early as the Bølling period in Ireland. The heathlands of Jutland were dominated by crowberry, later with an intrusion of increasing amounts of heather. All these plants were then reduced by the advancing forest, first birch and later pine. In England pine became dominant only in the southeastern half, birches elsewhere. Solifluction ceased and open grassy landscapes largely disappeared. In Germany the first oaks, elms and hazel began to appear by about 7000 B.C. The summers may already have been as warm as, or warmer than, today.

The Boreal period (c. 7000–6000 B.C.: corrected dates)

Temperatures continued to rise. The colder seasons of the year gradually became milder, though probably with some dry and frosty winters, and the summers became generally warmer than today. Pines reached Ireland and became abundant all over England and Denmark, and in these countries the trees of the mixed oak forest made their first appearance. But the most striking feature of the vegetation of this period was the great spread of hazel throughout western Europe, including Ireland, as well as in Germany and as far north as central Sweden (see fig. 16.21), far beyond its present limit (the hazel seems to have been overtaken in the next period by the advance of the great forest trees, an aspect of the vegetational succession rather than of further change of climate). In northeast Finland near 66°N 29°E the occurrence of sea buckthorn (REYNAUD 1976) points to an open vegetation and a somewhat maritime climate.

The Atlantic period (c. 6000–3000 B.C.: corrected dates)

This period spans most of the warmest postglacial times, and has also been variously known as the Postglacial Climatic Optimum or the mixed-oak-forest period in Europe, corresponding to the altithermal or hypsithermal in the American literature of the subject. The rate of general temperature rise eased off during the first 1000 years, and in Europe the summer temperatures may already have reached their maximum by 6000 B.C. In North America, especially the northernmost parts, temperatures presumably continued to become generally higher than before throughout the period of ice melt, i.e. until after 4000 B.C. In Europe the period was probably distinguished by the mildest winters of postglacial times and everywhere north of the Alps by more moisture than in the millennia immediately before and after. The rate of rise of world sea level also slackened (cf. fig. 13.27, see also Plate XI).

In the last 500 years or so of the period an advance of the glaciers in Europe, and probably changes that are registered in the vegetation of Europe and eastern North America (the sudden decline of the elm, ivy and lime) and an era of migration of the primitive agricultural peoples, signal a sharper oscillation (the *Piora oscillation*) towards cold climate than any for several thousand years previously. The spruce forest which had lately reached

Switzerland was widely replaced by open heaths and Alpine meadows. FRENZEL (1966, pp. 103–4) lists evidence of corresponding vegetation changes in North and South America and from Norway to East Africa. In Alaska the colder climate seems to have been drier than before, as was also the case in near-equatorial latitudes in Colombia and Kenya, whereas in Nevada (35–42°N) the fluctuation was towards more moisture.

In Denmark lime (linden, *Tilia*) is now recognized to have been the commonest tree in the forests of the Atlantic period; in England – particularly the Midlands, East Anglia and the southeast – and in Germany elms and limes bulked much larger than in any later period (ANDERSEN 1973, BIRKS *et al.* 1975).

Sub-Boreal times (*c.* 3000 to 1000–500 B.C.)

After the Piora oscillation the forests regained ground in Europe, so far as Man's increasing use of the land permitted. As the open areas were recolonized by trees, the beech came to far greater prominence than before in Denmark; oak, beech, hornbeam and fir in Germany, with spruce also increasing in the east; in England, alders, birches and, especially, oaks seem to have occupied a larger proportion of the woodlands than before. Oaks which grew in the English fenland during the warmest postglacial times, and have been found lying preserved in the peat, perhaps particularly from Sub-Boreal times, are reported to be often of remarkably large size, e.g. with trunks reaching a height of 27·5 m before the first branch (GODWIN 1956, p. 203).

The climate generally regained its former warmth, and may, in some centuries, have been a little warmer than at any time since Boreal times in Europe, but was apparently subject to recurrent fluctuations, particularly of rainfall, differentiating one century from another. Reference has already been made to these fluctuations affecting Europe (Chapter 13, pp. 210–12) and Egypt (Chapter 13, pp. 140–1). When the data are tabulated as in Table 13.12 (pp. 214–15), the operation of a 200-year periodicity seems to underlie the more important fluctuations. More variable, and presumably more meridional, circulation patterns than in Atlantic times seem an obvious interpretation.

The Sub-Atlantic period (from *c.* 1000–500 B.C. onwards)

Glacier advances, changes in the composition of the forests and retreat of the forest from its previous northern and upper limits, indicate significant cooling of world climates, its start being detectable in some places (e.g. Alaska, Chile, China) from as early as 1500 B.C. (e.g. GAMS 1937, HEUSSER 1966a, CHU KO-CHEN 1973). In Europe the most marked change seems to have been from 1200 to 700 B.C. By 700–500 B.C. prevailing temperatures must have been about 2°C lower than they had been half a millennium earlier and there was a great increase of wetness everywhere north of the Alps. It has been described as a period of mild winters and great windiness; cooling of the summers was presumably one of the most notable features as well.

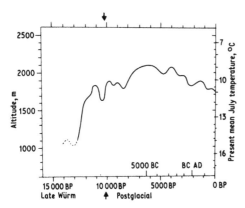

Fig. 16.6 Generalized graph of height changes of the upper tree line in the Alps and other temperate mountain regions over the last 15 000 years.

(*From* MARKGRAF (*1974*), *p. 71; reproduced by kind permission.*)

The secondary climatic optimum or Little Optimum (in the European Dark Ages and Early Middle Ages)

There was a gradual, fluctuating recovery of warmth and a tendency towards drier climate in Europe over the 1000 years after 600 B.C., particularly after 100 B.C., leading to a period of warmth and apparently of high sea level around A.D. 400. After some reversion to colder and wetter climates in the next three to four centuries, sharply renewed warming from about A.D. 800 led to an important warm epoch, which seems to have culminated around A.D. 900–1200 in Greenland and A.D. 1100–1300 in Europe. In these few centuries the climates in the countries concerned evidently became briefly nearly as warm as in the postglacial warmest times. This epoch and the cold regime of the Little Ice Age centuries which followed will be treated in Chapter 17.

The sources chiefly used in compiling this section were FIRBAS (1949), GODWIN (1956) and IVERSEN (1973). The names of the postglacial epochs go back to BLYTT and SERNANDER (see Chapter 13, pp. 79–80).

Temperatures and temperature fluctuations during postglacial prehistoric times

Reconstruction of the general atmospheric circulation patterns, which controlled the climatic regimes prevailing at different stages in postglacial times, depends largely on adequate knowledge of the geographical distributions of temperature in the summers and winters of the times concerned (see Chapter 13, pp. 40–4). Fig. 16.6 provides a convenient introductory summary of the sequence in central Europe: the author of this curve (MARKGRAF 1974) also reproduced curves, largely paralleling this one, representing different parts of the Alps, the Carpathians, Japan, the Rocky Mountains, New Guinea, Australia, New Zealand, East Africa and the Andes in Colombia.[1] The temperature scale

1. The distinctive departures from the course of the majority of these curves seem to be: (1) earliest progress of the warming is recorded in East Africa – attainment of half the total postglacial rise of tree

applied to these curves (as in fig. 16.6) indicates that summer temperatures in these regions were generally at least 4°C lower than now at the glacial maximum and 2°C higher than now in the warmest postglacial times. LAMARCHE's (1973) study of the upper tree line on the White Mountains, California, similarly indicates warm season temperatures about 2°C above those of today all through the warmest millennia, from before 5500 B.C. until about 2200 B.C. Other curves illustrating the temperature sequence have been given in Chapters 13 and 15 (see also Plate XII).

All the histories shown by proxy indicators of temperature, which have been established in sufficient detail, are marked by fluctuations superposed on the main trends. A spectral analysis of these fluctuations in deep sea sediments by PISIAS *et al.* (1973) stressed periods of about 380–400 years, 1300 and 2600 years – the first and last named being also present in the variations of radiocarbon incorporated from the atmosphere in terrestrial materials – and suggestions also of periods close to 1000 and 2000 years.[1]

There seems to be remarkably little difference in the timing of these fluctuations in different parts of the world, even where places are out of step as regards the progress of the longer-term trends. This point has been made in various ways by a number of authors. GEYH and ROHDE (1972) stress the differences of timing of the ('time transgressive') climate and vegetation changes as the Canadian ice sheet advanced towards the coast between 24 000 and 20 000 years ago, as well as in the case of the arrival of the first vegetation at places farther and farther north as the ice retreated (from southern U.S.A. 13 500 years ago to Alaska 11 700 years ago). Nevertheless there was general agreement between different regions of North America and Europe in indicating cold times (minimum vegetation) from 19 500 to 17 000 and around 14 500 years ago. WENDLAND and BRYSON (1974), concentrating on postglacial times, analysed the time distribution of some thousands of radiocarbon dates from all over the world which had been applied to vegetation (pollen zone) changes, to the bottom and tops of peat beds, to sea level changes and to discontinuities in human cultures. The results brought out a considerable number of globally synchronous discontinuities, in which the rapid environmental change generally preceded the change in human occupation by, on average, 50 years.

The major discontinuities were close to those already recognized as pollen zone boundaries in Europe (see fig. 13.13), the most frequent radiocarbon ages being about 10 800, 10 030, 9300, 8500, 5060 and 2760 B.P., but including also as major boundaries more recent botanical discontinuities around 1680 and 850 years ago. Other peaks, of lesser

---

line by 13 000 B.P.; and (2) earliest attainment of the actual peak warmth is suggested in New Guinea, around 7000 B.P. Latest attainment of the peak warmth seems to have occurred in the northern half of North America.

1. Approximately 1300 and 2500-year periods are also reported by BRAY (1970, 1972) from glacier fluctuations. A somewhat similar exploratory exercise carried out by KUTZBACH and BRYSON (1974) on

the longest meteorological instrument records, and on historical, botanical and oxygen isotope records going farther back in time, also stressed that the spectrum was very 'red', i.e. dominated by very long-term fluctuations with time scales of the order of 2000 years. But it indicated some additional contributions with time scales around 20–30, 50 or 50–70 and 100 years.

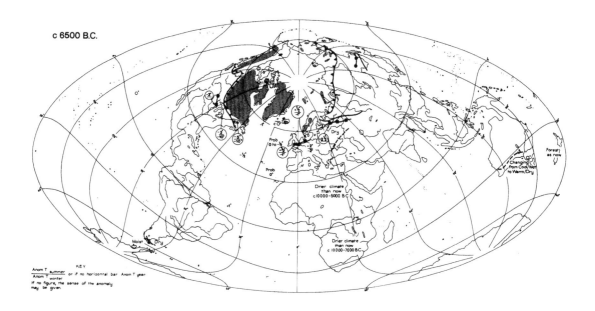

c 6500 B.C.

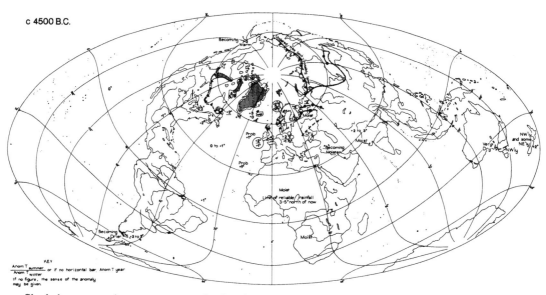

c 4500 B.C.

Shaded areas on these maps: extensive ice-sheets
Wavy line: permafrost limit
🔺...🔺...🔺   northern forest limit
🔲...🔲...🔲   northern limit of elms or oaks
⬤...⬤...⬤   northern limit of oaks
- - - - -   forest–grassland boundary

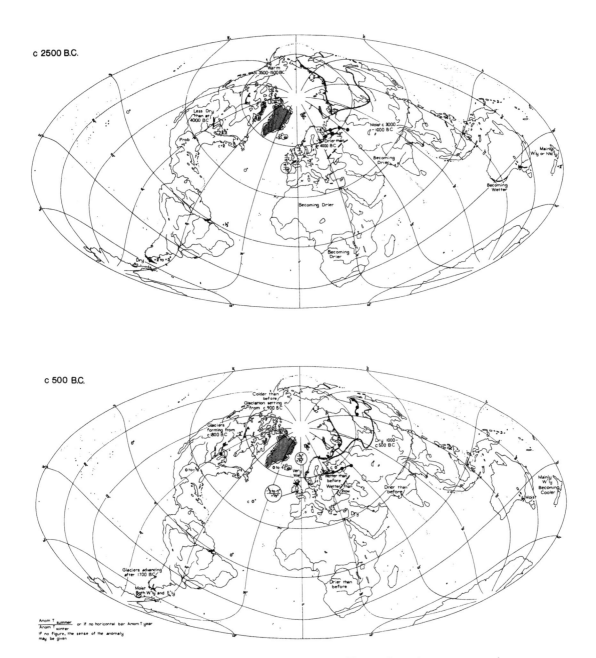

Fig. 16.7 World maps showing the geography and prevailing surface air temperatures in summer and winter about various dates as departures (°C) from the values prevailing in the first half of the present century (1900–50).

(*From* LAMB *et al.* (1966), *where all the sources used are included in the list of references.*)

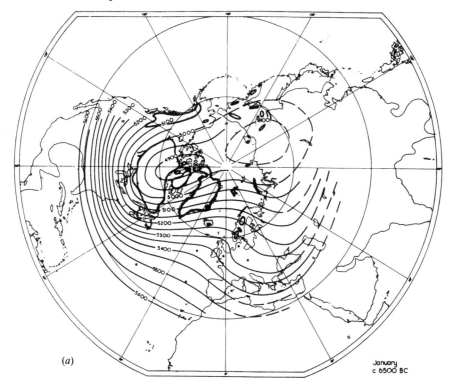

Fig. 16.8 Prevailing distributions of thickness of the 1000–500 mb layer, in metres, derived from the prevailing surface temperatures about 6500 B.C.
(*a*) January. (*b*) July.
(*Reproduced here by courtesy of the Royal Meteorological Society.*)

magnitude, seem to mark three oscillations within the 'Atlantic' period, about 7740, 6910 and 6050 years ago, and two in Sub-Boreal times, at 4240 and 3570 B.P. (all the dates here quoted refer to the environmental changes; the years B.P. are counted back from A.D. 1950, using the old standard radiocarbon half-life of 5570 years, as in fig. 13.13; they can be converted to approximate calendar ages by adding about 3% and applying the ^{14}C-age calibration indicated in fig. 13.11).

From all the curves one sees a clearly logical division of postglacial times into three sections:

(1) The time of warming – the *Vorwärmezeit* in FIRBAS's (1949) usage (cf. fig. 13.13).

(2) The warmest times or 'Climatic Optimum' – FIRBAS's *Wärmezeit*, often known in America as the 'hypsithermal' or 'altithermal'; attainment of the climax warmth was delayed in the northern half of North America by the residual ice.

(3) The time of cooling – the *Nachwärmezeit*, or 'Neoglacial', which has occupied the last 4000–5000 years, during which sharp cooling phases have been followed by

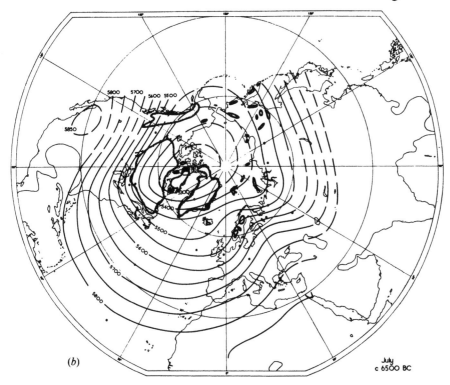

(b)   July
c 6500 BC

generally more gradual recoveries (without, however, regaining the temperature level of the *Wärmezeit*).

This simplest of all zonation schemes for pollen diagrams as well as for the climatic sequence was long ago proposed by VON POST (see comments by GODWIN 1956, pp. 27–8). The cool episodes (fluctuations) during (1) produced halts of the receding glaciers and ice sheets, and sometimes readvances, which on the whole failed to reach as far forward as previous advances. During (3), the glacier advances have on the whole pushed further forward in the later cool episodes,[1] though there were some cases of remarkable advances, locally the greatest in postglacial times, as early as 500 B.C.[2]

Figs. 16.7(a)–(d) present the temperature departures from modern values, summer and

1. This pattern of events has been elaborated in varying degrees of detail by BRAY (1970, 1971, 1972) from glaciers in all parts of the world.
2. Evidence of great glacier advances in both north and south Norway (GRIFFEY 1976) have been detected at that time and again a thousand years later (see also p. 427 regarding events in the Alps).

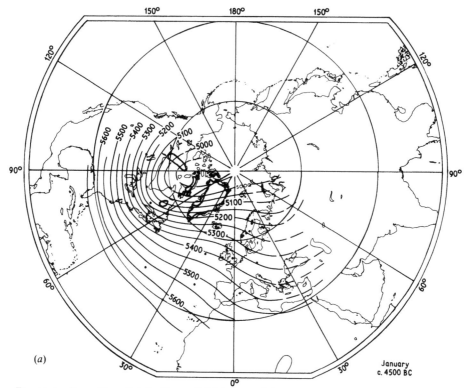

Fig. 16.9   Prevailing distributions of thickness of the 1000–500 mb layer, in metres, derived from the prevailing surface temperatures about 4500 B.C.
   (*Reproduced here by courtesy of the Royal Meteorological Society.*)

winter, derived from pollen analysis and vegetation boundary displacements on land, and from microfaunal analysis of ocean-bed deposits, which were used by the present author (LAMB *et al.* 1966) in arriving at the portrayal of prevailing atmospheric circulation patterns in the figures which follow. A few other salient facts which are sufficiently firmly established about the climatic regimes in different parts of the world at the relevant times are also inscribed on the maps in fig. 16.7. The dates represented by these maps have been readjusted since they were first published, to take account of the now better established calibration of radiocarbon ages, and may be taken as about 6500, 4500, 2500 and 500 B.C.

Reconstruction of the prevailing atmospheric circulation regimes

The thermal gradients and steering currents of the atmospheric circulation in the upper air, derived from these temperatures by the method used for fig. 13.7a, are here illustrated by the 1000–500 mb northern hemisphere thickness maps for the Januarys and Julys around 6500 B.C. (fig. 16.8(*a*) and (*b*)). The geography of ice, land and ocean of the time is clearly seen in the background to these circulation maps and in those in figs. 16.9 and 16.10. The dramatic

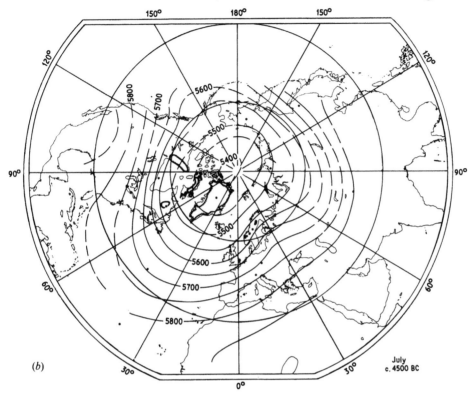

(b)

July
c. 4500 BC

thermal contrast at 6500 B.C. between glacial North America and the warmth of that period in Europe clearly resulted in a very strong tendency to steer the atmospheric disturbances emerging from North America near the ice limit north and northeastwards over the Atlantic, towards either side of Greenland and far away to the north of Europe. Hence the reportedly dry, warm regime over Europe in Boreal times. The contrast with the more zonal patterns of the time around 4500 B.C. (figs. 16.9(a) and (b)) is particularly marked, most of all in summer.

The corresponding patterns of mean sea level pressure (and implied surface winds, blowing counterclockwise around the low pressure, clockwise around the high pressure areas) are illustrated by the maps in figs. 16.10(a)–(h) for the Januarys and Julys of four epochs:

(1) Around 6500 B.C.: representing the Boreal regime, the early warm times, *Frühe Wärmezeit*, in Europe.

(2) Around 4500 B.C.: representing the Atlantic regime, the Postglacial Climatic Optimum, in Europe.

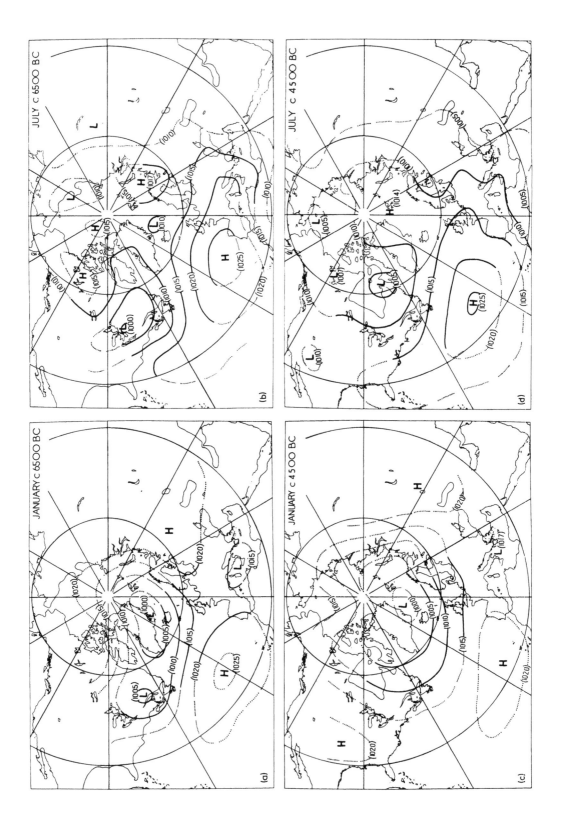

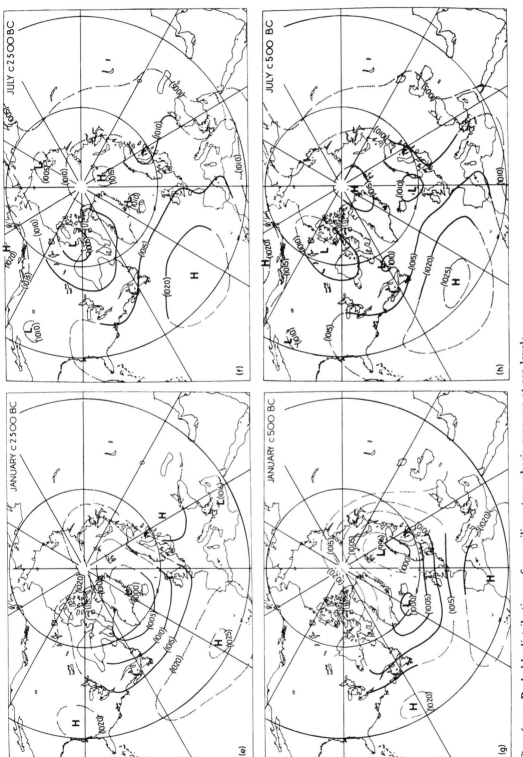

Fig. 16.10 Probable distributions of prevailing atmospheric pressure at sea level:
(*a*) Januarys (*b*) Julys in the seventh millennium B.C.
(*c*) Januarys (*d*) Julys in the fifth millennium B.C.
(*e*) Januarys (*f*) Julys in the third millennium B.C.
(*g*) Januarys (*h*) Julys in the first millennium B.C.
(*Derived by* LAMB *et al.* 1966. *Reproduced here by courtesy of the Royal Meteorological Society.*)

(3) Around 2500 B.C.: representing Sub-Boreal times, the later part of the warm time in Europe.

(4) Around 500 B.C.: representing the early part of Sub-Atlantic times.

The maps for 6500 B.C. indicate the steering of warm air far to the northeast over the Atlantic and prevalence of dry, anticyclonic conditions over Europe in summer and winter. This would be a situation conductive to warmer summers than now and winters in Europe that may often have been frosty but not liable to much snow. These maps readily explain the rapid opening up of Iceland and Greenland to warm air and to plants and birds, which mostly came from Europe (or in the case of western Greenland at this stage came from far south in North America).

Further implications of these maps seem to be as follows:

(1) Great vigour of the circulation in the earliest postglacial times, the difference from now being especially great in summer, in the North American and western Atlantic sector where the heating contrast between the still ice-covered region and the open lands exposed to stronger summer radiation than now must have ranked with modern autumn or winter situations. At the same stage strong southwest–northeast steering of the cyclonic activity must have been marked at all seasons, producing a northward displacement and extension of the anticyclonic area over the eastern Atlantic to Europe.

At some earlier stage, probably before 7000 B.C., the vigour and northeastward steering must have been still more pronounced, probably meaning that the anticyclonic regions over Azores-Biscay and northeast Europe were commonly joined. In the summers this pattern may have been so oriented that cold Arctic air commonly spread over parts of European Russia and northeast Siberia from the north, generating a zone of summer cyclogenesis and raininess somewhere between 40° and 60°N in that sector – though this was doubtless followed by a dry warm anticyclonic regime over most of the European sector, while the circulation patterns responsible for the cooler wetter tendency shifted farther east over northern Siberia and eventually weakened.

In connection with the suggestion of vigorous northeastward steering, it may be significant that in these earliest postglacial millennia the most prominent exotic pollens transported by the winds to southwest Greenland were pine, presumably from the southeastern United States, and birch and alder from Iceland. Pollens from Labrador (prominently spruce) came much later in postglacial times (from discussion in FREDSKILD 1972).

(2) From the stage represented by the maps for 6500 B.C. onwards, it seems that the circulation gradually became on the whole weaker, more zonally oriented and with its action centres farther north. Up to 4500 B.C., or after, these tendencies were accompanied by an opening out and eastward displacement of the North American

cold trough and an increase of wave length in the upper W'lies, which spread the surface W'lies – albeit somewhat anticyclonic W'lies – farther into Europe, and in winter possibly into northwest Siberia.

Cyclonic activity following zonal tracks was probably becoming more marked in high latitudes over North America, as the main thermal contrast was between the warm land areas and the only partly ice-free Arctic Ocean. The sea areas and sounds in that sector became wider and more extensive than now, owing to the land depression resulting from the former ice sheet.

(3) By 2500 B.C. shortening wave length in the upper W'lies and retrogression (westward shift) of the surface pressure pattern features over Europe is apparent, perhaps because the circulation had reached its weakest phase – especially the winter circulation. This would make the winters in Europe colder. In summer there are indications that the thermal contrast between North America, then at its warmest phase, and the partly open Arctic Ocean probably still gave rise to rather prominent cyclonic activity over northernmost Canada.

(4) By 500 B.C. renewed cooling of climates in the north had increased the thermal gradients again, generating a stronger circulation, longer wave length and more penetration of the W'lies over Europe. Northern, central and western Europe seem to have been much more exposed than before to cold air from the northwest in winter and summer alike. This pattern points to a very eastern position of the main North Atlantic centre of low pressure, presumably associated with notably long wave length in the upper W'lies. This seems reasonable if there had been substantial cooling over the highest latitudes, including the Arctic seas, whilst some regions in middle and lower latitudes (possibly the oceans?) remained rather warmer than now.

Present assessment of the reliability of these maps may be summed up as follows. The circulation development and steering patterns at 6500 B.C., particularly in July, must have been stronger than those of today and this probably means that the broadest-scale features of the maps for 6500 B.C. are reliable. The same is true, only in lesser degree, of the maps for 4500 B.C. The features of the maps for 2500 B.C. are much weaker; yet the sequence from 6500 B.C. to a much weaker circulation around 2500 B.C. is logical and the broad-scale consequences of it can be seen. At 500 B.C. the patterns appear stronger than at 2500 B.C. The indications of a broadly meridional (NW'ly or N'ly) flow over the eastern North Atlantic in summer and of a prevalent zonal (W'ly) wind regime over the eastern Atlantic and Europe in winter are the plainest implications of the maps for the time around 500 B.C.

If we use these maps to plot the changes with time of the mean latitudes of the subpolar low-pressure belt and of the subtropical high-pressure belt in the European sector (longitudes 0°E to 30°E) which characterized successive millennia, we obtain the curves seen in fig. 16.11. This diagram brings to light an obvious parallel between the latitude variations

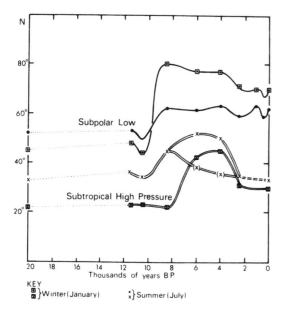

Fig. 16.11   Average latitudes of the subpolar low-pressure belt and subtropical high-pressure belt in European longitudes (0–30°E) and their variations over the last 20 000 years.

of these features and the changes of prevailing temperature.[1] These are only long-term mean positions that are reliably indicated by the data and methods of derivation used, however. Analysis of the century-to-century and shorter-term variations of circulation regime that can be established in outline within the last 1000 years shows that the varying incidence of blocking situations (with anticyclones in the zone 50–70°N) may introduce wide variations of the latitudes of highest and lowest pressures averaged over periods up to a century in length. Despite these reservations about the nature of the variability within each regime represented by the maps in fig. 16.10 and the diagram fig. 16.11, the changes of prevailing latitude indicated by the latter seem to explain many of the main features of the climatic sequence reported in different parts of the world.

SUZUKI (1975) has attempted a world survey and map of the (qualitative) evidence of the character of the rainfall departures in 25 areas of the globe in the warmest postglacial times from present-day conditions.

World survey of the climatic sequence of postglacial times

The postglacial history of moisture variations registered in the landscapes of Africa and Eurasia has been illustrated by various records of the lake levels in figs. 13.29–32 and by other indicators in figs. 13.36 and 37. Another record, more convenient for following the

---

1. The parallel is closest in winter, but is well marked in both seasons in the case of the subtropical high-pressure belt. The rather low latitude occupied by the subpolar lows in summer in this sector results from a shift of the entire circumpolar vortex towards this side of the hemisphere at the season when the Alaskan and northeast Asian land area becomes warm.

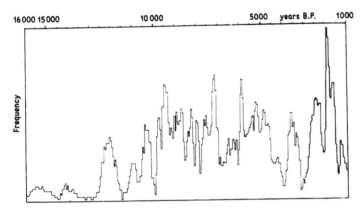

Fig. 16.12   Frequency distribution of all available radiocarbon dates from 16 000 to 1000 years ago for north Africa 15–30°N 5°W–30°E, taken as a proxy indicator of the amount of moisture available to produce organic material.

(*From* GEYH *and* JÄKEL (*1974*); *reproduced by kind permission.*)

moisture variations in postglacial times, in the Sahara, is given in fig. 16.12. It indicates that, on the whole, after a mainly dry period in the Sahara during the last millennia of the ice age, when the desert extended, or was displaced, far south of its present limit (see Chapter 15, p. 353) and the Nile is also known to have been very low (BUTZER and HANSEN 1968), moisture levels were high through the warmest postglacial times (see fig. 13.73 and text Chapter 13, p. 273).[1] Our maps suggest this was a regime which allowed the monsoon rains – associated with the equatorial system – to penetrate farther north than at other times. There seems to have been an abrupt decline of moisture to minima about 4200 and 3200 years ago, followed by a remarkable recovery and short-term peak of the curve in fig. 16.12 in the Roman era (at which time, however, the number of radiocarbon dates may be more a register of the peak of human activity). The history of earlier human occupation and animal presence in the Sahara, as indicated by rock drawings and Egyptian inscriptions (see p. 273) and worked out by BUTZER (1958) and others, corresponds well with the sequence and dates indicated by fig. 16.12.[2] The Nile floods declined after about 2800 B.C. (see fig. 13.37) and the fauna of the valley in upper Egypt lost many species (BUTZER 1958).

1. The River Nile record of the last 5000 years is surveyed in Chapter 13, and in Tables 30 and 31 in Appendix V.
2. Indications of moister conditions in west and northwest Africa in Carthaginian and Roman times than since from the Classical Roman literature are cited by PEJML (1962), who fails, however, to make any allowance for the persistence of a higher water table and more extensive oases in the Saharan region as a legacy of earlier climatic conditions. The account of the voyage in the fifth or sixth century B.C. by a Carthaginian squadron under Hanno along the Atlantic coast of Africa (see map in Appendix V) describes meeting with elephants in a coastal lagoon somewhere on the west coast of Morocco. A later account dated about 338 B.C. of Carthaginian settlements on the west coast of Africa speaks of the inhabitants living off milk and meat and producing

A partly similar history was found by OLAUSSON and OLSSON (1969) from examination of cores from the bed of the Gulf of Aden (12°N 44°E) – see p. 106 – where the Late Glacial part of the cores indicated dryness, but there was a great maximum of wind-blown sand in the Pre-Boreal (earliest postglacial) period, evidently marking a very windy regime. Correspondences with the varying spread of human settlement in the arid lands of the Near East have been noted on p. 134 (Chapter 13) and are demonstrated in the case of the Sahara desert by GEYH and JÄKEL (1974). North of the desert, however, cores from the bed of the western Mediterranean indicate an inverse climatic trend, with the warmest postglacial millennia as the driest time – as should be expected from the more northern position of the anticyclone belt in those times indicated in fig. 16.11.

There have been suggestions in recent years from various archaeologists, engineers and others, including meteorological theoreticians using mathematical modelling of the atmosphere, that the Sahara desert was created by Man destroying the vegetation through overgrazing by his animals. Such ideas ignore the development of the desert in Late Glacial times and during at least one phase in the warmest postglacial times (figs. 13.29–32). Old dunes formed of loose sand, now fixed by vegetation, indicate an extension of full desert conditions 500 km south of the Sahara – and a northward displacement of the Kalahari desert into the Congo basin – in the latter part of the last interglacial as well as in Late Glacial times. It seems that in periods of cooling climate, and in the final phases of the last ice age, these deserts spread somewhat towards the equator and were characterized by particularly great aridity; in the warmest millennia there was a dry regime extending poleward to latitudes 30–40° and beyond, while the monsoons penetrated farther north than now and increased moisture to near 20°N.

The dryness that characterized most of the Mediterranean region – especially its northern and western parts – in the warmest postglacial times extended northeastwards across the Near East to the Caspian Sea, which fell to a very low level (fig. 13.31), and to the Iranian (Persian) plateau (KRINSLEY 1970, p. 314). At 34–35°N, however, in northern Iraq (Mesopotamia), there is evidence of cultivation and irrigation works, at least at some times between about 5000 and 2000 B.C., extending 50 km north of the present limit of any possibility of such activity and of a density of human settlement that the land will not now support (J. L. OATES, personal communication, 17 May 1973). There is also evidence of moisture and human settlement in the present desert in Saudi Arabia. The archaeological

much wine from their vineyards, including some that was exported. Replacement of the species *Elephas Atlanticus* by the less moisture demanding *E. Africanus* can also be traced, first in the Sahara at the end of Neolithic times and later near the Atlantic and Mediterranean coasts. PLINY's writings in the first century A.D. indicate the distribution of elephants in North Africa at that time, including an extensive area at the southern foot of the Atlas Mountains, where numerous herds wandered in the forests and

emerged in winter to roam over rich pastures: they seem to have finally died out in the third century A.D., presumably through increasing aridity of the landscape. This trend is also indicated by the retreat of the cork oak (*Quercus suber*).

STRABO (54 B.C.–A.D. 21) also described how the tribes in southern parts of Roman Libya made journeys in northern parts of the desert in his day, fixing bags of water under their horses' stomachs.

excavations at Ur, Kish, Fara and Nineveh have given clear evidence of breaks in the stratification caused by flooding episodes, all dated between about 4000 and 3000–2400 B.C. In the fourth millennium B.C. there was a considerable spread of settlement in the Sinai peninsula, again indicating the exceptional moisture of those times in areas south and east of the Mediterranean.

In northwestern India a sequence similar to that of the southern fringe of the Sahara can be recognized, with the monsoon rains presumably spreading rather further north and northwest than now in the warmest postglacial times. The Rajasthan (Thar) desert had experienced great aridity in Late Glacial times: from around 8000 B.C. this was succeeded by an increasingly moist regime, with lakes in the area 27–29°N near 75°E which dried up again between 2000 and 1000 B.C. (SINGH *et al.* 1972–1974). In the Kashmir Valley, at 33–35°N 74–76°E, the earlier postglacial pine and cedar woods were succeeded by the warmth-and-moisture-demanding trees of the mixed oak forest and alder in mid postglacial times, followed by re-establishment of conifers and, after about 2000 B.C., deforestation by Man (SINGH 1963). The Harappa culture in the Indus Valley had its heyday between 2500 and 1750 ± 100 B.C. in the latter end of the moist period (RAMASWAMY 1968, SETH 1963, SINGH 1971), growing cereals in part of what is now the Thar desert. The main phases of the Harappan civilization seem, in fact, to have been disturbed by floods which, perhaps, caused the cultural breaks in its history (KIRK 1975). The fauna of the present Thar desert region in Harappan times included elephant, rhinoceros and water buffalo. And it must have been a legacy in the subsoil and in remnants of vegetation from that moist period that made possible the march of Alexander the Great across southwest Asia to the Indus between 330 and 323 B.C. On the return journey, Alexander's army crossed the Indus river in a flotilla of boats built of local timber (WADIA 1960).

In China also, during the time of the Yangshao culture, 5000–6000 years ago, the climate seems to have been warmer and wetter than now (CHU KO-CHEN 1973, p. 228; see also Chapter 13, p. 251, in this work).[1]

Farther north in Europe and Asia the history of moisture variation was different and seems related to the variation of strength and zonality of the general atmospheric circulation shown by the maps in fig. 16.10(*a*)–(*h*) in the middle latitudes belt. After the prevailingly anticyclonic patterns of the Boreal period, typified in figs. 16.10(*a*) and (*b*), the wind flow seems to have become increasingly W'ly and stronger, bringing more rainfall from the

---

1. Examination of corals and mollusk shells on raised beaches in Japan, radiocarbon dated 7000 to 5000 years ago, suggests (TAIRA 1975) that the Kuro Shiwo (warm North Pacific Current) had about the same northward extent as today but was between 2° and 8°C warmer than today. The smaller values of water temperature anomaly were apparently derived by species examination, the larger values by oxygen isotope measurements – a middle value of 5°C warmer than today may be realistic. Crustal movements in Japan and farther south are suggested by TAIRA as partly the cause of the subsequent change of the current and the prevailing temperatures in it, since the supposed explanation is that the warm water current reaching Japan in those warmest postglacial times was drawn from farther south, from the ocean areas just north of Papua/New Guinea that are still the warmest in the world – present-day mean 29–30°C).

Atlantic, particularly in the winters, across all northern Europe and on into Asia north of the great mountains. This was the so-called Atlantic period. After about 3000 B.C. the circulation seems to have become weaker, permitting some drier periods and colder winters in central and northern Europe and Asia. Meanwhile, in northern Turkey, among the mountains of Anatolia, the change was to moister conditions, with oak, pine and juniper gaining ground after 2000 B.C. at the expense of the previously dominant beech and fir (BEUG 1967), probably an early symptom of a return to more meridional circulation patterns with a winter trough in the upper W'lies near longitude 20–40°E.

Some related trends may be discerned in the southern hemisphere, which shared the same general temperature sequence through postglacial times as the northern hemisphere.

In southwestern Australia CHURCHILL's (1968) derivation of rainfall trends from variations in the ratio of the abundance of different species of eucalyptus trees (see Chapter 13, pp. 180–1) indicates a sequence (fig. 16.13) similar in some of its broadest features to that in northern Europe, though omitting or showing only in modest amplitude some of the variations that were strongly marked in the northern hemisphere. The climate from at least 4000 until 3000 B.C. seems to have been wetter than now. Thereafter it became increasingly

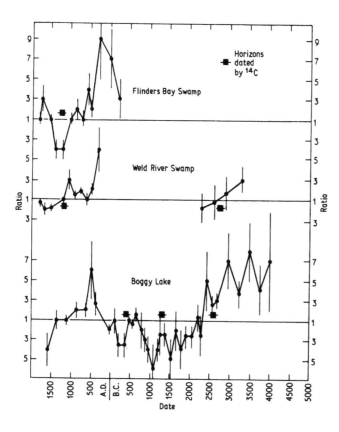

Fig. 16.13 Variations in the relative abundance of pollen of the Eucalyptus species *E. diversicolor* and *calophylla* in southwestern Australia since 4000 B.C. Ratios *E. diversicolor/calophylla* greater than 1 above the line: such periods imply greater rainfall.

(*From* CHURCHILL (*1968*); *reproduced by kind permission.*)

dry, was driest around 1200 B.C. and continued dry for some centuries after that.[1] Between about 500 B.C. and A.D. 500 it became increasingly wet, followed by increasingly dry conditions until A.D. 1200 or after (the medieval warm epoch or 'Little Optimum' in Europe and North America), but after A.D. 1500 the climate became wetter again until the present time.

In South America AUER's (1960) investigations of the vegetation history of Patagonia and Tierra del Fuego seem to imply a postglacial sequence of latitude shifts of the zones of the prevailing wind circulation resembling the sequence of such shifts over the North Atlantic and Europe:

(1) AUER identified a dry period in southernmost South America in early postglacial times, from about 7000 B.C., which he equated with the Boreal in Europe, when the forest was confined to the southwestern side of Tierra del Fuego, far west of its present limit. The arid zone in Chile extended farther south at that time than it does now and probably all the wind circulation zones were displaced towards high latitudes in that sector.

(2) During the following (Atlantic?) period the forest advanced eastward and about 3800 B.C. reached to about its present limit.[2]

(3) For some time after about 300 B.C. forest covered the whole of Tierra del Fuego. This implies a more frequent incidence of easterly winds than now with depressions quite frequently passing north of the island, in rather lower latitudes than commonly observed today.

(4) In fairly recent times the forest quickly receded to its present limit, with the eastern side of Tierra del Fuego becoming drier. The recession is still continuing, however, partly because of overgrazing.

In North America certain differences from the climatic history of the rest of the world resulted from the long time needed to melt the vast Laurentide ice sheet in its entirety (see Chapter 15). These differences were most important in the northern half of the continent and were probably at their greatest at the time of the readvance that produced the Cockburn and Cochrane moraines around 6500–6000 B.C. (fig. 16.5). As in all other climatic epochs, the significance of periods of great dominance of the W'ly winds differed from that seen in other continents (except South America), because it extended the area of dryness associated with the rain shadow of the Rocky Mountains. The timing of the major climatic shifts in Canada since the warmest postglacial time has been shown to correspond closely to that in northwestern Europe (NICHOLS 1967a).

1. BOWLER et al.'s assessment (1975) (see Chapter 15, p. 352 and references) based on lake levels in southern Australia generally agrees with this sequence. In New Guinea a fall of temperature about 3000 B.C. may be deduced from timber line and glacier changes. Since then there have been at least four glacier fluctuations, the timing of which agrees with the main fluctuations in other parts of the world.

2. This is based on a radiocarbon date corrected for the calibration adjustment now known to be applicable (see fig. 13.11).

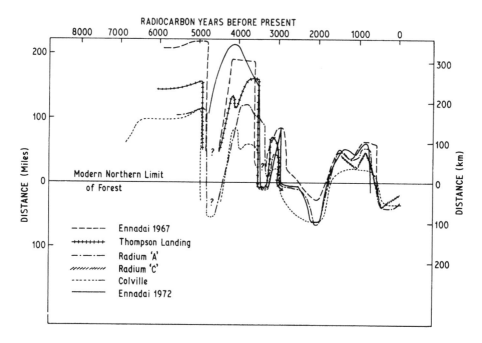

Fig. 16.14   Movements of the northern limit of forest at 100°w during postglacial times.

Distances measured north from the Cochrane moraines formed 8000–8500 years ago.

(*From* NICHOLS (*1967b*) *and later estimates: diagram kindly supplied by Dr H.* NICHOLS *for this book.*)

The spruce forest belt moved north from 6000 B.C. onwards, keeping close to the edge of the fast retreating ice, so that any tundra was at that time very limited in extent. By 4500 B.C. (corrected date) the forest edge was already north of its position today, and between about 4000 and 1500 B.C. this departure was as much as 250–300 km (fig. 16.14); summer temperatures in Keewatin (61°N 101°W) 3–3·5°C higher than today's are indicated (NICHOLS 1967b). Mixed hardwood oak forest (oak and other varieties) followed the conifer belt northwards. The succession over a range of 300 km from south to north in eastern Minnesota is shown in fig. 16.15. The northward advance of the southern woodlands from the southeastern United States – the assemblage includes the tulip tree, magnolia, black gum, sweet gum, red maple, hornbeam, hemlock spruce (*Tsuga*), hickory (*Carya*), and bald cypress (*Taxodium distichum*) as well as oak – was limited in the earlier postglacial millennia by the eastward spread of a dry area, evidently indicating increasing W'ly winds, producing an

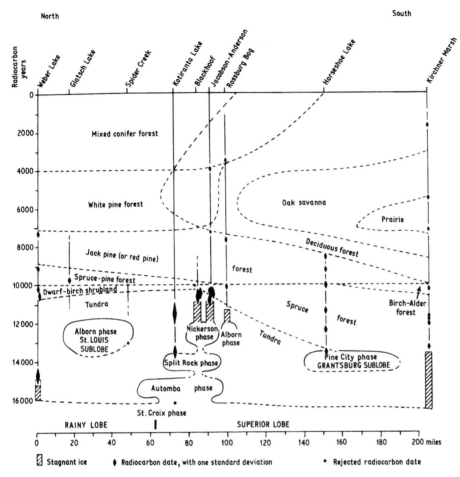

Fig. 16.15 Changes during the last 16 000 years in the positions of ice and vegetation boundaries along a north–south line in eastern Minnesota, near 93°w.
(*From* WRIGHT (*1971a*); *reproduced by kind permission.*)

extension of the 'prairie peninsula' (WRIGHT 1968), probably reaching 85°w in southern Michigan at its maximum about 7200 years ago. WRIGHT (1971b) regards that time as the culmination of the period of warmth and dryness. BRYSON (1966) has demonstrated the association between the prevalence of different airmass types and the vegetation boundaries in North America today (fig. 16.16) and on this basis suggests the airmass boundary positions of 8000 and 5000 to 3500 years ago shown in figs. 16.17 and 16.18.

In the southern high plains and the now arid southwestern areas of the United States, as in the Great Basin in western U.S.A., conditions during glacial times were clearly wetter than now, fulfilling in that area the old idea of a simple association of lower latitude pluvials with

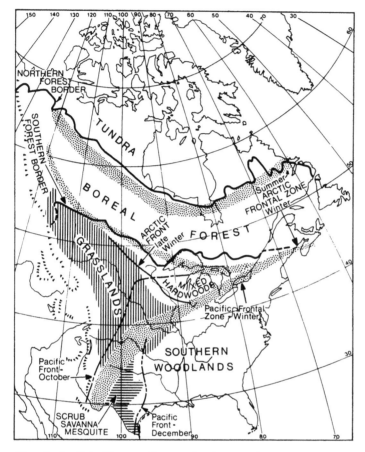

Fig. 16.16 Coincidences of vegetation boundaries and common
airmass boundaries over North America in the present climate.
(*From* B RYSON (*1966*); *reproduced by kind permission.*)

the high latitude glaciations. Forest covered the high plains in New Mexico between about
20 000 and 14 000 years ago. Conditions became drier as the postglacial regimes developed,
but there may have been another pluvial, or moisture maximum, in all the southern areas
during the warmest postglacial times. There have been fluctuations since, which also suggest
farther northward advance of the summer rains over the southwestern United States in the
warmer epochs. An example is mentioned on p. 415.

Since the warmest postglacial times the 'prairie peninsula' has undergone several
fluctuations, but has been on the whole rather less extensive and the axis of its eastern cusp a
little farther south. The belts of the various types of woodland have also generally retreated
somewhat southward. The retreat of the forest from the heights is registered in the case of the
White Mountains of California in fig. 16.19. The author of these curves suggests that the

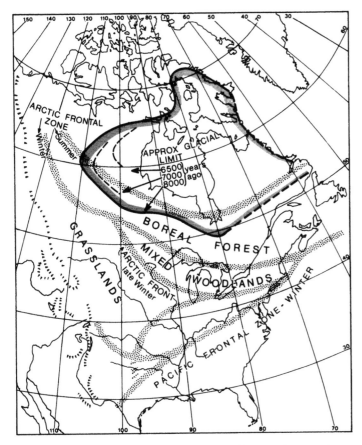

Fig. 16.17   Reconstructed frontal zone positions and vegetation belts over North America about 6500–6000 B.C.
  (*From* BRYSON *and* WENDLAND (*1967*); *reproduced by kind permission.*)

Campito Mountain upper tree line is lowered both by low temperatures and low rainfall, whereas the Sheep Mountain tree line is controlled by the temperatures of the summer half of the year. These differences are thought to account for the differences observed in the curves from these two mountains in the same group. There is, however, much in common between the histories on the two mountains. Both register the prevailing warmth prior to about 1500 B.C. There is even a suggestion from the Campito Mountain record that summer temperatures and rainfall in the area may have increased somewhat from 4000 to 2000 B.C. The lowering of the upper tree line in that area since has occurred in three sharp downward steps, around 1500 B.C., 600–400 B.C., and A.D. 1100–1500. The first and the last of these clearly represent temperature drops, perhaps also accompanied by increasing dryness, whereas the change around 500 B.C. in that area was evidently mostly a reduction of rainfall.

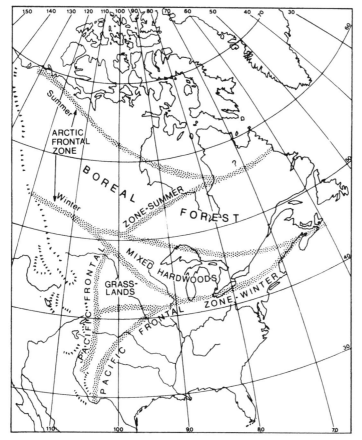

Fig. 16.18   Reconstructed frontal zone positions and vegetation
belts over North America about 3000 to 1500 B.C.
   (*From* BRYSON *and* WENDLAND (*1967*); *reproduced by kind
permission.*)

There is evidence of a scatter of dates from different parts of North America, as of
Europe, for the first main downward step of temperature from which there has been no full
recovery, marking the end of the warmest postglacial times. The dates for the onset of the
colder regime of the Sub-Atlantic or Neoglacial period range in both continents from about
1500 to 500 B.C. and probably mark a series of oscillations, some of which took more effect in
one area, some in another area. ADAM (1967) reports a date about 900 B.C. from the
Californian Sierra Nevada.

   The warmest postglacial regime did not reach northern Greenland until about 3000 B.C.,
and the incidence of driftwood at the head of the fjords there indicates that there was most
open water in the summers between about 2500 and 2000 B.C. (corrected dates), a time when

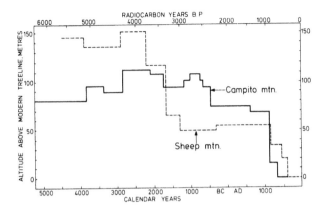

Fig. 16.19   Height of the upper tree line on two mountains in the White Mountains, California.

Minimum estimates of the upper limit based, in the earliest times, on isolated tree remnants.

(*From* LAMARCHE *(1973)*; *reproduced by kind permission.*)

there was also denser vegetation on the northern parts of the continents and presumably more precipitation in the summers than now (Plates Xa and b). The cooling which led to the present colder and drier regime seems to have taken place in abrupt stages about 1600 B.C., 100 B.C., and A.D. 1100. In west Greenland, near Jakobshavn (69°N), another cooling phase is indicated between about A.D. 400 and 600, for which there is some parallel in Europe (FREDSKILD 1967).

The changes of the vegetation limits in the northern hemisphere since the warmest postglacial times are displayed in a single map in fig. 16.20 (cf. also Plates IV, XI and XII). The net advance of the permafrost limit in northern Eurasia through the climatic fluctuations from that time to the present is also seen. The map also shows how the southern limit of forest, where broad-leafed and mostly deciduous forest gives way to steppe, has shifted in the last 4000 to 5000 years on the plains of Eurasia as well as North America. In general, the steppe has advanced, though the reasons for this are not altogether certain: human action and increased windiness may have combined to clear the trees and reduce soil moisture at the margin of the forest zone. The plains in the latitudes east of the Rockies and in European Russia and Siberia probably had a greater precipitation/evaporation ratio in glacial times, but the history of the Caspian Sea (figs. 13.31, 13.32) suggests that any legacy of that moisture had vanished by 5000 years ago. The water table may, however, have had a continued tendency to fall through most of postglacial time in the southern parts of these regions (as noted elsewhere in this chapter), as well as in the Sahara and central and southern Asia, with the wastage of the subsoil and aquifer moisture left over from the moister stages of the last glaciation and earlier postglacial times.

Fig. 16.21 reproduces a famous map, due to ANDERSSON (1902), in which the present limit of hazel in Sweden is compared with that of fossil finds from the postglacial warmest times. Hoards of fossil hazel nuts are found which were presumably collected by rodents and squirrels, and the nuts often show signs of having been opened by squirrels. ANDERSSON first deduced from this map that the summer climate of the time must have been about 2·4°C warmer than around the beginning of this century.

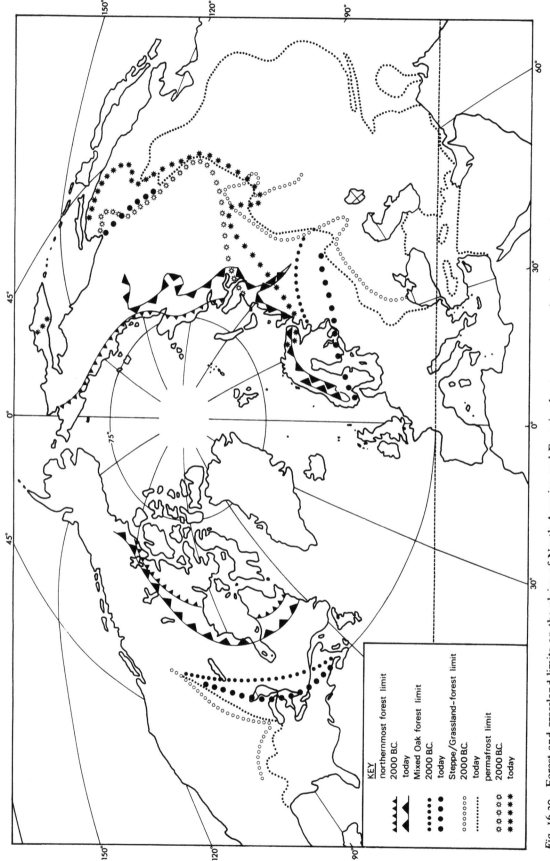

KEY

northernmost forest limit

2000 BC.

today

Mixed Oak forest limit

2000 BC.

today

Steppe/Grassland–forest limit

2000 BC.

today

permafrost limit

2000 BC.

today

Fig. 16.20 Forest and grassland limits on the plains of North America and Eurasia about 2000 B.C. and at the present day. (*Sources used include* FRENZEL (*1966, 1967*), NEJSTADT (*1957*) *and* NICHOLS (*1967a*).)

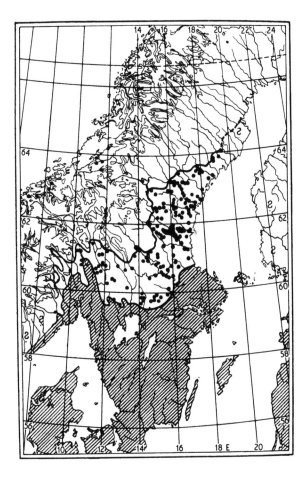

Fig. 16.21   Distribution of fossil finds (dots) of hazel in Sweden from the postglacial warmest times and its distribution in modern times (shaded). (*From* ANDERSSON *1902.*)

Note: There has been less change in the Norwegian coast region where hazel still grows today as far as 66–67°N.

In studying and interpreting the vicissitudes of climate and vegetation in the more intricate terrains near, and in, the mountains of Asia, Africa and the Americas, it is necessary to use indirect methods to elucidate the general circulation and climatic regimes. The task is simplified by the similarity of the main forest successions through postglacial time revealed by pollen analysis of the stratigraphy of deposits in most parts of the world. This indicates, with broadly similar datings, a climax of warmth followed by a return to a cooler climate and type of environment – for example, in the Alpine foothills of northern Italy (BEUG 1964), in Kashmir (SINGH 1963, VISHNU-MITTRE 1966, VISHNU-MITTRE and SHARMA 1966), in Japan (TSUKADA 1967), and in North and South America (HEUSSER 1966b, NICHOLS 1967a, 1970), as in northwest and central Europe (see, for example, AVERDIECK and DÖBLING 1959; FIRBAS 1949; GODWIN 1956; GODWIN, WALKER and WILLIS 1957).

Differences from sector to sector within one hemisphere

The regional climatic histories summarized in the previous section seem generally understandable in terms of latitude shifts of the main features of the global wind circulation,
very much as portrayed in fig. 16.11. To this must be added the variations that must be
supposed to have taken place in the total water content of the atmosphere, as the Earth's
surface, and particularly the oceans, warmed up and later began to cool down (see Volume 1,
p. 348). Some differences in the course and timing of the climatic changes in different
longitudes are, however, known to have affected middle and higher latitudes. These
presumably mark differences in the placing and amplitude of the waves in the upper W'lies.

NICHOLS (1967b) has demonstrated the parallelism of the climatic trends over the last
7000 years in northern Canada near 100°w and in northwestern and central Europe, and this
may be seen by comparing fig. 16.14 with fig. 13.4(c) and with fig. 16.6. This parallelism
seems about equally close when the variations in Greenland are compared with Europe,
either in terms of the oxygen isotope record from 77°N 56°w (figs. 13.16 and 13.18) or of the
dated readvances of the Greenland ice sheet (TEN BRINK and WEIDICK 1974), though the
main warming and cooling episodes in Greenland were sometimes as much as 200 years
earlier or later than in Europe. When the course of events in western North America (British
Columbia and Alaska) or in southern Chile (fig. 16.22) is compared with Europe, however, it
is clear that the early stages of the Neoglacial cooling from 1500 B.C. onwards were much
sharper near the Pacific than in Europe and that the warm periods in Europe around A.D. 400
and 900–1300 were poorly represented (if they were felt at all there). Central and eastern
longitudes in North America, the Canadian Arctic, and Greenland, as well as the western
half of Asia, seem to have shared with Europe the warmth of the Little Optimum in the early
Middle Ages, as did New Zealand.

In north Greenland (FREDSKILD 1969, 1973a, 1973b) the period between A.D. 400 and
1100 is distinguished by a greater downput of snow (and rain?) than since, presumably
indicating more cyclonic activity on tracks near 80°N on this side of the hemisphere in those
centuries.

The most striking difference so far noticed between different sectors of the northern
hemisphere is the coincidence in time of a period of cold climate in China and Japan with
the height of the early medieval warm epoch in Europe, Greenland and North America
east of the Rockies. The cold climate in the Far East was particularly marked in the twelfth
century A.D., when the mean blossoming date of the cherry trees in Kyoto was 10 days later
than the average of the last 1000 years and later than recorded in any other century (see Table
in Appendix V). This suggests an eccentric position of the circumpolar vortex over the
northern hemisphere in Europe's High Middle Ages, with the climatic zones persistently
displaced north over the Atlantic sector and south over the whole Pacific sector and the Far
East.

An antiphase relationship in the opposite sense is apparent during Europe's cold phases

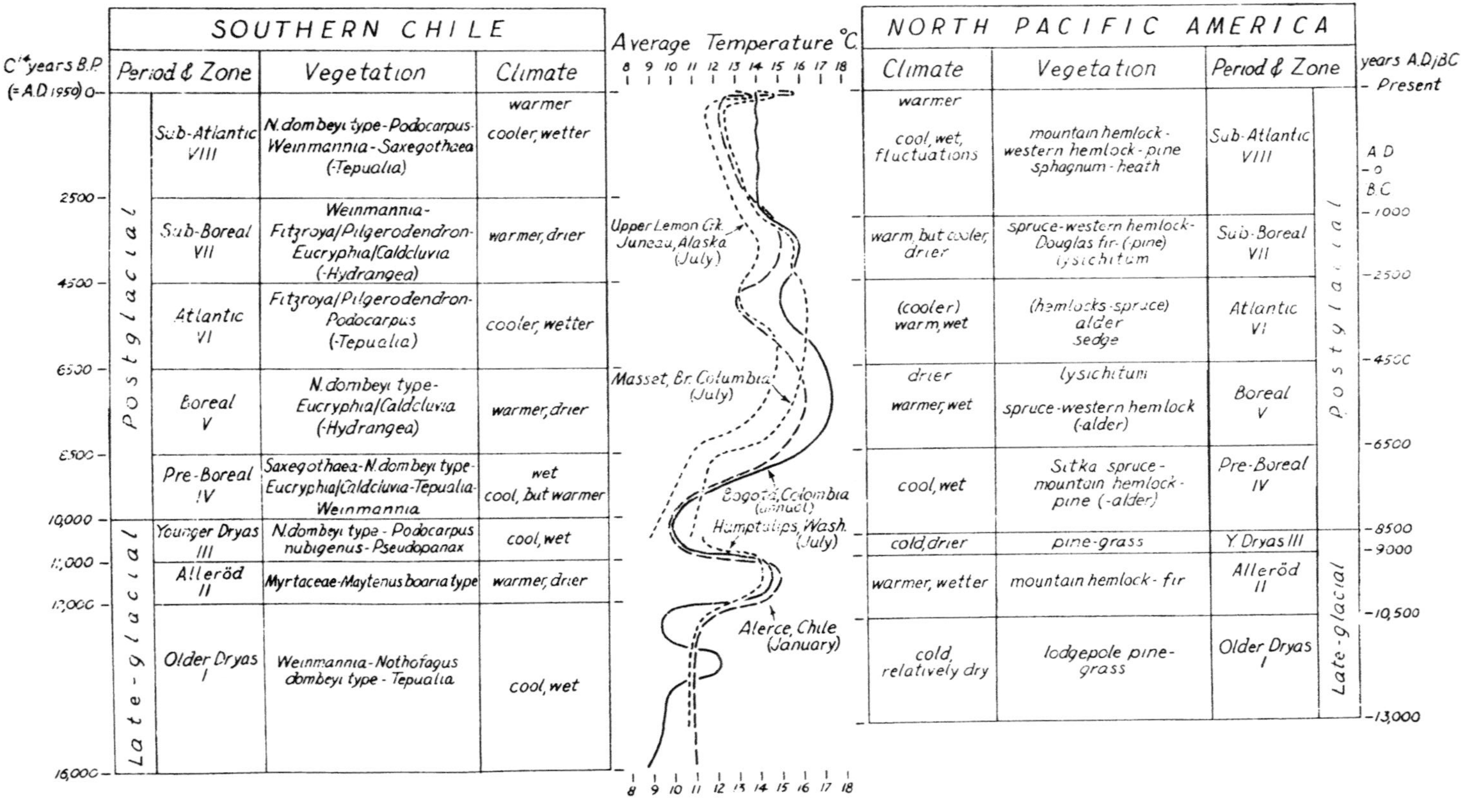

SOUTHERN CHILE

C¹⁴ years B.P. (=A.D.1950)	Period & Zone	Vegetation	Climate
	Sub-Atlantic VIII	N.dombeyi type-Podocarpus Weinmannia-Saxegothaea (-Tepualia)	warmer / cooler, wetter
2500	Sub-Boreal VII	Weinmannia-Fitzroya/Pilgerodendron Eucryphia/Caldcluvia (Hydrangea)	warmer, drier
4500	Atlantic VI	Fitzroya/Pilgerodendron Podocarpus (-Tepualia)	cooler, wetter
6500	Boreal V	N.dombeyi type-Eucryphia/Caldcluvia (Hydrangea)	warmer, drier
8500	Pre-Boreal IV	Saxegothaea-N.dombeyi type-Eucryphia/Caldcluvia-Tepualia Weinmannia	wet / cool, but warmer
10,000	Younger Dryas III	N.dombeyi type - Podocarpus nubigenus - Pseudopanax	cool, wet
11,000	Alleröd II	Myrtaceae-Maytenus boaria type	warmer, drier
12,000 / 16,000	Older Dryas I	Weinmannia-Nothofagus dombeyi type - Tepualia	cool, wet

(Late-glacial / Postglacial)

Average Temperature °C

(Curves: Upper Lemon Gk. Juneau, Alaska (July); Masset, Br.Columbia (July); Bogotá, Colombia (winter); Humptulips, Wash. (July); Alerce, Chile (January))

NORTH PACIFIC AMERICA

Climate	Vegetation	Period & Zone	years A.D./B.C. - Present
warmer / cool, wet fluctuations	mountain hemlock-western hemlock-pine sphagnum-heath	Sub-Atlantic VIII	A.D. 0 B.C. / -1000
warm but cooler, drier	spruce-western hemlock-Douglas fir-(pine) Lysichitum	Sub-Boreal VII	-2500
(cooler) warm, wet	(hemlocks-spruce) alder sedge	Atlantic VI	-4500
drier	Lysichitum		
warmer, wet	spruce-western hemlock (-alder)	Boreal V	-6500
cool, wet	Sitka spruce-mountain hemlock-pine (-alder)	Pre-Boreal IV	-8500
cold, drier	pine-grass	Y.Dryas III	-9000
warmer, wetter	mountain hemlock-fir	Alleröd II	-10,500
cold, relatively dry	lodgepole pine-grass	Older Dryas I	-13,000

(Late-glacial / Postglacial)

Fig. 16.22 Climate and vegetation changes over the last 16 000 years in Chile, Colombia, western U.S.A., Canada and Alaska compared. (*From* HEUSSER (1960a): *reproduced by kind permission.*)

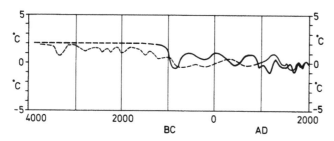

Fig. 16.23   The course of prevailing temperatures in China (bold line) and western Europe (thin line) over the last 5000 years.

(*Chinese data from* CHU KO-CHEN (*1973*). *Temperatures in central England from a curve by* G. MANLEY.)

around 600–200 B.C. and in the seventh and eighth centuries A.D. The temperature curves derived (fig. 16.23) for England and China show warm and cold phases occurring in opposition in the West and in the Far East at most times during nearly 2000 years between about 700 B.C. and A.D. 1200.[1] But the trends are broadly in step over the remainder of the last 5000 years, particularly as regards the onset of generally cooler regimes some time after 2000 B.C., the warmth in the thirteenth century A.D. and the sharp decline to the cold regime of the Little Ice Age, centred between A.D. 1550 and 1700, when colder climates than before were occurring simultaneously in and around the Atlantic sector and in the Far East.

This record of alternating warm and cold phases on opposite sides of the northern hemisphere, though only prominent over a limited section of the time surveyed in this chapter, may indicate a previously unsuspected manner of behaviour affecting the wave number 1 component of the circulation over the hemisphere. KING (1974) has suggested a long-continued westward drift of the one main trough in the upper W'lies (and implies that this trough circuits the hemisphere in about 700 years).

A case of similarly persistent, but abnormal, eccentricity of the circulation over the southern hemisphere may be noticed when the record of snow accumulation at Byrd station (80°S 120°W) in the Pacific sector over the last few centuries (fig. 7.23, Volume I, p. 303) is compared with the corresponding records for the South Pole and Southice (82°S 29°W) in the Atlantic sector (figs. 13.40(*a*), (*b*) in this volume).

Estimation of rainfall, run-off and stream flow

Most of what has been written in the preceding sections has had to do with the trends and fluctuations of temperature and of the large-scale atmospheric circulation. Some variations

---

1. The course of the temperatures in China over the last 2000 years shown in fig. 16.23 seems well verified by oxygen isotope measurements applied to tree rings in Yaku-sugi, Japan, performed by L. J. PANDOLFI, PAYTON and L. M. LIBBY (personal communication, 8 July 1975).

of apparent moistness of the climate have been noticed and explanations in terms of shifts of the prevailing axes of cyclonic activity suggested. It may also be important to estimate the actual amounts of rainfall, evaporation and run-off produced by the past climatic regimes. This is more difficult than the establishment of estimates of temperature.

It should be possible to establish past variations in the amount of snow accumulating on the ice caps in high latitudes as far back in time as the individual year layers can be recognized in pits and in cores taken from the ice, including recognition by oxygen isotope measurements revealing the yearly temperature cycle. Allowance must be made for compression by the overlying ice and for flow of the ice from higher up on the ice sheet. Estimates are available from Greenland ($77°N$ $56°W$) back to the tenth century A.D. (LANGWAY 1967, p. 97) and show higher rates of accumulation (about 42 grammes/cm^2/year) in the warm epochs of the present century and the tenth to thirteenth centuries A.D. than in the Little Ice Age (34–37 grammes/cm^2/year in the sixteenth and eighteenth centuries). It seems likely that accumulation on the high ice sheet in at least the northern half of Greenland (and on the ice sheet in Antarctica) was actually greatest in the warmest postglacial times. Nevertheless, Greenland's inland ice underwent shrinkage at that time and was withdrawing from the moraines, some of which now lie offshore, along the west coast, owing to greater surface ablation and run-off from the slopes, swifter flow of the ice itself and more icebergs calving and being lost from the ice-front in the fjords.

A method whereby the average rainfall in England and elsewhere in Europe in successive periods of a few decades or a century in length since the early Middle Ages can be estimated by calculations based on LAMB's summer and winter index values, which are related to frequencies of various types of reported weather, or by regression equations based on the prevailing temperatures, if these are known, has been explained in Chapter 13 (pp. 32–4). The latter method can be applied to the earlier postglacial climates wherever the temperatures – and, preferably, also the general character of the large-scale wind circulation patterns – are at least roughly known. The regression equations used must, of course, be appropriate to the place concerned. This is the basis of the rainfall estimates in Table 16.1. Unfortunately, the average high summer (partly convective, partly cyclonic) rainfall in England proved to be nearly independent of temperature and was better estimated from the medieval reports of wetness summarized in LAMB's summer wetness index. This item might also be estimated from the proximity or remoteness of the prevailing depression tracks. This procedure, where it can be used, is probably sounder than to estimate rainfall from former river and lake levels which express the balance between rainfall and evaporation (though, with care, it may be possible to use in reverse some of the procedure outlined in the following paragraphs to derive figures for rainfall from observed river levels and estimates of evaporation based on temperature and wind circulation characteristics). Table 16.1. gives the values used in the computations and the rainfall figures derived, indicating by brackets those considered most liable to be vitiated by the nature of the summer circulation, and leaving gaps where no basis for calculation exists. Some adjustment of the rainfall results indicated by this regression analysis, particularly as regards high summer rainfall, to take account of the prevailing pressure patterns derived for each epoch might be permissible. A control should be sought through some second approach, e.g. through botanical indicators that are sensitive to summer rainfall or to the difference between rainfall and evaporation.

Table 16.1   Derivation of average rainfall over England and Wales in different epochs, as percentage of 1916–50 averages

Dates (approx.)	Epoch by name	T_{JA}	T_{DJF}	T_y	R_y	R_{10}	R_{JA}	Adjusted T_y (see text)	Provisionally adopted R_y
		Temperatures in °C			Calculated (per cent)			°C	%
c. 7000 B.C.	Pre-Boreal	16·3*	3·2*	9·3	97	92		9·0	92–95
c. 4500 B.C.	Atlantic	17·8*	5·2*	10·7	111	(107)		11·0	110–115
c. 2500 B.C.	Sub-Boreal	16·8*	3·7*	9·7	(101)	95		10·0	100–105
900–450 B.C.	Sub-Atlantic onset	15·1*	4·7*	9·3	(97)	103			103–105
1150–1300	Little Optimum	16·3	4·2	10·2	103	106	85		
1550–1700	Little Ice Age	15·3	3·2	8·8	93	91	103		
1900–50	20th century warm decades	15·8	4·2	9·4	99	100	97		

* Values derived from anomalies indicated by botanical or other evidence.
 JA means July/August averages.
 The suffix 10 means averages for the other 10 months of the year.
 DJF means December/January/February averages.

In Table 16.1 only the regression analysis has been used; though adjusted values of prevailing yearly mean temperatures have been taken in the three earliest epochs because of probable differences in the seasonal distribution of temperature:

(1) At 7000 B.C. remaining features of the former ice age environment, including the dry Northsealand, suggest short hot summers and all the rest of the year colder than now.

(2) At 4500 and 2500 B.C., with little or no Arctic sea ice and the probability of a northward displacement of the zones of depression tracks and of subtropical anticyclones, short winters and lengthening of the warm season through spring and autumn are suggested.

These result in corresponding adjustments of the rainfall.

The rainfall figures derived in Table 16.1 range far less widely from the modern normal than BROOK's estimates (1949, pp. 297–9), which were based on lake levels in Ireland and central Europe, peat bog growth rates in northwest Europe and annual layers in the mud of Lake Saki in Crimea (see fig. 16.24, p. 408 and Tables 29(a)–(c) in Appendix V. BROOKS deduced for the Sub-Boreal a rainfall of only half (42–52 per cent) the present amount over the British Isles and central Europe and excesses of the order of 40 per cent in the British Isles around 4000 B.C. and for some centuries in the cold epoch about 500 B.C. It is hard to reconcile the large percentage changes of rainfall proposed by BROOKS with the small changes of prevailing temperatures in the air and sea derived from all types of evidence, as the changes in the amount of water vapour required on average to saturate the surface air in the neighbourhood of the British Isles would cover an overall range of only about 15 per cent (i.e. ±7–8 per cent) since 7000 B.C. Nevertheless, the values here put forward are complete and untested and based upon regression equations derived from British Isles data that could only be applied in rather closely similar regions. They cannot at this stage be regarded as more than indicators of the likely direction and order of magnitude of the differences of rainfall between different epochs.

A preliminary analysis of a similar kind carried out with appropriate regression equations for an area of the Russian plain between Kiev, Leningrad and Moscow indicates:

(1) That the range of variation of mean summer temperatures there within the last millennium was no greater (under 1°C), and possibly rather less, than in England; the range of variation of the mean winter temperatures (2–2·2°C) was almost twice as great as in England; the fluctuation from the early medieval warm epoch to the cold period of the Little Ice Age ran broadly parallel in Russia and England but with some minor differences of phase – the main period of dry summers may have fallen earlier in Russia, as in the Arctic (e.g. Greenland and Iceland), about A.D. 1000–1200 as compared with 1150–1300 in England; but it appears that the summers were again warm in Russia for 150 years after 1300 when the winter climate was becoming markedly colder.

(2) The range of variation of the yearly rainfall on the Russian plain, with deficiencies of the order of 30 per cent indicated for A.D. 1400–49 and 1600–99, was almost certainly greater than in Britain and probably explains the occasional drying up and apparent blocking of some river courses in the driest periods.

(3) Similar amplitude ratios for the temperature and rainfall changes in Russia and western Europe are indicated for 7000 and 4500 B.C. (average yearly rainfalls respectively about 90 per cent and 130 per cent of the modern values). Figures derived for rainfall around 2500 and 500 B.C. (respectively 116 and 97 per cent) are, however, hard to reconcile with the non-meteorological evidence; if they are

misleading, the error probably chiefly affects the figure for 2500 B.C. and may be due to more anticyclonic summers about that time.

The wetness or dryness of the ground, and the levels of lakes and rivers, depend not on rainfall alone but on the difference between rainfall and evaporation. The latter is principally a function of temperature, but also depends on wind and sunshine – hence also upon the seasonal distribution of temperature. These influences are taken into account in PENMAN's (1956) well-known formula (see Volume 1, p. 348) for potential evaporation, using saturation deficit and mean wind speed. It is doubtful whether their effect was sufficiently allowed for by BROOKS (1949).

An empirical expression derived by TURC (1958) from study of evaporation and run-off in drainage basins in all types of climate offers a means of calculating probable yearly evaporation (E_y) and the surplus water to be drained away yearly (D_y) in past climatic epochs, if the average yearly rainfall (R_y) and temperature (T_y) are known:

$$E_y = \frac{R_y}{\sqrt{0.9 + R_y^2/L^2}}$$

where $L = 300 + 25T_y + \frac{1}{20}T_y^3$. E and R are measured in mm and T in °C. When E_y has been evaluated, $D_y = R_y - E_y$.

The results of calculations with TURC's formula for the epochs here studied are summarized in Table 16.2.

These results must be regarded as raw and untested, though they indicate a sequence of changes from epoch to epoch in the amount of surplus water to be drained that is in general agreement with the Boreal dry, Atlantic moist, Sub-Boreal drier, Sub-Atlantic wet sequence indicated by bog stratigraphy in the British Isles, Scandinavia and central Europe. It is doubtful whether the rainfall and surplus water figures for the Sub-Atlantic are sufficient to account for the extreme wetness and growth rate of Tregaron bog in west Wales found by TURNER (1965) for the period between about 750 and 400 B.C.[1] Several possible adjustments may, however, account for the discrepancy. Margins of error in the radiocarbon dating admit of the possibility that the growth was spread over nearly twice as long as TURNER assumed and on the other hand the low evaporation rate indicated in our calculations by the

1. TURNER (1965) found by many radiocarbon dates that at Tregaron between 750 and 400 B.C. about 90 cm thickness of peat was formed, an average rate of 1 cm for every 4 years, whereas for much of the time between 1400 and 1000 B.C. the bog surface seems to have been dry and no growth taking place. Growth was much slower again afterwards and over the last 2000 years has averaged barely 1 cm/20 years. This low average may owe a great deal to prolonged periods with little or no bog growth in recent times and around A.D. 400 and 1200 – towards the latter time there was some recovery of the woodland, or at least scrub, on the heights and on the peat moors in Wales and Scotland.

There seems to be evidence that also in West Germany the period 800–400 B.C. was extraordinarily wet: between 3 and 3·5 m thickness of gravel was laid down on the river terraces in the Main Valley in that short time span (SCHIRMER 1972).

reduced summer temperatures may have caused difficulty over disposal of the surplus water in the drainage channels (TURC's formula takes no account of seasonal distribution). The bog growth probably implies, however, that the average summer rainfall around 700–450 B.C. was enhanced by frequent cyclonic W'ly situations with depressions passing near or over the British Isles: this would also make the average yearly total rainfall a few per cent higher than our estimates (Table 16.1) calculated by procedures that do not take account of this circulation character. Moreover, summer evaporation would be unlikely to increase by as much as the summer rainfall and might actually decrease a little.

Table 16.2   Evaporation and surplus water to be drained in different epochs in England and Wales, by TURC's formula (1958)

Dates (*approx.*)	Epoch by name	T_{JA}	T_{DJF}	T_y	R_y	E_y	D_y
		Temperatures in °C			(*Calculated: per cent of 1916–50 averages*)		
c. 7000 B.C.	Pre-Boreal	16·3	3·2	9·3 possibly 9·0	92–95	94–98	89–94
c. 4500 B.C.	Atlantic	17·8	5·2	10·7 possibly 11·0	110–115	108–114	112–116
c. 2500 B.C.	Sub-Boreal	16·8	3·7	9·7 possibly 10·0	100–105	102–104	100–106
900–450 B.C.	Sub-Atlantic (onset)	15·1	4·7	9·3	103–105	97	110–115
1150–1300	Little Optimum	16·3	4·2	10·2	103	104	102
1550–1700	Little Ice Age	15·3	3·2	8·8	93	94	92
1900–1950	20th century warm decades	15·8	4·2	9·4	(932 mm)	(497 mm)	(435 mm)

The figures in Table 16.2 for the Sub-Boreal and Atlantic epochs may require adjustment for increased frequency of anticyclonic influence, as compared with nowadays, favouring the higher estimates of evaporation (E_y) and lower estimates of surplus water (D_y) indicated. The ground condition in the Sub-Boreal can readily be envisaged from these figures as being in some centuries wetter and in others a little drier than in recent times, as suggested by botanical evidence; these alternations would be particularly readily brought about by variations in the frequency of dry anticyclonic weather in summer.

The estimates for prehistoric times presented in this section should be treated as 1000-year averages or at least as representing a broad average of conditions over many centuries.

However, we have noted elsewhere in this volume (e.g. Chapter 13, pp. 140, 210–13) evidence in river levels, lake bed varves and radiocarbon-dated changes of the rate of peat growth in bogs, of recurring fluctuations of rainfall from century to century, clearly indicated by field evidence in Europe and North Africa, which in Sub-Boreal times seems to have taken on a greater amplitude than before. Estimates of their amplitude are required and seem likely to

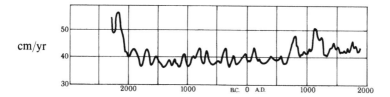

Fig. 16.24   Rainfall variations in the Crimea since about 2300 B.C. as indicated by the thickness of the yearly mud layers (varves) in the bottom deposit of Lake Saki, 45°7′N 33°33′E.
    (*From* SCHOSTAKOWITSCH (*1934*), *with the rainfall scale suggested by* BROOKS (*1949*).)

have to be derived from biological indicators (especially tree rings) and from the alternating cycles of aggradation and erosion of river beds, lake levels and so on.

The fluctuations from century to century

Fig. 16.24 shows an analysis, interpreted as a record of rainfall variations, of the remarkable series of yearly layers (varves) in the mud deposit on the bottom of a small salty lagoon on the west coast of the Crimea. The layers are presumed mainly due to the run-off from the land caused by the heavier rainstorms; but most of the layers are only a few millimetres thick, and some may have been missed in the counting. The earliest layer was counted as 2294 B.C., but could have been slightly earlier. A rainfall scale suggested by BROOKS (1949, p. 299) has been added to the diagram. This record suggests (*a*) that variations occurring 4–6 times in 1000 years are important, (*b*) that a few centuries between A.D. 800 and 1250 represented a temporary return to moister conditions rather as they were in that area before 2000 B.C. The fluctuations of rainfall or run-off in the Crimea appear often inverse to the changes in the prevalence of ground moisture found in northern, western and central Europe; this lake indicates dryness setting in about 2000 B.C. and no return to the former level except in the early Middle Ages. Particularly dry spells occurred around the twelfth, tenth and sixth centuries B.C. and a sharp return to drier conditions about A.D. 1250, both being times of marked wetness setting in (and renewal of the growth of the peat bogs) in western and northern Europe.
    The record of the Lake Saki mud layers is set forth in more detail in Tables 29(*a*)–(*c*) in

Appendix V, accepting the dating of the layers as correct, and a number of interesting and possibly important aspects appear. In general, the implied wetness variations show some parallel with those which are thought to have occurred in central Asia and are more often opposed to those over northern and central Europe. There are some runs of centuries in which the decade-to-decade variations within each century follow a similar pattern. In the

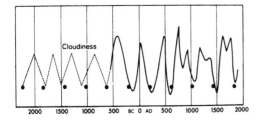

Fig. 16.25   Variations since 2300 B.C. in the cloudiness of the nights in latitudes between about 30 and 50°N in China, Egypt and Europe, indicated by LINK's (1958) analysis of the frequency of discoveries of comets.
Cloudiness increases as the curve rises.

first nine centuries A.D. the '40s–'50s and '90s–'00s seem to have been commonly wet (i.e. marked by thick mud layers) whereas the '10s–'20s and '70s were commonly dry. By the 900s A.D. the phase had changed, perhaps because fluctuations on a time scale of about 20 years ('bidecadal oscillation') had become more prominent. It also appears that the major changes of the regimes which persisted for centuries have been heralded by a few years (in some cases a single year) of remarkable wetness or dryness in the Crimea, presumably indicating extremely persistent occurrences of blocking of the W'lies (large stationary anticyclones or cyclones) in the sector of the northern hemisphere between 20 and 80°E. If so, this is a development which may be a valuable signal of climatic change, even though such a change did not always follow.

Another series covering the time from 2300 B.C. to the present is illustrated in fig. 16.25. LINK (1958) surveyed separately the available collections of reported discoveries of comets by observers in China and in western countries. The number of comet discoveries per century depends on:

(1) k, the number of comets actually within observable range of the Earth.

(2) C, the frequency of cloudless nights, so that the number of comets visible in practice is Ck, where C is expressed as a fraction.

(3) h, the human social or 'cultural' factor, which describes the observation skill of the existing state of civilization; so that, with h expressed as a fraction, hCk is the number of comets actually observed and of which records were kept.

When the number (*N*) of comet discoveries per century ($N = hCk$) is plotted on a graph against the date, the resulting curve shows a sequence of waves superposed on a generally upward trend. The waves are in remarkable agreement in China and the West from 500 B.C. onwards, when continuous curves for both areas can be drawn. The level of the curves represents the human skill: this and the upward slopes differ as long as the two communities were isolated from each other, being at first higher in China and after the ninth century A.D. higher in the West. Thus, the long-term mean level of the curve corresponds to the value of *h*, the human factor. LINK therefore plotted graphs of *N/h*, to eliminate the human factor, and obtained curves which showed only the waves. And since $N/h = Ck$, if *k* is in the long-term constant, these variations which were similar in China and the West must represent variations of night cloudiness in latitudes about 30–50°N. Groups of comet discoveries indicate the timing of the cloudiness minima back to 2300 B.C., though the amplitude of the swings cannot be assessed so far back. The curve suggests a fairly regular oscillation, a period length about 400 years; the detail after A.D. 600 points to the possibility of a superposed oscillation of period 200 years.

A statistical analysis of radiocarbon datings of organic matter buried in alluvial deposits in river flood plains all over the world has been carried out by KNOX (1973). The maxima of datable materials in the river gravels, etc., are surely an indicator of the frequency of flooding.[1]

Two points of general interest seem to emerge from the analysis:

(1) About eighteen centuries in the 12 000 years surveyed appeared as remarkable minima in the occurrence of washed-down, datable material. The intervals between these centuries (using calibrated radiocarbon dates) were on five occasions 400 years and six times 700–800 years.

(2) Study of the different regions separately indicated that maxima about 200, 400 or

---

1. It has usually been assumed that the maxima of datable material coincide with maxima in the frequency of flooding of the rivers. A closer study of the German finds of great oak trunks and branches dating from the postglacial optimum, buried in the gravels of the rivers Danube and Main (TROLL 1972), throws further light on the matter and suggests to FRENZEL (personal communication, 11 May 1974) that the trees were mainly felled by the rare floods in periods of prevailingly drier and more favourable climate, or possibly by the first onset of a wetter regime. The timbers that have been preserved were deeply buried by a period of rapid gravel deposition which quickly followed their demise.

It was also noted that the age distribution shown by 232 of these buried oaks was younger than that normally observed in an undisturbed oak wood where trees 600 and 700 years old would not be uncommon. 218 of the buried oaks were between 100 and 280 years old, and the oldest tree was only 420 years old.

These observations suggest that the centuries found by KNOX (1973) to be marked by striking minima of radiocarbon datings on vegetation remains in river gravels may correspond to the latter end of a long period of frequent flooding or the early part of an ensuing dry period (rather than marking the dry period itself as he supposed). The uncalibrated radiocarbon ages of the minima given by KNOX were 11 800, 11 050, 10 550, 9900, 9200, 8800, 8200, 7000, 6600, 5800, 4650, 4250, 3600, 3300, 2900, 2500, 1700 and 750 years B.P. This may explain the curious fact that most of the above dates were found by BRYSON *et al.* (1970) to represent *maxima* of radiocarbon datings of changes of regime indicated by the pollen record.

600 years apart were commonly not in phase, and were individually very far from being of equal importance, in the different regions.[1]

Shorter-term variability

The ranges of year-to-year and decade-to-decade differences of temperature all through the warmer millennia from 7000 B.C. and on until the present epoch probably did not vary by more than 50 per cent, being least in the most oceanic and greatest in the most continental regimes brought about by changes in the frequency of winds from the oceans. This estimate is suggested by what follows on p. 412 and in the case of England by Table 16.3 which presents the standard deviations of monthly mean temperature in regions rather more and rather less oceanic than England and at places near extensive glaciers and sea ice today. From Table 16.3 it seems likely that the ranges of year-to-year differences would be 2 to $2\frac{1}{2}$ times as great as now (most like the southwest Greenland and Novaya Zemlya situations) at a time when England was at once near to the extensive remains of the former north European ice sheet and to a large summer-warm, ice-free continental area and an ocean already warm for the latitude. This was the situation from 12 000 B.C., and perhaps earlier, until around 8000 B.C. Values nearly twice the modern standard deviation of temperature may have continued while England was still part of the continent. In the Full Glacial times before 12 000 B.C. the standard deviations of the monthly mean temperatures in England near the ice margin may have been as much as $2\frac{1}{2}$ to 3 times what they are today. Probably nowhere have the ranges of year-to-year variation changed since 8000 B.C. by much more than a factor of 2. All through the postglacial warm epoch the variability on these shorter time scales was presumably in most places similar to, or rather less than, today; in latitudes north of 70°N, in the absence of

1. The groupings in regions studied by KNOX were:
   North America west of 100°w, North America east
   of 100°w, Europe 10°w–40°E, and Australia.

   Frequencies of different intervals between epochs of maxima
   of ^{14}c-dated flood plain litter

Number of centuries apart	1	2	3	4	5	6	7	8	9	10	11	12
North America												
West of 100°w	0	7	4	7	2	0	1	4	0	1	0	0
East of 100°w	0	6	4	12	1	3	1	2	0	1	0	0
Europe												
10°w–40°E	0	8	5	7	1	5	1	1	0	0	0	1
Australia	0	2	1	7	2	1	0	1	0	1	0	0
All groups	0	23	14	33	6	9	3	7	0	3	0	1

permanent sea ice, the variability may have been much reduced, except in summer, and presumably became similar to the open oceans in the temperate zone today.

Table 16.3 Ratio of SD of monthly mean temperature to that of central England (period 1900–50 approx.)

	Ireland (eastern and inland districts)	Nether-lands–Rhineland	Bergen, Norway	Western Iceland	Western Greenland near 65°N	Novaya Zemlya (south island	Franz Josef's Land
January	0·8	1·5	1·1	1·2	2·1	2·4	3·0
April	0·9	1·2	1·0	1·4	1·5	2·2	1·7
July	0·7	1·1	1·2	0·8	1·2	1·8	0·3
October	0·8	1·2	0·9	1·2	0·8	2·0	1·8
Average	0·80	1·25	1·05	1·15	1·40	2·10	1·70

Note: The standard deviations of the monthly mean temperatures in central England in 1900–49, a period of somewhat oceanic climate, were: January 1·7°, April 1·2°, July 1·2°, October 1·1°C (these figures measure the variability of the mean temperature of the month named over a number of years).

An informative study bearing upon changes in the year-to-year variability of temperature and rainfall in postglacial and more recent, historical times has been published by FÜRST (1963), summarizing a number of authors' works on the observable differences in tree ring sequences shown by trees which grew near the same sites in central Europe in different epochs. The measured year-to-year variability was found in general to increase eastwards across central Europe, i.e. with increasing continentality (the one exception was a dry area in the lee of the Vosges Mountains which showed the great variability more generally found farther east). Measured variability shown by oaks at sites in northern Switzerland was 50% greater in Neolithic (Sub-Boreal) times, around and soon after 3000 B.C. (calibrated radiocarbon dates), than in the late Bronze Age between about 1600 and 1300 B.C. The variability shown by oaks in the area over the past hundred years was intermediate between these extremes. Oaks at sites in southwestern Germany (e.g. Spessart) and firs in Bavaria were used for a similar study of historical times and revealed increases of year-to-year variability of the order of 25–50% coinciding with the climatic deteriorations from about A.D. 1200 to the period 1280–1330, followed by a recovery (decline of variability that lasted from 1330 to 1380), and another great increase from the period A.D. 1550–1650 to 1650–1750. It is probably safe to associate these various occasions in the past when variability was greater than now with more meridional circulation character, as FÜRST suggested (see also corresponding evidence from floods of the Mississippi River and else-where in KNOX (1976)).

Comparison of the year-to-year variability of temperature in England (figs. 13.4(a), (b) and (c), Chapter 13 pp. 37–8) with the decade-to-decade and century-to-century variations makes very clear the greater statistical significance of the longer-term variations.

Climate and Man: early cultures and civilizations

Various instances have been given in Chapters 12 and 13 of the impact of climatic fluctuations and changes upon human society and of Man's reactions to them. Here we shall illustrate – mainly from Europe where there is most evidence – some general tendencies that corresponded to the successive climatic stages of postglacial times.

During the last ice age, primitive men living in groups made what was probably an easy living by hunting down the big grazing animals – reindeer, bison, mammoth, etc. – on the open grassy steppe and tundra in what is now France and elsewhere in Europe south of the ice. They recorded their life and the fauna they knew in the cave wall paintings in France (e.g. at Lascaux) and Spain (at Altamira). During the retreat of the ice, in Zone I (fig. 13.13), probably before 12 000 B.C., while the ice edge still crossed Denmark and northeast Germany, bands of reindeer hunters reached as far north as Hamburg.

With the rapid development of the postglacial warming, the rivers were swollen enormously, particularly in spring and summer, by the melting ice; gravels and sand were deposited in great quantities along the river courses, lakes formed and sometimes soon burst or quickly silted up. The landscape was changing rapidly. But the greatest change for the human population and the animals they hunted was the disappearance of the open plains, as the forest advanced north in Europe. Man seems to have adapted himself more successfully than the animals. The ranges of both moved northward in Europe, Asia and North America, but some species among the animals were lost, probably due to their reduction by Man.

Other great changes in the landscape were brought by the rise of sea level, proceeding over some thousands of years at the rate of 1 m a century. It seems likely that there was great loss of life, particularly among the fishing folk camped near the sea's edge and at the primitive industrial sites for making salt by evaporating sea water. It has even been suggested that in this way the end of the ice age was one of the times, rare in history, when the total population of mankind was significantly reduced.

In North America the change probably affected the early human inhabitants less drastically. The northernmost stands of forest during the ice age were never far from the ice-sheet margin; and as the ice retreated, though the forest spread north, from about 6000 B.C. onwards the tundra belt gradually widened somewhat and, farther south, the area of grassland spread farther east from the Rockies.

England had had some human inhabitants during the last ice age. Neanderthal men with their Mousterian culture survived at Creswell in Derbyshire until 50 000 years ago, when they were replaced by incoming *Homo sapiens* with an Upper Palaeolithic culture (WOOD 1963). Visiting groups of hunters also doubtless wandered north from France. Some of the Creswellian inhabitants seem to have lingered on until the arrival of the first postglacial settlers. These brought the Maglemosian culture from Denmark, evidence of the influx of peoples across the Northsealand plain. That plain was itself probably occupied by hunting and fishing peoples in early postglacial times. The site at Star Carr, Yorkshire, around 7400 B.C. was occupied by a few families of the Maglemos culture, using boats for fishing. And it seems that the first people to enter Norway in postglacial times moved north to favourable sites up the west and southeast coasts from Northsealand, perhaps as early as 8000–10 000 B.C.[1]

The next turning point in European prehistory in which, by its coincidence in timing, a climatic change seems to be involved is the spread of New Stone Age (Neolithic) culture – i.e. the first agriculture – over central and northern Europe. This development in the fourth millennium B.C. more or less accompanied the transition from the warm, moist climate of the so-called Atlantic period to the more variable, and on balance somewhat drier, Sub-Boreal regime. It seems likely that it was triggered in some way by the climatic break around 3500–3000 B.C. represented by the Piora oscillation, which ended the former regime, produced great glacier advances and must have severely disturbed settled human economies in regions of Europe and Asia that were marginal for either temperature or rainfall. How far the fall-off of the elm and lime trees should be attributed to the climatic break and how far to new forms of husbandry by Man (e.g. taking the new shoots for animals) is not yet fully agreed; but the widespread occurrence of the decline of those trees about the same time in Europe and North America – and suggestions of a similar event in the middle of previous interglacial periods – suggests that the cause was climatic.[2] The increased availability of openings in the forest, and Man's increased capacity to check the forest by fire and felling, and indirectly by grazing animals, probably led to a considerable increase in the numbers of mankind during Sub-Boreal times.[3]

The earliest beginnings of agriculture and the domestication of animals can be traced much farther back, perhaps to 9000 B.C., in some localities in southwestern Asia where the ancestors of wheat and barley grew wild and where there were wild goats, sheep, cattle and pigs (CHILDE 1954, 1958, WATERBOLK 1968). With the northward shift of the isotherms, and of the climatic belts and wind zones, in the millennia that followed between 8000 and 4000 B.C., and with the monsoon rains penetrating farther north than before, highly organized civilizations were gradually developed in a number of river valleys and alluvial plains in latitudes 20° to 40°N, where irrigation could be most easily arranged and provided seemingly reliable intensive cropping.

1. The origins of the much earlier Komsa culture at sites in north Norway (Finnmark) have not been explained with any certainty. Both French and Siberian affinities have been suggested. It may be from the last interglacial period (NORDHAGEN 1933).
   Other sources used in this section include CHILDE (1958) and CLARK and PIGGOTT (1965).
2. TAUBER (1965, pp. 54–61) surveys the evidence and speaks of a circumpolar reduction of the elms in middle latitudes around 3000 B.C.
3. This comment applies, of course, to Europe and Asia – not to other parts of the world that experienced no cultural change and where the change of climate was less or where a limited migration sufficed to obtain an environment similar to that which had been familiar.

The first agriculture was spreading northwards through New Mexico and Arizona about 4500–4000 B.C. (calibrated radiocarbon date) into the southwestern United States at a time when the climate in that region seems to have been becoming moister, presumably as the summer rains pressed farther north. It was probably only a minor addition to an economy still based mainly on hunting and collecting (WOODBURY 1961), and in the 2000 years of a cooler climatic period that followed (during the European Sub-Boreal) there is no further evidence of agriculture there.

Fig. 1 in Appendix V is a world map of the beginnings of civilization. The areas of origin of the civilizations that flourished between 4000 B.C. and the time of Christ are, except in the special case of Egypt and the lower Nile, areas of rather intricate geography, within which the sites then occupied range today from the natural domain of woodland to steppe or desert. Details of the changes that have taken place in the environment there could hardly be shown on a world map and have scarcely been probed so closely that valid regional maps of the former vegetation could yet be drawn. In at least three regions, however, there are unmistakable indications of formerly more extensive water supply and vegetation:

(1) The western and central Sahara, where rock drawings of a wide range of animals (BUTZER 1958), and finds of skeletons, including elephants (MONOD 1963), show that there was enough surface water for animals to pass across the terrain that is now desert and for at least some human occupation.

(2) The Rajasthan (Thar) desert areas of India and Pakistan, where rivers have disappeared (SINGH 1971) since the Indus Valley civilization and the cities of Mohenjodaro and Harappa (31°N 72$\frac{1}{2}$°E) were at their height before 2000 B.C. Also the other desert areas in southern Asia crossed by Alexander's army on expeditions in Iran and on the march to the Indus between 330 and 323 B.C.

(3) Sinkiang, particularly the Tarim basin in central Asia, near 36–42°N 80–90°E, was crossed by the ancient Silk Route which was used by trading caravans between China and the West in Roman times, when there was a chain of cities and settlements there and remnants of forest. Today it is largely a sandy desert (WADIA 1960, CHAPPELL 1970, 1971).

Other places, where conquest of the area by desert conditions within the last 2000 years suggests a continued trend towards natural desiccation, include Palmyra (34°40'N 38°0'E) (Plate XIII) and Petra (30°20'N 35°18'E), on the fringe of Palestine and the Syrian desert, and parts of the Anatolian plateau in central Turkey (CARPENTER 1966). It is unlikely that rainfall has been continuously decreasing for thousands of years but probable that it has been fluctuating with a net downward trend. The vegetation may, therefore, long have been in a subfossil condition, in which disturbance by Man might easily bring on its final demise. It is clear that the level of the water table in the arid zone in the Near East is still influenced by the supply of rain which fell in previous climatic regimes, particularly those of the last ice age. Certainly, the extent of the oases has been declining over the last 6000 years (BUTZER 1958).

A curious feature (see fig. 1 in Appendix V) is the spread of groups building stone circles as early as about 2000 B.C., and trading, evidently by a sea route through the western parts of the British Isles as far north as Orkney (59°N). This may suggest not only skill in navigation, but an era of more frequently quiet seas. In apparent agreement with this, there is now much evidence from radiocarbon dated pine stumps in the peat (H. H. BIRKS 1972 and personal communication, 5 July 1972, LAMB 1964, 1965, PENNINGTON, *et al.* 1972) that between about 5000 and 2000 B.C., forest grew much nearer to the open Atlantic coast of northwest Scotland than at any time since and also in parts of the Hebrides and the northern isles. The stone circles in Britain and the northern isles, some at least of which seem to have had uses as astronomical observatories (presumably for calendar fixing), may for this reason be connected with a climatic regime – implied by our maps for around 2500 B.C. – which offered more frequent clear skies for observing than nowadays (LAMB 1974) (it may have been necessary for Man to clear the forest and scrub sufficiently for these early observatories to be used).

The greatest postglacial spread of forest to the Atlantic fringe of Europe seems to have been before the times discussed above and apparently included some woods or thickets of birch and willow, and even oak, elm, hazel, and perhaps pine, in the Shetland Islands between about 5000 and 3500 B.C., according to a radiocarbon dated pollen stratigraphy in peat that is now below sea level (HOPPE and FRIES 1965). In the Orkney Islands only birch, hazel and pine are known, from macro remains, certainly to have been present, but they were extensive, even at the northwest coast (TRAILL 1868, MOAR 1969a). TRAILL quotes a dramatic account by a local resident at Otterswick (59°15'N 2°30'W) of how on Sanday in the northern part of the Orkney Islands, a prolonged NE gale in the winter of 1838 caused the sea to scour the sand from about 50 acres of the bay just above the low water mark, laying bare a forest floor of 'black moss' and fallen trees of up to 60 cm trunk width, all lying in the same direction from southwest to northeast. The same forest remains were then found to extend for 6·5 km across the shallow bay to Tuftsness. The forest had evidently grown close to the Atlantic shore of the time in a locality completely exposed to the ocean to the northwest. There is evidence of woods or thickets of similar composition in the Outer Hebrides, e.g. South Uist, so far undated by modern methods but thought to be from about the same period. And pines grew close to some of the most exposed parts of the Atlantic coast of the Scottish mainland in Wester Ross and Sutherland (LAMB 1964, 1965). Though there was some decline of these woodlands from as early as 3500 to 3000 B.C. onwards (MOAR 1965, 1969b), more rapid decline, which has been variously attributed to increased wetness, soil acidification and the beginnings of bog growth, possibly also to stronger winds, or to all these things, occurred between about 2600 and 1600 B.C.[1] (H. H. BIRKS, personal

1. A survey by H. H. BIRKS (1975) of the dates of pine stumps in the peat moors in various parts of Scotland, and evidence of the circumstances of the establishment, life and death of the trees, leads to the conclusion that there is no simple climatic explanation. The range of dates is too wide, although there is a great predominance of final dates between 4500 and 4000 years ago in the far north and west and as

communication, 5 July 1972, MOAR 1965, 1969b). GODWIN (1956, p. 338) records that, as late as the Bronze Age, Cornwall, Wales and Ireland were forest-clad right to their western shores and to higher altitudes on the hills than any present woods.

The farthest northern extensions of the forests in the continental sectors, in both North America and Eurasia, are dated later, between about 3000 and 1500 B.C. This may be attributed to the long aftermath of the ice age in those sectors, particularly the slow withdrawal of the cold sea from the regions of land surface that had been depressed by the former ice load. In latitudes near 80°N, radiocarbon dating of driftwood left on the now raised beaches of Spitsbergen and the Canadian archipelago indicates that there may have been most open water about 4000 B.C., though similarly open conditions continued until perhaps 1500 B.C. (BLAKE 1970). In the fjords of northernmost Greenland the period of most open water was not until 2500 to 2000 B.C. (FREDSKILD 1969). NICHOLS (1970) deduces from pollen diagrams for the Northwest Territories that the onset of cooler summers in northern Canada may be dated about 1500 B.C., in good agreement with the time of change indicated in Greenland and elsewhere. There was a warmer phase in both countries between about A.D. 900 and 1100–1200. Throughout the times which we have been considering, the spruce (*Picea abies*) seems to have been spreading westwards across Finland and Scandinavia at the expense of the earlier established trees (TALLANTIRE 1972a, 1972b). The advance of the spruce to dominance from the small pockets where it had existed from much earlier times in favourable sites throughout the region is probably to be attributed to this tree's competitive advantage in long cold winters, and therefore may be taken as indicating a fall in the general level of winter temperature. The spread of spruce forest from the east which betokens this, as derived from pollen diagrams from sites all over Fennoscandia, was not a smooth, continuous process but took place in distinct steps: (1) around 3400–3000 B.C. to the Finnish-Russian border, (2) around 2300–2100 B.C. to much of southern and central Finland, (3) between 1600 and 1300 B.C. a slight further advance, (4) 1000–400 B.C. across all central Sweden, (5) between A.D. 250 and 750 over southern Sweden and (6) between about A.D. 1100 and 1400 in Norway across Trøndelag to Trondheimsfjord. At each of these stages a renewed decline of the winter temperatures may be suspected.

Spruce also advanced into Switzerland from the east or southeast in stages, principally

---

far south as Rannoch Moor. A second group of dates in the Cairngorm and Galloway areas ranges between 7500 and 6000 B.P. The bog surfaces must have been relatively dry whenever the trees were established and grew. Death was commonly due to waterlogging, and failure of re-establishment of the trees could be attributed in the north, the west and southwest (Galloway) to the establishment of blanket bog plant communities, particularly with ling (*Calluna*), the fibrous peat formed by which inhibits the organisms (mycorrhiza) which help the uptake of

nutrients by tree seedlings from such phosphorus- and nitrogen-poor soils. Thus, failure of the pines in those parts of Scotland could be held due to a chain of circumstances set up by wetter conditions than before, though the effects of local drainage on the accumulating water would make the dates differ from one locality to another, and in some cases a single very wet season might change the situation irreversibly. Elsewhere in Scotland the balance may have been upset by Man, through forest clearance, burning and grazing leading to soil erosion, etc.

at times associated with intervals of colder climate, around 5000–4500 B.C. and at the time of the Piora oscillation, 3500–3000 B.C. (TALLANTIRE 1973).[1]

The vegetation changes discussed above probably owed little to human interference. But in Britain and central and southern Europe from Neolithic times (say 4000 B.C.) onwards, recession of the forest was increasingly due to Man (GODWIN 1956, p. 332 ff, TURNER 1965, NICHOLS 1967c, PENNINGTON 1970).

In southern England, particularly in the inland districts near the chalk downs and the Cotswolds and Mendip Hills, human activity has been considerable ever since Neolithic times. A closely argued paper by GODWIN and TANSLEY (1941) seems to have given a reliable picture of the vegetation with implications about the climate despite all the human interference. Oak and hazel were clearly present in some abundance near the chalk, and were the main species used as firewood by the Neolithic inhabitants around 3000 to 2000 B.C.; yew constituted 8 % of the charcoals analysed. But the evidence suggests that the trees were at the foot of the hills and on the slopes. The Neolithic people settled the chalk uplands, including Salisbury Plain, grew crops and kept sheep, and it seems likely, therefore, that the tops were already bare of trees. In the succeeding Bronze Age, the chalk plateau seems to have ceased to be used for settlement, though it was much used for grazing, for traffic and for burials. The implications of this archaeological evidence seem to be that the downs were already grassland and that the climate was drier than it had been earlier, in the Neolithic. The beautifully preserved form of the burial mounds could not have survived if their sites had been disturbed for long by the roots of woodland trees at any time since the barrows were made. There is more direct evidence of dryness of the Bronze Age climate in the digging at Wilsford, in the Salisbury Plain area, of a 33 m deep shaft, dated about the middle of the second millennium B.C., to 1·5–2 m below the present water table at its seasonal lowest (ASHBEE 1963, 1966). With the colder climates and increased wetness that set in from about 800 to 500 B.C. onwards, the early Iron Age peoples resettled the chalk uplands and tilled greater areas on the downs than ever before; it seems clear that the plateau remained largely bare of trees. Other archaeological evidence points to the same climatic sequence; wooden trackways were laid across the Somerset Levels (and other fens in the lowland areas of England), to keep open communications across the marshland when climatic conditions became wetter. The radiocarbon datings of these trackways are generally between about 3500 and 2500 B.C., and between 900 and 300 B.C. (see, for example, GODWIN and WILLIS 1959).

In northern England, also, the heights – even the summit plateau of the Pennines up to about 900 m above sea level – were occupied by settlements and defence forts in the warmest

---

1. According to MARKGRAF (1970, 1972), the spruce (*Picea*) migrating westwards across Europe had reached the upper Inn valley (11–12°E) in south Germany and Switzerland as the dominant tree already between 9000 and 8000 years ago, but the firs and larches reached the central Alps first and delayed the establishment of spruce there until various dates during about the last 5000 years. Its establishment west of the Rhine, at dates ranging in different places from 6000 to 600 years ago, may have depended on the clearings which Man made in the natural forest as well as on the occurrence of colder climatic phases.

postglacial times in the fifth millennium B.C. and again around 2000 B.C. (W. J. VARLEY, personal communication, 26 March 1973). Final abandonment of the summits after this second occupation period seems to have been associated with peat formation, when the inhabitants moved down to the drier oak and birch heaths on the lower slopes. Around 1000 B.C., or shortly after, a drier period seems to have made farming possible on the flat lowlands of east Yorkshire, about Holderness, but that too was later abandoned in a period of peat formation. There was some resettlement spreading up the slopes of the Pennines once more after 500 B.C.

Most European peat bogs show a sequence of recurrence surfaces in their stratigraphy (see Chapter 13, pp. 210–13), marking phases of wetter conditions when growth of the bog was renewed or accelerated after drier times during which the bog surface had become partially decomposed and eroded. Those dated around 2200, 1200 and 600 or 500 B.C. (as well as, in some places, 800 B.C. and A.D. 500–600) and A.D. 1250–1300 are most widely registered in the bogs in western, northern and central Europe. Among the various fluctuations at intervals of about 200 and 400 years, these are probably correctly identified as the main stages of sharp cooling and southward displacement of the most frequented storm tracks (expansion of the circumpolar vortex) which introduced the 'Neoglacial' climatic regimes of the *Nachwärmezeit*, or Sub-Atlantic, and of historical times.

During at least some of the stormier cooling stages, e.g. in about the last 1200 years before Christ and perhaps most of all between 1000 and 600 B.C., there seems to have been a rather general southward migration of peoples in Europe.[1] Thus, BROOKS (1949, p. 300) writes that the Federsee and other lakes in central Europe had been smaller than now in Neolithic times, culminating in a dry period about 2200–2000 B.C., followed by periods of catastrophic flooding (see also Chapter 13, p. 251, in this volume). Some of these floods may have been connected with the 'great eruption of Bronze Age peoples from the Hungarian plain, which probably occurred soon after 1300 B.C., and carried the Phrygians into Asia. . . . It is quite likely that this was the stimulus that caused the migration to a drier climate.' The lakes shrank to a minimum again in a dry period about 1000 B.C., during which the chief settlements were in moist places and agriculture was carried on above the forest level, even above Alpine passes which are now glaciated. 'Near the end of the Hallstatt period', BROOKS continues, following GAMS and NORDHAGEN (1923), 'about 500 B.C., the levels of the lakes rose suddenly; in the Bodensee (Lake Constance) the rise exceeded 30 feet (about 10 m). Most of the lake villages were destroyed, and settlement in the Alps reached a minimum . . . concentrated in the warmest valleys. . . . This climatic fluctuation had the appearance of a catastrophe.' GAMS (1937) confirms this statement, reporting a sharp sinking of the forest limit in the Alps in the course of the Hallstatt times; the formerly thriving mining industry and traffic in the high mountains ceased, and the lake settlements around the fringe of the Alps were finally abandoned. From the wettest period some recovery

---

1. There is also archaeological evidence of the adoption of thicker clothing in Greece after 1300 B.C.   (*Mycenaean Seminar*, Inst. Classical Studies, Univ. London, 12 March 1975).

followed and, according to BROOKS, the rainfall by Roman times was 'very little above the present ... in the first century A.D. the Bodensee was near its present level and roads were made across the bogs'.

KLEIN (1914, p. 45) notes that a careful reading of the Bible (e.g. Jeremiah 18:14; Proverbs 25:13) and other ancient Hebrew texts (see Chapter 12, p. 2, in this volume), reporting conditions broadly around 600 B.C., indicates more frequent snow than now and that it lay longer into the summer on the heights of Lebanon, possibly throughout the year.

A similar sequence can be traced in northwest Europe. BROOKS writes (1949, p. 302) of evidence of considerable traffic between Scandinavia and Ireland between 1200 and 1000 B.C., indicating a minimum of storminess. A time of eclipse for the northern peoples followed: 'in about 120 to 114 B.C. a period of especially great storminess in the North Sea (the Cimbrian flood) caused the wanderings of the Cimbri and Teutons' and pushed back the coast of Jutland and northwest Germany to the position in which the Roman geographer STRABO knew it. During the wet period, or periods, in the last millennium before Christ wooden trackways were laid across fens and marshlands in Somerset and elsewhere in England, presumably to keep open familiar routes, and were later abandoned in favour of the use of boats; around 250 B.C. lake villages were built at Glastonbury and Meare, possibly to take advantage for defensive purposes of the wet conditions then prevailing in the Somerset levels. There were shallow lakes, or 'meres', there again in the sixteenth century A.D. which have since disappeared.

Many aspects of the human history of the period of late medieval climatic decline towards increased wetness and the cold regime of the late sixteenth to seventeenth centuries A.D. show analogies with the events of the last millennium before Christ. Then, too, there was a time of eclipse for the northern peoples: the seat of government of Norway passed to Copenhagen, Scotland became increasingly subject to England, the dominance of the Hanseatic league replaced the earlier seafaring supremacy of the northern kingdoms, and the mining operations in the Alps which had reopened for a while in the high Middle Ages were once more abandoned. There were also substantial further losses of land to the North Sea by erosion of the low-lying coasts of the Netherlands, northwest Germany and Jutland (Denmark) with the frequent storms and high sea levels at the end of the medieval warm centuries.

The cultural gap in Greece between 1200 and 750 B.C.

The so-to-speak empty period in the history of Greece and the Aegean between 1200 and 750 B.C., and a somewhat similar blank in that area and, more particularly, in Turkey between A.D. 600 and 800, were picked out by CARPENTER (1966) as cases of some special interest in the impact of climatic events on human history. Both were attributed to drought in the areas concerned; but, as the areas were not identical in the two cases, the patterns which produced them may have been a little different. The earlier case, which coincided with the

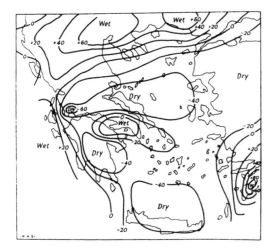

Fig. 16.26   Departures (%) of rainfall from normal in January 1955 over southeast Europe and the Aegean Sea, found to represent the fourth most prominent mode (eigenvector) of precipitation at the present day.

This corresponds well with the prevailing rainfall pattern in the twelfth century B.C. implied by CARPENTER's (1966) analysis.

(*From* BRYSON *et al. 1974. Reproduced by courtesy of* Antiquity *and my co-authors of the original paper.*)

decline of Mycenae, has been the subject of a study (BRYSON *et al.* 1974) that seems to throw some light on the atmospheric circulation behaviour involved.

CARPENTER noted that, unlike the history of aggression by which the Mycenaean power had replaced the Minoan civilization in Crete and the Aegean about 1400 B.C., the collapse of Mycenae around 1200 B.C. was accompanied by no evidence of invasion and about a century went by before the Dorians moved into what seems to have been a depopulated countryside. Indeed, Greece seems to have remained underpopulated until around 850 B.C. The decline of the Hittites about 1200 B.C. has similarly left no evidence of invasion of their homelands in central and southeastern Anatolia, whereas Egypt did suffer raids and threats of invasion at that time, including the arrival of bands of Hittites and other migrants from Syria and probably some Aegean islands as well as from Libya. CARPENTER's analysis suggests that Attica, Rhodes and the southwestern coasts of Turkey escaped depopulation at that time and were areas which probably, like Egypt, served as refuges for the wanderers from the drought-stricken lands in other parts of Greece and Turkey.

The mountainous topography of Greece and Turkey, and their positions in the boundary zone between regions of wet and dry climate, mean that intricate patterns of anomalously wet and dry weather commonly arise in individual months. BRYSON *et al.* (1974) report the results of eigenvector analyses of barometric pressure and wind patterns and rainfall distribution in individual months in the present century. It was found that the fourth eigenvector of the rainfall distributions (fig. 16.26) produced just the distribution of drought and immunity from drought implied by CARPENTER's analysis of the twelfth century B.C. described in the previous paragraph, as well as a very wet regime on the Hungarian plain and in the lower Danube consistent with the picture of central Europe about 1200 B.C. presented on p. 419. A classic case of this pattern was produced by the atmospheric circulation of the 1954–5 winter and particularly the month of January 1955

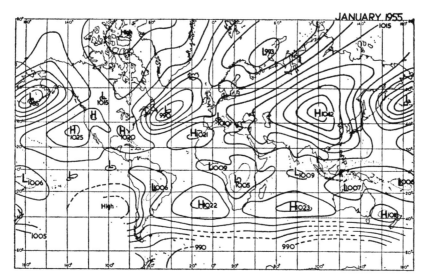

Fig. 16.27   Distribution of atmospheric pressure at sea level (mb) in January 1955, corresponding to the precipitation pattern in fig. 16.26.

(fig. 16.27). These results clearly suggest that circulation patterns resembling that which prevailed in January 1955 may have been frequent at the time of the decline of Mycenae and the Hittite domain around 1200 B.C. and may have continued rather prominent until about 850 B.C.

Chapter 17

Climate in historical times

When we come to civilized times, Man's increasing ability to modify the landscape – at first only by clearing limited areas of forest by fire, but ultimately turning over most of the temperate deciduous forest zone to cultivated grasses (food grains), draining or irrigating huge areas of swamp and near-desert, creating vast lakes by building great dams and regulating to some extent the levels of nearly all great lakes and rivers – means that, one by one, the environmental indicators which mainly registered climatic fluctuations in earlier times cease to do so. For that reason – as well as for more obvious ones – an attempt must be made to use the original reports of the weather prevailing, and related aspects of the landscape and effects on the human economy, which have come down to us from the observers who lived in the times concerned and of which numerous collections now exist (see Chapters 12 and 13 and the corresponding lists of references). Methods of analysis must take account of, and preferably be designed to eliminate the effect of, differences in the frequency of reporting or survival of the reports until our own times. Allowance must be made for the shortness of human memory in records such as 'the highest flood that ever was known', which unless measurements are given can be accepted only as meaning that there was a very high flood. Standards of comparison were simply not available to the reporter, when no long records of past observations were to hand for consultation. It is not surprising therefore that archaeologists, botanists, meteorologists and others, who have used this material, have sometimes been accused of arguing in a circle and suspected of calling in entirely hypothetical climatic changes – as a sort of *deus ex machina* – to explain the facts of history and shifts of vegetation limits which might more properly be ascribed to the works of Man, clearing and burning forests, grazing with goats, building and later neglecting irrigation works, and so on. In recent years, however, new techniques have provided new types of evidence – particularly measurements of the stable isotopes of oxygen, carbon, etc., in ice sheets, tree rings, lake sediments and other materials, and multivariate analysis of the responses to climate registered in such items – which, whatever the difficulties and uncertainties of interpreting such proxy data, are independent of the works of Man and of human bias in reporting. The use of such 'proxy' data offers the hope of verifying, or correcting, the analysis of the climatic sequence derived from original reports, as well as extending it to most parts of the Earth.

Through Classical times to the Middle Ages

Temperature

The probable course of prevailing temperatures in Europe and the Far East over the last 5000 years has been presented in fig. 16.22. In both regions the last few centuries before Christ register some general rise of temperature, representing a recovery from the coldest conditions of the onset of the Sub-Atlantic climatic period, which had culminated in great glacier advances in the Alps (HEUBERGER 1968) at various times between about 900 and 300 B.C. and apparently a lowered snow line in the high mountains in Lebanon and elsewhere in the Near East and in equatorial Africa. A number of severe winters had been reported in ancient Rome in that time (with mentions of the Tiber being frozen in 398, 396, 271 and 177 B.C.), and the Roman agricultural writers SASERNA (father and son) wrote that in the last century B.C. cultivation of the olive and the vine were spreading farther north in Italy where in the previous century winters had been too cold for transplants to survive (WARNER ALLEN 1961).[1] The next several centuries seem to have been an easier time, the reports of occasional 'severe' winters then being associated with flooding rather than ice on the Tiber. Some indication of a rather long immunity from great winters in Europe may be provided by the history of the bridge with many stone piers which APOLLODORUS of Damascus built in A.D. 106 for the Emperor TRAJAN across the Danube at the Iron Gates (Plate VI) into what is now Rumania. This bridge stood for almost 170 years without being carried away by ice and was reputedly destroyed by the Dacian tribes when the Romans withdrew from the area. The glaciers in the Alps seem to have been in retreat from about 300 B.C. to A.D. 400 (DELIBRIAS et al. 1975) and that they were for some time not so extensive as now is indicated by Roman gold mines high up in the Alps in the Sonnblick area (Austria), some of which are probably still under the ice since others have only lately come to light as the ice receded. Traffic over the Alpine passes continued even in winter time. The actual reports of winters in central and

1. The same climatic fluctuation may be the explanation of the fact that olive cultivation in Macedonia seems to have been sporadic and interrupted in Classical times, and again later. Macedonia even today marks the extreme northern limit of cultivation of the olive in the Aegean region (GREIG and TURNER 1974).

northwestern Europe that are recorded from those times indicate a few that were notable for snows in the British Isles (severe frosts being recorded there in A.D. 134, 173, 207, 221 and 231), but occasional great frosts that were prolonged and widespread over Europe seem not to come to prominence until between A.D. 359 and 565 and again between 664 and about A.D. 1000.

Periods of storminess

Another aspect of the centuries of colder climate around 500 B.C. in northwest Europe was evidently their storminess and the frequency of blowing sand near the coasts, the latter presumably owing to the lowered sea level that resulted from the build-up of glaciers. This has been referred to in Chapter 13 (pp. 119, 129); recent, well-dated evidence is reported by BAHNSON (1973) from Fuglsø Mose, a raised peat bog in eastern Jutland, which was found to contain layers of mineral dust attributed to distinct eras of frequent sandstorms, of which the greatest were between broadly about 600 B.C. and A.D. 0–100 and in the later Middle Ages. There was, perhaps, a final climax of the first of these epochs of marked storminess in the great North Sea storm, or storms, about 120–114 B.C. which altered the coasts of Jutland and northwest Germany in a great sea flood, the 'Cymbrian flood', which set off a migration of the Celtic (Cymbrian) and Teutonic peoples who had been living in those areas. Lesser episodes of the same kind appeared to date from about 2200 B.C. and A.D. 500.

Moisture variations

The centuries between about 750 and 500 B.C. and again between 350 and 100 B.C. seem also to have had a notably wet climate in northwest Europe (see references – e.g. pp. 210, 418–19 – to bog growth in Wales and on the Yorkshire lowland near Holderness, the laying of wooden trackways across ground that had become marshy, and ultimately the existence of lake villages on the Somerset levels in southwest England).

As regards moisture pattern, the opening centuries of the Roman era before about 200–100 B.C. may be regarded as continuing the conditions that existed around 500 B.C., accompanying generally cooler climates than now in extratropical latitudes. Fig. 17.1 shows schematically what is known, so far only in broad outline (i.e. without any knowledge of the shorter-term variations), of the prevailing level of Lake Rudolf (*c.* 3°N 36°E) and of the Nile floods in Egypt. In the early part of the times surveyed, through most of the Roman period, the diagram indicates that the summer monsoon rains in Ethiopia around 10°N which feed the seasonal high flood of the Nile were substantial and that the total rainfall near the equator which supplies Lake Rudolf was correspondingly meagre. All through Europe's warm dry time around 300–400 A.D. and the colder, wetter period which followed, and on into the early part of the early medieval warmth, the former conditions in these parts of

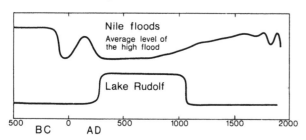

Fig. 17.1   Schematic diagram to show what is known of the general level of the Nile (annual high floods, from Egyptian records) and of Lake Rudolf in equatorial East Africa since 500 B.C.
    (*After data compiled by Ms.* SHARON NICHOLSON, *University of Wisconsin; reproduced by her kind permission.*)

Africa were apparently reversed. They returned to something like their former condition towards the end of the medieval warm epoch and the coldest times of the Little Ice Age centuries which followed (for more detail of the last 1400 years consult the upper curve in fig. 13.36). This history affirms a negative correlation between the long-term rainfall near the equator and 10°N over East Africa, but also indicates that the relationship of each to prevailing temperatures in Europe has varied: over a long period which led up to, lasted through, and followed Europe's rather cold climatic phase (or phases) between A.D. 500 and 900 this relationship was unlike that which characterized the cold climates around 500 B.C. and A.D. 1500–1900.

Fig. 17.2 traces the history of temperature variations in Europe near 50°N and of rainfall variations in Italy over 1300 to 1800 years prior to A.D. 1300, so far as these are indicated by the frequencies of manuscript reports of dry and wet summers, mild and severe winter weather and of great river floods and wet years in Italy.[1] The general impression of a rather colder climatic period in or about the third quarter of the first millennium A.D. receives some further elucidation in that the summer character ratios point to separate climaxes of cool summer climate (*a*) between about 500 and 700, (*b*) in the ninth century, and (*c*) in the eleventh century A.D. The balance of the winters (i.e. the ratio) seems to have been coldest in or around the third to fourth and the eighth centuries, and again between about 850 and 1000.[2]

1. The criteria used in the counts were similar to those which determined LAMB's summer and winter index values from A.D. 800–1100 to the present day (see Chapter 13). Statistically significant correlation coefficients, about +0·5, were found between the summer values used in fig. 17.2 for Britain and Germany, between the winter values for these two regions and between the latter and the frequency of wet years reported in Italy, encouraging belief that these indicators are each a useful approximation to reality.

2. There is some entirely independent confirmation of part of this sequence in a period of glacier advances in the Alps dated between about A.D. 400 and 750 (DELIBRIAS *et al.* 1975, RÖTHLISBERGER 1976, SCHNEEBELI 1976).
    Between 764 and 860 inclusive there seem to have been 6 to 10 very notably severe winters in Europe with the great rivers frozen and in a number of cases thick ice on the Black Sea, the Bosphorus and Dardanelles and the fringe of the Adriatic. In

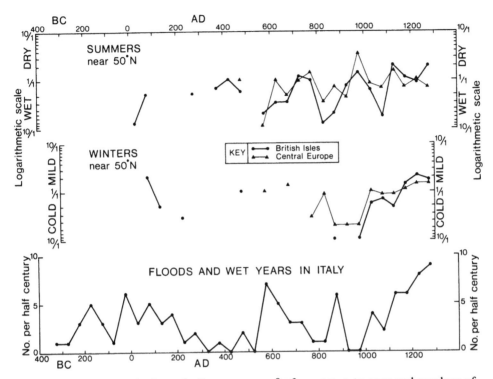

Fig. 17.2   Summers and winters in Europe near 50°N from A.D. 1 to 1300 and numbers of wet years and years with flooding in Italy from 350 B.C.

The character of the summers and winters is indicated by ratios of the numbers of years with reports indicating dry/wet and mild/cold respectively, by half centuries. The ratios are plotted on a logarithmic scale.

FREDSKILD (1967, p. 49) reports botanical evidence of a sharp change in climate in west Greenland near 69°N about A.D. 400, which probably corresponds to the downturn of temperatures and increase of moisture in Europe soon after that time.[1] In west Greenland the change apparently took the form of a turn to colder and drier conditions (more frequent E'ly winds?) than those which had supported a dwarf shrub heath before A.D. 400. Some time later (undated) moister conditions and dwarf shrubs returned again, followed ultimately by a

---

829 there was even ice on the Nile in Egypt. There is a suggestion of more frequent mild winter weather spreading east across Europe at times during the 800s and 900s, though there were a few more severe winters during those centuries and there was once again ice on the Nile in 1011 (data from HENNIG 1904). A diagnosis of the character of the atmospheric circulation change, based on fig. 13.3, suggests increasing wave length in the upper W'lies, the main depression tracks commonly passing well to the north of western and central Europe but frequent N'ly surface wind components in summer and winter

over eastern Europe, associated with a cold trough in the upper W'lies which ultimately moved still farther east and in the eleventh and twelfth centuries became confined to higher latitudes (i.e. became smaller in amplitude).

1. RÖTHLISBERGER (1976) and SCHNEEBELI (1976) have recently reported evidence of a sharp cooling also in the Alps setting in about A.D. 400, followed (possibly quite soon) by great advances of the glaciers, reaching maximum positions in the seventh century quite similar to those about 1820 (see also p. 379).

drier type of vegetation once more. More information and more complete dating of the later phases is promised by the ice cores taken from the inland ice in southern and central Greenland (DANSGAARD *et al.* 1975). A record from northwest Greenland is seen on p. 430.

There are some indications that the dryness that appears so marked in Italy (bottom curve in fig. 17.2) around A.D. 300 to 450 or after, and again around 800 and in the 900s, affected an extensive region of Europe and Asia between about 35° and 55°N. The Caspian Sea was at its lowest levels in the last 2000 years about A.D. 300 and 800, and these two epochs were noted by HUNTINGTON (1907) as times of abandonment of settlements all along the route through Chinese Turkestan that had connected China with the West (CHAPPELL 1971, p. 179). The level of the Caspian Sea was still generally low until the twelfth century (see Chapter 13, pp. 133–4). Droughts come rather notably to prominence after A.D. 350 among the scattered reports available from Britain and central Europe until about 484, immediately succeeded by a great predominance of flood reports in the next century or more, particularly in the 550s and between 575 and 590, which was a very wet period over much of Europe.[1] The most remarkable sequence of droughts in Britain and Germany was reported in the A.D. 900s, particularly in the 920s–40s and 980s–90s, and this seems to be verified in the tree ring series (see Appendix V, Table 22).[2]

The top and bottom curves in fig. 17.2 further indicate interesting differences between the regimes of the various warmer epochs in the centuries covered. The warm-summer periods around A.D. 300–500 and 700–800, and again in the 900s, were accompanied by general dryness and cold winters.[3] The warmth of the high Middle Ages in Europe between

1. A local example of the effect of these changes of prevailing moisture may be suspected in the fate of an early Saxon village in East Anglia, at West Stow near Bury St Edmunds. The settlement was established around A.D. 400 on a little sandy knoll near a river, surrounded by what is today mainly marshy ground, perhaps the only site in the district that was permitted in late Roman times. Evidence that it was then drier may be seen in the fact that in its heyday wheat, as well as barley, oats, rye and flax were grown. But by A.D. 650 the settlement was abandoned, the obvious supposition being that the ground was already becoming marshier.

2. It may also be noticed that the campaigns of Charles the Great between about A.D. 770 and 800, which ultimately reunited western Europe, seem to have been conducted in one of the drier periods, with a tendency to fine summers and cold winters.

CARPENTER (1966, pp. 14 and 78) draws attention to the evidence of widespread drought in southern Italy, Greece and Turkey, and also in Syria, in the seventh and eighth centuries A.D., with reports of migration to the coast and to the islands, leaving a depopulated countryside inland. This was the time of the decay of Ephesus, and depopulation of the area round about it, and of the abandonment of many

early Christian churches in Turkey and Syria together with the settlements in which they had been established. These observations suggest that the dryness of the seventh and eighth centuries was more extreme in the eastern Mediterranean area, and perhaps in North Africa, than it was in Italy (fig. 17.2) or the rest of Europe. If so, the rapid first spread of Islam would seem to have taken place in just those areas where the settled life was at that time most disturbed by drought.

Remains of abandoned irrigation works at a number of places in south Arabia (DAYTON 1975) bear witness to former times when there was more rainfall. Examples are the elaborate Mufija and Marib dams, with sluices for the release of surplus water, in the Najran area ($17\frac{1}{2}$°N 44°E) – site of an early Christian community – where wadis drain from the western mountains towards the desert to the east. The times of abandonment of successive irrigation systems in south Arabia have now been dated to the last centuries B.C., and around A.D. 300 and 600 – all, presumably, times of intensified drought: DAYTON writes of increasing desiccation towards A.D. 600 when there was drought of terrible severity and many Arab tribes emigrated from the area.

3. I am informed (personal communication, 28 April

A.D. 1000 and 1300 (most of all in the late twelfth and thirteenth centuries), and the more moderate warmth around the time of the first century A.D., was accompanied by mild wet winters: both these periods may have been generally moist in the Mediterranean with an incidence of thunderstorm rains in summer quite unlike the present day. Among the indications of this are the diary of CLAUDIUS PTOLEMAEUS of Alexandria about A.D. 120 and the existence of Roman bridges across Arabian rivers and wadis that are now dry (CROWN 1973, p. 329). An even more remarkable case is the five-arched medieval bridge at Palermo, Sicily (Plate V), built in A.D. 1113, to span a river which was then navigable but hardly exists today.[1]

Climatic history of the period from other sources

An independent indicator of the climatic record of the centuries we have been discussing is given by the series of deuterium (^{2}H) measurements on the wood material in dated tree rings of bristlecone pine in the White Mountains, California (near $37\frac{1}{2}°$N 118°W), from about A.D. 200 to the present century, which Dr IRVING FRIEDMAN has kindly made available to me (fig. 17.3). He explains that the variations are thought to correspond in a complex fashion to changes of temperature and air mass trajectories prevailing over the area (higher proportion of deuterium with higher temperatures and/or more Gulf of Mexico air and less air crossing the Rockies and Sierra Nevada – see FRIEDMAN and SMITH 1972). All the main periods of colder climate of which we have found evidence in Europe are shown in this Californian record also, but there seem to be differences in the relative severity of the various cold episodes in Europe and California. The relative magnitudes and times of culmination of the various warm epochs, including both the medieval and most recent ones also differ. One other point of interest is the pronounced cold episode indicated by the isotope record in California about A.D. 1300.[2] There were glacier advances in the Alps during the thirteenth century, but it seems certain that the main features of that time in Europe were a great

---

1975) by A. MAC GREGOR of the York Archaeo-logical Trust that there is evidence of the abundant use of skates (made of bone) in late Saxon and Viking times, evidence possibly tending to corroborate diagnosis of this as a period of rather continental-type climate, frequently anticyclonic over England, so that although the summers were fine enough to encourage vine growing there were commonly fine, frosty spells in winter. A. CARTER of the Norwich Survey has also told me that he regards the location of Pottergate, the first industrial section of Norwich, just *west* of the Saxon city centre, as possibly indicating a notable frequency at that time of light E'ly breezes – situations which would concentrate the pollution, especially in quiet winter weather.

1. Professor K. KREJCI-GRAF of the J. W. Goethe University, Frankfurt am Main has reported to me

(personal communication, 16 January 1964) that remnants of a cedar forest on the island of Flores in the Azores are buried by a thin layer of peat which formed from about A.D. 1000 onwards (from a ^{14}C date). Hence, the same medieval period may have been moist in the Azores also, presumably implying that the anticyclone belt was at that time frequently displaced somewhat to the north in connection with the dry, warm summers about latitude 50–55°N in Europe.

2. A somewhat different emphasis is given by the record of tree ring widths in fig. 13.61 (*b*) which more closely parallels the temperatures derived for Europe. This discrepancy between the two kinds of record from the bristlecone pines in California is not yet fully explained.

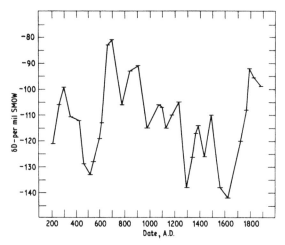

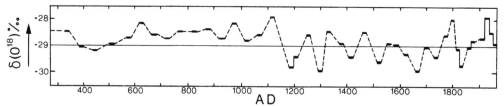

Fig. 17.3   A 'palaeotemperature' record for the White Mountains, California: measurements of the deuterium (a stable isotope of hydrogen (^{2}H)) proportion in the tree-ring dated wood of bristlecone pines by Dr I. FRIEDMAN – but see text p. 429 regarding alternative interpretations.

(*Kindly made available by Dr* FRIEDMAN.)

increase in the incidence of cyclonic activity, bringing severe wind storms and sea floods on the low-lying coasts. A generally high level of the prevailing temperatures in Europe seems well attested up to the 1280s, and probably until about 1310, by which time, however, a sharp increase of raininess was occurring, and it may be that upper air temperatures over Europe were falling more than the temperatures at lower levels.

The oxygen isotope record from north Greenland (fig. 17.4) is somewhat different again and in some respects agrees more nearly with the temperature history derived for Europe. Its most distinctive feature is the early cooling during the A.D. 1100s, which was probably a general feature of the high Arctic and was reflected in a sharp increase of the Arctic sea ice between Iceland and Greenland around A.D. 1200. A period of severe cold in high latitudes again in the late 1200s is indicated, as in California, and the increased temperature gradient between Greenland and the warmer parts of the Atlantic Ocean may well account for the storminess experienced in Europe.

An isotope 'palaeotemperature' record from the southern hemisphere (New Zealand) has been illustrated in fig. 13.19 and seems to agree more closely with Europe. A warm regime with the storm belt of the temperate and subpolar zone of the South Pacific mainly well away to the south would accord well with the increased range of Polynesian voyages in

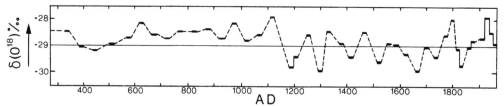

Fig. 17.4   Oxygen isotope variations measured in the ice in northwest Greenland (77°N 56°w) from about A.D. 300 to the present, dated by the ice flow method and believed to represent broadly the course of temperature variations.

(*Kindly made available by Dr* W. DANSGAARD.)

the early Middle Ages and the Maori settlement of New Zealand, which seems to have been broadly accompanied by extraordinarily extensive forest fires (see Chapter 13, pp. 255–6). HOLLOWAY (1954) has reported evidence from the forest composition of a warmer climate in South Island, New Zealand, between about A.D. 700 and 1400 than in the centuries immediately before and after. On the coast of East Antarctica, at Cape Hallett (72°S 170°E), a great modern penguin rookery seems from radiocarbon tests to have been first colonized between about A.D. 400 and 700, presumably during a phase of improving climate and to have been occupied ever since (HARRINGTON and MCKELLAR 1958). Earlier explorers of the Bunger Oasis (67°S 101°E) suggested a period of marked climatic improvement about 1000 years ago, since which there has been only a modest reversion.

BROOKS (1949, pp. 327, 354–6) adduced the high civilization reached by the Khmer empire centred on the lost city of Angkor, about 14°N in the jungle in Cambodia near the border of Siam (Thailand), and also by the Mayan people in what is now an area of dense forest on the Yucatan peninsula of southern Mexico, near 20°N, between A.D. 600 and 1100 or after, as evidence of sharp variations in the moisture of the climates prevailing in those latitudes. That the zone 10° to 20°N is liable to sharp moisture variations has been demonstrated by the events of the 1970s, particularly in Africa but also in other parts of the zone. The great building activity at Angkor, which seems to have reached a high point about A.D. 1000, and by the Mayans in Yucatan at various times – a first high cultural period soon after 100 B.C., their classic period now dated between A.D. 300 and 900, and, to some extent, a further period (characterized by despotic rule) between 1200 and 1450 – assuredly demanded a drier climate than that of recent times. After each period of decline the forest is believed to have advanced. But recently it has been suspected (Dr SHERRET S. CHASE, Botanical Museum, Harvard University, personal communication, 11 December 1974) that the Mayan civilization may at different times have been subjected to both extremes of moisture and, particularly around A.D. 790–810, may have been brought to the point of collapse by drought. R. A. BRYSON (personal communication, 18 December 1974) reports that examination of the classical Mayan culture shows intense preoccupation with rainfall, particularly the ensuring of adequate water supply: he further remarks that recent data show the rainfall in Merida (Yucatan) positively correlated with that in England and Ethiopia (Addis Ababa) and negatively correlated with that of the Egyptian Mediterranean coast (Alexandria).

The most obvious first reading of this suggested history of variations of the balance of rainfall and evaporation at these places in latitudes 14° to 20°N is that in the drier times the equatorial rain system (intertropical convergence zone) either failed to reach the area in the course of its seasonal migration north and south (monsoon failure) or made only a shortened seasonal sojourn so far north of the equator. This diagnosis fits well enough with the history of cold epochs farther north,[1] if dry periods in Yucatan (unlike northern Mexico) accompany

1. At the Galapagos Islands, on the equator in about the same longitude, near 90–95°w, the Little Ice Age period from the late seventeenth to the early nine- teenth century was wetter than the periods before or since, according to work by K. PEJML cited by SANCHEZ and KUTZBACH (1974).

cold periods farther north. The same can be seen to apply to Cambodia, when we remember that China and the Far East – unlike most of the rest of the world – underwent a cold climatic period around A.D. 1000.

A graph of the history of (estimated) population density in lakeside settlements in the driest part of the valley of Mexico (about 19°N) from 1200 B.C. to the arrival of the Spaniards about A.D. 1500, published by BLANTON (1972) also shows some points of correspondence to the supposed temperature and moisture sequence, in that population was particularly low (water shortage?) between 1200 and 700 B.C. and around A.D. 500. The third population minimum, around A.D. 1000–1100, cannot be explained in the same way; the evidence farther north in North America is unmistakably of a rather warm and moist regime with the climatic zones shifted north at that time. It seems logical to suppose that the valley of Mexico may have been exceptionally moist at that time with lakeside flooding.

A potentially valuable biological record bearing upon variations of ocean currents and upwelling off the coast of California about 35°N has been produced by SOUTAR and ISAACS (1969), by counting the fish scales which survive in varved deposits on the ocean bed in areas where the bottom conditions are anaerobic. Results of studying the deposits over the last 1850 years for the three species published appear to show:

(1) A great maximum abundance of the northern anchovy about A.D. 500 and a decline since to one fifth of that abundance, the latter level having been constant since A.D. 1600 or somewhat earlier.

(2) Fluctuations of abundance of Pacific hake and sardines, the former showing peaks about every 400 years, the last peak about A.D. 1750–1800.

Much less is known so far about the southern hemisphere. There is some evidence that the southern temperate and Antarctic zones had entered a colder period by 500–300 B.C. than in the preceding millennia. In southern New Zealand the evidence is consistent with prevailing westerly winds, but in Tierra del Fuego (about 54°S) spread of the forest to cover the whole island on both sides of the watershed at that time suggests quite frequent easterly winds and hence depression tracks commonly in rather lower latitudes than in any other period in postglacial times (AUER 1960, CRANWELL and VON POST 1936).

Sea level changes

Whatever doubts must be attached to attempts to establish the small variations of world sea level that presumably accompanied the climatic fluctuations of the last 2500 years, with their greatest range of world average surface temperature probably about 1–1·5°C (when the warmest and coldest centuries are compared), and the greatest duration of the warmer and colder phases only a few centuries, such evidence as we have points to a somewhat lower general sea level for some centuries after 500 B.C., a higher level during (and for some time after) the warmest periods in late Roman times and the Middle Ages and a lower level again in the seventeenth and nineteenth centuries A.D. (see Chapter 13, pp. 117–19, 257–8, for details).

Evidence of higher general sea level in the Early Middle Ages includes

(1) The watery state of the English Fenland, among the rush-and-willow-girt islands of which the Anglo-Saxon hero Hereward the Wake was able to hold out for ten years (1066–76) against William the Conqueror.

(2) The existence about A.D. 1000 (and reputedly also around A.D. 400) of an East Anglian inlet like a Danish fjord, in which the sea reached inland to Norwich (fig. 17.5).

(3) The ingulfing of a huge area of the Netherlands to form the Zuyder Zee through the agency of the storms in 1250–1 and 1287.

(4) The similar loss to the North Sea of great areas off the Danish and German coast.[1]

(5) The use by ships of sea channels which have since dwindled or entirely disappeared – e.g. Limfjord in northern Jutland as a passage from the North Sea to the Baltic avoiding the Skagerrak; another sheltered channel reputed, according to GEIKIE (1901), to have been used by Danish Viking ships on the south side of the Moray Firth in Scotland, joining Lossiemouth to Burghead Bay;[2] and the channel, known as the Minster Fleet, which separated the island of Thanet from the mainland of England. The history of Sandwich haven in Kent and the Minster Fleet parallels the change of sea level that would be expected to accompany the peak temperature level in the high Middle Ages and the declining temperatures thereafter. From the eleventh to the early sixteenth century the haven was big enough for 400 to 600 ships of the time, except around 1280 to 1320 when the waterway was blocked by walls built to check further inroads of the sea over the flatlands. Vessels sailed through to the Thames, and tolls were exacted. Later on there were difficulties over the shallowing of the water, at first between 1460 and the 1490s, then increasingly from 1537 onwards until the move in 1635 to establish a new haven between Sandwich and The Downs, just off the present east coast of Kent.

(6) From a close study of many sources, including botanical and geological work, BENNEMA (1954) concludes that sea level was high with major transgressions along the whole Belgian, Dutch and German coast around 1800 B.C., about the time of Christ, between A.D. 300 and 400, briefly about 900 and 1200 to 1300 or after, and that about A.D. 600–800 the mean water level was 60–80 cm lower than it is today.

Although all the indications of high sea level in the medieval warm epoch cited come

1. It was reported to the Council of Basel in the 1430s that sixty parishes with agricultural yield accounting for over half the revenues of the diocese of Slesvig (Schleswig) had been 'swallowed up by the salt sea' since the early Middle Ages, probably the worst loss having been on 16 January 1362 (an earlier storm in 1240 had started the series, separating the island of Nordstrand from the mainland of Slesvig for the first time). The sandy islands which were left offshore continued to give way in the greater storms, the most severe of which was that of 11–12 October 1634 all along the coast from Holland to Denmark, when the largest island was reduced to three fragments – Nordstrand, Nordstrandischmoor and Pelworm – and 6000 people drowned (STEENSTRUP 1907).

2. Spynie, 3 km north of Elgin and now 7 km from the sea at Lossiemouth, was a port as late as 1383.

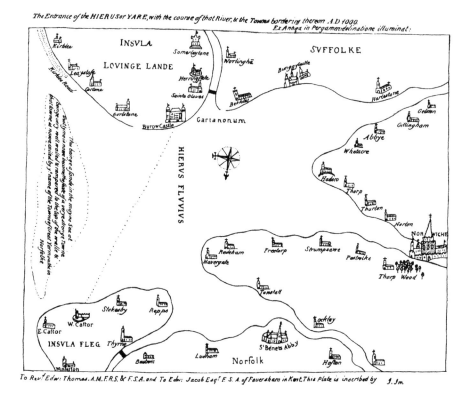

Fig. 17.5   Medieval map of the coast of East Anglia showing a fjord near Norwich: 'the Entrance of the Hierus or Yare with the course of that River and the Towns bordering thereon A.D. 1000'. The map was reprinted from a sixteenth century source in a booklet entitled *Remarks Upon the Garianonum of the Romans: The Site and Remains Fixed and Transcribed* by JOHN IVES, FRS (London, J. ROBINSON, 1803).

The representation of the architecture of the churches which mark the townships on this map seems to be sixteenth century. The map was used by W. CAMDEN (*Britannia* 1586) to show what was known in his time as the course of the coastline about A.D. 1000. This map makes the sea flood in 1087, reported around St Benet's Abbey, now about 20 km up the nearby flat river valley from the sea, understandable. Norwich itself was probably not much affected because the old city stands higher.

from places around the North Sea, which is a region subject to tectonic warping, they are so distributed that regions of progressive land rise as well as land sinking are represented. One must conclude therefore that a general, eustatic change of world sea level (of the order of 1 m) is indicated.

The same impression is given by the statistics of severe sea floods century by century in Table 13.4. Nevertheless, some severe floods and erosion of the coasts continued to occur in the fifteenth to nineteenth centuries, when sea level was presumably falling and ultimately became probably 7–15 cm lower than today.[1] Over the same time there was an increasing frequency, and severity, of storms of blowing sand (presumably sands laid bare and enabled to dry out by the retreating tides) which in some places near the coasts of northwest Europe overwhelmed substantial areas of fertile land between 1300 and 1800–50 (see Chapter 13, pp. 119, 129, 269–70). This combination of events points to an increasing tendency of the main cyclonic storms from the North Atlantic to approach the coast of Europe between latitudes 50° and 60°N as the climate cooled – the increased storminess offsetting any advantage from lower sea level.

The medieval warm epoch or Little Optimum

Evidence already cited at various places in this volume suggests that for a few centuries in the Middle Ages the climate in most parts of the world regained something approaching the warmth of the warmest postglacial times. The climax of this warm epoch was not quite contemporaneous everywhere, and the duration of the fairly stable warm regime seems nowhere to have exceeded 200 to 300 years. In the heartland of North America, as in European Russia and Greenland (see also DANSGAARD *et al.* 1975), the warmest times may be placed between about A.D. 950 and 1200.[2] In most of Europe the warmest period seems to have been between 1150 and about 1300, though with notable warmth also in the later 900s. In New Zealand (fig. 13.19) the peak may have been as late as 1200 to 1400. In southernmost South America the forest was receding rapidly to western aspects only, indicating more effective rain-shadow from the Andes, i.e. more predominance of W'ly winds, than in the previous 1500 years.

　　The conditions of that time are attested not only by the surviving manuscript reports of the weather, from which our analysis of the temperatures prevailing in England was derived, and from the fossil evidence of oxygen isotope measurements, tree rings and so on, but by such items as the following:

Europe south of 55°N

The northern limit of vineyards with a long history of cultivation lay some 300–500 km north of the limit of commercial vineyards in the present century (see Chapter 13, pp. 276–9). In

1. For example, the final loss of the great ports of Dunwich and Ravenser (or Ravenspur) on the east coast of England, in Suffolk and at the mouth of the Humber respectively, between the fourteenth and sixteenth centuries.
2. A year-by-year tree ring record from A.D. 885 to 1462 is preserved in the timbers with which the streets of Novgorod (58°N 31°E) were surfaced. This is available thanks to recent excavations, and is reproduced graphically in THOMPSON (1967). It has been used for dating the building of the medieval city, but seems not yet to have been examined for climatic implications. First impressions are of a rather general decline of ring widths after A.D. 1200.

many parts of England there are the characteristic traces (ridge and furrow) of medieval tillage far above anything attempted in the present century, even in wartime: up to 350 m above sea level on Dartmoor in the southwest and up to nearly 320 m on the fells in Durham and Northumberland in the north (cf. Plate XIV also footnote p. 455.[1] The tree line and the upper limits of various crops on the hills of central Europe were higher than today. Mining operations at high levels in the Alps (e.g. in the Hohe Tauern) which had been long abandoned were reopened, and water supply ducts were built to take water from points which were subsequently overrun by the glaciers and are in some cases still under the ice. The implications of some of these items as regards prevailing temperatures are assessed in Table

Table 17.1   Medieval temperature departures (warmest month of the year) from average values prevailing about 1900, implied by forest, agricultural and viticultural evidence

Displacements observed	Corresponding temperature anomaly
Upper limit of trees (or of particular tree species) on the Vosges, Black Forest and Riesengebirge (Sudetenland) subsequently fell by 100–200 m, mainly between A.D. 1300 and 1500 (FIRBAS and LOSERT 1949).	+0·7 to 1·4°C
Upper tree limit on Monte Rosa and in the Swiss Alps (Valais) fell after 1300 by 70 m (GAMS 1937).	+0·5°C
Upper limit of tillage in northern England (Northumberland) 150 m higher than in modern times.	+1·0°C
Upper limit of cultivation of various fruit and grain crops in the mountain districts of central Europe lowered after 1300 (GAMS 1937).	No figure given
Upper limit of vineyards in Baden, southwest Germany, lowered by 220 m after 1300–1430 (MÜLLER 1953).	+1·5 to 1·6°C
After 1550 retardation by 20 days (to mid October) of the average date of grape picking in the Vivarais district (45·7°N 4·8°E) of France near Lyon (ARAGO 1858) (calculated to give the same number of degree days above the 6°C threshold marking the beginning of the season of growth in early March).	+1·0°C*

* It is likely that the temperatures prevailing in the relatively warm first half of the sixteenth century were not as high as in the early Middle Ages. The situation in Baden is illustrated in fig. 13.71.

1. BERESFORD (1951) was the first to draw attention to this, and to the fact that there were complaints in 1234 that tillage was spreading so far up the hills that too little land was left higher up for grazing animals.

17.1, assuming a lapse rate of July temperature with height of about o·7°c/100 m (see also fig. 6.6, p. 236 in Volume 1, and Table 3 in Appendix V in this volume).

Northern Europe, Iceland and Greenland

The northern limit of grain crops allowed oats and barley to be grown in Iceland, though about A.D. 1200 – at the time of the first increase of the ice in the East Greenland Current – oats were being given up and the amount of barley declined by about half (G. S. BOULTON, unpublished report, March 1975, of pollen analysis and other investigations at a farm in south Iceland). Wheat was grown in the Trondheim district (Trøndelag) and the hardier grains were cultivated even north of Malangen ($69\frac{1}{2}°$N) in north Norway (Hålogaland) from OTTAR's time around A.D. 880 until some time after 1000, but had been given up by the fifteenth and sixteenth centuries (SOLVANG, 1942).

In central Norway the area of settlement, forest clearance and farming, which had been more or less static since early Iron Age times, spread 100–200 m farther up the valleys and hillsides – a height change that suggests a rise of prevailing summer temperatures by about 1°C – in the course of about two centuries, from around A.D. 800 to 1000, only to retreat just as decisively after A.D. 1300 (HOLMSEN 1961). It seems likely that in Scotland similar things were happening.[1] The early Middle Ages were certainly a period of high civilization, with the king (MALCOLM III), from his seat in Dunfermline, offering sanctuary from 1067 onwards to Anglo-Saxon exiles from England who settled in towns and villages all over Scotland (HENDERSON 1879). The next 200 years came to be regarded afterwards as a golden age in that country, with steadily increasing prosperity and, as elsewhere in Europe, the greatest period of building of cathedrals, churches and abbeys. In northernmost England (Redesdale, Northumberland), as also in north Wales (Hiraethog, Merionethshire), it is known that cultivation was extended up to remarkable heights above sea level; a similar extension up the glens in the Scottish Highlands has not been proved, but would go far to explain the unsettled state of that country through the centuries after A.D. 1300 when the climatic tendency everywhere was towards cooling.

1. The forests in Scotland were certainly more extensive around A.D. 1000 than in later centuries, when they were largely destroyed by human activities, though part of the decline (near the exposed coasts) seems to have been climatic (see pp. 466–7); they may indeed have been spreading in the earlier Middle Ages somewhat towards the more exposed places in the north of the country. HARTING (1880) indicates that wild boar continued to inhabit the Caledonian forests in some parts until some time later, as also in England where they were preserved for the sport of hunting in the royal forests. A few bears may have survived until about that time in Scotland. There were certainly still reindeer in the far north, which the Jarls of Orkney in the tenth century used to hunt along with the red deer every summer in the wilds of Caithness; these hunting visits continued (but possibly not the reindeer) until 1159. English travellers in the seventeenth and eighteenth centuries commented frequently on the desolateness of Scotland on account of the scarcity of trees at that time. The extinction of these animals was undoubtedly brought about by the hunting and the reduction of the forests, but the incidence of trouble from marauding wolves seems to have been related to the occurrence of severe winters, a fair number having been killed in various districts in Sutherland and Caithness between 1680 and 1700 and probably the last ones in Monadhliath mountains and the Speyside district in 1743.

The Viking colonies in west and southwest Greenland were able to bury their dead deep in soil that has since been permanently frozen. Corn pollen has been found in the turf at Erik the Red's original settlement, Brattahlid (now Qagsiarssuk), in the sheltered fjord country near Cape Farewell; and it is recorded in the *King's Mirror*, as late as the thirteenth century, that some of the wealthiest men in Greenland tried to grow grain there, though the majority of the inhabitants had never seen bread (corn growing – barley, oats and rye – has been resumed in both Greenland and Iceland in the present century). Study of the middens indicates that abundant cod were caught in the sea off western Greenland by the Old Norse colonists, and we have noted evidence (Chapter 13, pp. 183–4) that the water in the fjords was at least sometimes 4°C, or more, warmer than the present normal. It is not to be supposed that the wide Atlantic was that much warmer than today; the anomaly there was probably similar to that of the air temperatures prevailing in the British Isles and southern Norway, perhaps 1–1·5°C above the values that were normal around 1900. But there was certainly much less extensive ice on the Arctic seas, and ice was rarely seen in the eleventh and twelfth centuries on the sailing route around the Greenland coast south of 65°N. Expeditions are known to have reached Spitsbergen and points in northwest Greenland (Melville Bay as late as 1267) and the Canadian Arctic islands as far north as 79°N (JONES 1968, p. 295).[1]

Dr C. L. VEBÆK of the National Museum, Copenhagen (personal communication 1962), reports evidence from southwest Greenland of the climate becoming drier (more E'ly winds?) and the grass cover becoming thinner, from about A.D. 1300 onwards. Earlier, around A.D. 1000–1100, both archaeological and botanical evidence suggest that the pastures were richer than now (probably more W'ly winds and perhaps more moisture associated with higher sea temperatures).

North America

Archaeological evidence from the western Great Plains, the upper Mississippi valley and the arid southwest of the United States indicates a moister regime – presumably implying fewer W'ly winds – than now over North America east of the Rockies for some centuries prior to A.D. 1200, perhaps starting as early as A.D. 700. Peoples were moving northward and westward into Wisconsin by 700–800 and on into eastern Minnesota, agricultural settlers spreading up the valleys (GRIFFIN 1961, WOODBURY 1961). It was also a warm period generally, from New Mexico to northern Canada, where forest remnants between 25 and 100 km north of the present limit have been radiocarbon dated between about A.D. 880 and 1140 (BRYSON et al. 1965) and far to the north in Keewatin peat growth occurred for a time in or around the eleventh century A.D. (as it had done before 1500 B.C.) but ceased again around A.D. 1050 ± 75 years, presumably with the return of a colder drier climate (NICHOLS

1. If the recently discovered Vinland Map should finally prove authentic, it would appear that Greenland had been circumnavigated in Viking times either by the Old Norse navigators themselves or by Eskimos to the knowledge of the Norse colonists.

1970); but just at the warmest time, in the eleventh century, the southern plains seem to have suffered from drought (SMILEY 1961). As the reverse trend set in, the moist climate gradually retreated southwards and eastwards, and the people went with it. Pollen spectra from the middens of the settlements of the Mill Creek culture on the northern plains in Iowa that were abandoned from 1150–1200 onwards indicate a rapid decline of oak and rise of grass pollens in the twelfth century, and from the bones it is clear that the people went over abruptly to bison hunting and predominance of meat in their diet. About the same time conditions seem to have improved for agriculture – i.e. become moister – farther south in the dry 'Panhandle' region of Texas and Oklahoma (BRYSON *et al.* 1970).

Moisture distribution in Eurasia

There is much archaeological evidence that some centuries (around A.D. 1000–1200 or longer) in the Middle Ages were a moister period than now in the Mediterranean, with bigger and more permanent rivers and more frequent stream flow in the wadis of the African and Arabian deserts (see BUTZER and VITA FINZI references in Chapter 13). Tables 16.1 and 16.2 suggest that, apart from the dryness of the summers during much of the twelfth and thirteenth centuries, there may also have been somewhat increased rainfall in the British Isles and Europe north of the Alps; this could be at least partly attributed to the higher sea surface temperatures in the Atlantic leading to heavier, rather than more prolonged, rains and heavier thunderstorms in summer.

The Caspian Sea, after its low levels around A.D. 300 and 800, continued lower than now until about A.D. 1200 but then rose sharply over the next two to three centuries, being in the 1400s up to 8 m higher than now (see GUMILYOV and CHAPPELL references in Chapter 13). It is clear that a notably moister regime spread across central Asia, but this has to be regarded as part of the climate shift of the late Middle Ages towards the oncoming Little Ice Age. The earlier medieval warm epoch had been a relatively dry time in those regions.

North Africa

That the Middle Ages were a period of renewed moisture in North Africa, including the fringes of the Sahara, is indicated by (among other things) the writings of the great Arab geographers between the eleventh and fourteenth centuries (PEJML 1962). Journeys across the desert from North Africa to Ghana, Mali and Kufra are described. The desert seems to have had its northern limit about 27°N and the crossing of the uninhabited region took about two months, but even in that region there lived in 1352 'a large number of wild cattle which often approaches the caravan so that it can be hunted with bows and dogs'. There is a hint, however, of a drying tendency having set in again in the remark, in that journey account, that rearing of beef cattle had been given up in the Kufra region, where huge herds had formerly pastured in regions that had become desert.

Diagnosis of the atmospheric circulation character and the weather sequence

All these observations of the medieval period could probably be explained by a circulation pattern in which the northern hemisphere subtropical anticyclones were on the whole displaced somewhat to the north during the time of warm climate in Europe, Greenland and North America. During most of that time the middle latitudes westerly winds were presumably weaker and less prevalent in latitudes between 40° and 60°N than now or than they became from A.D. 1200–1300 onwards. While central and northern Europe were enjoying more frequent anticyclonic influence, and rather sluggish W'lies or more variable winds, the Mediterranean zone (including the Azores islands) may well have experienced more frequent cyclonic activity of various types – e.g. slow moving cut-off cyclones and khamsin depressions.

The sequence of prevailing summer and winter weather characteristics in Europe near 50°N from the British Isles to Russia over the twelve centuries since A.D. 800 can be studied in fig. 13.3. The sequence indicated by the summer wetness and winter severity indices (see Chapter 13, pp. 31–2) for the individual decades in England from 1100 to 1959 is illustrated in fig. 17.6. The index values for England, Germany and Russia are given in tables in Appendix V. A simple presentation of times of apparent excess of wet or dry summer months and mild or cold winter months decade by decade across Europe is given in fig. 17.7. In fig. 13.3 the trends of the indices have been smoothed by presenting running five-decade (half-century) mean values. This avoids excessive reliance on the accuracy of the individual decade values and makes it easier to see the general course of events. It stands out that between about A.D. 1000 and 1200, and especially in the twelfth century, all across Europe there was a high frequency of dry summers and of mild or, at any rate, moderate winters. As fig. 17.7 shows, all Europe near 50°N tended to undergo similar variations from decade to decade during that time. Similarly, between about 1550 and 1700 all across Europe there was a high frequency of wet summers and severe winters. From about 1200 onwards wetter summers

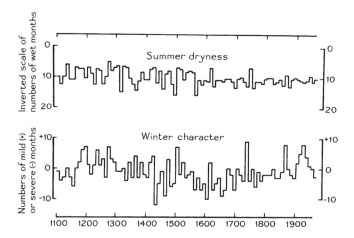

Fig. 17.6   Values of the indices of summer dryness/wetness and winter mildness/severity in England each decade from A.D. 1100 to 1969.

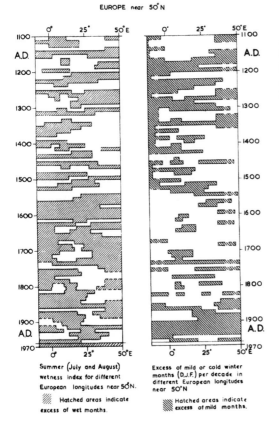

EUROPE near 50°N

Summer (July and August)
wetness index for different
European longitudes near 50°N.

▨ Hatched areas indicate
excess of wet months.

Excess of mild or cold winter
months (D.J.F.) per decade in
different European longitudes
near 50°N.

▧ Hatched areas indicate
excess of mild months.

Fig. 17.7 Excesses decade by decade of wet or dry summer months, mild or cold winter months in different longitudes in Europe near 50°N as shown by the decade values of the indices (Chapter 13, pp. 31–2) from A.D. 1100 to 1969.

and colder winters set in first in eastern Europe and, as the arrows on fig. 13.3 indicate, a region of this kind of anomaly seems to have progressed generally westwards across Europe between that time and about 1500.

Before A.D. 1000 one can trace signs of an eastward progress of similar anomalous weather tendencies, over the previous 200 years. A similar eastward movement is seen again in the period of climatic warming from 1700 to about 1900.

These movements are exactly what one might expect to accompany, during the periods of warming climate (A.D. 800–1000 and 1700–1900), a lengthening of the waves in the upper W'lies downstream from a more or less permanent ('anchored') warm ridge in the neighbourhood of the Rocky Mountains and to accompany a corresponding shortening of the wave length during the period of cooling climate (1200–1550). The experiment was therefore tried of measuring the apparent longitude displacements of the features identified in fig. 13.3. The results are summarized in Table 17.2 together with the apparent implications regarding other shifts in the general atmospheric circulation. It is of some interest to note that the changes of apparent wave length in the upper W'lies are so great as possibly to require not only a change in the prevailing latitudes of the main features of the circulation but some change of strength as well. Since, however, the possible range of long-term changes of energy of the general circulation must be small, and changes of latitude make big differences to the wave length (see fig. 3.5 in Volume 1, p. 89), the differences between the early medieval warm epoch and the colder climate that followed are likely to be largely a matter of latitude shifts.[1]

The data in Table 17.2 have been used to derive the maps of probable positions of features of the general atmospheric circulation prevailing in the eleventh and sixteenth centuries A.D. here presented as figs 17.8(a)–(c) and 17.9(a)–(c). Inspection of these maps suggests that they explain the known features of the climate of the times in Europe, North America and the Arctic satisfactorily but the trough positions indicated on figs. 17.8(a) and 17.9(a) may need adjustment over eastern Asia and the Pacific to accord with the evidence of colder conditions in China and Japan in the eleventh and (still more in the) twelfth centuries.[2]

1. It was found that:
 (1) In the summers of the early medieval warm epoch the longer wave length implied by the estimated upper ridge axis over or near western Europe would require either that the mainstream of the upper W'lies was 30% stronger than now or that the strongest flow was shifted 3–5° north of the modern normal. The likeliest solution seems to be that the main anomaly was a shift of latitude by about that amount. This would imply prevailing depression tracks over the eastern Atlantic about 65–67°N, as in the best European summers of the present century (e.g. 1933–5, 1941–3, 1945, 1947, 1949, 1959).
 (2) In the winters of the Early Middle Ages the evidence suggests a more northern position of the main depression track than now, perhaps allowing the same wave length but with a slightly weaker circulation than nowadays. If we assume that the commonest depression track was over the Barents Sea, owing to the retreat of the ice well to the north, there were probably many periods with just a two-wave pattern in the upper W'lies at the winter climax of circulation vigour. The rather cold winters experienced in Europe near 50°N in the eleventh and twelfth centuries could then be explained by frequent zonally extended anticyclones covering the continent between the Alps and latitude 60–65°N. These tendencies would probably imply a secondary maximum of rain-giving depressions in the Mediterranean, but less extreme cold either there or elsewhere in Europe than in those centuries in which the Arctic sea ice was extended and outbreaks of N'ly winds were common either

over the Norwegian-Greenland Sea or over Russia.
 (3) In the summers of the colder centuries, particularly between 1550 and 1700, it is implied that either the atmospheric flow was weakened by 30% or the mainstream was shifted south by 5° of latitude or more. Again, the likeliest solution (guided in this case by map analysis of the observed conditions around 1800) is a southward shift of the main flow and depression tracks over the eastern Atlantic by at least 4° or 5° and only a trivial, if any, weakening of the jetstream. There must also have been a change of the prevalent wave number: 5–6 waves in the main flow around the hemisphere, compared with 4–5 nowadays.
 (4) In the winters of the colder centuries, either a weakening of the main flow by 30% or a southward shift by 5° or more is implied. The main anomaly was probably a southward displacement: it is possible, indeed probable, that the atmospheric flow was actually stronger than now, though in a southern position, because of the strengthened thermal contrast between the snow and ice surfaces extending into middle latitudes and the warmth of the subtropical Atlantic and the Mediterranean.

2. If, as seems likely, there was in those centuries a prevalent eccentricity of the northern hemisphere circumpolar vortex, with the main flow in markedly lower latitudes over the Far East than over the Atlantic sector, there may have been a shorter wave length over the eastern sector and possibly therefore an extra wave on the east Asian-Pacific side of the hemisphere.

Table 17.2   Key parameters of the atmospheric circulation, North Atlantic sector

	Modern normal 1900–39 (1949–58 500 mb values)	*Extreme years in the early 1940s (partly after Scherhag)*	*Early Middle Ages warm epoch 1000–1200*	*Little Ice Age culminating period 1550–1700*
Summer (July)				
Latitude of strongest 500 mb flow	48°N	50°–52°N		
Departure from modern normal	—	+3°		
Longitude of European trough at 500 mb	10°–20°E	15°–20°E		0°–10°W*
Departure from modern normal	—	Slight +	+10	−15° to 30°
Wave length change implied (° longitude)	—	+6°	(+5° to 10°)	(−10° to 20°)†
Wave length (° longitude)	78°	84°	(83°–88°)	(65°–70°)
Latitude of depression track	62°N	65°N		
Departure from modern normal	—	+3°		
Winter (January)				
Latitude of strongest 500 mb flow	60°N	55°–60°N		
Departure from modern normal	—	−3°		
Longitude of European trough at 500 mb	40°–50°E	25°–33°E		10°–20°E*
Departure from modern normal	—	−15°	Slight	−20° to 30°
Wave length change implied (° longitude)	—	−10° to 15°		(−20°)
Wave length (° longitude)	120°–130°	110°		(100°–110°)
Latitude of depression track	68°N	65°N		
Departure from normal	—	−3°		

* Values read off charts for the period around 1800.
† Values in parentheses are derived estimates; all other values measured off actual charts and diagrams.

CHU KO-CHEN (1961, 1973) gives the details shown in Tables 17.3 and 17.4 for the Far East. Further details are given in Tables 24–28 in Appendix V.

Table 17.3   Average dates of blooming of the cherry trees in the royal gardens at Kyoto, Japan

Century (A.D.)	9	10	11	12	13	14	15	16	17	18	19	20
Date in April	11	12	18	24	15	18	13	18	12	*	12	14

* Records are lacking for the eighteenth century. A fuller version of this table is given in the Appendix.

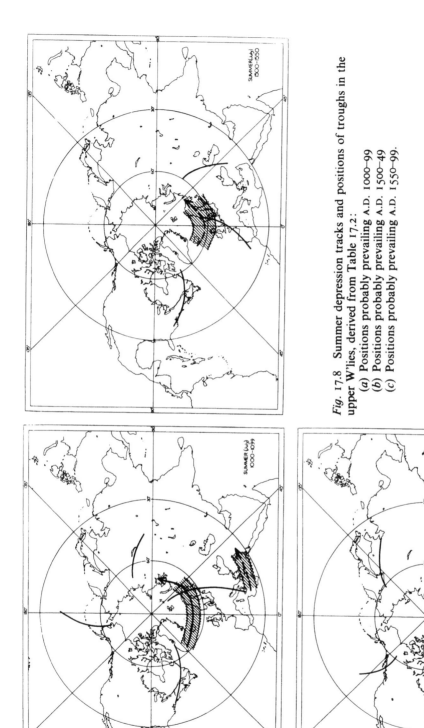

Fig. 17.8 Summer depression tracks and positions of troughs in the upper W'lies, derived from Table 17.2:

(*a*) Positions probably prevailing A.D. 1000–99
(*b*) Positions probably prevailing A.D. 1500–49
(*c*) Positions probably prevailing A.D. 1550–99.

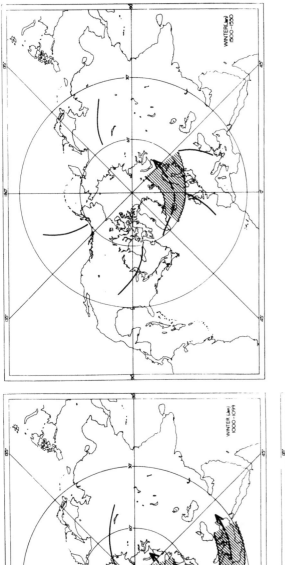

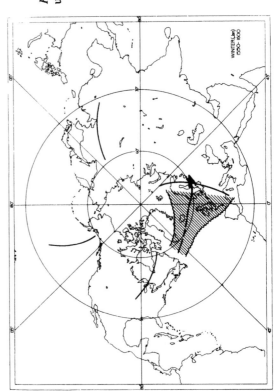

Fig. 17.9 Winter depression tracks and positions of troughs in the upper W'lies derived from Table 17.2:

(a) Positions probably prevailing A.D. 1000–99
(b) Positions probably prevailing A.D. 1500–49
(c) Positions probably prevailing A.D. 1550–99.

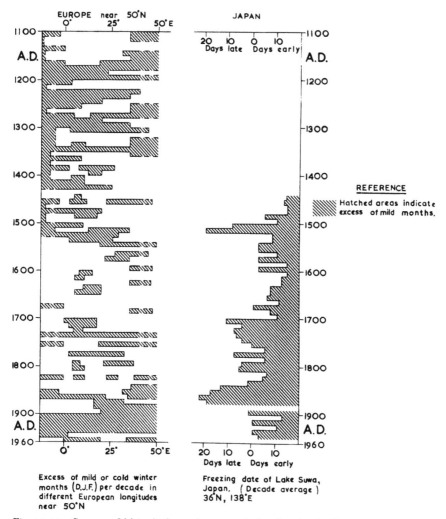

Fig. 17.10 Survey of historical weather records by decades in Europe and Japan.

Left-hand side: Predominance of winter mildness (shaded) or severity (clear) in different European longitudes near 50°N (as in fig. 17.7).

Right-hand side: Variations of average freezing date of a small lake in central Japan, decade by decade, from the overall mean date.

Europe: decades from A.D. 1100.

Japan: decades from A.D. 1440.

It is possible to view the figures in Tables 17.3 and 17.4 as implying an earlier onset in the Far East than in the West of the world climatic cooling which led to the cold centuries of the Little Ice Age; but this view seems inappropriate since we have no evidence of cooling of the

Table 17.4 Number of severe winters recorded in China

Century (A.D.)	6	7	8	9	10	11	12	13	14	15	16	17
Severe winters	19	11	9	19	11	16	24	25	35	10	14	21

Note: Regarding the criteria used for severe winters C H U K O-C H E N writes: 'We take only the extraordinarily severe winters.... They are easily traceable by freezing of rivers and lakes which in ordinary years go unfrozen, and by occurrence of snow and ice in tropical districts near sea level, where snow and ice are rarities anywhere in the world. For lakes we take the three biggest freshwater lakes in China, i.e. Lake Poyang with an area of 5100 km², Tungting with 4300 km² and Taihu with 3200 km².... They are all connected with the Yangtse River, Poyang and Tungting being... around latitude 29°N and Taihu at 31°–31°30′N. For rivers we take the River Huai in Kiangsu Province, at Yuji, and the River Han, the largest tributary of the Yangtse River at Hsiangyang.... For snowfall in tropical regions only district gazetteers in Kwantung Province and Kwangsi Chuang Autonomous Region south of the Tropic of Cancer were consulted....'

Arctic before about the 1190s. Moreover, the seventeenth century seems to have been the coldest century in Japan as in the West and was also among the coldest in China.

Fig. 17.10 traces the prevailing character of the winters, decade by decade, in Europe and Japan from the Middle Ages to recent times. There is clearly no reliable correlation between the prevailing temperatures indicated in Europe and the Far East, although the periods after 1550 till 1700 and from the late 1700s to the 1830s produced the greatest frequency of cold winters in both regions.

The temperatures prevailing in summer and winter and averaged for the whole year in England by half-centuries from A.D. 800 to 1950, derived from the summer and winter indices, have been presented in fig. 6.6 (Volume 1, p. 236; see also fig. 13.4(*b*) and Table 3 in Appendix V in this volume). The 50-year mean temperatures prevailing in England seem to have ranged from about 0·7°C below 1900–50 in the late seventeenth century (0·5° below in summer, 1·1° below in winter) to 0·7–0·8°C above 1900–50 in the twelfth and thirteenth centuries (almost 1°C in summer, but little difference in winter). Rainfall values, estimated from these indices, are displayed in fig. 6.7 (Volume 1, p. 237; see also Tables 16.1, 16.2 and Table 3 in Appendix V in this volume). Indications of the course of prevailing temperatures in other parts of the world from oxygen isotope work, tree ring measurements, and so on, have been given in a number of diagrams elsewhere in this volume and of the course of rainfall variations from lake and river levels, etc.[1]

1. An account, fascinating for its indications of detail of the sequence in the Alps, has been given by DELIBRIAS *et al.* (1975), based on radiocarbon datings and quotations from ancient documents. Particular attention was paid to dating the recumbent trunks and other remnants of trees (larches and Swiss pine), now buried in a moraine, the remains of a wood at 1650–1730 m above sea level pushed over by, and still partly under, the ice of the Unterergletscher at Grindelwald (Switzerland). The dates show that this wood grew mainly between A.D. 800 and 1280 ± 70 years: one tree at the lowest site examined gave a date as early as A.D. 640 ± 100 years. It was, however, always close to the glacier, which was subject then as in other times to fluctuations, and some of these around A.D. 640, 840, 1060 and 1240 (notice the 200-year intervals!) pushed over the nearest trees. After about 1280, however, the glacier advanced beyond the previous bounds of the period discussed and demolished the wood.

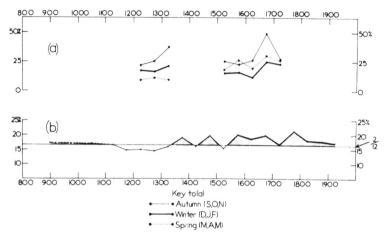

Fig. 17.11   Seasonal rainfall distribution in different half-centuries in England and Wales.

(a) Percentage of months described as wet in the autumns, winters and springs (snowy months have not been included as wet, because the equivalent rainfall is nearly always below normal.)

Data were insufficient before 1200 and between 1350 and 1500.

(b) Percentage of the year's total rainfall falling in high summer (July and August). The black area indicates more than two twelfths of the year's total.

Fig. 17.11 tells us something of the changes in the seasonal distribution of rainfall in England, from examination of the same historical records used in constructing the summer and winter indices. The noteworthy features are:

(1) Marked increases in the wetness of the autumns in two of the half-centuries covered, namely in the first stages of the climatic deterioration around, and especially just after, 1300 and in the coldest half-century of the Little Ice Age period, the late 1600s.

(2) The springs appear to have been more frequently described as wet all through the period 1500–1750 when the Arctic was particularly cold and the ice area extended (this may have been associated with more shower activity in spring because of a greater temperature difference from sea to air in NW'ly and N'ly windstreams over the open Atlantic).

(3) Much less change in the frequency of reporting wet winters than with the other seasons.

(4) A markedly greater frequency of reporting droughts in high summer in the early Middle Ages, and the high summer rainfall calculated from the indices constituting a much smaller fraction of the year's total then, than in any other centuries.

The increased wetness of the summers in England after about 1300 undoubtedly meant a general increase in the wetness of the ground – owing to reduction of the evaporation season – despite our indications (fig. 6.7 and Table 16.2) of a decreasing tendency of the annual total rainfall. A generally increased wetness of the ground after about 1300 is also marked by recurrence surfaces in the peat bogs of Europe. In England, and probably elsewhere in middle Europe, it was probably most marked in the fourteenth century, when the autumns were also often notably wet (see also p. 456).

The shift of climate in the Late Middle Ages

The general turn towards colder climates from A.D. 1200–1400 onwards, accompanied by shifts of the zones of most cyclonic activity as the polar cap and the circumpolar vortex expanded, and which in the seventeenth century seems to have produced a world-wide cold stage, is widely known as the climatic worsening (*Klima-Verschlechterung*) of the Late Middle Ages. Physical scientists tend to avoid such words, which confuse the climatic event with its consequences in the biological realm. Moreover, no climatic shift is likely to be either good or bad for all people, or all purposes, everywhere. But this particular climatic change probably was a change for the worse for most people in most places. The only obvious exceptions are the change towards more moisture for crops and pasture detected in the drier parts of the southern United States and, apparently, in the thirteenth century in central Asia. The increased strength and prevalence of the W'ly winds in middle latitudes, which was presumably responsible for this moisture transport to the innermost regions of Eurasia, had the opposite effect on the Great Plains and prairies of North America in the rain-shadow of the Rocky Mountains, where dryness increased and the agricultural tribes withdrew eastwards and southwards. It is possible also that in southern Europe – e.g. Italy – at some stages in the climatic shift there was some real advantage to human activity from shortening of the hot season. And, doubtless, the climatic shift caused changes in the balance of prosperity and power which advantaged some peoples at the expense of others, e.g. the gain to England and the Hanseatic League of the German and North Sea ports from the declining energy of competitors living farther to the north.

The physical conditions brought by the climatic change have already been indicated in many places in this work, particularly by diagrams showing the sequence of prevailing temperatures, also by figures indicating the increased raininess in England (figs. 6.7, 17.11) and central Europe. We have noted the increased wetness, marked by recurrence surfaces in peat bogs from Ireland to Sweden and northern Russia, and the increased storminess of the North Sea and around the other coasts of the British Isles. Another physical record (from about 1375 onwards) of this climatic change is the composite diagram of oxygen isotope ratio in the wood substance of the tree rings in oaks in two parts of western Germany, shown in fig. 17.12, calibrated in the later centuries against observed January-to-March temperatures in England. There is no mistaking the long-continued downward trend of temperature level from the fourteenth century to about 1700 indicated by these measurements of isotope ratio.

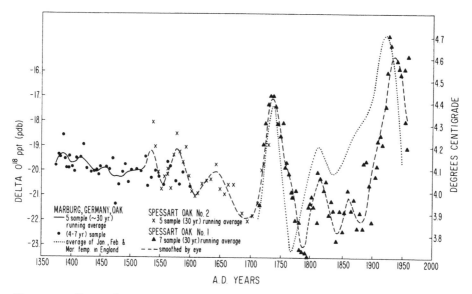

Fig. 17.12   Composite record made up of the oxygen isotope ratio anomaly measured in the (tree-ring-dated) wood of oak trees in two areas of Germany (Marburg: 50°49′N 8°45′E) and Spessart (around 50°N 9°30′E).

Internal consistency/continuity of the record is suggested by the overlap.

Calibration against the last 270 years of winter temperatures (average for January, February and March) observed in England is suggested by the dotted curve.

Note: The very low dot plotted about 1480 would fit much better with other data for the 1430s, and it seems that the overlap with Spessart would fit equally well if there were about a 50-year dating error in the Marburg tree.

(*Diagram kindly provided while measurements were still incomplete by Dr* LEONA M. LIBBY *and reproduced with her permission.*)

In the detail in available weather accounts of the time, the climatic decline is indicated by a fluctuating, but on the whole progressive trend towards wetter summers and more frequent severe winters in temperate Europe. This trend was first apparent in the far north (Iceland, Greenland) and in the east (Russia), though the increase of storminess in and around the North Sea was also marked, in the thirteenth century. It seemed a completely haphazard onset of harsher seasons to the inhabitants of Europe at that time but nonetheless it was clearly a time of storms and disasters more frequent than their fathers had known. After the great winter which gripped at least all central and eastern Europe and the eastern Mediterranean in 1010–11 there seem, despite mention of occasional cold winters in each century (and with possible exceptions in the cases of 1204–5, 1224–5 and 1268–9), to have been few, if any, more winters of comparable magnitude and extent for nearly 300 years. Because of the fluctuations which always occur from year to year, and from one short group of years to the next, superposed on any trend, the changing situation announced itself in Europe in a series of events: the great storms and sea floods in 1250 and 1251, which took a large area of Holland-Friesland and formed the northern half of the Zuyder Zee, preceded and followed by other great losses of land on the coasts of the North Sea, the southern Baltic and the Channel in 1216, 1240, 1287, 1304 and 1362;[1] the prolonged severity of the winter of 1306–7 from Britain and the Baltic to Italy, and more frequent severe spells about that time and later on, particularly in the winters of the early fifteenth century; but perhaps the worst shock of all was the wet cold summers of the decade starting 1310, particularly the unbroken series from 1313 to 1317. In 1315 there seem to have been harvest failures all over Europe, and over the following three years the effects of famine were widespread and dire (see Chapter 12, p. 7). There were, however, periods of recovery, or partial recovery, afterwards – e.g. notable returns to warmer and drier summers in Europe later, in the 1320s–30s and again in the 1380s, and nearly half a century of mostly more genial summers and winters in the early 1500s – which have doubtless made it difficult for historians to discern a trend.[2] Inspection of the summer and winter index values in Europe suggests that these fluctuations may be related to a 50-year periodicity in the incidence of blocking situations.[3]

The course of the climatic deterioration over five centuries from A.D. 1200 can quite well be traced by its effects under the following headings:

(1) Increasing spread of the Arctic sea ice into all the northernmost Atlantic and around Greenland, forcing the abandonment of the old sailing routes to Greenland which had been used from about A.D. 1000 to 1300 (see Chapter 12, p. 6).

(2) Advances of the inland ice and permafrost in Greenland and of the glaciers in Iceland, Norway and the Alps (see Chapter 12, pp. 8–10).

(3) Lowering of the tree line on the heights in central Europe (and in the Rocky Mountains) and more detailed studies of tree ring sequences.

1. There were records of earlier great floods of the North Sea coasts in 1015–17, 1087, 1099, 1113, 1212, 1218–19 and 1236, some of which caused enormous loss of life but which seem not to have resulted in great permanent inroads of the sea. One may perhaps deduce a continuing rise of sea level which was increasing the risk of permanent losses of land, but the greatest storms were probably those mentioned in the body of the paragraph, especially those of 1250, 1251 and 1362.

2. Thus, LADURIE (1971, p. 13) states: 'The receipts in wheat, wine and hay of the Abbey of St Denis near Paris, which represents a wide sampling of estates and vineyards, are as steady as a gyroscope between 1284 and 1342. ... During this time the hypothetical "deterioration of climate", if there was one, certainly had no negative consequence for the rich stable agriculture of the Paris region. It was the entirely non-meteorological series of disasters brought about by the plague, the English and the brigands after 1346–8 which make the hitherto impassive graphs for St Denis go hurtling downward.' (see, however, fig. 17.17 later in this Chapter).

3. A quasi-periodicity of approximately 50 years may be suspected from the dryness of the summers, presumably associated with maxima of anticyclonic influence near 50°N, indicated by our summer index values (cf. fig. 17.7 and Table 4 in Appendix V) for the 1130s, 1180s, 1230s–40s, 1270s–90s, 1320s–30s, 1380s, 1420s, 1470s, 1530s.

Surpluses of cold winter months, presumably due to north European blocking anticyclones, are indicated by the winter index values as culminating in the decades 1210–19 (more weakly), in the 1260s, 1310–19, 1370s, 1400–9, the 1430s and 1490s. M. K. MILES has pointed out (unpublished memorandum 1974) that EASTON's (1928) winter index values from A.D. 1200 also show a 45- to 50-year period.

Both sequences are lost in the climax phases of the Little Ice Age which followed, but may be picked up again in the observed maxima of blocking anticyclones during the 1780s, 1830s, 1880s and 1930s–40s.

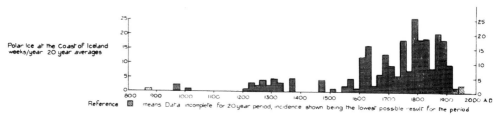

Fig. 17.13   Variations in the incidence of sea ice at the coast of Iceland since A.D. 860.
20-year averages, compiled by KOCH (1945) from Icelandic records and subsequently extended to 1960.

The curve may be used as an index of prevailing temperatures in Iceland and southeast Greenland.

(4) Increasing wetness of the ground and spread of lakes and marshes in many places in northern, western and central Europe (and, it is thought, all over northern Russia and Siberia), attended by swollen rivers and an increasing frequency of landslides and similar difficulties (see, for example, Chapter 13, pp. 132–3, 211, 272).

(5) Increasing frequency of the freezing of rivers and lakes (see Appendix V, Tables 6 and 18).

(6) Evidence of increasing severity of the windstorms and resulting sea floods and disasters by shifting sand in latitudes 50–60°N, particularly in the thirteenth century, and between about 1400 and 1450 and about 1530–1700 (the great North Sea storms of 1530, 1570, 1634 and 1694, the Hebridean storm of 1697 (see Chapter 13, p. 269) and the storm described by DANIEL DEFOE which passed across southern England in December 1703 seem to have been of a severity unmatched in the records from other times except perhaps by those mentioned above between 1212 and 1362).

(7) In the records of harvest failures, rising prices of wheat and bread, famines, tithes and tax remissions (sometimes with reasons given).

(8) In the records and archaeology of abandoning of crop growing, tillage and vineyards, abandoned farms and villages.

(9) In the incidence of disease and death among the human and animal populations, population changes and so on.

Some of these items will be enlarged on further in the following paragraphs. While the more general accounts of the European economic crisis of the late Middle Ages, and of the social disturbances, wars and raids, do not show where or how these climatic events had anything to do with it, in some cases it can be clearly established that they were the trigger.[1]

---

1. Perhaps the most obvious and specific examples are the loss of the Old Norse colony in Greenland, the decline of population in Iceland and upland Norway, and the migrations to England (East Anglia), Scotland and Wales of people made homeless by the coastal losses in Flanders and the Netherlands. The migration of the agricultural peoples in central North America (pp. 439, 449) supplies another example.

The history of temperature changes in Iceland and southeast Greenland may be deduced from the records of the ice brought by the East Greenland current to the coasts of Iceland (fig. 17.13).[1] It shows that the warmest times there were in the eleventh and twelfth centuries, together with an earlier period between 880 and 960 and some periods of recovery between 1340 and 1460. The increase of the sea ice about Greenland has already been partly documented in Chapter 12. Ice was increasingly encountered on the old sailing route from 1203 onwards, and by 1342 it had to be abandoned for one farther south. After 1410 there was no regular communication between Europe and any part of Greenland until the 1720s; the Old Norse colony died out, and no European ship ever got in through the ice even on the southwest coast after 1605. No one seems to have reached anywhere on the east coast of Greenland from the sea between 1476 and 1822; and after 1850 another deterioration severed communication and delayed the founding of any permanent European settlement – the first one was at Angmagssalik – until 1894. The graves of the Viking settlement at Herjolfsnes in southwest Greenland (NØRLUND 1924), including those of the fourteenth and fifteenth centuries, were dug deep into ground that was subsequently so predominantly frozen that only the shallower parts showed decay when unearthed in 1921. The clothes showed contact with European fashions as late as 1450–75, though burials in the more northern settlement near Godthåb (64°N), found a century later by JOHN DAVIS, showed no European influence except the Christian cross (BULL 1925, STEFANSSON 1943).

The changes of the climatic scene in Iceland have been noted in Chapters 12 and 13 and earlier in this chapter in terms of the glaciers and sea ice, the abandonment of corn growing and the numbers and stature of the population. Plate XV illustrates the growth of a glacier since the early Middle Ages. Excavations at the farmhouse site at Kvisker in southeast Iceland, about 20 km east of the farm in the picture and close to another tongue of the Vatnajökull ice sheet, and about 7 km south of the farms that were completely overrun by the ice about 1700 (Chapter 12, p. 10), have shown that the biggest of the successive farmhouses built on the same site was the first one, which stood there until A.D. 1090 – when there was a volcanic disaster – and the midden shows that its occupants enjoyed the richest food (including, presumably imported, oysters). When grain growing was given up in the thirteenth and fourteenth centuries, there was a shift of Iceland's population to the coasts and export of cod and whale- and fish-oil increased to pay for imports of grain. After somewhat easier times in the fifteenth and early sixteenth centuries, the sea ice situation near Iceland grew rapidly worse between 1575 and 1600. In 1695, the worst year, the drift ice surrounded the whole country except Snæfellsnaes on the west coast for many weeks. At most parts of the coast open water could not be seen from the highest mountains, and merchant vessels could not make their way to the harbours (KOCH 1945).

1. A proposed temperature curve for Iceland from A.D. 950 to 1950 derived by BERGTHORSSON (1962) in this way, seems unfortunately, but quite clearly, distorted by its author's overcorrection for supposed incompleteness of the records in the earlier centuries; as a result, the period 1920–60 appears much warmer than the early Middle Ages, a finding in conflict with most other evidence.

Myr ved Skage kirke, Overhalla v. B.Vorren

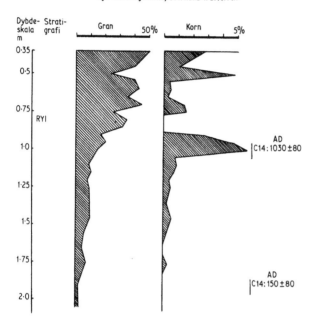

Fig. 17.14 Pollen diagram showing changes in the percentages of spruce (*gran*) and corn pollen over the last 2000 years indicated by the stratigraphy of a bog beside Skage church ($64\frac{1}{2}°$N 12°E) in Namdalen, central Norway.

(*Diagram constructed by* BRYNHILD VORREN *and reproduced by her kind permission.*)

For most of Europe the worst climatic shocks seem to have been those that came in the years 1310–19, in and about the 1430s and 1450s, in the 1560s and on to the end of that century, and in and around the 1690s. Much of the dislocation of society, decline of population and abandonment of farmland in those times has in the past been attributed to the Black Death of 1348–50 and subsequent outbreaks of plague. The troubles had, in fact, begun earlier. STEENSBERG (1951) reported records going back to 1334 of farms deserted, a phenomenon which became general in Denmark over the next 300–400 years. English visitors to a royal wedding in that country in 1406 reported much sodden uncultivated ground and that wheat was grown nowhere. Even where farmland was not totally abandoned, economic stringency was forcing the abandonment of the farmhouses and different families to share the same buildings (CHRISTENSEN 1938). All these changes were more general in Jutland than farther east in the Danish islands (GISSEL 1972). The phenomenon is now the subject of a continuing inter-Nordic historians' research project, the *Ødegårdsprojekt* (GISSEL 1972), and among its findings (GISSEL personal communication 28 March 1973) is that the beginnings can in very many cases be placed in the years of harvest failure around 1315.[1]

The regions in which villages and farmland were being deserted in the fourteenth and fifteenth centuries, and later, included England, Scotland and Germany, parts of the Low

1. E.g. 15 derelict churches reported in the diocese of Ribe, Jutland, around 1320. The islands farther east, notably Zealand, were much less severely affected than Jutland.

Countries, France and Hungary (FLOHN 1950), besides all the Scandinavian countries.[1] The sharp decline of corn growing just north of Trondheim, in Namdalen, in central Norway about A.D. 1250 seems to be registered in the pollen diagram in fig. 17.14 (rye is known to have been completely given up). There was a partial resumption of corn growing 200 years later, but spruce forest was increasingly taking over the area and some of the farms in the neighbourhood are known to have remained unoccupied (SANDNES 1971).

Examination of the history of deserted villages and abandoned cultivation in England (ALLISON, BERESFORD and HURST 1965, 1966) and Scotland (PARRY 1975) can carry the matter further and again demonstrates that the process had its origins long before the Black Death;[2] it even seems to have controlled the pattern of the plague's effect in devastating the countryside. Out of nearly fifty deserted village sites in Oxfordshire and thirty-four in Northamptonshire for which adequate population figures of taxpaying tenants at various dates have been published, only about 10 per cent were attributable to the Black Death. All appear to have suffered serious decline in the years of disastrous summers and famine between 1311 and 1319; at all events, the population figures in 1327 were down by a third on those for about 1280–1300. The decline by 1327 averages 67 per cent – that is, it was twice as great – in the case of those places that were later wiped out by the Black Death in 1350.[3] In

1. I am informed by Miss THELMA JEXLEV of Rigsarkivet, Copenhagen, that the northern parts of Sweden and Finland seem to have been immune, perhaps because they were more thinly settled and only the best ground had been taken up.
2. A close study by PARRY (1975) of the history of cultivation of oats and its abandonment on the heights of the Lammermuir Hills in southeast Scotland throws valuable light on the association of the phenomenon with climatic changes (cf. also p. 436).
    From a dense coverage of temperature data for the area in the late nineteenth century and other information available for that period it could be established that the upper limit of cultivation around 1860 coincided with an average summer accumulation of 1150°C-days above 4·4°C, a potential water surplus at the end of August of 30 mm (defined as the excess over the arly summer deficit) and a mean wind speed of 4·4 m/sec. Temperatures observed in years of harvest failure in 1782, 1799 and 1816 suggested that the accumulated warmth in those summers only reached 975°C-days at the 1860 cultivation limit. From these considerations and by comparison with other studies in Finland it was concluded that, in conditions of basic subsistence cultivation, oats might be grown to normal limits of 1050°C-days, 60 mm potential water surplus at the end of the summer and 6·2 m/sec mean wind speed, and that at this point very late or failed harvests would occur about 2 years in 7.
    The opening up of land for farming on the

heights of the Lammermuir Hills between A.D. 1150 and 1250 is documented in the records of the Scottish border abbeys, and 4890 hectares of subsequently abandoned cultivation still showing ridge and furrow can be identified by aerial survey. Its upper limit was near 400 m above sea level.
    Estimates of summer warmth and wetness based on our temperature and rainfall variations in England since the early Middle Ages indicate that the accumulated degree days in the summers around A.D. 1150–1250 exceeded 1050 at 400 m above sea level in the Lammermuir Hills and exceeded 1200 at 300 m above sea level where the frequency of failures of the oat crop would have been reduced to rather less than 1 year in 20. By A.D. 1350 the 50-year average frequency would have increased to 1 year in 5, by 1450 to 1 year in 3, and by 1700 to more than 1 year in 2. Moreover, available records of the dates at which farms were abandoned over these centuries show remarkable agreement with the estimated positions of the 1050°C-day limit and potential water surplus limit at the times concerned. Farms at even lower elevations on northeast aspects of the hill country were abandoned during the sixteenth century, presumably because of the increased frequency of N E'ly winds.
3. Of over 100 villages in Norfolk that disappeared after the Domesday Book survey of 1086, 34 had gone by 1316 and probably over half before the Black Death. Not more than one or two actually died out in the years just after 1348, so the Black Death played only a minor part in the depopulation of this

other words, there is evidence that they were already particularly bad places to live in, had been largely knocked out and were probably undernourished. The population figures for the other places concerned were actually higher in 1377 than before the Black Death, and in most cases the final decline came between about 1440 and 1480. JOHN ROUS of Warwick, who was born about 1430, wrote a history of England in 1485 in which he listed 58 villages in the Midlands that had tumbled down in his lifetime. The coincidence with the likely effects of the 1430s and 1450s in which there were four or five rainy summers per decade is suggestive. We should note at this point that it is even likely that the origin of the Black Death itself in China in 1333 was connected with the disturbed weather of the times, since it followed upon a time of extraordinarily great rains and river floods in 1332, which not only are alleged to have cost 7 million lives and destroyed human settlements and their sanitation arrangements but must have dislocated the habitats of wild life generally, including the plague-carrying rodents.[1]

Another aspect of the climatic change in England also seems to have been an increase in wetness (see also pp. 448–9), particularly in the early and mid fourteenth century and in some later decades – e.g. the 1450s, 1480s and 1490s. TREVELYAN (1928, p. 263) believes that the rivers of England were generally deeper and more navigable than now in the Middle Ages, but mentions it especially at the time of the Wars of the Roses (1455 to 1485). And in Cambridgeshire it is specifically mentioned in documents of the time that the cause of abandonment of some lands in the 1340s and 1350s was flooding.[2]

As in England and Denmark, so in upland Norway, according to Professor ANDREAS HOLMSEN (1961 and personal communication 28 September 1964), it seems that the climatic deterioration which was causing desertion of farmland and settlements at least as early as

---

part of England. But the countryside in the Breckland area around Brandon never recovered (ALLISON 1955, ZIEGLER 1969). During the tenth and eleventh centuries A.D., in the Danelaw period, and probably somewhat after, as also long before in Neolithic times, the Breckland seems to have been ploughland, much favoured because its light soil was easily tilled, and rich enough to support palaces and great religious houses in and around Thetford and Kenninghall. After the depopulation of the area in the later Middle Ages, the area was used for sheep, providing wool for the thriving East Anglian woollen industry of those times. But whether because of the grazing on the light soils, or perhaps an increase of windiness and evaporation contributed to drying this area more than before, the vegetation cover declined until the Breckland became known as near-desert and the diarist JOHN EVELYN described it in 1677 as an area that 'travelling sands have so damaged ... rouling from place to place ... like the sands of Libya ... quite overwhelmed some gentlemen's whole estates'. It was this situation that led to the first planting of pine trees as shelter belts in the following century.

I am indebted to Professor GORDON MANLEY (personal communication 18 January 1976) for the following quotation from an account by THOMAS WRIGHT, published in 1668 in one of the first volumes of the *Philosophical Transactions of the Royal Society*, London (No. 37, p. 722) of 'a sand flood at Downham in Suffolk': the origin of the sand is described as a warren in Lakenheath, 5 miles southwest by west of Downham, and WRIGHT added: 'The sands first broke loose 100 years since' and 'first reached Downham [actually Santon Downham] in Suffolk about 30–40 years ago'.

1. I am indebted to Dr M. BLOCH for drawing my attention to this information from the work of Dr J. NEEDHAM, F.R.S.
2. J. G. HURST (personal communication, 21 September 1966) suggests that the great activity in constructing moats around older manor houses in England in the fourteenth century may have been linked with generally increased wetness of the ground. The same factor may underlie an epidemic of collapses of great buildings – church towers, etc.

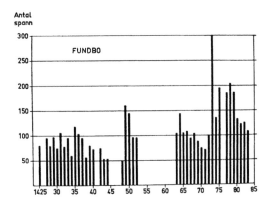

Fig. 17.15   Record of the receipts from corn tithes in the fifteenth century in the parish of Fundbo (near 59½°N 18°E), east of Uppsala, Sweden.

Note the difference in the yields registered by the tithes in the years about 1430–45 and in the years 1473–83.

(*From* Kultur-historisk leksikon for Nordisk Middelalder, **8**, *entry on Climate, 1963.*)

1320 may be the reason why areas depopulated by the Black Death remained empty for 200 years. In Sweden there seems to have been no abandonment of corn growing, and presumably the climate never became persistently unfavourable for it in the established agricultural areas; but in the fifteenth century there was an increased year-to-year variability and there were some runs of very bad years, particularly in the 1440s and 1450s. This period is registered by the records of corn tithe receipts in a parish in Uppland (fig. 17.15). In the later fifteenth century and on into the following century there was clearly an improvement. In north Sweden (Norrland) and in Finland there seems to have been a great expansion of the population, perhaps mainly in the sixteenth century (LARSSON 1972).

The 1430s seem to have brought a possibly unparalleled run of severe winters, or winters which included severe spells, in most parts of Europe (see tabulations in Appendix V, Tables 4, 5, 6 and 7). The British compiler (C. E. BRITTON 1937, p. 161) assumed that most of the reports from various parts of the British Isles must be erroneously dated and should refer to the same year. However, putting all the information on maps makes it clear that in Europe generally eight of the winters of that decade, and in England about seven, were either severe all through or included at least some prolonged frost.[1] Fig. 13.2 is a map of the information available for 1431–2. The difficulties of a Venetian ship, bound for Bruges (Brugge), but blown far out into the Atlantic by persistent E'ly winds to a point from where the survivors in an open boat were carried to north Norway, nicely complete the wind pattern around a great north European anticyclone, typical of prolonged severe weather. There were also many disturbed summers in western and central Europe in that century, particularly in 1400–9 and the 1430s and 1450s.

The period between about 1400 and 1480 stands out as the second great period of widespread *Wüstungen* (abandonment of cultivation and villages) in Europe, apparently exceeding the scope of those that took place between 1315 and 1360.[2] This time the weather

---

1. In the severe winters in the 1430s were some of the last occasions when wolves were troublesome in England.
2. FLOHN (1950) reports that between 20% and 60%

of all the previous villages in the various parts of western, central and northern Europe were abandoned in the Late Middle Ages, the climax of this phenomenon being in the time around 1450.

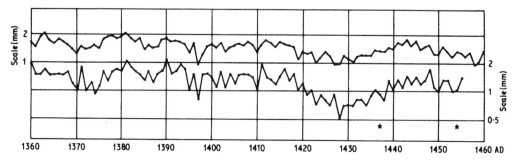

Fig. 17.16   A ring width series (semi-logarithmic scale) from eighteen oaks in southern England (upper curve) from about A.D. 1360 to 1460 and a series (lower curve) measured from an oak panel used for a portrait (of Elizabeth Woodville) in the Royal collection at Windsor Castle.

The stars mark two years with abnormally small vessels in the early wood.

These curves were previously dated four years earlier and there may still be some uncertainty within this range.

(*From* FLETCHER (*1975*); *reproduced by kind permission*.)

pattern seems to have taken the form of frequent severe spells of winter weather and great year-to-year variability of the summers and harvests; these and more directly reported aspects of the pattern suggest (unlike the 1310–19 period) an abnormally great frequency of blocking (with anticyclones over northern Europe in latitude 60–70°N) and meridional circulation, particularly in and around the 1430s and 1440s. There were also, as is well known, recurrences of the plague, though not on the scale of 1348–50, and how these were related to the weather is not clear.[1] The evidence of a considerable recovery in Iceland and, to a much less extent, in Greenland about this time may point to an enhanced frequency there too of anticyclones but also of S'ly winds over the eastern Atlantic sector.

The record of tree ring measurements from southern English oaks (upper curve, fig. 17.16) confirms the stressful nature of the years between about 1419 and 1459 – probably a combined effect of cold winters and springs and some dry summers – and the individual record (lower curve, fig. 17.16) from an oak panel used for a portrait is typical of a number of such records which showed an extreme effect in several of those years (the two years starred in this example) with much smaller vessel structure than normal in the earlywood formed during the period of most rapid growth in spring and early summer – a rare effect observed

1. Many of the seasons of this period in Britain and neighbouring countries were severe in a variety of ways and presumably taxing for health and sanitation arrangements. Thus, the outbreaks of plague in 1439 were preceded not only by a run of cold, and often severe, winters but by springs that were either wet or cold or both; the summers of 1433, 1434 and perhaps also 1436 were warm or very warm and 1435 was dry but cold, whereas the summers of 1432, 1437, 1438 and 1439 produced great rains, flooding and harvest failures, and most of the autumns of these years seem to have been wet. Other severe winters occurred in or about 1402–3 and (an outstanding case) 1406–7, whereas several in the next 20 years were very mild; and there were other great flood years in England in 1406, 1412 and 1413, and a famine in 1425.

with some frequency only in the rings dated 1348, 1390, 1437–9, 1444–5 and 1454–5 according to the dating of this series (FLETCHER 1975).

Further confirmation is provided by the great European famine of the years 1433–8 (ABEL 1943), preceded by the long cold winters and sometimes wet, cold springs of that decade, some of the summers being very hot and dry. W. G. HOSKINS has remarked that the years 1437–9 produced a famine in England second only to 1315–17. It is surely not fanciful to see some connection between these events and the social unrest of that century, as argued by FLOHN (1950). In England the Wars of the Roses, which went on for 30 years from 1455 to 1485, are usually alleged to have been largely a dynastic squabble which scarcely touched the common people, but TREVELYAN (1928, p. 257) mentions the probable effect of starvation on soldiers newly returned from France and the Hundred Years War, which ended in 1453, and used to ways of violence. BERESFORD's account (1954, pp. 158–67) indicates that the main period of depopulation of villages in England was probably between 1433 or soon after and 1485. In Scotland the fifteenth century seems to have seen the peak of the Highland troubles. BOECE (1536) mentions that in 1396 all the north of the country was heavily troubled by a clan war; there was great trouble again in 1411; and in 1433 the great central area of the Highlands between Dee and Spey which had been under the civilizing tutelage of the Earl of Mar (importer of good Hungarian horses) fell into poverty soon after the death of the earl and collapse of the Mar estate. The turmoil of the Highlands in 1436 led to the murder of the king, while on a hunting trip near Perth, and to the decision that in future the Scottish king must make Edinburgh castle their seat, because their safety could not be guaranteed at any of their previous capitals, in Dunfermline, Perth, Scone or Stirling (HENDERSON 1879).

In Denmark and Skåne difficulties associated with deepening crisis in agriculture seem to have increased during the fifteenth century. Depopulation of the countryside was accompanied by a drift to the towns, but by 1500 the towns too were affected (CHRISTENSEN 1938) – evidence, presumably, of a wider-ranging emigration. Farther south on the continent, Bohemia[1] was in turmoil in the 1420s and 1430s at a time when Russia is known to have suffered from famines which began with severe winters and summer droughts.[2] In Germany the fifteenth century growth of civic pride and mercantile splendour in the cities owed something to the community feeling engendered by defence of the towns and their

1. It is far from being the writer's meaning to ignore the religious and political ideas of independence and democracy about which the struggle of JAN HUS in Bohemia and other risings alluded to in this chapter were fought. But it is informative to note the coincidences in time with stresses produced by adverse weather and harvests that were disappointing or worse. These must surely always produce a disposition to unrest and are liable therefore to increase the readiness of many to join in conflict.

2. 1421–4 are all mentioned as famine years in Russia (BUCHINSKY 1957), so severe in 1423 that people emigrated to the west, and in 1424 another drought and epidemic of plague were added to the troubles. Again in 1436–7 frost killed the young crops and many emigrated from Russia to Germany because of the severity of the famine. In 1438 the famine about Smolensk was so severe that people 'died in the forests and along the roads ... wolves ate people and people ate people and small children'. Another great famine occurred in eastern Europe (Poland, Lithuania and Russia) in the next century beginning in 1568–9 and lasting until 1575.

people against a lawless countryside (FISHER 1936). From 1470 onwards was a time of more organized peasant revolts which culminated in the next century in the widespread rebellion in 1525.

Other confirmation of the trend of the climate and its effects may be seen in the records which show the decline of the English medieval vineyards (LAMB 1967). There had been not less than 50 vineyards in England south of about 53°N in the High Middle Ages (see map, fig. 13.74), of which a dozen or more sites were kept going for at least 100 years. They were mostly open vineyards, unprotected by walls, and some of them – e.g. in the Fenland of eastern England and in a nearly flat shallow valley at Tewkesbury – would be very liable to frosts in May in more recent times. The age of establishing new vineyards seems to have ended between 1270 and 1310, and the number then seems to have fallen greatly. It is recorded that the one at Abingdon (by the Thames) was doing quite well around 1380 (it was abandoned in 1412) and that there were yearly plantings in the vineyard at Honour Court, Windsor Castle, between 1377 and 1399. Two hundred years later all this was a thing of the past, and there was speculation as to why it had ceased.[1] At Ely the effort was continued to the bitter end, at least until 1469, when the accounts of year after year with nothing but sour, unripe juice ('verjuice') to sell suggest that little wine had been produced since a good year in 1319.[2] The annual values of wine sold before 1310–20 were sometimes 10 times as great as any mentioned after that.

Meanwhile on the coast, with an apparent increase since the twelfth century in the frequency of cyclonic storms in the latitude of the British Isles and the North Sea, and perhaps the beginning of a tendency for sea level to fall, there were important movements of wind-blown sand and deeply dug storm shingle. Morfa Harlech is a case in point where the fourteenth century port of Harlech on the west coast of Wales was closed by sand dunes, probably during the fifteenth century, when it ceased to be mentioned as a port. Other similar movements occurred in South Wales, in Kenfig sands some time after 1400, at Forvie on the east coast of Aberdeenshire in 1413, at the Culbin Sands south of Moray Firth culminating in 1694 (though with further episodes at Culbin frequent until 1704 and a previous occasion of severe sand blowing about 21 April 1663 (New Style); I am indebted to Mr S. M. Ross of Forres for these details), and in several stages dated between 1287 and about 1800 at Winchelsea and Dungeness on the Channel. Another such great storm one February in the seventeenth century permanently destroyed the sheltered natural harbour at Saksun in the

1. There has probably been at least one vineyard somewhere in the southern half of Britain at some time in each century since, but they have always been the pursuit of some enthusiast and have always been discontinued when the individual concerned died or retired from the scene. Like other families in later centuries, the Cecils established a vineyard at Hatfield House in 1610; a sheltered south-facing bank was chosen and high red brick walls were built all round it. Nevertheless, there is no record of any success with it, and the late Lord Salisbury described it (personal communication, 5 August 1959) as 'a rich man's fantaisie'. When Pepys visited it 50 years later, in July 1661, his only comments were on the coldness of the weather and the size of the gooseberries.

2. In southwest Germany (Baden) also this was a period of very poor vintages – there was no good wine year between 1451 and 1464.

northwest of the Faeroe Islands; great seas pouring in through the narrow entrance channel filled the mountain-girt haven with sand.

There was a temporary return to a warmer climate in Europe between about 1500 and 1540. This is clear alike from the direct reports of the weather at the time (figs. 13.3, 17.6, 17.7, 17.10) and from the records of harvests and vineyards, etc. The 1520s in England produced five bountiful harvests in a row, and there were six more from 1537 to 1542. The first such run was between 1492 and 1495, and there was a recurrence, at a slightly lower level, from 1566 to 1571 (see HOSKINS 1964, fig. 2) (in central Europe the German 1000-year tree ring series in oak and fir show an unequalled sequence of biennial oscillations through the years from 1529 to 1541, with narrower rings in 1530, 1532, 1534, 1536, 1538 and 1540 than in the years before and after (HUBER et al. 1964, 1969)).[1] But from about 1540–60 a further shift towards colder climate set in, soon to be accompanied by great advances of the glaciers and leading to the severest phases of the Little Ice Age. The sharpness of the change is shown in the daily observations in Zürich from 1546 to 1576 by the pastor W. HALLER, analysed by FLOHN (1949):[2] the relative frequency of snow in the winter precipitation from 1564 to 1576 was 63%, compared with 44% in the earlier years of the register. The severity of the winter 1564–5 in much of Europe undoubtedly surpassed anything since the 1430s, but it was followed by ten or more such winters in the next 175 years. Occasional wet, cold summers in the 1550s – harvest failures in England in 1551, 1555 and (an extreme case) in 1556 – marked an increasing incidence of such years, followed in 1594–8 by a sequence of wet summers which produced famines in England, and across Europe, ranking behind only those of 1313–18 and the 1430s. And once again there was a new wave throughout Europe of land going out of cultivation and villages and smaller settlements disappearing, though the picture is complicated more than in the previous cases by the activities of the wealthier landowners in enclosing land for other uses than formerly.[3] The century 1550–1650 produced only 24 years rated as good wine years in Baden as against 32 to 41 in the previous 100 years (FLOHN 1950).

The Little Ice Age: main phases

The period we are discussing has been dubbed 'The Little Ice Age' because not only in Europe but in most parts of the world the extent of snow and ice on land and sea seems to have attained a maximum as great as, or in most cases greater than, at any time since the last

1. The *Sägesignatur* (sawtooth structure) of the ring sequence of these years has become famous as an aid to dating. Other periods which show some approach to this illustration of quasibiennial oscillation appear in the tree ring series in 1058–65, 1147–55, 1204–10, 1264–71, 1441–51, 1566–77, 1774–82, and many shorter runs.
2. These daily observations in Zürich are printed *in extenso* in WOLF (1885).
3. The socially embittered history in England over many centuries, perhaps significantly between about 1300 and 1800, of wealthy, powerful and successful people enclosing land and converting it to other uses – through the first three centuries of this period commonly to sheep grazing – needs for completeness to be seen against the background of climatic change, abandonment of former tillage and spreading of wasteland recorded in this chapter.

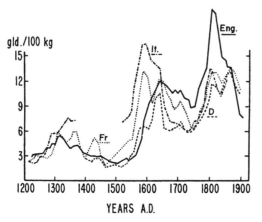

gld./100 kg

YEARS A.D.

Fig. 17.17   Wheat prices prevailing in England, France, the Netherlands and northern Italy (25-year means) from about 1200 to 1900.

Prices in Dutch guilders per 100 kg wheat, taken from *De Landbouw in Brabants, Westhoek in het midden van de achttiende Eeuw.*

Wageningen, Netherlands (H. Veenman & Zonen).

(*Reproduced from an unpublished paper,* 'Correlation of historic climate with historic prices *and wages', by* L. M. LIBBY, R *and* D *Associates, Santa Monica, California, 1974, and printed by her kind permission.*)

major ice age. Not surprisingly, the main climax seems to have come at different times in different regions: often the fluctuations within the centuries concerned seem to have been globally synchronous, but differed in relative amplitude in different parts of the world. On the whole, the culmination seems to have come earlier in the northern hemisphere, particularly in North America (see fig. 17.3), in the Arctic (fig. 17.4), and in China and Japan (fig. 17.10), and later in the southern hemisphere (fig. 13.19) where the maximum advance of the glaciers in Chile seems to have been in the eighteenth century and the greatest extent of ice on the Antarctic Ocean may have been as late as around 1900. There was, however, an important late climax of the Arctic sea ice around Iceland between 1780 and the 1830s, and many glaciers in the Alps reached their greatest extent towards 1850.[1]

Fig. 17.3 has given us a view of the succession of swings towards cold climate in North America, paralleled similarly by the oxygen isotope record for northern Greenland in fig. 17.4 and for New Zealand in fig. 13.19. An indication of a different kind for Europe is given by L. M. LIBBY's study of wheat prices in fig. 17.17. The changes in price register changes in the relation of wheat supply to expectation. In all the countries covered there are three main epochs of price rise: around 1300, 1600 and 1800; in all cases the biggest of these changes is a many-fold increase between 1550 and 1600–50, a phenomenon too widespread (and wrong in its timing) to be attributable to any of the rather localized wars of the period. The price rise around 1800 could be attributable largely to the interference of the Napoleonic wars with supplies and with trade, but the time does also coincide with the latest of the great periods of advance of the glaciers and of the Arctic sea ice about Iceland. It seems certain that the changes around 1300 and 1600 were largely climate-induced. It is surprising therefore that the 'bad' years in the 1400s make little impression on the price curves, except in France. Possibly the depopulation that had taken place in the previous century had so far reduced the

---

1. DELIBRIAS *et al.* (1975) produce documentary evidence that the advance of some of the glaciers in the Alps during the thirteenth century, particularly the Allalin glacier in Saastal, brought them around 1300 to a position not very far short of the positions occupied between 1589 and 1850.

pressure of demand upon supply from the available grain growing areas of Europe that the wheat price was less responsive to climatic fluctuations at that time.

With so many fluctuations towards colder and warmer runs of years discernible in these centuries, it is not surprising that different writers have set the limits of the Little Ice Age at different times. Times of onset around 1300–10, 1430 and 1560 have been commonly suggested. As times of ending 1700, 1850 and 1900 have been the commonest choices. It is reasonable to regard the time from about 1550 to 1700 as the main phase for most parts of the world and probable that the lowest values of 50-year, 100-year and 150-year mean surface air temperature in most places fell within that time. It is also reasonable to consider the whole sequence from about the time of Christ through the early medieval warm centuries and the cold climate that followed to our own times as an oscillation on the same time scale, and possibly of basically the same nature, as the Bølling and Allerød oscillations in Late Glacial times and the Piora oscillation and the Bronze Age to early Iron Age changes in the last 4000 years B.C.

The later stages of the Little Ice Age of recent centuries fell within the era for which meteorological instrument observations are available and will therefore be discussed in the next chapter. It is of interest to note here, however, that at the time (around 1800–50) of the last great maximum of the glaciers in the Alps, and of the sea ice around Iceland, the prevailing temperatures (most precisely documented in England but supported by the oxygen isotope records from America, New Zealand and, marginally, from Greenland) were not as low as they had been in the seventeenth century. It seems certain therefore that the advance of the glaciers around 1780–1850 owed more to increased precipitation and the advances between 1540 and 1700 owed more to lengthening of the snow season and shortening of the summer ablation period. The advances reported earlier around 1250–1300 were in these ways probably more analogous to those around 1800.

We know something of the character of the atmospheric circulation prevailing in the late sixteenth century, during the onset of the main phase of the Little Ice Age, from the observers living at that time. BROOKS (1949) quotes from an old book (*A Restitution of Decayed Intelligence*, by RICHARD VERSTEGAN) published in Antwerp in 1605 and in London in 1634: 'Old shippers of the Netherlands affirming, that they have often noted the voyage from Holland to Spaine to be shorter by a day and a halfe sayling than the voyage from Spayne to Holland.' The implication in this that the winds at that time over Europe between latitudes 35° and 60°N were commonly E'ly is borne out by the fifteen years of daily observations from October 1582 to April 1597, made by the Danish astronomer TYCHO BRAHE, on the island of Hven in The Sound (Øresund) (LA COUR 1876): these observations show that the commonest wind direction during those years was SE, while winds from all easterly points, and especially NE, were notably more frequent, and winds from S, SW and W less frequent, than in the present century. The total frequency of winds from all westerly points (SW, W, NW) was 40% against a total for all easterly points (SE, E, NE) of 37%. The maps in figs. 17.8(*c*) and 17.9(*c*) are samples derived for this period. Indirect confirmation of

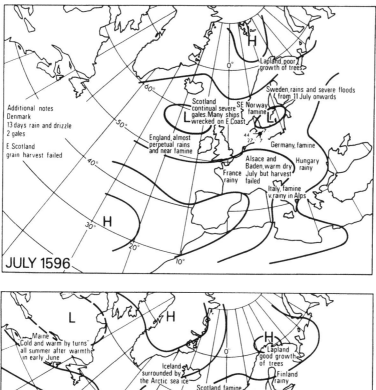

Fig. 17.18   Weather reported and suggested mean sea level pressure
and wind pattern.
    (*a*) July 1596.
    (*b*) July 1695.

the frequency of anticyclones in latitudes 60–70°N which these winds imply may be seen in
the warmth of these same decades indicated by tree ring series in northern Finland (as also in
Alaska) and by the oxygen isotope record from northern Greenland (fig. 17.4).

Northern Russia was also relatively warm and there was notably little ice on the Arctic coast in the sixteenth century and for some time after. Captain CHANCELLOR, seeking for the Northeast Passage to China, landed in 1553 on the shore of the White Sea, proceeded to Moscow and opened up diplomatic relations between England and Russia which were maintained by that route until Peter the Great built his new capital at St Petersburg (Leningrad) on the Baltic about 1700, by which time increasing ice was making the White Sea route difficult. In the sixteenth century, however, Russian sailings along the north coast as far east as the Gulf of Ob were common, and as late as 1620 a ship sailed past Cape Chelyuskin to the east side of the Taimyr peninsula before being trapped by the ice (this paragraph is from notes by J. E. CHAPPELL personal communication 3 May 1974, largely based on L. S. BERG).

By the first half of the next century the situation may already have been somewhat different, with evidence of more northerly wind components in Britain, Iceland and Norway and – from actual reports (LENKE 1960) – more W'ly winds (often cyclonic W'ly situations) over the German plain, though still interrupted by some years of markedly anticyclonic character.[1] All these characteristics, but probably with once more an added frequency of anticyclonic, E'ly and N'ly situations over all northern, western and central Europe, were evidently prominent in the second half of the seventeenth century.

Maps of the weather reported in July and tentative portrayal of a barometric pressure pattern that would account for it in two of the most extreme summers of the type commonly reported between 1550 and 1700 in Europe are shown in figs. 17.18(a) and (b).

Characteristics of this central period of the Little Ice Age may perhaps be summarized as follows:

(1) Low values of the long-term mean temperature at all seasons of the year by comparison with recent normals in most parts of the world (though some exceptions/different timing in high northern latitudes and in the southern hemisphere temperate and Antarctic zones have been mentioned).

(2) Enhanced variability of temperature from spell to spell, from year to year and on somewhat longer time scales. This is characteristic of regimes with frequent blocking, or meridional circulation, patterns in middle latitudes. Despite some record cold spells,[2] periods of great heat in England were also recorded[3] and

1. Writing in 1653, BACON considered the prevailing winds in England and the nearer parts of Europe to be N'ly in summer, W'ly in autumn, very variable directions in winter and E'ly in spring.
2. Examples of seasons of some special interest in England are the notably wet autumns and early winters of 1632, 1633 and 1634, each ending in a frosty period and in 1634–5 with a 6 or 7 week period of frost and snow; this was followed by such a mild fine autumn in 1635 that it was described as summerlike until about the end of November (modern calendar), and there are reports that 1636 was a year of continual SW'ly winds and mild or warm weather and a reputedly very hot summer. The great winter of 1607–8 in which the frosts killed many great trees (oaks, elms, ashes, etc.), splitting their trunks, is considered to have been unmatched in England for both length and severity except in A.D. 764, 1684 and 1740. The severest frost, as in 1963, was about 20 January (New Style). But there were a number of other very notable winters in or about the seventeenth century. In 1683–4 the ground is reported to

according to SAMUEL PEPYS's diary at least one run – from 1659 to 1662 – of three or four consecutive winters with no snow lying. And in the long record of the Baltic ice given by the dates of spring opening of the port of Riga (see Appendix V, Table 17 (i)) the two years with least ice (before 1975) and the year with most ice all fell in the same decade, the 1650s.

(3) Atmospheric circulation patterns associated with an expanded polar cap, expanded circumpolar vortex, frequent blocking and meridionality in middle latitudes of the northern hemisphere.

Consequences of the expanded circumpolar vortex were displacement of the most frequented depression tracks to somewhat lower latitudes than in other epochs (see figs. 17.8 and 17.9), frequent wetness of the summers in much of Europe, and wetness of the winters in parts of the Mediterranean and north Africa, while the winters in Europe north of the Alps (including the British Isles) were often snowy and sometimes drier than in the present century. In some decades in England the wet cool summers (1570s, 1590s, 1620s) were the more marked anomaly; in some decades the severity of the winters with one or more extremely severe winters stood out (1560s, 1600–9); and in others both summers and winters were generally cold (1580s, 1680s, 1690s) (see Plate XVI). Another consequence was the severity of the greatest windstorms on the coasts of Europe in the latitude of the British Isles and the southern Baltic (see p. 452).[1] This must be liable to happen at any time when the

---

have been frozen to a depth of 50–90 cm in various parts of England (in the southeast and near Manchester) and in southwest England (Somerset) wet ground was found to be frozen in places to a depth of 4 feet (1 m 20 cm); as in 1607–8, there was much damage to vegetation and wild life – birds died in great numbers. Yet in March the spring advanced rapidly. And then in 1708–9 the frosts were so severe in southern England and all neighbouring countries that the northernmost vineyard districts in France were permanently given up, and the limit of the industry settled at about its present position just south of Paris. The orange trees and olive trees were all killed in the south of France. In 1608 it is recorded that the sea froze 'as far as it ebbed on the east coast of Scotland and people walked out on the ice to ships in the Firth of Forth. Several winters of exceptional severity seem to have been experienced by early parties in New England and as far south as Virginia, and in the interior of North America, between 1603 and 1615; in one of those years it is reported that SAMUEL DE CHAMPLAIN's party found bearing ice on the edge of one of the Great Lakes as late as June. In 1684 there were belts of ice along the English and French coasts of the Channel and the coasts of Belgium and Holland; the ice extended about 5 km from the coast of England at Dymchurch and nearly 40 km from the coast of Holland. In 1709 the frost seems to have been relatively severer and more prolonged on the continent than in England, where the ice on the Thames was not thick and in the Lake District (northwest England) the lakes remained open; but there was ice on the harbours of Marseilles and Genoa. In 1740 the ice was strong enough for traffic on the Thames in London and on the biggest of the English lakes in the northwest, and the spring months that followed continued exceptionally cold. 1698 was another year in which there was ice on the coast of England (20 cm thick on the coast of Suffolk). See tabulations of winters in Appendix V Tables 5, 6 and 7.

3. MANLEY (1974) rates the four months July 1666, June 1676, July 1677 and July 1701, all with mean temperatures about 18°C or above, among the forty warmest months in the 315 years 1659–1974 in central England, and possibly amongst the top twenty-five – i.e. a proportion roughly as high as in the warmest half-centuries. The hot July of 1666 and a warm August, and prevailing dry weather since May of that year, immediately preceded the great fire of London, which started on 12 September 1666 (New Style calendar). See Appendix V, Table 8.

---

1. The cool summers and windiness of this period and the severity of the greatest storms may perhaps have been the reason why woods that had survived until this time near the exposed northern and northwes-

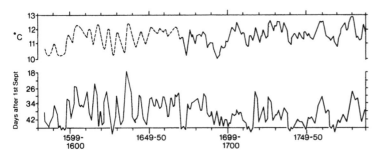

Fig. 17.19   Wine harvest dates in France (upper curve) and suggested mean temperature of the months March to August in central England (lower curve).

Two-year running means developed by Professor GORDON MANLEY from vintage data by E. LEROY LADURIE.

(From MANLEY *(1965); reproduced by kind permission.)*

mainstream of the circumpolar vortex shifts far enough south to steer the main travelling cyclones towards this part of the coast of Europe; but in the Little Ice Age there were probably times when the jetstream, in its southern position, was exceptionally strong owing to the enhanced thermal gradient between the subtropical Atlantic and the southward penetrating cold air. It might be supposed that another consequence of extreme sharpening of this frontal zone in similar situations would be enhanced tornado activity in Europe and on the Great Plains of the United States (though it is hardly possible to verify this point).

The variability of seasonal temperature from one group of years to the next in the seventeenth century is illustrated by the sequence of 2-year mean temperatures of the springs and summers in England (fig. 17.19) estimated by MANLEY (1965) from the correlation between the central England temperature series and the wine harvest dates in France and Luxemburg.

Fig. 17.20 presents estimates from isotope work of the probable course of the long-term mean winter temperatures (January to March) in central England from 1530, to compare with the temperature of the warmer seasons in the previous figure.

FLOHN (1950) arrived at estimates of the mean winter temperatures in the late sixteenth century by applying a statistical method to the descriptive reports of the weather of the time in central European longitudes. The proportion of winters described as mild to the total mentioned as either mild or cold in HENNIG's compilation (1904) is 35 %. This proportion

---

tern coasts of Scotland were dying. According to the *Laxdaela Saga* ships had been built in the northern-most county of Caithness around A.D. 950, pre-sumably indicating more local timber and larger trees there than now. The Earl of Cromertie (1710) reported as a familiar sight the woods standing dead around 1650 near the northwest coast and then after

some decades being blown over, soon overgrown by moss and sinking into the peat.

Maps compiled by the present author indicate the paths of frequent cyclonic depression centres passing over or near Scotland between about 56° and 62°N in 80–85 % of the summers between 1586 and 1605 and again between 1688 and 1700.

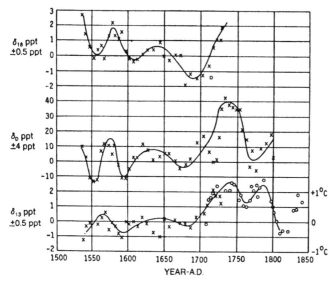

Fig. 17.20   Isotope ratios $^{18}O/^{16}O$, $^{2}H/H$ and $^{13}C/^{12}C$
measured in an oak tree in Germany (Spessart forest area)
from 1530 to 1800, calibrated against central England
average temperatures for the late winter and early spring
(January to March) from 1700 and used as estimates of the
temperatures of that time of year in England back to 1530.
   Thirty-year means derived by L I B B Y and P A N D O L F I
(1974).
   (*Reproduced by kind permission of Dr* L. M. L I B B Y.)

can be reproduced in the winters from 1775 onwards, by classing as mild all those winters
with a temperature deviation amounting to $\geq +2°$c from the overall average and classing as
cold all those with a deviation of $\leq -1\cdot2°$c. Applying this argument to the winters of the
period 1560 to 1599 indicates that the mean winter temperature was $1\cdot3°$c lower than in the
first half of that century. A similar basis of calculation, using the proportion of snowfall
reports to the total number of entries of precipitation in winter at Hven in the Danish Sound,
indicates that the mean temperature of the winters in Denmark between 1582 and 1597 was
$1\cdot5°$c lower than in the period 1886–1925.

   Also farther south in Europe climatic stress made itself felt in the period about 1550–
1700. The French historian L A D U R I E (1956–7, 1966, 1972) began his interest in climatic
history when, in studying the incidence of shortages and famines in the upland districts of
southern France (Languedoc) in this period, the original documents made clear the
connections with severe winters and wet summers.[1] Similarly, in Spain, it was a time of

1. P O S T (1973) notes that in the middle two decades of
   the seventeenth century major revolts occurred in
   many parts of Europe – in England, France, the

   Netherlands, Portugal, Catalonia and Naples, while
   there were peasant riots in central and eastern
   Europe – and in China the Ming dynasty collapsed.

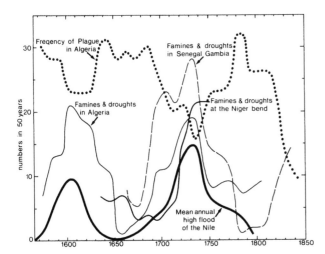

Fig. 17.21   Variations in the incidence of plague in Algeria and of droughts and famines in territories at the fringe of the Sahara, together with the heights of the Nile floods, from 1565 to the nineteenth century.

   (*Data compiled by Ms.* SHARON NICHOLSON *of the University of Wisconsin from original sources and kindly supplied for this book.*)

deepening agricultural crisis, which has commonly been ascribed by historians to a supposed tendency to increasing aridity. I am informed by Professor R. DAVIS of Leicester University (personal communications, 21 September, 27 October and 13 December 1971) that the evidence is rather of great variability of the summers in the sixteenth and seventeenth centuries in Spain, with recurrent phases of dryness but also some wetter periods, making agricultural yields more uncertain, and that there were some severe winters around 1600 and 1680 and perhaps some severe winter storms in those centuries. The whole Mediterranean area was in economic decline in the seventeenth century, but there is no firm evidence of any general increase of aridity.

   An increased frequency of cyclonic influence in the northern part of the Mediterranean area in the seventeenth century as compared with earlier and later periods is indicated by the high levels of the lakes at Fucine (near 46°N 11°E) and Trasimeno (near 43°N 12°E) in Italy.

   The question of moisture variations in the Mediterranean may be taken further by data from Africa. Fig. 17.21 indicates the variations during this period of items registered in the historical records of territories at the northern and southern fringes of the Sahara, which make possible some deductions about how the climate varied in Africa (I am greatly indebted to SHARON NICHOLSON for the supply of this recently gathered information, 7 April 1975).[1] The occurrences of plague in Algeria may probably be taken as marking moist periods near the northern edge of the Sahara. On this interpretation, there were considerable fluctuations of rainfall in this zone, with drier periods at the times of coldest climate in Europe around 1600 and the 1690s but also during Europe's remarkable warm climate phase

In central Europe the Thirty Years War could be blamed for upheavals and desolation over wide areas, but its whole course took place in a time of rather cold and more than usually variable climate, with late harvests and major advances of the Alpine glaciers. Who can say that the preconditions of these very widespread upheavals in the social structure itself had more to do with causing them than the climatic vicissitudes of the time?

1. See NICHOLSON (1976) for further details.

in the 1730s. The famine and drought frequency curves in fig. 17.21 all relate to areas essentially within the Sahara region, at its northern and southern fringes. These show a variation on about the same time scale as, and inverse to, that of the plagues. They confirm dry periods there also at the times for which we deduce most dryness in Algeria. This parallelism between the rainfall fluctuations at the north and south fringes of the Sahara is surprising and contrasts with the experience during the Sahel zone droughts of the 1960s and 1970s, when it is clear that there was a general southward shift of the climatic zones, suggesting in the seventeenth century a difference of pattern which is not yet understood.[1] The same may be said of the dryness in these regions which accompanied the warmth of the 1730s in much of Europe; evidently the prevailing pattern of atmospheric circulation over the zone south of the Sahara was not the same as in the twentieth century warmest epoch (1933–52). The varying heights of the annual flood of the River Nile in Egypt, which depend on the summer rains over Ethiopia, as seen in fig. 17.21, go clearly in parallel with the frequency of droughts in the Sahel zone (10–17°N) – i.e. at about the same latitude as the Blue Nile catchment in Ethiopia in the western sector of the Sahara, west of 0–10°E; this relationship (high Nile: Sahel drought in west Africa) differs sharply from recent years and suggests a more meridional pattern in those latitudes, with differences from east to west Africa not seen in the present century.[2]

A curve closely resembling the variations in the heights of the Nile flood in fig. 17.21 is given by MALEY (1973) for the level of Lake Chad (14°E) in a central Saharan longitude and at the same latitude as Ethiopia. At present, we can only place these newly acquired facts on record and not attempt any complete explanation. Lake Chad is registered by MALEY as high (about 286 m above sea level) in the ninth century and again around 1600 and in the eighteenth century, but low (about 281–2 m) from A.D. 1000–1500, again briefly around 1650, 1820–50 and 1900–40. It rose to 284 m for a time between about 1870 and 1890 and to 283 m about 1960.

Opposition, as in the present century, between the variations of rainfall in equatorial Africa and in latitudes near 10–20°N (Chad, Ethiopia, etc.) may perhaps be traced over the longer time-scale climatic fluctuation of the last 700–1000 years. Arab records of the thirteenth century A.D. (when Lake Chad was low) indicate denser human settlement on the

1. What seems to be indicated is a more extreme meridional distortion of the prevailing northern hemisphere atmospheric circulation – a marked ridge and trough distortion penetrating farther equatorward – in the Little Ice Age of the seventeenth and eighteenth centuries than at any time in recent years.

2. The report of a missionary, MANOEL DE ALMEIDA, travelling in Ethiopia about A.D. 1628 of a big river Machy (or Meki) draining a great lake Zuaj (or Ziway, whose present water surface is 1636 m above sea level) in the Galla basin (near 8°N 39°E) northeastwards to join the River Awash, i.e. overflowing the present watershed at 1670 m –

mentioned, though not accepted, by GROVE et al. (1975) as a drainage pattern operating in such recent times – seems at least unambiguous evidence of a much wetter regime at that latitude in east Africa at that time, when there was permanent snow in the heights of Ethiopia. It is known that lake levels in that area were falling rapidly between about 1880 and 1895 and that at some previous time the lakes in the basin had all been joined together to overflow by the route mentioned, but radiocarbon dates so far obtained all point to dates for this 5000–10 000 or more years ago. However, changes of lake level up to 8 m above and 1 m below present in quite recent decades are registered by belts of dead trees.

coastal lowland of East Africa in Kenya, between 2° and 5°s, and the decline of these settlements in the sixteenth and seventeenth centuries (Dr E. B. MARTIN, Nairobi, personal communication, 28 December 1974).[1] It is conjectured that the likeliest explanation of the decline is failure of the wells in a drier climatic regime.

Other anomalies elsewhere contemporaneous with the Little Ice Age in Europe have already been mentioned in other chapters, notably the lowered snow line on the mountains of Ethiopia in the seventeenth century[2] and the frosts in China between 1646 and 1676 which caused the abandonment of orange growing in Kiangsi province.

The long records (see Table 44 in Appendix V) of rainfall near 33°s in Chile, since A.D. 1535, show a history with a remarkable inverse parallelism to the advances of the Arctic sea ice south towards Iceland (fig. 7.22, p. 302, in Volume 1). This suggests a broad tendency, through the Little Ice Age period and since, for all the climatic zones in the American and Atlantic-European-African sectors to move south or north at the same time. This is an item which can be further investigated by study of the yearly ice layers in the ice sheets, e.g. in Antarctica (fig. 7.23). Some study of rain and drought records in the southernmost part of Africa since 1642 was reported by HUTCHINS (1889), though his deduction of an approximately 9-year cyclical tendency is hardly relevant to our present discussion.

Returning briefly to the British Isles before ending this chapter, we find in Scotland particularly convincing evidence that the seventeenth century brought the harshest phases of the Little Ice Age in that country. It was in 1612 that the government of King James VI of Scotland, who had acceded as King James I in 1603 to power in England, used that power to establish in Ulster a 'Plantation' of his Scottish subjects to relieve the suffering arising from the continual famines in Scotland and to reduce their frequency. And so began the long troubled history of Northern Ireland. Yet in the last decade of the seventeenth century came the worst famine of all in Scotland, the 'ill years of King William's reign', when there were 7 years of harvest failure out of 8 between 1693 and 1700 in all the upland parishes of Scotland where more people (and a bigger proportion of the people) died of starvation at that time than in the Black Death of 1348–50 (see Plate XVII, also Chapter 12, pp. 10–11). The 1690s seem to have brought the severest phase of this climatic period in most of western and

1. The early Portuguese records indicate that the denser settlement continued in the fifteenth and early sixteenth centuries. The timing of the change is thus verified as coinciding with the climatic change in the mid sixteenth century in Europe.

2. Early travellers' tales (not known to be confirmed as yet by any report of botanical evidence still present) that there were oakwoods – presumably the cork oak *Quercus suber* – growing within the present territory of Mauritania in the late seventeenth century A.D. and houses built of wood may be evidence of a cooler, moister climate than now there also. BRYSON (personal communication, April 1974) associates both the abandonment of the great city of Fatepur Sikri in India in 1588, only sixteen years after it was founded, and the collapse of the empire of Mali in west Africa in 1591 with increasing incidence of drought, so that the climatic change implied by the report from Mauretania may have set in a century earlier, at the same time as the changes in Europe. Empires had flourished in the Mali region from the eighth to the thirteenth centuries A.D.

GROEBEN'S account (1694) of his voyage to Guinea in 1682–3 is, however, explicit about the desert nature of the land south of Cape Blanco (21°N) where 'neither greenery nor grass but only here and there a prickly bush' grew (quoted by PEJML 1962).

northern Europe, as well as in Iceland.[1] All through the Little Ice Age period, from 1550 onwards, Scotland developed and maintained trade with the Baltic as her 'emergency granary' for years of need, particularly with Danzig, through which passed the exports from the Polish grainlands, and later increasingly from Königsberg and Riga, shipping Russian grain, and from Swedish ports. There were Scottish colonies in the Baltic ports connected with the trade (LYTHE 1955, SMOUT 1963). Details of the trade and its fluctuations are traceable in the registers of the tolls levied by the Danish administration of the time on ships passing through the Sound (BANG 1906, 1922, FRIIS 1925)[2] and some in the port records of Dundee (LYTHE 1955). Despite the usual availability in that century of surpluses in the east European grainlands, however, there were some years – e.g. in Poland in 1685, 1698 – in which the harvest failed there too, fortunately not always the same years as in Scotland and apparently less often than in the west.

Finally, it is appropriate – and may be important – to note that the best estimates of population changes in most parts of the world other than the Ottoman empire, China and the present territories of the U.S.S.R. (where the temperature and moisture trends may have been different), compiled by historians unconcerned with climate, so nearly parallel the course of the gross-scale climatic shifts discussed in this chapter and the next that they might be alleged to register the Little Ice Age development. The figures, assembled by CROSBY (1972), are indicated in Table 17.5.

Table 17.5   World population estimates (in millions) by continents and regions

	1650	1750	1800	1850	1900	1950
Africa	100	95	90	95	120	198
Latin America	12	11	19	33	63	162
North America	1	1	6	26	81	168
India	100–125		120	130	300	500
Asia (total including India but excluding present U.S.S.R.)	327	475	597	741	915	1320
Europe with all present U.S.S.R.	103	144	192	274	423	593
Oceania	2	2	2	2	6	13
World total	545	728	906	1171	1608	2454

1. Contemporary discussion of the matter led some to enquire 'by how many degrees the general earthquake some years since hath put back the terrestrial globe towards the North, which is the reason (till another shall shake it into its place again) that we shall have no more ripe grapes' (quoted here from *Calendar of State Papers, Domestic, of the Reign of William III, 1698*, edited by E. BATESON (London, H.M.S.O., 1933)).

2. The original registers are kept in Rigsarkivet, Copenhagen. The details may perhaps be used to derive weather information.

Europe apart from Russia and the Balkans probably decreased in total population from 1650 to 1700 or after. The populations of Iceland (see Chapter 13, p. 265) and Scotland are known to have declined, and the same probably applied to the other countries north of the Alps, except perhaps England. CROSBY (1972) specifically mentions that the population of France was declining during the first decades of the eighteenth century – and, by implication, for some long time before that – due to the incidence of harvest failures and that the situation in the south was improved by the introduction of maize. Likewise in Italy the population was declining from 1650 to 1700. In India, too, the total was approximately static all through those times and on until about 1800. Moreover, the regions – Turkey, the Russian territories and China – that were exceptional in enjoying population growth during the seventeenth and early eighteenth century may also have been exceptional climatically. It is known (see Chapter 13, p. 136) from direct reports of the time that western and northwestern Siberia enjoyed exceptional warmth in the 1690s, the very same decade that marked the extreme for cold in western Europe. It also seems likely from the reports of lake levels and the level of the Caspian Sea that some of the arid regions of Asia were better watered by the same climatic phase.

Clearly no climatic cause of the population changes can be positively identified except in the case of Europe. (In the Americas, in the sixteenth century, decimation of the previous native population by diseases resulting from the first contact with people from the 'Old World' was by far the main cause.) But if the population estimates in Table 17.5 are a sound guide the phenomenon of population decline was on a nearly global scale and must surely point to the working of some universal influence in the physical environment. It may be significant that our population estimates are taken from a work that does not mention climatic change and are clearly therefore not subject to bias in favour of this as the cause. Yet the population declines in the northern countries, particularly Iceland and Scotland, can be seen to have been due to the effects of cold climate (cold summers above all) and in southern Europe to crops ruined by rainy cool summers. In Africa, and elsewhere in lower latitudes, if climate was the cause, it presumably operated mainly through droughts ruining crops and pastures and causing wells to run dry. The population changes may in this case therefore be indirect evidence of the pattern of incidence of drought.

Chapter 18

Climate since instrument records began

The latest part of the history of climate is told by the regular reports of meteorological observers with instruments. As noted in Chapter 13, most of the basic meteorological instruments had been invented by A.D. 1700 and then, or soon after, were coming into regular use for daily weather observations at places in Europe and eastern North America. The longest observation series that have been carefully homogenized take us back to that time (see references in Chapter 13).

Development of the instruments and their exposure so as to get representative readings presented problems which took many decades to sort out, and extension of the network of observations to the whole Earth had to wait until our own century (see Chapter 13). In the fullest sense, coverage was not complete until the era of satellites, i.e. since 1960. Fuller treatment of the instrument and observation problems has been given by VON RUDLOFF (1967) for Europe; reference should also be made to MITCHELL *et al.* (1966, especially pp. 7–13), LAMB and JOHNSON (1966) where the equivalents of many ancient units of measurement are given, and for the specific problems of temperature and rainfall observations to various works by MANLEY (e.g. 1953, 1959, 1974) and CRADDOCK (1976). By 1700–50 some of these instruments were sufficiently reliable and reasonably exposed for it to be possible to derive monthly values that probably represent a good approximation to the true values of mean pressure and air temperature. The best barometers could be considered reliable instruments from some time earlier and their exposure (which presents little problem), instrument temperature, and so on, if recorded, allow their readings to be used. In a few cases this takes the record back into the seventeenth century (e.g. in MANLEY's 'central

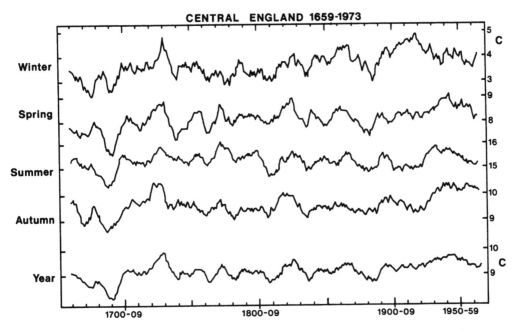

Fig. 18.1   Average temperatures in central England by seasons and for the whole year from 1659 to 1973.

Ten-year running means.

(Adapted from MANLEY *(1974); reproduced by kind permission.)*

England' series) monthly mean temperatures that are probably reliable to within about 0·3°C go back to 1680 and within 1°C to some time earlier. (Other very long records are listed in Chapter 13, pp. 24–5; the series for Berlin could be extended back to 1697 with the seventy years of observations made by the KIRCH family in that city.) Representative rainfall records are much more difficult to achieve because of the magnitude of the effects of varying the exposure of raingauges and the real differences of rainfall over quite short distances. Doubts inevitably linger therefore over the homogeneity of the longest rainfall records, and confidence can only be attached to values from before about 1800–50 in so far as a number of places show much the same variations and trends and where reasons – in terms of general wind circulation and temperature levels prevailing, etc. – can be seen for the differences from recent times. In all cases, particularly in the case of rainfall, the most nearly homogeneous long series of observation values are likely to be district or regional averages – averages, that is, of several good observation records within some region of reasonably uniform terrain and uniform response of its rainfall regime to any changes in the prevailing winds. Tabulations of year-by-year seasonal values for England and the eastern United States are given in Appendix V, Tables 33 and 34; other rainfall histories are given in Tables 35–44.

Air temperature

Fig. 18.1 shows the history of temperatures prevailing in central England from the mid seventeenth century to 1973 in terms of the 10-year running averages (see also the tabulated year-by-year values for each season in Appendix V, Table 10; a tabulation for eastern North America from 1738 is given in Appendix V, Table 11).[1] From about 1700 to around 1945 it is a one-way story, a trend towards greater warmth that was interrupted only by various shorter-term fluctuations. In consequence, the temperatures prevailing between about 1920 and 1960, and especially between 1933 and 1952, represented a recovery to something near the level which seems to have prevailed for a longer time in the early Middle Ages. This also means that the average temperatures of that period were probably unmatched over any period of comparable length in the intervening centuries and must be taken as untypical of the longer record of climate in historical times. The long-term trends stand out more clearly in figs. 13.4(b), 13.19 and 13.61(a). From the parallelism with the curves representing widely separated regions of the Earth, and the near agreement between the annual curve in fig. 18.1 and the sequence of 5-year mean temperatures averaged over the whole world from 1870 to 1969 in fig. 18.2, it seems that the changes in England of temperature averaged over the whole

1. Since this chapter was written a series of mean temperatures of the winters (December, January and February) in the Netherlands from 1634 to 1975 has been derived by Professor J. DE VRIES of the University of California at Berkeley, using the records of dates of freezing of the canals for the early years of the series. The mean values and standard deviations of the winter temperature in different sections of the record were found to be as in the Table in the next column:

	Mean temperature	Standard deviation
1634–98	1·43°C	1·99°
1699–1757	2·50°C	1·55°
1758–1839	1·61°C	1·93°
1840–1939	2·32°C	1·56°
1940–75	2·16°C	1·98°

Mean temperatures derived for the seventeenth century winters in the Netherlands were as follows:

Decade beginning	Year									
	0	1	2	3	4	5	6	7	8	9
1630					0·2	−0·2	2·2	(4·2)	1·4	3·0
1640	1·7	1·9				3·2	−0·6	−0·2	(4·2)	0·6
1650	2·6	1·6	1·8	1·9	3·4	−0·2	0·4	1·8	−1·0	2·6
1660	−0·5	(4·2)	(4·2)	−2·6	(4·2)	0·2	2·2	0·6	2·8	1·9
1670	−0·3	(4·2)	−2·1	1·4	1·8	(4·2)	(4·2)	0·4	1·8	−2·1
1680	(4·2)	−1·7	3·0	3·4	−0·9	1·2	(4·2)	2·2	1·5	1·0
1690	2·4	−1·3	−1·3	3·3	1·9	−1·7	2·8	−2·5	−1·3	(4·2)

Winters for which the value 4·2° was derived have been put in brackets, as this figure indicates no closing of the canals by ice and a possibility that the true mean temperature was even higher. Even though such very mild winters (like the very cold and severe winters) were commoner in the seventeenth century than in other periods, a little calculation indicates that the average winter temperature for the period 1634–98 quoted by DE VRIES (above) could not be more than 0·1°C too low. The present writer suggests that 1·5°C be adopted as the value for that period.

year and taking about 10–20 years together have probably been a fair approximation to the average for the whole Earth.

The representativeness of the changes in England over the last three centuries, and probably over most of postglacial time, is judged to be a fact related to the position of England, in the middle latitudes zone of prevailing westerly winds, at the eastern limit of an ocean where winds and ocean currents freely and continually effect heat exchanges between all latitudes.[1] It is not contradicted by the observation that the cooling of climates setting in after the 1940s (fig. 18.3) took effect almost a decade sooner in the Arctic – and in tongues extending from the Arctic towards the heart of the northern continents – than it did in much of Europe and the U.S.A. Its amplitude was also much greater in the Arctic and in northern parts of the continental interiors than elsewhere. The amplitude of the early twentieth

1. What the changes of temperature noted in this chapter have meant for many practical concerns in one sample country in middle latitudes may be illustrated by the variations in England of the length of the growing season and of the heating fuel requirement as well as the frequency of snow-covered ground (see Table 18.7, p. 524).

M ANLEY (1957) studied the central England temperature series from 1698 to 1955, and provided a diagram showing the accumulated month-degrees below 56°F (13·4°C) each year, as an indicator of the likely fuel demand for indoor heating. This points to the demand in the coldest individual year being just double that in the warmest year and the demand in the coldest decade since 1700 being 20% greater than in the warmest decade.

The frequencies of very warm and very cold months, and of 'summer days' (maximum temperature over 25°C) and freezing days (maximum under 0°C) in England have changed correspondingly (see fig. 18.31 and Tables 6–9 in Appendix V).

Changes in the length of the growing season in the lowlands of central England are sampled in Table 18.1 (see also footnote p. 535 and further details in L AMB 1965, 1967(a)).

Table 18.1   Lengths of the growing season in central England

Period	1870–95	1930–49	1950–9
Average length of the growing season (temperatures above 5·5°C) at Oxford (range of the decade averages)	255–65 days	270–5 days	265 days
Shortest growing seasons at Oxford (shortest each decade)	205–25 days	237–43 days	226 days

It seems likely that in the coldest decades of the Little Ice Age period the average growing season was shortened by 3 weeks to a month by comparison with 1930–49 in the English lowlands and by a rather greater amount on the upland farms in northern England and Scotland.

B RYSON (1974) has given some figures for Iceland which demonstrate the direr effects in a country near the thermal limit of agriculture. The mainstay of Icelandic farming is grass and the animals raised on it. Since there is always ample moisture, the yield is approximately proportional to the number of growth days when the temperature is above 5·5°C, apart from a few anomalous years when the pastures were poisoned by ash deposit – some-times including fluoride – from the volcanoes. In the warm years in Iceland in the late 1950s with mean temperature for April to October 7·6°C the average hay yield was 4·33 metric tons/hectare (with application of 2·83 kg of nitrogen fertilizer/hectare). In 1966 and 1967 the April to October temperature average was 6·8°C, and the hay yield averaged 3·22 metric tons/hectare (despite application of 4·83 kg of nitrogen fertilizer/hectare).

The remarkably abundant historical records for Iceland show only 12 famine years in a 525-year span before 1500, 5 of them in the relatively cold period in Iceland between 1250 and 1390. Between 1600 and 1804 there were 34 famine years.

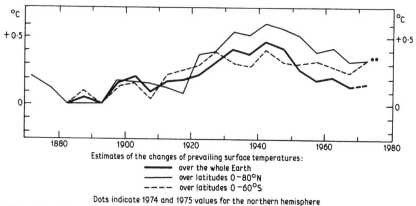

Estimates of the changes of prevailing surface temperatures:
— over the whole Earth
— over latitudes 0−80°N
− − − over latitudes 0 −60°S

Dots indicate 1974 and 1975 values for the northern hemisphere

Fig. 18.2 Average temperature over the whole Earth: changes since 1870–74.

Successive 5-year means expressed as departures from the means for 1880–84.

Best estimates based on calculations by MITCHELL (1961), extended to 1965–69 by calculations at the National Center for Atmospheric Research at Boulder, Colorado and for the northern hemisphere to 1970–73 by W. BRINKMANN at the University of Wisconsin, Madison. Dots show the corresponding northern hemisphere values for the individual years 1974 and 1975 derived from information published in the *Berliner Wetterkarte*. The curves for the whole Earth and for latitudes 0–60°s have been provisionally extended to 1970–73 on an *ad hoc* basis by reference to work by ANGELL and KORSHOVER of N.O.A.A., Washington D.C. on the temperatures in the upper troposphere. (Margins of error of the estimates must be presumed greatest in the extensive ocean areas, especially 0–60°s; but repetitions of the calculations in different laboratories indicate that the main warming from the 1880s and the cooling from the 1940s are beyond doubt.)

century warming of climates had been greatest in just the same regions where the subsequent cooling has been great (the maps of decade-to-decade changes from 1900 to the 1960s given by KIRCH (1966) document this precisely). In both the warming up to around 1940 and the cooling since, the change began to be registered soonest in the Arctic, especially where the inner Arctic Ocean region is open to communication with the Atlantic and indeed along an axis from southwest Greenland and the Labrador Sea, through the Barents Sea to the northernmost tip of Asia, and from there southeastwards into Asia as well as northeastwards across the pole to Canada. And these were the regions where it was strongest.[1] From fig. 13.3, and from what is known of the timing of the early medieval warm epoch and the subsequent cooling, both described in Chapter 17, one suspects that those major climatic changes affecting several centuries were also registered soonest and most strongly in about the same regions.

1. Both changes were strongest about Franz Josefs Land (82°N 53°E), where anomalies of monthly mean temperature exceeding 10°C in the sense of the prevailing temperature trend became common. In the cooling case in the 1960s this was commonly the locus of an extra trough – i.e. a third lobe – in the circumpolar vortex in high latitudes (See also pp. 281, 385, in Volume 1).

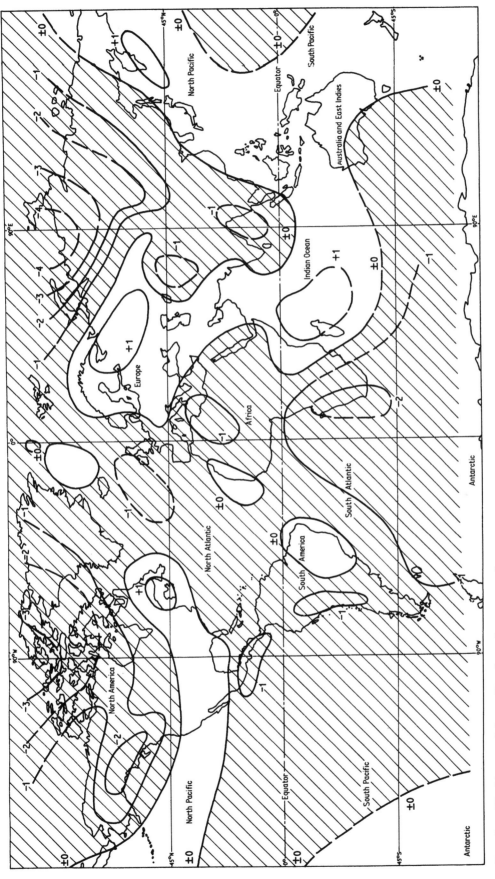

Fig. 18.3 World map of the cooling from the 1940s to the 1950s. Isopleths (°F) temperature change: 1951–60 averages minus 1941–50 averages. Shaded areas becoming colder. (*From* MITCHELL *(1963); reproduced by kind permission of the author and UNESCO.*)

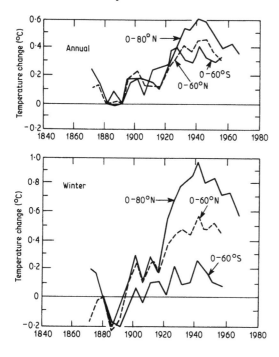

Fig. 18.4   Average temperatures over each hemisphere from 1870–4 to 1955–9.

Successive 5-year means estimated as departures (°C) from the mean for 1880–4.

Note the big increase of amplitude introduced by including the zone between 60 and 80°N despite its small area; temperature changes averaged for this latitude zone appear to be about four times as big as the average for the whole Earth and ten times as big as for zones near the equator.

(*From* MITCHELL (*1961*); *reproduced by kind permission.*)

The overall range of the changes of prevailing temperature during the last 1000 years in England has been surveyed on pp. 436–8, 447 in Chapter 17. For Japan (Osaka region) YAMAMOTO(1971), using a statistical approach based on the freezing of lakes and rivers and the yields of the rice crop and dates of disappearance of snow on the mountains, has derived mean January temperatures around 1770–1830 1·6°C below the early to mid twentieth century level and in summer 1·7°C below. The winters of that time – in the latter part of the Little Ice Age – seem to have been notably snowy on the western (Japan Sea) side and the summers rainier than in this century, except in Hokkaido, the northern island (see, however, Tables 24–26 in Appendix V).

In fig. 18.4 we see that the sequence of temperature changes from 1880 to 1960 was much the same over the northern and southern hemispheres, though the range was somewhat greater over the northern hemisphere, particularly in winter and particularly in the Arctic. As both WILLETT (1950) and MITCHELL (1961) noted, however, there has been a tendency for the temperature changes south of about latitude 40°s to go in antiphase to the rest of the world. Closer analysis (e.g. in the maps given by MITCHELL (1961)) indicates that the opposite temperature trend has in each case been by far strongest in one sector of the higher latitudes of the southern hemisphere: around South America and the South Atlantic during the early twentieth century global warming and around New Zealand and the South Pacific during the global cooling since 1950. Thus, the antiphase phenomenon has to do with an eccentricity of the southern hemisphere's circumpolar vortex and a shift of all the climatic

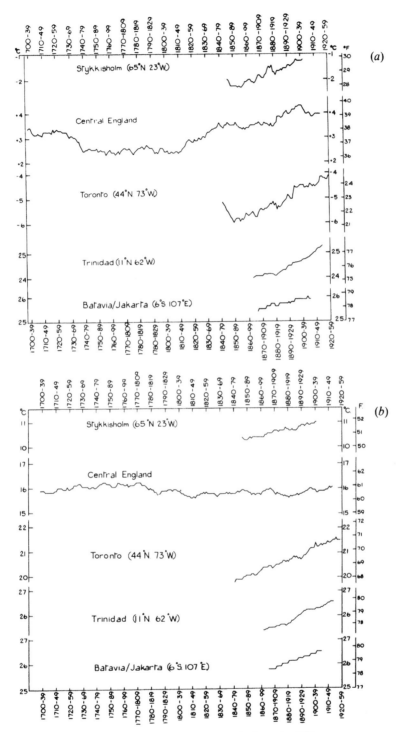

Fig. 18.5   Average temperatures at selected places in different latitudes:
Forty-year running means.
(*a*) January.
(*b*) July.

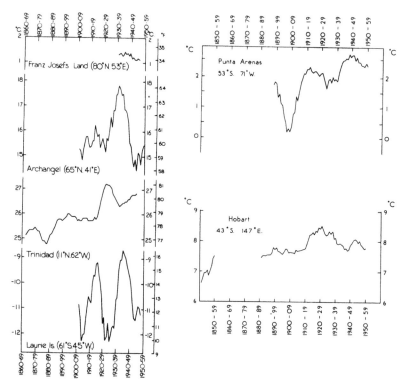

Fig. 18.6   Average temperatures at selected places in different latitudes
in July.
    Ten-year running means.

zones of both hemispheres towards the south or the north in the sector concerned (see
Volume 1, pp. 302–3, and fig. 8.11).[1]

    Fig. 18.5 displays sample long temperature records for a few places between latitudes
66°N and 6°S.[2] The general experience of warming of climates over about 100 years up to
some time in the first half of the present century is further illustrated by the curves in fig. 18.6,
where it is seen that the decade means quite commonly show large-amplitude superposed
fluctuations such as are also a feature of fig. 18.1. An enormously greater selection of such
curves – and others for rainfall and pressure changes too – for places all over the world up to

1. An extreme example is the drop of 10-year mean July
(mid winter) temperatures at Punta Arenas (53°s
71°w) in southernmost Chile to approximately 0°c
around 1900, just at the start of the great warming of
the Arctic and when the ice was receding from
Iceland. In most decades before and since Punta
Arenas had mean July temperatures about +2 to
+2·5°C. The coldest July there, in 1900, with a mean

temperature of −2°C surely means that the Antarctic
sea ice had advanced to quite near the southernmost
coasts of South America.
2. Other curves of mean temperature – individual year
averages from 1811 to 1910 for the northern hemisphere temperate zone and for the tropics and decade
averages from the 1750s to the 1950s for Europe – are
presented in figs. 10.18 and 10.23 in Volume 1.

about 1940 is given by LYSGAARD (1949). As fig. 18.1 and most corresponding curves show, the record of temperature change appears as a history of overall warming from the seventeenth century to the early or middle part of the present century. The closer detail in fig. 18.1, which is certainly representative for most of Europe, and perhaps for global average conditions, indicates very rapid warming from the late 1600s to the 1730s – when full twentieth century warmth was attained, or slightly exceeded, for just one decade – followed by a sharp setback, ushered in by the great winter of 1739–40.[1] There was then little real change from 1740 to about 1900, apart from some individual colder decades in the late eighteenth and nineteenth centuries, often coinciding with the prevalence of great volcanic dust veils in the atmosphere (see LAMB 1970 and Chapter 10 in Volume I), and a 20-year period of rather more sustained warmth (though several severe winters also) in the 1820s and 1830s. Apart from this latter period of continental-type climate in England, associated with frequent anticyclones near 50°N, cold winters were generally commonest in those decades when the prevailing temperature was low.

The incidence of severe (and notably mild) winter months, and of very warm (and notably cool) summer months in England is indicated by Table 11 in Appendix V. As seen in fig. 18.5, there was a long period from around 1720 to 1810 when summers tended to be warmer (but not statistically significantly warmer) in England than in the present century. This effect seems to have been exaggerated in some central European records, probably because the instrument sites were in city centres, to the point where even the average temperature for the year in the late eighteenth century appeared higher than at any time since. DRONIA (1967) believes that observations made within the area of increasing warmth due to the growth of cities and industrial areas has similarly exaggerated the accepted analyses of strength and duration of the twentieth century warmth (see Volume I, Chapter 11). Despite this caution, however, it is clear (figs. 18.1, 2 and 4) that from some time around 1890–1900 a sharp warming of world climates set in, which in some ways paralleled the changes that took place about 200 years earlier.

The changes of circulation regime observed since 1940–50 have brought a lowering of the overall average temperature (fig. 18.2) to about the level of 1900–20, this cooling having probably been longer sustained than any other period of downward trend since the examples known in England (fig. 18.1) of the decades 1660–90s and 1730s to about 1770. The global regime changes in this latest period will be analysed later in the chapter.

Because of the almost unbroken one-way (warming) long-term trend of world climate since instrument measurements began there is an obvious need to extend the climatic record farther back in time by the use of various kinds of proxy data (ice cores, isotope measurements, tree rings, lake varves, etc.) and by other studies such as are described in this book. It is clearly now possible to acquire increasing knowledge of the long periods of

---

1. Effects of the warmth of the 1720s and 1730s on health, social life and even on the numbers of the population, as well as on the crops grown, in Scotland and Sweden have been noted in Chapters 12 and 13.

cooling climates between A.D. 1200 and 1600 and between approximately 1200 and 600 B.C.; by the use of some of these kinds of proxy data – and in the late Middle Ages historical reports of the weather – it will be possible even to study the year-by-year sequence.

Periodogram and spectral analyses of the series of monthly mean temperature values in England over the last 290 years suggest that quasiperiodicities of 20–25 years and about 45–55 years play a modest but significant part in the variations in at least several months of the year. Periods of these lengths were also found to be among the most prominent in tree ring series in pines near the Norwegian coast near 68°N (OTTESTAD 1942) and are suggested by long records of the fisheries and sea temperatures. A 55-year periodicity was suggested by BEVERIDGE (1921) from his series of wheat prices in England since the early Middle Ages. One effect of the fluctuation of about 20 years is that the temperature records in England and the Netherlands show that from the 1720s to 1819, from the 1830s to 1880s, and from the 1930s to 1970s all the even-numbered decades have had colder winters than the decades immediately before and after.

GRAY (1974, 1975) has used the freezing dates of Lake Suwa, Japan (see fig. 17.10 and Table 25(*b*) in Appendix V), to derive the probable mean temperature (D J F) of each winter in Tokyo (140 km to the east) from 1443 to the present century.[1] Comparison of the individual winters of the last 200 years in Japan with the temperature of the same winter in western and central Europe and in the eastern U.S.A. (Philadelphia) produced correlation coefficients that were essentially zero, but the present author noticed the following signs of relationship in the case of winters which tended to either extreme (temperature deviations exceeding $1\cdot5\,\sigma$):

(1) All England's severe Januarys in the last 300 years were in winters with normal, or late, or no freezing of Lake Suwa.
(2) All winters with very early freezing of Lake Suwa (before 22 December New Style calendar) gave normal or mild Januarys in England.
(3) Among winters with very late or no freezing of Lake Suwa (after 9 February) very mild and very severe winters in England were both rather strongly represented.

Despite the absence of any general relationship between the individual winters in these widely separated sectors of the northern hemisphere, GRAY reports that the successive half-century mean values of Tokyo winter temperature from 1450–99 to 1900–49 gave a correlation coefficient $r = \pm0\cdot64$ (significant at the 5 % level) with the winter temperatures in central England derived from our winter severity index (see Table 4 in Appendix V). Thus, there seems to be some coherence between the long-term trends in Japan and Europe. Cross-correlation of the decade values through the last five centuries of winter temperature indicated for Japan and other sectors of the middle latitudes of the northern hemisphere at various lags produced the interesting results shown in Table 18.2 (GRAY 1975).

---

1. Series of dates of freeze-up and break-up of the ice on the estuaries in the Hudson's Bay region of Canada from 1714 to 1871 are also now available (MOODIE and CATCHPOLE 1975).

Table 18.2   Correlation of decade values of mean winter temperature DJF around the northern hemisphere with the decade values derived for Japan (Tokyo)

	Years lag giving maximum correlation coefficient	°Longitude west of Tokyo	Rate of progress westward (°Longitude
China	50	30	6·0°/decade
Russian plain, near 35°E	180	105	5·8°/decade
Germany, near 12°E	200	122	6·1°/decade
England	220	140	6·4°/decade
Eastern U.S.A. (Philadelphia) since 1738	350	220	6·3°/decade

The right-hand column of this table gives rather strong evidence that some disturbance affecting the prevailing winter temperature makes steady progress westward around the hemisphere, so as to complete one circuit in about 600 years. This is presumably an aspect, unrecognized before, of the behaviour of wave number 1 which expresses the eccentricity of the circumpolar vortex.[1] This phenomenon, however, accounts for less than half of the variance of winter temperatures and its range differs in different regions, being smaller in Tokyo (0·3°C) than in Philadelphia (0·8°C). An important component of the remaining variance must be the world-wide fluctuation of temperature which produced the warmth of the early Middle Ages and the cold of the Little Ice Age about five centuries later.

Not only changes of the mean temperature level but also changes of the year-to-year variability must be important. WALLÉN (1953) found that variance of summer temperatures in Sweden in the past 200 years has tended to be greatest in periods of meridional circulation character. Similarly, variance of the winter temperatures in western Europe tends to be greatest in periods when W'ly winds are least prevalent. The changes are illustrated in Table 18.3. Note how the standard deviation increased in the periods of warmer summers and colder winters (see also footnote p. 476).

Another noteworthy aspect of the temperature changes observed is the decrease – at all seasons of the year – of the south-to-north temperature gradient across middle latitudes which accompanied the secular warming of climates. This is illustrated in figs. 18.7(*a*) and (*b*). It is an obvious consequence of the fact that the warming was strongest in high northern latitudes. It might also be an expected result of the increasing vigour of the general atmospheric circulation and its heat transport. But it raises a question regarding how that increase of energy of the atmospheric circulation was supplied. Some implications of this

1. KING (1974) may be credited with first suggesting that there is a component of the circumpolar vortex which makes continual westward progress around the hemisphere (and which he suggests is linked with a westward drift of an anomaly of the Earth's magnetic field).

It may be noticed that the time required for one circuit of this feature around the hemisphere roughly coincides with the spacing between the outbreaks of raininess marked by recurrence surfaces in the peat bogs of northwest Europe around 1200 and 600 B.C. and A.D. 500–600 and 1200–1300 (see p. 211).

Table 18.3   Means and standard deviations of temperature in different periods

Summer (JJA) temperatures in Stockholm	Period average °C	Standard deviation °C	January temperatures in central England	Period average °C	Standard deviation °C
1770–89	16·7	1·25	1795–1824	2·6	2·26
			1879–98	2·9	2·26
1906–25	15·2	0·94	1906–35	4·2	1·45
1928–47	16·5	1·23	1938–63	3·1	2·10

question were discussed on pp. 95 and 306 in Volume 1. Investigations there referred to make it clear that the trends of upper air temperature are not necessarily the same as at the surface. In particular, it seems from observation of the reverse changes since 1950 that the middle and upper troposphere over subtropical latitudes was more intensely warmed by subsidence in the anticyclones there during the warm epoch of the early twentieth century, while the temperatures prevailing in those layers of the atmosphere over the Arctic (and high latitudes generally) were not much changed. Hence, in those layers the south–north temperature difference may actually have increased with the secular warming up to 1900–50. Two other factors must be mentioned to explain the decline of the latitude gradient of surface temperature in middle latitudes during the warming epoch, as seen in fig. 18.7:

(1) Increasing vigour of the wind circulation is very destructive of surface temperature inversions and hence to the development of very low temperatures over ice and snow surfaces in the northern regions.
(2) With the secular warming, the mainstream of the wind circulation was tending to be displaced northwards, snow and ice cover on land and sea was becoming less frequent near the northern United States, and hence the locus of strongest temperature gradient moved generally well away north of Charleston.

Atmospheric circulation changes[1]

These have been studied by reconstructing monthly mean maps of barometric pressure at sea level for each January and each July over as much of the world as the data available allow (the method used and test of reliability have been described in Chapter 13, pp. 41–4). The map series for the Januarys and Julys from 1750 have been published (LAMB and JOHNSON 1966). A tentative analysis, extending the series back to 1680, based on isolated barometric readings and more widespread data on wind direction frequencies, temperatures prevailing and weather character, has been carried out and tested in a similar manner to the maps for

1. Tabulations of various circulation indices and wind flow histories are given in Appendix V, Tables 48–50.

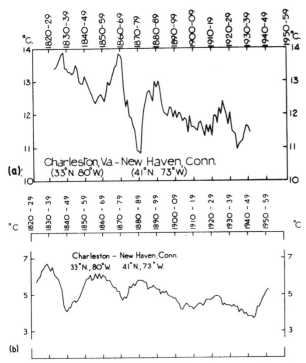

Fig. 18.7   South-to-north differences of mean surface air temperature at 33° to 41°N over North America.

    Ten-year running means.

    (*a*) January.   (*b*) July.

the later years in the eighteenth and nineteenth centuries.[1] These maps represent the most direct possible use of the observation material available to reconstruct the character of the atmospheric circulation prevailing during the most extreme phase of the Little Ice Age.

    The results of these studies may be surveyed in the maps in figs. 18.8, 9, 10 and 11. Figs. 18.8(*a*) and (*b*) compare the mean barometric pressure in January 1790–1829 with that in 1900–39, the latter being the period with most frequent W'ly winds and mild winters in Europe in the record.[2] The Iceland low and the Azores high appear on both maps, but the pressure differences are much weaker on the map for the earlier period and the continental

1. A putative isobaric analysis of maps for ten individual Januarys and ten individual Julys in the late nineteenth century (1880s–90s) was drawn on the basis of the same kinds and amount of data as were available for the 1680s and 1690s, and these analyses were then compared with the analysis of complete data, including ships' reports. The patterns were found reliable over the region shown in fig. 18.10.
2. The use of 40-year averages reduces the random errors of the mean to an acceptable margin (root mean square error < 2·5 mb in January, < 1·0 mb in

July) and thus makes it possible to print a reliable analysis right across the North Atlantic. The isobaric pattern is cut off at limits determined by this error margin. Mercator maps are used because in the later years the analysis extends to both hemispheres. The growth of the observation network and the area of the globe that can be covered by an isobaric analysis showing the mean pressure distribution in individual Januarys and Julys within these prescribed limits of error is shown by maps for the 1750s, 1850s and 1950s in Appendix V, fig. 3.

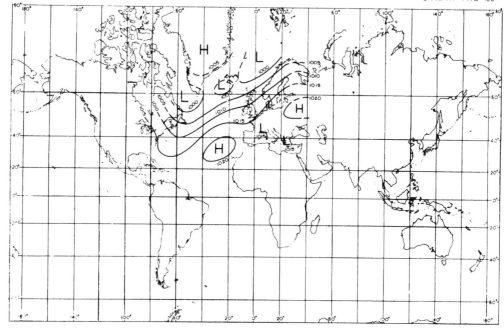

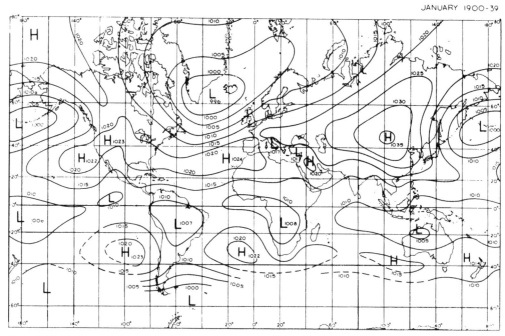

Fig. 18.8 Average mean sea level pressure in January.
Isobars at 5 mb intervals.
(*a*) 1790–1829.
(*b*) 1900–39.

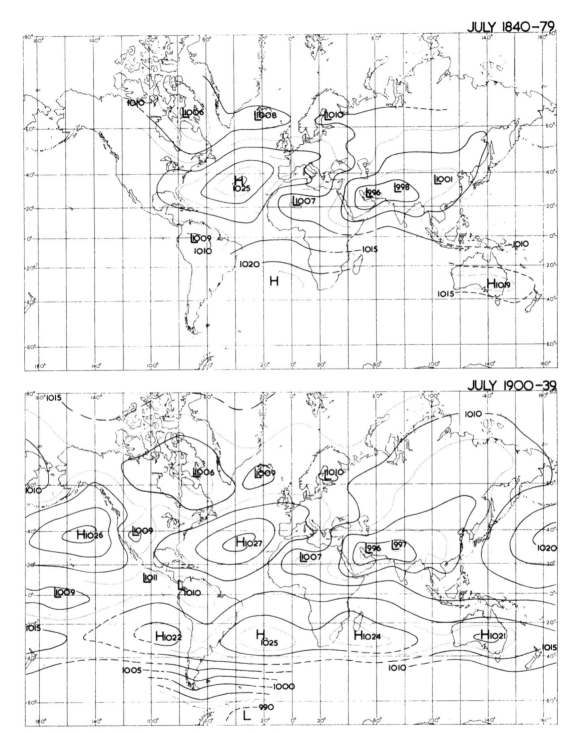

Fig. 18.9 Average mean sea level pressure in July.
Isobars at 5 mb intervals.
(*a*) 1840–79.
(*b*) 1900–39.

anticyclone is more prominent and displaced somewhat to the north over eastern Europe. These differences explain why the winds over Europe, although still mainly W'ly, were more variable in direction – and sometimes E'ly – in the winters around 1800 than in 1900–39. And because the transport of oceanic air from the west was less frequent, it is not surprising that the average winter temperatures were lower around 1800. Figs. 18.9(a) and (b) similarly show mean barometric pressure in July for the two 40-year periods which differed most from each other in the time over which maps can be reconstructed right across the Atlantic, in this case 1840–79 and 1900–39 compared. These July maps, with the characteristically weaker pressure gradients of the northern hemisphere summer season, differ from each other in less obvious ways than was the case with the Januarys. Yet there were significantly more cool Julys in Europe in the mid nineteenth century period. The difference is associated with a slightly more southern position of the axis of high pressure over Europe and the Mediterranean, a more WNW'ly alignment of windstream and isobars over the British Isles and central Europe and less anticyclonic curvature (implying less anticyclonic influence).

Figs. 18.10(a)–(d) show what it is possible to reconstruct by the methods described of the prevailing wind and pressure distribution pattern (the actual pressure values are not to be relied on) in the Januarys and Julys of the coldest decade at the end of the seventeenth century and the remarkably warm decade, the 1730s. And, for comparison, figs. 18.11(a)–(d) show the mean conditions in those decades of the last 200 years in which the Januarys and Julys in Europe were most generally either warm or cold, including the warmest decades of the present century.

The apparently unique features of the maps for the 1690s (figs. 18.10(a), (c)), when compared with maps for later decades are at the western and eastern extremities of the mapable area, namely the low latitude of the subpolar low-pressure belt (which corresponds to the main concentration of travelling depression centres) over the western-central Atlantic and the suggestions of a centre of low pressure somewhere in the European Russia sector (probably near Novaya Zemlya in winter and over northern Russia in July). Although the maps are least trustworthy at their edges, these features would go far to explain the cold, unsettled weather conditions observed over eastern North America and western Europe at the time and probably also accord well with the advanced position of the limit of the Arctic sea ice. The southern position of the low pressure trough right across the eastern Atlantic and European longitudes shown on the July map for the 1690s was more or less matched in the 1760s and 1810–19, but the truncation of the Azores anticyclone and prevalence of NW'ly isobars over central Europe appears much more marked in the Julys of the 1690s. The most remarkable feature of all, however, is the appearance on the mean January map for the 1690s of a centre of high pressure over Scandinavia and a great meridional ridge between 0° and 30°E, implying resultant mean N'ly winds over Finland and north Russia. In no later decade analysed was the frequency of blocking so great as to produce a Scandinavian centre of high pressure in the decade mean for January. And in no individual January since 1714 has a long narrow north–south ridge in about this longitude, linked with the Azores anticyclone,

Fig. 18.10 Results of averaging the putative pressure distribution at mean sea level on maps of observed winds, weather and a few atmospheric pressure and temperature means.

(*a*) Januarys 1690–9.
(*b*) Januarys 1730–9.
(*c*) Julys 1690–9.
(*d*) Julys 1730–9.
(*Observations compiled and maps for the individual years analysed by* H. H. LAMB.)

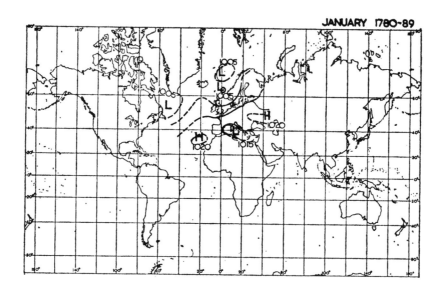

JANUARY 1780-89

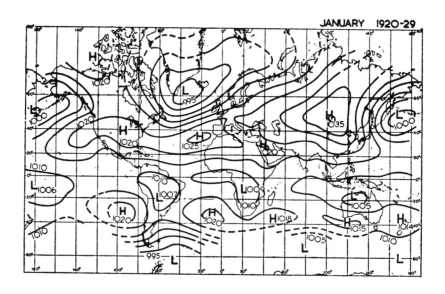

JANUARY 1920-29

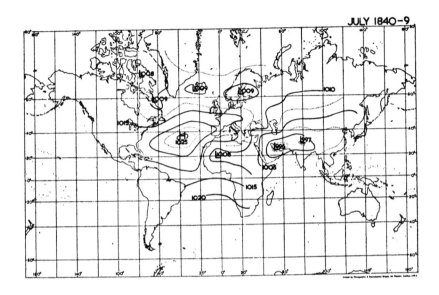

JULY 1840-9

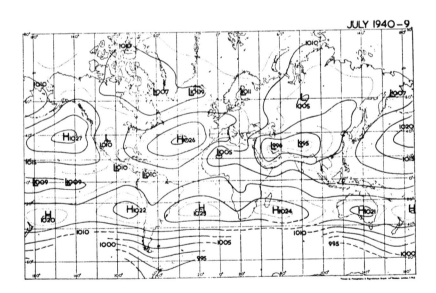

JULY 1940-9

Fig. 18.11   Average mean sea level pressure.
    (*a*) Januarys 1780–9 (cold in Europe).
    (*b*) Januarys 1920–9 (mild in Europe).
    (*c*) Julys 1840–9 (cool in Europe).
    (*d*) Julys 1940–9 (warm in Europe).

appeared as a feature of the monthly mean chart, though it had been common in the previous 35 years.[1]

The analysis of the maps shows that despite the differences between the circulation patterns prevailing in the cold 1690s and in the warmth of the 1730s there were some features in common between them which distinguish a regime unlike the present century. The most obvious example of this is the extent to which Europe was apparently covered by anticyclonic (high pressure) patterns in every decade between 1680 and 1750. A concomitant of that (which unfortunately lies in the less guaranteeable edge region of our maps) is the low latitude of the main cyclonic (low pressure) region over the western Atlantic.

The maps (fig. 18.11) illustrating the most different regimes of later times indicate that most mild winters in Europe have been associated with a strong circulation, as in the 1920s, giving vigorous flow of W'ly and SW'ly winds (though a few of the most extreme cases have been S'ly and rather anticyclonic, as in 1796, 1973 and 1974). All the decades with many warm summers showed markedly anticyclonic patterns over western and central Europe, as in the 1940s. The decades with more cold winters showed relatively weak circulation patterns, indicating much more variable directions of the wind flow and variable positions of the controlling pressure systems. The cool summers, as in the mean for 1840–79 (fig. 18.9(a)) and for the 1690s (fig. 18.10(c)) showed the main low pressure axis near 60°N across the east Atlantic and Europe, a more southern position of the high pressure axis and a net northerly component added to the prevailing west wind flow over Europe.

A survey of the changing strength of the main surface windstreams over the globe from about 1800–50 to around 1960 has been given in fig. 7.1(a) in Volume 1 and of the upper

---

1. The experimental map analyses begin in 1680. Cases of this long linked north-south meridional ridge appear on the January maps for 1680, 1691, 1693, 1694, 1698, 1705, 1710, 1712, 1713 and 1714. The longitude of the ridge axis at 60–65°N ranged in different cases from 0° to nearly 30°E. A somewhat similar feature, but with a southwest-northwest orientation and its axis lying across central or southern Britain and Scandinavia, appears on the January maps for 1729 and 1731, as in 1681, 1704, 1707, 1711 and 1715, but with a further lateral displacement of the ridge axis to the southeast could describe many mild winters of more recent times when SW'ly winds prevailed over Britain and northern Europe. The significance of the extreme north-south orientation of the European high pressure ridge in the period around 1680–1714 presumably lies in its pointing to an extreme meridional circulation form and prevailing short wave length in the upper W'lies and in introducing a short-track flow of direct Arctic air from the north over Europe.

Other Januarys of the period showed a Scandinavian anticyclone separated from the Azores anticyclone by a belt of low pressure across Biscay or farther south to the Mediterranean, as in known E'ly

wind winters in Britain and central Europe in more recent times. Examples appear on the January maps of 1684, 1689, 1695, 1697, 1706, 1709, 1716, 1717, 1718, 1740, 1742 and 1744.

Some straight W'ly Januarys also occurred in Britain and central Europe even at this extreme phase of the Little Ice Age, though perhaps with the main cyclone centres keeping south of 67°N. Examples were 1682, 1683, 1686, 1690, 1696, 1699, 1700, 1701, a less marked case in 1702 and then, after a long gap, 1734, 1735, 1737 and not again till 1754. The gaps were marked by much high pressure near 50°N and some SW'ly or S'ly winds over England, but also some cyclonic cases. Between 1750 and 1900 the Januarys with winds from between south and west heavily predominant over Europe over a wide belt between latitudes 50° and 65°N were: 1754, 1756, 1759, 1761, 1762, 1764, 1777, 1779, 1790, 1791, 1793, 1796, 1798, 1806, 1817, 1818, 1819, 1822, 1834, 1836, 1839, 1846, 1849, 1851, 1852, 1859, 1863, 1864, 1866, 1869, 1872, 1873, 1874, 1882, 1884, 1890, 1894, 1899.

For the patterns of the Januarys and Julys after 1750, the reader is referred to the maps given by LAMB and JOHNSON (1966).

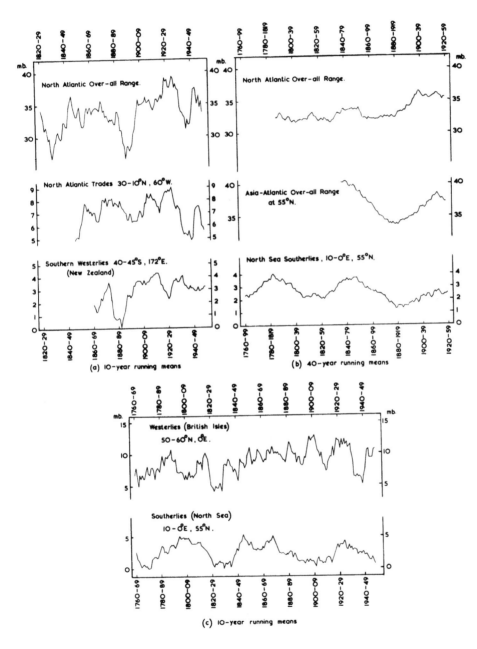

Fig. 18.12 Pressure difference indices of the strength of various windstreams (particularly zonal and meridional components) in January since about 1760.

W'lies in winter over the North Atlantic and Europe from 1880 to 1963 in fig. 7.9.[1] A further view of the matter is given here in figs. 18.12 and 18.13. Other examples are given in LAMB and JOHNSON (1959, 1961). From these surveys the most general features are:

(1) A rising trend of the pressure differences which control the strength and prevalence of most of the great zonal windstreams (North Atlantic westerlies and Trade Winds, southern hemisphere W'lies over most sectors, in January and July and the Indian southwest monsoon current) to attain in each case a main maximum some time between 1900 and 1950, followed by some decline and a falling level of successive shorter-term maxima.

(2) Superposed fluctuations of substantial amplitude affecting the zonal currents on time scales of the order of 20 years.[2]

(3) No apparent trend in the pressure differences which control meridional windstreams but signs of repeating fluctuations on time scales of the order of 50–70 years (and probably around 20–25 years as well).

A further glimpse of these fluctuations as they affect the incidence of blocking anticyclones over northern Europe is seen in the pressure record for Trondheim (fig. 18.14).

The changes of latitude of the main features of the global pressure distribution (North Atlantic subpolar low pressure and subtropical anticyclone, the intertropical low pressure axis, and the southern hemisphere subtropical anticyclones in various sectors) in January and July have been illustrated in figs. 7.1(b) and 7.21 in Volume 1 and will not be further illustrated here. Some further data are given in LAMB and JOHNSON (1961, 1966). The most interesting aspects seem to be:

(1) Northward displacement of the 40-year mean position of the Iceland low (subpolar pressure minimum in the North Atlantic) by some $1\frac{1}{2}°$ to 3° of latitude between the late eighteenth century and the early part of the twentieth century. The change of latitude in July is greater than in January and in July apparently rather closely parallels the withdrawal of the Arctic sea ice, including the temporary reversal of that situation in the last quarter of the nineteenth century. But the renewed southward tendency of the Iceland low since about 1940 (since about 1930 in the winters) preceded the fall of temperatures prevailing in the inner Arctic by some years and preceded the significant readvance of the sea ice by about two decades. As the change of circulation pattern and strength also preceded any significant increase

1. The selection of indices to express the variations of strength of the mean circulation, identification of the most representative ones and the correlation coefficients connecting those that represent different parts of the global atmospheric circulation, are discussed in LAMB and JOHNSON (1959, pp. 112–120; 1961, pp. 371–380).

2. These fluctuations seem to affect the middle latitudes W'lies of both hemispheres about equally, though nothing can yet be said about phase relationships between the two hemispheres or between different sectors of each (see also Chapter 20, pp. 680–1).

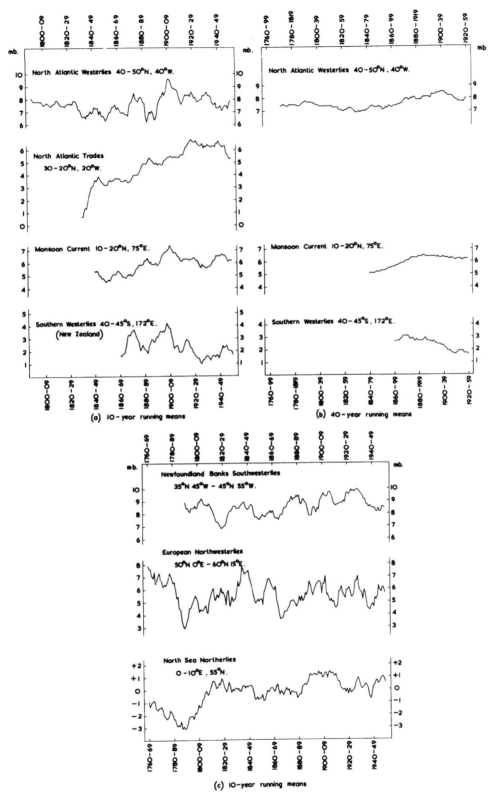

Fig. 18.13 Pressure difference indices of the strength of various windstreams (particularly zonal and meridional components) in July since about 1760.

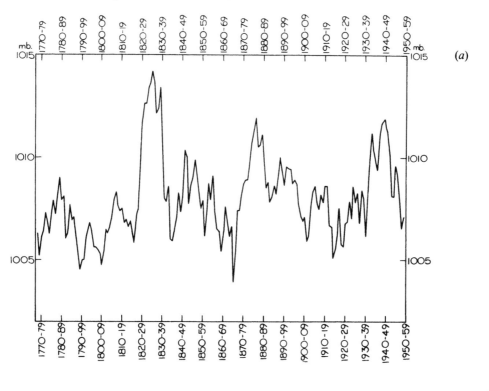

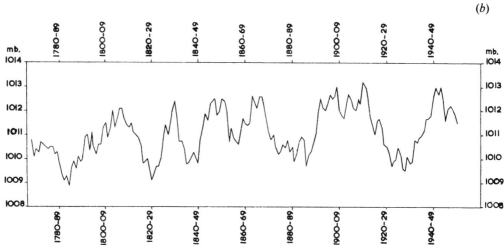

Fig. 18.14　Atmospheric pressure (mean sea level values) at Trondheim (63°N 11°E). Ten-year running means from 1762.
(*a*) January.
(*b*) July.

of volcanic dust in the atmosphere, and the case for increased significance of Man-made dust just at that time is hard to make out, it may be that the likeliest origin is a long-term solar fluctuation (see fig. 2.3(a) and pp. 23–4 in Volume 1).

(2) The 40-year mean positions of the intertropical convergence trough in the Indian Ocean and Australian sectors, and the subtropical anticyclones farther south in those sectors, moved about 3–4° of latitude south in January and north in July between about 1880 and 1925. These movements were what might be expected if the latitude of these features responded to the changing strength of the monsoon windstreams (winter N'lies and summer S'lies) over and near East Asia, with the general intensification of the global wind circulation. Both tendencies seem to have reversed in more recent decades as the circulation weakened.

(3) There is no lasting consistency of parallel movements of the main pressure systems in the Atlantic sector; though, on the whole, the Iceland low and Azores high have moved north and south together, their latitude separation has varied, being probably greatest in the decades when the circulation indices (pressure differences) had their greatest values. The latitude of the intertropical trough in the Atlantic sector, between 40° and 10°w, has shown a movement more like that of the South Atlantic anticyclone, so far as the latter has been adequately diagnosed, both having moved 1–2° of latitude north from around 1910, or earlier, to 1940.

(4) No general tendency appears for expansion or contraction of the circumpolar vortex over either hemisphere, in summer or winter, to accompany increase or decrease of the general strength of the main windstreams.

Study of the secular shifts of longitude of the prevailing trough and ridge positions in the upper W'lies and corresponding features of the atmospheric circulation near sea level is also informative. Long histories, as always, have to be illustrated by features of the surface

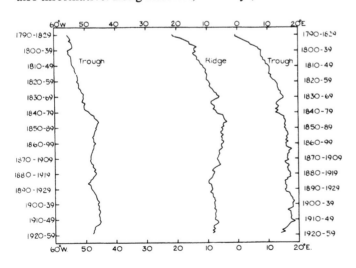

Fig. 18.15 Longitudes of the semi-permanent pressure troughs and ridge (positions of lowest and highest average pressure at mean sea level) at 45°N in the Atlantic sector since about 1800: January. Forty-year running means.

weather maps. The illustrations here in figs. 18.15, 18.16 and 18.17 show the longitude positions of features at 45°N, related to cyclogenesis east of the Canadian upper cold trough and downstream features of the circumpolar vortex, in the Atlantic and European sectors in January and July, and features of similar relevance at 45°S in the mid winter month of July (when the analysis is best established so far south). It is clear from these diagrams that the positions of the key features over the North Atlantic and Europe moved east and their spacing increased as the general circulation strengthened to its maximum in the early part of the twentieth century. These are the changes that would be expected if the wave length in the upper W'lies were increasing as the circulation strengthened east of some feature more or less anchored by the Rocky Mountains. At 45°S a similar increase of spacing and eastward shift is observed in fig. 18.17 to a brief maximum in the 1870s and a sustained maximum between 1900 and 1940. Over the Southern Ocean at 45°S the most nearly anchored feature appears to be the low-pressure trough around 110°E, east of the upper cold trough that is maintained by the protrusion of the Antarctic continent and ice surface into the southern Indian Ocean sector. Over the Southern Ocean, as in the northern hemisphere, the maxima of longitude spacing (and easternmost positions of the low pressure in the southwest Pacific) seem to correspond to maxima of the circulation strength (zonal indices). Hence, trough/ridge diagrams such as here illustrated, and related diagrams like fig. 13.3, may be used for rough measurements of the variations of circulation strength in times before barometric pressure observations were available and over parts of the world where the network is inadequate today.

Since the foregoing circulation index diagrams all relate to January or July, it is of interest to note that the frequency of W'ly situations over the British Isles over the whole year

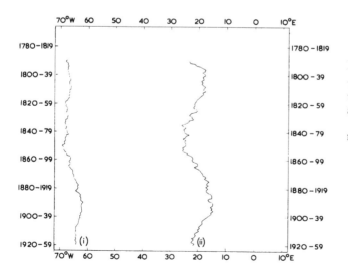

Fig. 18.16   Longitudes of the semi-permanent pressure trough and ridge at 45°N in the Atlantic sector since about 1800: July.
Forty-year running means.
(i) Trough,
(ii) ridge.

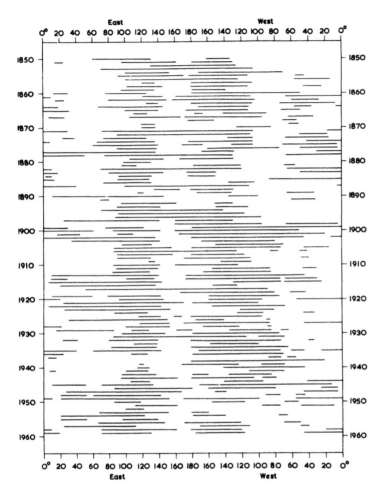

Fig. 18.17   Trough positions at 45°s in July since 1850.
  Longitude ranges with monthly mean pressure < 1010 mb
each July are shown by the horizontal lines.

(fig. 18.18)[1] has varied in a way that fairly closely parallels the indexes of strength of the North Atlantic westerlies and Trade Winds, particularly in January (fig. 18.12). The degree of parallelism with the course of global mean temperature (fig. 18.2) suggests that the number of W'ly days over the British Isles is an index of global significance. Other items which seem to show a parallel history (positive correlation) with this index include the northward penetration of the monsoon over west Africa (see fig. 18.36, p. 544), and therefore the amount of rain along the southern fringe of the Sahara (Sahel) (WINSTANLEY 1973), and the decade averages of accumulation of snow at the South Pole (LAMB 1967b). A much longer

1. See also Appendix V tables on pp. 642 and 647–8.

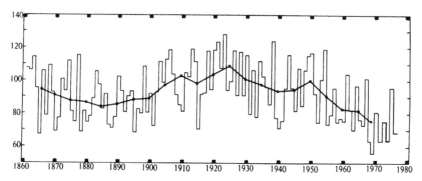

Fig. 18.18   Number of days each year of general W'ly type over the British
Isles: 1861 to 1975. (The number of W'ly type days in 1976 was 59.)
Individual years and 10-year means at 5-year intervals.

history in terms of the average yearly frequency of SW'ly surface winds in England since A.D.
1340 is shown in fig. 6.8 in Volume 1.[1]

The relationships mentioned and others which also seem significant, including a
number illustrated in what follows, have in most cases not yet been examined statistically as
fully as is to be desired. In the case of long-distance relationships it rather commonly appears
that these affect the longer-term changes more strongly than, and sometimes in a different
sense from, the short-term variations. Thus the temperature anomalies that characterize
individual summers and winters in northwest Scotland and north Norway, and still more so
Iceland and Greenland, are often unlike those ruling over central Europe and even the
southeastern half of Britain in the same season.[2] In some cases it may be possible to
demonstrate a significant negative correlation. Yet the long-term fluctuations represented by
10- or 20-year means (and still more by 50- or 100-year means) may be found to show
significant positive correlation (cf. also the relationship between Japanese and European
winter temperatures mentioned on pp. 484–5).

It seems likely that many other teleconnections within the global wind circulation
remain to be discovered. The changes of the global mean wind circulation during the last 300

1. It may be of more than passing interest that in the
years 1968–76 the average number of W'ly days (69·4)
and the numbers in the extreme years (1968, 1969,
1971, 1973 and 1976) (see fig. 18.18) have fallen to
levels which had not occurred since 1860. The only
previous match for the figures of these years so far
discovered is about 190 years earlier, in 1781–5, when
the frequencies of the other wind direction patterns
over the British Isles were also similar to 1968–75 (see
KINGTON 1975). (Analysis of the year 1785 just
completed, however, shows an altogether lower level
of frequency of W'ly days, totalling 45, perhaps an
effect of the exceptional volcanic dust veil over the
Earth after the two great eruptions in 1784 (see
Chapter 10 in volume 1).) Fig. 6.8 suggests that

corresponding sharp declines of the prevailing W'lies
may have occurred at about 200-year intervals since
1340.
2. An out-of-phase relationship between the temperatures characterizing individual winters in west
Greenland and in northern and central Europe was
first noted by HEINRICH WILHELM DOVE in his
remarkable work 'Über die geographische Ver-
breitung gleichartiger Witterungserscheinungen',
published in the *Abhandlungen der Kgl. Akad. Wiss.*
(Berlin 1838), pp. 287–415. See also LOEWE 1937).
Nevertheless, when the average conditions of different centuries – still more of different millennia –
are compared the Greenland and European records
are more nearly parallel.

years surveyed in this section also provide a framework against which it is possible to interpret changes in the incidence of rainfall, snowfall, great storms and other aspects of the climate. These will be reviewed in the following paragraphs.

The equatorial rain-belt seems to have been generally more active in the nineteenth century than since. It is possible, therefore, that slightly greater world cloud cover contributed to lessening the global intake of solar radiation, though it does not seem likely that the change from this cause was as great as that resulting from the greater extent of snow and ice. Reduction of the net radiation supply over a long period by at least 0·5 to 1 % may have to be invoked (LAMB 1963, p. 141) to account for the coldness of the climate around A.D. 1800, whether this reduction was due to the frequent volcanic dust veils at that time, or to a variation of the sun's output at source, or to both these causes.

Incidence of gales, tropical storms, tornadoes

The change in geographical distribution of gale frequency (Beaufort force $\geqslant 8$) in the Atlantic and European sector of the Arctic from the 1930s to the late 1940s, mapped in figs. 7.15(a) and (b) in Volume 1 (p. 284), conforms well to what should be expected from the southward shift of the belt of lowest pressure by 2–3° latitude between the two epochs. The frequency of gales everywhere in that sector north of 70°N decreased, figures for the areas of maximum frequency near Jan Mayen and the Greenland coast fell from 10 % to 6 % of the observations and in the Barents Sea from 6 % to 3 %, while the frequencies near Iceland increased from around 3–4 % to over 6 % of the observations in the later years. PETTERSSEN (1949) had reported an opposite change in the frequencies of cyclone centres over the previous decades, with a great increase from 1900–19 to 1920–39 in the numbers passing through Denmark Strait and reaching the northern part of the Greenland Sea and Norwegian Sea area, while the numbers over the eastern Atlantic south and east of Iceland decreased. In conformity with this, places on the Greenland coast, Jan Mayen island and Spitsbergen had a sharp maximum of gale frequency in or about the late 1920s and 1930s, frequencies at the beginning of the century having been mostly similar to those in the late 1940s.[1]

Farther south in middle latitudes in the Atlantic and European sector, despite the wealth of high-grade observing stations, there are not many long anemometer records where the instrument has not been either altered or affected by changes of exposure due to new buildings and the growth of trees, etc. There seems to have been not much change in the

---

1. Despite the evidence in this paragraph, and later, of how many secular changes of windiness can be explained by latitude shifts of the zone of maximum incidence of cyclonic activity – and, therefore, can be associated with expansion or contraction of the circumpolar vortex – the present writer believes there must also occur important variations of gale frequency at places in middle and high latitudes (parti- cularly near the mountainous north–south aligned coasts of Greenland, Novaya Zemlya and the Ross, Bellingshausen and Weddell Seas, and probably of British Columbia, Norway, Chile and New Zealand) associated with changes in the *longitudes* most frequented by slow moving cyclone centres and upper troughs.

frequency of gales in Shetland apart from the lower frequencies reported between about 1915 and 1933 than before or since, though the record is suspect before 1929 because of instrument and site changes and almost certainly exaggerates the variation. South of 60°N, however, there seem to be reliable indications of increasing gale frequency in the British Isles from the 1940s to the 1960s or later, in good agreement with the increasing frequency of cyclone centres in these latitudes.[1] The gale frequency at Tiree in the Hebrides, at Holyhead, Anglesey and Shoeburyness in the Thames Estuary, more than doubled over these decades.

The frequencies of tropical cyclones in nearly all the regions of the world that are liable to them, and the frequencies of those that develop to hurricane intensity, are also found to have undergone a sequence of changes that seem to be related both to the sea surface temperature changes in the world's warmest ocean areas (see Volume 1, p. 405) and to the changes in the general wind circulation. The magnitude of the changes has not only attained statistically significant levels but has been of serious concern to the insurance of risks on land and sea.[2] In the case of the tropical cyclones over the North Atlantic (see Table 18.4), though the minimum in the period 1907–31 must be associated with the fact that the ocean surface temperature in low latitudes was often below the apparently critical value of about 26·5°C in those years, it seems that the frequency of development of these storms has generally fallen off in those years or periods when the middle latitudes W'lies have been most strongly and persistently developed.

Table 18.4   Numbers of tropical cyclones of all intensities in the Atlantic and Caribbean area (5-year averages)

Years	Average yearly number	Years	Average yearly number
1886–90	9·2	1931–5	11·6
1891–5	8·8	1936–40	9·2
1896–1900	6·6	1941–5	9·4
1901–5	6·8	1946–50	10·0
1906–10	7·4	1951–5	10·8
1911–15	4·0	1956–60	8·8
1916–20	5·8	1961–5	8·4
1921–5	5·4	1966–70	9·2
1926–30	5·6	1971–3	(7·7)

Sources: Mainly from CRY *et al.* (1959). See also GENTRY (1963), MILTON (1974).

1. Compare the references in Chapter 17 to increased frequency of gales in the fifties latitudes in the Late Middle Ages and the particular storminess of the late sixteenth and seventeenth centuries – a more extreme example of the same effect.

2. The changes of frequency of hurricanes in the North Atlantic from 1887–96 to 1897–1931 and to 1932–55 appear to be statistically significant at the 0·1 % level (Student's *t* test).

(a)

(b)

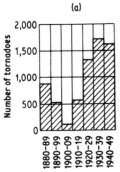

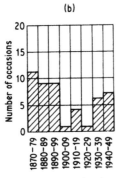

Fig. 18.19   Numbers of tornadoes reported in each decade: (a) in the United States from 1880, (b) in the British Isles from 1870.

(*From* LAMB (*1957*); *main source used for American figures* BROWN *and* ROBERTS (*1935*). *Reproduced by kind permission of the Comptroller of Her Majesty's Stationery Office, London.*)

The number of tropical storms developing may have been more nearly constant over the western North Pacific – the most prolific region – than over the other areas.

More extensive tables of tropical storm frequencies in those parts of the world where reporting has been most complete are given in Tables 45–47 in Appendix V. In some areas, where the network of observing posts has improved in recent years, questions have naturally arisen as to whether that is the reason why greater numbers of storms have been reported. There is no reasonable doubt, however, that in the most consistently surveyed areas there was a decrease of the frequency of these storms towards the end of the last century and a recovery of their numbers between about 1930 and 1950 that was in some areas gradual but in others (especially in the North Atlantic) startlingly abrupt.

The abruptness of the changes has perhaps been the more general experience (see also the low frequencies in the 1870s, 1890s and 1920s in the southwest Indian Ocean, recorded in Table 47 in Appendix V) and probably points to the effect of some threshold condition. Whether or not this threshold is primarily a matter of sea surface temperature in low latitudes, there is no doubt that changes in the patterns of the global wind circulation, and particularly the meridional amplitude and the longitude positions of the troughs in the upper westerlies, have affected the frequency with which the paths of tropical storms recurve towards higher latitudes and the point in their tracks at which this change of direction has taken place. Thus, the increased meridionality of the circulation in middle latitudes in the late 1950s and after was associated with tropical storms occasionally recurving northwards over the interior of the United States on tracks not observed earlier in the century; other characteristics of the mean circulation in these recent decades seem also to have increased the frequently of hurricanes which failed to recurve at all and continued westwards through the Gulf of Mexico. Increases in these same decades of tropical cyclones affecting the Solomon Islands in 7–10°S near 160°E and of others passing through the Timor Sea and affecting northernmost Australia may be associated with the same tendency for some storms to continue westwards on long tracks in low latitudes, perhaps controlled by extended development of a subtropical anticyclone and of the tropical E'lies in a rather low latitude.

The incidence of another type of violent storm, the tornado, appears to show a strong secular variation which largely parallels that of the hurricanes in the North Atlantic. Fig.

18.19 shows the frequencies of tornadoes reported in the United States and in the British Isles by decades from 1870 or 1880 to the 1940s (the survey illustrated is discontinued at that point because new techniques of observation are believed responsible for the very great increases in the numbers of tornadoes reported since 1950 – see COURT 1970, p. 35). In the case of these storms it seems likely that the main physical variables accounting for changes in their incidence are things that affect the sharpness of the thermal contrast (and hence the wind shear) at fronts and the moisture content and thermal/convective instability of the uplifted air. These things are likely to increase with the meridionality of the large-scale wind circulation pattern and the temperature of the nearest warm ocean areas. Both these items varied broadly in the right sense over the decades surveyed in figs. 18.19(a) and (b) and seem capable of accounting for the minimum tornado incidence in the early decades of this century.

Sea temperatures

Histories of sea surface temperature in widely scattered parts of the world, in some cases from the 1880s, have been given in fig. 10.11 in Volume 1. The places for which records are available – whether at the equator or near 40–50°N – show a rather general warming of the overall average sea surface temperature by about 0·5 to 1·0°C from the beginning of the record, or in some cases from a minimum between 1915 and 1925, to around 1940 to 1950.[1] Near south Iceland the rise from 1910–19 to 1940–9 was 2·1°C, and near the coast of Norway in 60–70°N it ranged from 1·1 to 1·8°C increasing northwards (BROWN 1963). Maps of mean sea surface temperature in the Atlantic Ocean, averaged over periods of 40 years or longer back to 1780, can be presented and are illustrated in figs. 8.9 and 10.12 in Volume 1 by surveys from different sets of observations which tend to confirm each other. The ocean generally north of 45–50°N, in latitudes 0–20°N, and south of 30–40°s, appears to have been colder in the 1780–1850 period than in this century. Averaged over the whole Atlantic Ocean between 55°N and 40°s, the sea surface was probably cooler by about 0·3°C than in the 1920s and 1930s or by 0·8 to 1·0°C than around 1950.[2] The departures showed a strong

1. The reversal of trend since 1940–50 may have been greater than currently accepted sea surface temperature figures indicate because of the general adoption since about the same time of water temperatures measured in the intake to the ships' engines instead of by the former canvas bucket method. Investigation by SAUR (1963) of the results obtained by the two methods indicated that, on average, the intake temperatures were 0·7°C higher.

2. We have illustrated in fig. 15.13 an experiment carried out by WEXLER (1956) in which he moved the mean winter insolation isopleths over North America southwards by the amount required to reduce the radiation received at each point by 20% and then moved the isopleths of 1000–700 mb thick-

ness in the atmosphere south by the same amount. He presented the result illustrated as a theoretical picture of the atmospheric circulation pattern that should prevail under a reduction of the incoming solar radiation by this amount. The main difference from the regime observed in the present century is a deepening and sharpening of the upper cold trough over North America. This result was exactly the nature of the change from modern conditions observed on the actual maps for the period about A.D. 1800 constructed by the present author (LAMB and JOHNSON 1959, 1966) though WEXLER's case was much greater in amplitude. If we apply WEXLER's argument in reverse, and make the obviously crude assumption that the whole explanation of the shifts

geographical pattern with maximum cooling in high latitudes and off the American seabord in 40–50°N, where the cold Labrador Current evidently extended farther south. And, as we shall see, the supply of polar water to the East Iceland Current evidently increased and it advanced strongly towards the Faeroe Islands. In the central Atlantic, in subtropical latitudes, the water was reported as warmer than now (see remarks about variations of the Gulf Stream and North Atlantic Drift on p. 335 in Volume 1).[1]

Few sea temperature records either for individual places or individual ships before about 1850 have come to light. Special interest may therefore be attached to an isolated record (WEST 1970) from northern waters in the summer of 1789 (which was a pleasantly normal summer in western and northern Europe with a rather warm May and August). Sea and air temperatures were carefully measured by JAMES WRIGHT, surgeon and botanist on JOHN THOMAS STANLEY's expedition from Scotland (Leith) to the Faeroe Islands, Iceland and Denmark between May and September 1789. Departures of the sea surface temperature values he observed from the U.K. Meterological Office Atlas averages for 1887–99 and 1921–38 for the relevant month were as follows (see also fig. 18.20):

Between the Firth of Forth and Orkney	−0·3 to −0·5°C
Faeroes area	−1·1°C
Between Faeroes and southeast Iceland	−1·0 to −1·6°C
Near Westman Islands	zero
Near Reykjavik	0 to +1·5°C
Off south Norway	−0·5°C
Danish Sound (Øresund)	+0·3°C

of sea surface temperature in the Little Ice Age period lay in a deficiency of the incoming radiation, the southward displacement of the 1000–500 mb thickness isopleths over the North Atlantic and eastern North America derived by LAMB (1963) about 1800 is found to correspond to a reduction of the incoming radiation in summer and winter of the order of 5%. Extending the examination to the detail of different regions of the ocean, we find that the figure of supposed radiation deficit for the North Atlantic is influenced rather strongly by the sea surface temperature anomaly in the fringe regions where the cold currents were clearly extended. The average radiation deficit indicated for the whole Atlantic Ocean between 50°N and 40°S is between 0·7 and 1·5%. This figure may be too low because the most extensive region of colder surface north of 50°N has been excluded. We may guess that the anomalies in the great Pacific Ocean were smaller than in the Atlantic, yet it seems that an average deficiency of the incoming solar radiation of rather over 1%, maintained over some long period (possibly of the order of a couple of centuries) to allow an approximate equilibrium to be attained, is a possible explanation of the conditions observed around 1800.

1. A similar but much weaker tendency for positive temperature anomaly within the same area at the climax of the last major ice age, 17 000–20 000 years ago, has been independently deduced by A. McINTYRE from the biological evidence in the ocean-bed sediment (see figs. 15.21(a) and (b)).

A reconstruction of yearly values (estimates) of sea surface temperature from A.D. 1611 to 1964 in the upwelling region in the Pacific off southern California has been produced by DOUGLAS (1974), using multiple linear regression analysis to exploit the apparent correlation between sea surface temperatures in the region and tree ring data for five sites in southern California. The results indicate a general warming, amounting to about 2°C, from the mid seventeenth century to the period between about 1830 and 1930, but with notably cold decades around 1710, in the 1880s, around 1900 and in the 1950s, and notable warmth in 1611–12, the 1680s, 1770s, 1830s, around 1870 and from about 1910 to 1930. The range between the warmest and coldest individual years of the whole sequence appears to approach 7°C.

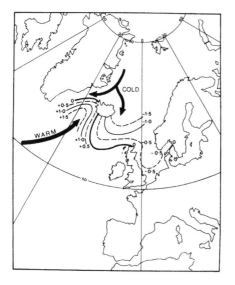

Fig. 18.20   Sea surface temperatures observed in the summer of A.D. 1789.

The values are departures (°C) from modern (1921–38) averages.

The chief import of these values seems to be that the East Iceland cold current was stronger than it has usually been in this century and was transporting cold water towards the Faeroes, while the Irminger Current was probably active in bringing warm water from the south towards Iceland.[1] Meanwhile, quiet summer weather allowed the water to become quite warm in the Baltic and near Denmark.

The longest continuous sea temperature history seems to be that put together (with some use of proxy data from land stations in Denmark) by DIETRICH and KALLE (1957) for the North Sea near the coast of Denmark from 1849 (fig. 18.21). It shows the climatic warming since the middle of the nineteenth century to the 1940s and early 1950s, amounting to about 0·7°C, taking place in two rapid stages around 1900 and around 1930. There is a levelling off, but no minimum as in some other areas, in the 1920s. This may be compared with the record indicated by travelling ships' observations in the outer Biscay area of the North Atlantic 45–50°N 5–10°W from 1854 to the early 1970s in fig. 18.21(*b*) (see also Table 12 in Appendix V), where the range of the 19-year averages is about 0·8°C, but the timing differs and may be more typical of the world ocean. Another long record – the longest that is entirely a record of observations of the water in one small area – is that from the Faeroe Islands from 1867 to 1970, homogenized by Dr P. M. KELLY of the Climatic Research Unit, Norwich, from overlapping records of water temperature measurements at Mykines in the west and Torshavn on the southeast side of the islands. The overall average value for 1867 to 1970 was 7·7°C. This record of conditions in the open Atlantic again differs from that in the partly landlocked European waters, and is in some respects more similar to the outer Biscay

1. There had been sea ice reported near parts of the   was common at that epoch.
   coast of Iceland from January till early June 1789, as

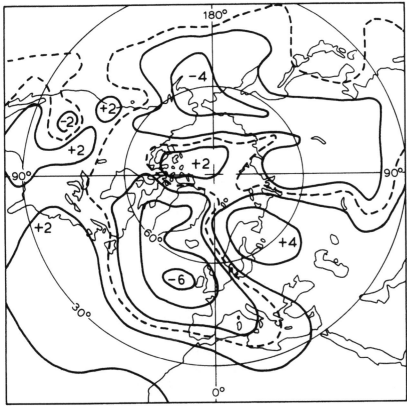

Fig. 18.21 Sea surface temperature histories (19-year running means).

(*a*) In the eastern North Sea, from 1849. Observations at Horns Rev lightship, near 55°30′N, 7°30′E, expressed as departures (°C) from the 1901–30 average (approximately 9·2°C). The early part of the record, before 1880, makes use of observations at land stations in Denmark In the absence of later data for Horns Rev, we may note that mean sea surface temperatures at Weather Ship P in the German Bight in late summer (September) fell by 0·8°C from 1934–53 to 1955–74.

(*Adapted from a diagram given by* DIETRICH *and* KALLE *1957*.)

(*b*) In the Atlantic outside the Bay of Biscay, from 1854. Approximate averages for the 'square' 45–50°N 5–10°W indicated by travelling ships (see Appendix V, Table 12).

(*Adapted from* COLEBROOK *1975*.)

record, though in all cases the earliest years were nearly the coldest and conditions became colder again after 1955 p. 533). At the Faeroes the years 1882–4 brought warm water, and there was a 12- to 15-year period of warmth from 1890, followed by about 20 years of generally colder water centred around 1917. The warmest water was observed between 1940 and 1957, the peak being in 1951 (average 8·9°), and by the late 1960s values were back to the lowest level which had prevailed earlier in the record. The difference between the means of the warmest and coldest individual years was about 2·4°C.[1]

The Faeroe Islands lie in a peculiarly sensitive region of the ocean for registering climatic changes. In warm periods, with generally strong, or northward-displaced, circulation features in atmosphere and ocean, these islands lie continually in the broad sweep of the main arm of the Gulf Stream–North Atlantic Drift as it approaches the entry to the Arctic via the eastern Norwegian Sea. In colder periods, when the North Atlantic Drift weakens or its main arms pass either altogether farther south or towards southwest Greenland, southwest Iceland and the coasts of Europe mainly south of 50–60°N, a tongue of the polar water from the East Iceland branch of the East Greenland Current is liable to approach the Faeroe Islands from the north.[2] The East Iceland cold current itself seems to be a somewhat occasional feature, prominent only in cold years.[3] In this respect the region is not fully typical of the world ocean but so placed as to register the largest amplitude shifts of the water current boundary in the North Atlantic (cf. the discussion of the ice age development in Chapter 15, p. 328, and figs. 15.21(a) and (b)). The timing of the fluctuations of this boundary may not fully coincide with the fluctuations of overall mean surface temperature in the world ocean, although some general correspondence may be expected. An illustration of such differences of timing is provided by a 90-year history of sea surface temperatures in the summer half-year off west and south Greenland (HERMANN *et al.* 1973), where water temperatures were low from the 1870s to about 1920 and the lowest 5-year mean occurred about the latter year, followed by a sharp rise of 1·7 to 1·8°C to the highest 5-year means in 1926–32. The 5-year means there remained generally 1 to 1·5°C above the previous level (except briefly around 1940) until 1960–4, but fell abruptly back to the 1870s level in the years 1966–72 (these details should be compared with the sea temperature histories in other latitudes, in fig. 10.11 in Volume I, summarized above on pp. 506–7).

Since the abundance of cod appears to be limited by the 2°C water temperature isotherm, a much longer history of sea temperature in the region about the Faeroe Islands

---

1. The difference between the warmest 5-year means (8·2° around 1955, 8·1° around 1895) and the coldest 5-year means (7·2° around 1869, 1917 and 1968) in the period of this record was about 1°C – i.e. twice as great as the variations of prevailing air temperatures over England and central Europe.

   A secular maximum of sea surface temperatures appears to have been registered about the following dates, differing in the different ocean regions of the world:

Southern North Sea and Baltic	1945–9
Western North Atlantic	1951–5
Eastern North Atlantic	About 1935–9 and 1955–9
Off west Greenland	1960–4

(See also the graphical presentations for other ocean areas in Volume I, p. 405.)

2. The map (fig. 8.10) and text on p. 335 of Volume I are interesting in this connection.

3. A clear case of the advance of these prongs of the polar water was observed (PICKETT and ATHEY 1968) in a survey with an airborne radiation thermometer on 9 April 1968, showing the warm North Atlantic Drift water sharply limited along a line from southeast Iceland (where a tongue of the Arctic sea ice was advancing) via 65°N 11°W to 63½°N 6½°W and thence south towards the Faeroe Islands. Immediately south and west of this line water surface temperatures were between 5° and 7°; immediately the other side of the line they were between 2° and 3°C.

may be detected from the records of the fisheries. There were many years during the Little Ice Age period when the Faeroese cod fishery failed, and SVABO recounted from his visit to the islands in 1781–2 verbal reports that the fishery did not fail so often before about 1600–50[1] (even the famous bird colonies of the islands seemed to have been partly deserted after that time, so it is probable that many fish which the birds feed on had moved elsewhere). Failures of the Faeroe Islands fishery were reported in 1626 and 1629, when many of the inhabitants died of hunger, and in most years between about 1675 and 1685 (when the taxes were reduced by 50 % because of total failure). Again in 1715 total failure was reported. Between 1709 and 1788 there were only four years (1748–50 and 1757) when there was a significant surplus allowing export of over 36 tons of fish; between 1789 and 1828 there were eleven such years, and between then and 1856 only one year (1839) in which the export failed to exceed 36 tons – over ten times that amount was more usual.[2]

Table 18.5 Yearly export of fish from the Faeroe Islands by the government handling monopoly
  Units: Voger (1 vog = 36 pounds)
  (Data from SVABO 1959)

Decade	Year									
	0	1	2	3	4	5	6	7	8	9
1720s										603
1730s	373	159	0	0	0	20	106	1011	40	40
1740s	267	0	0	0	420	112	0	436	4702	2421
1750s	3860	1084	50	360	0	0	100	2636	1404	0
1760s	0	0	588	931	0	801	1600	966	420	110
1770s	0	520	1092	0	0	0	124	136	460	33
1780s	0	0								

The arrivals of schools of whales in Faeroese waters seem to have varied in the opposite sense to the fisheries, the whale catches between 1717 and 1739 constituting the famous period of this activity. How far this can be taken as an indicator of the approach of the polar water from the north has not been determined.

Another indication of the more extreme variations of winds, currents and ice in the same regions is probably to be seen in a number of recorded instances of isolated Eskimos in their

1. I have not found any report that the fisheries off the main inhabited regions in south and west Iceland failed during most of that time, but L. KRISTJANSSON (1971) reports total failure of the Iceland fisheries between 1685 and 1704, even off the southwest of the country. The first hints of impoverishment of the Faeroe Islands fisheries come in 1615–24 in the form of complaints of unfair encroachment of English and Scottish fishing boats. The nadir of the fisheries was evidently between 1670 and 1790. In the most extreme year 1695 (see p. 583)

failure of the fisheries also in Norway, from Stavanger to the far north, suggests that the polar water may have spread across the surface of the entire Norwegian Sea.

2. TAIT and MARTIN (1964) report evidence that the haddock fishery about the Faeroe Islands and the herring fishery in the Minch (Hebrides area) are sensitive to the occasional, and relatively minor, intrusions of polar watermasses into the lower levels of the Faeroe-Shetland Channel, which have occurred in this century.

kayaks arriving in Faeroese and Scottish waters, four known instances between 1682 and 1700 and one that reached the river at Aberdeen as late as 1843. It has been supposed (G. MANLEY, personal communication, 28 January 1962) that these cases arose when large areas of pack ice, from which the Eskimo had been fishing, advanced or broke away and drifted far beyond the normal limit.

If the outflow of polar water with the East Greenland Current, and especially with its East Iceland branch, is as greatly variable as we have indicated in this section, and as seems to be substantiated by oceanographic observations off northeast Iceland in the 1960s (MALMBERG 1969), it is appropriate to ask where the extra water comes from. Between A.D. 1300 and 1500 (as mentioned in an earlier chapter), when the East Iceland Current was also stronger and transporting ice farther south than had been the case in several previous centuries, fishery records for the Norwegian coast indicate a broad current of the warm North Atlantic Drift water pushing well to the north; so that at that time there seems to have been a strengthening of the whole gyre in the East Greenland Sea, and there is evidence of a similar situation at times in the 1960s. But after 1500 the warm current by the Norwegian coast seems to have been mostly weaker than before (except about the 1730s) and the supply of North Atlantic Drift water to the inner Arctic must have partly failed. That the Irminger Current branch of it, approaching southwest Iceland from the south or southeast, seems to have been fairly strong hardly affects the issue because of the evident dominance of the ice-bearing polar current about other parts of the Iceland coast and at times as far south as the Faeroes region. The scientific voyage in 1789, quoted on p. 507, indicated that the Norwegian Sea was colder than nowadays across its whole width. It seems necessary to suppose that in the period 1500–1800 the compensating supply of water to the polar region was at least partly dependent upon an increased inflow through the Bering Strait from the Pacific (see pp. 528, 533).[1]

1. Since this chapter was written a memorandum by STOLLE (1975) has appeared which presents a useful brief summary of the history of the Gulf Stream-North Atlantic Drift from 1577 to 1974. The observations used start with FROBISHER's account of his voyage from the Orkney Islands in search of a Northwest Passage and of the great quantities of driftwood which he met, whole fir trees rooted up, it was presumed in Newfoundland, and carried by the current setting towards the east and then towards Iceland. They also include Benjamin FRANKLIN's rediscovery of the Gulf Stream, the results of his own observations and other surveys instigated by him, following a complaint by the Board of Customs in Boston in 1769 that mailships from England (Falmouth) to New York took 14 days longer than ships on another course to Rhode Island (farther north). Many original maps are reproduced in STOLLE's history, the most important sources of the maps and observations being apparently BLAGDEN (1781), FRANKLIN (1786), STRICKLANDER (1802),

TRUXTUN (1794), BECHER (1843, 1852), KOHL (1868) and the observations carried out by the Woods Hole Oceanographic Institution in the last 40 years.

This intermittent observation series shows that already in 1577 the Gulf Stream must have turned east from the American coast on a more southern track than has been usual in this century, although a strongly marked branch of it was reaching Iceland. By the mid eighteenth century the warm current was still farther south, turning east near Nantucket Island at 41–42°N and thence east-southeast from a point about 62°W, though at 45–50°W the current branched northeast towards Iceland and southeast towards the Azores. The southernmost position and a set of 120° towards mid ocean was shown by the publication of 1794 and is confirmed by the sea temperature observations between 1791 and 1819. There was some theorizing at that time as to why the current turned farther south than it was believed to have done earlier. The surveys from this period on,

We may perhaps summarize our survey of the ocean surface temperature variations in the last three centuries by concluding that the changes indicated have generally been about as great in magnitude as the change of prevailing surface air temperature over the globe since 1870 and, probably, similar to the change in England over the three centuries. There are also indications of lag in the sea surface temperatures, which have registered their highest (lowest) values at the end of the period of high (low) air temperatures – e.g. the climax of warmth in the ocean around 1950–5 and of cold in the eighteenth century. Against the indications of general worldwide warming of the ocean surface over the last 100 years by between 0·5 and 1·0°C, and perhaps 1·5°C since around 1700, we see that 5-year means of sea surface temperature in regions affected by the shifting boundary of the polar water – between Iceland and the Faeroes, off south and west Greenland, and probably off eastern North America – have had a range of 1·0 to 1·8°C within the last 100 years and general lowering from average twentieth century values by as much as 4 to 5°C in these regions may be implied by the fisheries and ice data of the most extreme runs of twenty to thirty years in the Faeroes/Iceland region between about 1670 and 1770.[1] These observations seem to make sense of the rather numerous reports of travellers between about 1600 and 1700 (see p. 525) that there was always snow on the tops of the Cairngorm Mountains in eastern Scotland, implying that 50- to 100-year mean air temperatures in the northern half, or more, of Scotland (and probably in south Norway also) in that period were about 1·5°C below the 1931–60 average, an anomaly twice as great as in England.[2]

which were completed with bottle drift and temperature observations (cf. fig. 8.10 in Volume I), substantiate a return of the current to more northern positions by the 1850s and after (see remark by SABINE quoted on p. 335, Volume I). The observations in recent decades point to a further northward shift of the Gulf Stream limit off the American seaboard, and a more northeastward set from there, up to 1942. Between 1947 and 1950–60 the observations concentrated on the immediate area of the Gulf Stream limit off that coast suggest that the northward trend continued, but from some time between 1950 and 1960 and the latest year available the tendency appears to have been for the turning point to shift farther south again.

W. S. BROEKER reported evidence from Arctic ocean bed cores in the Chuckchi Sea (in discussion) at the WMO/IAMAP Norwich Symposium on Long-Term Climatic Fluctuations, August 1975, that the amount of water passing northwards through the Bering Strait had varied and was significantly greater about 200 years ago, in the Little Ice Age period, than since.

1. Cf. the ranges we have indicated for changes of prevailing air temperature in figs. 18.1, 18.2 and Tables in the Appendix. Five-year means of air temperature in central England probably rose by 1·5 to 1·8°C from the 1690s to the 1730s and to the

warmest years in the present century. Half-century averages in England seem to have fallen by 1·5°C from the medieval warm period to the late seventeenth century and to have risen by 0·7 to 0·8°C to 1900–50.

2. The following observation which seems to require a lowering of the overall prevailing temperatures in Scotland in the 17th century by fully 1·5°C for its explanation has been brought to my attention by Professor GORDON MANLEY: a report by Sir GEORGE MACKENZY, found in *Hutton's Abridgment of the Philosophical Transactions of the Royal Society*, Vol. II, 1672–1685, includes these words: ... 'There is another little lake in Straglash [presumably Strathglass] at Glencannich on land belonging to one Chisholm ... the lake is in a bottom between the tops of a very high hill, so that the bottom itself is very high. This lake never wants [lacks] ice on it in the middle, even in the hottest summer, tho' it thaws near the edges ...' 'Our famous Lake Ness never freezes...' The tarn referred to seems almost certain to be Loch Tuill Bhearnach, which is 500 m long by 200 m across and 750 m above sea level, between the peaks Braigh a'Choire Bhig and Carn nan Gobhar, which rise to about 1000 m. The loch (near 57°22′N 5°02′W) has steep mountain walls on most sides of it, particularly to the west and south. (Footnote continues on p. 514.)

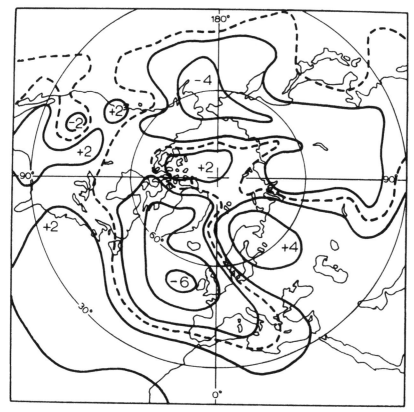

Fig. 18.22   Average observed differences in mb of atmospheric pressure
at mean sea level over the northern hemisphere prevailing in winters
(December, January, February) in which the central north Pacific
(30–45°N 155–175°w) sea surface (November to January) was cold, minus
winters in which the same ocean area was warm.
   (*After* FARMER (*1973*); *reproduced by kind permission.*)

Figs. 10.7 and 10.8 in Volume 1 indicate the feedback effect of marked sea surface
temperature anomalies in the region of the North Atlantic Drift and Labrador Current
boundary on the Newfoundland Banks upon the atmospheric circulation in succeeding
months over the northern hemisphere, and particularly over the eastern North Atlantic and
Europe, such that cold water in that region generally favoured blocking anticyclones over

The report by Sir GEORGE MACKENZY, entitled 'On a Storm and some Lakes in Scotland', was
communicated by Dr JAMES GREGORY and appeared originally in *Phil. Trans*, Vol. X, issue No.
114, p. 307, 1675. Significantly greater lowering of
the temperature in Scotland than in England during
the Little Ice Age seems also to be required to explain
the wetness of large tracts of Scotland noted in
Chapter 13, p. 135.

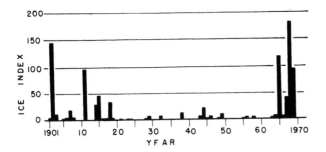

Fig. 18.23 Sea ice at Iceland:
KOCH's index – duration in
weeks multiplied by the number
of sections of the coast affected.
    (*After* KOCH (*1945*), *updated
to 1969 by* SCHELL (*1974*);
reproduced by kind permission.)

northern Europe or Iceland and in some months over northern Canada also. Fig. 18.22 demonstrates the comparable effect of a cold sea surface anomaly in the region of the corresponding boundary of the warm Kuro Shiwo current in the central North Pacific (30–45°N 155–175°w). Once again the tendency of the cold water cases is to be accompanied by high pressure in high latitudes, particularly in certain sectors, but also by above-normal pressure in the region of southward-displaced subtropical anticyclones.

LAMB and JOHNSON (1959) noted that the difference of overall annual mean sea surface temperature in the Atlantic between latitudes 20° and 60°N was about 15°C for 1900–29 compared with 16°C in the 1880s and 16·5°C in 1940–59. This may have something to do with the minimum of many forms of convective activity about 1900–30, e.g. in the incidence of tornadoes in the British Isles and of tropical hurricanes over the North Atlantic as well as the average yields of rain per rain day in showery weather in the British Isles.

Sea ice

A survey of the record over more than the last 1000 years of sea ice affecting the coasts of Iceland has been presented in fig. 17.13. Year-by-year records of the ice in the western North Atlantic off Newfoundland since 1880, off west Greenland since 1820, at the coast of Iceland since 1600, and in the Baltic since 1530 are given in Tables 13–17 in Appendix V, as well as some notes with dates of exceptional positions of ice in the North Atlantic.[1] A plot of an alternative index (due to KOCH 1945 and updated by SCHELL 1974) of the ice at the Iceland coast in the present century is displayed in fig. 18.23.

All these indicators show the minimum of ice in the earlier part of the present century and the increase in the 1960s or, in the case of the western Atlantic, since 1970. Years in earlier times in which KOCH's estimates of the index in fig. 18.23 (which multiplies the duration by the extent of the ice along the coast) exceeded 150 were A.D. 1233, 1258, 1261,

1. Ice also occurs along the continental coasts of the North Sea, particularly over the shallower waters off Jutland, in the German Bight and off the Dutch coast, in any spells of severe winter weather, as well as in Pegwell Bay, east Kent, and (rarely) in the estuaries and firths of the eastern side of England and Scotland. In 1947 and 1963 the belts of ice were some kilometres wide in all these areas, and loose drift ice was encountered in the middle of the North Sea.

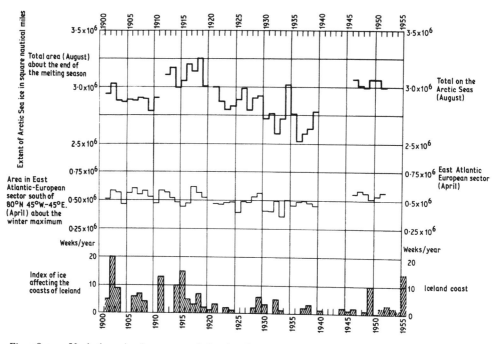

Fig. 18.24 Variations in the extent of the Arctic sea ice at the end of the winter and end of the summer, 1901–55.

Area measurements in the upper two graphs were made by planimeter on the monthly maps published in *Isforholdene i de Arktiske Have*, Copenhagen.

1470, 1615, 1639, 1695, 1817, 1821, 1835, 1840, 1866, 1881 and 1888. Variations of the incidence of Arctic sea ice in mid latitude regions, where it obtrudes more particularly during the spring and early summer loosening and drift of the pack-ice, present a somewhat confusing picture, since seasons with excessive ice there may be the result either of excessive production farther north during one or more previous winters or of winds or warm weather causing an exceptional break-up and clearance of ice from the source regions in the Arctic. Some indication of the discrepancy is obtained by comparing the upper graphs and the lower ones in fig. 18.24; moreover, there was a period of severe ice on the southeast coast of Greenland even in the spring of 1938 owing to the exceptional rapidity of the outflow of ice from near the North Pole (see fig. 7.4 in Volume I). The late summer of that year saw the most extensive open water ever known north of the coast of Asia. Nevertheless, these various records all indicate the decline of the Arctic ice cover from the early years of the century.

In general, periods of vigorous atmospheric circulation penetrating the Arctic are conducive to decline of the ice, both because of the enhanced heat transport and because stormy winds from any direction tend either to break up and disperse the ice or to compact it into narrow belts. Periods of more frequent calm weather are likely to favour growth of the ice. During such periods the ice in the central Arctic moves only slowly and may circulate several times around the Arctic Basin before emerging to melt in lower latitudes (see fig. 7.4 in Volume 1). Declining values of many atmospheric circulation indices from the mid 1930s onwards and again 20–30 years later seem to have marked a change which allowed the ice on the Arctic Ocean to begin to grow and thicken once more. The climate in the central Arctic became significantly colder (particularly in terms of colder summers and a shortened melting season) from 1961 onwards, but it was not until the mid 1960s that a substantial increase of ice extent affected the seas about Greenland and Iceland and not until 1971 onwards off Labrador and Newfoundland.

Other maps and diagrams displaying the variations of the Arctic ice are found in figs. 7.3, 8.11 and 8.12. The last-named indicates that, as with sea surface temperature, the Faeroe Islands lie in the path of variations of exceptional range in the extent of the ice. Twice in the 1880s (once in 1888) fishing vessels putting out from the islands met an ice barrier at no great distance (in one case 20 sea miles) west of Mykines and had to turn back. Also in 1888 ice extending from north to south just east of the islands could be seen from the hill above Torshavn.[1] A 1000-year survey of the ice is seen in fig. 10.22 in Volume 1.

Table 18.6 summarizes the Arctic sea ice information from recent centuries in a way that facilitates comparisons between different sectors around the hemisphere. The simple numerical indices for the ice off Newfoundland, Greenland and Iceland are the same as those given for individual years in Tables in the Appendix and from the same sources. The percentage figures quoted for other areas are percentages of the available ships' reports (ARMSTRONG 1958) which indicate ice in the late summer months of July and August – 'slight' or 'moderate' ice or 'partial cover' counted as half cover, 'unnavigable ice' counted as full cover – figures based on less than ten observations being given in brackets. Most figures, even in the earliest years, are based on over twenty observations.

The striking things about Table 18.6 are:

(1) The indications of rather extensive open water east of Greenland as far as Novaya Zemlya or beyond in the years immediately preceding the severe European winters of the early 1600s, the 1740s, 1880s and 1940s (and as happened again before the beginning of those of the 1960s).[2]

1. I am indebted to Professor BØRGE FRISTRUP of Copenhagen (personal communication, 13 November 1959) for information that on one occasion '200–300 years ago a polar bear landed on the Faeroe Islands from the ice and was killed there'. The exact spot is said to be still remembered in the locality and has been named after the incident.

2. In all these cases other groups of severe winters had occurred in Europe about half a century earlier – in the 1560s, 1690s, late 1820s–40s, and 1890s – with a temporary predominance of E'ly winds. Much less is known about the ice situation in these other cases, though the years 1817–20 (and possibly after) were

certainly much more open in the Spitsbergen sector than for a long time before that. It seems likely that in some of these cases, as in the 1930s and 1940s, prevalence of blocking anticyclones over northern Europe may have meant S'ly wind components over the Norwegian Sea, producing extensive open water. In the 1830s and 1890s this is known to have occurred (Table 18.5); much inner Arctic ice was simultaneously expelled, though this depends also on a rather fine point of prevailing wind directions between Spitsbergen and northern Greenland. There is some possibility that extensive open water (and anomalously high water surface temperatures) in the

Table 18.6   Summary of sea ice extent on northern waters in different decades since 1590

Years	Off Newfoundland south of 48°N (Groissmayr's index)	Off West Greenland (after Koch)	Iceland coast (after Koch)	Barents Sea west of Novaya Zemlya 68–76°N 45–60°E	Western Kara Sea 70–76°N 60–70°E	Eastern Kara Sea 73–78°N 70–90°E	Laptev Sea 72–77°N 110–140°E	East Siberian Sea 69–72°N 160–180°E	Chuckchi Sea and Behring Strait 66–70°N 180–170°W	Years
1594–97			0	15%						1594–97
1734–41			2·3		20%		(50%)			1734–41
1780–89			12·7							1780–89
1790–99			8·4							1790–99
1800–09			9·6							1800–09
1810–19			9·0							1810–19
1819–24				33%						1819–24
1820–29		5·7	8·9							1820–29
1830–39		3·9	8·4							1830–39
1840–49		1·5	3·2							1840–49
1850–59		5·2	7·2							1850–59
1860–69		5·1	12·9		18%					1860–69
1870–79		5·0	7·7	5%	48%	(25%)			(17%)	1870–79
1880–89	4·85	7·4	12·3	21%	45%				20%	1880–89
1890–99	4·96	6·2	6·9	10%	25%	(33%)				1890–99
1900–09	5·22	3·2	5·1	21%	48%				20%	1900–09
1910–19	4·71	6·4	5·5	20%	37%	33%	5%	29%	34%	1910–19
1920–29	5·09	1·5	1·5	7%	43%	43%	(22%)	44%	37%	1920–29
1930–39	5·06	1·7	1·4	3%	38%	36%	31%	40%	37%	1930–39
1940–49	4·16		1·5	8%	41%	27%	15%	(31%)	33%	1940–49
1950–59	3·40		2·0							1950–59
1960–69	3·18		8·0							1960–69

(2) That the variations in the eastern Arctic, north of Siberia and the Pacific, in the present century were partly out of phase with those in the Atlantic sector.

Not nearly so much detail is known about the variations of the Antarctic ice on the Southern Ocean and no long series of index values are available. Nevertheless several useful

Norwegian Sea and Barents Sea significantly favour the occurrence of anticyclones over northern Europe in winter.

Figs. 13.3 and 17.7 suggest that the increase of winter severity in Europe between 1350 and 1550 became established in eastern and central Europe before it was noticeable in the west and while there was still extensive warm water in the Norwegian and Barents Seas. One might perhaps deduce a change in radiation conditions affecting the heart of the Eurasian continent (and presumably the polar region) more rapidly than the Atlantic Ocean could respond.

analyses and collections of observations from the voyages of the 1770s onwards exist (e.g. TOWSON 1859, FINDLAY 1884, 1920, and AURROUSSEAU 1958) and data for more recent years have been compiled by Mr J. HEAP at the Scott Polar Research Institute, Cambridge.

The regions south of about latitude 40°s do not seem to have fully shared the cold epoch of about 1550–1800 or after, which was so marked in the northern hemisphere. The farthest south positions reached by the early voyagers between COOK's expedition in 1773 and BISCOE's in 1831 in various sectors were all 1–2° of latitude south of the normal limit of pack-ice at the end of the melting season in recent years. When the frailty of the wooden ships of those times is allowed for, it seems clear that the southern sea ice tended to be somewhat less extensive then than since. A deterioration followed, the ice increasing on the whole until 1900 or rather later, with some recession since. There were several groups of extraordinarily bad years, 1832–4, 1840, 1844, 1854–7, 1888–90, 1892–6, 1898–9, 1904–7, 1929–31, with large numbers of great icebergs in temperate latitudes in many sectors – in extreme cases reaching the latitude of the River Plate in the western South Atlantic and to near the Cape of Good Hope. Taking acount of the circumstantial evidence (cf. air temperatures at Punta Arenas, fig. 18.6(b)), it seems probable that most of the years 1888 to 1907 were bad ice years around Cape Horn and that reports are sometimes lacking because ships avoided the risk by keeping to Magellan Strait. Accounts are known to exist of ships about the turn of the century finding only 100 sea miles of sailing room south of the Horn, and such reports probably spread as a warning amongst mariners. Another period of severe ice conditions, primarily in the Bellingshausen Sea, in 1958–60 was of a different order and produced no more than a slight advance of pack-ice north of the South Shetlands. Some years when most of the Weddell Sea was clear of ice in the summers followed in the 1960s (as it had been in 1823).

The causes of the phenomenal ice years in the nineteenth century and the beginning of the present century contain an element of mystery, for ships' reports consistently allude to extraordinary heights and horizontal extents for the vast masses of ice concerned. One great ice island sighted by many ships between 40° and 44°s in the South Atlantic between December 1854 and April 1855 was 100 miles long, in the form of a hook, with a bay 40 miles across (in which an emigrant ship was lost and others nearly trapped). The occurrence of other ice islands of this magnitude suggests the breaking adrift of parts of the great coastal ice shelves, perhaps mostly from the great bights in the Pacific sector of the Antarctic coast, where for instance the Bellingshausen Sea (70°s) was found unusually clear of ice in 1893. Tidal waves (tsunami) from earthquakes, subglacial volcanic activity and surges in the flow of ice rivers in the Antarctic ice sheet, caused by rising temperature or internal mechanical processes (ROBIN 1955), are among the possibilities; though it is known that the tidal waves from the Chilean earthquakes of May 1960 had their greatest amplitude along (westward) paths normal to the Andean range and produced little effect on the Antarctic ice.

Although we have found suggestions that the southern hemisphere W'lies, like the northern hemisphere circulation, were increasing in average intensity from at least the mid nineteenth century to around 1900 or later, the changes of latitude of the southern systems appear quite different from those of the northern. The long periods of dryness in central Chile in the eighteenth century, and in southern Australia down to the 1840s, suggests that the southern depression tracks were at that time far south. This development may itself have been sufficient to confine the Antarctic pack-ice to more southern limits and, through this and the heat transport achieved by the travelling depressions, to cause a generally positive temperature anomaly in the eighteenth century in the zone south of 40°s.[1] Our charts show that, by contrast, around 1890–1900 in the Chilean sector the Antarctic depression tracks, both summer and winter, were unusually far north. These displacements have clearly been in parallel with the trend of the Antarctic ice limits discussed in the preceding paragraphs. The unlike trends of the extent of Antarctic sea ice and of temperature and ice in the northern hemisphere are illustrated in figs. 8.11 and 7.22 in Volume 1.

Precipitation

Of all the elements of climate affected by changes in the patterns of flow and intensity of the prevailing atmospheric circulation, none is more obvious than precipitation. The effects are very clearly seen (especially when analysed in terms of percentages), despite the intricacy of local rainfall distribution owing to its dependence on even small-scale orography and the haphazard incidence of convective rainstorms over flat terrain. They amount in some regions to large percentage changes which can be very serious for agriculture and the human economy. The relationships of important rainfall changes to the global atmospheric circulation characteristics of the early twentieth century period of general warming were first pointed out by KRAUS in a valuable series of papers (1954, 1955a, 1955b, 1956, 1958, 1959). They represent in part diversion of the rain-giving depressions to other paths and partly changes of orographic exposure and shelter associated with changed frequencies of winds from various directions.

Fig. 18.25 surveys the changes of average yearly downput of rain and snow decade by decade over the longest sound records available up to the 1950s in sample areas all over the world. It shows (in the upper five graphs in the left-hand column and the bottom two in the right-hand column) how the amounts generally increased in all those places in middle latitudes in both hemispheres which are exposed to moisture transport from the oceans by

---

1. Interesting light is shed on the reverse effects of the sea ice on cyclonic development and the general circulation over the southern hemisphere in a brief study of the regular seasonal changes by SCHWERDFEGER et al. (1974). Interest arose from the fact that the belt of lowest monthly mean sea level pressure around the Antarctic exhibits a half-yearly variation, being farther south in spring and autumn than in summer or winter, whereas the sea ice undergoes a simple yearly variation reaching farthest north in spring and retreating to nearest the Antarctic continent in late summer to early autumn. By counting over 2000 cyclonic vortices on satellite cloud photograph mosaics, it was found that cyclonic vortices were consistently more frequent in the lower latitudes north of 50°s in spring than in autumn, and the reverse was the case south of 50°s, the figures having a high level of statistical significance. It was also clear that the maximum frequencies of cyclonic vortices occurred in longitudes just east of the prominences of the ice outside the Ross and Weddell Seas, and that these maxima were most pronounced in the southern spring.

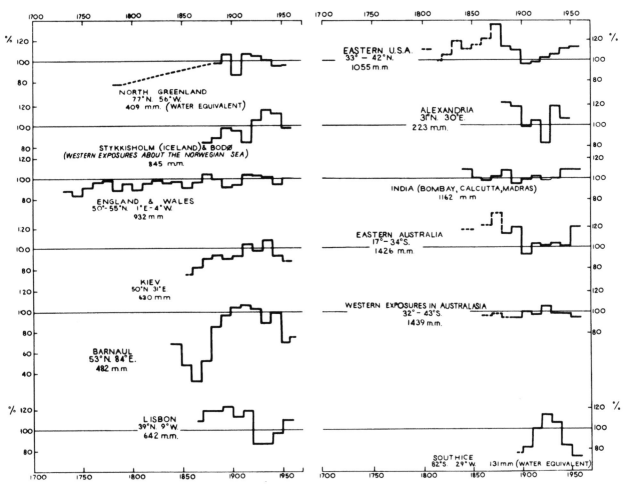

Fig. 18.25 Precipitation changes registered by successive decade averages at selected places in widely separated parts of the world.

Expressed as % of the 1900–39 average yearly total for each place, which is quoted in mm.

the prevailing W'ly winds, and in high latitudes, over the long period – from the nineteenth century or earlier up to 1900–50 – during which the general atmospheric circulation intensity was increasing. All these graphs show a maximum in the first half of the twentieth century, about the same time as the maximum strength registered by the indices of the wind circulation, and a tendency to decline since. The amplitude of the changes (all measured in per cent to make them comparable) increased towards the arid regions in continental interiors and towards high latitudes. The greatest percentage changes occurred in central Asia, at the eastern extremity of the great plain that extends across Europe from the Atlantic, and in the Arctic and Antarctic. In the central Asian region the Soviet grainlands were affected – as the pastures used by the nomadic tribesmen must have been affected by such changes there in former times. The changes measured in the snow and ice layers on the polar ice-caps must obviously be attributed partly, perhaps even largely, to changes in the moisture content of the airstreams, as the ocean surfaces became warmer or colder, as well as to changes in the vigour of the winds responsible for the moisture transport. Both causes operate in the other areas as well, though at the places farthest from the open oceans the strength and direction of the winds are likely to be most important.

Where mountain ranges or the crest of an ice sheet lie athwart the wind flow, the trend of precipitation changes differs in their lee (see Chapter 13, p. 147). For this reason, North America east of the Rocky Mountains and Argentina in the lee effect of the Andes suffered from a dry regime in those decades when the westerlies were strongest (e.g. the 'Dust Bowl' years of the early 1930s in the U.S. Middle West and in the first decade of the century in Argentina).[1] Fig. 18.25 shows this different trend in the eastern U.S.A. and similarly in eastern Australia. And in the zone affected by the intensified, and most prevalent, subtropical anticyclones during the decades of strongest circulation (e.g. Lisbon, Alexandria) the rainfall was also low.

Figs. 13.40(a) and (b) have shown the history of snow accumulation at the South Pole and in northern Greenland in more detail, revealing a remarkable degree of parallelism between the two areas in 90°s and 77°N and with the increasing strength of the general atmospheric circulation (e.g. as shown by the North Atlantic and Southern Ocean indices in fig. 7.1(a) in Volume 1) up to its maximum around 1925 ± 25 years and with the subsequent decline. Another element in the causation of the variations observed in northwest Greenland is probable, namely the occurrence during the warmest periods of climate in the early Middle Ages and during the first half of the twentieth century of more frequent cyclonic activity – especially in summer and early autumn – travelling along paths around the fringe of the polar ice in such high latitudes as to give more W'ly (i.e. upslope) winds than at other times. A similar diversion of the South Pacific cyclonic activity to higher latitudes, giving increased downput of snow as far as 80°s in West Antarctica, in the seventeenth and eighteenth centuries is apparent from fig. 7.23 in Volume 1 and has been discussed on p. 520 above.

In New Zealand DE LISLE (1956) found different secular variations of rainfall in the different seasons of the year, which could be related to the strength of the W'lies and the latitude of the subtropical anticyclones.

In summary, the long-term changes of precipitation amounts and their global distribution which we observe – whether in the course of seasonal change, and the spells of weather which last up to several weeks, or from century to century and from glacial to interglacial periods – are mainly produced by

1. Rainfall in Buenos Aires in the 1940s was 20% greater than the total for 1902–11.

(1) Changes in prevailing moisture content of the winds (associated with the temperature changes but also affected by windiness over the ocean regions where the moisture is picked up).

(2) Changes in the strength and distance scale of the main windstreams and, hence, of the moisture transport.

(3) Diversion of the moisture-bearing winds and cyclonic activity onto other paths.

(4) Orographic exposure to different winds.

These physical causes seen at work in the data reviewed in this section, and plainly illustrated in fig. 18.24, justify a final note about the question of reliability of the rainfall data alluded to on p. 475. Doubts and disbelief have sometimes been expressed about the lower rainfall totals indicated by the eighteenth century European measurements. Adjustments certainly must be made for differences of exposure, particularly for the underreading of gauges exposed on roofs. Yet, the most recent review, by CRADDOCK (1976), of the early raingauge records in England produces the result that the downput measured in those early gauges between 1725 and 1750 was about 5% below the longer-term average of those same records.[1] It seems to the present writer that the evidence for believing that there was a rather general increase of rainfall from that time to the present century is strengthened by the parallelism registered by representative measurements in the middle and high latitudes of both hemispheres and the general continuance of the trend up to 1925 ± 25 years. Moreover, the fact, noticed earlier in this chapter (in a footnote on p. 494) that the mean circulation maps between 1702 and 1754 indicate two long gaps when for 32 years and 17 years respectively no single January showed a prevailing westerly circulation pattern over the British Isles provides an independent reason to expect lower rainfall totals.

1. Observations maintained by THOMAS BARKER, F.R.S. from 1736 to 1798 at Lyndon, Rutland, in the East Midlands of England, with much care and shrewd comment, indicated the increase from the first to the second half of the eighteenth century. He wrote to GILBERT WHITE of Selborne that the mean rain of any place cannot be ascertained until a person has measured it for a very long time. 'If I had only measured the rain for the first four years 1740 to 1743, I should have said that the mean rain at Lyndon was 16½ inches; if from 1740 to 1750, 18½ inches. The mean rain before 1763 was 20¼; from 1763 and since 25½; from 1770 to 1780, 26. If only 1773, 1774 and 1775 had been measured, Lyndon mean rain would have been called 32 inches.' Thus, there were two 3- to 4-year periods, the mean for that in the 1770s being nearly double that in the 1740s, and the mean for the whole decade from 1770 exceeded that from 1740 by 41%.

Mr J. M. CRADDOCK, who has examined the earlier English rainfall records with great care, writes (personal communication, 14 November 1975) that 'there is no evidence on the rainfall regimes before 1760 to rival in cogency BARKER's record' and that it 'is unique not only for its length [over 60 years], but for the hard facts [given] about the site, etc., not to mention BARKER's own wide-awake attitude'. Nevertheless, CRADDOCK believes that the whole BARKER record needs some upward adjustment because of the exposure of the gauge on a wall, 9 feet above the ground, where two walls joined.

There is no doubt about the extraordinary dryness BARKER established in the 1740s and of the increase of rainfall which built up over the following decades. The general dryness of the earlier part of the eighteenth century is in keeping with what can be established about the prevailing wind circulation and the colder seas. So H. R. MILL's suggestion that the raingauge readings need an upward adjustment of as much as 9% to correct for their exposure seems unlikely.

Snow in middle latitudes

The changes of prevailing temperature and of frequency of the westerly winds in middle latitudes noted in this chapter have big effects on the frequency of snow-cover in places where it is a seasonal and marginal event and upon its duration on the mountains (this also affects the formation and growth of glaciers, though the amount of snow which falls in the winter and spring and the duration and warmth of the summer melting season are of most account). The effect in lowland areas in Europe is illustrated in Table 18.7; the periods covered exclude the great winter of 1962–3 and other cold winters up to 1969–70. After a further increase in the 1960s there was once more an abrupt decline to very low frequencies from 1970–1 to 1974–5.

Table 18.7   Average number of days per winter with snow lying and (in brackets) percentage of winters with 15 days or more

Place and first year of observation	1870–1 to 1894–5	1895–6 to 1937–8	1938–9 to 1961–2
Lerwick, Shetland, from 1917		10·3 (23%)	24·6 (83%)
Tynemouth, northeast coast of England (data before 1921 derived mostly from a Sunderland record which goes back to 1857)	10·6 (32%)	7·4 (16%)	10·9 (33%)
Cambridge, from 1916		7·7 (14%)	12·4 (33%)
Ross-on-Wye, southwest Midlands of England, from 1916		5·4 (9%)	9·7 (21%)
Freiburg im Breisgau, southwest Germany, from 1884	38·9 (100%)	23·2 (70%)	31·8 (92%)

The effect may also be traced in the average dates of occurrence of the first and last snows of the winter. Thus, in Tokyo for which long records of this exist (ARAKAWA 1956) the average date of first snow-cover between 1870 and 1910, which was the mildest period there, was 22 January and for 1910–50 the average was 11 January. Similarly, 1962 brought the first November snowfall in southeast England for several decades and snow has fallen at some time in many Novembers since. There has been a corresponding increase in the frequency of snow in April from about the same year.

Changes in the duration of snow on the mountains in the British Isles are not well established but seem from comparison of the accounts of travellers in Scotland in the seventeenth and eighteenth centuries with modern experience to have been very significant.[1] It is reasonable to conclude from these reports that the snowbeds surviving the summers in the shaded gullies on the Scottish mountains were more numerous, widespread and persistent in the Little Ice Age period than in this century. Indeed, there is evidence from the distribution and measured diameters (ages) of the lichens present in the corries in the Cairngorms today (SUGDEN 1974) that a few of them may have been occupied by glacier ice as late as the 1740s.[2] It seems possible also that there were periods of some decades duration, and perhaps most of the time from about 1570 to 1770, when the summit plateau of the Cairngorms, in the eastern Highlands, was almost always covered with snow.[3]

Variations within the eighteenth century are revealed by the observations noted by Pastor J. J. SPRÜNGLI of the occurrences of new snow in summer between mid May and mid September on the mountains in Switzerland below about the 2300 m level (PFISTER 1974). Averages appear as follows:

1759–64	6 times a summer
1766–9	14 times a summer
1770–4	15 times a summer
1775–9	12 times a summer
1780–4	7 times a summer

MANLEY (W.M.O. Norwich Symposium on Long-Term Climatic Fluctuations 1975) has pointed out that, because of the greater lapse rate of temperature with height in unsettled, cold cyclonic weather, the increase with height in the number of days with snow-cover ranges from 1 day for 8 m in a mild winter to 1 day for 5 m in cold unsettled winters and springs. Since observations in Norway and the Alps indicate that the firn limit (above which glaciers persist) occurs at about the level where representative open ground is snow-covered

1. Witness the report by JOHN TAYLOR (1618) who visited Deeside about 1610 and wrote of the Cairngorms (the tops of which are mainly 1220–1310 m above sea level): 'for you must understand that the oldest man alive never saw but the snow on the tops of divers of these hills, both in summer as well as in winter.' THOMAS PENNANT (1771) wrote of the same hills: 'naked summits of surprising height ... many of them topped with perpetual snow.' He also reported on Ben Wyvis (summit 1045 m), 57·6°N, in Ross-shire: 'snow lies in the chasms of Ben Wyvis in the form of a glaciere throughout the year.' There is also a reference in the Statistical Account of Scotland (Sir JOHN SINCLAIR, Vol. 1, 1791) to a report from the 1740s that the summit of Ben Uaish (= Ben Wyvis?) 'is always covered with snow'.

Semi-permanent snowbeds mostly in north and east facing corries and gullies are a feature of Ben Nevis (top 1343 m above sea level) and the Cairngorms today but not of Ben Wyvis; it is unusual for all of them to melt, but probably rare for any

individual one to go unmelted for more than a decade or two. The snowbeds surviving the summers in the gullies at about 1100 m on the precipitous north face of Ben Nevis were known in 1630 and reported again in 1760, 1787 and 1880s (MANLEY 1969, see also MANLEY 1949, 1971).

HUTCHINSON (in his History of Cumberland, 1794) reported that within the previous 70 years snow had never lain all year on Crossfell (893 m), the highest top in the Pennines in northernmost England; though the remark may suggest that this had been sometimes near to being the case.

2. MANLEY suggests (personal communication, May 1976) that it may have been firn rather than having attained the density of ice.

3. JAMES WRIGHT's diary of his expedition in the Faeroe Islands (WEST 1970) in the generally rather warm and pleasant summer of 1789 reported a deep, extensive snowbank on a hilltop there at about 550 m above sea level on 20 June.

for 270–280 days, a fall of only 1·3°C in the overall mean yearly temperature in the Scottish Lowlands, similar to that shown by the central England temperature record in the Little Ice Age, combined with the cold unstable type of winter weather that was common, would have sufficed to bring the snow line down to 1200 m on the mountains and to allow glacier ice to form in some shaded corries. Our analysis in a previous section of this chapter (p. 513) of the sea temperature and ice variations around the Faeroe Islands suggests, furthermore, that ocean current changes may have caused a somewhat greater lowering of prevailing temperatures below twentieth century values in Scotland than in England in the seventeenth and early eighteenth century: perhaps as much as 1·5°C lower than today for the lowest 50-year mean, and rather over 2°C for the lowest 5-year mean, at least in the northern half of Scotland.

In 1933 the twentieth century warming brought about for the first certainly known time complete melting away of all the snowbeds in Scotland. This occurred in 3 years each decade from the 1930s to the 1950s, and there is one rather uncertain case in the 1960s, but it has not happened since then. The years recorded (up to 1976 inclusive) are: 1933, 1935, 1938, 1940 (uncertain), 1945, 1949 (probably), 1953, 1958, 1959, 1969 (almost certain, just towards the end of the warmest October in the 300-year central England temperature record).

Survey of the Scottish snowbeds has been undertaken regularly each summer since the mid 1960s by Mr P. C. SPINK of Ulceby, Lincolnshire, and reported in yearly letters in *Weather*. Less complete coverage, but with some reports for each month, is given in the yearly issues of the official *Snow Survey of Great Britain* (London, Meteorological Office). I am indebted to Mr SPINK for the reports on the latest years.

Studies by IVES (1962), BRADLEY and MILLER (1972) and BRADLEY (1973) indicate even greater changes in the extent of snow and ice-cover on Baffin Island. Geomorphological markers and rock lichen development indicate that extensive tracts of north-central Baffin Island were 70% buried beneath permanent ice and snow between 200 and 400 years ago, where only 2% was so covered by 1960. By 1970, after a decade of summers averaging 2·1°C colder than before 1960 and the other months (September to May) averaging 2°C warmer than before, accompanied by marked increase of easterly winds off the sea and snowfall, snowbanks were found to have regained the size they had about 1930 and two new incipient glaciers were found. Upper air temperature measurements at the Canadian Arctic stations between 1964 and 1972 showed the average height of the freezing level in July lower by 276 m than in 1955–63. Tree ring studies in the Canadian north suggest that temperatures in the interior of the Yukon in the eighteenth century and after were a little over 1°C lower than in the last few decades (RAMPTON 1971), an anomaly smaller than that in Iceland or than is probable in the other places near the northernmost Atlantic which we have discussed.[1]

The Meighen Island ice cap (in the Arctic Ocean near 80°N 100°W), which seems to have formed 3000–4500 years ago, i.e. since the warmest postglacial times, and which dwindled from about 2000 to about 660 years B.P., registered the coldest period in its history – with

---

1. This estimate seems too small, however, by comparison with the Canadian and United States cities farther south. It may be that the tree ring series did not go back to the coldest period in the Yukon.

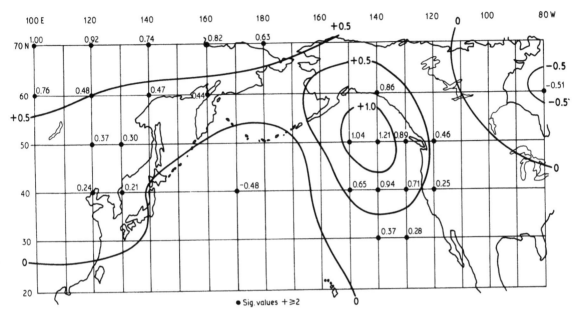

Fig. 18.26   Proposed average mean sea level pressure prevailing in winter (December, January and February) 1700–1899: departures in mb from the 1900–39 winter average.
Derived by H. C. FRITTS and co-workers from tree-ring data in the western United States.
(*Map supplied by Professor* H. C. FRITTS *and reproduced by his kind permission.*)

little summer melting – between about 560 and 390 years B.P. (i.e. A.D. 1390–1560 approximately). A period of marked growth of the ice cap then set in and continued till about 1870, when it had its greatest thickness (KOERNER and PATERSON 1974).

Further indications of the climate of the Little Ice Age period
from the era of instrument measurements

Professor H. C. FRITTS (personal communication, 3 January 1975) has kindly supplied the map in fig. 18.26 of proposed anomalies of overall winter-average mean sea level pressure for the years 1700–1899 across the Pacific Ocean sector of the northern hemisphere in comparison with 1900–40. The pattern, which needs verifying (calibrating) by reference to any available independent markers of prevailing winter circulation or weather over that, time, was derived from tree ring measurements in western North America by the method of FRITTS *et al.* (1971) (see also Chapter 13, pp. 225–7, and fig. 13.66 in this volume). Its chief features – though rather weak, doubtless because so many different years and decades were averaged together – seem highly probable in the light of what we know about the Little Ice Age climate; the pressure differences indicate an added component of northerly winds over all the western half and central longitudes of North America and a continuation of the implied cold airstream as NW'ly and W'ly surface winds towards the Atlantic seabord. A slighter tendency for increased NE'ly and N'ly winds over east Asia and Japan also is included. Much enhanced S'ly winds in mid Pacific must have made for continual warmth of the water in the Kuro Shiwo current boundary area, recognized first by NAMIAS (1963) and illustrated here in fig. 18.22 as an indicator for anomalous effects on the atmospheric circulation all around the hemisphere – a situation which favours blocking anticyclones in the Iceland and northeast Atlantic sector as well as over the northeast Pacific.

The anomalous wind flow over the northernmost Pacific in the average pattern of the winters of the period 1700–1899, indicated in fig. 18.26, also seems likely to have favoured some additional flow of water through the Bering Strait from the Pacific into the Arctic basin. This looks like some support for the entirely independent suggestion on pp. 512–13 that increased supply of water by this route was required to explain the oceanographic situation in the Greenland–Norwegian Sea.

In studies of the earliest weather observation network covering the eastern half of the present United States WAHL (1968) and WAHL and LAWSON (1970) found the entire eastern half of the country colder in all seasons of the year in the 1830s and 1840s than in 1931–60. The average departure for the 1830s was least in winter (a fraction of a degree, but ranged up to 1·1°C in spring, nearly 2°C in summer and almost 2·5°C in autumn), the greatest negative anomalies being in all seasons near the Great Lakes and extending south and southwest from there. Precipitation excesses of up to 30% in summer and late autumn, and up to 20% in the other seasons, along 600–1000 km-wide bands south and southwest (in summer south and southeast) of the Great Lakes point to depression tracks south of the usual twentieth century positions. Other areas were drier than in the recent past. In the 1850s and 1860s the colder climate conditions continued over the eastern half of the United States, while the western half (from which reports were then available for the first time) was distinctly warmer and wetter than in the first half of the present century.

The 1830s were a decade of unusual circulation character over Europe also, having an abnormally great frequency of anticyclones with their centres or axes near 50°N (see LAMB 1963), a frequency of this probably unmatched in the times since A.D. 1700 or earlier. Conditions observed in Europe, with milder years in the 1850s and 1860s and a high frequency of W'ly winds (fig. 18.18), and the beginning of the century-long recession of the glaciers (see Plate XVIIIa and b), indicate that the Little Ice Age regime was already ending then. Indeed, the circulation characteristics observed in Europe during the 1830s (LAMB 1963, LAMB and JOHNSON 1966), with a high frequency of anticyclones near 50°N and often vigorous W'lies farther north, led to an alternation of very mild and very cold winter months in Britain and central Europe and apparently led to a great diminution of the Arctic sea ice affecting Greenland and Iceland in the following decade or more.[1]

1. Severe months in Europe especially in 1829–30 and early 1838. Particularly mild in the Decembers of 1831–4 and 1837 and in the whole 1833–4 winter, but also very mild in the Februarys of 1833–5. The Januarys of 1831, 1833, 1835, 1837 and 1839 were cold or very cold.

Changes since the early twentieth century warming

General note

The trend of the indices (figs. 7.1(a), 18.12, 18.13) which had registered the very long-continued general increase of strength of the mainstreams of the global circulation of the atmosphere seems to have reversed as early as 1925 ± 15 years, and successive short-term peaks thereafter have been lower than the ones before (see also Table 48 in Appendix V). Since this downward trend of circulation strength began some decades before any significant renewal of the amount of volcanic dust in the atmosphere, its causes must be sought elsewhere. Some aspects of the sequel over the decades since show so much similarity with the events of about 200 and 400 years earlier (see, for example, fig. 6.8 in Volume 1) that association with the supposed 200-year cycle in solar activity (deduced from the several-thousand-year-long record of radiocarbon in the atmosphere) may seem the likeliest ultimate cause (see also the downward tendency, amounting already to 2 % or 3 % before the volcanic eruptions of 1963–8, of the measured intensity of the solar beam, displayed in fig. 2.3(a) in Volume 1). The fact that the warm climax conditions, prolonged over almost half a century before 1960, were apparently more extreme than any since the early Middle Ages suggests, however, that another much longer-term cyclic change was also involved.

Temperature

Air temperatures prevailing over most of the globe (fig. 18.2) reached their maximum about the early 1940s, towards the end of the decades of strongest atmospheric circulation, and their subsequent fall to the levels maintained from 1965 to 1975, or after, seems likely to be the longest-continued downward trend of this item since 1700.[1] Prevailing sea surface temperatures may have reached their maximum about a decade later than the air temperatures; fig. 10.11 shows the maximum in many areas around 1935–45, but the general maximum in the North Atlantic Ocean Weather Ships region between latitudes 33° and 66°N was probably as late as 1951–5 (cf. also fig. 18.21).

As always, the onset of change in the air and sea temperature regimes shows a geographical pattern. HARVEY (1965) noted that the highest water temperatures in the top 200 m of the Barents Sea at 70–72°N 33°E, north of the Kola peninsula, appeared to have been reached already in 1935–9 and despite a secondary maximum about 1951–5 had fallen away 0.5°C by 1960. Fig. 18.3 maps the change of air temperatures to the 1950s, showing the

---

1. DRONIA (1967) has suggested (see Volume 1, p. 517) that the increasing warmth of the cities where many of the longest-established observing sites are may have prolonged the registration of the warm climate and made the maximum of global air temperature appear some years later than more representative observations would place it.

Although the wind circulation pattern produced a general tendency for the 1950s to be warmer than the 1940s in the eastern United States and most of Europe, including European Russia, the warming trend which had been interrupted in all those regions in the 1940s was replaced by sharp cooling in all three regions in the 1960s (see Plates XIX and XX).

greatest cooling in the Arctic and the northernmost interior regions of the northern continents while some other regions were still becoming warmer. A rather sharp step downward in the general temperature level prevailing over the northern hemisphere, by an average of 0·6°C for the whole atmosphere up to the 18 km level, seems to have occurred between 1958 and 1963, according to a survey by STARR and OORT (1973), who also explored its geographical distribution and distribution with height; greatest cooling was found throughout the depth of the troposphere (a) over the Arctic, and (b) over latitudes between 5° and 22°N (though not comparable with the changes in the Arctic surface inversion layer), and at most heights north of 60° and south of 40°N. There was also strong cooling in the stratosphere north of about 35°N and in the uppermost levels of the troposphere farther south (continuing at least as far as 12°S). A 5-year change of this magnitude must, however, be seen as part of some fairly short-term fluctuation and cannot be representative of any much longer trend (since it would lead to an ice age within a few decades). In fact, ANGELL and KORSHOVER (1975) report that between 1963 and 1973 the global atmosphere between the 1000 and 50 mb levels (i.e. up to about 21 km) on balance cooled no more, a slight further cooling by 0·2°C over the northern hemisphere being probably a little more than offset by a 0·4°C warming over the southern hemisphere. However, the trends at the Earth's surface are not well represented by the variations averaged through such great depths of the atmosphere. Surface temperatures averaged over the globe had remained almost steady between 1958 and 1963 and underwent a renewed fall in about the next 5 years thereafter.

Dr JILL WILLIAMS reports (personal communication, October 1975), on the basis of work carried out jointly with H. VAN LOON at the National Center for Atmospheric Research, Boulder, Colorado, that estimation of the temperature changes over the wide ocean areas by systematic use of the heat transport indicated by the wind patterns suggests that the apparent downward trend of temperatures since the 1940s has been interrupted both by shorter-term rises and by sector-to-sector differences of trend over periods up to several years at a time, largely associated with wave number 1, i.e. with switches of eccentricity of the circumpolar vortex sometimes towards one sector, sometimes another.[1] Warming of the wide Pacific sector between 1950 and 1964 may even have outweighed the net cooling of the rest of the northern hemisphere during those years. Changes of the common positions adopted by the major waves in the W'lies have similar effects. The magnitude of these shorter-term and regional changes exceeds that of the resultant global (or hemispheric) trend, which because of the sparsity of data in the great ocean spaces cannot be computed with so much confidence. In fact, one of the reasons for believing in the correctness of the

---

1. A secular rise of overall mean temperature in New Zealand by about 0·5°C, sustained for 20 years, 1953–74, has been shown by TRENBERTH (1975) to be associated with the frequency of a circulation pattern giving warm air transport over New Zealand while South Australia and Tasmania cooled slightly. (See also pp. 535–6.)

Another new point of interest in this analysis of the circulation regimes that prevailed over New Zealand in different spans of up to a couple of decades was that the variance spectra showed different time scales of shorter-term variation (5-month to 5-year quasi-periodicities) within the dominance of the different regimes.

global temperature trend as commonly presented (e.g. in fig. 18.2) may be its parallelism with such independent data as the frequency of W'lys winds over the British Isles and the downput of snow at the South Pole.

The geography of surface temperature changes has continued to show the greatest cooling from the 1940 level in the Arctic. Tables 18.8 and 18.9 show this in terms of the numbers of months each year that have been respectively warmer or colder than the 1931–60 average over most of the region north of 70°N.

Table 18.8   Numbers of months each year warmer than the 1931–60 average over most of the Arctic north of 70°N

	Year									
Decade	0	1	2	3	4	5	6	7	8	9
1950s							4	4	1	7
1960s	4	1	0	0	0	1–2	1	1–2	0	2
1970s	0–1	1–2	1–2	0–1	1–2	0–1	1–2			

Table 18.9   Number of months each year colder than the 1931–60 average over most of the Arctic north of 70°N

	Year									
Decade	0	1	2	3	4	5	6	7	8	9
1950s							2	3	6–7	3
1960s	5	8	6	9	11	7–8	8–11	7–9	10–12	6
1970s	8–9	7–9	7–8	7	6	8–10	6–8			

The data for Tables 18.8 and 18.9 were compiled from the monthly maps published in *Die Grosswetterlagen Europas* (Offenbach, Deutscher Wetterdienst) and in the *Beilagen zur Berliner Wetterkarte* (Berlin, Inst. f. Met. der Freien Univ.).

The colder conditions prevailing in the central Arctic after 1961 are demonstrated in graphical presentations of various data by FLOHN and RODEWALD (1975).

Changes over the past twenty-five years similarly treated for the North Sea coast of Germany (DRONIA 1974) show a great predominance of cold years in air and sea since 1961 and of rainfall and wind strengths above the previous averages in that region.

Surveys of air and water temperatures, and of the circulation in the Arctic in the 1960s, and comparisons with the preceding decades, by KIRCH (1966) and RODEWALD (1972), bring out a number of interesting points, some of which may also be relevant to the understanding of other cold periods. The biggest fall of temperature (see fig. 18.27) was near Franz Josefs Land (80°N 53°E), where the overall average temperature for the whole year in

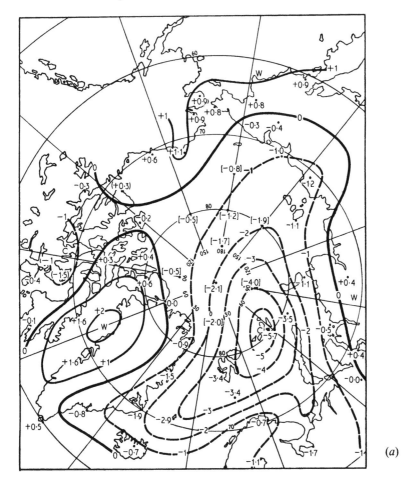

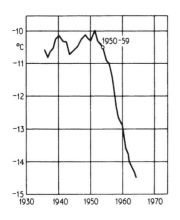

Fig. 18.27   (a) Change of average air temperature (°C) for the four winter months December–March in the Arctic: values for 1961–70 minus 1951–60. Note the general cooling, greatest near the Norwegian Sea and European sector, but warming near the Bering Strait and northernmost Pacific, also off Arctic Canada and west Greenland.

(b) Average temperature for the year at Franz Josefs Land (80°N 53°E). 10-year running means.

(*From* RODEWALD (*1972*); *reproduced by kind permission.*)

each 10-year period from the beginning of the permanent station record in the late 1920s until 1960 had varied between -10.8 and $-10.0°C$; the value for 1960–9 was $-14.5°C$, and the Januarys of that decade averaged more than $8°C$ colder than in the 1950s. In most of the colder months of the year there was a remarkable frequency in the 1960s of monthly mean temperatures over $10°C$ below the average of the previous 30 years, an anomaly never indicated anywhere in the north Russian sector in the previous 80 years and apparently associated with the development of a third cold cyclonic centre throughout the depth of the troposphere, additional to the previously familiar upper cold troughs over northeast Canada and northeast Siberia (see also pp. 279–281, 385 in Volume I). KIRCH's maps show that the cooling at the surface in the 1960s was greatest in just the same places where the warming in the early decades of this century had been greatest and was of about the same magnitude. Of particular interest is the change in overall average mean sea level pressure values from the 1950s to the 1960s, showing a rise of over 4 mb over an area between Iceland and southeast Greenland and smaller rises over a belt from northeast Greenland to the Yukon and Alaska, while pressure was more than 2 mb lower than in the 1950s over the Kara Sea, the Taimyr peninsula and the northeasternmost part of Asia (the 1950s had already shown some change in the same sense from the previous decades). There was thus an increased flow of air right across the Arctic towards the Norwegian Sea, which must have caused an increased wind drift of the water northward through the Bering Strait and of the Arctic pack ice towards the East Greenland exit (cf. p. 512).

MALMBERG (1969) reports the results of yearly oceanographic surveys between Langanes, northeast Iceland, and Jan Mayen which show the resulting watermass changes in that region. Water samples taken in the summer cruises in the top 200 m of the ocean there between 1948 and 1958 showed salinities only between 34·72 and 34·92‰, in the range typical for the Irminger Current water when it reaches around the north of Iceland and approaching the typical North Atlantic Drift salinity of $\geqslant 35‰$ in that region. Water temperatures were between $-1°$ and $+2°C$. By contrast, the samples taken in the years 1964–9 showed much more variable and mostly lower salinities, sometimes as low as 33·6‰ and averaging about 34·5‰, and temperatures ranging down to $-1·8°C$, both items indicating common presence of the polar water. Negative anomalies, generally of both items, though weaker, continued in an unbroken series until 1974, and in 1975 a stronger negative temperature anomaly developed again.

A downward tendency of water temperatures was noted farther south in the Atlantic from about 1953 onwards and was registered in the subsurface waters also as far west as the coast of Maine. By contrast, the Irminger Current remained as strong as, or stronger than, before and transported ever warmer water towards southwest Iceland, south Greenland and the west coast of South Greenland till as late as 1964 (I.C.N.A.F. 1967). Thereafter, however, those regions also registered cooling. By 1968–72 the 5-year mean water temperature averaged for all the nine Ocean Weather Ship positions in the North Atlantic between 33° and 66°N was 0·56°C colder than in 1951–5 (RODEWALD 1973). Superposed on these long-

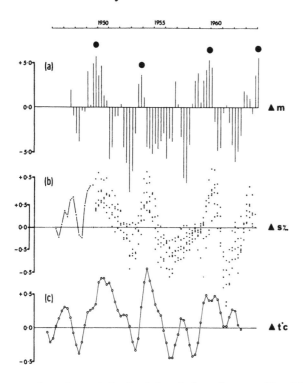

Fig. 18.28   Four- to six-year fluctuations of southerliness of the winds, and of salinity of the North Sea and temperature of the Barents Sea waters.

Running 9-month means of
(*a*) British Isles meridional index anomaly (m positive means more S'ly wind).
(*b*) Surface salinity anomaly in the German Bight.
(*c*) Water temperature anomaly in the surface to 200 m depth layer in the Barents Sea on the Kola meridian.

(*From* DICKSON (*1971*); *reproduced by kind permission.*)

continued trends in the Atlantic (see also p. 508), which are more comprehensively reviewed in DICKSON and LAMB (1972), DICKSON (1971) has reported a recurring 4- to 6-year fluctuation (see fig. 18.28), which has strongly marked effects on temperatures and salinities in European waters from the North Sea to the Barents Sea and is associated with variations in the frequency of S'ly wind components over the eastern Atlantic.

A similar 5- to 6-year oscillation has been found to affect sea surface temperatures in the North Pacific in the region of the Kuro Shiwo (North Pacific Current), having a half wave length of 7000–8000 km and propagating eastwards slowly at a rate of about 5 km/day (WHITE 1975). Blocking development over the North Pacific sector in autumn and winter seems linked to these variations. The quasibiennial oscillation has been demonstrated by RODEWALD (1974) as affecting not only the temperatures and rainfall, particularly in the summers, in Germany, as in northwest Europe generally, and in the eighteenth, nineteenth and twentieth centuries alike, but also affecting the sea surface temperatures in the Baltic, the North Sea and the North Atlantic outside the Bay of Biscay. A difference between the odd and even years, averaging 0·7°C by the end of the summer (September), was found in the southern North Sea and southern Baltic over periods of 25–30 years and of 1°C at Ocean Weather Ship K (45°N 16°W).

It is necessary to stress these changes in the higher latitudes of the northern hemisphere not just because of their magnitude which has had great effects upon the fisheries, and on farming and forestry near the northern and upper limits of agriculture, but because they seem to provide the clearest clue to the understanding of much that has been happening elsewhere over the globe to the prevailing temperatures and to the patterns of the atmospheric circulation and the distribution of rainfall.[1] It is likely that the more fundamental cause of the weakened global atmospheric circulation in the last decade or two and the more restricted latitude scale of the HADLEY circulation – and the related latitudes of the subtropical anticyclones (and other features of the circulation) which we shall note below – will be found in some failure of the energy supply in low latitudes. But the strongest (and therefore most easily established) effects on temperature are – apparently as in other, earlier climatic fluctuations – observed in the highest northern latitudes. Moreover, the resulting increase in the Arctic ice and snow-cover causes a further loss of total energy receipt and therefore tends to amplify the change, particularly in the far north, and to modify the circulation pattern through its effects upon the location and strength of the thermal gradients.

Farther south in the northern hemisphere no fully worked out series of latitude averages of surface air and sea temperature up to recent years are known to the writer, but it seems likely that there has been a slight ($\leqslant 0.1°$C) further rise of the temperatures prevailing near the tropic and a fall near the equator.[2]

In the southern hemisphere no complete survey of surface temperatures is known by which to extend MITCHELL's graphs (in fig. 18.4) after 1960 or to compare with ANGELL and KORSHOVER's (1975) analysis of the upper air temperatures. Many reports, however, indicate a rising trend of annual mean surface temperatures, particularly south of 30°s, but extending to about 10°s in the 1960s in South America (SANCHEZ and KUTZBACH 1974).[3] The average for that continent south of 10°s in the 1960s appears to be about 0.1 to 0.2°C warmer than in 1930–60. There has been a continuous warming trend since the 1930s in all

---

1. Examples of effects on the fisheries and location of fish stocks in the North Atlantic and the North Sea have been given in Chapter 13, pp. 189–191, and in fig. 13.50. Cod at the northern limit of their range off west Greenland were abundant there while the water was warming, i.e. until 1964, but decreased thereafter, whereas in the later part of that decade they were increasing towards, and probably extending, the southern limit of their range. And it seems safe to deduce from the abundance of the fisheries along the Norwegian coast far to the north between A.D. 1300 and 1500 that the water remained quite warm during that time, but may have become colder later, when the herring disappeared from the coast.

   A 40-year farm record from 160–170 m above sea level in northern England, in Swaledale, Yorkshire (kindly supplied by Mr E. POTTS, now of Northallerton) illustrates what the temperature and other changes since 1950 have meant for farming in England: despite introduction of the combine har-

vester about 1950 the harvests in 1960–73 were completed on average 9 days later than those of the previous 20 years. The date of the spring sowing in 1970–3 was almost as late as in the 1940s, which were affected by several severe winters; but in the 1960s, despite colder winters and springs, the sowing was particularly early – the earliest group of years in the 40-year record – presumably because the ground dried early. (It seems clear that the early spring sowings in the 1950s and 1960s, which were generally a week to ten days earlier than in the other decades, were associated with reduced winter rainfall.) The effect of the twentieth century changes upon the frequency of hot summer days in East Anglia is seen in Table 9 in Appendix V.

2. Tentative conclusions based on incomplete temperature series, supported by consideration of cloudiness and rainfall distribution.

3. A possible reason for this is suggested on pp. 544–5.

parts of New Zealand, where (in marked contrast to the northern hemisphere and lower latitudes in the southern hemisphere) 1900–35 was the coldest time in the country's recorded history (SALINGER and GUNN 1975);[1] 20-year means at Auckland and Christchurch rose by about 1°C from around 1915–35 to 1950–70. The Antarctic sea ice also seems to have decreased significantly over about the same period and, perhaps in consequence, there has also been warming at those places around the fringe of Antarctica where there are long enough temperature records.[2] At Little America, at 77°S on the ice barrier at the head of the Ross Sea, WEXLER (1959) noted that the rising trend of annual mean temperature seemed to have been continuous since 1911 and the warming amounted already to 2·6°C by 1958. Other sectors of the Antarctic coast showed much less change than this, and the state of the glaciers around most of East Antarctica suggested that any change was hardly significant up to that time.[3] Despite the evidence cited of general warming in the southern hemisphere, particularly strong where a former accumulation of sea ice has diminished, and strongest in some regions (e.g. New Zealand and the Weddell Sea) just in those recent decades when the northern hemisphere has been cooling, increased incidence of meridional circulation patterns there, as in the northern hemisphere, has produced more frequent winter snowfalls in the 1960s and 1970s on the high ground in South Africa, Tasmania and Australia.[4]

In 1970–1 an abrupt change took place in the northern hemisphere which produced an unbroken run of five very mild winters in Europe and some southern and eastern parts of the U.S.A. and the Far East (cf. fig. 18.29), while most of the rest of the hemisphere – particularly in the Arctic and low latitudes and over the oceans – had a predominance of cold winters. The coldness of the other seasons of the year (also in Europe) meant that the overall average temperature for the year over the northern hemisphere continued at the level of the 1960s, about 0·2°C below the 1930–60 average.[5] The explanation of this change seems to lie in the fact that in 1970–1 the Canadian sector of the Arctic became for the first time fully involved in the Arctic cooling and quickly became so cold that the thermal gradient between the subtropical Atlantic with the Gulf Stream water and Arctic Canada was greatly strengthened.[6] This revived the intensity of the atmospheric circulation over the Atlantic sector,

1. The warming has been accompanied by an increase of rainfall in North Island which has raised lake levels above a previous peak attained about 1905 to position which appear to be the highest for several centuries past (J. HEALY in *J. Roy. Soc. N.Z.*, **5** (1), 77–100; 1975).
2. According to survey of satellite photograph mosaics by G. KUKLA the ice area, has, however, increased after 1967 (verbal report at CLIMAP meeting, Norwich, May 1973).
3. 5-year mean temperatures in the South Orkney Islands (average of Laurie Island and Signy Island, near 61°S 45°W), the longest continuous record south of 60°S (graphed by LIMBERT 1974), indicate a downward trend from values between −4° and −5°C in the years 1903 to 1925 to −6°C around 1930, then a sharp rise to a peak around −2·5°C about 1955,

followed by an undulating decline to values between −3° and −4°C in the years 1964 to 1973.
4. Including, in the 1960s, the first time snow was ever observed in Queensland.
   Snow all over upland South Africa and Southwest Africa on 18 to 20 June 1964, with drifts many feet deep, was the heaviest and most widespread in that country since 1895.
5. This figure was again confirmed for 1975, the latest complete year available (*Beilage zur Berliner Wetterkarte:* Nordhemisphärische Wetterlage im Jahr 1975).
6. About 1972 there was a run of 19 consecutive months all colder than the previous averages in the Canadian north. The difference of mean surface air temperature between Miami, Florida and Goose, Labrador, in the Januarys of 1972–5 was mostly between 8° and

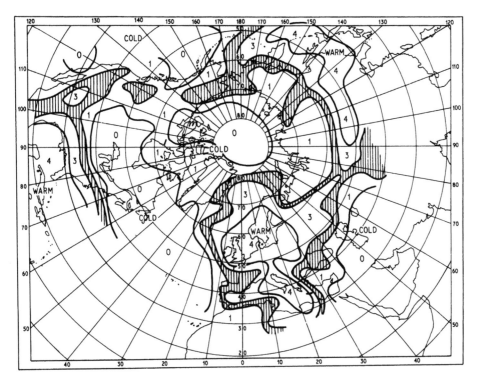

Fig. 18.29   Distribution of mild winters over the northern hemisphere 1971–4.
The figures give the number of those winters that were milder than
the 1931–60 average in each area.
    (*After* WRIGHT *1975*.)

particularly in winter, deepening the Iceland low, producing frequent S'ly and SW'ly winds
over Europe and in the Norwegian Sea and causing the ice to recede once more in the Iceland
and Barents Sea sectors. The warmth does not seem, however, to have penetrated the inner
Arctic to the same extent as in the early twentieth century warming, and it is not clear that the
shift of the cold centre of the northern hemisphere circulation to the Canadian and Pacific
side in the early 1970s is in itself a mechanism capable of bringing about a renewal of the
general warming of the Arctic and of the hemisphere.[1] At the same time frequent cold NW'ly
winds intensified the Labrador Current and brought more ice into the western and
northwestern Atlantic (see Table 11 in Appendix V). Some places in Newfoundland and in
the interior of Canada farther north experienced in these years the lowest temperatures ever
recorded there. KUKLA and KUKLA report (1974) that the net effect was that the area of

12°C greater than in the 30-year average for 1931–60
and in some cases exceeded 45°C (RODEWALD
1975).

1. Soviet shipping operations reported the heaviest sea
ice for over 20 years past in July–August 1975 on the

Northern Sea Route in the Kara Sea area and,
particularly, off the coast of northeast Siberia.
    In 1976 the cold centre of the circulation
returned to the Siberian side of the Arctic.

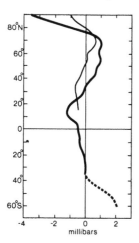

Fig. 18.30   Departures of average atmospheric pressure at mean sea level in different latitudes in recent years from the average for the first half of the century.
    Thin line = 1951–66.
    Bold line = 1970–2.

snow and ice surface integrated over the year over the northern hemisphere, which had increased 3% since the satellite surveys of it first became possible in 1967, underwent a sharp further increase in 1971. In 1972 and 1973 it was 12% more than in 1967 and, despite a decreasing trend in 1973–4, steadied in 1974 at 11% above the 1967 extent. Probably because of the intensification of the winter circulation, the increase in these years has been mainly in a lengthening of the snow season in spring and autumn and not any greater area of snow in mid winter.

General circulation and rainfall distribution

The changes in the general circulation over the northern hemisphere since the end of the early twentieth century warming, and the last great peak of W'ly type frequency (fig. 18.18) in the British Isles in 1948–54, have been (1) a weakening,[1] and shortening of wave length, in the mainstream of the W'lies over both hemispheres (see, for instance, the top and bottom curves in fig. 7.1(*a*) in Volume 1, and Table 48 in Appendix V of this volume), (2) a drift of the mainstream of the upper W'lies and of the subtropical anticyclones towards rather lower latitudes,[2] (3) an initial expansion of the north polar high pressure regions shown by the average map for 1951–66 (fig. 7.2 in Volume 1); this was followed (4) by a great increase of cyclonic activity near the

1. Monthly values of the sea level zonal index for the sector of the northern hemisphere 0–180°w computed by P. W. QUADE (University of Wisconsin, unpublished M.Sc. thesis, 1968) in over half of all months 1951–65 were in the bottom tercile of the frequency distribution that ruled from 1909–38. The change was greatest in winter and summer, no January and no July after 1950 ranking as high index according to the 1909–38 distribution.

2. This showed itself, according to unpublished work by WAHL, in a southward displacement of the

northern hemisphere W wind maximum at 700 mb in most months of the year, amounting to 4–5° of latitude in July, August, February and March, when 1965–72 is compared with 1951–8, and similarly the frequency of zonal mean W winds at 700 mb in low latitudes (25°N) exceeding 8 m/sec increased from about 35 to 55 days/year. A shift of about the same amount took place in the mean latitude of the 5100 m 500 mb height, and a corresponding expansion of the area of the Arctic cap described by it, according to T. ASAKURA.

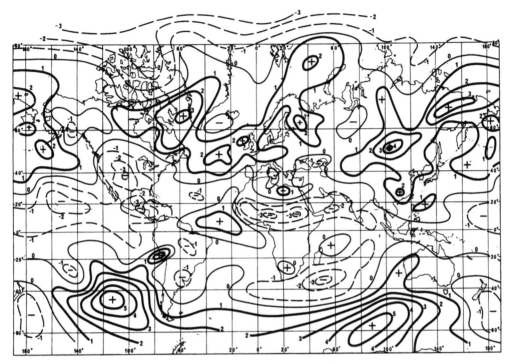

Fig. 18.31   World map of the departures of average atmospheric pressure at mean sea level for 1970–2 from the average for the first half of the century.

pole, presumably fed by cyclonic systems generated in the zone of increased thermal gradient around the fringe of the cooling Arctic, and accompanied by a great increase in the frequency of blocking anticyclones and a rise of the average pressure level over the zone 40–70°N (figs. 18.30 and 18.31). The curves in fig. 18.30 are strongly affected by the often very high pressures in blocking anticyclones over middle latitudes and the subpolar zone. The effects of equatorward displacement of the subtropical anticyclones can be seen in the map in fig. 18.31 over Egypt and Libya, southern China and the Philippines, and in the southern hemisphere over Peru, South Africa and Madagascar. There is indeed a remarkable symmetry in the latitude distribution of the anomalies north and south of the equator, seen both in the map and, more clearly, in fig. 18.30.

The distribution of anomalies in pressure distribution, which show some continuity of development from the 1950s to the 1970s (cf. figs. 7.2 and 18.31), certainly explains some of the changes of upper air temperature shown in figs. 7.25, 7.26 in Volume 1 and by STARR and OORT (1973) and described on p. 530 above. In particular, the observation of greatest fall of upper air (mid tropospheric) temperature over subtropical latitudes can be attributed to the weaker, or more intermittent, development of the subtropical anticyclones, and less subsidence, there in the last decade or two. And the fact that upper air temperatures have

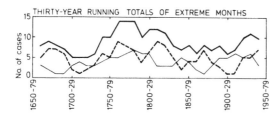

Fig. 18.32 Frequencies of extremely warm and
extremely cold (freezing) months in central
England: 30-year totals.

Thin line = months with mean temperature
over 17·5°C.

Broken line = months with mean temperature
below 0·5°C.

Full (bold) line = combined totals of both
types of extreme month.

fallen much less than surface temperatures over the sub-Arctic zone can be attributed to the more frequent anticyclones and subsidence of the mid tropospheric air over that zone.

These circulation changes made themselves apparent in significant changes in the world distribution of rainfall and the frequencies of droughts and floods in different regions from about 1960 onwards. Since these aspects affect the main agricultural and industrialized regions in middle latitudes and the heavily populated regions with difficult food production problems in low latitudes, the changes of the rainfall pattern and the range of year-to-year variability seem more important to mankind than the slight fall of overall average temperature since 1950.[1] We shall refer to these matters again in Chapters 19 and 20. Since blocking anticyclones tend to dominate somewhat different longitudes from one year to the next, and sometimes from one spell to the next, climatic periods with frequent blocking in middle latitudes tend to show an enhanced year-to-year variability (also from spell to spell, lasting a few weeks, and from one group of a few successive years to another). With the anticyclone in one position a given country may experience a long spell of warm S'ly winds or of drought, while another sector in the same latitude zone has prolonged cold N'ly winds or wet weather; and in the following year, or a few weeks or a few years later, the same country may have the opposite experience.[2]

1. It may be, however, that the temperature fall in high latitudes is of very great direct significance to the world's food supply, particularly in view of the increased incidence of meridional circulation patterns. The greatest threat probably lies in damage to the Canadian wheat crop and to corn in the United States Middle West from early frosts and snow, as in September 1973 and 1974. In higher latitudes the risk is greater. There was a temporary cover of 5 cm of snow in central Finland as early as 24 August in 1973 – an event unparalleled earlier in this century.

2. Thus there have been years of drought in the eastern United States and in Australia, and years of great floods in both countries, particularly in the United States Middle West, in 1970–4, where lake and river levels have been higher than since the 1840s. Also, between 1960 and 1975 drought-produced failures or shortfalls of the harvests in India and Soviet central Asia about four times in each country, and about twice in China, have caused such massive purchases of North American grain that the world's food reserves were run down and in the worst case in

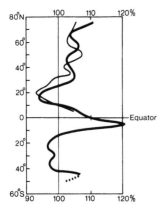

Fig. 18.33   Mean yearly rainfall averaged by latitude zones in recent years, as per cent of the 1931–60 averages.
          Thin line = 1960–9.
          Bold line = 1970–2.

Periods of frequent blocking are similarly associated with an enhanced liability to both hot and cold spells and with temperature extremes. Fig. 18.32 shows how the incidence of extreme months with either great heat waves or persistent freezing temperatures in England has varied since A.D. 1660 (see Tables 6, 7 and 8 in Appendix V). The 30-year totals of months with mean temperatures above 17·5°C and below +0·5°C, added together, are highly significantly negatively correlated with the frequency of SW'ly surface winds in England. The same is true of the extreme cold months alone correlated with the frequency of SW'ly surface winds. No statistically significant result was found in the case of the very warm months, though there was an appearance of a weak negative correlation with the frequency of the SW'ly surface winds. In other words, both extremes occurred more notably during periods of frequent blocking of the W'lies. Such times could be said to have a more continental character. Only in the late 1600s and around 1810–40 and the 1880s was the frequency of the cold extremes (freezing months) high while months of extreme warmth remained rare. The reverse situation, with extreme warm months relatively frequent and extreme cold months rare, prevailed only between about 1900 and the 1930s, if we overlook the briefer instance around 1710 to the 1730s.

1972 the world price of wheat doubled within a few months; in 1975 another drought cut the total Soviet wheat harvest even more severely, reportedly by more than 30%.

    Analysis by A. B. PITTOCK (personal communication, August 1975) of the variations of Australian rainfall during the present century has shown that in the south and east of the continent it is strongly dependent on the northward penetration of the prevailing W'lies of the Southern Ocean and the latitude of the subtropical high pressure maximum. All over eastern Australia, east of the mountains, however, and in the centre of the continent, the rainfall is also significantly correlated with an index of the Southern Oscillation (Table 49 in Appendix V), which at least in part expresses a tilt of the circumpolar vortex towards this or that sector (and is not significantly correlated with the general latitude of the W'lies).

    It also appears that the more sluggish general wind circulation (lower zonal index values and less frequent W'ly zonal flow patterns in middle latitudes) after about 1960 resulted in more frequent slow-moving cyclones and slow-moving rains – hence for example, a much enhanced frequency of heavy 24- to 48-hour rains in lowland districts of the British Isles.

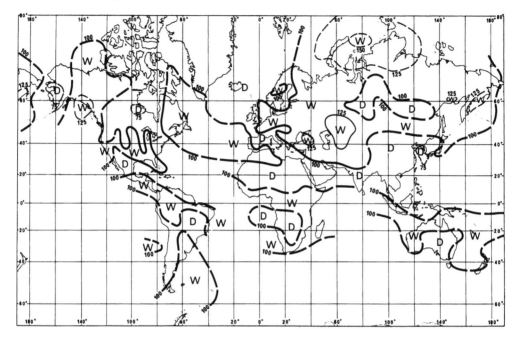

Fig. 18.34   World map of rainfall distribution 1960–9 as per cent of the 1931–60 averages.

The years 1962–75 have seen many temperature records in the northern hemisphere middle latitudes of both high and low extremes which could only be matched, if at all, about 200 to 220 years earlier.[1] It is worth noting that in the long record of the Baltic ice at Riga

1. As examples, we may cite the 1962–3 winter, which was the coldest in England since 1740, the 1963–4 winter which was the driest in England since 1743 and brought frosts to the shores of the Persian Gulf, the great Baltic ice winter of 1965–6 when also the Russian Arctic port of Murmansk was reached by the Arctic sea ice for the first time, and the winters of 1973 to 1975 when there was almost no ice on the Baltic and roses bloomed in Copenhagen till late January. The same period produced some very cold summers in Europe but also the highest temperatures ever recorded in northern Finland and north Russia, reaching 33°C in places in late July 1972.

There were striking results of the same behaviour – i.e. of blocking systems in different longitudes – on a shorter time scale in 1975:

(1) Widespread snowfalls in England as far south as Essex on 2 June 1975.
(2) The most sustained heat wave for many decades past over most of western and northern Europe, including England, giving the highest average temperatures for a whole week ever recorded at many places in the Netherlands and Denmark (24·8°C in

Copenhagen area at Lille Dyrehavegård), 4–11 August 1975.

(3) The earliest autumn snow lying in Belgium and the Netherlands in the present century on 13 October 1975.

FLOHN and RODEWALD (1975) also note the following changes in the incidence of extremes comparing periods of frequent blocking, rather as in recent years, with the prevalence of zonal circulation forms from the late 1890s to the early 1930s. Of the ten highest and lowest values of rainfall and temperature for the four seasons of the year at Philadelphia, the years 1930–67 contributed 39 cases, the years 1890–1929 only 26, and the period 1780–1819 53 cases. Similarly with the extreme individual days in the well-homogenized temperature data for Munich, the highest numbers of freezing days (maximum temperature $\leqslant$ 0°C) were 83 in 1784–5, 80 in 1829–30, 70 in 1962–3 and the lowest numbers (i.e. the other extreme) occurred in about the same epochs of frequent blocking: 7 in 1821–2, 8 in 1833–4, 11 in 1974–5, 16 in 1942–3. It was much the same with summer days (maximum temperature $\geqslant$ 25°C).

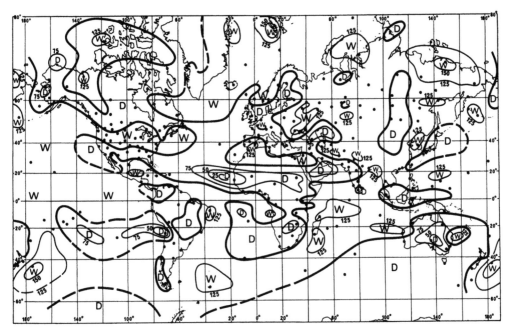

Fig. 18.35   World map of rainfall distribution 1970–2 as per cent of the 1931–60 averages.

since A.D. 1530 (see Table 17(*a*) in Appendix V) the two most open years and the year with the longest ice season of all (before 1975) occurred in the same decade, the 1650s.

The changes in the average global distribution of rainfall since the earlier decades of the century are shown in figs. 18.33, 18.34 and 18.35.[1] As in figs. 18.30 and 18.31, there is a good deal of symmetry north and south of the equatorial zone. The drought tendency in recent years has been nowhere more extreme than in the Cape Verde Islands at 17°N and St Helena at 16°s (St Helena rainfall data in HENRY 1974). All these figures show considerable continuity of development in the 1960s and 1970s as regards latitude distribution of rainfall and some other features, and the differences from the previous 30 years (or more) are highly significant both in the conventional statistical sense and in their effects on agriculture and water supply. The main features of the change may be described as:

(1) Greater concentration of the equatorial rains near the equator – marking reduced seasonal wandering of the system north and south of the equator – and consequently greater downput near the equator (or just south of it) in comparison with the preceding decades.

1. Decade-by-decade maps of rainfall over the British Isles from 1820–9 to 1920–9 were published by GLASSPOOLE (1933), but the region covered is too small to indicate the character of the wind circulation except in the more extreme decades: viz. the 1850s for general dryness, the 1890s and 1900–9 similarly except for much of northern Britain, the 1870s for general wetness especially in the east, and the 1920s for general wetness except in some extreme eastern districts.

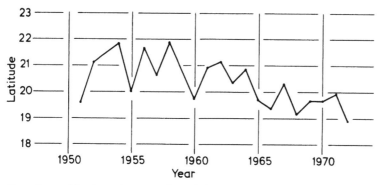

Fig. 18.36   Northward penetration of the monsoon rains each year in the Sahel zone of Africa, the southern fringe of the Sahara: 1951 to 1972.

The farther north the rains penetrate, the greater the rainfall at places between latitudes 10° and 20°N

(*In* BRYSON (*1973*) *and* WAHL *and* BRYSON (*1975*); *reproduced by kind permission.*)

(2) Reduced rainfall – failures of the monsoons – over zones near 10–20°N and 12–20°S (and farther south).

(3) Much smaller changes over middle latitudes,[1,2] where the most significant feature of these latest decades has been the very awkward type of variability from year to year, associated with the behaviour of blocking systems and meridional circulation patterns, described above.

(4) Substantially increased downput (mainly snow) in the highest northern latitudes.[3]

The sharp maximum of rainfall increase shown by this analysis (fig. 18.33) near 5°S, indicating the concentration of the equatorial rains within a narrower range than formerly about that latitude, may be taken as confirming the equatorward shift of the subtropical anticyclones over recent decades previously referred to. In as much as this shift has been more pronounced in the northern hemisphere and seems to have kept the intertropical convergence farther south than before, it seems possible that the slight warming trend in the

1. Average flow of the rivers in Germany 1960–73 was reported to be about 12% less than in the first 60 years of the century. It is not known to the author how much of this change may be due to increased use of the water for industry and irrigation.
2. See, however, GRAY (1976).
3. This aspect suggests that the situation in the 1960s and 1970s, shown by figs. 18.2 and 18.30–35, could legitimately be put forward as a glacial onset pattern. It may be important to observe that there appear to be a number of characteristics in common between this and most earlier global climatic cooling episodes – e.g. an increase in the frequency of N'ly and NW'ly winds, and of storms from these directions, affecting the British Isles, the North Sea and much of northern Europe. It may be that all climatic cooling episodes are alike in being marked by decline of the middle latitudes W'lies (perhaps particularly in the British Isles sector), reduced zonal circulation index values and increased meridional circulation character. Hence all major episodes of this kind would have some appearance of a glacial-onset regime.

The question of whether a lasting increase of glaciation and permanent shift of the climatic belts results from any given one of these episodes must depend critically on the radiation available during the recovery phase of the 200-year and other, shorter-term, fluctuations.

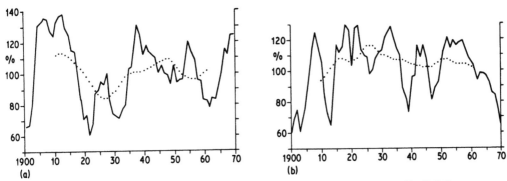

Fig. 18.37   Rainfall variations since 1900 across north Africa, Arabia and India (full lines 5-year means, dots 20-year running means).
(*a*) Winter + spring rainfall (average of 14 stations) in the zone 35–20°N.
(*b*) Summer monsoon rainfall (average of 8 stations) in the zone 30–15°N.
Expressed as percentages of the 1931–60 averages.
(*From* WINSTANLEY (*1973*); *reproduced by kind permission.*)

southern hemisphere since 1960 may be attributable to a decrease of the net northward transport of heat by winds and ocean currents across the equator.[1] This probably affects the heat transport by the summer monsoon winds crossing the equator and by the South Atlantic Equatorial Current in the ocean, more of which will have been turned south at the 'nose' of Brazil. More latent heat of condensation will also have been released in the atmosphere by the equatorial rain system just south of the equator.

Examples of the consequences of these features include a number of serious items besides the extremes of cold and warmth, drought and flood associated with the occurrences of blocking in middle latitudes. The greater yield of the equatorial rains since 1961 over Africa between the equator and 12°S led to abrupt rises of the levels of the great lakes there, drowning harbour installations and much land near the lake shores (LAMB 1966). Lake Victoria ($\frac{1}{2}$°N–3°S), which is as extensive as the southern North Sea, rose by 1·5–2 m (see fig. 13.35) to levels not previously remembered, though subsequently found in old records to have occurred between about 1876 and 1892, and Lake Tanganyika (4–9°S) rose about 3 m also between 1961 and 1964 to a level which had not occurred since the 1880s. Both lakes have remained significantly higher than in the first sixty years of this century. But far more serious were the droughts in the zones to the north and south, which have been less regularly reached by the equatorial (monsoon) rains than formerly. Fig. 18.36 shows the change from the early 1950s to the 1970s in the northernmost limit reached by the rains in the southern fringe of the Sahara, the region known as the Sahel.[2] In the drought in that zone of Africa in

1. NEWELL *et al.* (1974) have found from their thorough examination of global data for the years 1957–64 that in the atmosphere there was a slight resultant flow of heat southward across the equator. No similarly reliable assessment for earlier years is known to the writer.

2. A more or less similar decline of rainfall was noted in corresponding latitudes south of the equator (see fig. 18.33), e.g. at St Helena where 1973 appears to have been the driest year in the present century.

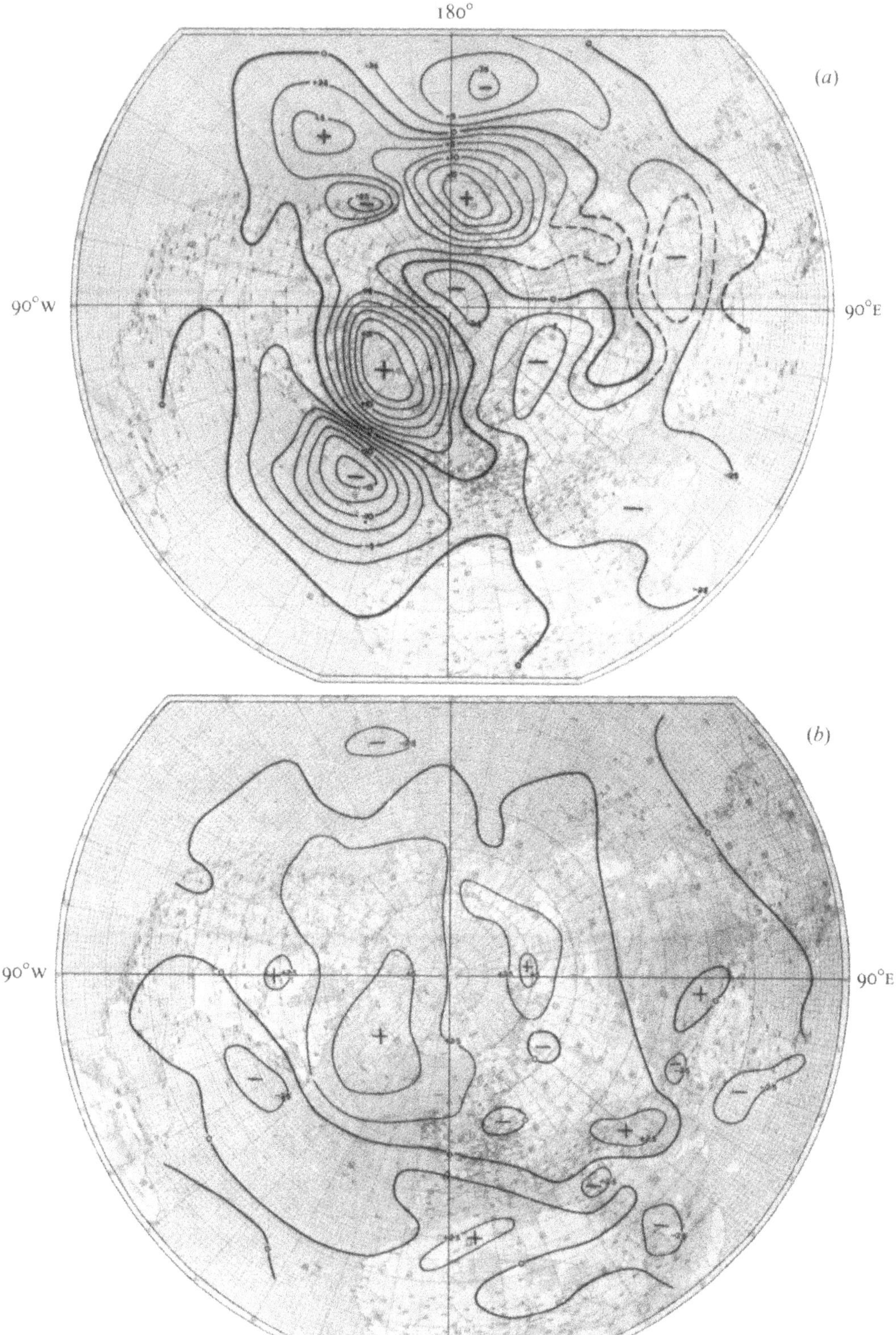

(a)

$90°\text{W}$

$90°\text{E}$

(b)

$90°\text{W}$

$90°\text{E}$

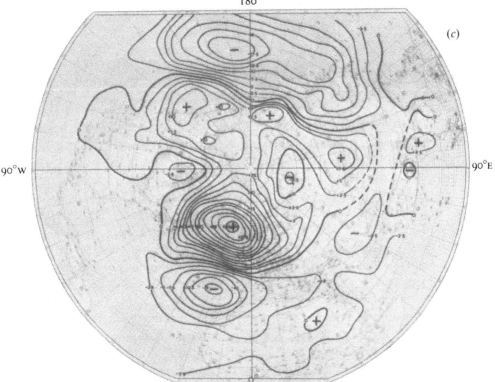

Fig. 18.38 Distribution over the northern hemisphere of departures (in millibars) of mean sea level barometric pressure from the 1900–39 average.

(a) December 1961.
(b) July 1962.
(c) January 1963.

1972 and 1973 between 200 000 and 400 000 people died and much greater numbers of cattle. A great southward trek was started from lands abandoned in west Africa, and in Ethiopia the disturbances that followed led to the collapse of the age-old imperial regime. WINSTANLEY (1973) has presented the rainfall records since 1900 in fig. 18.37 and pointed to the fact that the summer monsoon variations in this zone (fig. 18.37(b)) just south of the desert appear to parallel those of the frequency of westerly type weather in the British Isles (fig. 18.18).[1] The winter rainfall variations in the zone along the northern fringe of the Sahara desert show an inverse relation to those in the Sahel and the frequency of the W'lies in 50–60°N. Examination by LANDSBERG (1975) of the longest available rainfall records in the Sahel zone of Africa since the 1880s has shown that the frequency distribution is bimodal, suggestive of two alternative regimes. It is clear, however, from figs. 18.36 and 18.37 that, apart from the short-term fluctuations, the tendency to the dry regime has been becoming

1. The history of rainfall in Central America, and the Caribbean islands and coasts between latitudes about 5° and 20°N since 1920, presented by HASTENRATH (1975), shows a strong parallelism with this record of rainfall in the Sahel. By contrast, the Seychelles islands near $4\frac{1}{2}$°S $55\frac{1}{2}$°E in the Indian Ocean have experienced an increase of rainfall roughly paralleling the rise in the level of Lake Victoria (D. ASPIN in *Weather*, **31** (1976), 47–56).

increasingly dominant since the climax strength of the general atmospheric circulation around 1925.[1]

In those parts of northern and northwestern India near the limit reached by the summer monsoon, BRYSON (1973) has noted a corresponding effect, scarcely less threatening to the inhabitants than the 6-year drought from 1968 to 1973 in the West African Sahel. In the first quarter of this century there was a severe drought in north and northwest India every third or fourth year. Then 'as the Earth warmed up and the circumpolar vortex contracted, the monsoon rains penetrated regularly in northern India and drought frequency declined' to 2 in 36 years between 1925 and 1960. But since about 1960, 'with the cooling of the Earth and the southward movement of the subtropical high pressure areas, drought frequency has been increasing' again and the probability may now be more than once a decade. BRYSON adds that if a drought frequency like that which prevailed at the beginning of the century were to occur now, with India's population increased fourfold, or in the future, the human and political consequences would be enormous.

An aspect of the changes in the global atmospheric circulation since the period of strongest zonal circulation in the first half of the century that is not yet understood is the extent to which the distribution of anomalous high and low pressure has been similar at all times of the year. This is illustrated by the northern hemisphere maps in figs. 18.38 (a), (b) and (c), relating to two winter months and one summer month in the 1960s, when it was first noticed. The features are stronger in the winter cases, but their positions are essentially the same in both seasons, regardless of the very different thermal patterns and the circulation development to be expected therefrom. Other examples could be presented from subsequent years. We see in these maps persistent tendencies of the atmospheric circulation, particularly of the locations of blocking development, which characterize the secular change within the run of years examined. Such persistence from season to season suggests either an extraterrestrial cause or one connected with events in low latitudes where seasonal changes are slight.

The most basic aspect of climatic changes is probably to be found in relatively small amplitude changes in the energy absorbed and the temperatures prevailing in the oceans in low latitudes, but because these affect the energy and scale of the atmospheric and ocean circulation they result in much bigger changes in high latitudes. This means that the changes are most easily determined in high latitudes. The amplification in high latitudes is, however, important in its own right because of the changes in the extent of ice and snow and, hence, the albedo and the total heat intake. This in turn affects the location of the strongest thermal gradients, and so the jetstream, and the development and steering of cyclones and anticyclones and their heat transport, ultimately producing large amplitude changes in the

1. In east Africa at this latitude – in Ethiopia, for example – the short-term fluctuations of rainfall on time scales up to 10–12 years, seem to bulk larger in comparison to the long-term trend than is the case in west Africa (cf. WOOD and LOVETT 1974).

ranges of the main ocean currents, and hence in sea surface temperatures prevailing in some regions, in high and middle latitudes.

The logic of these patterns of variation would appear to be that on a uniform Earth indicators of the global condition should be identified in terms of temperature in low and high latitudes, though the generally bigger changes in high latitudes would be the easier to work with; in terms of wind and ocean circulation the relevant indices should be mainly in middle latitudes. In practice, however, rather big sea surface temperature variations do occur near the equator in the Pacific Ocean and we also find that there is one mode of circulation variation, the Southern Oscillation (see Volume 1, Chapter 6, pp. 241–50), which spans nearly all latitudes.

As regards causation of the climatic changes from A.D. 1600 to the 1970s, SCHNEIDER and MASS (1975), while admitting evidence for effects of great dust-producing volcanic eruptions and Man's output of carbon dioxide, were able to simulate most of the supposed course of global mean surface air temperature over the past 370 years by use of a formula expressing variations of the solar energy available in terms of sunspot numbers, using KONDRATIEV and NIKOLSKY's relationship which gives maximum solar energy at sunspot number 80 (see Volume 1, p. 18). SCHNEIDER regards this as no more than 'a demonstration that it is feasible that these external factors may well have contributed to the shape of the observed record', suggesting that high priority should be given to obtaining high precision extraterrestrial measurements of variations of solar beam intensity.

I. R. YAMAMOTO and M. HOSHIAI (personal communication, August 1975) have demonstrated that the surface temperature anomalies in the northern hemisphere following the volcanic dust-veil-producing eruptions in the years 1952 to 1970, regardless of where the eruption occurred, were in all cases greatest in high latitudes 50–90°N.

Note: A useful compendium of data in graphical form on the changes since 1940 in temperatures, precipitation, sunshine, cloudiness, snowfall and pressure in different parts of Canada is given by THOMAS (1975). Many further details about the climate of Europe since about 1670 are given by VON RUDLOFF (1967).

Appendix to Part III

V Supplementary tables and figures

Figures

Tables

554

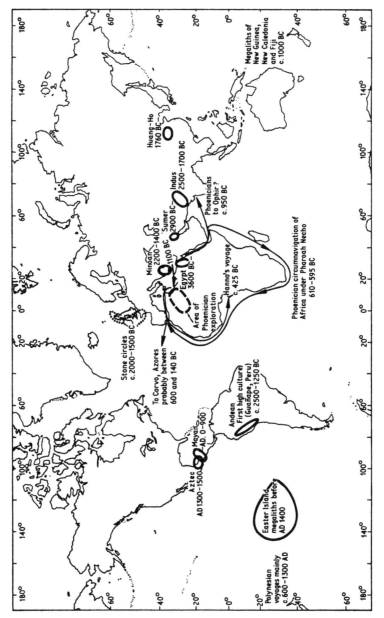

Fig. App.V.1 World map of early civilizations and early voyages by sea.

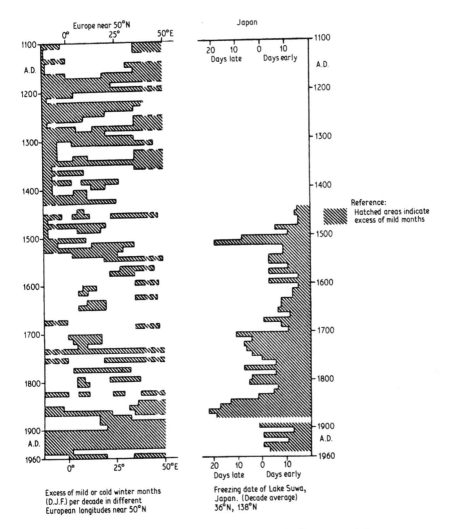

Fig. **App. V.2** Indices of winter mildness/severity in Europe and Japan compared, decade by decade, since 1440. Survey of historical records by decades: European winter mildness/severity index compared with variations in the freezing date of a small lake in Japan, decade by decade, from the overall mean date (15 January).

Note: The frequencies of various kinds of bad summer weather (continual rains, floods and storms) in Japan, and of famines in that country, by centuries or shorter periods since A.D. 1300, as plotted by T. YAMAMOTO ('A review of climatic change in Japan', pp. 223–31 in *Proc. WMO/IAMAP Symposium on Long-Term Climatic Fluctuations, Norwich, 18–23 August 1975*; Geneva, WMO No. 421), seems to show an inverse parallelism to the frequency of SW'ly surface winds in England.

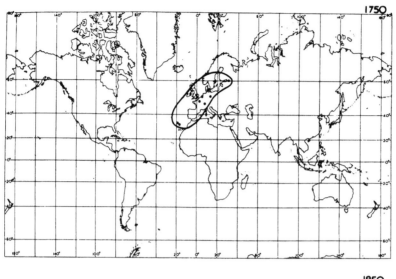

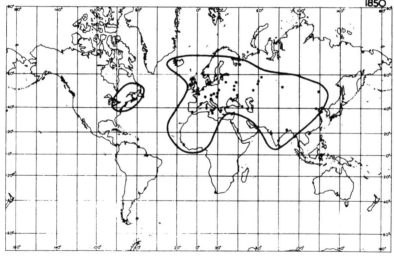

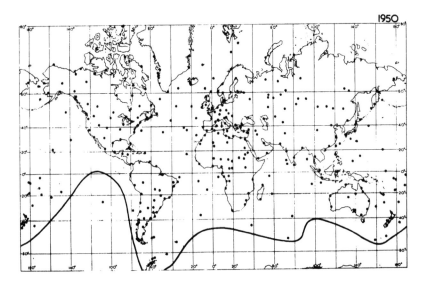

Fig. App. V.3 Growth of the network of observations, including barometric pressure, over the world and (bold line) limits of the area for which reliable 10-year mean isobars can be drawn.

(*a*) 1750–9.
(*b*) 1850–9.
(*c*) 1950–9.

Table App. V.1   Geological eras and their subdivisions, with time scale

Era	System or Period	Series or Epoch	Notes on climate, fauna and flora	Radiometric age
CENOZOIC (Kainozoic)	Quaternary	Holocene or Recent (post-glacial)		
				c. 10 000 years ago
		Pleisto-cene	Glaciations and Interglacial ages alternating. *Homo sapiens* emerged about mid last glacial (40 000 years ago). Earliest forms of Man in early Pleistocene.	
				1–3 million years ago
	Tertiary	Pliocene	Becoming colder. Large carnivorous animals.	
				c. 10 million years ago
		Miocene	Warm to moderate. Whales, apes and grazing animals.	
				c. 25 million years ago
		Oligocene	Warm. First monkeys and man-like apes.	
				c. 40 million years ago
		Eocene	Moderate becoming warm.	
				c. 60 million years ago
		Palaeocene	Becoming warmer. Early horses, elephants, etc. Most modern genera of flora present, but world distribution different.	
				c. 70 million years ago
MESOZOIC	Cretaceous		Some evidence of a cold, but not glacial phase towards end. Ichthyosaurus became extinct; many other extinctions about end of Cretaceous to early Tertiary including among phytoplankton, marine turtles, dinosaurs.	
				c. 130 million years ago
	Jurassic		First birds and small mammals. Dinosaurs' heyday. First crocodiles and flying reptiles.	
				c. 180 million years ago
	Triassic		First dionosaurs, ichthyosaurus, etc.	

Table App. V.1 – continued

Era	System or Period	Series or Epoch	Notes on climate, fauna and flora	Radiometric age
				c. 230 million years ago
PALAEOZOIC	Permian		Glacial at first, later moderate. Conifers abundant.	
				c. 270 million years ago
	Carboniferous		Warm at first, becoming glacial in some regions. First winged insects. First spiders. Climax of seed ferns and clubmoss trees; great coal forests. First coniferous trees.	
				c. 350 million years ago
	Devonian		Moderate, becoming warm. First insects. First trees. Fish abundant.	
				c. 400 million years ago
	Silurian		Warm. First known land plants and animals.	
				c. 440 million years ago
	Ordovician		Moderate to warm. First corals. First fishes.	
				c. 500 million years ago
	Cambrian		Climates becoming warm. First foraminifera, starfish, snails (mollusks), bivalve shells.	
				c. 600 million years ago
PRECAMBRIAN	Proterozoic (first period with life forms)		First algae, radiolaria and worms. Evidence of some ice ages, perhaps severe enough to cover the whole Earth.	
	Azoic or Archaean		Before the dawn of life.	
				> 3000 million years ago

Table App. V.2   Mean temperatures for each latitude in the present century: reduced to sea level
(After HANN-SÜRING 1940, with adjustments* in high latitudes)

| Latitude | Air temperatures (°C) | | | | | Sea surface temperature |
	Jan.	Apr.	July	Oct.	Year	
90°N	−38	−28	−1	−22	−22	−1·7
85°	−34	−27	0	−20	−21	−1.7
80°	−31	−25	0	−18	−19	−1.7
75°	−29	−20	3	−14	−16	−1·2
70°	−26·3	−14·0	7·3	−9·3	−10·7	0·7
65°	−23·0	−7·3	12·4	−4·1	−5·8	3·1
60°	−16·1	−2·8	14·1	0·3	−1·1	4·8
55°	−10·9	1·8	15·7	2·9	2·3	6·1
50°	−7·1	5·2	18·1	6·9	5·8	7·9
45°	−1·7	10·4	20·9	11·5	9·8	10·8
40°	5·0	13·1	24·0	15·7	14·1	14·1
35°	9·6	17·0	25·8	18·9	17·2	18·3
30°	14·5	20·1	27·3	21·8	20·4	21·3
25°	18·7	23·2	27·9	24·6	23·6	23·7
20°	21·8	25·2	28·0	26·4	25·3	25·4
15°	24·0	26·7	27·9	27·0	26·3	26·6
10°	25·8	27·2	26·9	26·9	26·7	27·2
5°N	26·3	26·8	26·2	26·3	26·4	27·4
Equator	26·4	26·6	25·6	26·5	26·2	27·1
5°S	26·4	26·5	24·9	26·0	25·8	26·4
10°	26·3	25·9	23·9	25·7	25·3	25·8
15°	25·9	25·2	22·3	24·4	24·4	25·1
20°	25·4	24·0	20·0	22·8	22·9	24·0
25°	24·3	21·8	17·5	20·6	20·9	22·0
30°	21·9	18·7	14·7	18·0	18·4	19·5
35°	18·7	15·2	11·8	15·3	15·2	17·0
40°	15·6	12·5	9·0	11·7	11·9	13·3
45°	12·3	8·0	6·2	8·0	8·8	9·9
50°	8	6	4	5	5	6·5
55°	5	3	1	2	2	3·5
60°	2·5	−1	−5	−4	−2	0·5
65°	0	−5	−13	−8	−7	−1·0
70°	−2	−13	−20	−15	−13	−1·3
75°	−6	−24	−30	−21	−21	−1·7
78°S	−7	−27	−32	−22	−22	—

* The adjustments are based on the tabulated data in SCHWERDTFEGER (1970) and VOWINCKEL and ORVIG (1970).

Table App. V.3 Prevailing temperatures (°C) in central England and rainfalls (% of 1916–50 averages) in England and Wales

Temperatures from 1680: Averages taken from MANLEY's (1959, 1961) homogenized records.

Temperatures before 1680. Averages derived from decade values of the winter mildness/severity index and the summer wetness/dryness index (p. 32) using regression equations based on comparisons with observed values since 1680–1740. Bracketed values have been adjusted for systematic departures from the regression line indicated by wind circulation characteristics. A separate column gives winter values adjusted on other meteorological considerations explained in the text, so that the mildest decades between 1150 and 1300 have temperatures equalling those of the 1920s and 1930s. However, the balance of *all* the evidence, meteorological and non-meteorological, appears to favour the still higher values indicated by the thin line in fig. 6.6 (Volume I, p. 236). Summer temperatures adjusted to fit certain botanical indications are given in a separate column.

Rainfalls from 1740: Averages taken from NICHOLAS and GLASSPOOLE (1931) and Meteorological Office records.

Rainfalls before 1740: Averages derived from decade values of the summer wetness/dryness index and from the adjusted average values of annual mean and winter temperature, as explained in the text, using regression equations.

Period	Temperatures					Rainfalls			High Summer	Period
	Winter (DJF)	Winter (DJF) adjusted for probable under-reporting of mild winters in medieval times	High Summer (JA)	High Summer (JA) adjusted to meet certain botanical considerations	Year based on the adjusted values given for winter and summer	Year	High Summer (JA)	Sept. to June	Rain as % of the year's rain	
800–1000	3.5	3.5	15.9	15.9	9.2	96	93	97	17	800–1000
1000–1100	3.7	3.7	(16.2)	16.2	9.4	97–98	94	98	17	1000–1100
1100–1150	3.5	3.5	(16.2)	16.5	9.6	98–100	93	102	17	1100–1150
1150–1200	3.9	4.2	(16.3)	16.7	10.2	100–106	86	107	15	1150–1200
1200–1250	3.8	4.1	(16.3)	16.7	10.1	100–104	86	105	15	1200–1250
1250–1300	3.9	4.2	(16.3)	16.7	10.2	100–106	84	107	14	1250–1300
1300–1350	3.6	3.8	15.9	16.2	9.8	98–101	89	102	16	1300–1350
1350–1400	3.6	3.8	15.7	15.9	9.5	97–98	105	96	19	1350–1400
1400–1450	3.4	3.4	15.8	15.8	9.1	95	88	97	16	1400–1450
1450–1500	3.5	3.5	15.6	15.6	9.0	95	106	93	20	1450–1500
1500–1550	3.8	3.8	15.9	15.9	9.3	97	84	99	15	1500–1550
1550–1600	3.2	3.2	(15.3)	15.3	8.8	93–94	106	91	20	1550–1600
1600–1650	3.2	3.2	(15.4)	15.4	8.8	93–94	99	92	19	1600–1650
1650–1700	3.1	3.1	(15.3)	15.3	8.7	92–93	104	90	20	1650–1700
1700–1750	3.7	3.7	15.9	15.9	9.24	96–97	92	98	17	1700–1750
1750–1800	3.4	3.4	15.9	15.9	9.06	94	109	91	21	1750–1800
1800–1850	3.5	3.5	15.6	15.6	9.12	96	97	96	18	1800–1850
1850–1900	3.8	3.8	15.7	15.7	9.12	97	98	97	18	1850–1900
1900–1950	4.2	4.2	15.8	15.8	9.41	99	97	100	17	1900–1950

Table App. V.4 Decade values of the indices of (*a*) summer wetness/dryness and (*b*) winter mildness/severity from 1100 to the 1960s for different European longitudes near 50°N

(*a*) High Summer Wetness/Dryness Index (Europe 50°N). Each July or August with unmistakable evidence of frequent rains counted 1, an unremarkable July or August ½, a plainly dry month 0. Totals per decade.

Decade	England c. 0°E	Germany c. 12°E	Russia c. 35°E	Decade	England c. 0°E	Germany c. 12°E	Russia c. 35°E
1100–09	10	10½	9½	1440s	8	8½	8½
1110–19	12½	9½	8½	1450s	14	16	10
1120s	10	11½	7½	1460s	9	11½	12
1130s	6	9½	9½	1470s	8	9	8½
1140s	11	10½	10½	1480s	12½	11½	9½
1150s	11	7½	10½	1490s	16	10½	12½
1160s	7	10½	6½	1500–09	7	8	9½
1170s	7½	11½	10	1510–19	7½	10½	10
1180s	7½	5	10½	1520s	11	13	9
1190s	10½	12½	9½	1530s	8½	6	7½
1200–09	7	8½	10½	1540s	8	9½	13
1210–19	9	10½	12½	1550s	8½	6½	11
1220s	12½	10½	10	1560s	16	14½	12
1230s	7½	10½	9	1570s	11	15	10
1240s	8	7½	9½	1580s	12½	17	11½
1250s	12½	12½	10½	1590s	11	10½	11½
1260s	10	9½	10½	1600–09	11½	12½	13½
1270s	5	6½	11½	1610–19	7	7	8½
1280s	8	7½	12	1620s	13	14½	10½
1290s	7	7½	8½	1630s	11	6	11½
1300–09	6½	10½	11½	1640s	11	9½	7½
1310–19	15	14	9½	1650s	11½	11	10
1320s	6½	9½	8½	1660s	10½	12½	10½
1330s	7	11½	5½	1670s	10½	8½	10½
1340s	10½	13	7½	1680s	11½	10½	11
1350s	12½	11½	8½	1690s	13½	12	13½
1360s	14½	13	8½	1700–09	10½	9½	10½
1370s	11½	4	7½	1710–19	10	8½	10½
1380s	8½	5	9	1720s	11½	8	13½
1390s	11	10½	9½	1730s	10	10½	13½
1400–09	11	17	7½	1740s	7	6½	10½
1410–19	7½	11½	10	1750s	13½	12	10
1420s	7	2	9½	1760s	11½	12	9
1430s	12	10½	8½				

Decade	England c. 0°E	Germany c. 12°E	Russia c. 35°E	Decade	England c. 0°E	Germany c. 12°E	Russia c. 35°E
1770s	$11\frac{1}{2}$	11	6	1870s	13	$17\frac{1}{2}$	12
1780s	12	10	10	1880s	$10\frac{1}{2}$	$13\frac{1}{2}$	$10\frac{1}{2}$
1790s	$10\frac{1}{2}$	$12\frac{1}{2}$	10	1890s	10	$12\frac{1}{2}$	7
1800–09	10	12	11	1900–09	$10\frac{1}{2}$	$11\frac{1}{2}$	$7\frac{1}{2}$
1810–19	10	$14\frac{1}{2}$	8	1910–19	11	12	14
1820s	12	$16\frac{1}{2}$	9	1920s	$11\frac{1}{2}$	8	$7\frac{1}{2}$
1830s	$11\frac{1}{2}$	$15\frac{1}{2}$	9	1930s	10	$10\frac{1}{2}$	$10\frac{1}{2}$
1840s	$10\frac{1}{2}$	$14\frac{1}{2}$	10	1940s	$9\frac{1}{2}$	$6\frac{1}{2}$	9
1850s	$12\frac{1}{2}$	$12\frac{1}{2}$	$10\frac{1}{2}$	1950s	$11\frac{1}{2}$	$10\frac{1}{2}$	$7\frac{1}{2}$
1860s	9	$10\frac{1}{2}$	10	1960s	$10\frac{1}{2}$	10	$11\frac{1}{2}$

The index was compared with rainfall records to continue the series to the present century. Best results appeared to be obtained by counting Julys and Augusts as wet or dry if they fell within the highest or lowest quintiles of the present century's records. The figures for Germany are based partly on comparisons with records compiled in connection with vine harvests as well as on actual weather records. Perhaps because of the abundance of information about central Europe more doubt and difficulty arise over the index figures, which are often a result of compromise, in this series than in those for Britain and Russia.

(*b*) Winter Mildness/Severity Index (Europe near 50°N). The number of unmistakably mild months (Decembers, Januarys and Februarys only) minus the numbers of unmistakably severe months (D, J, F only) per decade.

Decade	England c. 0°E	Germany c. 12°E	Russia c. 35°E	Decade	England c. 0°E	Germany c. 12°E	Russia c. 35°E
1100–09	−1	−2	0	1460s	−9	−6	−6
1110–19	−4	−5	0	1470s	+3	+1	−6
1120s	−3	−3	−6	1480s	−6	+1	−10
1130s	0	−3	−6	1490s	−5	−12	−8
1140s	−3	−9	+2	1500–09	+7	−3	−6
1150s	−6	−3	0	1510–19	−2	+2	0
1160s	0	−2	+6	1520s	+2	+11	−6
1170s	+2	+2	+4	1530s	−3	+6	−6
1180s	+6	+5	−6	1540s	−2	−1	+6
1190s	+7	+5	+6	1550s	−1	−3	−12
1200–09	+1	−3	−6	1560s	−7	−3	+4
1210–19	−2	−5	−12	1570s	−3	−1	0
1220s	+1	+6	+2	1580s	−7	−5	+12
1230s	+6	+1	0	1590s	−5	−5	0
1240s	+1	+3	−6	1600–09	−10	0	−6
1250s	+3	−5	0	1610–19	+2	−3	−10
1260s	−3	−1	0	1620s	−7	−3	0
1270s	+7	0	+6	1630s	−5	+2	−10
1280s	+3	+3	−18	1640s	−3	+5	−36
1290s	+2	+5	0	1650s	−9	−7	−24
1300–09	−1	+1	+6	1660s	−7	−3	−14
1310–19	−1	−10	−6	1670s	0	−6	−24
1320s	0	−1	0	1680s	−2	−7	0
1330s	−4	−1	0	1690s	−7	−5	−6
1340s	+2	+4	0	1700–09	0	+1	−6
1350s	−3	−4	0	1710–19	−2	+1	0
1360s	+4	−2	−6	1720s	−4	−1	−12
1370s	−3	−5	−12	1730s	+9	+3	+6
1380s	+2	+11	−8	1740s	−4	−1	−6
1390s	−2	+2	−6	1750s	0	−5	+10
1400–09	−3	−2	−22	1760s	−6	0	−12
1410–19	+2	−1	−10	1770s	−1	+4	−6
1420s	+4	+7	−6	1780s	−4	−4	−20
1430s	−12	−18	−10	1790s	−2	−2	0
1440s	−5	−2	−28	1800–09	−2	−1	+1
1450s	−1	−3	+8	1810–19	−2	−3	−1

Decade	England c. 0°E	Germany c. 12°E	Russia c. 35°E	Decade	England c. 0°E	Germany c. 12°E	Russia c. 35°E
1820s	0	0	0	1900–09	+2	0	+2
1830s	−3	−2	−1	1910–19	+5	+6	+1
1840s	−3	−2	+1	1920s	+8	+6	+2
				1930s	+4	+4	+5
1850s	−1	−2	+3	1940s	+1	−1	−1
1860s	+8	+6	+1				
1870s	0	0	−3	1950s	0	+1	+2
1880s	−2	0	+1	1960s	−3	−1	−3
1890s	−3	−1	+1	1970s			

Table App. V.5 EASTON's (1928) rating of the winters in western Europe
Winters are identified by the year in which the January fell.

(*a*) Winters rated as the *mildest* in each century (with index rating in brackets)

584 (80)

808 (80)

1172 (80)

1218 (80), 1249 (82), 1289 (90)

1301 (79), 1365 (82), 1368 (82)

1405 (79), 1421 (79), 1427 (82), 1478 (90)

1507 (82), 1515 (79), 1529 (84), 1576 (79)

1609 (79), 1617 (82), 1661 (79)

1702 (79), 1703 (82), 1717 (79), 1708 (82), 1725 (82), 1764 (75), 1796 (84)

1807 (76), 1822 (78), 1834 (89), 1846 (81), 1869 (79)

(*b*) Winters rated as the *coldest* in each century (with index rating in brackets)

296 (10)

359 (10)

401 (21), 411 (10), 432 (21), 462 (10)

545 (21), 554 (10), 566 (21), 593 (21)

695 (10)

760 (21), 764 (10)

822 (10), 845 (21), 856 (21), 860 (10), 874 (21), 881 (21)

913 (10), 928 (21), 940 (21), 975 (21)

1033 (21), 1044 (21), 1068 (21), 1074 (10), 1077 (10)

1125 (10), 1143 (21), 1150 (10), 1179 (21)

1205 (10), 1210 (17), 1211 (25), 1217 (21), 1219 (17), 1225 (10), 1236 (10), 1270 (17), 1276 (25), 1282 (25)

1303 (21), 1306 (10), 1316 (17), 1361 (25), 1363 (21), 1364 (10), 1372 (25), 1394 (21), 1399 (21)

1408 (4), 1423 (10), 1432 (17), 1435 (5), 1443 (17), 1458 (17), 1465 (21), 1469 (25), 1481 (17), 1491 (25)

1508 (25), 1511 (21), 1514 (10), 1544 (21), 1546 (21), 1548 (25), 1565 (4), 1569 (17), 1571 (10), 1573 (17), 1587 (21), 1591 (21), 1595 (10)

1608 (4), 1621 (10), 1622 (25), 1656 (17), 1658 (10), 1663 (25), 1667 (21), 1672 (21), 1674 (25), 1677 (21), 1684 (17), 1695 (21)

1709 (4), 1716 (21), 1729 (25), 1740 (8), 1742 (25), 1757 (24), 1763 (24), 1784 (11), 1789 (10), 1795 (11), 1799 (21)

1823 (25), 1830 (4), 1838 (20), 1841 (24), 1845 (20), 1871 (19), 1880 (12), 1891 (8), 1895 (16).

Scale: ≥85 means the warmest class
   76–84 warm
   61–75 mild
   55–60 rather mild
   45–54 normal
   39–44 rather cold
   26–38 cold
   14–25 severe
   6–13 very severe
   ≤5 great winters

EASTON's index values, defined as above, and applied to the winters from early times onwards, have been superseded by many more precise indicators, including the various homogenized thermometer records from recent centuries. Those for winters assigned to the extreme classes are likely to remain valid, though not necessarily a complete list as regards nearly extreme cases.

There are some curious features of the distribution in the tables that should perhaps be explored further with more adequate data:

(1) The warm winter occurrences appear nearly twice as frequent in the first halves (21 cases) as in the second halves (11 cases) of the centuries covered.*

(2) The warm winter ratings appear to be particularly clustered in the decades 00 to the twenties (18 cases) and sixties to eighties (10 cases), the remaining four decades having only a few (6 in all).

(3) The severe winter ratings appear more evenly spread, though 72% of them were in the even numbered decades.

(4) Some more sharply focused clusters may be suspected, e.g. 16 out of the 107 winters rated as severe occurred in the years numbered '63±2 and '95±2.

(5) The frequency of both extremes (high incidence of blocking?) in the years '05 to '09.

Presumably these tendencies, if verified as statistically real, are expressions of oscillations with periods close to 100, 50 and 20 years.

* Cf. the indication of raininess of the summers in the second halves of the centuries noticeable in fig. 6.7, p. 237, in Volume 1.

Table App. V.6   Cold winters in England: Provisional list 1400–1659

Winters with weather descriptions similar to those in later years given in Table App. V.7. **Bold** face indicates that the River Thames froze over in London,* probably requiring that the mean temperature of one month was below +0·5°C over much of southern England. Winters are identified by the year in which the January fell.	*Number per decade*
1403, 1404, **1408**	3
	0
1423 (possibly also 1429)	1–2
1432, 1433, 1434, **1435**, 1436, 1439 (possibly also 1431 and 1437)	6–8
1443 (possibly also 1440 and 1441)	1–3
1458	1
1460, 1464, 1465, 1469	4
1470 (possibly also 1477)	1–2
1481 (possibly also 1487)	1–2
1491, 1492, 1496	3
1506 (possibly also 1508)	1–2
1511 or 1512, **1514** (possibly also 1517)	2–3
	0
1534, **1537** (possibly also 1536)	2–3
1546 (possibly also 1541 and 1548 or 1549)	1–3
1554	1
1561, **1565**, 1569 (probably also 1564)	3–4
1571, 1573 (possibly also 1570 and 1578)	2–4
1584, 1586, 1587 (possibly also 1580)	3–4
1595 (probably also 1590 and 1599)	1–3
1600, **1608** (possibly also 1601, 1603, 1605)	2–5
1615 (possibly also 1612 and 1618)	1–3
1621 (possibly also 1623, 1624, 1625 and 1628)	1–5
1635, 1637 (possibly also 1631)	2–3
1649 (possibly also 1644, 1645 and 1646)	1–4
1655, 1658 (possibly also 1656)	2–3

* *Main source:* WALFORD, E. (1887) *Frost Fairs on the Thames*, London (Wyman – privately printed *Opuscula* No. 12, issued to the Members of the Sette of Old Volumes – available in the British Museum).

Table App. V.7   Cold winters in England since 1659

Winters with mean temperature below +3°C (*December, January and February*) in central England. *Long-term average* +4°C: *the winters below 3°C make up the coldest quartile.* *Winters are identified by the year in which the January fell.*	Number per decade	Freezing months (*mean temperature below* +0·5°C)
1660, 1663, 1665, 1667	4	Jan. 1660 (0°C), Jan. 1667 (0°)
1670, 1672, 1674, 1675, 1677, 1678, 1679	7	Feb. 1670 (approx 0°), Dec. 1676 (−0·5°)
1681, 1684, 1685, 1689	4	Jan. 1684 (−3°) and Feb. 1684 (−1°)
1691, 1692, 1694, 1695, 1697, 1698	6	Feb. 1692 (0°), Jan. 1694 (0°), Jan. 1695 (−1°), Jan. 1698 (0°)
1709	1	Jan. 1709 (−1·5°)
1716	1	Jan. 1716 (−2·0°)
1729	1	
1731	1	
1740, 1741, 1746	3	Jan. 1740 (−2·8°) and Feb. 1740 (−1·6°)
1755, 1757	2	Jan. 1757 (0·3°)
1760, 1763, 1765, 1766, 1767	5	Jan. 1763 (−0·8°), Feb. 1765 (0·4°), Jan. 1767 (0·1°)
1771, 1772, 1774, 1776, 1777, 1778	6	Jan. 1776 (−1·6°)
1780, 1784, 1785, 1789	4	Jan. 1780 (−0·9°), Jan. (−0·6°) and Dec. (0·3°) 1784, Feb. 1785 (0·4°), Dec. 1788 (−0·3°)
1792, 1795, 1797, 1799	4	Jan. 1795 (−3·1°), Dec. 1796 (−0·3°)
1800, 1802, 1803, 1805, 1808	5	
1814, 1816	2	Jan. 1814 (−2·9°), Jan. 1815 (0·3°)
1820, 1823, 1827	3	Jan. 1820 (−0·3°), Jan. 1823 (−0·1°) Jan. 1826 (0·4°), Jan. 1829 (0·3°)
1830, 1831, 1838	3	Jan. 1830 (−0·2°), Jan. (−1·5°) and Feb. (0·4°) 1838

Table App. V.7 – continued

Winters with mean temperature below +3°C (*December, January and February*) in central England.
    Long-term average +4°C: *the winters below 3°C make up the coldest quartile.*
    Winters are identified by the year in which the January fell.
	Number per decade	Freezing months (*mean temperature below* +0.5°C

1841, 1845, 1847	3	Dec. 1844 (0·4°)
1855	1	Feb. 1855 (−1·7°)
1860, 1861, 1865	3	
1871, 1875, 1879	3	Dec. 1874 (−0·2°), Dec. 1878 (−0·3°), Jan. 1879 (−0·7°)
1880, 1881, 1886, 1887, 1888	5	Jan. 1881 (−1·5°)
1891, 1893, 1895	3	Dec. 1890 (−0·8°), Jan. (0·2°) and Feb. (−1·8°) 1895
	0	
1917	1	
1929	1	Feb. 1929 (0·4°)
	0	
1940, 1941, 1942, 1947	4	Jan. 1940 (−1·4°), Feb. 1942 (0·1°), Jan. 1945 (0·4°), Feb. 1947 (−1·9°)
1951, 1956	2	Feb. 1956 (−0·2°)
1963	1	Jan. (−2·1°) and Feb. (−0·7°) 1963

Later winters in which the Thames froze over in London were

   1663, 1666
   1677
   1684
   1695
   1709
   1716
   1740
   1768   (river 'more or less frozen over')
   1776   omitted by WALFORD
   1785   (river 'more or less frozen over')
   1795   ? omitted by WALFORD
   1814

The river has never been frozen over in London since the replacement of London Bridge, the building of the embankments and channelling of the tributary streams in pipes in the early-mid nineteenth century, but considerable amounts of broken ice were reported in 1838, 1895 and 1929.

Table App. V.8 Hot summers in England since 1659

Summers with mean temperature 16°C or over (June, July and August) in central England	Number per decade	Hot months (mean temperature above 17·5°C)
1666, 1667, 1669	3	July 1666 (18°C)
1676, 1679	2	June 1676 (18°), July 1677 (18°)
1704, 1706, 1707	3	July 1701 (18·3°)
1718, 1719	2	July 1714 (18·0°), July 1719 (18·0°)
1726, 1727, 1728	3	
1731, 1733, 1736	3	July 1733 (18·3°), Aug. 1736 (17·8°)
1747	1	
1759	1	July 1757 (18·4°), July 1759 (18·2°)
1760, 1762	2	July 1762 (17·8°)
1772, 1775, 1778, 1779	4	July (17·9°) and Aug. (17·6°) 1779
1780, 1781, 1783	3	Aug. 1780 (17·6°), July 1783 (18·8°)
1794, 1798	2	July 1793 (17·6°), July 1794 (18·1°)
1800, 1801, 1807, 1808	4	July 1800 (17·7°), July 1803 (17·6°), July 1808 (18·4°)
1818	1	July 1818 (18·2°)
1822, 1826	2	July (17·9°) and Aug. (17·6°) 1826
1831, 1834, 1835, 1837	4	
1846	1	June 1846 (18·2°)
1857, 1859	2	July 1852 (18·7°), July 1859 (18·3°)
1868	1	July 1868 (18·3°)
1870, 1876, 1878	3	
1887	1	
1893, 1899	2	Aug. 1899 (17·8°)
None 1900–09	0	July 1900 (17·7°), July 1901 (18·0°)
1911	1	July (18·2°) and Aug. (18·2°) 1911
1921	1	July 1921 (18·5°)
1933, 1934, 1935	3	July (17·8°) and Aug. (17·6°) 1933, July 1934 (18·2°)
1947, 1949	2	Aug. 1947 (18·6°)
1955, 1959	2	July (17·7°) and Aug. (18·1°) 1955
None 1960–69	0	
1975, 1976	(2)	Aug. 1975 (18·6°), July 1976 (18·4°)

Table App. V.9   Number of days each year with maximum temperature over 26·5°C (approx. 80°F) at Belstead (52·0°N 1·1°E) near Ipswich, England

Decade	Year										Decade average
	0	1	2	3	4	5	6	7	8	9	
1900–09			5	7	12	12	13	0	5	4	7·3
1910–19	1	33	7	3	6	2	3	8	2	9	7·4
1920s	0	13	3	16	1	8	7	0	8	11	6·7
1930s	6	1	10	19	8	20	4	6	9	6	8·9
1940s	8	14	9	9	10	7	2	26	10	19	11·4
1950s	8	1	8	3	0	9	2	7	2	17	5·7
1960s	2	7	0	0	4	0	0	3	2	3	2·1
1970s	8	3	1	6	0	15	21				

Table App. V.10   Seasonal and yearly mean temperatures in central England (°C) since 1659
From the monthly figures provided by MANLEY (1974).
N.B.: The December in the winter means is the December of the previous year.

Year	Winter DJF	Spring MAM	Summer JJA	Autumn SON	Av. for calendar year J–D	Year	Winter DJF	Spring MAM	Summer JJA	Autumn SON	Av. for calendar year J–D
1659	(3·5)	8·0	15·0	9·3	8·8						
1660	(2·0)	8·7	15·0	9·7	9·1	1680	3·2	7·7	14·7	10·5	8·8
1661	5·0	8·3	14·7	10·7	9·1	1681	1·0	7·5	15·0	10·7	8·7
1662	5·7	8·3	15·0	10·0	9·5	1682	3·7	7·5	14·5	9·3	9·0
1663	1·7	7·3	14·7	10·0	8·6	1683	3·8	8·8	15·0	8·0	8·5
1664	4·7	8·0	15·7	9·3	9·3	1684	−1·2	7·5	15·5	8·7	7·9
1665	2·0	7·3	15·0	9·3	8·3	1685	2·7	8·7	14·3	10·0	9·1
1666	3·7	8·3	16·7	10·3	9·8	1686	6·3	9·3	15·5	9·5	10·2
1667	2·3	6·3	16·0	9·3	8·7	1687	4·7	7·3	14·5	9·3	8·8
1668	4·3	7·7	15·3	10·0	9·5	1688	3·7	6·5	14·5	7·7	7·8
1669	3·3	7·7	16·0	10·0	9·0	1689	2·7	8·0	14·3	8·7	8·5
1670	1·9	8·0	15·3	10·0	8·9	1690	4·3	7·3	14·7	9·3	8·9
1671	3·5	8·0	15·0	9·3	9·1	1691	2·2	7·2	15·2	8·3	8·0
1672	2·3	8·0	15·0	9·7	8·9	1692	1·8	6·8	14·5	7·7	7·7
1673	3·7	7·8	15·2	8·0	8·7	1693	3·8	6·2	14·8	9·2	8·4
1674	2·5	(6·6)	13·7	8·7	8·2	1694	2·7	6·7	13·7	7·8	7·7
1675	2·8	6·8	13·7	7·7	7·9	1695	0·7	6·0	13·2	8·7	7·2
1676	5·0	7·7	16·8	7·5	8·8	1696	4·7	6·5	14·7	8·7	8·5
1677	2·0	8·2	15·3	8·7	8·7	1697	1·3	8·0	14·3	8·3	8·0
1678	(1·8)	7·2	15·3	9·8	8·5	1698	1·0	6·5	14·0	8·7	7·7
1679	1·0	7·8	16·2	9·2	8·8	1699	3·4	6·8	15·5	9·4	8·8

Table App. V.10 – continued

Year	Winter DJF	Spring MAM	Summer JJA	Autumn SON	Av. for calendar year J–D	Year	Winter DJF	Spring MAM	Summer JJA	Autumn SON	Av. for calendar year J–D
1700	3·3	7·3	14·5	8·9	9·1	1740	−0·4	6·3	14·3	7·5	6·8
1701	3·2	6·1	16·2	9·5	8·7	1741	2·8	6·9	15·8	11·2	9·2
1702	5·1	7·4	14·8	9·6	9·3	1742	3·1	7·1	15·5	8·6	8·4
1703	3·5	8·6	15·4	8·5	9·1	1743	3·4	8·0	15·8	10·8	9·8
1704	3·5	8·4	16·0	8·9	9·1	1744	3·1	7·4	15·4	9·7	8·8
1705	3·4	7·8	15·1	8·2	8·7	1745	3·2	7·8	14·4	10·1	8·8
1706	3·8	9·2	16·1	9·9	9·8	1746	2·2	7·6	15·3	8·4	8·6
1707	3·7	8·2	16·3	9·7	9·4	1747	4·8	7·6	16·6	10·2	9·8
1708	4·5	8·8	15·3	10·2	9·7	1748	3·2	6·2	15·3	10·2	8·8
1709	1·2	8·0	15·2	10·3	8·7	1749	5·0	8·1	14·9	10·2	9·4
1710	3·2	8·0	15·0	10·3	9·5	1750	5·1	8·9	15·5	9·5	9·7
1711	4·8	8·7	15·3	9·8	9·4	1751	3·2	7·5	14·9	8·4	8·4
1712	3·8	7·8	15·3	9·5	9·1	1752	3·1	7·6	15·4	10·3	9·2
1713	4·2	6·8	14·2	9·3	8·6	1753	3·3	8·5	15·2	9·2	9·0
1714	4·5	7·5	15·8	9·8	9·4	1754	3·5	7·4	14·7	10·0	8·8
1715	4·3	9·0	14·8	10·3	9·4	1755	2·3	7·8	15·1	8·9	8·6
1716	0·8	8·0	15·0	9·2	8·4	1756	4·3	7·3	14·9	9·0	8·8
1717	3·3	7·7	14·8	9·5	9·0	1757	2·4	7·9	15·9	9·5	8·8
1718	3·2	8·0	16·2	9·8	9·3	1758	3·2	8·7	15·1	8·6	9·0
1719	3·8	7·8	16·7	9·8	9·5	1759	5·2	8·9	16·5	9·8	10·0
1720	4·0	7·5	14·7	9·5	9·1	1760	2·7	9·2	16·0	10·2	9·8
1721	3·8	7·2	15·2	9·7	8·9	1761	5·8	9·4	15·5	9·9	10·0
1722	4·5	8·0	14·7	10·4	9·3	1762	4·4	8·9	16·7	8·7	9·6
1723	3·1	9·4	15·3	10·6	9·8	1763	2·6	8·2	15·1	9·1	8·9
1724	5·2	7·8	15·5	9·4	9·3	1764	4·6	7·8	15·1	8·6	8·7
1725	3·7	8·0	13·1	9·7	8·7	1765	2·7	8·0	14·9	8·8	8·5
1726	3·1	8·7	16·0	10·3	9·3	1766	1·4	7·3	15·3	9·9	8·6
1727	3·7	9·3	16·2	10·0	9·9	1767	2·9	7·3	14·4	10·1	8·7
1728	3·3	9·3	16·4	9·7	9·5	1768	3·0	8·3	15·2	8·8	8·9
1729	1·7	6·7	15·9	11·6	9·3	1769	3·3	8·0	14·8	8·9	8·8
1730	4·6	9·1	15·2	11·8	10·1	1770	4·4	6·0	14·7	9·4	8·5
1731	2·5	8·3	16·2	11·8	9·8	1771	2·6	6·9	14·8	9·2	8·6
1732	4·7	8·8	15·7	10·6	9·7	1772	2·9	7·0	16·4	10·6	9·2
1733	5·0	9·0	16·5	9·5	10·4	1773	3·8	8·4	15·9	9·2	9·2
1734	6·1	9·5	15·5	9·3	9·8	1774	2·9	8·6	15·6	9·2	9·1
1735	4·1	8·5	14·8	10·3	9·6	1775	4·7	9·5	16·4	9·5	10·1
1736	5·0	8·7	16·6	10·6	10·3	1776	2·2	8·9	15·2	9·8	9·0
1737	5·6	9·1	15·7	9·7	9·9	1777	2·9	8·6	14·9	10·6	9·1
1738	4·7	8·9	15·5	9·7	9·8	1778	2·6	7·8	16·5	8·8	9·3
1739	5·6	8·0	15·3	8·8	9·2	1779	5·6	9·7	16·6	10·6	10·4

Table App. V.10 – continued

Year	Winter DJF	Spring MAM	Summer JJA	Autumn SON	Av. for calendar year J–D	Year	Winter DJF	Spring MAM	Summer JJA	Autumn SON	Av. for calendar year J–D
1780	1·4	9·0	16·2	9·7	9·1	1820	1·4	8·3	14·7	8·7	8·6
1781	3·4	9·3	17·0	10·4	10·2	1821	3·5	8·2	14·5	11·3	9·5
1782	4·2	6·1	14·9	7·7	8·0	1822	5·8	9·6	16·0	10·4	10·1
1783	3·2	7·9	16·5	9·6	9·3	1823	1·5	8·0	13·6	9·3	8·4
1784	1·2	7·3	14·3	9·4	7·8	1824	4·6	7·6	14·8	10·1	9·3
1785	1·4	7·3	15·4	9·3	8·6	1825	4·3	8·6	15·9	10·4	9·7
1786	3·0	7·1	15·4	7·5	8·2	1826	3·8	8·8	17·6	9·7	10·1
1787	4·1	8·6	15·1	9·0	9·3	1827	2·7	8·9	15·2	10·7	9·5
1788	3·8	8·9	15·7	9·8	9·2	1828	5·7	9·1	15·6	10·6	10·3
1789	2·1	7·3	15·3	8·8	8·9	1829	4·0	7·8	14·8	8·0	8·2
1790	5·7	8·1	15·0	9·5	9·4	1830	1·1	9·5	14·2	9·7	8·7
1791	4·4	8·9	15·3	9·5	9·3	1831	2·7	9·3	16·3	10·7	10·1
1792	2·6	8·7	15·1	9·2	9·2	1832	4·1	8·4	15·5	10·1	9·5
1793	3·9	7·1	15·4	9·7	9·1	1833	4·0	8·9	14·9	9·6	9·5
1794	4·8	9·5	16·4	9·4	9·9	1834	6·5	9·3	16·2	10·4	10·4
1795	0·5	7·5	15·0	10·7	8·7	1835	4·7	8·6	16·1	9·6	9·6
1796	6·2	8·2	14·8	9·1	9·0	1836	3·4	8·0	15·1	8·5	8·8
1797	2·6	7·7	15·6	8·5	9·0	1837	3·8	5·6	16·0	9·4	8·8
1798	4·1	9·5	16·5	9·4	9·6	1838	1·4	7·2	15·0	9·0	8·1
1799	2·0	6·1	14·6	8·9	7·9	1839	3·6	6·9	14·6	9·6	8·7
1800	2·1	8·5	16·1	9·5	9·3	1840	3·8	8·3	14·6	8·1	8·5
1801	4·2	9·0	16·0	9·7	9·8	1841	1·6	9·3	13·8	9·1	8·7
1802	2·3	8·4	14·8	9·7	8·9	1842	3·1	8·5	15·7	8·6	9·2
1803	2·9	8·6	15·8	8·6	9·1	1843	4·4	8·2	14·3	9·3	9·1
1804	4·4	8·3	15·9	10·5	9·6	1844	4·3	8·4	14·5	9·5	8·6
1805	2·8	8·2	15·2	9·2	8·9	1845	1·5	6·7	14·2	9·2	8·3
1806	4·0	8·0	15·5	10·6	9·8	1846	5·8	8·7	17·1	10·4	10·2
1807	4·4	7·5	16·1	8·3	8·7	1847	1·7	8·2	15·5	10·0	9·2
1808	2·4	7·6	16·6	8·6	8·8	1848	4·1	9·3	14·6	9·4	9·4
1809	3·3	8·1	14·5	9·2	8·9	1849	5·1	8·2	15·0	9·7	9·3
1810	3·3	7·4	14·8	9·7	8·8	1850	3·5	7·9	15·4	9·2	9·1
1811	3·1	9·6	14·9	11·2	9·7	1851	5·0	7·9	14·8	8·9	9·1
1812	3·7	6·6	13·8	9·1	8·2	1852	4·8	8·0	15·9	9·5	9·8
1813	3·1	8·7	14·4	8·3	8·7	1853	4·5	7·3	14·6	9·2	8·3
1814	0·4	7·2	14·3	8·5	7·7	1854	3·1	8·7	14·6	9·6	9·3
1815	3·7	9·3	14·8	9·0	9·1	1855	1·9	6·4	15·3	9·4	8·0
1816	2·4	6·8	13·4	8·7	7·8	1856	3·8	7·3	15·2	9·3	9·2
1817	4·7	7·3	14·3	9·6	8·9	1857	3·8	8·0	16·5	11·0	10·1
1818	3·2	7·6	16·6	11·6	9·8	1858	4·2	7·8	15·8	9·5	9·2
1819	4·1	9·0	15·7	8·9	9·2	1859	5·1	8·8	16·4	9·1	9·6

Table App. V.10 – continued

Year	Winter DJF	Spring MAM	Summer JJA	Autumn SON	Av. for calendar year J–D	Year	Winter DJF	Spring MAM	Summer JJA	Autumn SON	Av. for calendar year J–D
1860	2·3	7·4	13·5	8·5	7·9	1900	3·1	7·4	15·8	10·2	9·6
1861	2·7	8·1	15·2	9·7	9·1	1901	4·3	8·1	15·8	9·5	9·1
1862	4·3	8·9	13·8	8·8	9·2	1902	3·2	7·7	14·3	9·7	8·8
1863	5·7	8·6	14·8	9·6	9·7	1903	5·3	8·2	14·2	10·0	9·3
1864	3·7	8·6	14·4	9·5	8·9	1904	3·6	8·0	15·2	9·1	9·0
1865	2·7	8·7	15·8	10·9	9·7	1905	4·2	8·3	15·5	8·1	9·1
1866	5·3	7·8	15·2	10·1	9·7	1906	4·4	7·6	15·6	10·7	9·4
1867	4·7	7·9	15·1	9·2	9·0	1907	3·1	8·1	13·6	10·0	8·8
1868	4·5	9·7	16·9	9·2	10·4	1908	4·1	7·6	14·9	10·7	9·3
1869	6·8	7·8	15·3	10·0	9·6	1909	3·4	7·8	13·9	9·0	8·6
1870	3·0	8·5	16·1	9·0	9·0	1910	4·2	8·2	14·7	8·8	9·2
1871	2·4	9·1	15·1	8·6	9·1	1911	5·0	8·5	17·0	9·8	10·1
1872	5·2	8·2	15·5	9·5	9·7	1912	5·1	9·4	14·3	8·5	9·4
1873	4·1	7·7	15·3	8·9	9·0	1913	5·3	8·6	14·7	11·1	9·8
1874	4·9	8·8	15·4	9·9	9·3	1914	5·2	8·9	15·5	10·1	9·9
1875	2·8	8·7	15·0	9·7	9·4	1915	4·3	8·0	14·8	8·4	8·9
1876	4·1	7·3	16·0	10·0	9·5	1916	5·5	7·7	14·5	10·1	9·2
1877	5·9	7·0	15·0	9·2	9·2	1917	1·5	7·1	15·5	9·8	8·5
1878	5·0	8·7	16·0	9·0	9·2	1918	4·2	8·5	14·9	8·9	9·5
1879	0·7	6·4	13·7	8·5	7·4	1919	3·9	8·1	14·6	7·8	8·5
1880	2·5	8·2	15·2	9·0	9·1	1920	5·6	9·1	14·0	10·1	9·6
1881	2·3	8·1	14·6	9·6	8·6	1921	5·4	9·0	16·2	10·5	10·4
1882	5·1	9·1	14·4	9·2	9·4	1922	4·9	7·6	13·7	8·8	8·7
1883	4·8	6·9	14·6	9·6	9·1	1923	5·7	7·8	15·1	8·5	9·0
1884	5·5	8·3	15·9	9·7	9·8	1924	3·9	7·5	14·4	10·2	9·3
1885	4·4	7·0	14·6	8·5	8·6	1925	5·8	8·0	15·7	8·5	9·2
1886	2·4	7·4	15·1	10·5	8·7	1926	4·7	8·6	15·6	9·5	9·7
1887	2·7	6·3	16·1	7·8	8·3	1927	4·2	8·8	14·7	9·7	9·3
1888	2·5	6·7	13·7	9·3	8·2	1928	4·4	8·6	14·8	10·2	9·6
1889	3·7	8·2	15·1	9·4	9·0	1929	1·7	8·1	14·9	10·5	9·0
1890	4·0	8·3	14·0	9·9	8·7	1930	4·6	8·1	15·4	10·1	9·4
1891	1·5	6·5	14·6	9·7	8·5	1931	3·8	7·8	14·7	9·4	9·0
1892	3·3	7·2	14·3	8·6	8·2	1932	4·8	7·4	15·8	9·4	9·4
1893	2·9	10·2	16·5	9·3	9·9	1933	4·1	9·4	17·0	10·2	9·8
1894	4·4	8·5	14·5	9·6	9·3	1934	3·2	8·0	16·2	10·4	10·0
1895	1·2	8·6	15·3	10·0	8·7	1935	6·1	8·2	16·3	10·0	9·7
1896	4·4	9·2	15·6	8·1	9·3	1936	3·0	8·3	15·4	9·8	9·3
1897	3·8	7·9	15·9	9·9	9·4	1937	5·4	8·3	15·7	9·6	9·6
1898	5·7	7·7	15·1	11·2	10·1	1938	4·6	9·1	15·3	11·2	10·2
1899	5·8	7·6	16·9	10·2	9·7	1939	4·7	8·7	15·4	10·4	9·7

Table App. V.10 – continued

Year	Winter DJF	Spring MAM	Summer JJA	Autumn SON	Av. for calendar year J–D	Year	Winter DJF	Spring MAM	Summer JJA	Autumn SON	Av. for calendar year J–D
1940	1·5	9·1	15·7	9·8	9·1	1960	4·6	9·4	15·4	10·2	9·7
1941	2·6	7·0	15·7	10·5	9·1	1961	4·9	9·7	15·0	10·7	9·9
1942	2·2	8·5	15·5	9·6	9·1	1962	3·6	6·9	14·4	9·5	8·6
1943	5·9	9·6	15·6	10·1	10·1	1963	−0·3	8·4	14·8	10·7	8·5
1944	4·3	8·9	15·7	9·3	9·6	1964	3·5	8·8	15·1	10·1	9·4
1945	3·7	10·1	15·7	11·2	10·3	1965	3·3	8·3	14·5	9·3	8·9
1946	4·5	8·6	14·7	10·6	9·4	1966	4·3	8·2	15·0	9·8	9·4
1947	1·1	8·6	17·0	10·9	9·6	1967	5·1	8·4	15·5	9·9	9·6
1948	5·1	9·6	14·8	10·4	10·0	1968	3·5	8·1	15·1	11·0	9·3
1949	5·6	8·8	16·5	11·5	10·6	1969	3·2	7·3	15·7	10·8	9·3
1950	5·1	8·8	15·9	9·4	9·4	1970	3·3	7·8	15·9	11·0	9·6
1951	2·9	7·0	15·0	10·7	9·3	1971	4·4	8·1	15·0	10·4	9·7
1952	3·9	9·9	15·7	7·9	9·1	1972	4·9	8·4	14·2	9·5	9·2
1953	3·5	8·5	15·4	10·7	9·8	1973	4·9	8·2	15·6	9·8	9·5
1954	4·1	8·2	14·0	10·5	9·2	1974	5·3	7·6	14·2	8·7	9·2
1955	3·5	7·4	16·5	10·1	9·3	1975	6·3	7·6	16·8	9·8	9·9
1956	2·9	8·3	14·1	9·9	8·8	1976	5·0	8·2	17·5	10·1	9·9
1957	5·5	9·5	15·6	9·9	10·0	1977	3·2				
1958	4·2	7·4	15·3	10·8	9·4	1978					
1959	3·6	9·8	16·6	11·5	10·5	1979					

The sources and nature of the original observation records, and the methods and problems of derivation of the temperature values quoted, have been described by MANLEY (1974 and in a number of earlier papers). The figures in the above tabulation are averages of the monthly values given by MANLEY. For the years before 1680 the monthly values were generally estimated only to the nearest whole degree Celsius, and it is possible that some of the seasonal means are in error by up to 1°C; the most difficult and therefore doubtful cases are indicated by brackets.

Currently, since the unavoidable closure of observing stations which had long been used, the average temperatures for central England are taken as the arithmetic mean of the values for Bedford (Cardington), Birmingham (Elmdon) and Manchester (Ringway), a combination which appeared to give the same values as in this table in the earlier, overlap years examined.

Table App. V.11   Seasonal and yearly mean temperatures in the eastern United States, reduced to Philadelphia equivalent values, from 1738

From figures provided by LANDSBERG *et al.* 1968.

Year	Winter DJF	Spring MAM	Summer JJA	Autumn SON	Av. for calendar year J–D	Year	Winter DJF	Spring MAM	Summer JJA	Autumn SON	Av. for calendar year J–D
1738		13·8	24·0	13·3	13·6	1780		10·8	24·6	13·7	
1739	1·6	12·7	23·4	13·1	12·6	1781	2·6	12·1	24·3	12·9	13·1
1740	2·2	12·0	23·6	12·7	12·2	1782	0·8	12·2	24·2	13·5	12·6
1741	−1·1					1783	2·6	12·6	24·3	12·9	12·8
1742		12·6	24·0	12·7	12·6	1784	−3·8	10·8	23·7	13·2	10·9
1743	1·5	12·6	23·3	13·3	12·8	1785	−2·1	9·0	23·7	12·7	11·2
1744	1·9	12·9	23·1	13·3	13·6	1786	−1·1	11·3	23·3	12·8	11·6
1745	1·9	13·6	24·6	14·6	12·7	1787	−1·8	11·6	22·8	13·3	12·2
1746	2·5	12·7	24·7	14·2	13·3	1788	−0·5	11·6	24·1	14·8	12·5
1747	2·7	12·6	23·6	13·6	13·1	1789	−0·8	10·9	24·1	13·2	12·4
1748	1·7	12·4	23·4	13·6	12·8	1790	1·7	11·1	23·8	13·8	12·3
1749	2·4	12·3	22·8	12·1	12·0	1791	−1·2	12·8	24·2	13·4	12·7
1750	1·4	12·6	23·2	13·2	12·8	1792	−0·5	12·7	23·4	13·8	12·7
1751	3·0	13·2	23·1	12·9	12·8	1793	1·4	12·7	24·6	14·1	13·3
1752	1·6	12·3	23·1	13·9	12·9	1794	0·9	11·8	24·1	13·9	13·3
1753	3·6	12·0	23·0	13·6	12·9	1795	2·0	12·4	24·3	14·5	13·2
1754	4·1	12·2	22·9	14·1	13·3	1796	2·6	11·4	24·0	13·5	12·4
1755	3·1	11·1	22·4	13·1	12·0	1797	0·4	11·4	24·0	13·0	12·3
1756	3·8	12·0	23·0	13·8	13·5	1798	−0·1	12·3	24·6	13·9	12·7
1757	2·1	12·0	22·9	13·8	12·9	1799	0·0	10·9	24·3	13·9	12·4
1758	2·4	10·7	22·9	13·6	12·1	1800	0·6	11·9	24·5	13·9	12·7
1759	1·1	10·3	23·5	13·8	12·1	1801	2·2	12·5	24·2	14·6	13·1
1760	2·2	11·8	23·8	13·3	12·7	1802	2·8	11·8	24·7	14·8	13·2
1761	1·3	12·3	24·3	13·3	12·6	1803	2·2	11·7	24·8	13·7	12·9
1762	0·8	11·1	23·5	12·3	12·1	1804	1·3	11·6	24·4	14·1	12·4
1763	2·2	10·9	22·2	12·7	11·9	1805	−0·1	12·9	24·9	13·8	13·3
1764	3·2		21·9	12·6		1806	3·0	10·3	23·7	13·8	12·6
1765	2·5	12·2	22·3	13·2	12·4	1807	−0·2	10·4	23·8	13·3	12·5
1766	3·3	11·5	22·8	13·5	12·6	1808	2·1	11·7	24·1	13·7	12·8
1767	2·8	10·9	23·2	13·0	12·1	1809	−0·3	11·0	23·1	13·6	12·4
1768	3·3	9·8	22·3	12·8	11·5	1810	2·4	11·3	23·3	13·4	12·3
1769	3·0	10·3	22·6	12·7	11·6	1811	0·2	11·7	23·7	14·1	13·6
1770	2·2	10·0	22·7	13·0	11·7	1812	−0·8	9·4	22·8	12·8	11·3
1771	3·2	11·2	22·6	13·4	11·7	1813	−0·7	10·4	23·7	13·6	12·1
1772	2·2	8·8	22·2	13·2	11·8	1814	0·3	11·7	23·1	13·6	12·2
1773	3·2	11·7	22·7	13·5	12·6	1815	−1·2	10·7	23·2	13·0	11·5
1774	2·9	11·4	22·4	13·7	12·1	1816	−0·3	9·7	22·1	13·1	11·2
1775	2·5	11·9	22·6	13·7	12·4	1817	−1·3	10·3	23·1	13·6	11·6
1776		10·6			11·9	1818	−1·1	10·3	24·1	13·8	11·8
1777		9·9			10·6	1819	1·6	10·1	24·2	14·4	12·7
1778											
1779											

Table App. V.11 – continued

Year	Winter DJF	Spring MAM	Summer JJA	Autumn SON	Av. for calendar year J–D	Year	Winter DJF	Spring MAM	Summer JJA	Autumn SON	Av. for calendar year J–D
1820	0·1	11·2	24·4	13·6	12·3	1865	−0·4	12·3	24·3	14·1	12·6
1821	−0·5	10·7	23·9	13·8	12·2	1866	0·7	11·2	23·8	14·0	12·2
1822	−0·5	12·2	23·9	14·7	12·6	1867	0·1	10·6	23·7	13·9	12·0
1823	−0·1	11·2	23·8	13·0	12·2	1868	−1·8	10·2	24·3	13·3	11·6
1824	1·8	11·1	23·3	13·8	12·8	1869	1·1	10·3	23·6	13·1	12·1
1825	2·1	12·3	24·6	13·6	12·9	1870	1·7	10·8	24·8	14·5	12·8
1826	1·3	11·8	23·9	13·7	12·7	1871	0·6	12·4	23·8	12·9	12·2
1827	0·4	12·0	23·4	13·2	12·3	1872	−0·3	10·2	24·4	13·2	11·5
1828	3·5	11·7	24·5	13·9	13·6	1873	−0·9	9·9	23·2	12·7	11·5
1829	0·3	11·1	23·6	13·0	12·1	1874	1·9	9·8	23·3	13·7	12·1
1830	1·7	12·4	24·0	14·9	13·3	1875	−1·1	9·6	23·2	12·5	11·1
1831	−0·1	12·4	24·6	14·0	12·3	1876	2·1	10·2	24·4	12·8	12·1
1832	−0·9	10·8	22·9	13·2	12·2	1877	−0·3	10·5	23·9	14·2	12·6
1833	1·6	11·6	22·5	13·1	12·2	1878	2·5	12·3	23·4	14·1	12·8
1834	1·6	11·7	23·6	13·1	12·5	1879	−0·4	10·9	23·3	14·0	12·2
1835	−0·6	10·8	23·8	13·6	11·9	1880	3·8	11·9	23·7	13·2	12·7
1836	−1·8	10·3	22·2	12·1	10·7	1881	−1·2	10·7	23·1	15·4	12·6
1837	−1·2	10·0	22·6	13·1	11·4	1882	2·9	10·3	23·4	14·1	12·4
1838	0·4	10·3	24·2	12·6	11·8	1883	0·6	10·1	23·3	13·6	12·1
1839	−0·1	11·5	22·8	13·3	12·1	1884	0·7	10·8	23·0	14·4	12·4
1840	0·4	12·1	23·4	13·3	12·2	1885	−0·8	9·3	23·2	13·4	11·5
1841	0·2	10·4	23·8	13·2	12·0	1886	0·1	11·1	22·8	14·1	11·9
1842	1·7	11·4	23·2	12·7	12·2	1887	0·4	10·8	23·7	13·3	12·2
1843	0·0	9·7	23·4	13·1	11·6	1888	0·3	9·3	23·3	13·1	11·7
1844	−0·1	12·6	23·3	13·4	12·2	1889	1·4	11·8	23·0	13·5	12·6
1845	1·3	11·8	24·0	13·8	12·4	1890	4·9	10·8	23·2	13·7	12·7
1846	−0·3	12·0	23·4	14·6	12·4	1891	1·7	10·8	22·9	13·7	12·6
1847	1·0	10·4	23·6	13·8	12·2	1892	2·2	10·3	23·9	13·4	12·1
1848	1·9	11·9	23·5	13·0	12·3	1893	−1·0	10·4	23·4	13·6	11·8
1849	0·3	10·7	23·5	14·0	11·9	1894	1·3	12·0	23·6	13·8	12·7
1850	1·9	9·9	23·5	13·8	12·2	1895	−0·4	10·8	23·6	13·9	12·1
1851	1·5	11·2	23·3	13·6	12·3	1896	1·0	11·3	23·4	13·9	12·3
1852	−0·4	10·7	23·4	13·7	12·1	1897	0·7	11·4	22·9	14·1	12·4
1853	2·5	11·7	23·8	13·7	12·4	1898	1·9	11·6	24·1	14·3	12·8
1854	0·6	11·3	23·9	14·2	12·3	1899	0·1	11·3	23·9	14·1	12·4
1855	−0·3	10·8	23·4	14·0	12·1	1900	1·3	10·6	24·6	15·4	13·0
1856	−1·2	10·0	23·8	13·7	11·4	1901	0·2	10·7	24·2	13·4	12·1
1857	−0·1	9·8	23·2	13·6	11·7	1902	−0·1	11·9	22·9	14·6	12·3
1858	1·8	10·6	23·7	13·4	12·1	1903	0·8	12·7	22·2	13·5	12·2
1859	1·2	11·9	22·9	13·3	12·1	1904	−1·4	10·8	22·9	13·2	11·4
1860	0·3	11·8	23·4	13·9	12·3	1905	−1·7	11·5	23·3	13·7	12·1
1861	0·8	11·0	23·6	14·2	12·4	1906	2·1	10·8	23·8	14·3	12·6
1862	0·7	11·0	23·5	14·1	12·2	1907	−0·1	10·3	22·8	13·6	11·9
1863	1·9	10·6	24·0	13·8	12·3	1908	1·0	11·9	23·7	14·3	12·7
1864	0·9	11·9	24·4	13·4	12·5	1909	2·4	10·8	23·3	13·9	12·4

Table App. V.11 – continued

Year	Winter DJF	Spring MAM	Summer JJA	Autumn SON	Av. for calendar year J–D	Year	Winter DJF	Spring MAM	Summer JJA	Autumn SON	Av. for calendar year J–D
1910	0·4	12·6	23·3	14·1	12·5	1945	−0·1	13·2	23·3	14·4	12·7
1911	0·7	11·3	23·7	13·8	12·8	1946	0·6	12·4	22·7	15·1	13·0
1912	0·0	11·2	23·1	14·5	12·3	1947	1·8	10·8	23·5	14·6	12·4
1913	3·2	12·1	23·7	14·3	13·2	1948	−0·4	11·6	23·6	14·6	12·6
1914	0·9	11·1	23·4	14·2	12·2	1949	3·9	12·0	24·6	14·3	13·6
1915	1·9	11·2	22·8	14·8	12·7	1950	6·3	10·6	23·2	14·1	12·6
1916	1·4	10·4	23·2	14·1	12·3	1951	1·9	11·8	23·5	13·9	12·9
1917	0·9	10·4	23·6	12·7	11·6	1952	3·0	11·5	24·3	13·9	13·1
1918	−2·0	12·3	23·4	14·1	12·5	1953	3·2	12·1	23·8	14·5	13·4
1919	2·9	11·8	23·6	14·7	12·9	1954	2·8	11·5	23·4	14·4	12·8
1920	−1·0	10·8	23·1	14·6	12·2	1955	1·3	12·3	24·2	13·9	12·7
1921	2·6	13·4	23·7	14·7	13·6	1956	0·9	10·3	23·2	14·0	12·4
1922	1·3	12·3	23·5	14·6	12·9	1957	2·3	11·9	23·7	14·2	12·9
1923	0·5	10·9	23·4	14·1	12·6	1958	0·8	11·2	23·1	13·9	11·9
1924	2·2	10·4	23·2	13·7	12·1	1959	−0·2	12·0	23·9	14·6	12·9
1925	1·8	11·8	23·8	13·6	12·8	1960	2·3	10·8	23·3	14·2	12·1
1926	1·1	9·8	22·3	13·8	11·9	1961	−0·5	10·9	23·5	14·9	12·4
1927	1·5	11·5	23·5	15·1	12·8	1962	0·6	11·9	23·2	13·3	12·1
1928	2·0	10·7	22·9	14·2	12·6	1963	−1·2	11·8	23·3	14·3	12·1
1929	1·8	12·3	24·0	14·1	12·7	1964	0·1	12·0	23·3	13·8	12·6
1930	2·4	11·6	24·1	14·6	13·1	1965	0·9	11·3	23·1	13·8	12·3
1931	1·7	10·9	23·7	15·9	13·6	1966	1·2	11·1	23·7	13·7	12·3
1932	4·9	10·9	23·9	14·3	13·3	1967	1·3	10·7	23·0	12·7	12·0
1933	3·7	11·9	23·9	14·0	13·1	1968					
1934	−0·3	11·3	23·8	14·4	12·4	1969					
1935	0·5	11·5	23·0	14·2	12·4	1970					
1936	−1·1	11·9	23·8	14·0	12·6	1971					
1937	3·6	11·0	24·2	13·6	12·8	1972					
1938	1·8	12·2	23·9	14·4	13·2	1973					
1939	2·2	11·4	24·0	13·9	12·8	1974					
1940	0·0	10·0	23·2	13·4	11·8	1975					
1941	1·1	11·6	23·5	15·1	12·8						
1942	1·1	12·6	23·6	14·4	12·7						
1943	0·6	10·7	24·4	13·4	12·4						
1944	1·2	11·3	24·0	13·9	12·6						

Table App. V.12   Mean sea surface temperatures (°C) averaged over the 'square' 45–50°N 5–10°w in the eastern North Atlantic, yearly from 1854

Decade beginning	Year										Decade average
	0	1	2	3	4	5	6	7	8	9	
1850					13·19	13·10	13·85	13·78	13·75	14·00	(13·61)
1860	13·02	13·35	13·44	13·36	13·81	14·01	13·48	14·10	13·99	13·85	13·64
1870	13·74	13·50	13·66	13·57	13·82	13·77	13·64	13·58	13·79	13·20	13·63
1880	13·66	14·10	13·76	13·58	14·07	13·33	13·69	13·61	13·25	13·39	13·64
1890	13·45	13·29	13·71	14·41	13·47	14·04	14·02	13·95	14·16	14·25	13·87
1900	13·74	13·28	13·28	13·34	13·37	13·26	13·53	13·27	13·63	13·31	13·40
1910	13·30	13·75	13·31	13·45	13·53	13·54	13·59	13·68	13·61	13·60	13·54
1920	13·45	14·35	13·35	13·14	13·30	13·33	14·02	13·58	13·87	13·83	13·62
1930	13·52	13·43	13·73	14·01	13·68	13·51	13·87	14·25	13·99	13·71	13·77
1940	13·96	13·71	13·82	13·97	13·82	13·88	13·80	14·25	14·42	14·99	14·06
1950	14·45	13·88	14·18	14·41	14·06	14·68	14·00	14·25	14·15	14·67	14·27
1960	14·40	14·63	14·34	13·90	14·20	13·81	14·28	14·01	14·05	14·18	14·18
1970	14·08	14·10	13·41	14·02							(13·90)

Source: Monthly and yearly values kindly supplied by the Institute for Marine Environmental Research, Plymouth, from marine data decks held by the National Climatic Center, Asheville, North Carolina and the U.K. Meteorological Office.

The averages for the whole year in this outer Biscay region seem much influenced by the warmth developed above the thermocline in the warmest months of the year.

Table App. V.13   Ice in the western North Atlantic, off Newfoundland, since 1880

The longest records of ice in the western North Atlantic for correlation and comparison with weather data are derived from the yearly counts of the number of icebergs which pass south of latitude 48°N to affect the shipping lanes.

The iceberg counts are nearly always quoted in terms of an index, devised by SMITH (1931), which provides a scale from −5 to +5. Key points on the scale appear to be as follows:

Iceberg count	Index value	Iceberg count	Index value
0	−5	400	+0·2
50	−2·5	700	+2·0
100	−1·8	1000	+3·2
200	−1·0	1250	+4·0

The greatest number of bergs ever reported to have passed south of 48°N was in 1972: 1594 bergs. The previous biggest total was 1351 in 1929. In 1974 there were 1387. There were also over 1000 also in 1909 and 1912. In 1966 there were none at all. In 1958 only one iceberg passed south of 48°N, though in the previous year there had been 931. Other years with very low totals were 1940 and 1941 with 2 each.

Full details are believed to be given in the yearly *U.S. Coast Guard Bulletins: Reports of the International Ice Patrol Service in the North Atlantic Ocean.*

The totals before 1946 were based on ships' reports; from 1946 onwards they are based mainly on aircraft reconnaissances. The totals for the years 1943–5 are thought somewhat unreliable because of less than normal shipping and inexperienced observers.

Iceberg index off Newfoundland, yearly values

Decade	Year										Decade average
	0	1	2	3	4	5	6	7	8	9	
1880	−0·1	−2·4	+1·3	−0·1	+1·6	+2·6	−0·8	+0·2	−0·5	−1·3	+0·05
1890	+3·8	−1·7	−0·8	−0·4	+1·3	−1·8	−1·0	+1·3	+0·3	+0·6	+0·16
1900	−1·8	−1·8	−2·3	+2·5	−0·7	+2·6	−0·1	+1·6	−1·0	+3·8	+0·28
1910	−2·0	−0·2	+3·8	+0·9	+2·0	+0·6	−2·0	−2·3	−1·1	−0·6	−0·09
1920	+0·3	+2·0	+1·1	−0·7	−2·8	−1·5	−0·5	−0·2	+0·8	+4·2	+0·27
1930	+0·4	−2·0	+0·7	−1·0	+0·9	+2·8	−2·6	+0·4	+1·5	+2·6	+0·37
1940	−4·7	−4·7	−2·4	(+2·6)	(+2·0)	(+3·9)	+0·1	−2·0	+1·1	−2·2	−0·63
1950	+0·3	−4·7	−3·0	−2·0	−0·6	−2·0	−1·8	+2·9	−4·7	+1·6	−1·40
1960	−0·8	−1·5	−1·6	−2·6	−0·3	−1·8	−4·8	+0·2	−1·0	−2·0	−1·62
1970	−1·7	−1·7	+4·7	+2·6	+4·5	−1·8	−1·3				

SCHELL (1962) reports a correlation coefficient $r = +0.47$ between the yearly iceberg counts south of 48°N off Newfoundland and the mean atmospheric pressure difference between Belle Isle (off the northernmost tip of Newfoundland) and Ivigtut (southwest Greenland) for December to March in the preceding winter. Thus strong NW'ly winds between Labrador and Greenland in the winter, as when the Iceland low is large and intense, tend to produce a great number of icebergs in the western Atlantic in the spring and summer following. The mean pressure distribution in the winters before light ice seasons in the western Atlantic shows a weak development of the Iceland low pressure region and NE'ly winds on the Labrador coast.

Apart from these year-to-year differences, the trend towards fewer icebergs, culminating (so far) in the very low average for the 1960s, has been reasonably described as a delayed response to the twentieth century climatic warming, probably chiefly due to a reduced rate of calving of icebergs from the retreating fronts of glaciers and ice sheets, though the weakening of the Iceland low in the 1960s played a part. The period average yearly counts of bergs south of 48°N were:

1880–1926	480
1901–35	442
1936–70	294
1962–71	148

See also GROISSMAYR 1939, MILES 1974, RODEWALD 1971, 1974, for further aspects.

Table App. V.14 KOCH's (1945) index of the greatest extent of the Arctic drift ice (*Storis*) along the west coast of Greenland after rounding Cape Farewell, from 1821

Scale 0–10:

1	Cape Farewell (60°N)	7	Godthåb (64°N)
2	Nanortalik	8	Sukkertoppen (66°N)
3	Julianehåb (61°N)	9	Holstensborg (68°N)
4	Ivigtut	10	Egedesminde (69°N)
5	Frederikshåb (62°N)	11	North of Egedesminde
6	Fiskernæsset (63°N)		

Decade	Year										Decade average
	0	1	2	3	4	5	6	7	8	9	
1820s		8	9	1	7	8	0	8	9	7	6·3
1830s	2	2	1	0	8	4	6	9	6	1	3·9
1840s	9	1	0	1	1	1	1	0	1	0	1·5
1850s	6	0	0	1	2	9	9	9	7	9	5·2
1860s	8	0	8	5	0	2	9	9	1	9	5·1
1870s	9	8	2	2	2	0	8	5	9	5	5·0
1880s	6	11	9	8	0	6	8	10	11	5	7·4
1890s	6	9	9	5	5	7	8	6	6	1	6·2
1900–9	5	2	4	2	1	1	8	7	1	1	3·2
1910–19	1	9	6	5	7	8	6	6	9	7	6·4
1920s	1	6	0	2	1	2	0	0	1	2	1·5
1930s	1	0	6	1	0	0	0	1	8	0	1·7

Table App. V.15   Number of weeks each year with drift ice at the coast of Ireland, from 1600

The ice seasons in the table below are counted from October of the previous year to September of the year of the entry.

The data below are mainly taken from KOCH (1945). For a few years in the eighteenth century for which THORARINSSON (1956) had discovered reports after KOCH's work was done THORARINSSON's figures have been used (agreement between the latter's compilation and KOCH's is generally good). The figures for the years from 1940 onwards were compiled from reports in *Veðrattan Islændi*, Reykjavik, and the official records of the *Veðurstofan Islændi* (Iceland Weather Bureau) kindly supplied by Dr HLYNUR SIGTRYGGSSON and FLOSI SIGURDSSON in January 1977.

The table begins at the year 1600, because KOCH was not satisfied about the completeness of the reports available for the fifteenth and sixteenth centuries, when a few years with much ice in the Denmark Strait are known which KOCH was not able to include in his reports of ice affecting the Iceland coast. Those years between 1600 and 1780 for which no reports of drift ice at the coast of Iceland are known have been left blank in the table. It is probably safe to assume in most cases that there was little ice in those years, but it may not be safe to assume that there was none. The worst of all years were in the 1690s, particularly 1695 when the ice surrounded the whole country with the exception of one point on the west coast (Snaefellsnaes); in most places open water could not be seen from the highest mountains, and ships could not approach the country."

Decade	Year										Decade average
	0	1	2	3	4	5	6	7	8	9	
1600		20			7	10			9		
1610	22		10			18		11	7	5	
1620		7	11		17	17	9		23	7	
1630				24					26	18	
1640									17		
1650								2			
1660		7		7	7	5					
1670	2		9		15						
1680	2			14	19	22		7			
1690					18	27		18		13	
1700		9		7		17	10		24		(6·7)
1710	19				24		17		5		(6·5)
1720							4		11	25	(4·0)
1730			15	7							(2·2)
1740		23	8		1	24				20	(7·6)
1750	38	11			26	30	24	15		15	(15·9)
1760					10		12	13			(3·5)
1770	14	7		11	5	6		7		7	(5·7)
1780	0	12	22	7	17	13	17	10	10	19	12·7
1790	0	16	24	0	9	4	18	0	9	4	8·4

Table App. V.15 – continued

Decade	Year										Decade average
	0	1	2	3	4	5	6	7	8	9	
1800	0	15	30	14	4	0	3	25	5	0	9·6
1810	0	27	10	0	0	3	15	27	2	6	8·7
1820	2	20	17	2	0	4	0	24	3	17	8·9
1830	5	6	3	0	12	23	3	23	5	4	8·4
1840	23	0	0	3	0	0	3	0	3	0	3·2
1850	6	0	4	3	3	19	5	12	7	13	7·2
1860	14	0	3	16	0	6	29	16	20	25	12·9
1870	18	5	11	10	19	0	12	6	12	2	9·5
1880	6	20	19	10	0	5	15	20	29	0	12·4
1890	5	13	22	9	0	7	7	0	5	1	6·9
1900	0	5	20	9	0	0	6	7	4	0	5·1
1910	0	13	0	0	10	15	5	3	7	2	5·5
1920	1	3	0	2	1	0	0	0	2	6	1·5
1930	3	0	5	1	0	0	0	2	3	0	1·4
1940	0	0	0	7	10	0	4	0	2	7	3·0
1950	0	0	0	0	0	2	7	0	1	0	1·0
1960	0	0	0	4	4	23	2	11	27	22	9·3
1970	13	13	0	0	2	9					

Table App. V.16 Exceptional positions of ice in the North Atlantic

Notes by the late Commander J. HENNESSY from records in the Marine Division of the United Kingdom Meteorological Office and in (1) *Marine Observer*, Vol. XVI (1939), 60–1, and map at end of the April issue 1939, (2) *North Sea Pilot*, Vol. I (1949).

14.1.1836	60°55′N 5°50′W 2 bergs (1), (2). This is a report by Capt. James Ross, H.M.S. *Cove*, who regarded it as very exceptional at that time. The 2 bergs were judged to be over 70 feet (22 m) high and one about a mile in diameter. Ross feared they might be the cause of many wrecks, being 'such an unusual sight'.
2.9.1883	35°40′N 30°00′W. Lump of ice, sighted by the barque *Olivette* (1).
15.10.1883	37°N 18°W. Piece of ice, sighted by S.S. *Elenora* (1).
25.6.1886	48°40′N 15°22′W. Large berg, sighted by the brig *Blanch* (1).
In 1888	20 miles west of the Faeroes, an ice barrier turned back fishing boats from Mykines and Suðuroy, bound for Iceland. It is also known that ice was seen to the east from the high ground above Tórshavn in that year.
July 1890	48°53′N 24°11′W. Last remnants of berg (1).
Sept. 1895	36°35′N 71°36′W. Southeast of New York. 2 bergs 30 feet high, 300–400 feet long, also much field ice (1).
July 1902	56°30′N 6°30′W. West side of the island of Mull. Piece of ice 40–50 feet long, 15 feet wide and 2 feet 6 inches above the water (1).
19.9.1906	54°20′N 22°00′W. Small berg 20 × 6 feet. Report by S.S. *Lord Landsdowne* (1).
June 1907	80 miles west of Fastnet, *c.* 51°N 11½°W. Berg sighted by the barque *Silverstream* (1).
7.8.1908	50°31′N 18°55′W. 2 pieces of ice 10 feet square and 15 feet square. Report by S.S. *Caronia* (1).
27.8.1912	42°30′N 15°26′W. Piece of ice 50 feet square, 4 feet out of water. Report by S.S. *Lux* (1).
16.4.1926	61°03′N 10°30′W. Floating ice about 40 feet long and 3 feet high. Seen by trawler *Orizaba* (1), (2).
23.10.1927	30 miles east-southeast of Out Skerries, Shetland, *c.* 60¼°N 0°W. Piece of ice 100 feet long, 6 feet high. Seen by the trawler *Grecian Empire* (2).
16.7.1933	52°32′N 22°00′W. Small piece of ice about 25 feet long, 12 feet wide. Seen by S.S. *Rein*.
1.12.1959	59°00′N 19°00′W. Ocean Weather Ship I reported sighting a berg.

Note:

5.10.1976	58°28′N 0°15½′E. Several large bergy bits were reported close to an oil rig, presumed to be the remains of a berg driven away from the East Greenland ice-belt by N'ly winds about 7 to 9.9.1976.

Table App. V.17   Baltic ice variations

The three tables below give data on the Baltic ice from different observation points. Examination of the tables shows that they represent some distinctively different aspects of the climatic regime affecting Europe. The data were collected by BETIN and PREOBAZENSKY (1959) and by SPEERSCHNEIDER (1915, 1927).

The date of opening of the port of Riga (*a*) shows least variability, evidently being related to the spring warming of the Eurasian continent.

Lack of correlation in some years between the maximum area the ice attained during the winter (*b*) and the persistence of ice into spring at Riga (*a*) probably indicates that the former has much to do with the character of the winter in northern Scandinavia as well as in the southern Baltic.

The variability is greatest (in proportion to the mean value) in the case of the duration of ice at the entrance to the Baltic, in Danish waters (*c*). Here the conditions are more complex than at Riga or in the inner Baltic, being affected as well by variations in the transport of warm, saline water from the Atlantic and the North Sea.

(*a*) Date of final opening of the port of Riga (57°N 24°E) counted from 1 January = 1, 1 February = 32, etc. (New Style calendar throughout)

Decade	Year										Decade averages
	0	1	2	3	4	5	6	7	8	9	
1530	113									85	
1540	105		104								
1550			98				100	110	81		
1560			101	97	119	107	73	88	114		(99·9)
1570		88	101				97	86	104	91	(93·0)
1580	101	97	117	103	81	70	94	117	95	86	96·1
1590	113	106	91	106	94	107	95	116	86		(101·6)
1600		111	91							90	
1610			99			97	102	76	87	111	(95·4)
1620		103	104	94			100				
1630											
1640				120						101	
1650	87	96	33	34						122	(74·4)
1660			49					108			
1670											
1680								88		107	
1690			109								
1700										106	
1710	106	96	114	83	98	92	112	101	91	101	99·4
1720	104	99	99	72	82	100	100	112	91	100	95·9
1730	105	113	95	94	91	68	99	79	91	78	91·3
1740	116	107	113	95	100	99	100	100	108	110	104·8

Table App. V.17. – continued

Decade	Year										Decade averages
	0	1	2	3	4	5	6	7	8	9	
1750	63	86	91	81	103	101	79	90	110	74	87·8
1760	118	79	100	108	91	90	96	90	109	81	96·2
1770	101	116	94	101	94	93	98	105	96	79	97·7
1780	79	101	95	103	107	111	97	100	104	110	100·7
1790	82	85	95	98	79	97	108	92	92	107	93·5
1800	102	85	79	96	104	104	101	105	114	111	100·1
1810	116	100	116	88	102	97	106	92	116	92	102·5
1820	102	95	62	94	78	102	95	95	103	108	93·4
1830	101	97	94	96	67	71	75	108	111	121	94·1
1840	104	100	96	92	109	108	82	111	82	110	99·4
1850	107	96	104	110	105	102	103	95	95	82	99·9
1860	102	91	100	83	85	103	96	112	99	79	95·0
1870	104	107	92	91	92	110	93	98	94	97	97·8
1880	99	107	70	109	73	85	92	85	95	101	91·6
1890	77	101	96	95	86	101	66	95	94	97	90·8
1900	106	95	85	60	102	92	99	102	102	102	94·5
1910	60	97	81	76	36	100	91	100	95	100	83·6
1920	69	78	74	91	96	60	97	72	96	111	84·4
1930	71	112	97	85	82	84	72	84	65	86	83·8
1940	101	106	104	79	69	85	88	90	92	87	90·1
1950	84	94	104	92	98	106	111	88	105		(98·0)

Period averages

	Number of years with information	*Coverage*	*Opening date*
1550–79	17 years	57%	97·5
1580–99	19 years	95%	98·8
1600–49	15 years	30%	(99·0)
1650–69	7 years	35%	(75·6)
1680–1709	4 years	13%	(102·5)
1710–49	40 years	100%	97·9
1750–99	50 years	100%	95·2
1800–49	50 years	100%	97·9
1850–99	50 years	100%	95·0
1900–49	50 years	100%	87·3

N.B.: The earliest and latest years occurred in the same decade in the seventeenth century: 1652 (2 February) and 1659 (2 May).

(b) Greatest area attained by the ice on the Baltic Sea ($1000 \times km^2$), winters denoted by the year in which the January falls

Decade	Year										Decade averages
	0	1	2	3	4	5	6	7	8	9	
1720	130	130	78	220	78	79	360	132	220	281	171
1730	79	279	220	182	131	80	221	131	286	217	183
1740	420	115	105	90	109	310	245	400	300	317	241
1750	84	380	240	110	420	330	100	381	360	100	250
1760	380	146	170	124	90	180	230	324	320	90	205
1770	360	330	344	100	291	281	210	240	200	110	247
1780	289	237	300	330	380	380	380	280	320	420	330
1790	70	80	260	100	100	330	156	152	142	420	181
1800	400	136	220	400	320	400	251	140	280	420	297
1810	320	220	240	300	400	160	300	90	268	80	238
1820	370	260	76	180	84	96	160	250	281	380	214
1830	420	328	105	90	233	70	395	172	420	330	256
1840	362	377	80	90	383	396	180	280	130	182	246
1850	330	180	173	305	127	304	410	215	90	80	221
1860	180	315	337	90	80	360	104	420	412	132	243
1870	178	420	126	90	88	340	412	420	72	140	229
1880	125	420	81	337	127	167	198	74	420	327	228
1890	81	126	252	420	81	282	143	180	140	183	189
1900	330	180	360	92	176	134	85	139	235	181	191
1910	81	110	161	118	127	183	330	400	161	150	182
1920	151	125	260	330	280	89	382	126	172	390	231
1930	58	175	148	148	110	90	148	161	70	61	117
1940	420	371	420	84	64	109	215	420	201	62	237
1950	113	149	120	153	272	156	396	165			(191)

Note: In 1966 the Baltic once again became almost completely ice-covered, and only nine years later in the winter of 1975 the Baltic remained almost totally ice-free.

(c) Number of days with ice in the Danish Sound (Øresund), winters denoted by the year in which the January falls

Decade	Year										Decade averages
	0	I	2	3	4	5	6	7	8	9	
1760				45	0	61	12	46	62	7	(33) 7 yr
1770	96	98	45	0	31	—	48	45	12	0	(42) 9 yr
1780	14	0	—	—	106	57	59	0	43	134	(52) 8 yr
1790	0	0	40	0	0	75	0	0	0	135	25·0
1800	109	0	11	60	48	76	0	0	—	60	(40) 9 yr
1810	—	58	5	57	125	—	—	0	0	0	(35) 7 yr
1820	—	—	0	—	0	0	—	26	—	59	(17) 5 yr
1830	125	43	0	0	0	0	0	0	117	11	29·6
1840	—	44	0	0	39	61	0	5	18	—	(21) 8 yr
1850	14	0	0	29	0	85	—	13	20	0	(18) 9 yr
1860	39	49	36	0	0	75	0	62	41	0	30·2
1870	63	75	0	0	0	44	74	12	0	46	31·4
1880	70	80	0	16	0	0	27	0	124	46	36·3
1890	0	34	8	70	0	55	0	29	0	0	19·6
1900	57	23	I	0	0	0	0	51	0	57	18·9
1910	0	0	36	0	0	0	0	61	20	0	11·7
1920	I	0	36	14	103	0	0				(22) 7 yr

Period averages: percentage of years for which the information shows ice was present in the Sound on at least 1 day:

1763–99	65%
1800–49	55%
1850–99	55%
1900–26	45%

Table App. V.18   Winters in which it is recorded that Bodensee (47°N 8–9°E) froze over
Winters identified by the year in which the January falls.

1216
1277
1326 (perhaps only the northern part, at Überlingen)
1370 (perhaps only at Konstanz, though this probably means the wide part)
1378
(1422 or) 1423 (perhaps only the northern part, at Überlingen)
1435 (perhaps only the northern part, at Überlingen, but probably the whole lake)
1460 (perhaps only the northern part, at Überlingen)
1465
1517 (perhaps only the northern part, at Überlingen)
1565
1567 or 1569
1571
1573 or 1574 (Überlingen arm and parts of Bodensee frozen and bearing transport at Christmas 1572
        or 1573)
1587
1603
1608
1672
1684
1695
1740
1830
1880 (Limited areas only)
1963

N.B.: In many years around 1800 the apparently abnormally high level of the water of Bodensee is reported. The lake may have frozen over more readily in the Middle Ages when notably low water levels were often reported.

Table App. V.19 Mean tree ring widths (bristlecone pine) near the upper tree line Campito
Mountain, White Mountains, California (37°30′N 118°12′W)
  20-year means from 3431 B.C. to A.D. 1969.
  Kindly supplied by Professor V. C. LaMarche,
  Laboratory of Tree Ring Research, University of Arizona, Tucson.
  Mean ring widths in hundredths of a millimetre.
  ( )=Values based on samples of less than 5 tree radii.
  (Entry under "median date" 3421 B.C. is the average for
  3431–3412 B.C. and so on.)
  Regarding reliability of the dates see
  Chapter 13, p. 54.

Century ending	Median year					Average for the
	81	61	41	21	01	
3400 B.C.				(63)	(49)	(56)
3300	(58)	(70)	(56)	(55)	58	(59)
3200	50	47	49	42	46	47
3100	49	53	45	46	41	47
3000	38	38	35	35	40	37
2900	39	35	35	31	34	35
2800	48	41	34	36	29	38
2700	48	66	49	55	62	56
2600	55	54	56	50	56	54
2500	52	44	43	41	41	44
2400	36	40	39	42	46	41
2300	47	52	51	44	45	48
2200	46	38	46	42	43	43
2100	44	44	47	56	49	48
2000	51	45	41	47	46	46
1900	54	46	43	42	46	46
1800	51	40	43	44	46	45
1700	44	39	50	46	43	44
1600	49	43	37	43	38	42
1500	47	45	47	54	59	50
1400	53	48	53	55	45	51
1300	47	53	53	54	57	53
1200	61	49	43	44	41	48
1100	35	37	35	38	35	36
1000	33	33	33	34	37	34
900	36	27	31	36	40	34
800	36	37	47	35	35	38
700	34	38	42	39	37	38

Table App. V.19 – continued

Century ending	Median year					Average for the century
	81	61	41	21	01	
600	36	36	37	32	35	35
500	32	30	30	34	32	32
400	27	30	49	37	33	35
300	33	38	39	38	43	38
200	44	38	38	50	41	42
100	40	49	48	48	46	46
I B.C.	41	42	37	42	55	43

Century beginning	Median year					Average for the century
	10	30	50	70	90	
I A.D.		50	54	47	51	49
100	40	41	39	43	44	41
200	44	44	48	57	47	48
300	35	43	38	45	39	40
400	42	45	41	36	37	40
500	41	42	35	33	(28)	36
600	(45)	(27)	36	46	44	40
700	29	39	38	41	44	38
800	33	35	29	33	24	31
900	24	21	26	28	33	26
1000	34	38	44	35	45	39
1100	61	39	36	60	66	52
1200	55	48	44	43	40	46
1300	44	56	32	33	40	41
1400	40	38	38	21	29	33
1500	33	50	40	43	36	40
1600	37	30	35	40	27	34
1700	28	37	45	40	38	38
1800	45	34	35	42	50	41
1900	46	52	62	57		(54)

Values based on 5–9 tree radii 3311–2672 B.C., thereafter generally 10–21 tree radii until A.D. 100 and 7–11 tree radii in later times.

Ring widths in this series probably vary with warm season temperature in eastern California. However, the potential importance of precipitation variations remains to be investigated.

Table App. V.20 Tree ring index values for northern Finland (Lapland) 68–70°N 22–29°E, yearly from A.D. 1181 to 1960

The values given below were kindly supplied by Dr G. SIRÉN. The index used (SIRÉN 1961) is log M_g, where M_g is the geometric mean of the ring widths from the individual trees (this index was preferred to the arithmetic mean, which follows a nearly parallel course except for higher peaks in the years of strongest growth).

Decade	Year									
	0	1	2	3	4	5	6	7	8	9
1180		−176	−310	−303	−214	−132	−175	−155	−104	−27
1190	−35	−61	−30	−64	−76	−47	−59	+45	+66	−27
1200	−34	−18	+90	123	58	67	41	61	73	169
1210	240	215	278	249	152	169	215	127	16	123
1220	58	34	−1	20	24	80	211	79	−5	−16
1230	−149	−175	−109	−171	−188	−360	−332	−413	−549	−566
1240	−446	−384	−389	−376	−647	−234	−262	−312	−110	−316
1250	−309	−298	−197	−256	−75	−141	−140	−38	−200	−93
1260	−9	−6	−121	−45	−36	+70	+13	+27	−121	−148
1270	−100	−117	−158	−38	−80	−51	+17	−5	+39	−44
1280	−119	+89	171	194	222	233	236	329	316	271
1290	+385	322	386	412	464	368	397	385	283	114
1300	+257	129	43	−57	−251	−293	−19	−53	+51	243
1310	+324	172	118	197	242	135	68	50	140	119
1320	+93	174	158	178	199	48	49	79	−44	+66
1330	+20	−115	−81	−98	−18	+10	22	119	106	99
1340	+52	117	−5	−21	−48	−64	−55	−18	−183	−91
1350	+27	232	222	175	286	247	253	296	228	213
1360	+225	196	135	54	−69	+117	44	75	59	−17
1370	−43	124	142	184	325	257	269	354	345	375
1380	+361	312	114	210	185	139	98	101	−11	−96
1390	−25	+111	+44	−52	+14	6	70	−103	−162	−116
1400	−290	−225	−168	−34	+79	169	164	56	158	164
1410	+108	214	62	109	101	42	−26	+118	112	194
1420	+82	−126	−144	−166	+9	+1	−127	−148	−101	−49
1430	−205	−14	+90	233	263	103	−47	−138	−73	−111
1440	−97	−120	−56	−77	−181	−99	−153	−256	−474	−366
1450	−239	−414	−331	−498	−633	−551	−667	−711	−522	−523
1460	−523	−487	−321	−386	−300	−305	−229	−199	−150	−88
1470	−28	−77	−137	−108	−88	−114	−93	−124	−180	+3
1480	+59	+31	−142	−168	−150	−123	−202	−113	+80	+90
1490	+64	150	123	206	191	65	159	174	141	156

Table App. V.20 – continued

Decade	Year									
	0	1	2	3	4	5	6	7	8	9
1500	+178	126	154	35	119	97	108	158	98	67
1510	+37	115	143	93	−8	+63	−55	+84	98	50
1520	+76	+37	−24	+52	−54	−51	+80	−51	−19	−195
1530	−101	−150	−126	−81	−124	−22	−28	−56	−96	−53
1540	+40	−7	−27	−43	−13	−8	+84	143	159	1
1550	+12	45	−8	+48	61	133	115	80	132	156
1560	+204	222	202	165	175	181	188	180	245	236
1570	+215	130	171	152	25	175	164	130	69	177
1580	+43	68	96	163	136	189	158	66	−62	+23
1590	−67	−22	+41	−70	+54	−76	−101	−80	−115	−85
1600	−120	−497	−317	−214	−304	−509	−439	−719	−680	−696
1610	−667	−524	−459	−436	−402	−521	−530	−274	−317	−274
1620	−304	−133	−166	−179	−112	−76	−32	−111	−27	+60
1630	−71	−53	+156	46	98	22	72	108	89	137
1640	+116	−150	−196	−18	−281	−286	−114	−138	−107	−97
1650	−102	+4	79	174	231	335	283	301	365	335
1660	+473	383	379	333	392	409	207	150	278	119
1670	+75	+87	−10	−34	−95	−317	−364	−281	−275	−192
1680	−569	−324	−45	−6	+94	123	268	237	176	430
1690	+428	441	470	512	503	78	−306	−111	−49	−82
1700	−86	+16	138	119	209	214	82	194	148	−154
1710	+60	+77	−19	+33	83	188	−9	−45	+81	−207
1720	−52	−122	−90	−127	−100	+9	−2	+139	97	243
1730	+257	57	79	87	−349	+70	179	161	251	313
1740	+91	112	216	233	248	204	342	79	206	222
1750	+219	167	297	340	387	442	487	423	367	471
1760	590	500	485	393	250	308	369	212	201	−162
1770	+44	32	90	35	182	106	96	183	51	−28
1780	+132	−70	−25	−48	−139	+145	−153	−95	−4	−33
1790	−310	−236	−125	−297	−263	−240	−168	−104	−133	+49
1800	−146	−187	−87	−156	+45	−6	−415	−101	+17	−46
1810	−172	−290	−375	−494	−351	−300	−247	−241	−95	−95
1820	−247	−271	−191	+48	116	6	227	270	198	269
1830	+269	200	136	71	18	−104	−118	−514	−277	−379
1840	−250	−420	−533	−453	−349	−186	−209	−227	−264	−117
1850	−100	−28	−39	−60	+10	83	17	18	72	40
1860	−16	+21	−22	−59	+49	+50	−63	−9	+8	+19
1870	−3	−106	−100	+82	−43	−83	+95	+87	+13	−39
1880	−86	−102	+31	31	37	119	153	40	−67	+57
1890	+83	−10	−246	−138	+9	−14	−38	−165	−4	−141

Table App. V.20 – continued

Decade	Year									
	0	1	2	3	4	5	6	7	8	9
1900	−327	−157	−414	−642	−482	−492	−490	−522	−479	−488
1910	−694	−733	−373	−323	−124	−102	−67	−182	−81	−2
1920	−1	+101	183	146	129	225	−76	+83	−76	−207
1930	+227	231	79	130	256	228	166	422	343	223
1940	+128	339	246	173	244	281	131	128	106	191
1950	+197	79	132	330	354	196	80	281	48	163
1950	+402									
1970										
1980										
1990										

Table App. V.21 German oak tree ring widths from west of the Rhine, near Trier (due to HOLLSTEIN 1965)

This series is generally based on between 10 and 36 sample measurements each year but on only 1–9 in the years before A.D. 910 and 4–8 in the period between A.D. 1060 and 1129.

Decade	Year										Average
	0	1	2	3	4	5	6	7	8	9	
820			152	161	115	143	186	190	175	147	(159)
830	144	204	185	165	138	121	174	231	148	191	170
840	181	201	216	162	199	178	187	173	172	163	183
850	168	128	130	177	153	132	148	156	137	165	149
860	133	145	138	128	135	135	156	149	127	160	141
870	141	159	143	151	80	136	134	144	147	152	139
880	128	144	138	112	150	156	163	139	149	145	142
890	119	135	154	154	159	139	139	158	110	108	137
900	141	141	117	141	118	143	132	109	117	149	131
910	95	130	126	106	94	107	78	89	79	109	101
920	82	117	88	92	91	89	82	117	81	109	95
930	83	120	118	104	117	91	115	102	131	129	111
940	131	123	153	95	100	77	100	113	129	128	115
950	144	124	134	128	107	123	111	116	112	138	124
960	115	130	107	118	99	114	114	129	124	148	120
970	119	117	124	124	91	132	118	120	104	129	118
980	100	98	119	108	101	144	119	111	108	129	114
990	86	113	102	108	97	89	105	98	95	132	103

Table App. V.21 – continued

| Decade | Year | | | | | | | | | | Average |
	0	1	2	3	4	5	6	7	8	9	
1000	94	124	116	133	109	116	110	109	115	121	115
1010	110	115	139	138	95	122	111	114	85	114	114
1020	99	139	116	131	127	113	131	128	106	104	119
1030	148	118	119	127	116	118	105	144	114	103	121
1040	123	110	114	92	78	115	113	118	100	119	108
1050	109	110	133	98	122	123	129	127	114	93	116
1060	129	110	113	145	100	130	145	138	162	132	130
1070	107	111	118	166	122	131	121	134	108	139	126
1080	139	99	171	132	116	168	138	166	158	164	145
1090	109	163	189	180	180	164	160	187	168	170	167
1100	218	168	119	145	149	132	170	150	182	174	161
1110	174	118	193	148	196	184	256	288	216	248	202
1120	198	165	264	262	186	169	189	159	184	153	193
1130	211	188	187	142	180	178	174	135	178	180	175
1140	232	195	164	209	182	206	168	167	149	186	186
1150	147	158	121	186	169	155	184	159	175	155	161
1160	182	175	131	101	66	60	80	60	131	130	112
1170	115	149	142	154	156	148	119	81	126	124	131
1180	141	127	142	114	86	96	109	136	88	129	117
1190	130	138	102	143	128	147	133	111	140	121	129
1200	110	160	153	160	144	120	138	89	129	128	133
1210	148	131	92	118	108	116	115	86	108	137	116
1220	115	134	106	125	124	121	118	106	103	142	119
1230	125	108	95	75	83	100	84	113	109	108	100
1240	133	120	118	132	87	135	110	138	122	140	123
1250	135	133	86	114	92	122	121	85	111	99	110
1260	102	91	73	81	119	117	138	101	128	114	106
1270	77	118	93	110	108	134	86	116	72	95	101
1280	116	98	108	141	109	116	138	75	92	81	107
1290	101	95	115	101	87	111	130	98	133	106	108
1300	111	104	118	90	91	86	88	99	79	119	99
1310	96	93	125	109	121	109	127	126	113	113	113
1320	85	117	113	94	99	89	72	107	106	108	99
1330	90	94	86	83	93	103	79	98	97	88	91
1340	74	93	99	102	77	102	101	79	90	115	93
1350	84	100	86	75	107	85	102	84	126	132	102
1360	87	90	133	131	124	110	158	158	140	175	131
1370	157	130	131	132	150	133	143	130	143	126	137
1380	134	122	107	135	132	103	142	116	137	127	125
1390	116	126	121	84	89	86	72	82	129	128	93

Decade	Year 0	1	2	3	4	5	6	7	8	9	Average
1400	117	135	156	161	131	141	134	122	130	144	137
1410	166	152	145	140	139	119	144	94	111	96	131
1420	80	119	72	85	104	89	128	131	130	144	108
1430	143	123	120	112	88	132	152	143	126	135	127
1440	138	120	99	120	113	119	107	119	90	137	116
1450	114	137	99	108	136	132	145	135	125	99	123
1460	129	107	91	110	78	122	105	126	154	120	114
1470	145	131	158	129	165	163	120	131	98	147	139
1480	150	149	137	138	188	139	144	179	150	135	151
1490	140	110	120	119	128	130	136	145	95	150	127
1500	117	143	138	89	73	120	84	119	109	131	112
1510	111	108	124	97	106	133	110	69	93	95	105
1520	95	112	100	103	110	98	128	111	152	137	114
1530	99	155	84	132	85	132	91	116	77	118	109
1540	100	111	124	116	136	128	108	93	109	100	113
1550	97	108	83	99	85	125	81	86	91	85	94
1560	131	124	144	120	122	125	96	87	120	101	117
1570	125	107	95	99	76	124	93	113	97	120	105
1580	118	129	99	86	136	124	136	117	94	131	117
1590	92	107	125	130	105	104	90	116	115	96	108
1600	113	88	93	69	97	96	117	148	118	118	106
1610	98	98	120	146	125	100	101	130	115	98	113
1620	83	73	91	124	103	130	99	139	93	101	104
1630	97	101	136	118	91	86	59	85	99	76	95
1640	93	90	100	103	75	84	82	82	119	113	94
1650	109	78	63	78	100	115	100	101	106	80	93
1660	98	122	97	108	124	99	105	76	76	95	100
1670	81	102	91	129	93	100	85	105	104	117	101
1680	112	81	118	121	93	94	118	122	147	132	114
1690	137	154	132	158	137	114	92	80	115	116	123
1700	128	133	123	116	150	108	145	132	130	82	125
1710	80	86	132	141	113	137	117	135	110	93	114
1720	121	110	137	103	122	155	133	188	120	137	133
1730	147	105	133	133	140	162	130	151	162	136	140
1740	147	98	116	89	65	87	137	130	109	136	111
1750	125	133	161	108	168	123	152	119	87	96	127
1760	85	139	104	149	130	107	132	119	153	145	126
1770	154	118	101	167	166	162	137	139	114	115	137
1780	103	90	96	124	100	107	79	100	172	170	114
1790	130	128	148	107	137	135	148	147	135	108	132

Table App. V.21 – continued

Decade	Year										Average
	0	1	2	3	4	5	6	7	8	9	
1800	110	128	88	108	118	122	97	107	86	134	110
1810	110	122	95	109	104	104	129	124	107	91	109
1820	111	126	90	136	130	93	112	111	118	128	115
1830	109	117	85	95	124	99	100	113	106	108	106
1840	112	100	115	94	84	93	77	105	104	120	100
1850	121	121	105	109	70	111	95	93	72	102	100
1860	108	125	132	118	119	102	119	115	98	105	114
1870	64	115	101	102	96	127	89	121	119	118	105
1880	82	96	104	95	102	95	93	89	107	106	97
1890	106	104	89	71	130	137	123	126	154	125	117
1900	128	133	129	136	134	100	68	76	73	70	105
1910	116	111	127	125	133	98	135	124	128	100	120
1920	113	88	123	103	130	105	108	115	105	106	110
1930	107	131	130	105	97	99	112	97	114	87	108
1940	89	84	63	90	88	80	108	80	97	120	90
1950	108	117	104	124	119	123	84	67	98	97	104
1960	103	105	88	100	85	—	—	—	—	—	(96)

The nature of the weather dependence of this series has not been investigated at the time of writing. It appears more complex than in the case of the Spessart oak series in Table App. V.22. In this Trier series decades with mild winters and warm summers which also gave plenty of rain seem to have produced the most growth; colder or drier summers or winters probably restricted growth. The ring widths seem to register the effects of more than one year.

Table App. V.22 German oak tree ring average widths from *Quercus petraea* trees in the Spessart forest area (near 50°N 9°30′E)

See B. HUBER and V. GIERTZ-SIEBENLIST (1969) Unsere tausendjährige Eichen-Jahrringchronologie durchschnittlich 57 (10–150) fach belegt, *Sitz-berichte I: Biol., Mineral., Erdkunde & verwandte Wiss.*, **178** (1–4), 37–42, Vienna (Österreichische Akad. der Wiss.).

Extended series of values here printed, kindly made available by H. GRÜNHAGEN, Lemgo and the Forstbotanisches Institut, Munich.

This series is based on between 10 and 150 sample measurements in the individual years.

Decade	Year										Average
	0	1	2	3	4	5	6	7	8	9	
840	56	61	62	55	58	53	56	62	61	55	57.9
850	42	43	44	60	49	40	48	52	46	58	47.2
860	48	51	46	40	44	45	44	46	40	50	45.4
870	48	69	63	53	27	52	54	53	57	60	53.7
880	49	51	49	39	47	47	50	45	45	48	47.0
890	37	41	47	45	48	42	43	48	36	32	41.9
900	40	39	34	44	36	42	36	30	30	36	36.7
910	27	32	33	31	27	29	21	22	22	27	27.1
920	22	32	24	24	24	25	22	28	23	28	25.2
930	23	29	33	31	34	24	27	27	28	30	28.6
940	28	30	32	29	24	20	23	29	34	40	28.9
950	48	41	41	39	34	36	34	35	33	37	37.8
960	34	36	31	30	28	28	27	32	34	35	31.5
970	32	30	30	30	23	31	30	30	27	32	29.5
980	24	25	26	27	27	35	32	30	24	29	27.9
990	20	28	27	29	27	24	28	26	23	31	26.3
1000	25	28	29	34	30	30	26	24	29	31	28.6
1010	28	25	44	39	37	34	32	35	25	31	33.0
1020	33	51	46	53	52	36	51	52	40	43	45.7
1030	58	39	38	43	50	52	45	62	40	36	46.3
1040	53	42	32	34	30	48	40	38	30	35	38.2
1050	54	56	70	54	60	66	72	60	54	42	58.8
1060	53	45	41	47	36	51	54	58	69	60	51.4
1070	39	49	52	67	48	55	52	49	49	62	52.2
1080	68	46	69	65	60	65	65	60	71	78	64.7
1090	62	65	65	69	67	57	61	61	61	55	62.3

Table App. V.22 – continued

Decade	Year										Average
	0	1	2	3	4	5	6	7	8	9	
1100	65	64	36	36	37	33	42	34	42	48	43·7
1110	54	45	46	46	52	49	56	55	58	48	50·9
1120	58	48	55	58	50	40	39	38	40	36	46·2
1130	46	41	44	36	50	52	57	46	39	43	45·4
1140	63	60	48	61	57	61	58	58	49	56	57·1
1150	50	50	43	54	51	55	60	53	56	48	52·0
1160	58	60	46	44	36	28	28	24	42	50	41·6
1170	44	49	46	46	40	41	40	32	39	42	41·9
1180	46	48	48	53	45	54	55	54	38	47	48·8
1190	48	46	37	44	41	50	43	41	36	42	42·8
1200	32	46	40	42	38	30	39	32	36	36	37·1
1210	39	38	34	36	35	32	36	28	35	38	35·1
1220	36	43	38	37	38	37	41	34	33	42	37·9
1230	36	31	31	32	29	34	31	40	40	46	35·0
1240	50	40	44	42	40	49	46	54	44	49	45·8
1250	50	48	34	40	32	46	45	38	40	36	40·9
1260	45	40	34	38	40	38	44	31	36	42	38·8
1270	30	46	34	36	37	40	34	40	30	36	36·3
1280	40	36	43	42	36	34	40	30	35	36	37·2
1290	39	42	40	40	37	40	44	41	38	35	39·6
1300	32	34	40	37	34	40	32	39	43	48	37·9
1310	48	36	45	48	51	53	51	50	44	42	46·8
1320	34	48	51	46	42	42	30	37	43	52	42·5
1330	44	38	43	44	41	50	41	44	41	38	42·4
1340	33	42	42	44	38	43	42	40	39	42	40·5
1350	42	41	39	35	40	43	45	44	47	49	42·5
1360	40	44	52	48	48	44	44	35	39	44	43·8
1370	50	45	43	44	54	52	64	58	56	55	52·1
1380	58	60	64	62	68	62	75	69	66	65	64·9
1390	58	54	52	31	29	36	42	38	58	51	44·9
1400	54	56	55	54	52	52	51	41	44	45	50·4
1410	56	52	52	52	52	50	50	38	46	41	48·9
1420	34	49	36	37	47	45	47	50	54	52	45·1
1430	50	49	46	48	38	42	50	44	43	44	45·4
1440	43	42	36	42	40	42	34	38	37	42	39·6

Table App. V.22 – continued

Decade	Year										Average
	0	1	2	3	4	5	6	7	8	9	
1450	38	43	37	35	38	38	46	41	40	32	38·8
1460	37	32	26	29	24	32	32	34	39	36	32·1
1470	36	34	38	38	46	50	38	34	37	36	38·7
1480	41	45	40	40	43	44	45	52	47	47	44·4
1490	46	37	38	40	41	48	52	53	46	66	46·7
1500	60	58	53	33	29	49	48	55	50	59	49·4
1510	52	50	52	49	47	60	51	32	40	45	47·8
1520	39	46	40	40	46	37	46	46	56	53	45·4
1530	44	53	34	50	35	41	31	43	30	45	40·6
1540	36	47	50	38	42	52	43	38	41	40	42·7
1550	34	38	36	44	37	50	40	40	38	40	39·7
1560	50	43	54	46	44	45	38	32	40	33	42·5
1570	30	32	31	38	32	36	32	34	33	38	33·6
1580	38	41	33	30	36	38	39	33	27	29	34·4
1590	28	30	34	36	30	28	28	34	34	25	30·7
1600	30	26	25	24	26	28	32	39	33	33	29·6
1610	30	31	29	36	36	28	27	32	31	28	30·8
1620	26	23	25	30	27	30	26	30	26	25	26·8
1630	27	25	28	26	22	20	17	21	28	28	24·2
1640	31	30	32	33	26	26	30	33	34	31	30·6
1650	32	28	27	26	30	36	30	28	32	31	30·0
1660	30	34	37	32	35	32	37	30	30	30	32·7
1670	26	32	38	59	42	39	38	45	44	54	41·7
1680	58	42	45	48	38	34	43	52	51	48	45·9
1690	42	46	41	42	40	36	34	27	37	38	38·3
1700	40	48	46	44	52	46	48	46	46	36	45·2
1710	44	47	52	61	53	49	48	46	36	36	47·2
1720	48	50	46	42	43	50	42	57	48	46	47·2
1730	52	44	44	50	51	50	42	51	52	50	48·6
1740	42	37	36	32	29	28	36	40	37	42	35·9
1750	42	42	44	32	41	43	46	38	31	29	38·8
1760	32	44	30	44	40	36	41	42	45	40	39·4
1770	43	42	40	42	50	48	48	45	49	43	45·0
1780	44	40	42	50	36	38	34	39	46	48	41·7
1790	36	40	48	39	41	42	50	51	46	39	43·2

Table App. V.22 – continued

Decade	Year										Average
	0	1	2	3	4	5	6	7	8	9	
1800	36	50	44	48	47	50	44	54	47	50	47·0
1810	47	53	50	48	44	38	48	47	46	44	46·5
1820	44	46	38	46	45	41	43	43	46	52	44·4
1830	46	48	42	40	49	36	38	38	34	39	41·0
1840	36	31	36	41	40	40	40	38	36	40	37·8
1850	44	40	38	45	36	44	41	35	29	34	38·6
1860	36	37	38	37	40	34	38	40	34	41	37·5
1870	30	32	28	30	30	36	32	29	35	37	31·9
1880	34	32	32	32	40	33	35	31	31	30	33·0
1890	31	36	31	28	34	32	33	30	31	32	31·8
1900	33	32	32	32	34	28	28	29	26	26	30·0
1910	30	32	34	33	36	29	34	32	34	30	32·4
1920	30	27	30	33	33	25	28	35	33	30	30·4
1930	32	40	38	36	28	31	31	30	31	26	32·3
1940	24	18	14	19	21	18	24	17	20	22	19·7

The nature of the responses to weather in this tree ring series does not seem to have been rigorously investigated. The narrow rings prevailing in much of the tenth century, particularly the 920s and 990s, have been tentatively attributed to droughts and summer warmth. In the case of the 990s this is almost certainly true. The decline of ring widths after 1740 is attributed to the severe winter of that year followed by a number of dry summers (and, indeed, dry years) in that decade, and the very narrow rings in the 1940s may perhaps be similarly explained.

Agreement between this series and that for the oaks near Trier, west of the Rhine, may be described as fair. In parts of both series, particularly in the 1530s, the quasibiennial oscillation is well marked.

Shorter tree ring series for oaks on the north German plain are reported to show less agreement.

Table App. V.23   A tree ring series from Java, from A.D. 1514

Measurements on *Tectona grandis* trees growing in Java from A.D. 1514 onwards, originally published by BERLAGE (1931), have been converted by DE BOER (1951) using a simple mathematical formulation to the index values given in table (*a*) below. The index is intended to give a series suited to investigation of both short-term and long-term weather variations.

The ring width index value W for year y is given by

$$W = 100 \frac{w_y - \Omega_y}{\Omega_y}$$

where w_y is the average of the ring widths measured for that year and Ω_y is the value for year y read off a smooth quadratic curve fitted to the growth of the trees over whole time span from 1514 to the present century. This curve expresses the radial growth of the trees varying according to their age (decreasing as the age of the trees increases).

Investigation by DE BOER of the short-term variations shown by the departures of each year's ring width from a 13-year weighted mean centred on that year showed the nature of the response of these trees to weather by the following correlation coefficients linking ring widths with the meteorological observations at Jakarta (Batavia) from 1864 to 1929: with cloudiness +0·49, number of raindays in the year +0·48, amount of rainfall +0·10, yearly mean temperature −0·26, number of dry months −0·43.

(*a*)

Decade	Year										Decade average
	0	1	2	3	4	5	6	7	8	9	
1510					2	−36	−20	−13	−14	12	(−11·5)
1520	25	18	−10	−4	25	31	20	20	16	6	14·7
1530	25	24	19	22	40	20	22	15	7	6	20·0
1540	18	−4	22	−7	−27	−32	3	−19	−23	−24	−9·3
1550	−9	−31	−32	−18	20	−54	−10	11	−7	−44	−17·4
1560	40	15	−57	−16	−34	−21	−19	−31	−25	−17	−16·5
1570	−21	−9	−13	−5	28	16	21	15	31	−32	3·1
1580	−45	65	41	34	10	46	26	1	32	59	26·9
1590	39	65	5	9	8	13	10	40	−16	−12	14·1
1600	−8	−31	42	−44	16	−11	−36	−22	−29	−29	−15·2
1610	−23	−33	−22	−14	−18	−20	−35	12	−25	2	−17·6
1620	−5	1	5	−10	0	−13	−24	4	12	15	−1·5
1630	50	5	1	−21	−60	−8	−10	−51	−43	30	−10·7
1640	−33	35	44	−26	−41	−17	−35	−27	−29	−34	−16·3
1650	−53	−50	−30	−50	−10	−15	−3	−23	−10	−33	−27·7
1660	−33	−10	−24	−40	−61	−67	−15	−21	−22	−43	−33·6
1670	−17	−13	13	−27	−21	−29	−5	1	−7	−24	−12·9
1680	4	28	40	58	−34	4	−1	−16	32	14	12·9
1690	−17	9	45	48	3	−2	22	5	−8	−6	9·9

Table App. V.23 – continued

Decade	Year										Decade average
	0	1	2	3	4	5	6	7	8	9	
1700	−23	6	3	−27	−33	−2	−8	−19	−11	0	−11·4
1710	−22	−42	−30	−22	−12	0	15	−17	−22	−37	−18·9
1720	−31	1	2	−34	−25	−47	−45	−57	−41	−57	−33·4
1730	−41	−32	−24	29	−11	32	31	−56	0	12	−5·0
1740	30	26	22	−6	−7	53	−7	24	−27	28	13·6
1750	15	33	12	31	−26	0	17	−8	9	8	9·1
1760	16	18	−16	31	1	29	48	27	54	16	22·6
1770	12	22	23	31	47	38	51	49	30	70	37·3
1780	87	45	16	23	48	53	21	33	85	8	41·9
1790	−3	30	40	20	26	62	−4	22	−9	−26	15·8
1800	−14	33	31	−18	0	10	−4	−3	−12	21	4·4
1810	−13	0	0	7	−30	−25	−36	−52	−41	−29	−21·9
1820	15	5	−16	29	8	−5	68	−24	69	40	18·9
1830	45	47	16	52	5	24	44	5	9	11	25·8
1840	55	18	−13	30	8	−7	−26	−6	−38	−15	0·6
1850	1	2	8	−26	1	−52	−10	10	31	18	−1·7
1860	−15	43	−3	40	2	17	−6	2	−11	41	11·0
1870	27	14	−9	−6	18	16	−3	−13	6	52	10·2
1880	24	41	−6	0	47	−13	−4	54	−2	31	17·2
1890	−6	9	20	9	17	1	−30	−16	−12	−18	−2·6
1900	33	−32	−32	−18	37	−9	13	−13	−4	0	−2·5
1910	29	−20	−28	−46	−46	−29	0	−22	−42	−38	−24·2
1920	−14	−33	0	−10	−25	−67	−44	−26	−11	−56	−28·6

The unadjusted ring widths are given in table (*b*). The average number of measurements on which each year's ring width is based varied from 10 to 15 between 1514 and 1760 and was generally between 30 and 80 after 1825.

(*b*)

Decade	Year									
	0	1	2	3	4	5	6	7	8	9
1510					33·57	20.61	24·93	26·58	25·53	32·46
1520	35·40	32·52	24·09	25·08	31·98	32·61	29·10	28·20	26·46	23·64
1530	27·00	26·01	24·21	24·12	27·00	22·50	22·14	20·22	18·36	17·52
1540	18·93	15·00	18·36	13·65	10·29	9·27	13·65	10·32	9·45	8·94

Table App. V.23 – continued

Decade	Year									
	0	1	2	3	4	5	6	7	8	9
1550	10·14	7·38	6·93	8·01	11·16	4·08	7·56	8·91	7·11	4·14
1560	4·32	8·10	2·97	5·58	4·32	5·07	5·07	4·26	4·56	4·98
1570	4·68	5·37	5·07	5·52	7·38	6·66	6·93	6·57	7·44	3·84
1580	3·12	9·30	7·95	7·53	6·15	8·13	6·99	5·58	7·29	8·73
1590	7·62	9·00	5·73	5·94	5·85	6·12	5·91	7·53	4·50	4·68
1600	4·92	3·66	3·06	2·94	6·09	4·65	3·33	4·08	3·69	3·66
1610	3·99	3·42	3·99	4·38	4·17	4·05	3·27	5·64	3·75	5·13
1620	4·74	5·04	5·22	4·44	4·92	4·29	3·72	5·07	5·46	5·61
1630	7·26	5·07	4·86	3·81	1·89	4·41	4·26	2·34	2·70	6·12
1640	3·15	6·30	6·72	3·42	2·73	3·84	3·00	3·36	3·24	3·00
1650	2·13	2·25	3·15	2·25	4·05	3·78	4·32	3·42	3·99	2·97
1660	2·97	3·93	3·33	2·61	1·68	1·41	3·69	3·39	3·33	2·43
1670	3·54	3·69	4·80	3·09	3·30	2·97	3·96	4·23	3·87	3·15
1680	4·29	5·25	5·76	6·45	2·67	4·23	3·99	3·39	5·31	4·59
1690	3·33	4·33	5·79	5·88	4·05	3·87	4·80	4·11	3·60	3·63
1700	2·79	4·08	3·93	2·79	2·55	3·75	3·48	3·06	4·17	3·75
1710	2·91	2·16	2·61	2·91	3·24	3·70	4·22	3·04	2·84	2·30
1720	2·52	3·66	3·70	2·38	2·70	1·90	1·96	1·54	2·08	1·52
1730	2·08	2·38	2·66	4·50	3·10	4·56	4·52	1·52	3·44	3·84
1740	4·42	4·28	4·14	3·18	3·12	5·12	3·12	4·12	2·42	4·24
1750	3·78	4·36	3·68	4·30	2·42	3·26	3·78	2·98	3·52	3·46
1760	3·70	3·76	2·66	4·14	3·20	4·06	4·64	3·96	4·82	3·60
1770	3·46	3·76	3·78	4·02	4·50	4·20	4·60	4·52	3·92	5·12
1780	5·62	4·36	3·46	3·66	4·40	4·52	3·56	3·92	5·42	3·14
1790	2·82	3·78	4·06	3·48	3·64	4·66	2·74	3·48	2·58	2·10
1800	2·44	3·74	3·68	2·30	2·80	3·06	2·68	2·68	2·44	3·34
1810	2·40	2·74	2·74	2·90	1·90	2·02	1·74	1·30	1·58	1·90
1820	3·08	2·78	2·22	3·40	2·84	2·50	4·39	1·98	4·40	3·64
1830	3·76	3·79	2·98	3·90	2·70	3·17	3·66	2·67	2·77	2·80
1840	3·90	2·97	2·18	3·25	2·70	2·31	1·83	2·32	1·52	2·10
1850	2·47	2·51	2·64	1·81	2·45	1·16	2·17	2·64	3·15	2·82
1860	2·04	3·41	2·32	3·34	2·41	2·78	2·23	2·40	2·09	3·29
1870	2·95	2·65	2·10	2·18	2·73	2·68	2·23	1·99	2·43	3·47
1880	2·82	3·21	2·13	2·25	3·33	1·95	2·15	3·44	2·20	2·93
1890	2·09	2·43	2·67	2·40	2·58	2·23	1·55	1·85	1·93	1·78
1900	2·89	1·48	1·48	1·79	2·95	1·96	2·43	1·86	2·07	2·15
1910	2·76	1·72	1·55	1·14	1·14	1·51	2·12	1·64	1·22	1·31
1920	1·80	1·40	2·09	1·89	1·57	0·68	1·15	1·53	2·29	0·91

Table App. V.24   Blossoming dates of the cherry trees at Kyoto, Japan, since A.D. 812

The dates, a function of the spring temperatures up to the date mentioned, were recorded for the yearly cherry blossom festival of the emperors and governors at Kyoto, the old capital of Japan. ARAKAWA (1955b) writes: 'Even if not a few of the old records leave the reader in doubt whether the reference is to *Prunus yedoensis*, *Prunis serrulata* or *Prunus subhirtella* and whether the reference is to efflorescent date, date at their best or some other stage, there still remains such a large number of certain records of blooming dates of the cherry blossoms that the epochs of the climatic changes are very definitely manifest.'

All dates corrected to the modern (Gregorian) calendar.

(*a*) Each year.

Decade beginning	Year									
	0	1	2	3	4	5	6	7	8	9
810			1 Apr			15 Apr				
830		6 Apr								
850		18 Apr		14 Apr						
860					9 Apr		16 Apr			
910		4 Apr						5 Apr		
920							8 Apr			
940		19 Apr								17 Apr
950								22 Apr	19 Apr	
960		28 Mar		4 Apr		14 Apr	8 Apr	13 Apr		
970				18 Apr	17 Apr					
1000							10 Apr			
1010							20 Apr		23 Apr	
1020										26 Apr
1040		13 Apr								
1100						24 Apr				
1120								9 Apr		
1170										13 Apr
1180					24 Apr					
1230	12 Apr									
1240							29 Mar	16 Apr		
1260				30 Apr						
1270									23 Apr	
1280						15 Apr	22 Apr			

Table App. V.24 – continued

Decade beginning	Year 0	1	2	3	4	5	6	7	8	9
1290						6 Apr				
1300			6 Apr					14 Apr		
1310							12 Apr	12 Apr		
1320			16 Apr	4 May			17 Apr			
1330		22 Apr	24 Apr							
1340					11 Apr		17 Apr	19 Apr		
1360							22 Apr			
1420			19 Apr		8 Apr		9 Apr			
1430	18 Apr		13 Apr							24 Apr
1440				17 Apr						
1450	10 Apr	13 Apr	1 Apr				15 Apr	15 Apr	22 Apr	23 Apr
1460	21 Apr		16 Apr	1 Apr	13 Apr	8 Apr	15 Apr		10 Apr	
1470									1 May	
1480		16 Apr			6 Apr	4 Apr	9 Apr	3 Apr		
1490						4 Apr			7 Apr	21 Apr
1500	7 Apr	13 Apr	12 May			16 Apr	16 Apr	18 Apr	20 Apr	12 Apr
1510	20 Apr	21 Apr	18 Apr	20 Apr	15 Apr			16 Apr	15 Apr	21 Apr
1520	14 Apr	17 Apr	13 Apr		18 Apr	14 Apr	8 May	15 Apr	15 Apr	17 Apr
1530	17 Apr	13 Apr	17 Apr							
1540					19 Apr					
1560	12 Apr			11 Apr		29 Apr	10 Apr			
1570										15 Apr
1580	10 Apr					15 Apr			18 Apr	
1590					19 Apr			24 Apr		
1600				13 Apr	9 Apr	15 Apr	16 Apr			18 Apr
1610			27 Mar	16 Apr						
1620			16 Apr							
1630				8 Apr						
1650		17 Apr								
1840							7 Apr	17 Apr		
1850		12 Apr		12 Apr						
1860				14 Apr						

Source: ARAKAWA (1955b), based on TAGUCHI (1939).

(b) Frequencies of recorded blossoming dates of the cherry trees at Kyoto, Japan, in various ranges of calendar dates (modern calendar)

Century A.D.	Before 1 April	1–10 April	11–20 April	21–30 April	After 30 April	Average
9th	0	3	4	0	0	11 April
10th	1	5	7	1	0	12 April
11th	0	1	2	2	0	18 April
12th	0	1	1	2	0	24 April
13th	1	1	3	3	0	15 April
14th	0	1	8	3	1	18 April
15th	0	13	11	5	1	13 April
16th	0	3	31	4	2	17 April
17th	1	2	7	0	0	12 April
19th	0	1	4	0	0	12 April
1917–53 Prunus yedoensis efflorescent	3	22	11	0	0	7 April
Prunus yedoensis at its best	0	11	21	3	0	14 April

Table App. V.25(a)   Freezing dates of Lake Suwa, in central Japan, 36°N 138°E, each winter since the later Middle Ages

Winters identified by the year in which the January fell.

D = December (of the previous year), J = January, F = February, M = March.

Dates in brackets are approximate, but can be accepted as meaning that the lake was frozen by the date mentioned.

Modern (Gregorian) calendar throughout.

Decade beginning	Year									
	0	1	2	3	4	5	6	7	8	9
1390									18D	
1440					7J	23D	31D	1J	30D	7J
1450	12J	7J	22J	28D	2J	4J	31D	4J	Open?	5J
1460	20D	9J	14J	29D	9J	10J	11D	20J	11J	10J
1470	2F	26D	22D	24D	3J	1J	31D	6J	17J	16J
1480	5J	3J	15J	24J	31D	1J	10J	11J	5F	20J
1490	24D	9F	5J	1J	5J	8J	3J	22J	11D	13J

Table App. V.25(*a*) – continued

Decade beginning	Year									
	0	1	2	3	4	5	6	7	8	9
1500	18D	27D	15J	11J	15D	31D	Open	20D	Open	Open
1510	Open	Open	Open	Open	Open	Open	1J	5J	5F	29D
1520	3J	14J	26J	1J	1J	11J	14D	14J	16J	27D
1530	6J	1J	2J	9J	21D	16J	21D	5F	13J	18D
1540	13F	23D	3J	15J	4J	13J	1F	Open	5J	13J
1550	7J	4J	3F	25D	Froze	8J	Open	2J	24J	31D
1560	2J	5F	10J	4J	12J	6J	30D	8J	4J	25D
1570	22D	26J	22D	16J	30D	7F	25J	1J	5J	28J
1580	13J	11J	29J	17D	18D	28D	14J	28D	23D	14J
1590	31D	13J	13J	28D	4M		24J	25J	13D	20D
1600	21J	19D	29J	3J	31D	14J	18D	23D	24J	5J
1610	18D	31D	22D	25J	9J	21D	22J	22D	2J	10J
1620	10J	8D	7J	12J	24J	22J	8J	9J	8J	21D
1630	20J	22D	24J	18J	9J	28D	2J	11J	31D	21J
1640	19J	18J	15D	25J	5J	1J	12J	9F	4J	29D
1650	5J	14J	13J	6J	13J	7J	10J	10J	24J	28D
1660	10J	11J	18J	1J	10J	26D	28D	5J	10J	16D
1670	25J	6J	Open	26D	22J	18J	21J	5J	2F	5J
1680	28D	28D	21J		(4J)	(8J)	(9J)	(4J)	(10J)	(15J)
1690	(4J)	(5J)	(19J)	(27D)	(7J)	(9J)	(7J)	29D	(31D)	(6J)
1700	(2J)	(18J)	(9F)	Open	(6F)	(31J)	(25J)	(31D)	(4J)	(7J)
1710	(4F)		(6J)	(5J)	(27D)	(28J)	(17J)	(26J)	(26J)	(16J)
1720	(12J)	(23J)	(5F)	(27J)	(4J)	Open	(8J)	(2J)	(28J)	(23J)
1730	(15J)	(1J)	(26J)	(8J)	(30J)	(5J)	(4J)	(9J)	Open	(4J)
1740	Nearly open	(16J)	16J	(4J)	(4F)	(29D)	Nearly open	8J	27J	13J
1750	23J	2J	29D	7J	(24J)	1F	1J	5J	(20J)	16J
1760	Open	31D	26D	12J	(12J)	(9J)	(22J)	(16J)	(17J)	18J
1770	9J	Open	Froze	(16J)	(21J)	29D	28J	(19J)	(11J)	(7F)
1780	(15J)	(15J)	(24J)	(20J)	(16J)	(10D)	(20J)	(15J)	(2J)	1J
1790	10J	30D	Froze	28D	Open	25J	(17J)	2J	6J	10J
1800	Open	6J	Froze	Nearly open	17J	Open	3F	5J	11J	7J
1810	27D	4J	4J	26D	13J	(6J)	31D	21J	15J	(30J)
1820	16J	6J	10J	22J	Froze	Open?	Froze	Froze	Froze	8J
1830	16J	13J	9J	(16J)	7F	13J	5J	11J	6F	28J
1840	13J	24J	7J	Open	17J	Open	28D	10J	(12F)	(2F)

Table App. V.25(*a*) – continued

Decade beginning	Year									
	0	1	2	3	4	5	6	7	8	9
1850	(25J)	5F	(28J)	2F	Nearly open	6F	12J	(4F)	8F	23J
1860	(25J)	30J	(4F)	21J	17J		Open	Open	Open	Open
1870	21J	23J	18J	Froze	Froze	Froze	Froze	Froze	Froze	Froze
1880	Open	Froze	Froze	Froze	Froze	Froze	Froze	Froze	Froze	Froze
1890	Open	Froze	Froze	(19J)	(28J)	(26J)	(12F)	(25J)	26D	18J
1900	9J	19J	31D	15J	22D	25D	16J	30D	24D	2J
1910	27D	3J	4J	2J	8J Nearly open	10J	Open	(21F)	23D	6J
1920	26J	8J	5J	20D	24J	15J	3J	26D	4J	31D
1930	9J	11J	Open	14J	6J	7F	21D	13J	4J	4J
1940	17J	2F	20J	6J	9J	3J	27D	4J	20D	Open
1950	12J	2J	13J	5J	25J					

Table adapted from ARAKAWA (1954), who considers the data from the period 1682 to 1923 inferior to those of earlier and later times.

GRAY (1974) has shown that the mean winter (December to February) temperatures at Tokyo, near sea level, 145 km to the east of Lake Suwa which lies at a height of over 500 m above sea level in the centre of the island, can be estimated to a reasonable approximation from the correlation ($r = +0.65$) with the freezing date of Lake Suwa and has published the yearly values from 1443 to 1953 so derived. The decade means appear to be as in Table (*b*) below. Brackets mark values for decades in which information on the freezing dates was inadequate or incomplete.

Table App. V.25(*b*)   Prevailing winter (DJF) temperatures in Tokyo: decade means from A.D. 1443

Century beginning	Decade									
	00–09	10–19	20–29	30–39	40–49	50–59	60–69	70–79	80–89	90–99
From freezing dates of Lake Suwa:										
1400					(3·9)	4·1	4·1	3·9	4·2	4·0
1500	4·4	4·7	4·0	4·1	4·2	4·3	4·0	4·3	4·0	4·2
1600	4·0	4·0	4·1	4·1	4·1	4·2	4·0	4·3	4·1	4·0
1700	4·6	4·4	4·5	4·4	4·4	4·3	4·1	4·4	4·1	4·3
1800	4·4	4·1	4·4	4·4	4·6	4·8	4·9	(4·7)	(5·3)	(4·1)
1900	3·9	4·3	4·0	4·4	4·2	(4·5)				
From observations in Tokyo:										
1800								(4·4)	3·6	4·2
1900	4·0	4·1	4·0	4·3	4·4	(4·8)				

Table App. V.26   Dates of first snow cover in Tokyo since 1632

Winters identified by the year in which the January fell.
N = November (of the previous year), D = December (of the previous year),
J = January, F = February, M = March, — = No data.

Decade beginning	Year									
	0	1	2	3	4	5	6	7	8	9
1630	—	—	—	6J	—	—	—	—	16D	—
1640	2F	—	28N	—	—	—	—	—	10J	—
1650	—	—	—	—	—	—	—	9J	4J	20J
1660	—	—	14D	—	—	16D	—	—	—	—
1670	12F	—	—	31D	—	—	—	—	—	17J
1680	21D	27J	8J	—	—	—	8D	—	—	—
1690	28D	7D	4J	15J	—	29J	—	—	—	—
1700	24N	—	—	—	—	1J	2F	—	—	3J
1710	—	18J	—	—	—	12J	12J	31J	7F	17F
1720	3D	14J	18J	—	17D	21J	—	17J	—	17J
1730	19D	—	31D	29J	1J	—	—	—	24D	25D
1740	2J	—	—	31D	22J	13D	—	—	16J	30J
1750	—	—	27D	21D	—	29J	—	30D	—	—
1760	—	8J	28D	10J	25J	5F	10D	25J	19J	13D
1770	—	4J	22J	18D	23J	22J	23J	21J	25J	31D
1780	12D	—	8D	5J	—	—	30D	5J	19D	—
1790	11J	16J	—	31D	—	14J	21D	26J	23J	29D
1800	—	29J	14J	15N	12J	26J	4F	—	—	4J
1810	4D	20D	9J	—	12J	18D	—	—	27D	6D
1820	—	—	—	—	—	22J	—	7J	—	—
1850	—	—	—	14J	—	—	—	—	6D	—
1860	29D	27D	—	24D	—	—	—	17F	—	—
1870	7F	—	—	—	—	—	—	26J	31J	12J
1880	22J	16F	2F	12J	17J	30D	30J	6J	1F	3J
1890	11F	31J	12J	29J	9M	18J	13F	9J	23J	25J
1900	18D	25F	22J	31J	7F	13D	29D	2J	7J	9J
1910	11F	15J	3F	28D	16D	23D	7F	2J	28J	11D
1920	29J	7D	15J	24J	27J	10J	9J	19D	25J	28J
1930	12J	1J	10F	6J	3F	5D	25J	21D	30J	18J
1940	26D	17D	12J	31J	5J	13D	23D	24J	17D	14M
1950	21D	10J	3J	7J	23J	6J				

Source: ARAKAWA (1956)

1632–1870	Mean date	6·5 January
	Earliest	15 November 1802
	Latest	17 February 1719 and 1867
	Standard deviation	19·4 days
1876–1955	Mean date	15·2 January
	Earliest	5 December 1934
	Latest	14 March 1949
	Standard deviation	20·5 days

Table App. V.27   Severe winters in the Yangtse river basin, China (near 30°N), with freezing of some or all of the large lakes and tributary rivers

Years A.D.

Lake Tungting,	112½°E	(Tu)
Lake Poyang,	116°E	(Po)
Lake Taihu,	121°E	(Ta)
River Han		(Ha)
River Huai		(Hu)

After CHU KO-CHEN (1973)

1111 Ta

1219 Hu

1329 Ta

1353 Ta

1416 Ha, 1449 Ha

1454 Ta Hu, 1476 Ta, 1493 Ha

1503 Ta, 1510 Tu, 1513 Ta Po Tu, 1519 Ha, 1529 Ha

1550 Hu, 1564 Hu, 1568 Ta, 1570 Po, 1578 Ta

1619 Hu, 1620 Ha, 1621 Tu Ha, 1640 Hu

1653 Tu Ha Hu, 1654 Ta, 1660 Ha, 1665 Ta, 1670 Po Ha Hu, 1671 Hu, 1683 Ta,
    1690 Tu Ha Hu, 1691 Ha

1700 Ta, 1715 Hu, 1720 Hu

1761 Ta, 1790 Tu

1830 Ha, 1840 Po, 1845 Hu

1861 Ta Po, 1865 Po Ha, 1871 Ha, 1877 Ta Tu Ha, 1886 Ha, 1893 Ta, 1899 Ha

---

Table App. V.28   Years with snowfall and frosts on low ground in south China (Kwangsi and Kwangtung), 22–25°N

Years A.D.
After CHU KO-CHEN (1973)

1245

1415, 1449

1506, 1512, 1522, 1532, 1536, 1537, 1547, 1549

1578

1602, 1606, 1621, 1635, 1636

1654, 1655, 1656, 1681, 1682, 1683, 1684

1711, 1713, 1721, 1729, 1737, 1742

1757, 1758, 1763, 1768, 1781

1824, 1831, 1832, 1835, 1840, 1846

1854, 1856, 1862, 1864, 1871, 1872, 1878, 1882, 1893

Table App. V.29 (*a*) Mean thickness (mm) of the yearly mud layers in Lake Saki, Crimea (45°07′N 33°33′E), by decades

Century	Decade (years)										Century average	Standard deviation of the decade values
	00–91	90–81	80–71	70–61	60–51	50–41	40–31	30–21	20–11	10–01		
B.C.												
23rd	(25)	33·1	24·0	37·1	19·6	15·2	16·7	27·3	24·6	32·0	25·5	7·14
22nd	23·3	25·3	32·0	19·3	22·0	35·8	18·8	20·1	15·5	18·0	23·0	6·43
21st	24·4	21·5	16·0	14·4	14·7	14·7	12·6	19·0	16·9	15·4	17·0	3·64
20th	11·4	10·7	11·0	15·3	17·5	10·8	17·1	11·9	17·1	18·6	14·1	3·26
19th	14·2	14·2	16·2	9·3	11·5	17·4	16·1	15·1	9·3	11·5	13·5	2·91
18th	11·5	10·0	8·9	7·8	9·9	12·0	9·0	11·1	12·7	25·3	11·8	5·00
17th	13·3	11·3	14·2	14·7	10·7	17·4	7·4	9·9	9·5	12·7	12·1	2·93
16th	9·0	10·0	12·2	15·1	6·2	11·5	17·2	13·2	15·8	9·1	11·9	3·46
15th	9·2	10·3	11·7	11·9	11·8	8·0	10·0	11·2	10·2	8·0	10·2	1·47
14th	10·8	8·7	8·6	8·7	10·7	11·2	9·6	11·1	14·3	12·9	10·7	1·88
13th	12·3	10·3	8·8	8·5	12·8	9·2	12·2	12·1	13·6	15·0	11·5	2·18
12th	9·9	9·3	12·0	10·8	9·4	11·1	9·7	8·6	8·9	9·0	9·9	1·10
11th	15·1	10·4	11·9	13·4	16·0	18·4	9·7	12·9	13·7	8·2	13·0	3·08
10th	8·2	8·8	10·1	9·1	8·8	11·3	10·7	9·5	14·9	12·1	10·3	2·02
9th	12·9	18·3	10·6	13·2	12·7	13·4	11·9	10·2	16·0	17·7	13·7	2·78
8th	19·6	16·3	10·9	12·9	13·0	11·7	12·3	10·3	9·5	10·4	12·7	3·10
7th	14·4	12·4	15·1	12·1	11·6	11·2	12·7	10·5	12·4	11·0	12·3	1·46
6th	7·6	11·2	10·5	10·1	9·6	8·7	10·1	13·6	10·5	11·3	10·3	1·61
5th	9·6	13·1	12·1	13·2	10·9	12·3	8·9	9·7	10·7	9·8	11·0	1·55
4th	11·0	9·4	14·9	12·6	12·7	17·8	21·0	5·4	12·9	9·1	12·7	4·47
3rd	7·9	9·4	8·7	12·4	11·6	13·1	14·7	12·8	15·6	11·9	11·8	2·51
2nd	10·1	10·2	12·7	9·3	9·6	13·0	11·0	16·9	8·7	11·2	11·3	2·42
1st	9·0	15·2	15·6	13·2	10·5	12·4	13·7	15·2	16·0	13·8	13·5	2·29

Table App. V.29 – continued

Century beginning	Decade (years)										Century average	Standard deviation of the decade values
	00–09	10–19	20–29	30–39	40–49	50–59	60–69	70–79	80–89	90–99		
A.D.												
1	9·8	9·1	10·7	11·2	12·4	11·5	14·0	8·2	11·3	13·5	11·2	1·83
100	10·5	14·3	13·5	16·1	21·6	17·1	10·4	9·7	11·9	11·6	13·7	3·73
200	13·3	11·7	12·4	12·4	15·1	9·7	9·1	12·6	9·2	13·0	11·9	1·95
300	12·0	9·8	9·6	10·4	14·7	12·1	9·4	12·5	8·9	13·1	11·3	1·91
400	13·8	12·4	15·3	12·5	12·4	11·9	12·8	11·6	13·4	11·7	12·8	1·13
500	12·7	8·6	12·6	13·1	11·9	14·7	13·0	11·2	14·5	13·2	12·5	1·74
600	11·7	10·2	10·8	10·5	9·1	11·8	11·0	10·9	13·4	10·0	10·9	1·17
700	11·3	10·4	12·7	11·9	13·8	13·5	12·8	11·4	12·7	20·6	13·1	2·83
800	37·9	14·0	13·4	15·7	17·5	19·4	13·7	16·6	14·5	10·1	17·3	7·68
900	10·4	17·5	17·7	13·2	15·2	11·8	17·9	13·0	12·8	15·5	14·5	2·65
1000	12·6	21·7	12·4	18·4	16·1	12·6	12·3	15·6	16·4	17·8	15·6	3·16
1100	17·7	27·6	18·6	18·8	24·6	24·9	21·1	20·1	20·8	18·7	21·3	3·31
1200	21·7	13·6	17·4	19·6	25·2	18·2	15·7	20·2	12·2	13·4	17·7	4·11
1300	13·1	14·8	11·5	13·4	12·8	12·8	12·2	14·6	18·6	19·9	14·4	2·77
1400	14·7	12·9	11·6	17·8	20·8	14·7	22·4	13·0	15·7	10·1	15·4	3·94
1500	14·5	23·7	13·2	13·9	13·5	17·6	15·0	13·9	14·3	12·1	15·2	3·32
1600	17·1	17·1	14·2	17·0	10·7	13·7	19·0	12·9	14·1	12·6	14·8	2·59
1700	21·9	13·9	12·3	10·3	21·9	21·8	19·9	18·5	11·0	21·7	17·3	4·89
1800	13·7	13·7	18·6	17·3	10·8	12·2	15·1	20·4	19·8	(24·8)	16·6	4·33
1900												

There is some interest in the successions of individual yearly values about the times of the major changes of regime indicated by the above Table – e.g. in the twenty-second, nineteenth and perhaps the ninth centuries B.C. and around A.D. 800 and 1280 – which, if we may follow C. E. P. BROOKS in reading the Lake Saki mud layer thicknesses more or less directly as a measure of rainfall, seem to be heralded *either* by one (or several, occasional) extremely wet year(s) *or* by a run of 5 to 10 extremely dry years, and an increased decade-to-decade variability, in the Lake Saki region. Some of these periods are detailed in the following short tables (*b*).

Both these types of deviation may be presumed to indicate extremes of persistent 'blocking'.

(b) Yearly mud layer thickness (mm) in Lake Saki, selected periods

Decade ending	Year									
	9	8	7	6	5	4	3	2	1	0
2190 B.C.	11	14	22	34	40	26	12	18	20	40
2180	30	18	34	24	26	21	16	20	24	22
2170	16	20	**110**	38	20	18	16	25	35	18
2160	20	26	20	11	20	18	16	14	30	10
2150	14	12	14	20	40	20	22	28	40	**230**
2140	12	20	18	15	12	12	18	11	10	12
2130	18	20	28	18	20	12	15	13	32	10
2120	45	10	45	10	20	8	28	11	14	20
2110	10	12	20	16	8	17	18	18	16	18
2100	20	12	22	13	11	10	14	28	32	30
1890 B.C.	11	16	17	10	14	18	12	14	9	11
1880	19	12	12	18	22	10	12	10	16	10
1870	20	10	11	16	26	25	9	11	24	**10**
1860	**9**	**10**	**5**	**5**	9	6	**10**	11	18	12
1850	10	10	12	15	20	10	10	11	5	14
1840	10	12	6	11	9	35	40	20	16	30
1830	40	14	18	15	4	6	13	11	10	9
1820	8	38	22	10	20	15	**10**	**10**	9	**4**
1810	**8**	**6**	**9**	**8**	**10**	**10**	**10**	16	12	11
1800	10	12	11	10	11	13	18	9	10	20
890 B.C.	25	10	9	2	11	4	6	36	17	5
880	**110**	6	5	6	11	11	15	6	8	9
870	16	9	12	6	8	11	16	11	8	9
860	13	10	28	16	14	9	11	16	6	10
850	10	10	10	20	11	10	6	10	30	18

Bold print marks extreme deviations – individual very thick mud layers (wet years) and runs of thin layers (dry years).

Table APP. V.29 – continued

Decade beginning	Year									
	0	1	2	3	4	5	6	7	8	9
780 A.D.	10	10	12	12	10	18	9	10	16	20
790	20	50	30	13	10	13	9	20	11	30
800	18	22	18	20	16	220	20	15	14	16
810	12	18	12	16	16	8	12	16	14	16
820	20	10	16	10	9	11	16	8	10	24
830	10	15	20	20	14	10	18	24	16	10
840	8	12	14	22	33	20	16	18	14	18
1250	24	16	18	12	12	30	20	12	16	22
1260	12	15	18	14	12	20	10	24	20	12
1270	26	32	22	8	12	26	4	8	12	52
1280	14	10	8	10	8	10	12	14	12	24
1290	20	20	4	14	12	20	8	10	14	12

The following table (c) gives a *complete* list of the years with extraordinarily thick mud layers, exceeding three times the currently prevailing (100-year) mean thickness.

(c) Extreme years for mud layer thickness in Lake Saki

Year	2204 B.C.	2177	2150	2087	1843	1839 B.C.	
Thickness (mm)	85	110	230	56	40	40	
Year	1705 B.C.	1668	1648	1042	889	86 B.C.	
Thickness (mm)	90	39	40	40	110	50	
Year	A.D. 140	424	440	791	805	1117	
Thickness (mm)	90	42	40	50	220	76	
Year	A.D. (1249)	1279	1392	1432	1464	1759	(1873)
Thickness (mm)	(46)	52	70	45	50	110	(45)

(d) The following were the longest and most notable runs of successive years with layer thicknesses ⩽ 10 mm:

1870–1863 B.C.	(av. 8·0 mm)	1119–1115 B.C.	(6·4 mm)	A.D. 72–81	(6·9 mm)
1823–1813	(8·5 mm)	1108–1104	(7·0 mm)	100–104	(6·2 mm)
1774–1763	(7·5 mm)	1073–1069	(7·4 mm)	174–182	(7·8 mm)
1640–1631	(7.4 mm)	1004–1000	(4·0 mm)	194–198	(8·8 mm)
1614–1607	(6·1 mm)	935–928	(7·7 mm)	253–261	(7·1 mm)
1572–1567	(7.7 mm)	722–718	(7·2 mm)	313–318	(8·0 mm)
1561–1551	(6·5 mm)	575–570	(7·0 mm)	362–369	(7·9 mm)
1497–1493	(7·0 mm)	393–389	(9·0 mm)	462–466	(9·2 mm)
1449–1444	(6·8 mm)	330–320	(6·9 mm)	470–475	(9·0 mm)
1408–1404	(6·8 mm)	305–298	(9·3 mm)	512–516	(7·4 mm)
1370–1363	(8·0 mm)	296–290	(6·7 mm)	902–907	(8·0 mm)
1334–1330	(6·6 mm)	135–131	(7·2 mm)	1281–1285	(9·2 mm)
1250–1243	(8·4 mm)			1734–1740	(6·6 mm)
1190–1185	(8·7 mm)			1799–1803	(7·0 mm)

Figures in brackets give the average yearly thickness in the periods quoted.

Table App. V.30   Low stage of the River Nile at Cairo

Decade averages of the minimum gauge readings (converted to meters), as given by TOUSSOUN (1925).

Century beginning	Decade										Century average
	00–09	10–19	20–29	30–39	40–49	50–59	60–69	70–79	80–89	90–99	
A.D.											
600			(11·2)	12·07	11·66	11·39	11·62	11·31	11·26	11·33	11·48
700	11·48	11·53	11·29	10·81	10·71	11·08	10·05	10·14	10·17	10·59	10·79
800	11·13	11·70	11·35	11·82	10·88	11·59	12·50	11·10	11·19	11·98	11·52
900	11·72	11·66	11·75	11·44	10·51	11·73	11·33	11·41	11·09	10·93	11·36
1000	11·35	10·84	11·56	11·32	12·33	11·47	11·81	11·28	11·96	11·81	11·57
1100	12·92	12·80	13·13	12·11	11·68	12·34	11·96	12·08	12·19	12·01	12·32
1200	11·07	11·28	10·93	11·29	11·49	11·51	11·66	12·42	11·16	11·27	11·41
1300	11·03	10·63	11·45	11·26	11·59	11·67	12·13	11·94	12·34	12·07	11·61
1400	10·44	11·60	11·50	12·45	12·81	12·56	12·51	(12·50)	(12·97)	(12·41)	(12·17)
1500	(12·4)	(12·8)	(12·7)	(9·4)	(9·0)				(9·5)	11·37	(10·4)
1600	11·09	11·14	(11·4)	(11·9)		(11·9)	(11·7)				(11·6)
1700	(11·6)	11·64	11·45	11·90	(11·5)	11·33	11·49	(12·5)	(11·9)	(11·9)	(11·7)
1800	(11·35)	(11·0)	(11·1)	(11·9)	12·29	12·28	12·97	12·85	13·01	13·79	(12·25)
1900	For later years consult the Nile flow data in *World Weather Records*										

Bracketed values appear to have been based on incomplete data.

It is thought that reliance can hardly be put upon the figures given by TOUSSOUN for the years after 1890, presumably owing to allowance needed for some extra alteration (silting?) of the bed of the river. The sequence of figures through the nineteenth century up to the 1880s is in reasonable accord with other information, but whereas TOUSSOUN's figures show a rapid and continuing rise after 1890 the flow of the river in fact declined sharply (see Chapter 13, p. 139).

Table App. V.31 Height of the yearly Nile flood at Cairo

Decade averages of the maximum gauge readings (converted to metres), as given by TOUSSOUN (1925).

Century beginning	Decade										Century average
	00–09	10–19	20–29	30–39	40–49	50–59	60–69	70–79	80–89	90–99	
A.D.											
600			17·44	17·76	17·54	17·36	17·68	17·49	17·19	17·43	17·49
700	17·44	17·56	17·66	17·38	17·62	17·55	17·23	17·29	17·23	17·21	17·42
800	17·49	17·46	17·69	17·10	17·14	17·43	17·69	17·47	17·45	17·49	17·44
900	17·36	17·64	17·60	17·47	17·19	17·44	17·20	17·57	17·41	17·56	17·44
1000	17·42	17·62	17·46	17·58	17·77	17·59	17·51	17·56	17·66	17·76	17·59
1100	17·76	17·85	17·73	17·76	17·56	18·00	17·72	17·73	17·79	17·83	17·77
1200	17·44	17·56	17·56	17·51	17·64	17·87	17·75	17·80	17·74	17·45	17·63
1300	17·64	17·65	17·71	17·74	17·83	18·04	18·05	17·85	18·11	18·05	17·87
1400	18·06	18·26	18·18	18·47	18·37	18·06	18·12		(18·0)	(17·6)	(18·1)
1500	(18·15)	18·25	(18·1)						(18·7)	18·88	(18·5)
1600	18·68	18·77	(19·7)								
1700	(18·89)	(18·02)	19·09	19·37	19·65	19·36	19·26	19·32	18·79	18·63	19·04
1800		(19·1)	(18·76)	19·63	19·36	19·71	20·00	19·36	19·68		(19·45)
1900	For later years consult the Nile flow data in *World Weather Records*.										

Remarks as for Table App. V.30 giving the low stage of the Nile. See Chapter 13 as regards allowance needed for progressive silting of the river bed.

Some details of the earliest records of the River Nile between about 3000 and 1200 B.C. are given in Chapter 13 on pp. 140–1. Further work by BELL (1975) on ancient texts has produced more information on both the annual flood levels and the annual low water stage of the river since about 2000 B.C.

Some years with exceptional high floods did occur between 1991 and about 1570 B.C. (i.e. during and after the Middle Kingdom time), and there was an abnormally great year-to-year variability (particularly around 1870 B.C.). It is not clear whether the Middle Kingdom high flood years represented a return to the pattern of early Neolithic times before 3000 B.C., with the monsoon reaching farther north than now, or resembled the much later Little Ice Age patterns with occasional 'cloudbursts' only – though it seems likely that big floods of the Nile necessarily involved the monsoon. BELL deduces prevalence of low floods in the years about 1768–1740 B.C. and perhaps again in the mid 1600s B.C. (though this last suggestion is only based on the fact that Egypt's fortunes were low at that time). One last great northward penetration of the monsoon, with a rainy season in Nubia 'so that the hills glistened' is recorded in 683 B.C. There may have been another flood of similar magnitude in one year about a century earlier.

Abnormally prolonged high levels of the River Nile were recorded in A.D. 1371 and in 1817 and 1818. An account of A.D. 1835 indicates that showers were expected four or five times a year at Thebes and that once every eight or ten years cloudbursts filled the torrent beds from the mountains: characteristically these events are produced by intense thunderstorms in warm, moist air from the Red Sea. Such an event occurred in December 1843 at Asswan.

Table App. V.32 Seasonal and yearly rainfall in southeast England, at Kew (near London), from 1697, as percentages of the 1916–50 averages

From figures given by WALES-SMITH (1971).

Winters include the December of the previous year.

The monthly values of Kew rainfall gave a correlation coefficient $r = +0.789$ when compared with the England and Wales series from 1727 to 1971 (significant at the 1% level). They may therefore be used to extend the latter series farther back in time, usually correctly indicating whether any particular season was wet, middling or dry (an overlap with the England and Wales rainfall table shows how the two compare; where discrepancies do occur this may be due either to heavier rain on the western hills or thundery rainfall in the southeast).

The months which comprise the seasons for the values given in this table for the years 1697–1749 are the months of the Old Style calendar.

Year	Winter DJF	Spring MAM	Summer JJA	Autumn SON	Calendar year J–D	Year	Winter DJF	Spring MAM	Summer JJA	Autumn SON	Calendar year J–D
1697	(34)	58	86	80	71	1725	37	101	117	75	83
1698	81	127	112	137	111	1726	106	77	119	108	112
1699	78	51	67	77	69	1727	138	127	58	58	88
1700	65	69	68	121	86	1728	107	149	169	91	122
1701	127	63	75	79	85	1729	43	83	82	110	85
1702	99	85	85	98	93	1730	62	118	96	91	89
1703	96	164	112	103	109	1731	42	33	95	51	57
1704	31	116	66	67	72	1732	65	140	57	71	82
1705	48	55	67	78	77	1733	61	87	115	42	79
1706	151	111	111	112	111	1734	92	126	95	73	102
1707	67	49	81	88	74	1735	143	107	104	78	96
1708	107	108	108	41	87	1736	130	81	92	82	104
1709	113	122	121	122	121	1737	117	113	120	124	110
1710	95	85	84	84	84	1738	72	120	101	144	111
1711	97	109	107	108	107	1739	139	139	90	136	128
1712	106	109	108	109	108	1740	49	91	87	86	87
1713	169	88	127	54	106	1741	108	73	71	124	82
1714	52	62	58	45	51	1742	79	62	64	109	79
1715	42	131	194	104	118	1743	39	99	84	46	68
1716	48	71	51	98	72	1744	64	128	73	175	106
1717	66	114	89	119	95	1745	28	124	130	104	104
1718	65	104	56	119	94	1746	122	99	99	74	96
1719	90	81	89	70	74	1747	115	57	82	75	103
1720	68	118	126	103	107	1748	142	116	76	47	84
1721	70	104	95	131	98	1749	192	84	53	71	87
1722	43	95	120	60	78						
1723	65	45	67	49	61						
1724	99	79	112	66	87						

Table App. V.33 Seasonal and yearly rainfall averages for England and Wales from 1727, as percentages of the 1916–50 mean values listed at the end

Averages of the monthly percentages indicated by Nicholas and Glasspoole (1931) and by Meteorological Office data since then.

Note: This 'England and Wales' series has been criticized in several respects – in particular that Wales is probably underrepresented and contributed no data in the eighteenth century – but at the time of writing it is the best general indication for England that has been published. It is at least partly vindicated by the high correlation coefficient (see Table App. V.32 heading) connecting it with the Kew record, which was not available when it was constructed; this, however, also suggests that it is more representative of the eastern and inland districts than of the west.

Another criticism relates to the fact that most, if not all, of the eighteenth century raingauges of which accounts exist were exposed either on a roof or on the top of a high wall, sometimes at the junction of two walls. All such exposures obtain a significantly smaller catch of rain than modern gauge-exposures near the ground. The authors of the 'England and Wales' series below have not made clear whether they made any adjustment for this. It seems unlikely, however, that the deficiency of rainfall apparent in this series in the eighteenth century, and particularly in the years before 1750, is entirely – and perhaps it is not to any significant extent – to be explained in this way. The apparent increasing trend of rainfall indicated continued from that time right up to 1910, although some raingauges were installed on the ground from 1770 onwards, and seems to have been paralleled by an increasing trend in much of Europe and western-central Asia through the nineteenth century, as well as in the measured accumulation of snow on the ice cap in northern Greenland. In principle, therefore, it seems likely to be real. Moreover, the wind circulation patterns over Britain derived for the eighteenth century decades mentioned suggest a higher frequency of anticyclonic conditions than in this century.

Winters are dated by the year in which the January fell; the December is that of the year before. The dates are those of the modern (Gregorian) calendar.

Year	Winter DJF	Spring MAM	Sum-mer JJA	Autumn SON	Calendar year J–D	Year	Winter DJF	Spring MAM	Sum-mer JJA	Autumn SON	Calendar year J–D
1727	(110)	134	90	56	95	1745	48	137	130	83	98
1728	123	135	101	85	106	1746	93	84	97	87	86
1729	53	95	85	151	98	1747	100	101	72	85	99
						1748	84	120	115	41	81
1730	67	110	93	86	79	1749	132	97	83	51	84
1731	47	40	103	59	63						
1732	75	130	62	83	89	1750	74	81	66	68	71
1733	77	129	117	52	80	1751	84	151	122	91	107
1734	108	128	114	76	112	1752	91	98	132	35	88
						1753	103	71	111	85	101
1735	132	103	118	85	95	1754	92	88	129	45	87
1736	117	82	126	81	106						
1737	96	98	112	83	92	1755	82	101	111	109	92
1738	69	79	89	70	71	1756	60	137	155	60	95
1739	104	102	104	70	95	1757	52	127	127	63	95
						1758	90	68	143	75	94
1740	39	50	102	86	78	1759	55	108	126	83	84
1741	90	26	63	119	66						
1742	55	63	75	108	77						
1743	36	79	84	61	68						
1744	55	111	92	135	94						

Table App. V.33 – continued

Year	Winter DJF	Spring MAM	Sum- mer JJA	Autumn SON	Calendar year J–D	Year	Winter DJF	Spring MAM	Sum- mer JJA	Autumn SON	Calendar year J–D
1760	83	46	113	128	98	1795	74	80	104	95	90
1761	77	70	118	116	92	1796	102	79	81	79	83
1762	87	66	84	118	87	1797	52	104	140	111	105
1763	61	88	181	96	118	1798	87	60	102	108	86
1764	159	78	123	79	103	1799	85	83	127	123	101
1765	67	120	75	98	85	1800	48	115	34	135	90
1766	49	99	106	77	79	1801	75	86	82	126	95
1767	103	92	105	86	93	1802	95	57	126	73	86
1768	109	68	164	149	129	1803	79	79	87	63	82
1769	89	65	120	85	85	1804	115	106	91	87	92
1770	71	100	105	130	103	1805	77	84	108	46	80
1771	78	63	77	93	79	1806	104	65	110	93	102
1772	102	90	116	137	104	1807	98	93	79	102	84
1773	67	115	76	138	106	1808	53	81	97	109	87
1774	117	96	115	88	97	1809	101	76	111	72	92
1775	99	63	144	127	107	1810	74	86	104	105	99
1776	91	58	134	91	93	1811	104	107	94	90	92
1777	59	101	127	91	91	1812	101	117	107	98	96
1778	59	86	102	118	101	1813	72	96	75	103	86
1779	59	76	114	110	91	1814	55	94	103	94	98
1780	83	95	53	108	71	1815	93	123	95	91	92
1781	63	55	105	84	79	1816	88	99	119	99	102
1782	72	169	123	99	110	1817	106	86	149	51	99
1783	89	83	112	96	91	1818	117	171	37	103	94
1784	59	99	133	59	86	1819	103	102	81	83	97
1785	59	27	109	113	83	1820	88	97	96	81	86
1786	81	72	97	111	94	1821	56	126	85	131	109
1787	82	92	115	103	101	1822	101	104	106	107	95
1788	93	51	97	60	63	1823	103	89	122	99	110
1789	83	92	134	112	111	1824	86	113	103	147	114
1790	76	76	97	85	90	1825	94	106	72	111	92
1791	140	76	93	101	102	1826	102	68	54	107	80
1792	88	149	123	109	115	1827	68	123	76	113	101
1793	110	77	80	83	84	1828	125	104	148	80	111
1794	90	95	67	140	97	1829	67	85	168	94	93

Table App. V.33 – continued

Year	Winter DJF	Spring MAM	Sum- mer JJA	Autumn SON	Calendar year J–D	Year	Winter DJF	Spring MAM	Sum- mer JJA	Autumn SON	Calendar year J–D
1830	56	107	136	105	101	1865	101	88	116	108	107
1831	88	104	107	127	110	1866	116	81	114	114	107
1832	74	100	129	93	97	1867	103	122	93	68	93
1833	121	79	100	87	101	1868	93	93	61	92	98
1834	135	57	133	56	85	1869	152	109	48	110	98
1835	81	106	71	132	93	1870	95	68	56	99	79
1836	68	120	102	133	109	1871	88	90	110	107	93
1837	104	58	92	78	85	1872	118	116	140	145	139
1838	70	81	115	97	86	1873	120	89	98	84	86
1839	72	77	148	133	111	1874	68	69	76	115	91
1840	105	56	89	106	85	1875	106	63	126	148	111
1841	65	101	131	156	121	1876	83	107	88	124	113
1842	87	90	90	113	91	1877	164	129	122	113	122
1843	84	131	117	95	101	1878	81	122	124	109	106
1844	80	61	92	100	84	1879	106	99	186	67	105
1845	52	86	120	86	93	1880	63	80	121	140	109
1846	95	121	105	103	101	1881	112	92	135	96	105
1847	63	102	74	99	92	1882	97	117	141	131	123
1848	132	118	157	112	122	1883	136	74	105	129	103
1849	92	102	77	97	91	1884	95	82	82	65	85
1850	85	91	102	81	86	1885	113	101	67	136	98
1851	84	104	107	60	83	1886	73	139	81	111	113
1852	104	55	166	167	137	1887	93	73	51	90	72
1853	106	83	136	94	94	1888	61	105	131	84	95
1854	61	66	85	66	74	1889	72	131	86	85	91
1855	58	78	128	88	85	1890	77	87	117	80	87
1856	75	99	91	78	87	1891	38	96	131	115	107
1857	82	100	108	101	92	1892	97	72	115	112	91
1858	41	94	84	75	76	1893	93	38	82	81	80
1859	76	100	103	113	97	1894	114	93	117	99	104
1860	97	111	169	93	115	1895	76	81	105	92	93
1861	89	81	106	98	87	1896	57	75	90	122	90
1862	62	148	108	91	100	1897	115	122	103	80	98
1863	79	62	106	109	88	1898	79	101	77	85	84
1864	65	94	51	92	74	1899	114	102	63	93	91

Table App. V.33 – continued

Year	Winter DJF	Spring MAM	Summer JJA	Autumn SON	Calendar year J–D	Year	Winter DJF	Spring MAM	Summer JJA	Autumn SON	Calendar year J–D
1900	140	65	112	84	105	1935	138	86	92	155	109
1901	85	92	88	73	85	1936	123	79	126	97	105
1902	88	93	98	74	81	1937	153	135	68	75	106
1903	92	136	133	140	124	1938	95	49	101	112	95
1904	114	93	85	56	86	1939	129	88	116	111	109
1905	61	111	107	87	83	1940	90	98	62	124	98
1906	97	89	75	103	98	1941	113	106	98	68	93
1907	70	110	105	91	96	1942	73	112	74	80	91
1908	90	128	94	67	88	1943	117	74	90	91	90
1909	57	116	120	94	101	1944	71	57	99	139	97
1910	139	90	119	85	109	1945	96	84	103	71	90
1911	99	78	74	100	91	1946	112	85	138	120	114
1912	140	101	186	84	121	1947	94	170	73	54	89
1913	104	150	57	99	95	1948	118	83	119	79	103
1914	77	107	98	82	105	1949	77	94	52	115	85
1915	178	77	97	70	106	1950	117	89	119	123	110
1916	155	122	98	110	110	1951	121	150	92	112	120
1917	75	99	138	91	95	1952	82	111	87	113	97
1918	79	82	92	126	103	1953	73	89	109	85	82
1919	126	118	79	75	101	1954	79	92	136	138	117
1920	118	155	113	71	105	1955	96	103	75	70	85
1921	79	70	57	59	68	1956	99	51	145	72	94
1922	117	99	115	65	102	1957	115	69	112	102	97
1923	142	100	81	117	109	1958	123	89	147	101	114
1924	86	127	122	120	116	1959	82	95	69	75	88
1925	134	106	69	104	102	1960	147	77	123	168	130
1926	119	85	106	109	99	1961	120	83	85	98	97
1927	82	99	155	124	119	1962	94	99	81	83	87
1928	127	89	117	103	111	1963	56	129	116	106	94
1929	68	52	83	121	97	1964	36	125	89	56	78
1930	129	108	110	127	113	1965	81	111	113	112	110
1931	108	108	147	89	105	1966	142	114	127	99	113
1932	55	145	85	116	100	1967	107	131	83	129	109
1933	91	91	68	82	78	1968	82	109	127	124	108
1934	46	104	77	80	92	1969	101	139	96	73	101

Table App. V.33 – continued

Year	Winter DJF	Spring MAM	Sum- mer JJA	Autumn SON	Calendar year J–D	Year	Winter DJF	Spring MAM	Sum- mer JJA	Autumn SON	Calendar year J–D
1970	114	99	90	110	100	1976	62	67	35	155	83
1971	77	96	128	70	91	1977	146				
1972	89	121	86	66	97	1916–50					
1973	76	95	103	77	84	Averages					
1974	118	56	112	143	108	(mm)	252	184	221	269	926
1975	86	114	61	85	85						

Table App. V.34 Seasonal and yearly rainfall (mm) in the eastern United States, reduced to Philadelphia equivalent values, from 1738

Total downput, including rainfall equivalent of snow.

From figures provided by Landsberg *et al.* (1968).

Year	Winter DJF	Spring MAM	Sum- mer JJA	Autumn SON	Calendar year J–D	Year	Winter DJF	Spring MAM	Sum- mer JJA	Autumn SON	Calendar year J–D
1738		277	307	246	1072	1760	211	239	307	244	1059
1739	229	282	371	244	1125	1761	221	251	302	236	1069
						1762	226	236	292	231	1041
1740	234	274	302	236	1041	1763	211	251	312	234	1029
1741	234	284	307	244	1079	1764	211	262	320	239	1067
1742	221	284	292	236	1029						
1743	224	274	297	239	1041	1765	213	267	317	234	1072
1744	226	269	315	239	1069	1766	185	264	340	226	1029
						1767	188	267	297	246	1049
1745	249	277	343	234	1074	1768	208	188	325	231	1026
1746	224	269	323	234	1039	1769	203	234	254	239	1003
1747	231	262	297	249	1057						
1748	224	267	315	249	1079	1770	203	208	356	241	1044
1749	234	292	328	234	1087	1771	218	312	290	229	1062
						1772	203	241	340	274	1077
1750	216	272	358	231	1072	1773	196	218	269	221	1008
1751	246	254	391	236	1095	1774	211	244	297	239	1029
1752	211	254	315	236	1044						
1753	221	262	345	257	1077	1775	178			239	
1754	226	231	361	231	1067	1776				218	
						1777	188	229	320		1019
1755	224	264	310	234	1041	1778	198	221	335		1026
1756	206	254	302	234	1016	1779	201				
1757	226	259	307	229	1052						
1758	229	239	381	231	1054						
1759	221	246	351	249	1052						

Table App. V.34 – continued

Year	Winter DJF	Spring MAM	Sum- mer JJA	Autumn SON	Calendar year J–D	Year	Winter DJF	Spring MAM	Sum- mer JJA	Autumn SON	Calandar year J–D
1780						1815	201	269	338	239	1064
1781						1816	254	198	244	246	1011
1782		287	295	236	1041	1817	173	188	384	224	1047
1783	216	196	394	274	1079	1818	180	330	267	229	1031
1784	221	284	363	277	1123	1819	173	300	272	206	1001
1785	249	302	323	279	1123	1820	224	267	315	246	1074
1786	224	297	300	229	1064	1821	239	251	267	279	1077
1787	236	257	300	259	1090	1822	211	178	277	229	935
1788	234	269	302	246	1069	1823	198	300	282	239	1072
1789	226	257	305	246	1034	1824	292	254	287	216	1036
1790	224	259	279	241	1008	1825	208	206	328	206	1019
1791	231	264	345	257	1128	1826	203	216	371	234	1044
1792		262	249	239	998	1827	231	284	328	279	1107
1793	196	259	231	224	1006	1828	231	290	287	274	1039
1794	203	196	290	211	1001	1829	231	274	315	234	1057
1795	188	297	351	241	1107	1830	170	264	264	272	1097
1796	216	249	292	229	1039	1831	290	282	320	262	1095
1797	216	284	320	234	1067	1832	234	277	297	216	1062
1798	211	277	305	234	1047	1833	234	244	305	290	1107
1799	206	264	333	234	1085	1834	213	244	317	234	1021
1800	213	302	295	244	1057	1835	208	259	333	203	1047
1801	224	284	325	221	1047	1836	295	216	302	234	1049
1802	201	274	302	226	1029	1837	224	302	305	201	1087
1803	226	211	315	211	1026	1838	216	262	295	305	1095
1804	224	259	264	249	1044	1839	170	264	307	213	1077
1805	262	213	259	241	1031	1840	251	287	333	259	1082
1806	188	170	328	249	1016	1841	234	292	340	246	1138
1807	284	295	312	211	1054	1842	226	279	368	221	1092
1808	224	300	328	231	1067	1843	216	287	310	251	1064
1809	241	234	357	231	1090	1844	206	246	302	254	1031
1810	216	229	351	241	1039	1845	246	221	323	236	1052
1811	216	244	328	241	1074	1846	246	262	295	239	1077
1812	295	196	345	236	1054	1847	262	236	320	267	1123
1813	201	351	333	244	1115	1848	224	267	284	231	1031
1814	246	290	378	234	1087	1849	196	267	292	251	1036

Table App. V.34 – continued

Year	Winter DJF	Spring MAM	Sum-mer JJA	Autumn SON	Calendar year J–D	Year	Winter DJF	Spring MAM	Sum-mer JJA	Autumn SON	Calandar year J–D
1850	239	310	340	267	1158	1885	259	201	323	249	1029
1851	246	295	274	231	1016	1886	269	284	290	221	1041
1852	193	290	312	234	1079	1887	251	231	353	213	1062
1853	264	297	323	257	1090	1888	269	300	300	290	1113
1854	216	340	251	236	1052	1889	239	282	378	292	1141
1855	262	218	335	251	1092	1890	196	290	300	257	1052
1856	241	257	335	221	1034	1891	295	257	315	213	1064
1857	221	317	361	206	1128	1892	236	267	295	218	1016
1858	249	251	323	229	1067	1893	239	292	292	229	1059
1859	300	310	351	269	1176	1894	221	259	246	267	1036
1860	224	213	333	262	1074	1895	221	267	282	226	991
1861	244	297	279	262	1107	1896	229	241	292	216	1001
1862	269	239	323	269	1110	1897	191	262	386	221	1092
1863	249	302	325	231	1133	1898	262	274	330	267	1102
1864	224	328	254	251	1087	1899	257	254	297	229	1031
1865	262	317	302	254	1151	1900	226	244	277	241	1029
1866	249	249	310	251	1077	1901	155	356	325	216	1090
1867	231	295	445	221	1168	1902	305	246	302	264	1110
1868	201	325	274	292	1158	1903	290	231	363	221	1072
1869	262	279	290	284	1115	1904	198	254	307	244	1034
1870	305	290	284	224	1057	1905	213	229	345	229	1041
1871	198	300	343	249	1115	1906	208	312	361	226	1110
1872	160	249	391	267	1100	1907	213	257	282	297	1090
1873	257	282	351	269	1153	1908	259	272	312	183	1016
1874	229	292	330	234	1087	1909	241	251	282	208	1019
1875	196	254	345	241	1059	1910	279	221	292	221	1006
1876	208	333	310	279	1110	1911	183	208	366	257	1095
1877	183	274	315	305	1087	1912	208	335	274	234	1077
1878	244	287	340	231	1110	1913	226	307	277	259	1085
1879	244	264	371	185	1049	1914	226	257	282	185	1029
1880	236	241	330	211	1019	1915	335	211	333	193	1082
1881	292	272	284	213	1029	1916	244	262	310	201	1008
1882	292	249	277	277	1079	1917	211	259	317	226	1026
1883	239	231	297	229	1036	1918	206	277	302	216	1024
1884	300	272	338	198	1077	1919	236	295	338	236	1102

Table App. V.34 – continued

Year	Winter DJF	Spring MAM	Summer JJA	Autumn SON	Calendar year J–D	Year	Winter DJF	Spring MAM	Summer JJA	Autumn SON	Calandar year J–D
1920	226	282	333	229	1113	1945	226	254	363	254	1115
1921	236	254	312	216	1019	1946	224	262	351	198	1026
1922	191	267	340	183	1013	1947	193	297	317	244	1059
1923	249	241	284	234	1041	1948	236	333	325	226	1118
1924	264	302	315	216	1067	1949	300	257	269	216	1024
1925	211	234	292	234	1006	1950	221	244	317	196	1052
1926	229	208	305	257	1067	1951	246	282	282	257	1069
1927	216	218	351	267	1077	1952	257	351	328	221	1100
1928	234	257	363	206	1052	1953	257	394	290	231	1118
1929	211	297	272	246	1047	1954	206	287	279	246	1016
1930	208	234	282	203	970	1955	196	239	386	254	1036
1931	183	284	338	191	1026	1956	198	272	305	249	1077
1932	221	269	292	274	1074	1957	208	251	251	236	1008
1933	198	343	335	229	1128	1958	302	315	338	239	1110
1934	224	277	290	264	1059	1959	168	236	356	241	1054
1935	241	208	302	262	1054	1960	244	251	328	254	1072
1936	244	264	320	213	1072	1961	221	312	305	208	1054
1937	297	267	338	257	1077	1962	229	244	328	241	1057
1938	193	234	351	269	1087	1963	193	241	287	257	1001
1939	269	264	302	216	1059	1964	231	229	262	196	980
1940	185	343	297	241	1074	1965	241	218	274	203	955
1941	206	206	333	185	1001	1966	218	234	246	282	1039
1942	213	262	373	236	1090	1967	168	267	335	183	1013
1943	229	284	272	241	1013	1968					
1944	180	282	262	277	1049	1969					

Table App. V.35   Dates of establishment of the Indian summer monsoon and of the Bai-U rains at Tokyo

Counted from 1 May = 1, 1 June = 32, etc.

Year	Dates of onset of Indian monsoon			Bai-U at Tokyo $(35\frac{1}{2}°N)$	Year	Dates of onset of Indian monsoon			Bai-U at Tokyo $(35\frac{1}{2}°N)$
	Travancore (Cochin: 10°N)	Kolaba (18–19°N)	Delhi (28$\frac{1}{2}$°N)			Travancore (Cochin: 10°N)	Kolaba (18–19°N)	Delhi (28$\frac{1}{2}$°N)	
1891	27	52		32	1920	27	37	54	− 1
1892	22	31		22	1921	32	43	74	30
1893	22	41		17	1922	25	43	55	12
1894	32	38		7	1923	35	44	69	14
					1924	31	43	74	19
1895	39	46		25					
1896	30	32		19	1925	27	29	49	10
1897	30	38		41	1926	28	41	73	12
1898	33	39		41	1927	23	41	61	53
1899	23	41		22	1928	31	38	63	24
					1929	29	37	59	12
1900	37	40		35					
1901	32	38	69	4	1930	21	40	61	12
1902	31	43	51	12	1931	23	45	64	33
1903	39	43	74	19	1932	14	34	61	12
1904	29	39	65	46	1933	22	32	52	17
					1934	37	41	50	28
1905	37	41	72	41					
1906	34	39	56	39	1935	41	45	72	26
1907	31	42	81	35	1936	20	32	53	15
1908	39	42	62	27	1937	34	43	57	32
1909	32	34	59	11	1938	32	35	53	46
					1939	37	40	56	1
1910	28	34	60	28					
1911	34	35	73	18	1940	38	49	60	36
1912	35	43	69	22	1941	23	47	71	32
1913	24	38	56	25	1942	35	44	56	37
1914	28	44	64	28	1943	12	21	66	14
					1944	29	41	66	44
1915	32	49	64	8					
1916	26	32	70	17	1945	32	44	59	18
1917	26	36	63	34	1946	29	33	60	17
1918	7	25	68	17	1947	31	39	78	29
1919	16	37	65	40	1948	25	39	70	32
					1949	23	25	70	23

Table App. V.35 – continued

Year	Dates of onset of Indian monsoon			Bai-U at Tokyo $(35\frac{1}{2}°N)$	Year	Dates of onset of Indian monsoon			Bai-U at Tokyo $(35\frac{1}{2}°N)$
	Travan-core (Cochin: 10°N)	Kolaba (18–19°N)	Delhi $(28\frac{1}{2}°N)$			Travan-core (Cochin: 10°N)	Kolaba (18–19°N)	Delhi $(28\frac{1}{2}°N)$	
1950	26	41	70	24	1965	(32)			
1951	(32)				1966	(33)			
1952	(19)				1967				
1953	(37)				1968				
1954	(27)				1969				
1955	(28)				Aver-				
1956	(20)				age				
1957	(31)				1891–				
1958	(41)				1950	29	39	63	24
1959	(31)				Stan-				
1960	(16)				dard				
1961	(17)				devia-				
1962	(15)				tion				
1963	(31)				(days)	7·0	6·0	7·8	11·6
1964	(33)								

Sources: ARAKAWA (1959), BHULLAR (1952), RAMDAS *et al.* (1954), de la MOTHE and WRIGHT (1969).

Table App. V.36   Dates of onset of the dry season in western Java, 1915–40
Counted from 1 April = 1, 1 May = 31, etc.

Decade	Year									
	0	1	2	3	4	5	6	7	8	9
1910						44	18	27	12	38
1920	18	16	36	42	30	−2	20	34	22	1
1930	30	36	36	37	19	19	27	44	44	37
1940	39									

Average										
1915–40	28				Standard deviation 12·5 days					

Source: ARAKAWA (1959).

Table App. V.37   Yearly monsoon season (June–September) rainfall (mm) at Bombay (Kolaba) 18°54′N 72°49′E, 11·3 m above sea level, from 1817
On average 95% of the year's rain falls in these months.

Decade beginning	Year										Decade average
	0	1	2	3	4	5	6	7	8	9	
1810								2631	2008	1979	
1820	1964	2098	2850	1567	803	1835	1946	2035	2936	1651	1969
1830	1825	2526	1865	1808	1691	1580	2223	1653	1290	1870	1833
1840	1550	1734	2306	1499	1661	1390	2193	1737	1615	2875	1856
1850	1146	2424	1721	1589	1849	1044	1617	1217	1461	1944	1601
1860	1491	1889	1822	1899	1145	1782	1971	1450	1574	2233	1726
1870	1562	859	1914	1754	2086	2127	1270	1628	2672	1416	1729
1880	1695	1735	1725	1993	1815	1634	2450	2308	1386	1601	1834
1890	1604	1925	2348	1491	1612	1619	2206	2004	1858	870	1754
1900	1756	1883	1787	1794	833	821	1426	2534	1339	1809	1598
1910	1610	1172	1253	1743	2065	1870	2041	1964	584	1689	1599
1920	1001	2179	1769	1968	1314	1206	1734	1561	2164	1351	1625
1930	2221	2077	1626	1411	1772	2042	1415	1737	1890	1372	1756
1940	2300	867	2069	1809	1855	2248	1804	1820	1645	2697	1911
1950	1522	1667	1551	2291	3423	2079	2445	1705	3310	2103	2210
1960	1833										
1970											

Source: World Weather Records.

Table App. V.38   Yearly rainfall totals (mm) at Calcutta (Alipore), 22°32′N 88°24′E, 6·4 m above sea level, from 1829

On average (1829–1920) 91 % of the year's rain falls in May to October, 75 % between June and September.

Decade beginning	Year										Decade average
	0	1	2	3	4	5	6	7	8	9	
1820										1522	
1830	1607	1445	1288	1538	1746	2172		1107	1346	1650	1544
1840	1509	1530	1933	1634	1876	1546	1941	1864	1491	1791	1711
1850	1937	1630	2068	1323	1688	1787	1631	1752	1518	1744	1708
1860	1336	2266	1866	1553	2139	1564	1670	1847	2324	1575	1814
1870	1530	2370	1297	1150	1562	1521	2038	1609	1487	1105	1567
1880	1764	1767	1681	1335	1590	1695	1656	1475	1770	1460	1619
1890	1556	1169	1186	2165	1236	1000	1352	1482	1512	1827	1449
1900	2269	1781	1582	1375	1605	1772	1453	1362	2051	1834	1708
1910	1454	1101	1443	2193	1454	1675	2103	1795	1483	1618	1632
1920	1620	1437	2111	1435	1296	1505	1912	1164	1998	1515	1599
1930	1521	1661	1598	2080	1382	909	1784	1915	1209	1989	1605
1940	1575	2221	1493	1572	1480	1365	1645	1702	1714	1651	1642
1950	1666	1582	1547	1252	1213	1384	1753	1295	1177	2006	1487
1960	1626										
1970											

Source: World Weather Records.

Table App. V.39   Yearly rainfall (total downput of rain and snow, etc.) in mm at Seoul, Korea ($37\frac{1}{2}°$N 127°E), from 1771

On average about 77% of the year's rain falls in the months May to September.

Decade beginning	Year										Decade average
	0	1	2	3	4	5	6	7	8	9	
1770		1130		1387	870		1044	867	693	1617	(1087)
1780	1632	1458	1396	1301	2213	865	1029	2852	1809	1993	1655
1790	1407	2479	1872	1296	1532	911	2164	1356	1432	1524	1597
1800	2118	1319	1929	1222	1759	2333	2752	1175	1418	1481	1751
1810	1965	1733	1634	1777	1589	1569	2324	2219	1757	1692	1826
1820	1821	3186	1278	1594	1834	1045	1392	2516	2572	1432	1867
1830	1506	1525	2744	2172	1129	1745	1026	1272	1329	3220	1767
1840	1336	1355	1605	1804	1453	1881	2329	2670	1401	1683	1752
1850	1800	2226	1582	1144	1875	1754	1749	2415	1368	1578	1749
1860	896	1182	1590	1681	1600	2330	1302	1823	2228	1143	1577
1870	1343	1959	1555	1801	2103	2308	879	2160	1477	3148	1873
1880	1483	1387	1157	1164	1684	1475	1364	758	843	944	1226
1890	1105	1654	1138	1204		1058	1315	1832	1301	1097	1300
1900	960	685	1233	1557	800	1512	1037	862	1066	827	1054
1910	1021	1067	1129	917	1263	1578	1710	982	1132	1196	1199
1920	1674	972	1553	1195	1001	1934	1833	1270	1054	1130	1362
1930	1627	1388	941	1401	1245	1202	1532	1063	1059	639	1210
1940	2135	1094	1131	647	1089	1562	1508	1496	1475	634	1277
1950				1382	1436	1231	1701	1220	1325	1371	(1380)
1960	1188										

Source: Seoul Monthly Precipitation Records (1770–1960). Seoul (Central Meteorological Office, Republic of Korea), 1961.

Note: It seems possible that the annual values of rainfall at Seoul have a strong negative correlation with the zonal index for W'ly winds in the east Asian sector and could be used to indicate probable values of the latter.

Table App. V.40   Yearly rainfall totals (mm) Alice Springs, central Australia, 23°48's 133°53'E, 587 m above sea level, from 1874

On average (1873–1924) 78% of the year's rain falls in the months October to April.

Decade beginning	Year										Decade average
	0	1	2	3	4	5	6	7	8	9	
1870					234	399	159	518	276	691	(379)
1880	168	163	271	146	137	437	297	263	256	174	231
1890	365	265	214	155	506	360	265	145	260	166	270
1900	147	196	138	405	335	244	305	249	448	205	267
1910	464	180	227	214	267	110	346	230	107	297	244
1920	726	538	325	370	137	232	214	199	60	142	294
1930	281	160	237	257	152	155	264	231	216	419	237
1940	271	264	245	263	304	270	299	487	269	330	300
1950	383	154	264	351	251	267	228	318	186	177	258
1960	156										
1970											

Source: World Weather Records.

Table App. V.41 Yearly rainfall totals (mm) at Adelaide, South Australia, 34°56's 138°35'E, 43 m above sea level, from 1839

On average (1839–1924) 78% of the year's rain falls in the months April to October.

Decade beginning	Year										Decade average
	0	1	2	3	4	5	6	7	8	9	
1830										504	
1840	615	456	516	437	429	478	683	701	501	646	546
1850	497	809	697	688	390	588	633	537	547	377	576
1860	500	610	555	601	502	394	511	484	508	374	504
1870	606	591	576	533	438	742	341	634	561	526	555
1880	571	457	399	680	476	404	366	653	370	784	516
1890	655	356	547	546	528	541	385	392	527	479	496
1900	551	457	407	647	516	566	673	452	624	703	560
1910	625	406	497	462	289	492	715	734	442	437	510
1920	678	575	589	757	595	419	564	430	494	445	555
1930	461	565	636	562	514	596	491	584	489	592	549
1940	410	573	646	453	435	453	548	556	544	463	508
1950	408	648	507	509	425	623	691	426	446	289	497
1960	587										
1970											

Source: World Weather Records.

Table App. V.42   Yearly rainfall totals (mm) at Cape Town, South Africa, 33°56′s 18°29′E, 12·2 m above sea level, from 1838

On average (1841–1924) 68 % of the year's rain falls between May and September, 84 % between April and October.

Decade beginning	Year										Decade average
	0	1	2	3	4	5	6	7	8	9	
1830									621	444	
1840		636	667	630	477	531	571	568	591	625	(588)
1850	850	515	589	539	509	623	495	560	616	933	623
1860	740	647	813	650	480	474	488	583	507	821	620
1870	712	511	745	604	665	653	677	904	1042	476	699
1880	450	650	744	814	719	709	706	586	916	787	708
1890	669	770	1039	595	558	588	470	510	605	680	648
1900	540	652	857	760	808	766	515	505	628	610	664
1910	548	696	559	609	636	660	538	643	543	497	593
1920	685	656	485	708	478	620	504	482	423	496	554
1930	418	485	603	459	486	557	513	712	639	491	536
1940	614	808	693	546	849	637	573	574	586	521	640
1950	676	713	629	616	836	609	734	856	488	607	676
1960	428										
1970											

Source: World Weather Records.

Table App. V.43 Yearly rainfall totals (mm) at Buenos Aires, Argentina, 34°36's 58°22'w, 25 m above sea level, from 1861

Changes of site in the 1930s are not thought to have made a very material difference.

Decade beginning	Year										Decade average
	0	1	2	3	4	5	6	7	8	9	
1860		584	1061	685	744	775	865	583	1145	1171	(846)
1870	837	748	778	778	957	937	872	994	1131	631	866
1880	901	1046	949	1150	1105	1029	915	708	1089	1278	1017
1890	831	954	701	547	881	1454	759	845	1004	1021	900
1900	2023	891	790	1042	793	1060	771	685	760	799	961
1910	666	1230	1504	1137	1741	928	504	854	767	1389	1072
1920	935	934	1193	976	654	1239	865	967	988	706	946
1930	1257	848	962	918	1054	766	1290	914	857	1115	998
1940	1432	945	746	858	1189	1227	1414	770	1007	710	1030
1950	886	1086	976	1276	931	886	853	946	1157	1858	1085
1960	932										
1970											

Source: World Weather Records.

Table App. V.44   Yearly rainfall in central Chile, since 1535

The figures in this table express the classification published by T A U L I S (1934) on the basis of an exhaustive survey of official documents, relating to winter rainfall in the Santiago district of Chile, rain which comes essentially from the travelling cyclonic depressions of the southern temperate zone when they reach farthest north. Equivalent yearly rainfall totals established for the years after 1849 when raingauges were available are:

Classification	*Year's rainfall* (mm)
1 = Very dry	< 100
	(or the rain coming only at the end of the winter)
2 = Dry	100–200
3 = Normal	200–500
4 = Wet	> 500
	(but falling on a small number of days)
5 = Very wet	> 500
	(well distributed throughout the winter)

The sources used begin with the conquest of Chile by the Spanish in 1535 and include reports about expeditions affected by swollen rivers in wet seasons and fording rivers even to the south of Santiago in dry winters, enactments about irrigation and water supply, reports of damage by flooded rivers to water mills or of no water for their working, harvests lost through lack of water, epidemics of plague and prayers for rain, etc., from a great miscellany of chronicles, memoirs and correspondence

Decade	*Year*										*Decade average*
	0	1	2	3	4	5	6	7	8	9	
1530						5					
1540		3	3	3	3	5			3	4	(3·4)
1550	3	4	4	3	3	3	3	3	3	3	3·2
1560	4							3	3	3	
1570					3	5	4	3	3	3	(3·5)
1580	3	3	5	3	3	3	3	3	3	3	3·2
1590	3	3	3	3	3	3					(3·0)
1600					3	3	3	3	4	3	(3·2)
1610	5	3	3	3	3	3	3	3	3	4	3·3
1620	2	3	3	3	3	3	3	2	3	1	2·6
1630	3	3	3	3	3	3	3	1	2	2	2·6
1640	2	3	3	3	2	3	3	3	5	4	3·1
1650	1	4	3	3	3	3	4	1	3	3	2·8
1660	3	2	3	3	3	1	3	3	3	3	2·7
1670	3	3	3	3	3	3	2	3	2	1	2·6
1680	3	3	3	1	4	3	3	4	4	4	3·2
1690	3	3	3	3	3	3	4	2	5	4	3·3

Table App. V.44 – continued

Decade	Year										Decade average
	0	1	2	3	4	5	6	7	8	9	
1700	3	3	3	3	3	3	4	3	3	3	3·1
1710	3	3	3	3	3	3	3	3	1	3	2·8
1720	3	3	3	3	4	3	3	3	3	3	3·1
1730	3	3	3	1	3	3	3	3	3	3	2·8
1740	3	3	3	3	3	4	3	5	3	4	3·4
1750	3	3	5	3	3	3	3	3	3	3	3·2
1760	3	3	3	3	3	5	3	3	3	4	3·3
1770	3	2	1	1	1	1	2	2	1	2	1·6
1780	2	2	2	1	5	3	3	3	3	3	2·7
1790	3	3	1	1	3	3	3	3	2	3	2·5
1800	3	2	2	2	2	3	3	3	3	3	2·6
1810	3	3	3	3	3	3	3	3	5	3	3·2
1820	4	4	4	2	2	3	3	2	5	4	3·3
1830	5	2	2	3	5	3	3	3	4	3	3·3
1840	3	3	4	3	5	3	5	3	3	2	3·4
1850	3	4	5	3	3	3	4	4	3	4	3·6
1860	3	4	3	3	1	5	3	2	3	5	3·2
1870	2	3	3	2	3	3	3	2	5	3	2·9
1880	2	5	3	3	3	3	3	1	4	5	3·2
1890	3	3	5	3	3	3	3	3	3	3	3·2
1900	5	5	3	4	2	5	5	3	3	3	3·8
1910	2	3	2	3	3	5	3	3	3	3	3·0
1920	5	3	3	3	3	1	3	5	3	3	3·2
1930	3	3	3	3	4	3	3	3	3	3	3·1
1940	3	5	3	3	3	3	2	3	3	3	3·1
1950	3	3	3	5	3	2	3	3	3	3	3·1
1960	2	3	3	3	3	3	3	2	1	2	2·5
1970	3	3	5	2							

Table App. V.45   Numbers of tropical hurricanes in the North Atlantic each year affecting coastal states of the U.S.A., from 1886

Decade beginning	Year										Decade average
	0	1	2	3	4	5	6	7	8	9	
1880							8	10	5	5	(7·0)
1890	1	8	5	10	5	2	6	2	4	5	4·8
1900	3	3	3	8	2	1	6	0	5	4	3·5
1910	3	3	4	3	0	4	11	2	3	1	3·4
1920	4	4	2	3	5	1	8	4	4	3	3·8
1930	2	2	6	10	6	5	7	6	3	3	5·0
1940	4	4	4	5	8	4	3	6	7	8	5·3
1950	11	8	6	6	8	9	4	3	7	8	7·0
1960	4	8	3	7	6	4	7	6	4	10	5·9
1970	3	5	3	4							(3·7)

Mainly from CRY *et al.* (1959) and more recent data published in the *Monthly Weather Review*, Washington.

Table App. V.46   Numbers of tropical storms (typhoons) reported in the western North Pacific each year, from 1884

Decade beginning	Year										Decade average
	0	1	2	3	4	5	6	7	8	9	
1880					13	7	14	20	13	12	(13·2)
1890	12	15	20	17	11	14	16	12	17	17	15·1
1900	16	7	14	15	16	20	17	20	15	22	16·2
1910	26	25	27	18	23	26	18	15	20	24	22·2
1920	19	25	23	20	24	20	16	21	20	18	20·6
1930	20	18	24	27	27	21	21	22	28	38	24·6
1940	39	27	26	32	19	33	23	38	30	30	29·7
1950	27	25	33	34	16	17	22	17	26	21	23·8
1960	30	30	31	27	43	40	28	31	24	15	29·9
1970	22	31									

Source: Tropical Cyclone Data 993, Data Processing Division, USAFETAC.
National Climatic Center NOAA, Environmental Data Service, Asheville, North Carolina, U.S.A.

Table App. V.47 Numbers of tropical cyclones reported in various regions, from the 1850s

Region	Decade averages												
	1850s	1860s	1870s	1880s	1890s	1900–9	1910–19	1920–9	1930–9	1940–9	1950–9	1960–9	1970–9
North Atlantic (U.S. seabord)[1,2]				7·0	4·8	3·5	3·4	3·8	5·0	5·3	7·0	5·9	(3·7)
Bay of Bengal (estimated from MILTON's (1974) graph)					6·7	6·3	4·8	7·5	8·2	9·0	7·5		
Japan (from 1912)							17·9	15·8	12·8	17·0	16·6		
SW Indian Ocean[3]	9·3	9·6	few	7·2	4·7	5·7	7·0	4·3	7·1	about 5	>6[4]		
N Australia–Timor Sea 10–15°S 120–145°E							0·7	1·2	1·0	1·6	2·5	3·7	
NW Australia 15–20°S 115–120°E							0	0·5	0·9	1·8	2·0	1·9	
E Australia N of 20°s, E of 135°E							3·4	1·8	2·1	3·5	4·6	4·1	
E Australia S of 20°s, E of 135°E							2·4	1·6	1·6	2·8	3·6	3·6	

(1) 1880s average for 1886–9 only.
(2) 1970s average for 1970–3.
(3) 1850s average for 1856–9, 1860s for 1860–8, 1870s for 1874–9.
(4) According to some estimates the numbers in the southwestern Indian Ocean (Mauritius region) in the 1950s and 1960s averaged 8–9 per year in both decades.

Table App. V.48   Zonal circulation index 50–60°N, Atlantic/European sector 60°w–60°E

(*a*) Mean winter (DJF) values of the geostrophic west wind component (m/sec) at the 500 mb level, from 1882

Values kindly supplied by Dr H. TRENKLE, Deutscher Wetterdienst.

Decade beginning	Year										Decade average
	0	1	2	3	4	5	6	7	8	9	
1880			13·3	11·0	12·9	11·4	11·4	11·9	11·3	10·5	(11·7)*
1890	11·5	10·1	11·5	11·8	12·7	9·8	10·8	11·5	11·2	13·5	11·4
1900	10·0	12·4	13·0	12·4	11·8	13·2	13·4	12·4	12·6	11·5	12·3
1910	13·1	11·0	11·7	12·1	12·5	11·5	12·2	11·0	10·5	12·2	11·8
1920	13·0	11·8	12·0	12·2	10·5	13·3	12·0	11·2	11·5	9·3	11·7
1930	10·7	11·8	11·4	10·2	11·0	11·5	12·6	12·2	10·1	11·4	11·3
1940	8·0	10·8	9·7	11·6	12·3	11·3	11·2	8·3	11·5	13·9	10·9
1950	12·1	11·3	13·4	10·4	10·0	9·7	8·6	11·6	13·2	9·9	11·0
1960	9·0	11·6	11·7	6·8	6·5	10·2	8·9	10·3	10·7	5·8	9·1
1970	9·5	8·8	7·9	12·2	11·8	13·3	10·5				(10·0)

* Inclusion of the two severe 'blocking' winters 1880 and 1881 to complete the decade would give a mean zonal index for the 1880–9 decade in the range 11·0 to 11·4.

In this table the winters are identified by the year in which the January fell.

Values for the years and decades before the development of the radio-sonde observation network were derived by the work described by TRENKLE (1956) and must be considered as a first approximation only.

(*b*) Monthly and yearly mean values of the geostrophic west wind component (m/sec) at the 500 mb level, from 1881

Values kindly supplied by Dr H. TRENKLE from charts prepared in the Deutscher Wetterdienst.

Year	J	F	M	A	M	J	J	A	S	O	N	D	Average
1881	9·9	8·4	7·4							9·5	14·0	13·5	
1882	14·0	12·4	13·4							9·9	13·0	10·4	
1883	11·9	10·8	7·2							13·7	13·8	13·4	
1884	15·1	10·3	6·1							11·8	10·8	14·5	
1885	9·4	10·4	9·0							12·1	12·9	14·4	
1886	12·1	7·4	8·8							9·4	12·2	14·8	
1887	10·9	10·1	8·8							9·3	11·3	13·8	
1888	11·1	8·9	11·9							12·0	11·8	11·7	
1889	9·1	10·8	10·2							9·2	12·3	13·1	
1890	14·9	6·6	10·7							13·4	11·6	9·9	
1891	10·0	10·5	10·2							9·9	10·8	14·7	
1892	11·0	8·9	8·3							9·9	11·5	11·8	
1893	9·4	14·1	11·1							13·3	12·6	13·4	
1894	11·8	13·0	10·3							11·0	13·2	14·2	

Table App. V.48 – continued

Year	J	F	M	A	M	J	J	A	S	O	N	D	Average
1895	8·2	7·0	10·0							11·3	12·5	11·7	
1896	11·6	9·2	10·6							9·5	9·6	12·6	
1897	8·3	13·7	12·6							11·0	12·0	10·1	
1898	12·6	11·0	8·0							10·9	12·0	16·0	
1899	12·8	11·8	8·7							13·6	13·8	7·7	
1900	11·2	11·1	6·3							15·3	10·0	16·8	
1901	11·4	9·1	7·3							10·2	14·5	13·2	
1902	17·0	8·9	11·1							12·3	10·1	12·1	
1903	12·1	12·6	10·3							15·5	12·9	10·4	
1904	12·4	12·7	5·8							11·2	14·4	15·0	
1905	13·5	11·2	9·2							9·7	11·2	14·7	
1906	14·8	10·6	10·4							11·3	12·2	13·3	
1907	13·3	10·6	11·1							10·3	12·1	12·8	
1908	12·7	12·3	8·0							10·0	12·1	12·2	
1909	13·3	8·9	7·8							13·2	11·6	14·4	
1910	13·8	11·0	13·0							11·4	9·8	11·6	
1911	10·2	11·3	6·9							11·6	13·6	11·4	
1912	11·2	12·4	12·6							12·2	13·3	16·2	
1913	10·2	9·9	12·2							10·3	14·0	14·5	
1914	11·4	11·6	15·3							8·6	13·6	13·6	
1915	11·9	9·1	7·2							7·9	8·9	13·5	
1916	12·5	10·6	6·8							12·5	10·1	13·7	
1917	8·6	10·7	7·8							9·4	12·4	10·4	
1918	10·2	11·0	6·1							10·8	11·9	14·8	
1919	9·4	12·4	10·2							9·6	10·1	14·5	
1920	12·4	12·2	10·4							7·9	11·3	10·2	
1921	14·9	10·3	11·7							12·0	8·7	15·1	
1922	9·0	12·0	9·2							8·6	12·7	14·9	
1923	13·2	8·4	7·3							16·1	12·9	11·9	
1924	10·2	9·5	10·6							10·8	10·8	13·8	
1925	14·2	11·8	8·9							12·8	11·1	15·0	
1926	10·3	10·8	10·7							10·9	10·6	12·2	
1927	11·4	10·0	8·9							12·2	11·5	8·3	
1928	13·9	12·3	6·4							12·0	12·1	13·0	
1929	8·0	7·0	7·9							15·0	12·6	13·3	
1930	10·1	8·6	13·2							14·0	15·2	11·7	
1931	12·8	11·0	7·2							11·1	9·4	13·0	
1932	13·2	8·1	7·3							14·4	11·2	12·3	
1933	10·0	8·3	9·8							11·0	8·8	9·4	
1934	12·0	11·7	11·0							15·2	10·4	11·8	
1935	9·4	13·3	9·3							13·7	9·3	13·4	
1936	15·4	9·1	6·1							12·5	13·2	13·8	
1937	8·9	14·0	9·1							8·6	9·6	8·4	
1938	13·3	8·5	12·8							13·7	14·2	9·5	
1939	12·8	11·8	7·5							9·1	13·5	10·4	

Table App. V.48 – continued

Year	J	F	M	A	M	J	J	A	S	O	N	D	Average
1940	7·0	9·3	12·9							9·3	13·2	12·2	
1941	9·2	11·1	9·5							12·4	9·7	13·5	
1942	9·0	6·6	6·5							14·2	12·0	13·0	
1943	8·8	13·1	8·7							12·0	12·3	12·1	
1944	15·6	9·1	8·4							10·0	12·5	10·2	
1945	9·6	14·0	10·2							11·0	9·2	11·1	
1946	11·4	11·0	7·1							7·5	11·7	11·7	
1947	8·2	5·1	11·5							10·9	11·9	11·5	
1948	13·5	9·4	10·3	11·1	5·8	7·0	7·9	8·9	12·3	15·2	10·3	8·7	10·0
1949	17·9	15·1	8·6	12·1	9·0	6·5	7·6	9·3	6·5	13·2	10·4	15·7	11·2
1950	8·6	12·0	8·9	10·6	7·3	9·3	8·3	9·1	15·6	10·9	10·9	10·0	10·3
1951	12·3	11·5	9·5	11·2	5·8	5·6	11·2	8·1	9·7	8·7	12·6	17·2	10·3
1952	14·7	8·3	4·7	9·8	5·6	9·6	9·9	11·0	9·9	11·0	7·6	10·2	9·4
1953	12·4	8·7	10·6	5·5	5·3	6·0	9·6	9·9	11·4	11·7	14·8	11·4	9·8
1954	8·2	10·4	6·8	9·1	3·4	8·8	8·7	8·2	14·0	15·2	15·4	14·2	10·2
1955	9·1	6·3	7·4	8·1	8·9	8·0	8·6·	8·1	12·2	11·7	8·0	12·6	9·1
1956	12·9	0·2	4·9	7·6	10·5	8·8	7·3	10·4	10·1	15·4	12·7	13·3	9·5
1957	13·4	8·0	6·2	8·6	6·4	7·0	6·6	10·0	10·8	18·3	9·7	16·2	9·8
1958	13·6	9·7	8·5	7·6	11·7	4·5	5·7	7·8	7·9	13·0	9·8	11·0	9·2
1959	9·4	9·2	7·1	8·4	7·6	8·0	8·9	8·2	11·8	10·9	10·3	11·2	9·2
1960	7·5	8·3	3·7	7·3	7·1	8·8	7·3	5·6	10·4	6·8	10·0	14·7	8·1
1961	9·5	11·8	14·0	8·5	7·6	12·1	9·2	13·7	12·1	10·6	12·6	7·6	10·8
1962	16·2	11·2	6·2	10·0	6·3	11·7	6·6	14·7	10·6	13·1	10·6	11·7	10·8
1963	4·5	4·7	8·0	5·4	10·3	6·2	8·6	8·9	14·1	15·9	11·3	6·9	8·7
1964	7·9	5·4	4·6	9·2	7·9	9·3	11·0	6·8	13·0	10·9	12·9	14·7	9·5
1965	11·1	4·9	5·8	8·5	7·1	8·5	6·7	11·7	10·5	11·7	8·7	13·3	9·0
1966	4·6	8·8	13·7	5·9	9·7	7·3	9·4	7·1	14·9	10·4	12·5	12·7	9·8
1967	8·6	9·6	15·6	10·1	6·6	10·6	9·6	10·8	8·8	17·1	11·8	13·0	11·1
1968	11·6	7·6	14·3	8·1	6·6	7·7	6·2	5·6	8·5	9·9	7·4	6·9	8·2
1969	5·6	5·0	5·7	9·8	6·8	10·0	10·0	6·4	14·5	13·7	15·9	9·3	9·4
1970	5·8	13·3	7·7	10·3	8·9	7·2	11·7	7·6	12·2	12·6	14·9	7·3	10·0
1971	8·5	10·7	6·1	9·2	9·8	7·5	8·0	10·8	10·9	14·6	14·7	12·5	10·3
1972	6·8	4·5	10·8	10·8	9·1	11·7	7·0	8·6	8·7	10·9	15·9	12·2	9·8
1973	9·5	14·8	9·6	7·4	7·5	8·5	7·7	9·6	12·2	12·1	13·4	16·0	10·9
1974	10·0	9·3	6·1	4·2	5·7	6·2	9·2	9·0	12·5	10·3	12·8	15·9	9·3
1975	17·1	6·9	6·4	9·0	6·4	9·7	8·7	8·5	13·1	8·5	11·5	17·1	10·2
1976	16·2	5·4	10·9	9·2	7·5	10·7	6·0	8·0	8·1	9·8	10·4	6·7	9·1
Average values 1948–75	10·7	8·7	8·0	7·9	7·6	8·2	8·5	9·1	11·4	12·3	11·8	12·3	9·7

Same footnotes as for previous table apply.

Table App. V.49 WRIGHT's (1975) index of the Southern Oscillation, from 1860

This index, unlike previous representations of the Southern Oscillation, is based on barometric pressure values alone. It is a weighted average (different weightings for different times of the year) of the mean sea level pressures observed at up to eight places in the tropical and subtropical latitudes, currently Adelaide, Apia (Samoa), Bombay, Cape Town, Darwin and Santiago (Chile), but in earlier years sometimes using Batavia (now Jakarta) and Honolulu.

Positive values (particularly the higher values) of the index indicate a strong 'Walker Circulation' (high pressure over the eastern Pacific, low pressure over the Indian Ocean), the 'transverse' circulation between eastern and western longitudes.

The northern winter seasons (DJF) are dated by the year in which the January fell.

Year	DJF	MAM	JJA	SON	Year	DJF	MAM	JJA	SON
1860	0·69	0·26	0·19	−0·06	1890	2·19	1·21	1·11	0·53
1861	0·83	1·17	1·42	0·15	1891	−0·32	−0·87	−0·97	−0·42
1862	1·17	1·11	0·98	1·64	1892	0·09	1·54	0·93	1·64
1863	1·91	1·86	1·33	0·35	1893	0·88	1·00	2·06	1·22
1864	−0·01	−1·01	−1·28	−0·64	1894	0·81	0·66	0·72	0·91
1865	−1·28	−0·03	−0·62	0·08	1895	−0·08	0·66	−0·18	0·02
1866	−0·19	−0·05	−0·33	−0·05	1896	−0·01	0·24	−0·81	−1·48
1867	−1·39	−0·15	0·69	0·48	1897	−0·67	−0·12	0·11	0·72
1868	−1·88	−1·58	−1·99	−1·47	1898	0·30	1·58	0·18	0·65
1869	−1·94	−1·15	−0·13	0·41	1899	1·14	−0·16	−1·68	−1·59
1870	2·43	1·17	0·76	1·00	1900	−1·29	−1·16	−0·73	−0·22
1871	0·76	0·20	0·37	0·40	1901	−0·58	−0·40	−0·62	−0·35
1872	0·88	0·84	1·39	1·42	1902	−0·00	−1·14	−2·05	−1·26
1873	1·99	0·52	0·92	−0·62	1903	−1·57	−0·97	0·64	0·83
1874	−1·88	0·34	1·26	1·41	1904	0·76	0·77	−0·68	−0·87
1875	0·31	0·93	0·25	0·24	1905	−2·26	−2·02	−1·09	−0·57
1876	0·71	0·29	−1·34	−1·50	1906	−1·48	−0·50	0·72	0·33
1877	−1·81	−2·01	−4·18	−3·41	1907	0·32	−0·15	−0·39	−0·34
1878	−1·37	−1·26	0·58	1·95	1908	−0·14	0·62	−0·01	0·65
1879	0·92	0·24	1·05	0·46	1909	0·42	0·86	1·68	0·95
1880	1·47	−0·36	−0·81	−0·92	1910	0·91	1·84	1·21	1·04
1881	−1·63	−1·68	−1·02	−0·27	1911	0·84	−0·94	−1·00	−1·06
1882	0·62	0·65	1·23	0·52	1912	−0·84	−0·96	0·27	−0·05
1883	0·29	0·81	0·30	−0·20	1913	−0·19	0·47	−0·83	−1·05
1884	−0·62	−0·38	−0·39	−0·38	1914	−1·84	−1·25	−1·63	−1·92
1885	−0·81	−0·15	−1·40	−1·00	1915	−1·08	−1·09	1·30	1·09
1886	−0·86	0·70	1·28	1·62	1916	0·76	1·32	2·85	2·38
1887	0·59	0·45	−0·11	−0·61	1917	2·33	1·89	2·64	2·45
1888	−0·89	−0·91	−0·68	−1·50	1918	0·99	−0·53	−0·94	−1·05
1889	−1·42	−1·02	1·18	1·29	1919	−0·98	−1·01	−0·39	−0·28

Table App. V.49 – continued

Year	DJF	MAM	JJA	SON	Year	DJF	MAM	JJA	SON
1920	−0·55	0·18	0·93	0·23	1950	1·02	1·05	1·30	0·74
1921	1·34	1·31	1·26	0·22	1951	0·09	0·59	−0·12	0·09
1922	0·40	0·76	0·55	0·58	1952	−0·03	0·56	0·70	0·19
1923	0·78	0·43	0·58	−0·69	1953	0·45	0·30	−0·10	0·22
1924	−0·11	0·72	0·63	1·16	1954	−0·17	0·56	0·68	0·74
1925	0·60	0·97	−0·68	−1·31	1955	0·39	2·10	1·50	2·13
1926	−1·88	−1·67	0·08	0·63	1956	1·47	1·75	2·14	1·03
1927	0·72	0·90	0·04	−0·28	1957	0·45	−0·43	−0·78	−0·95
1928	−0·65	0·43	−0·01	0·50	1958	−1·01	−0·60	0·26	0·51
1929	0·20	0·52	−0·67	−0·31	1959	−0·28	−0·11	−0·16	0·34
1930	0·24	−0·88	−1·38	−1·06	1960	0·80	0·52	0·77	0·67
1931	−0·74	−0·76	0·33	0·59	1961	0·17	0·59	0·78	1·60
1932	−0·32	−0·27	0·41	0·56	1962	1·30	1·44	1·20	0·74
1933	0·53	0·60	0·81	1·26	1963	0·65	0·53	0·01	−0·97
1934	1·48	0·53	−0·21	0·07	1964	−1·12	0·28	1·12	0·62
1935	0·11	0·20	−0·12	0·06	1965	−0·22	−0·17	−1·59	−0·70
1936	−0·46	−0·16	0·30	0·05	1966	−0·89	−0·35	−0·39	0·32
1937	0·93	0·58	0·00	0·31	1967	0·74	0·25	0·09	0·27
1938	1·10	1·12	1·12	0·83	1968	0·04	0·27	1·05	−0·52
1939	1·03	0·52	−0·14	−0·66	1969	0·46	−0·22	−0·46	−0·60
1940	−0·65	−1·22	−2·00	−1·95	1970	−0·21	−0·04	0·90	1·34
1941	−1·04	−1·74	−1·41	−1·28	1971	1·60	1·63	0·83	1·64
1942	−0·64	−0·26	1·10	0·81	1972	1·06	−0·05	−2·23	−1·25
1943	0·60	0·06	0·26	0·32	1973	−0·74	0·60	1·04	1·57
1944	−0·47	−0·42	−0·41	0·12	1974	2·16	1·71	1·30	1·11
1945	1·44	0·42	−0·14	0·19	1975	0·04			
1946	0·12	0·07	−0·34	−0·32	1976				
1947	1·07	0·15	0·28	0·86					
1948	0·14	0·17	0·28	0·50					
1949	−0·09	−0·04	0·53	1·16					

Table App. V.50 History of the winds

(*a*) Gradient wind direction over London

Apparent decade averages (numbers of days/year) from daily observations of surface wind adjusted to a veer of 45° with height.

	W	NW	N	NE	E	SE	S	SW	N+NW+NE
1723–9	103	42	26	25	40	25	16	35	93
1730–9	123	34	62	15	40	23	33	26	111
1740–9	101	32	56	15	59	17	29	12	103
1750–9	114	38	38	30	51	17	19	30	106
1760–9	104	38	33	40	39	16	17	39	111
1770–9	71	122	39	40	38	22	17	10	201
1780–9	68	64	25	23	37	57	17	30	112
1790–9	69	68	8	4	16	88	33	70	80
1800–9	59	58	16	40	41	40	16	38	114
1810–9	80	57	37	34	42	32	19	50	128
1820–9	75	70	26	30	36	50	17	41	126
1830–9	70	61	21	30	36	39	14	42	112
1840–9	79	40	24	37	37	30	14	42	101
1850–9	99	28	21	39	42	24	16	40	88
1860–9	124	40	20	30	38	30	19	35	90
1870–9	103	42	16	31	37	33	19	38	89

The figures must be treated as provisional. It cannot be foreseen when it will be possible to carry out a thorough critical appraisal, though it should be possible, after sufficient research, to present a fully homogenized series from 1669 to the 1970s. It is clear, however, that there are some variations from decade to decade in the figures given above that are so strong that they can hardly be modified significantly by further research and must be accepted as broadly true.

(b) Frequencies of different circulation pattern weather types over the British Isles[1]
   Decade averages (numbers of days/year)

	W	NW	N	E	S	A	C	NW + N
1781–5	64	27	32	37	26	93	75	59
1861–9	98	21	25	27	27	95	62	46
1870–9	87	18	27	30	31	86	73	45
1880–9	85	19	27	30	36	93	62	46
1890–9	88	20	28	26	31	101	59	48
1900–9	97	15	28	26	33	93	62	43
1910–9	99	13	29	28	28	89	68	42
1920–9	109	16	25	26	34	79	65	41
1930–9	98	24	27	27	30	90	65	51
1940–9	95	19	24	30	31	97	57	43
1950–9	90	19	31	26	31	93	62	51
1960–9	80	19	30	34	31	85	70	49
1970–5	75	25	28	28	31	101	62	53

A = Anticyclonic
W = Westerly   NW = Northwesterly   N = Northerly
E = Easterly   S = Southerly
For definitions see Volume 1, pp. 264–6.

The remarkable fall-off in the frequency of westerly type situations in recent years has been most directly seen in the frequency of westerly winds in the southernmost parts of England. Thus, for example, at Thorney Island on the south coast during the first nine months of 1971 surface winds from SW, W or NW were reported at 37 % of the observations (previous normal 54 %) and winds from N, NE or E were reported at 43 % of the observations (previous normal 30 %).

[1] Full details of the classification are given in *British Isles weather types and a register of the daily sequence of circulation patterns 1861–1971* by H. H. LAMB: *Geophysical Memoir* No. 116. London (Her Majesty's Stationery Office for Meteorological Office) 1972.

(c) Overall averages and standard deviations (days) for each weather type over the British Isles 1861–1973, in different months of the year and for the whole year

	W		NW		N		E		S		A		C	
	Mean	SD	Mean	SD	Mean	SD	Mean	SD	Mean	SD	Mean	SD	Mean	SD
J	10·2	5·3	1·1	1·4	1·9	2·2	1·9	2·3	3·8	3·1	6·5	4·1	4·4	3·0
F	7·5	4·9	1·2	1·6	1·8	2·0	2·5	3·0	2·8	2·6	7·0	4·5	4·2	3·4
M	7·0	5·0	1·4	1·7	2·5	2·4	3·5	3·5	2·9	2·6	7·5	4·5	5·0	3·4
A	5·6	3·2	1·7	1·7	2·8	2·1	3·4	2·7	2·3	1·8	8·1	4·2	5·0	3·2
M	4·7	3·1	1·2	1·3	3·2	2·5	3·6	2·6	2·6	1·9	8·6	3·8	5·7	3·5
J	6·6	3·4	1·5	1·7	2·8	2·3	2·1	2·1	1·7	1·5	8·7	4·2	5·4	3·2
J	8·1	4·3	2·2	1·9	2·3	2·0	1·2	1·4	1·5	1·6	7·6	4·6	6·8	3·8
A	8·6	4·4	1·7	1·5	2·0	1·9	1·4	1·6	1·8	1·7	7·2	4·3	7·2	4·1
S	8·2	4·5	1·2	1·4	1·9	2·0	1·7	1·7	2·5	2·0	8·6	4·8	4·8	3·1
O	8·3	4·9	1·2	1·3	2·1	2·3	2·4	2·6	3·1	2·8	7·6	4·3	5·1	3·4
N	7·8	4·2	1·4	1·7	2·4	2·3	2·1	2·2	2·7	2·6	7·0	4·2	5·4	3·3
D	10·1	4·7	1·8	2·0	1·7	1·8	1·8	2·1	2·9	2·4	6·6	3·9	5·0	3·5
Year	92·7	16·7	17·5	6·5	27·3	7·9	27·8	8·2	30·5	7·6	91·0	17·3	64·1	12·1

(d) Provisional numbers of Januarys per decade with net wind components (estimated isobar direction from different quarters over the region between London and Copenhagen, centred about Amsterdam

Decade	N	E	S	W	A	C	Decade	N	E	S	W	A	C
1550s	2	1	1	9	—	—	1730s	3	1	7	9	2	1
1560s	1	5	3	3	—	—	1740s	3	3	6	6	6	2
1570s	4	4	1	6	—	—							
1580s	5	5	2	3	—	—	1750s	2	2	5	5	2	—
(−81, 82)							1760s	1	3	7	3	5	1
1590s	4	3	5	6	—	—	1770s	1	1	3	5	1	4
(−98)							1780s	2	—	5	7	2	—
							1790s	—	2	9	7	5	1
1600–09	1	6	—	3	—	—							
(−04)							1800–09	—	3	9	5	2	1
1610–19	1	1	2	8	—	—	1810–19	—	2	6	6	2	1
(−18)							1820s	2	2	3	5	5	1
1620s	3	4	4	3	—	—	1830s	1	2	2	7	5	—
(−20)							1840s	—	3	8	6	—	—
1630s	4	4	5	6	—	—							
1640s	2	3	4	7	—	—	1850s	1	1	8	7	2	1
							1860s	—	—	9	8	2	2
1650s	3	4	—	3	—	—	1870s	2	1	7	8	1	—
(−51, 52)							1880s	1	2	4	6	5	—
1660s	2	3	3	4	—	—	1890s	—	1	2	7	4	1
(−65, 66)													
1670s	3	2	5	8	—	—	1900–09	3	—	3	10	3	—
1680s	8	2	2	8	—	—	1910–19	—	2	3	5	3	1
1690s	6	4	4	6	—	—	1920s	1	1	4	7	1	—
							1930s	—	—	7	8	2	3
1700–09	5	4	5	6	—	—	1940s	2	3	3	6	1	1
1710–19	5	4	5	5	—	—	1950s	2	—	2	7	3	—
1720s	1	1	9	8	—	—	1960s	1	1	4	8	1	1

A means anticyclonic central region.
C means cyclonic central region.
1580s (−81, 82) means 1580s omitting 1581 and 1582, for which no estimate was possible.

(e) Provisional numbers of Julys per decade with net wind components (estimated isobar direction) from different quarters over the region between London and Copenhagen, centred about Amsterdam

Decade	N	E	S	W	A	C	Decade	N	E	S	W	A	C
1550s	2	0	6	10	3	3	1720s	5	1	5	8	0	0
1560s	6	1	2	9	1	1	1730s	5	1	3	9	2	0
1570s	1	0	2	9	2	2	1740s	2	2	5	8	5	0
(−79)							1750s	1	0	5	10	3	0
1580s	2	1	6	6	0	1	1760s	2	0	3	9	4	0
(−81, 82)							1770s	1	0	4	8	4	0
1590s	3	0	5	9	0	1	1780s	1	0	8	10	7	0
(−97)							1790s	0	0	9	9	3	0
1600–9	3	2	3	4	4	1	1800–09	1	2	6	6	6	2
(−00, 09)							1810–19	4	0	1	9	3	4
1610–19	2	0	2	3	2	0	1820s	6	1	3	8	2	3
(−10, 11, 13, 14, 17, 18)							1830s	3	1	1	8	2	4
1620s	5	0	2	9	1	3	1840s	4	0	3	10	0	2
(−20)													
1630s	3	0	4	10	0	1	1850s	2	0	4	9	4	1
1640s	5	0	4	10	1	2	1860s	4	1	2	9	4	3
1650s	1	0	3	8	3	0	1870s	4	0	5	10	2	2
(−59)							1880s	5	1	3	9	3	4
1660s	0	0	3	6	0	1	1890s	8	0	1	9	3	1
(−60, 62, 64, 65)							1900–09	6	1	1	7	5	0
1670s	5	0	3	10	0	0	1910–19	6	2	1	8	1	3
1680s	(4)	0	(4)	10	0	0	1920s	2	0	3	8	2	2
Observer poorly exposed for NW winds in the							1930s	4	0	5	10	4	5
1680s							1940s	5	0	3	10	4	1
1690s	7	2	3	7	0	2							
							1950s	4	1	3	7	4	5
1700–09	4	2	4	8	0	0	1960s	6	0	0	8	3	1
1710–19	4	2	6	8	1	0							

Table App. V.51   Conversion factors: energy and power units

The following various units are commonly used in discussing energy problems, particularly the increasing prospects of some future impact of artificially generated energy output upon global climate. The list below shows the relationships between them.

(The mechanical equivalent of heat, 1 joule, is defined by J: 1 joule $= 10^7$ ergs.)

1 gramme calorie	$= 4 \cdot 1868$ joules
1 watt	$= 1$ joule/sec $= 14 \cdot 33$ g cal/min
1 langley	$= 1$ g cal/cm^2/min $= 0 \cdot 07$ watts/cm^2

Hence the mean energy supply by the sun at the Earth's distance, averaged over the whole Earth and over the year, taken as 720 g cal/cm^2/day (see Volume 1, p. 55), amounts to about $1 \cdot 8 \times 10^{17}$ watts for the whole Earth,* of which about 70% is absorbed either at the Earth's surface or in the atmosphere. Note also the larger units sometimes used:

1 megawatt	$= 10^6$ watts
1 terawatt	$= 10^{12}$ watts
1 exawatt	$= 10^{18}$ watts

---

* The Man-made contribution to the energy balance (all forms including CO_2 effect), at present effectively about $1 \cdot 8 \times 10^{14}$ watts, may, it has been suggested, reach 4×10^{14} watts by A.D. 2000 or soon after and is likely to be concentrated over a small fraction of the Earth's surface (further consideration of this has been given by FLOHN 1973 and WASHINGTON 1972).

Part I V

The future

Chapter 19

Man-made climatic changes

To determine how we should address ourselves to the problems of foresight in regard to climate, we must estimate

(1) The future course of natural climatic fluctuation (and the range of events which accompanies each regime).

(2) The impact of human activities on local, regional and perhaps ultimately on global climate.

Both types of climatic change are also liable to affect each other.

Man's impact on the climate began 5000 to 9000 years ago as soon as he was able to alter the environment by burning and felling forest and tilling the Earth.[1] Also in those river valleys – the cradles of ancient civilizations – in arid regions where he learnt to irrigate the soil and produce crops, i.e. vegetation, where previously all but the river was dry, the climate was in some measure altered. But the increase of windspeed in the little forest clearings and the lowering of albedo and raising of relative humidity where crops replaced desert would hardly have any significance beyond the fields. Over irrigated land the incoming energy from sun and sky is partly used in evaporation, and the air temperature rises less in the course of the day in consequence. The latent heat passed into the air in such localities is later released as

---

1. In some parts of the world this particular alteration of the Earth's natural surface is still proceeding, only faster than ever. The southernmost province of Brazil, Rio Grande do Sul (28–34°s), was still 19% forest covered in 1900; by 1971 the proportion was 2%. It is also estimated that the Amazon jungle itself may have been largely destroyed by A.D. 2000.

feelable heat when condensation occurs, in cloud and precipitation processes, commonly after the air has travelled some thousands of kilometres. Thus, there is a moderating effect on maximum temperatures just over the fields, but little effect concentrated enough to be significant anywhere else.

The most extensive change wrought by Man prior to our own times was the gradual conversion of most of the temperate forest zone to crops, i.e. an artificial steppe or savanna. But, though increased windiness and changed evaporation over deforested ground changes the precipitation regime slightly, according to DROZDOV and GRIGORIEVA (1963), the effect is relatively small where the forest is replaced by other types of vegetation (cited by BUDYKO 1974, p. 463).

Thus, until the industrial revolution, and (as we shall see) probably until the present century, Man has had little effect on the climate except on a very local scale. And up to the middle of the present century, outside the limits of urbanized and industrialized areas and away from artificial lakes and irrigated areas, Man's effects on the climate have probably in no case – apart, perhaps, from the output of carbon dioxide – been significant in comparison with the magnitude of the natural fluctuations of climate with which this volume has so far been concerned.

Anxieties are now, however, reasonably felt, and widely expressed, about the possible imminence of global effects owing to the increasing energy at Man's disposal, the increasing variety of pollutants, and his rapidly increasing technical power. Certain grandiose schemes for modifying world climate deliberately, e.g. inducing a permanent warming by abolishing the Arctic ice, with the aim of increasing the total cultivable area of the globe, have been discussed from time to time and assessments made of the energy cost and and technological development required to achieve them. Aside from these aspects, a basic objection to all such plans is that no changed pattern of climate (which either Man could devise or Nature might produce) would be an improvement for all human needs everywhere. All would be bound to produce distress, and probably disasters, in some regions. None could properly, or advisably, be pursued without the broadest international agreement, and all would be liable to entail some movement of populations. Moreover, it cannot be pretended that the consequences of deliberate action would all be predictable.

Schemes have also been published for deliberate actions to change the global pattern of climate as a strategy of war. And already it is reported (*The Times*, London, 24 June 1975) that talks have begun on a Soviet proposal for a draft treaty, submitted to the United Nations, under which the nations would forswear attempts to manipulate the weather or climate as an instrument of war.

Scales of human intervention

Examples of Man's impact on weather and climate may be classified in various ways: e.g.

(1) Unintended, and often undesirable, side effects of Man's activities, so far mainly

local in extent, many now, however, rapidly increasing in scale and some perhaps ultimately calling for countermeasures.

(2) Planned attempts to modify weather or climate so as to moderate the effects of their extremes on various human activities, or to shift the balance of Nature in a way favourable to Man, or for other reasons.

Most discussion prior to about 1960 centred on planning for deliberate action on various scales, but since that time awareness both of the limitations of planning and prediction in this field and of the possible imminence and scale of unintended side effects has caused most attention to be switched to consideration of these and attempts to estimate them by every available modern refinement of measurement, monitoring and theoretical modelling.

Various types of deliberate intervention are already established, however. The scale and the aims of human intervention may range from extremely local – the artificial climate in a garden frame, a greenhouse or a cattleshed might logicaly be included – to global.

The planting of shelter belts of trees to protect the light soils from blowing, which was among the agricultural improvements introduced in Norfolk in the eighteenth century, may be reckoned the first deliberate attempt, other than the many-thousand-years-old practice of irrigation, to modify the climate on a local scale. Shelter belts were planted later in the same century near the coast of Scotland, in Morayshire, where the Culbin Sands disaster had occurred in the windstorms of the late 1600s, and have since been established in many other areas, e.g. near the Danish coast and more recently in Norway as far north as Finnmark, on the Lüneburger Heide and (perhaps most extensively) on the plains of Russia. The benefits are found to include an increase of soil moisture attributable to the reduced evaporation where wind speeds are curbed.

Within the present century various other local, and mostly short-term, measures for weather protection have been introduced. These range from special localized heat sources for fog clearance on airfields, and smoke sources for frost protection of orchards and vineyards, to 'cloud seeding' – with silver iodide or 'dry ice' (solid carbon dioxide at very low temperature) to provide condensation nuclei for ice crystals and encourage the transfer of supercooled water droplets in the clouds to the ice nuclei – with the object of inducing rain or averting a risk of hail damage (by preventing the growth of large hail within the cloud). A useful review of these techniques, their economics and effectiveness, is given by MAUNDER (1970) with references to the extensive literature on such practices. None of these measures, however, can be thought to have any large-scale or lasting effect significant for the climate.

Proposals for large-scale modification of the climate of wide regions or of the globe have so far not proceeded beyond the research and discussion stage, though several meetings of experts have been held. Such proposals were among the items considered by the Conference on Modification of Climate, organized by the Voejkov Central Geophysical Observatory and the Institutes of Applied Physics and Geography of the U.S.S.R. Academy of Sciences at Moscow, 25–28 April 1961 (GALTSOV 1961) and by the Panel on Weather and Climate Modification of the U.S. National Academy of Sciences in 1966.

Large-scale schemes for climate modification

To many people the prospect of Man's meddling with the large-scale climate of the Earth seems impious. And, to the extent that this attitude is based on apprehension of the likelihood of conflicting interests and of the limits of predictability of the actual effects of climate, it is well that it is so. Certainly, there is an obligation to proceed humbly and with caution. Since any climatic change or fluctuation must introduce a new variability of agricultural yields in some areas, and world population has now reached a level at which the food reserves built up over the past 30 years have been largely exhausted following partial harvest failures in the Soviet grainlands in 1972 and in India in 1972 and 1974, it is hard to see how any contrived shift of the large-scale climatic pattern – even one aimed at increasing the total cultivable area of the world – could be achieved without some local or regional disasters. Nevertheless, BERGERON (1965) has already warned that by the year 2000 weather modification may have come to seem a more important aim than weather forecasting.

Because of the still vastly greater energy exchanges involved in climate and the global wind circulation, and even even in day-by-day local weather processes, than the energy at Man's disposal, human intervention is only likely to be effective when it is applied strategically, i.e. at sensitive points where the balance between two alternative modes of atmospheric or ocean circulation is nearly neutral. In some cases, where it might be possible for Man to tip the balance, the ensuing change of regime might nevertheless be irreversible. Because of this risk, and the risk of unforeseen side effects, however much theoretical research had been done beforehand, it would be essential to proceed in small stages, allowing many years for observation and assessment of the full range of effects of each stage. And it would be preferable first to make pilot studies affecting a limited, but similar, terrain in any case where a relevant test could be devised.

In what follows space only permits the briefest description of a few of the apparently most important schemes that have been proposed.

Creation of inland seas in Africa

BERGERON (1960) proposed at the Conference on Physics of Precipitation held at Woods Hole Oceanographic Institution, Massachusetts, in 1959 that it should be practicable to increase the rainfall of the semi-arid (Sahel–Sudan) zone of North Africa, the savanna and the steppe, by injecting extra humidity on a great scale into the monsoon air. The airmass which forms the African southwest monsoon is secluded in the lower layers between two tropical fronts that are continually renewed at the limit of penetration of fresh airmasses from other latitudes in the northern and southern hemispheres (see fig. 19.1). Thus, the humidity in the monsoon airmass, which is precipitated in rain from cumulonimbus clouds developed in all the region where this airmass is deep enough for their growth (and is then evaporated and precipitated again and again within the realm of the same airmass), cannot

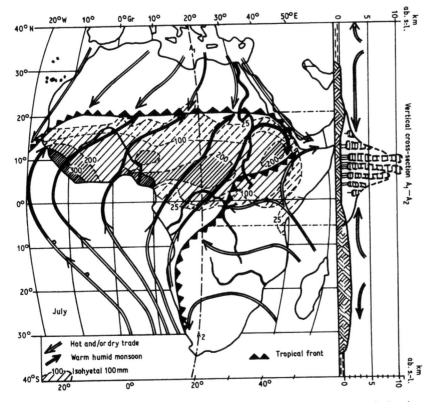

Fig. 19.1 BERGERON's (1960) schematic picture of the general circulation in
the lower troposphere and rainfall distribution over Africa in July.
    The barbed lines mark the fronts (see text p. 658).
    (Kindly supplied by Professor T. BERGERON *for reproduction here.)*

leak away from Africa on any great scale, apart from what ultimately escapes with the upper
winds to the east of Ethiopia.[1] Such a scheme, like the small-scale operation of 'rain making'
by cloud seeding, is to a large extent open to the criticism that its effectiveness must be limited
to places – in this case to the zone between the two tropical fronts – where clouds and a good
deal of humidity are already present, but it is true that the injection of extra humidity will
facilitate further, and more extensive, cloud development through the release of latent heat of
condensation realizing the conditional instability of the air for vertical convection.

    BERGERON's scheme might be brought into operation by the plan originally conceived
by the German engineer H. SÖRGEL in the 1930s. Suitably massive damming of the Congo
River at the Stanley Gorge near Kinshasa (Leopoldville) (at a point where in the Cretaceous
era the river broke through the rocky rim of an earlier lake) should in the course of a few
years cause the great inland basin of Zaire, between 15° and 26°E 4°N–5°S, to fill with water to

1. See the vertical section diagram in the right-hand portion of fig. 19.1.

a level sufficient to turn back the tributary river Ubanga, so that it would flow north to join the Shari River and fill Lake Chad once more into an inland sea about 2 million km² in extent (a map is given by ADABASHEV 1966). The two water bodies together would cover an area of about 3 million km² or 10% of the African continent. The meteorologically important aspect of the plan is that this great expanse of water would lie broadly within the regions that are for much of the year covered by the cloudmasses and vertical air circulation of the equatorial rain system, so that the extra water available should be continually recycled through the vegetation growth in the zone and the atmospheric circulation over it (by contrast, the water lost by evaporation from the new lake at Asswan under the usually cloudless skies at that latitude must mainly be transported far beyond the confines of Africa before it falls to the Earth again).

Creation of a Siberian Sea

According to FLOHN (1963), the proposal of the Russian engineer M. DAVIDOV to dam the rivers Ob and Yenesei (Jenesei) should produce an expanse of water of about 250 000 km² in northwest Siberia. From this sea, it is suggested, water could be channelled through the Turgai region to arrest the drying up of the Caspian and Aral Seas and water the arid steppes thereabouts while still allowing enough flow over the dams to maintain navigation on the lower reaches of the Ob and Yenesei (see, however, p. 324 in Volume 1 and pp. 671–4 later in this chapter).

Removal of the ice on the Arctic Ocean

Much research has been devoted in the Soviet Union to the possibility of removing the Arctic sea ice and the probable effects of such action on the climates of the northern landmasses, as well as to the question of whether the ice would form again (e.g. BUDYKO 1966, 1969, 1974, DORONIN 1968, RAKIPOVA 1962, 1966). This may be considered the biggest project proposed for deliberate modification of climate. The object would be to raise the temperature of northern lands and so to open up great regions to human settlement and cultivation.

Melting the ice by blackening its surface with soot deposited from aircraft during an Arctic summer was the first method proposed, but estimates pointed to a number of flights and an expense that was still unattainable.

An alternative plan proposed by the Russian engineer P. M. BORISOV (1962), and much publicized, is to dam the Bering Strait, blocking off the present flow of water from the Pacific Ocean into the Arctic and instead pumping 500 km³ water/day from the Arctic to the Pacific. BORISOV's supposition is that thereby more, and warmer, Atlantic water would be drawn into the central Arctic than now reaches there. FEDEROV (1962), however, considers the concept unsound; Atlantic water would be unlikely to flow straight through the Arctic, and other workers have suggested that it would be mainly the vast reserve of cold deep water in the Arctic that would pass through the pumps into the Pacific.

Opinions – and verdicts from calculation – differ sharply on whether the Arctic ice would reform once it had been destroyed. BUDYKO (1974) follows DORONIN (1968) in considering that melting of the ice would leave a fresh water layer on the ocean surface, that it would remain there because of the stable stratification, and that it would be easy for the ice to form again in that layer. In another work BUDYKO (1969), reviewing his own and other model calculations, found that the best estimate of the average surface temperature that would prevail in the polar region in winter with an open Arctic Ocean is about $+1°c$. This value differs from the freezing point of sea water by less than the margin of error of the calculations. Moreover, the year-to-year variability about such a value suggests that within a few years the ice might reform. Another model calculation by SELLERS (1969) in the United States suggests that the ultimate effect of disappearance of the ice, without any primary change in the energy available over other latitudes, would be to raise the overall average temperatures prevailing north of 70°N by no more than 7°c, a result which implies that the ice would soon reappear.

Other, more serious, objections to the scheme concern side effects on the distribution of precipitation. Most calculations indicate that, with an open ice-free Arctic Ocean, more rain and snow than now would fall over the central Arctic and the surrounding lands in high latitudes, possibly leading to the onset of glaciation as snow accumulated on the mountainous regions and existing ice caps, while the northern hemisphere land areas in general, farther from the Arctic coasts, would suffer from a drier regime than now. This conclusion was first indicated by DROZDOV (1966) empirically, from consideration of the observations of the warmest and most ice-free years in the present century (see figs. 19.7(a) and (b) later in this chapter). Much the same is also implied by a computation by NEWSON (1973), with a dynamical climate model developed in the U.K. Meteorological Office, which showed that winter temperatures prevailing in latitudes 50–60°N over Europe and Asia and 25–50°N over North America with an ice-free Arctic would be lower than now by 6–8°c in some inner continental areas, presumably owing to a much increased frequency of anticyclone development over the continents between 50° and 70°N.

Other proposals

Other schemes have been suggested such as diversion or regulation of the flow of the Gulf Stream and the corresponding warm current, the Kuro Shiwo, in the North Pacific. One scheme is to control a possibly useful branch of the Kuro Shiwo on the Asian side of the Japanese islands by a giant sluice at the mouth of the Tatarsk Strait, acting as a one-way valve to allow the flood tide through and check the ebb so as to increase the warm water flow into the Okhotsk Sea and reduce the ice there (FLETCHER 1969). And in North America there is a proposal – the North American Water and Power Alliance – for diverting water southward from the great rivers of northwest Canada and Alaska to irrigate the dry western regions of the United States and parts of Mexico, similar to the Russian proposals about the Siberian rivers and the watering of central Asia (see Volume 1, p. 324, and later in this chapter).

Of all the large schemes mentioned in this section BERGERON's is the proposal that seems soundest, in that it takes most strategic advantage of the natural tendencies of the energy pattern and organization of the global wind circulation. For this very reason it can be criticized as the least radical plan, in that it seeks only to extend and enlarge somewhat the already existing rainfall distribution, though it may do so on a sufficiently great scale (in a warm climate zone where the moisture carrying capacity of the air is very great) to make a very significant contribution to food production. Moreover, BERGERON has suggested that the project should be preceded by smaller-scale tests in partly similar airmass situations in southeast Australia, southwest Norway or elsewhere. All the other proposals listed are open to serious objections because of the likelihood of undesirable and very wide-ranging side effects, some of which might well nullify the main objectives of the schemes. These unwanted side effects will be discussed at the appropriate places in the next sections.

Inadvertent modification of the climate

Since about 1950 many studies have been undertaken, and a great variety of evidence has been presented, showing that modern Man-made modifications of the environment may affect climate not only locally but also on a regional, or even a global, scale. The observed variations of average temperature over the world since 1880 have sometimes been interpreted as Man-made effects, despite the fact that similar – and also greater – variations occurred before the industrial era, attributable only to natural influences such as we have discussed in this book. Several recent reports (e.g. WORLD METEOROLOGICAL ORGANIZATION 1972, NATIONAL ACADEMY OF SCIENCES 1975) have concluded that Man-made effects on regional or world climate may now be just detectable, but must be expected to wax in importance, even allowing for the buffering and damping influence of the oceans with their great storage capacity for carbon dioxide and for heat, etc.

The reports of the SCEP Conference held in 1970 in the U.S.A. and of the SMIC Conference held in 1971 near Stockholm, and the work by MATTHEWS et al. (1971), have made available a mass of valuable reference data and information about the processes at work and observable effects of Man's activities. Shorter summaries by LANDSBERG (1970, 1971), MITCHELL (1971, 1973) and SAWYER (1971) also give useful bibliographies.

Some account of the observations of, and physical processes involved in, Man's effects on climate has already been given in the present work (Volume 1, pp. 44–7, 435–9, 517–19). The following paragraphs briefly review the present state of knowledge regarding the effects of various activities of Man and their by-products, and present some indications from monitoring different kinds of pollution.

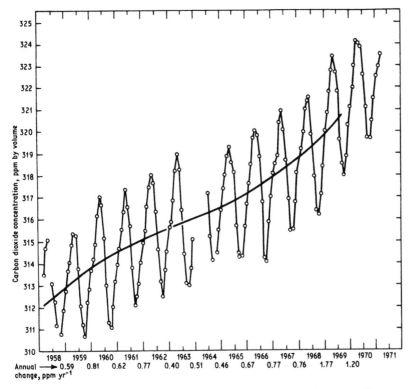

Fig. 19.2   Mean monthly observed CO_2 concentration in the atmosphere at
Mauna Loa peak (3398 m), Hawaii, from 1958 to 1971.
   (Units: parts per million by volume.)
   (*From data largely due to* C. D. K EELING; *diagram originally published
in the SMIC report and reproduced here by kind permission of the M.I.T.
Press.*)

Particularly neat examples of Man's inadvertent modification of the weather (and,
through their frequency, of the climate) on a very localized scale are the fogs generated by
input of water vapour, released by exhausts and in flues, etc., into the very cold surface air on
airfields in the Arctic and along motorways and railways (particularly where steam engines
are used) in temperate zone countries in winter. This may be described as pollution of the
atmosphere by excess water.

Carbon dioxide

The increase of CO_2 in the atmosphere from the burning of fossil fuels is estimated to be
now (1974–5) 13 % above the amount present in 1850. When considered in conjunction with
estimates of the amount of fossil carbon burnt, it appears that only 50–75 % of the carbon

dioxide produced has stayed in the atmosphere, the rest having gone into the oceans and into the biosphere.

Fig. 19.2 shows the results of monitoring the CO_2 content in the air at the mountain-top observatory on Mauna Loa, Hawaii (19°N 3398 m asl), remote from all local sources of Man-made pollution. The pronounced seasonal variation marks the diminution of the carbon dioxide in the atmosphere each northern hemisphere spring and summer, produced by the growth of the vegetation (which is mostly in the northern hemisphere). But, apart from the seasonal swings which cover a range of 6 or 7 parts per million, there has been a net increase of CO_2 from year to year throughout the period surveyed. The rate of increase is not constant; it declined a little in the early 1960s, increased from 1968 to 1970, and is reported to have declined again towards the mid 1970s. The reasons for these apparently cyclic changes, repeating themselves over periods of about 10 years within the time so far surveyed, are not yet understood (but may have to do with the vertical circulation of the oceans and changes in the rate of absorption in the sea). Monitoring is now carried on at other points also, including the South Pole. The carbon dioxide is quite well mixed throughout the world's atmosphere; it appears to be most at the most northerly observing station and least at the South Pole, though the overall difference averages only 1 part per million.[1] The difference implies a lag of 18 months before the amount observed at the South Pole receives the concentration produced by the CO_2 output which occurs mainly in the northern hemisphere.

Projections of the carbon dioxide in the atmosphere, based on extrapolation of Man's fuel consumption, suggest (conservative estimates) 22% above the 1850 level by 1990 and 32% above that level by A.D. 2000.

From its radiation absorption properties it is certain that the most direct effect of increasing carbon dioxide in the atmosphere is to raise the temperature (see Volume 1, p. 45), but assessments of the amount of the temperature rise that should be expected from a given CO_2 increase differ. One of the most widely accepted results of the use of mathematical models of the atmosphere has been the computation by MANABE and WETHERALD (1967) of the effects of introducing increasing amounts of carbon dioxide into the atmosphere, using a model in which not only the radiation exchanges within the atmosphere but also changes in the convective activity were taken into account. Relative humidity and the amount of cloud cover were assumed constant. This model indicated that doubling the carbon dioxide content would raise the equilibrium temperature at the surface of the Earth by 1·9°C (earlier, it had been suggested that doubling the CO_2 would raise the global mean surface temperature by 3·6°C, and that the increase of this gas since the industrial revolution might account for the whole of the twentieth century warming of world climates). However, as noted in Volume 1, water vapour is an effective absorber of most of the same wave lengths which CO_2 absorbs; and BUDYKO (1974) cites work of KONDRATIEV and NIILISK (1963) which suggests from consideration of the effect of the atmospheric water vapour on the

1. The seasonal range is also greatest at high northern latitudes, reaching 9 parts per million.

absorption of long-wave radiation that the changes which have occurred in the amount of CO_2 in the atmosphere may have had only a small effect on temperature.

The complexity and uncertainties that remain in this matter are underlined by a suggestion of SIEMERLING (1974) that the increase of carbon dioxide that has resulted in the sea, by increasing the acidity of the water, causes some of the calcium compounds and other plant nutrients which would otherwise settle to the ocean bed to dissolve. Hence, it is alleged, the phytoplankton populations are becoming denser, increasing the albedo of the ocean surface and thereby tending to cool the water. However, it seems unlikely that world observational coverage is yet sufficient to assert that there has been in fact any such general increase of phytoplankton density. The effect of a slight (1 or 2 per cent?) increase in the carbonic acid ions in the surface water may tend in that direction; but the observed cooling by 0·5°C (and the associated later spring flush of the phytoplankton in the surface waters at

Table 19.1   Estimated global production of particulate matter
(millions of metric tons/year)

Estimates about 1970	All sizes			Particles <5 μm diameter		
	Natural	Man-made or Man-induced	Total	Natural	Man-made or Man-induced	Total
Produced directly as particles:						
Sea salt	1000	—	1000	500	—	500
Wind-blown dust	200	300	500	100	150	250
Volcanic	250	—	250	25	—	25
Forest fires and slash/burn agriculture	35	100	135	5	60	65
Cosmic dust	10	—	10	<1	—	<1
Industrial plants	—	80	80	—	22	22
Other human activities	—	30	30	—	8	8
Subtotals	1495	510	2005	630	240	870
Particles formed from gases:						
Sulphates	420	220	640	335	200	535
Nitrates	75	40	115	60	35	95
Hydrocarbons	75	15	90	75	15	90
Subtotals	570	275	845	470	250	720
Grand Totals	2065	785	2850	1100	490	1590
	72%	28%	100%	69%	31%	100%

least of the North Atlantic) surely tends to produce the opposite result. SIEMERLING'S suggestion, however, undoubtedly indicates another item which should be monitored and which must be assessed before any final verdict on the total effect of the global increase of carbon dioxide can be reached.

On balance, the effect of increased carbon dioxide on climate is almost certainly in the direction of warming but is probably much smaller than the estimates which have commonly been accepted.

BRYSON (1974) also suggests that, because of its effect on the vertical lapse rate of temperature – and hence decreasing the thermal (or convective) stability – in the atmosphere, increasing carbon dioxide may shift the latitude of the subtropical anticyclones equatorward (cf. fig. 3.20, p. 120, in Volume 1) and hence affect the world distribution of rainfall directly.

Particulate matter (dust, smoke, etc.) in the atmosphere

The atmosphere's total burden of dust, smoke and other particles about 1970 has been estimated (MITCHELL 1973) at about 4×10^7 tons, of which about 1×10^7 tons may have come directly or indirectly from Man's activities. The history of this item is much less well established than that of CO_2, but the Man-made and Man-induced contribution is probably increasing, perhaps by as much as 4% yearly, which would lead to a concentration 60% above the 1970 level by A.D. 2000. MITCHELL (1973) gives the broad estimates shown in Table 19.1 (p. 665).

Man-made particles are estimated to be 10 times as numerous in the atmosphere over the northern hemisphere as over the southern hemisphere. Note (from Table 19.1) the large contribution of particles from agricultural practices as compared with industry: much of this is attributed to the smoke particles from the burning of great areas of grass (savanna fires) each year in the dry season. There must also be an increased uptake of mineral dust in dry areas where the natural vegetation has been destroyed. The estimates (due to FLOHN) in Table 19.2 of the changes of the total particle injection relative to the 1970 figure are taken from the W.M.O. Report (1972).

According to MITCHELL, about 90% of the particle load on average is within the troposphere. The remaining average figure of 10% in the stratosphere is primarily volcanic

Table 19.2   Estimated yearly total particle injection into the atmosphere (relative to 1970 = 100%)

Year	Man-made	Total all sources
1880	25%	69%
2000 (projected)	158%	124%

dust and sulphate particles in the JUNGE aerosol layer at about 20–25 km, also largely of volcanic origin; this item varies by up to 100 times from one group of years to another, depending on volcanic activity.

The effect on climate of the dust in the lower atmosphere is hard to assess, and there is as yet no firm consensus of opinion or conclusion about it. The effect on radiation passing through the atmosphere depends on the particle sizes and varies according to the wave length. Particle sizes between 0·1 and 5 µm are certainly the most important because of their considerable abundance and relatively long residence, provided they are not washed out by rain (average residences of 10 days in the lowest 1·5 km and up to 30 days in the upper troposphere are suggested, compared with a year or two for the volcanic dust particles in the stratosphere). Such particles are much more effective in intercepting solar short-wave radiation than the outgoing long-wave radiation from the Earth. Thus the Earth's albedo is increased. The net effect in any area must depend on whether the dust is lighter or darker (has higher or lower albedo) than the ground beneath. RASOOL and SCHNEIDER (1971) have indicated that increasing the suspended aerosol content of the atmosphere by a factor of 4 should lower the mean temperature of the surface of the Earth by 3·5°C due to back-scattering of the short-wave solar radiation while leaving the passage of the Earth's long-wave radiation almost unaffected. This is certainly the best estimate available. BRYSON (1968) believes it to be the most important agent in the recent cooling of world climate. However, there are other effects which must enter into any complete assessment. Cloudiness may be increased through some of the particles acting as condensation nuclei, and this would tend to raise the temperatures prevailing in the higher latitudes though it would cool the lower latitudes. Also the dust itself absorbs some of the radiation and thereby tends to warm the air layer in which it resides, an effect instanced by IDSO (1973) in the surface air over Arizona and reported also by DAVITAYA (1969) in the U.S.S.R. For this reason Man-made dust in the lower atmosphere cannot have the same climatic effect on the surface air as volcanic dust in the stratosphere. Also, because of the limited residence time in the lower atmosphere, Man-made dust does not spread so uniformly over the Earth as either the carbon dioxide or volcanic dust veils in the stratosphere.

Pending more adequate observational evidence, it seems safe to conclude that the effect of Man-made dust in the lower atmosphere is mainly in the direction of cooling, at least in all the lower latitudes, and there is probably a tendency to moderate the diurnal and annual extremes of temperatures. There may also be some effect on precipitation.

The effect of water vapour discharged by aircraft: artificial cirrus cloud

There is no doubt that the frequency of condensation trails near major air routes has increased the amount of cirrus cloud there. This is illustrated by fig. 19.3 from a point near the busy transcontinental airways in the United States. But the percentage of the Earth's surface so affected is still small, and evidence of any significant effect on global cloudiness is

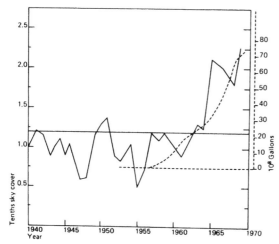

Fig. 19.3   The change of average high-cloud (Cirrus) cover at Denver, Colorado.

Yearly values from 1940 onwards.

The dashed line shows the growth of jet fuel consumption by United States domestic commercial aircraft, paralleling the increase of cirrus since 1956.

(*From* MACHTA *and* CARPENTER (*1971*); *reproduced by kind permission of the authors and M.I.T. Press.*)

still inconclusive.[1] It must be important to make use of satellite technology to establish (for the first time) how much the Earth's total cloudiness varies from year to year both from natural and Man-made causes. It is supposed on theoretical grounds (MACHTA 1971) that an increase of water vapour in the stratosphere will tend to warm the climate at the Earth's surface.

Pollution of the stratosphere

The (still minute) water vapour content of the stratosphere seems to have increased significantly over recent years (fig. 19.4), but this is thought to have been a natural variation. If supersonic aircraft and rocket exhausts add H_2O to the stratosphere in the future, perhaps to 3–3·5 parts per million by 1990, there may be a significant effect on world temperature. The effect will be a balance between that due to some increase of albedo of the Earth, if the frequency of thin clouds at the mesopause level (80 km) increases, and the absorption and re-radiation of the Earth's long-wave radiation. It is generally supposed that the net result would be a tendency towards warming at the Earth's surface.

The amount of ozone in the stratosphere also increased during the 1960s (fig. 19.5). This too is supposedly a natural fluctuation, though its causes and mechanism are as yet unknown.

Other forms of pollution of the stratosphere, due to Man, have been discussed in Volume 1 (p. 439). A W.M.O. Executive Committee Panel on the Meteorological Aspects of Air Pollution about 1970 expressed concern lest chemical constituents of jet aircraft exhausts

---

1. HUNT (personal communication, August 1975) reports estimates from modelling experiments that increasing cirrus cloud cover globally by 20% might be expected to raise the global average surface temperature by 0·5 to 0·8°C. The U.K. Meteorological Office general circulation model, with radiative transfer processes within the atmosphere accounted for, indicated a decrease of middle and low cloud when the cirrus cover was increased and hence the suggested rise of global mean surface temperature.

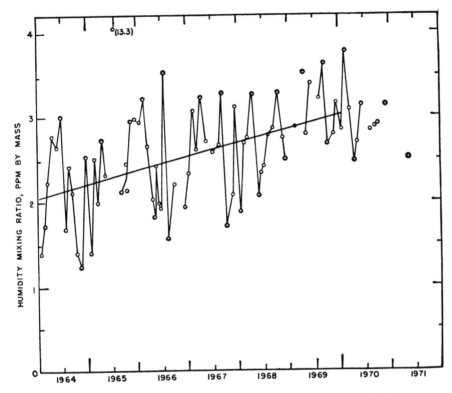

Fig. 19.4   Monthly values of the concentration of water vapour in the lower stratosphere over the Chesapeake Bay station near Washington, D.C., from 1964 to 1970.
(Humidity mixing ratio in parts per million by mass.)
(*Data due to* MASTENBROOK *published in* MATTHEWS *et al.* (*1971*); *reproduced by kind permission of M.I.T. Press.*)

might cause the destruction of the stratospheric ozone layer.[1] There is certainly no evidence of that happening, at least in the years surveyed in fig. 19.5, and no specific process has been alleged whereby the ozone would be likely to be destroyed. But since ozone is highly active chemically as an oxidizing agent, the fear of what might happen if the volume of stratospheric flying increased greatly and introduced significantly more foreign constituents into the atmosphere at those levels is not unreasonable. Caution and careful observation, preferably continuous monitoring of the stratospheric ozone, are clearly advisable.

Human actions affecting the conservation of water and input of water vapour at the surface

Under this heading should be considered the possible effects of afforestation and of forest destruction, so far mainly in middle and higher latitudes but perhaps destined to include the

1. Chlorofluorocarbons, popularly known as freons, used as a propellant in aerosol sprays are also believed capable of destroying the ozone in the stratosphere.

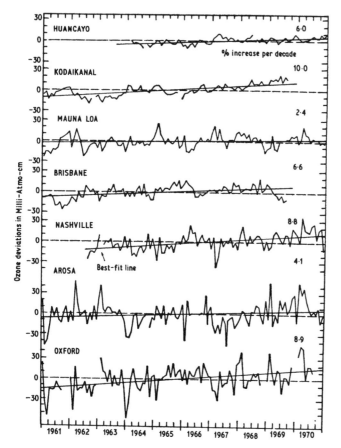

Fig. 19.5   Total ozone amounts observed in the column of the atmosphere over selected stations in many parts of the world, showing the increasing trend in the 1960s.

(*From SMIC Report; data by* KOMHYR *and others; reproduced here by kind permission of M.I.T. Press.*)

disappearance of the Amazon and Congo jungles (tropical forests) by the end of the century. To date there has been surprisingly little serious effort to establish the effects of these changes. Consequently no convincing observational evidence can be presented here. It seems likely, however, that where one type of vegetation is replaced by another any effect on climate is small.

Much more serious, certainly, is the effect of destruction of the vegetation in dry areas, where any plant cover may be difficult to restore and the bare soil may be blown away by the wind. This seems to have happened in the Breckland area of East Anglia in the late Middle Ages (see Chapter 17), though the situation was gradually recouped from the eighteenth century onward (and more particularly since 1920) by a scientifically planned and pursued policy of afforestation and the planting of shelter belts. CHARNEY (1975) has given theoretical justification for the view that overgrazing in the Sahel region of the Sahara's southern fringe may have increased the natural tendency to desertification. Owing to the high albedo of desert sands, estimated by CHARNEY from satellite measurements as averaging 35%, destruction of the vegetation means an increased loss of heat from the surface, which encourages sinking motion (anticyclonic subsidence) in the atmosphere over the regions of bare surface – a 'positive feedback' mechanism which tends to intensify and perpetuate the dryness.

No less serious are the possible side effects that have to be considered in connection with schemes to divert the flow of the great rivers which enter the polar basin to water the arid lands in middle latitudes in North America and Asia. These are illustrated here by the Soviet scheme for central Asia (fig. 19.6).

There is no doubt that the water emergency which threatens central Asia is real. Current use of the two central Asian rivers Amu Darya and Syr Daria to irrigate 5 500 000 hectares results in half the water being lost by evaporation, etc., before it reaches the Aral Sea, the level of which is reported to have dropped 1·5 m in recent years in consequence. There are 34 000 000 hectares of fertile land in the region which, if irrigated, could yield cotton, rice, sugar beet, fruit and pasture and which probably must be brought into cultivation within the next 50 years to meet the needs of an increasing population. Current and projected needs would, it is estimated, cause the total disappearance of the Aral Sea unless further sources of water can be arranged.

The level of the Caspian Sea is also dropping: it is reportedly now a continuous fall, at least partly owing to the use of the water from the rivers which supply it – mainly the Volga – for irrigation. Total inflow from the rivers into the Caspian Sea over the 20 years 1946–65 is estimated as 900 km³ less than it would have been but for this use of the water (ADABASHEV 1966, p. 101). The Caspian is receiving 30 km³ too little water yearly to balance the evaporation from its surface, and this deficit is expected to increase by 50–70% by 1985. The decreasing volume of water in the Caspian Sea, which has no outlet and is shallow, particularly in its northern part, is increasing its salinity and threatens the valuable fishing industry there, the main source of supply of the sturgeon from which the famous caviar comes.

The salt content of the Sea of Azov, described as the most productive fishing basin in the world, is also likely to increase as water from the Black Sea flows in to make up the declining flow from the rivers Don and Kuban due to industry and irrigation.

All over the wide continental interior region discussed in these paragraphs the water table is falling. The ambitious Russian scheme to solve the problem – which might be the first success in providing extra water on anything like the scale concentrated in a large region where the natural supply falls short of modern needs – has been the subject of research for 30 years and, it is reported, will be proceeded with only gradually, with a careful watch on its actual impact and with due regard to its possible effect on other peoples. It raises difficult questions (outlined on p. 324 in Volume I), because it appears to touch a peculiarly sensitive spot in the existing climatic regime of the northern hemisphere. The layer of low salinity water which occupies the top 100–200 m of the Arctic Ocean – remaining insulated by its low density from the deep convection in the salt water beneath –

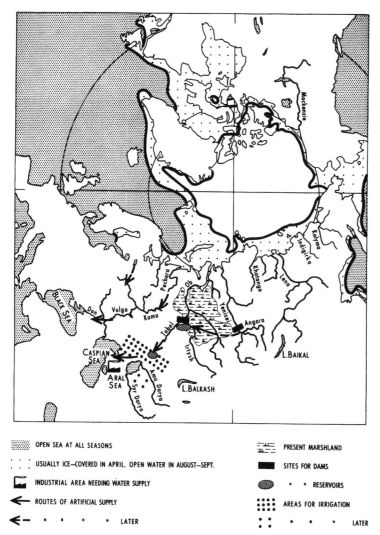

OPEN SEA AT ALL SEASONS

USUALLY ICE—COVERED IN APRIL, OPEN WATER IN AUGUST—SEPT.

INDUSTRIAL AREA NEEDING WATER SUPPLY

ROUTES OF ARTIFICIAL SUPPLY

◄— " " " " LATER

PRESENT MARSHLAND

SITES FOR DAMS

• " " RESERVOIRS

AREAS FOR IRRIGATION

" " " LATER

Fig. 19.6   Map showing the places concerned in the Russian engineering project for diverting the flow of the rivers Pechora, Ob and Yenesei and draining the west Siberian marshes to water the arid lands about the Caspian and Aral Seas.

The average limits of the Arctic pack ice about the end of April (seasonal maximum) and about the end of August (seasonal minimum) over the years 1911–56 are also shown.

is the medium in which the pack-ice at present covering the polar ocean is formed. This layer is maintained by:

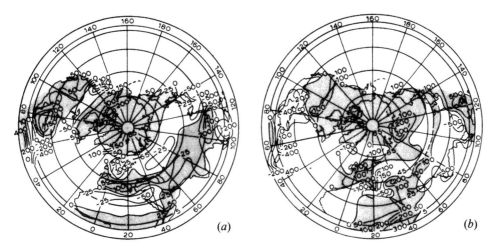

Fig. 19.7   Average precipitation anomaly in millimetres in seasons when the Arctic area in the sector concerned was $\geq 5°c$ warmer than normal
(*a*) November to March.
(*b*) April to October.
(*After* DROZDOV *1966*.)

(1) Freshwater from the great rivers of Asia and Canada, altogether estimated as averaging 5×10^3 km³/year of which the Russian and Siberian rivers contribute $3 \cdot 2 \times 10^3$ km³, the rivers involved in the scheme supplying 1 to $1 \cdot 2 \times 10^3$ km³/year.

(2) Downput of rain and snow over the polar ocean, estimated as contributing $0 \cdot 5$ to $0 \cdot 9 \times 10^3$ km³/year under the present regime.

(3) Alternate melting and freezing of ice in the Arctic Ocean, some of it derived from the ice sheets on Greenland and other smaller islands.

(4) Inflow of Pacific Ocean water through the Bering Strait (salinity about 32‰ compared to 37‰ in the water derived from the Atlantic). This contribution has been variously estimated at 10 to 30×10^3 km³/year and may well vary within about this range from one period of years to another.

Although the Pacific water is the major contribution to the upper layer, it seems likely that the actual surface water in which the ice forms (and in which salinities around 20‰ and less are observed in the seasonally open areas of the Arctic Ocean in summer) is much more largely contributed by the rivers than these figures suggest.[1] Tapping off any large proportion of the river flow might therefore substantially increase the area of salt water at the surface in the Arctic and reduce the ice correspondingly. In the regions affected the

1. The pack-ice is generally up to 3 to 4 m thick.

prevailing surface temperatures, which are now in the neighbourhood of $-20°$C to $-25°$C in winter and $-10°$C averaged over the year, would be raised to around $0°$C.

So a great change in the distribution of heating of the atmosphere would tend to change the position and alignment of the main thermal gradients in the northern hemisphere and, with them, the jetstreams and the development and steering of cyclonic activity. In this connection, the results (figs. 19.7(a) and (b)) of an empirical study by DROZDOV (1966) of the changes of precipitation totals over different parts of the northern hemisphere in seasons with temperature anomalies exceeding $+5°$C in the Arctic have to be considered. Note particularly the dryness indicated over great regions of the northern continents in middle latitudes, including the population centres and the central Asian region itself which the scheme for diverting the great rivers is designed to water. Yet the cases of $5°$C anomaly represent a modest estimate of the change, especially in winter, if the Arctic were to become an open ocean.[1] It appears, however, from DROZDOV's maps that an increase of precipitation directly over the Arctic Ocean might tend to replenish the freshwater component in the surface layer and make it easier for the ice to reform. Presumably a balance would ultimately be attained with less net reduction of the Arctic ice than at first appeared, but perhaps also with some awkward shifts of the climatic regime including a rather drier regime in the continental interiors.

Finally, in this section we must notice another implication of irrigation. According to BUDYKO (1974) the total area of lands now irrigated is approximately 2 million km^2 or 0.4% of the Earth's surface (1.3% of the land surface). Irrigation of arid regions decreases their albedo, perhaps on average by 10% (probably a conservative estimate). The corresponding change of the albedo of the entire Earth-atmosphere system due to the present extent of irrigation is estimated at 0.03%. But since, according to BUDYKO's calculations, a change of the Earth's albedo by 1% should alter the average temperature near the Earth's surface by $2.3°$C, the present use of irrigation may be maintaining the global mean temperature at a level nearly $0.1°$C higher than it would otherwise be. This figure is a small, but not negligible, contribution to compare with the climatic changes observed over the last 100 to 300 years (discussed in Chapter 18).

Man's direct output of heat (thermal pollution)

Artificially generated heat is significant so far only on a local scale, in the urban heat islands (Volume 1, pp. 517–19), where the surface air temperature in the city centre may average up to $2°$C higher than in the surrounding country, and in certain industrialized reaches of rivers,

---

1. DROZDOV's study – based on contemporary correlations between the temperatures prevailing in an individual winter or summer about Greenland, Spitsbergen, the north coast of Siberia, Wrangel Island, Barrow (Alaska), etc., and precipitation over the rest of the northern hemisphere – and the argument outlined in this section are described at rather great length in LAMB (1971). NEWSON's study (1973) with a dynamical model (see comment on p. 661 is also relevant in the connections here discussed.

lakes and inlets of the sea.[1] Annual average temperatures in such waters (near power stations, steelworks, oil refineries, etc., are occasionally 5° to 10°C above those prevailing in neighbouring unaffected water.[2]

According to FLOHN (1971), the best estimate of Man's present (around 1970) total output of heat, mainly from the burning of fossil fuels, is 0·03 langleys/day or 0·11 ly/day averaged over the land areas. The latter figure is approximately the same as the average flux of geothermal heat from the Earth's interior and, as such, is of no significance globally. However, estimates for the area of concentrated population and industry in western Europe between the Ruhr, Antwerp and Namur (500 000 km²) suggest an over-the-year average of 17 ly/day, and in the great centres such as Hamburg and Paris 100 ly/day; for Manhattan (central New York) the figure is 250 ly/day. Present average rates for Montreal are believed to be 117 ly/day in summer and 313 ly/day in winter. These are, essentially point sources of heat. They may be compared, however, with the world average values of solar radiation available at the outer limit of the Earth's atmosphere, 720 ly/day, and of radiation actually absorbed at the Earth's surface, 200–250 ly/day. The extreme observed natural input of heat from the surface into the overlying atmosphere is represented by figures in the range 1400–2200 ly/day in fast-moving Arctic airstreams over the warm waters of the northeastern Atlantic and Gulf of Alaska respectively.

Forward estimates of Man's energy use and corresponding output of waste heat into the atmosphere and oceans are precarious, being subject to many sorts of human, political and economic uncertainties, as well as the limits of raw materials on a finite Earth. Extrapolation based on any assumptions other than collapse of the economy or deliberate maintenance of near-zero growth, however, indicates that there are likely to be climatic problems over disposal of the increasing output of waste heat (as well as those not altogether dissimilar problems attending the disposal of nuclear wastes). An official United Nations report (*World Energy Supplies*, Statistical Papers, Series J: United Nations Statistical Office, Department of Economic and Social Affairs, New York) estimated by carrying forward the present growth of demand in countries all over the world that the total output of thermal pollution (waste heat) would rise (with a growth rate of the order of 5% per year) from $5·5 \times 10^6$ megawatts in 1970 to $9·6 \times 10^6$ MW in 1980 and 31·8 MW in A.D. 2000, a roughly sixfold increase over the 30 years. A further twelve- to fifteenfold increase is assumed by BUDYKO, DROZDOV and JUDIN (1966) on the basis of continuity of development up to A.D. 2050. By that time, on the given assumptions, the output of Man-made heat averaged over all the land areas would be of the order of 10% of the present average absorption of incoming radiation at the surface (HAEFELE 1974a, 1974b); as the waste heat output is at present many hundred

---

1. There is also a significant effect on precipitation, which seems to be increased by up to 10–15% over the centre of the heat island of a great city and as far as 80 km downwind.

2. In some localities exotic faunas have been established in the artificially warm water.

times larger than average in certain industrial areas and city centres, these would presumably have become uninhabitable.[1] Also, because of the incipient melting of the great land-based ice sheets in the Arctic and Antarctic, world sea level would begin to rise more substantially.

WEINBERG and HAMMOND (1970) predicted that Man's heat output would 'ultimately'.reach 3×10^{14} watts with a population of 20 000 million by about the middle of the next century. This heat output is equivalent to about 4·1 langleys/day averaged over all the land areas, or 13 ly/day over all the major population concentrations foreseen (and 78 ly/day over substantial areas within them). A map used by WASHINGTON (1972) – to explore the implications of this heat input for climate and the general wind circulation with a theoretical model – placed the following population concentrations:

(1) Over all eastern, southern and southwestern U.S.A. and central America, extending into limited coastal strips south to Peru and east through Venezuela.
(2) Over all the eastern third of South America between the equator and 40°s, right across the continent in 31–35°s and over Chile from there to 47°s.
(3) Over all Europe south of 60°N and northwest Africa, north of the Atlas Mountains.
(4) From 20° to 60°N in the sector 30–50°E and in Siberia between 50 and 60°N to 95°E.
(5) Persia (Iran) and Iraq, Afghanistan and the Indian subcontinent.
(6) All southeast and east Asia between the equator and 50°N, including China and Japan.
(7) New Zealand and a rather narrow strip of southern, southeastern and southwestern Australia.
(8) South Africa.
(9) Ethiopia and the Kenyan highlands.
(10) A limited coastal strip of West Africa north of the equator.

WASHINGTON acknowledges that his conclusion that the expected levels of Man's thermal energy production over the next 75 years entail rather small modification of the Earth-atmosphere heat balance is rendered doubtful by the serious shortcoming that the model assumed no change of ocean surface temperature.

BUDYKO (e.g. 1962 and in GALTZOV 1961) has suggested many times that Man's output of heat is likely to become the most serious cause of change of climate within about 100 years, and has mentioned that countermeasures such as the deliberate creation and maintenance of an artificial dust veil to reduce the receipt of radiation from the sun may have to be undertaken. Others (e.g. SCHNEIDER and DENNETT 1975) have pointed out in the same connection the advantage of tapping natural energy flows such as wind and water

---

1. Assuming that already by the year 2020 the world is inhabited by 10 000 million people with an average standard of living and energy use equal to the present United States level (10 kW per person), the output of energy would, according to SCHNEIDER (1974), amount to a few tenths of 1 per cent of the solar energy receipt and might be expected to raise the world temperature average by several tenths of one degree.

movement, which would not add heat that otherwise would not have been put into the environment.

Summary conclusions

The various ways in which Man has polluted the atmosphere and disturbed the environment so far have probably not individually affected the general level of prevailing world temperature by more than 1–2 tenths of a degree Celsius. Because some tend towards warming, others towards cooling the Earth,[1] their overall effect on world temperature has probably been very small, though anxiety about the continuing carbon dioxide increase continues.

Complementary to these conclusions, a theoretical modelling study by SCHNEIDER and MASS (1975) gives fresh ground for believing that the observed changes of global average surface air temperature from 1880 to 1970 could be rather well explained by the combined effects of volcanic activity and solar variations. These items therefore demand further study and continuous monitoring.

The projections of the various items due to human activity, and their effects both positive and negative on world temperature, into the future, however, commonly tend towards exponential increase. Forward estimates of their net effect are therefore unstable, but it appears likely that the increasing output of Man-made heat will gain the upper hand in the next century unless stronger controls are instituted than any that are believed to be yet contemplated (or unless there is a catastrophic breakdown of the international economy).

1. A computer modelling study by POTTER et al. (1975) at the Lawrence Livermore Laboratory, California suggests that the net effect of removal of the tropical rain forests (e.g. the Amazon forest) by Man, to increase the arable acreage, but in the process increasing surface albedo, should be a global cooling (of the order of 0·2 to 0·3C) as well as weakened atmospheric circulation, leading to reduced precipitation in the higher latitudes 45–85°N and 40–60°S.

Chapter 20

Approaches to the problem of forecasting

In 1969 one of the last acts of the World Meteorological Organization's Commission for Climatology before its untimely abolition was to establish a working group to review the increasing demand for forecasts of future climate and the possible approaches to meeting that demand in a sound scientific manner. The group reported to the Commission for Atmospheric Sciences, but the report (WORLD METEOROLOGICAL ORGANIZATION 1972) has not been published. That report has, however, guided the writing of this and the previous chapter. In the meantime, anxieties about the impact of climatic fluctuations upon the reliable year-by-year production of enough food to feed the world's rapidly growing population (and the need for a strategy for maintaining food reserves) have been sharpened by the object lessons of climatic impact in 1972 (and, to a lesser extent, in 1974 and 1975). These anxieties underlie the publication in the United States of recommendations for action (NATIONAL SCIENCE FOUNDATION 1974, NATIONAL ACADEMY OF SCIENCES 1975; see also COLLIS 1975). Serious attention to 'applications of meteorology to mitigation of undesirable environmental, social and economic consequences of climatic fluctuations' in future years has been urged by the W.M.O. Working Group on Climatic Fluctuations and Man (Chairman G. A. McKAY) (unpublished, Report of First Session in Geneva, January 1975). And on 15 April 1974 the United States Secretary of State HENRY KISSINGER stressed the increasing urgency of investigation of climatic changes in a speech to the United Nations General Assembly in New York and called on the International Council of Scientific Unions and the World Meteorological Organization to offer guidelines for international action.

The need for advice on future climate and for research to provide the understanding on which it must be based

The demand for climatic forecasts over periods from one season to many years ahead arises from agriculture and forestry (to guide planting policy), from the fuel and power industries (atomic energy, oil, natural gas, electricity – including hydroelectricity), from horticulture and the honey and vine trades and all those concerned in promoting new or marginal crops (e.g. extending the growing of maize in England and farther north in the U.S.A.), from civil engineers (concerned with the design size and provision of dams for irrigation and hydroelectricity), from water supply authorities, from the insurance industry (concerned principally with the incidence of tropical hurricanes and other severe storms, also losses due to exceptional droughts and floods), in connection with sea level changes affecting the probability of sea floods and the building of coast defences, from ship design interests, from the fisheries industry, from international trading interests, from the winter sports and tourist industries, in connection with proposals for the disposal of radioactive waste materials in the Antarctic ice sheet or in the ocean deeps, and from government and international agencies concerned with all these things. The greatest stakes of all have been involved in the stresses with which Dr KISSINGER was concerned, in some cases amounting to great human disasters with direct political consequences, arising from floods, famines and droughts (as in Bangladesh and the Sahel zone of Africa and Ethiopia in the 1970s, not to mention the Soviet and Indian grain purchases in the West and their upset to world prices).[1]

The vulnerability of the international economy – and therefore of international order – to climatic fluctuations is enormously increased (a) by the present size and growth of the world's population[2] and (b) to a less, but still very important, extent by the rationalization of

---

1. The world price of wheat doubled within a few months at the end of 1972 following harvest failures in India and the Soviet Union. Reserves of grain (which represents 70% of the world's food consumption) fell from their peak of 95 days' supply around 1960 to 69 days by 1970, dropped to 37 days in 1973 and to around 19 days' supply in early 1976.

2. The best estimates of world population history are believed to be (based on *Encyclopaedia Britannica* and United Nations Statistical Office):

About 250 million	at the time of Christ
500 million	in A.D. 1650
1094 million	in 1850
2070 million	in 1930
About 4000 million	in 1975

Extrapolation suggests that a population of 28 000 million might be present by A.D. 2070, but there are some grounds for supposing that the growth may level off and stabilize the population at between 8 and 12×10^9 in the middle of the next century.

The growth rate of 2% yearly over the past 25 years has been more or less matched by the growth of agricultural production thanks to technological advance – mechanization, nitrogen fertilizers, pesticides and new varieties of seed. Yields of wheat (per hectare sown) approximately doubled in Europe and North America between 1948 and 1970, but made much smaller advance elsewhere (McKAY and BAIER 1974). There is most room for increasing the yields of various crops in the potentially highly productive soils and climates of some of the underdeveloped countries in the warm zones of the Earth, but traditional farming practices are an obstacle, and it must be remembered that the modern high yields in the advanced countries are very costly in terms of energy use (oil); in 1974 in the United States it is estimated that it took 2 calories of fuel used directly on the farms to produce 1 calorie of food, and when transport and storage are included the ratio may have increased to 12 to 1. There is not much additional land which could be converted to grain

agriculture and international trade arrangements, which are leading extensive regions to concentrate on the one crop to which their climate and soil appear best suited; this is a return, on a large scale, to the one crop economies which underlay the historic famines of the Middle Ages. This aspect of the situation may even be aggravated by the 'wonder crops' developed by the plant-breeding efforts of the 'green revolution' since the 1960s, in so far as new strains are finely attuned to give a very high yield under just a narrow range of expected climatic conditions. The contributions of progress in technology, farm equipment and the use of energy (see footnote above) to increased yields have masked the fact that the turn to cooler moister summers in the late 1950s to early 1970s in the United States Middle West also made an important contribution to the increase. Moreover, the world total grain production in 1972 is believed to have been the second highest ever achieved up to that date (second only to 1971) in spite of the fact that weather caused yields to fall seriously below expectation in India and in the Soviet grainlands.

The scale of the impact of such events on international affairs is noted in the *Report of the Ad Hoc Panel on the Present Interglacial* (NATIONAL SCIENCE FOUNDATION 1974):

> In the late 1950s and 1960s food grains were in surplus in the United States and desperately short in India . . . as a result of unfavourable crop weather in India. . . . In 1966 and 1967, the United States exported one-fifth of its total wheat crop to India to forestall famine. . . . Since the United States could not make use of the rupees (which had been credited to an American account in Indian banks over all those years) without seriously dislocating the Indian economy, it reached an agreement with India in December 1973 by which it reverted $2·2 billion in rupees back to India for use in development programs.

Poor monsoons in India produced further serious shortfalls (and threats of famine overcome in a similar way) in 1972 and 1974. Weather in the U.S.S.R. in 1972 – severe frosts and inadequate snow cover in January in European Russia followed by prolonged heat and dry weather in spring and summer, with only 20–50% of the usual rainfall over a vast area including also the central Asian grainlands – caused the total Soviet grain production to be 12% (i.e. 22 million metric tons) below expectation. One quarter of the 1972 United States wheat crop was sold to the Soviet Union. That autumn, as in 1963 and 1965 for the same reason, massive Soviet grain purchases from Western countries were made. China had been in a similar situation in 1960. Only North America and, to a much lesser extent, Australia and Argentina now produce in most years an exportable surplus of grain.[1] But among the

---

production either in the cooler or in arid countries without great cost.

Another item in the equation is the growth of affluence over recent decades, leading to increased diversion of grain to the production of meat; because of this the demand for grain has been increasing at a rate of 4–6%, i.e. much faster than the global population and outstripping the development of higher yields.

1. On a limited sample of sixteen years 1949–64, WAUGH (1967) (see also THOMPSON 1969) reports statistically significant positive correlation coefficients between wheat yields in the same year in the U.S.A. and U.S.S.R. ($r = +0·47$) and in the U.S.A.

cases of large-scale climate-induced stress in the past must be listed the famous drought of the Dust Bowl years 1933–6 in the United States Middle West, when total U.S. wheat production was only three-quarters of that in the preceding and following years. Rather similar drought tendency in that region has been observed to recur at intervals of close to 22 years, threatening severely the exportable surplus for relief of difficulties in other countries.

An approximately 20- to 22-year ('bidecadal') cycle, affecting the latitude of the subtropical anticyclones and the frequency of blocking, is suggested by a number of phenomena (see the analysis of power spectrum and periodogram results, Volume 1, p. 242) and, if understood better in terms of mapped sequences of the circulation evolution, might be of use in forecasting. Among the phenomena cited in this connection are (a) the clusters of very mild winters in Europe in the 1910s, 1930s, around 1950 and in the early 1970s and the colder winters in nearly all the even numbered decades of the last three centuries,[1] (b) the droughts (usually accompanied by excessive summer warmth) that affected grain yields in the United States Middle West in the 1890s, 1910s, 1930s and 1950s (THOMPSON 1975). This sequence is traceable in Minnesota back through several further decades to droughts in the 1830s and 1850s, but in a time with more persistent tendency to drought in the Great Plains in the early Middle Ages there is evidence of a run of not less than 20 years of drought. Suggestions of some parallelism of experience in the grainlands of Argentina, Australia and the U.S.S.R. have been made.

Among the examples of climatic stress which should be remembered from the past are:

(1) The great Irish Potato Famine of the years 1845–8, with intermittent recurrences to 1851, caused by multiplication of the potato blight fungus (which had lately arrived in Europe from diseased tubers in a ship from North America) in a series of summers with suitably warm and moist weather. The folk-emigration, largely to the United States (but also to Britain), that followed halved the population of Ireland (which has never recovered) and left a lasting legacy of bitterness against England and the Anglo-Irish upper class who were blamed for many of the associated misfortunes.

(2) The comparable Scottish famines produced by the cold wet summers and harvest failures in many years between 1550 and 1700 (but particularly between 1693 and 1700), which caused a massive death rate and depopulation of the upland villages

---

and Australia ($r = +0.77$), which if representative of longer runs of years must emphasize the world food difficulties and increase the case for carrying reserves to tide over several bad harvest years.

In most of the United States corn (maize) and soya bean belt summer temperatures (and hence evaporation) are often somewhat too high for optimal production; yields improve in the cooler years. In the northern part of the Canadian prairies, however, wheat yields are cut by the colder summers and occurrences of early frosts.

Optimum temperatures for corn and soya bean given sufficient moisture are stated to be 32–35°C, and for most other grains in North America are 25°C.

Most grains and grasses will germinate at temperatures as low as 4°C, but corn and soya beans not below 8°C. Growth of the grasses (and of peas) usually ceases with temperatures above 37°C (or less), but with corn and soya beans at about 45°C.

1. My attention was drawn to this by Lt-Col. W. G. VAN VOSS of Ryswyk in the Netherlands.

and parishes and led to the establishment of the original 'plantation' of a colony of Scots in northern Ireland in 1612 and influenced the Union of Scotland with England in 1707.

(3) The late medieval *Wüstungen* and depopulation of many areas in northern Europe, referred to in Chapter 17, the decline of Iceland (and medieval Norway) and the dying out of the Old Norse colony in Greenland.

(4) The horrors (reported in detail by BUCHINSKY 1957), due both to harvest failures and harsh winters, suffered in eastern Europe, particularly Russia, in the late Middle Ages, also referred to in Chapter 17.

It is surprising that little study seems so far to have been made of the economic costs of the commonly occurring minor climatic fluctuations and of their comparison with the probable costs of research and organization of an international climate monitoring and advisory capacity. Data on the economic impact of weather on most kinds of human activity are still scarce (see, however, MAUNDER 1970), except in the case of electric power demand.[1] The big values involved are, however, illustrated by one relatively small-scale example (PITTOCK 1972): the difference in total farm incomes from the dairy produce of 750 000 cows in the South Auckland province of New Zealand between a wet January and a dry January is about two million New Zealand dollars. ELLISON (1970) rates the potential benefit to agriculture and industry in Britain that might be reaped from reliable forecasts for one season ahead at £40 million yearly (or perhaps more realistically at £4 million when allowance is made for lack of full use of such a forecasting service).

Other reasons for the need to develop a climatic forecasting capability are provided by questions with which policy-makers and the public increasingly confront science, ranging from the likely side effects on climate of this or that human activity to the implications of the increased climatic variability observed in recent years and the possibility of recognizing the onset of any major climatic shift (up to, and including, the end of the present interglacial).

It may be asked what Man can do about current or expected climatic changes, even if reliable warning could be given. The answers lie in the adaptation of planning and designs (of structures, etc.) to obtain the maximum prospective economic advantage and the minimizing of risks, in the choice of crops and planting limits, in plant-breeding to provide varieties with suitable tolerances, and in developing food supply and trading arrangements that are fortified by suitable reserves and emergency alternatives.

Forecasting: aims, methods and strategy

The atmosphere's existing store of heat and potential energy is used up in the development of winds and weather, including all the smaller-scale convective processes and eddies, and in

---

1. More electric power is now used in every major American city for cooling systems in an average summer than for heating in winter (J. D. McQUIGG, unpublished communication 1971).

driving the ocean currents, over a period of about 5 days. That is also roughly the normal life span of any one of the stable, more or less stationary, anticyclone cells associated with the prevailing steering of the travelling atmospheric disturbances in middle and higher latitudes and within the Trade Wind zone and of the run of weather associated with that steering pattern. The replenishment of the energy (and moisture) supply, which will control the weather after the present store is exhausted and the existing weather systems have decayed, depends to an important extent on the patterns of cloud and ice and snow, of parched or waterlogged ground, and of upwelling cold water or stagnant warm water maintained at the surface of the oceans, by the weather of the coming days. The problems of forecasting the weather or climate on all periods beyond 3 to 5 days have, therefore, this in common that they are in all cases concerned with foreseeing the development (and steering) of circulation systems which do not now exist and with identifying any persistent, or long-term continuously developing, features which will constrain and guide the future weather systems to a significant extent (i.e. within certain limits).

Forecasts of the weather beyond 3 to 5 days ahead therefore differ from shorter-range forecasts in not dealing with extrapolation of the movement and development of particular, known, cyclones and anticyclones, and attempting to specify their individual positions at this or that precise time, but must take a more statistical form (quite similar, however, to the treatment of showers and thunderstorms in an individual daily forecast).

The desiderata of forecasting the weather of seasons ahead and the climate of the coming years may be set down as:

(1) To approximate the central values or averages of temperature, rainfall and other items which specify the prevailing character of the weather.
(2) To indicate the frequencies (or probabilities) of events such as days with rain, hail, snow or thunder, frost or fog, great gales, long frosts or droughts.
(3) To indicate the probable duration of temperatures and rainfalls above or below such threshold values as are critical for various human purposes.

Practice must, however, be guided by what it is possible to do with a satisfactory rate of success. Most forecasts issued by various authorities so far for a month or a season ahead have attempted only category 1 of these objectives, commonly specifying the expected temperatures within quintile classes (identified as 'very warm', 'warm', 'average', 'cold' or 'very cold') and rainfall in tercile classes (as 'wet', 'normal' or 'dry'). It is probably the case at present (1976) that nowhere has the reference material needed to make forecasts in categories 2 and 3 yet been organized, even though it is widely suspected that the frequency, and perhaps the duration of rain over such periods ahead may be less difficult to indicate than rainfall amount.

For reliance to be placed upon any series of forecasts, or advice about the future, particularly in the early stages of developing a scientific capacity to make forecasts, and where very great prospects of gain or loss to the community are involved in the outcome, several requirements must be met:

(1) The methods used in deriving the forecast must be open and should be expressed in terms understandable to the recipients, so that those who have to bear the risk may decide for themselves how much they can (or should) venture upon it.

(2) This means that a statistical statement is most appropriate, of the form that in similar circumstances in the past a given outcome x followed in p_x cases and alternative outcomes y and z followed in p_y and p_z cases (p_x, p_y and p_z should be expressed as actual numbers of cases as well as percentages of all the relevant instances surveyed).[1]

(3) The identification of 'similar circumstances' as the basis of selection leading to the type of statement mentioned under item 2 actually demands understanding of the physical processes at work. Without such understanding the selection may be wrong, and the statements made (although seemingly objective and confidence-inspiring) may be downright misleading.

We may classify the possible scientific methods of forecasting or obtaining guidance about the weather and climate over long periods ahead in the following categories:

(1) The acceptance of some past period of years as representing the 'normal' climate or, it is hoped, as sufficiently representative of the conditions ruling in the planning period.[2] Climatic tables are then provided to present the statistics of weather observed in the past reference period and statistical routines are used to derive estimates of the probability, frequency or return period of rare and extreme events from examination of the frequency distribution of values observed within the span of past years used.

(2) The recognition that climatic fluctuations do occur, and are continually occurring, with a wide range of time scales and amplitudes, including some 'preferred' or characteristic time scales, but assuming that in the present state of knowledge the

---

1. The value of statements such as those envisaged under item 2 depends upon the correctness of the identification of relevant precedent cases.

2. By agreement adopted by the former International Meteorological Organization at its meeting in Warsaw in 1935 recent 30-year averages of weather observations are defined as climatic 'normals'. The original standard period so adopted was 1901–30; later the period 1931–60 was substituted, and it is planned to change the datum period every ten years, using the figures for 1941–70, 1951–80, and so on, as soon as they are available. The practice may be convenient from some points of view but can be, and

undoubtedly has been, very misleading. Both the 30-year periods so far used were in fact highly abnormal, being among the warmest (and 1931–60 probably *the* warmest) runs of three decades for some hundreds of years past. Moreover, it is the antithesis of the object of a datum to keep changing it. Furthermore, it can be shown that, to correspond with human memory of the past, and to provide a minimum error estimate of conditions over the next year or two ahead (see e.g. COURT 1967–8), averages of the last 10 to 15 years provide better guidance than 30-year averages; though to predict the next decade mean it appears best to use the data of at least 100 antecedent years.

timing of the onset of climatic shifts must be treated as random. This makes possible a number of actuarial statements such as the following:[1]

(a) The probability of onset within the next 1000 years of transition to a glacial climate associated with the 100 000-year glacial-interglacial cycle is about 2%.

(b) The probability of a shift within the next 100 years associated with the approximately 2000-year 'Neoglacial' (or it might be called 'intraglacial') cycle is about 8% (within the next 1000 years it is about 50%).

(c) The probability of one or more climatic shifts occurring within the next 10 years associated with those observed in the past on time scales of about 100 to 200 years is about 10–20% (within the next 100 years it is about 60–80%).

(3) The adoption of extrapolation of identified cyclic, or quasiperiodic, recurrences and the assumption that their effects are additive and not seriously complicated by interactions between them.[2] Various degrees of sophistication are possible. Some examples of climatic forecasts that have been attempted (see later in this chapter and refer to Appendix V for details) have interpreted time series of, for instance, mean summer temperature values as consisting essentially of just one, two or three sinusoidal oscillations. Others have resolved the variations of the prevailing barometric pressure and wind fields over the northern hemisphere into up to 50 orthogonal (mutually independent) components, and, proceeding then to analyse the quasiperiodicities in the variability of each of those components which (over some past reference period of 60 to 100 years) contributed more than a very small percentage of the variance of the whole, have extrapolated them some years or decades into the future.

(4) The appeal to analogous cases in the past ('analogue method'). This has been much used in monthly and seasonal forecasting, recognition of the essential character of the analogy being ideally extended to all the significant physical controls of the circulation development in atmosphere and ocean. The same argument could in principle be applied to forecasting from one year to many decades ahead, on the basis of the starting situation and its emergence over some time previously being essentially the same as the analogous cases in the past in terms of the most important long-term cyclic processes. The argument is insecure, however, and the forecast correspondingly without confidence, unless and until the nature of those processes is known and kept under constant watch.

---

1. The examples here given are specified in the National Science Foundation Ad Hoc Panel's report (1974).

FLOHN (1974), who attaches more weight to WILSON's theory of the critical role of surges of the Antarctic ice sheet in initiating glaciations in both hemispheres, and identifies five occurrences during the last 100 000 years of rapid hemispheric or global coolings of about 5°C within about a century, suggests a probability of about 1% for such an event to start within the next 50 years.

2. The case against the use of raw extrapolation of cycles for forecasting without understanding their physical nature and origins, even in the case of the well attested quasibiennial and 11- to 12-year cycles, has been well put by FENNER (1975).

(5) Combined physical and statistical handling of long-term weather processes. This is the necessary extension of (3) and (4) for application to forecasting over any length more than a week or two ahead. The essential point is to identify the causes, or causative influences and complexes, which trigger and control the characteristic successions of large-scale circulation development and processes over the weeks, months or years that follow. This approach is applicable alike to those processes that express themselves as cyclic or quasiperiodic evolutions (e.g. corresponding to the seasonal changes of the radiation budget or the long-term variations of tidal force) and to those that are started by the creation of a great volcanic dust veil or a solar disturbance. Examples have been given in Volume 1 of forecasting rules based on this type of understanding that have been found useful in seasonal weather forecasting (so far mainly in Europe and India, though presumably the same method could be applied to other parts of the world). Nevertheless, it must be stressed that recognition (at least in outline) of the physical processes at work is essential if the forecasting rules (and the statistics used to derive them) are to have any validity.[1]

(6) The use of elaborate physical and dynamical models for forward calculations of the tendencies of the large-scale circulation of the atmosphere and oceans and of the climatic elements such as radiation exchanges, temperature, rainfall and atmospheric content of moisture and pollutants. The contribution and outlook for this approach has been discussed in Chapter 13. The fine review of the status and prospects of modelling climate and climatic change by GATES (1975) is valuable not only in allowing comparison of the results of different models, and their ability to represent the real climatic mean conditions, but also leads to discussion of the very important problem of establishing the 'noise levels' – the standard deviations and higher moment statistics – of each model and hence the statistical significance of results indicated (for example, for the response to this or that environmental change). It is the present writer's view that the main (i.e. most realistic and useful) contributions to be expected in the foreseeable future from this approach are:

(a) Towards identification and understanding of the natural physical processes that produce changes of weather and climate on the time scales we are discussing and estimation of the magnitudes of their effects and interactions (feedback processes).

(b) Identification and understanding of the spread of pollutants in the atmosphere and ocean and the effects of these and other human activities (including

1. There are grounds for thinking that the rapid production of a huge additional amount of statistical reference material over the first years since the introduction of the giant computers into weather services, around 1960–5, has caused some deterioration (at least in some centres) in the practice of monthly and seasonal weather forecasting, owing to the proliferation of unsoundly based statistics and the failure to recognize redundant data which add nothing to the argument for this or that sequel, because they are no more than other symptoms of the same (unrecognized) large-scale process.

deliberate schemes for modification of climate), their time scales and magnitudes and their interactions with the natural processes of climate and the circulation of atmosphere and oceans.

(*c*) Estimates of conditions to be expected far into the future, beyond, say, 50 years.

Finally, in the pursuit of improved understanding and assessment of the probable future development, the following points are important:

(1) The extension of knowledge of the past record of climate and understanding of its behaviour in the past and present has been greatly helped, and may be further helped, by the extent to which the global wind circulation behaves as an entity and by the discovery of single items which apparently serve as rather simple indicators of the global condition. The frequency of W'ly days in the British Isles, the northward penetration of the monsoon rains in Africa, the southern limit of extension of the winter (upper) W'lies in the Arabian Sea and Indian sector (see JOSEPH 1975),[1] the sea surface temperatures in the equatorial Pacific and in the northern margin areas of the Kuro Shiwo and the Gulf Stream, the extent of the Arctic and Antarctic sea ice, and the downput of snow on the ice sheets in high latitudes, as well as the more obvious zonal and meridional indices of the circulation, may be examples of such indicators.

(2) The range of the seasonal changes which take place each year exceeds that of all but the greatest glacial/interglacial climatic shifts; the range of the year-to-year variations exceeds most longer-term ones (though not by the amount statistically expected on the assumption that no systematic long-term processes occur). Investigation of the physical links between various aspects of the seasonal and year-to-year changes may therefore throw light on the working of physical processes and linkages in the long-term climatic changes.[2]

1. Investigation of the period between 1964 and 1974 with full upper air data (up to the 50 mb level) showed an approximately 3-year recurrent oscillation over the east African, south Asian and Indian Ocean sector, most clearly seen in the prevailing latitude of the upper westerlies at the 150 mb level and in their latitude in the winter seasons. There has also been a net southward shift of the upper W'lies over the sector 30–110°E over the years investigated (and a lowering of 500 mb heights since 1957–64). Failures of the Indian summer monsoon and breaks in the monsoon occurred in those years (1965, 1966, 1968, 1972 and 1974, with severe failures in 1965 and 1972) when the upper W'lies in that sector were farthest south. In the 1965–6 and 1974–5 winters the northern hemisphere upper W'lies extended south to the equator and spread over into the southern hemisphere between longitudes 30° and 65°E. The monsoon rainfall was about or above average in

those years (1964, 1967, 1970, 1973) when the W'lies were displaced north and strong upper E'lies (at 150 mb) had prevailed over equatorial areas of south Asia in December–March.

2. An example of this type of reasoning is KUKLA's (1975) study of the closeness of the correspondence at different times of the year and in different parts of the Earth between (*a*) the seasonal changes of the radiation absorbed at ground level (dependent on solar radiation available and local albedo, neglecting contributions from heat transport and cloudiness and radiation exchanges in the atmosphere) and (*b*) the prevailing surface air temperature. The analysis showed that the change of irradiation is greatest between September and November, associated with the advance of snow cover over the northern continents and the seasonal break-up of pack-ice on the Southern Ocean (both of which take place more rapidly than the corresponding changes in the op-

(3) In studying the power spectra of the variations of the elements of weather and climate, and the possibilities of forecasting them within acceptable margins of error, the concept of 'noise' is useful. By analogy with the study of sound, the name 'noise' is given to variations in which no orderly elements are detected – i.e. to random variations. 'White noise' is the name (increasingly given in climatic time series analysis) applied to a variance spectrum in which there is no concentration of amplitude either in any narrow frequency band or over any broader range of frequencies. Many climatic time series have the appearance of what is called 'red noise', i.e. (whether or not there are any orderly elements such as oscillations of fairly constant frequency) there is a general increase of variance per frequency unit towards the low-frequency (long time-scale) end of the spectrum.

The most obvious implication of this for forecasting is a persistence tendency. It does, however, appear that the reality of a number of very long-term oscillations of roughly constant period lengths, notably including those of about 100 000, 40 000, 20 000 and 2000 years, and perhaps others of about 700–1000, 350–400, 180–200 and 90–100 years, can be established and that these contribute the appearance of trends and red noise in shorter runs of data.

All advice about future climate, over whatever period ahead, and whatever methods are used, must be qualified by the provisos that (a) no great sequence of volcanic explosions producing world-wide dust veils, and (b) no unforeseeable increase in the effect of Man's activities take place and alter the situation. Partly for these reasons, SCHNEIDER (1974) recommends the actuarial approach in wording forecast statements and that it is important that action be taken on

(1) Examination of past climate, including use of proxy data, to try to estimate the probabilities of droughts, floods and temperature extremes, etc., in different regions, and the prospects of droughts or rains (and good or bad crops) occurring simultaneously in different regions.

(2) Use of mathematical models to intensify study of the question of human effects on climate.

(2) Development of a world food security plan and strategies to forestall and relieve climate-induced disasters.

---

posite season). The response of the locally prevailing and world-average surface temperatures is also quickest and greatest at that season. KUKLA notes that the cyclic changes of September and October mean global total of the solar radiation at the top of the atmosphere calculated from the (MILANKOVITCH) astronomical variables of the Earth's orbital arrangement (see Volume 1, pp. 30–7, 492–3) showed the last maximum of irradiation in the northern hemisphere autumns 6000 years ago and the last minimum 17 000 years ago, in closer agreement with the times of postglacial highest and glacial lowest world temperatures than the radiation curves for any other season of the year. It seems justified to conclude that the northern hemisphere autumn and the northern part of the interiors of the great northern continents are the time and place of greatest sensitivity of the climatic regime to changes in the incoming radiation available.

(The analysis shows that the next minimum of the September and October radiation will occur about 4000 years hence.)

(4) Securing international agreement of all nations to refrain from unilateral action to modify the climate.

(5) Provision for compensation and relief in case of unforeseen damaging or disastrous consequences to some communities of attempts to modify the climate.

Historical note on the development of weather forecasting towards longer periods

The official weather services in most of the leading countries of the world were set up in the period 1855–75. Admiral FITZROY, the first Director of the British Meteorological Office, began a series of daily weather forecasts, based on extrapolation of the movement of storms seen on his maps of telegraphically collected observation reports, in 1855. Criticism led to withdrawal of the service for some years from the 1860s, but the issue of gale warnings to coastal fishing boats was valued, and continued, and by the 1880s is reported to have already reached the standard which has since been maintained. The first International Meteorological Congress met in Vienna in 1873 and started the international collaboration on weather observation networks and the exchange of messages, as well as an organization for the development of meteorology. This was followed after very few years by the observations of TEISSERENC DE BORT (1881) on the role of the semipermanent subtropical anticyclones, or 'centres of action', and in spells of unusual weather type the occurrence of slow-moving anticyclones in abnormal positions elsewhere, in controlling the weather development over wide areas. By the 1920s the behaviour of anticyclones was well enough known to warrant a service of fine-spell warnings for farmers wishing to be notified whenever two or three days without rain could be foreseen for haymaking and harvesting. This was apparently first introduced by the Danish Meteorological Institute in 1923, and similar services soon came to be operated in other countries. In the 1930s a success rate of about 80–85 % was claimed for the routine 24-hour weather forecasts in European countries based on 'synoptic weather map' methods; the further outlooks were less successful, and were only extended beyond the second day when a newly formed anticyclone dominated the scene.

The dawn of more modern methods was heralded by L. F. RICHARDSON's demonstration in 1922 that the future development of the wind field over 24 hours or so could be calculated by the laws of dynamics and thermodynamics from the existing situation, though the time taken (some weeks with a team of assistants) and the quality of the results made the method impractical at the time. Progress had to await extension of the regular observation coverage to three dimensions by the introduction of the radio-sonde network, the stationing of weather ships in the oceans of the northern hemisphere and the coming of the electronic computer to speed up the calculations – all in the late 1940s. A 10-level model introduced in the U.K. Meteorological Office (BUSHBY and TIMPSON 1967) incorporates effects of humidity into the calculations for the first time and is intended to provide broadly quantitative indications of expected rainfall amount. It is further hoped that the development on satellites of improved radiation sensing techniques will in the coming years make available satisfactory temperature observations at the surface, and at various levels in the atmosphere, for as close a network of points over all the oceans as may be required.

The greatest improvement in forecast accuracy which has so far been achieved by these methods seems to be in the forecasts for 2 to 3 days ahead. Forecasting beyond this introduces new problems, which make it still doubtful whether the range that can be satisfactorily dealt with by these methods will ever be extended beyond perhaps 5 to 7 days. It is a fundamental difficulty that the atmosphere uses up its initial store of energy in about 5 days. Experiments to test the reliability of numerical prediction from the fundamental physical and dynamical equations, reported by LORENZ (1969), indicate that it is just the largest-scale systems, and therefore the broad character of the weather, that are likely to be predictable within tolerable limits of error over the longest periods ahead for which any useful forecasting capacity can be developed. This has, in fact, long been recognized as the appropriate aim of long-range (monthly and seasonal) weather forecasting research, which has developed by quite other methods that have achieved some demonstrated skill.

Long-range meteorology, appropriately named *Grosswetterkunde* (the science of large-scale weather) by one of its pioneers, FRANZ BAUR, has developed over more than half a century along somewhat different lines from the dynamical meteorology that serves the aims of short-range weather forecasting. The dynamics of the atmosphere make a necessary contribution, though only as one of many aspects that must be considered. Equally fundamental are the great changes in the heating pattern that occur with the changing seasons, the varying heat and moisture exchanges between the oceans and the atmosphere, variations in the thermal characteristics of the Earth's surface due to snow and ice, variations in the transparency of the atmosphere due to cloudiness and volcanic dust veils, and the apparent small variations in the energy output of the sun itself.

A philosophy which has underlain much long-range weather forecasting research is that, if we can understand the relationships between the seasonal progression of the global atmospheric circulation and the great changes of the heating pattern that develop in the round of the average year (LAMB 1973), we can go on to trace the connections between the anomalies of that heating pattern, which distinguish each individual year, and the particular course of the windstreams and the steering of weather systems that make that year differ from the average.

The first scientific forecasting along these lines was instituted by Sir GILBERT WALKER, then Director General of Observatories in India, in 1908. On the results of an extraordinarily extensive probe for correlation coefficients, he proceeded to forecast the total rainfall of the summer monsoon season over all India each year in dependence on the amount of snow on the mountains in northwest India at the end of May, the mean barometric pressure in May in Mauritius, the mean pressure at three places in South America in April and May and the May rainfall in Zanzibar and the Seychelles. Over the first 13 years, from 1909 to 1921, excess or deficiency of the monsoon was correctly forecast 9 times; and in the 8 years when the calculated rainfall averaged over India departed by more than 25 mm (1 inch) from the 'normal' the sign of the departure was a correct prediction 7 times. Later, WALKER extended his studies and published a historic series of papers on 'World Weather' (see references in Chapter 6, Volume 1).

In 1912 MULTANOVSKY started a series of long-range forecasts in Tsarist Russia based on tracking the paths of anticyclones. His methods seem to have continued in use more or less to the present day. A report published in the 1960s noted that the method continued in use until the 1950s, when it was abandoned in favour of a supposedly more modern method with more theoretical content, only to be resumed after a few years when the latter method had failed to give as good results. A new contribution from Russia, published (with inadequate explanation of the groundwork) by BORISOVA and RUDIČEVA in 1968 (see Volume 1, p. 297), claims to recognize forerunners of the prevailing circulation character of the next one or two seasons, occurring as brief interruptions around certain common dates in the preceding seasons. This deserves further study, as it has given some apparent success when applied without modification to the British Isles.

BAUR in Germany began to build on WALKER's ideas in the 1920s, constructing multiple correlation tables connecting the values of certain indices of the 'action centres' of the atmospheric circulation over the Atlantic and Europe with subsequent weather. Ten-day forecasts were issued for parts of Germany in the summers of the 1930s. Later he was able to identify many persistences and recurrent processes; he also increasingly made use of statistically established relationships to the phase of the sunspot cycle (and to the incidence of individual solar disturbances) at a time when this topic was severely out of fashion in many other countries. In 1929 he had been appointed to establish a Long-Range Weather Forecasting Research Institute under the Ministry of Agriculture in Prussia, which unhappily later became the victim of political events and was finally dissolved in 1945. Nevertheless, BAUR's work trained a whole generation of German meteorologists in the basic concepts of large-scale and long-range meteorology and put this branch of science on the map by providing many physical insights, and establishing a set of 'physical-statistical forecasting rules', a classification of large-scale circulation patterns (*Grosswetterlagen*) over Europe and the surrounding seas, and a calendar of 'singularities' or characteristic seasonal weather episodes which tend to recur close to the same date in many years and certain abnormalities in which in any given year may serve as indicators of the character of the season that follows.[1] The seasonal forecasts, mainly for the summers and winters in central Europe, which he personally issued in the 1950s, though restricted to cases where his physically based statistical reasoning gave positive indications, were independently assessed as over 90% successful. This is the highest success rate so far achieved by any series of long-range forecasts. Temperature and rainfall values were indicated within broad ranges (possibly too broad, however, for many practical requirements).

Much of BAUR's philosophy on the subject and some of his actual 'rules' have been

1. Note the similarity with the ideas and technique later
   developed by BORISOVA and RUDIČEVA.

incorporated into the research and practice of long-range weather forecasting in Britain, France and other countries (this is apparent in the examples of forecast rules given in Chapter 7, Volume 1, of the present work). Monthly forecasts were first published in Britain in 1963. 'Thirty-day forecasts', together with forecast maps covering the entire northern hemisphere, had been issued by NAMIAS in the United States from some years earlier. These were based on statistical treatment, averaging and extrapolation procedures applied to indices of circulation strength as well as to the pressure distribution and features seen on upper air map analyses; as in BAUR's work, the treatment was based on physical and dynamical considerations.

A valuable further development has been the recognition of statistically significant associations between persistent anomalies of sea surface temperature in certain areas of the great oceans and the patterns of development of the global wind circulation (see Chapter 10, Volume 1). Although the anomalies in the sea are a response to the climate above the ocean surface over some previous period, we see here a feedback which has strong influence on the weather that follows over the hemisphere for some months, and sometimes for several years, afterwards. Because of the great well of heat which the oceans represent, and their mechanical inertia, the role of the oceans is likely to appear in most ways more important than that of the atmosphere in maintaining changes of regime on all the longer time scales of climate (a point stressed in NATIONAL ACADEMY OF SCIENCES 1975, p. 25).

Monthly and seasonal (3-month) forecasts, issued every month or season as a routine, and based on the techniques developed by this research have been found broadly successful on about 60–75% of the occasions in different years. As such, there is no doubt that they should show a net economic return when any long series of practical decisions in agriculture, or other activities, takes account of them; although, with any series of long-term forecasts issued at such intervals, crises of confidence are likely to arise from time to time when two or three forecast failures occur in succession.

The present writer's experience of forecasting by such methods since 1956 indicates (LAMB in WORLD METEOROLOGICAL ORGANIZATION 1972) that:

(1) Most forecasting success has been obtained in cases where the physical process underlying the months-long sequence of connected circulation developments seems to be at least recognized in outline and where the indicator rules all give the same verdict. There is some evidence also that the mean time over which a valid forecast can be given is longer by 1 to 3 months in these cases than in the general run of forecasts.

(2) Many forecast failures can be attributed to cases where undue weight was probably given to a number of forecast 'rules' which gave the same verdict because of an unrecognized repetitiveness of the rules concerned – i.e. the preconditions stipulated by the different rules expressed only different parts of one and the same large-scale process.

It has to be expected that the same lessons will apply to attempts to forecast the prevailing climate over still longer periods ahead.

Towards climate forecasting

The longest periods ahead for which any such forecast rules or indicators as those mentioned in the last section, and illustrated in Chapter 7 in Volume 1, have so far (1976) been developed are believed to be 6 to 9 months (in the case of the Indian southwest monsoon and the winter weather in Europe). Possibilities of forecasting over periods from half a year to 20 years ahead have received far too little research effort in relation to their importance in connection with economic decisions and planning.[1] Clearly any valid associations between the large-scale atmospheric circulation and weather pattern development and the 11-year and longer quasiperiodic variations ('cyclical' elements) of solar activity, any predominant patterns and time scales of development following known volcanic eruptions, and any authentic persistences and cyclical tendencies in the behaviour of the atmosphere and oceans, contain within them (and could be used to provide) first indications of the developments to be expected over the coming years.

The obvious problems which need to be solved are:

(1) To establish the physical nature of the causation of such associations, persistences and cyclical tendencies and the characteristic magnitudes and durations of the effects of each.

(2) To trace the interactions and feedback processes which cause modulation of the amplitude, course and duration of the individual cycles. Only then can these be appropriately monitored and interpreted so that the course of the current evolution may be correctly foreseen.

(3) To improve knowledge and forecasting capacity in relation to all potentially significant external influences, viz. solar and cosmic variables, the operation of tidal

. Some research directed at forecasting the character of seasons about one year ahead is, however, reported since 1973 in the Institute for Environment Studies, University of Wisconsin, Madison (under R. A. BRYSON and T. STARR) and in the U.K. Meteorological Office (RATCLIFFE 1975).

Anxiety about the impact of climatic fluctuations on the world food supply and economic problems led the United States to establish (with effect from 15 November 1974) a Center for Climatic and Environmental Assessment, under the National Oceanic and Atmospheric Administration, at Columbia, Missouri. The current global weather patterns and development of crops will be kept under constant review, and a Model Development Division will seek to promote better understanding of relationships between crop yield and previous weather.

An interesting contribution to the problem of forecasting a few seasons ahead has appeared since this chapter was written. R. A. EBDON (1975) has produced maps which reveal a hemisphere-wide pattern of statistically significant differences – over the 20 years for which stratospheric maps are available – of prevailing mean sea level pressure distribution in mid summer and mid winter according as the quasibiennial oscillation finds the equatorial stratospheric winds at the 30 mb level in the strongly W'ly or strongly E'ly phase. The E'ly phase appears to favour weakening, or reversal, of the surface W'lies in middle latitudes of the northern hemisphere, both in winter and summer, particularly in the east Atlantic and European sectors but also in other parts of the hemisphere.

variations, the Earth's wobble and things affecting the rotation of the Earth (length of day), volcanic events, and all effects of the works of Man.

Also the need for continual efforts to improve in detail and in length our knowledge of the past record of climate – particularly the repeats of cyclical processes – must be stressed, because this is vital both to the resting of theoretical models and to the framing of sound actuarial statements regarding the range and probability of items of future climatic behaviour.

It is broadly agreed that the framework of a climatic forecast at the present time must take account of the following:

(1) Since the present-day climate is much warmer than the average of the last million years, and since the climate of the first half of the present century was significantly warmer than for many centuries past, and that warmth has not been fully maintained since 1950–5, the 'law of averages' suggests a trend towards cooler conditions.

(2) If all the apparently significant quasiperiodic variations were accepted as real, approximately sinusoidal and constant in period, and non-interactive, forecasts could be made on the basis of provisional estimates of their magnitudes. These have been assessed by the Ad Hoc Panel on the Present Interglacial, as shown in Table 20.1.

(3) The possible impact of future volcanic activity which is always in the direction of cooling.[1]

(4) Some estimate of the future impact of human activity. The prevailing opinion at present (unlike that around 1955–60) seems to be that this is currently working in the direction of cooling, but there is a consensus that (even given a great diminution in the output of carbon dioxide as fossil fuels approach exhaustion or are progressively given up in favour of nuclear fuel and the use of solar energy) Man's effect will be in the direction of increasingly powerful warming starting about A.D. 2000–2050.

On the long time-scale effects produced by the Earth's orbital variations, the probable future development can be calculated and can indeed be read off fig. 15.2(b). The same calculations have been presented in the rearranged diagram here reproduced as fig. 20.1, designed to show the degree of correspondence between the present situation and that about the end of the warmest part of the last (Eemian) interglacial about 115 000 years ago (placed in the upper panel of the diagram immediately above the present epoch in the lower panel). There is no doubt about the qualitative resemblance, but the radiation changes which followed 115 000 years ago were much greater than any expected in the next several thousand years. That is because the ellipticity of the Earth's orbit has remained slight over the last

---

1. At present this is unpredictable, but there is some evidence of quasiperiodic components of its vari-   ation, including one of about 200 years length.

Table 20.1   Estimated characteristics of principal climatic fluctuations, assumed quasiperiodic

(After NATIONAL SCIENCE FOUNDATION 1974)

Characteristic wave length in years (approx.)	Estimated average range (double amplitude) in °C	Estimated time of last temperature maximum (years ago)	Present phase: temperature level	Rates of temperature change (assuming smooth sinusoidal shape of curve) in °C/year	
				Fastest $2\pi A/P$	Around 1970s
P	$2A$				
100 000	8·0	10 000	Very high	±0·00025	−0·00015
20 000	3·0	8 000	High	±0·00045	−0·0003
2 000*	2·0	1 750	Middle	±0·0025	+0·0024
200*	0·5	75	High	±0·0075	−0·0053
100*	0·5	35	Middle	±0·015	−0·0121
			Totals	±0·0257	−0·0154

Note: The physical origins and statistical significance of quasiperiodicities of lengths around 200, 400 and 2000 years (also those of around 2 to 3 and 20–22 years and many other lengths) are not nearly so clear or firmly established as those around 20 000, 40 000 and 100 000 years (see Chapter 15, p. 312). Nevertheless, fluctuations of about 200 years in length have been indicated in a considerable diversity of climatic data (see Volume 1, Chapter 6), and in the isotope measurements in the Greenland ice sheet stratigraphy, as well as in the fossil evidence of the quantity of radioactive carbon in the atmosphere (and, curiously, also in volcanic eruption data). PISIAS *et al.* (1973) report similarly on fluctuations of about 380 years length. Provisionally, all the fluctuations with which this footnote is concerned may be taken to be of solar origin and this is generally supposed to be the origin also of fluctuations of around 80 to 100 years in length.

100 000 years and is expected to continue slight for the next 100 000 years, a situation calculated to have last occurred between 1·8 and 1·6 million years ago. On that ground MITCHELL (1972) doubts whether the climatic changes that lie ahead will be so severe as those that marked the last several interglacial-glacial oscillations. However, the severest development of the last ice age in fact set in twice within the last 70 000 years, particularly around 70 000–60 000 years ago and from 30 000 years ago onwards, a 40 000-year spacing which suggests association with the variations of obliquity of the Earth's axis. Fig. 20.1 shows that there is some quantitative as well as qualitative resemblance between the present situation and those about 68 000 and 40 000 years ago. Other workers (see, for example, KUKLA and MATTHEWS 1972) have concluded from the parallel course of vegetation history in postglacial times and previous interglacials that the present situation is that of an interglacial in its late temperate stage – the third of the four typical stages in WEST's (1968, p. 303) scheme – and that successive future climatic oscillations on time scales from centuries to millennia superposed on the declining tendency of the radiation (northern summer temperatures) regime will bring stepwise advances of the natural vegetation boundaries

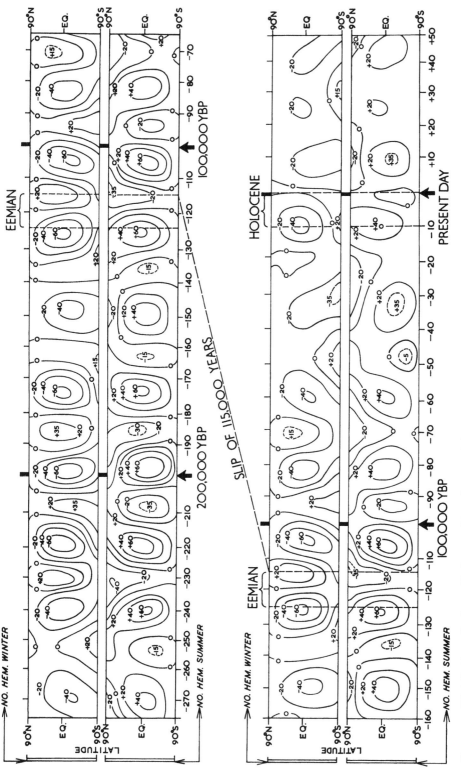

Fig. 20.1 Distribution of solar radiation reaching the top of the Earth's atmosphere in the two halves of the year, by latitude, from pole to pole.

Upper panels = from 275 000 years ago to 65 000 years ago.

Lower panels = from 160 000 years ago to 50 000 years in the future.

Units: calories/cm²/day, expressed as departures from present values at each latitude.

The upper and lower panels have been placed so as to show the correspondence between the present situation after 10 000 years of the postglacial (Holocene) and the situation 115 000 years ago, at the end of the warmest part of the last (Eemian) interglacial.

(*From* MITCHELL (*1972*); *reproduced by kind permission. Calculations by* A. D. VERNEKAR – *see Chapter 15.*)

equatorward. Parallels in the last two interglacials for the mid postglacial elm decline are among the details which seem to support this diagnosis. On all these grounds it seems justified to expect a glacial climax situation about 12 000 years hence. KUKLA's (1975) suggestion that it is the radiation budget in September–October that is most critical would place this temperature minimum only 4000 years in the future.[1]

On the shorter term, the analysis presented in this section indicates that the present tendency of the natural climate is towards a further, but fluctuating, cooling and no likely recovery of the temperature levels prevailing in 1930–55 until around A.D. 2100 (though this date may be brought forward by the effects of Man's waste heat). BRYSON (1975) has usefully pointed out on the basis of KUTZBACH's examination of the 1000-year record of the incidence of sea ice about Iceland, and estimates of prevailing temperature derived by BERGTHORSSON therefrom, that coolings such as that since 1945 have never in the past millennium continued for less than 40 years, nor has the climate returned to its previous state in less than 70 years. On the other hand, the operation of the approximately 2000-year variations should moderate the implied temperature minimum in the coming decades and suggests that it is unlikely to equal the level of A.D. 1650–1700 or around 1800.

The apparent implications of the recent cooling, and perhaps of most other known cases of climatic cooling trend in the past, for the world distribution of the downput of rain and snow and the frequency of monsoon failures in India are pointed to in Chapter 18 (pp. 538–48) and have been illustrated by BRYSON (1973).

It must be borne in mind that the assumptions on which this framework of the climatic outlook for the future is built are crude – e.g. the adopted constancy of period lengths, the absence of interactions between the different components of the variability – and therefore probably the assumption of sinusoidal shape of the basic oscillations is somewhat unrealistic. Because of the pressing need of climatic foresight for some projects, and also as a result of the challenge presented by the problem to individual research workers and institutes, a number of scientifically based forecasts of the natural climate have been attempted already. Those known to the W.M.O. Working Group on Climatic Fluctuations which reported in 1972 and a few independently derived ones collected since are listed briefly in Appendix VI.

In most cases, the forecasts attempted so far have been based on a single line of scientific reasoning, assumed dependence on just one controlling variable (often solar activity) or the extrapolation of a single time series, and may therefore be classed as 'one-track-mind forecasts'. Methods which take account of all variables which are likely to be relevant should be preferred. Nevertheless, despite the wide variety of physical and statistical approaches

---

1. An analysis by BRAY (1971) of the effect of the superposed oscillations of period length (according to BRAY) around 2400 to 2600 years, attributed to a recurring solar variation, is interesting. The next fluctuation comparable to the Little Ice Age climax of A.D. 1500–1900 and the cold oscillation between 900 and 100 B.C. must be expected to reach its climax about A.D. 4300 according to BRAY (or possibly between A.D. 3500 and 3900), and may be cool enough to develop some sub-Arctic continental ice sheets and mid-latitude mountain glaciation.

involved, there is a remarkable degree of agreement between the forecasts for the next 30 years collected in Appendix VI. The list can hardly be complete, but it includes all the most important statements that were known to the Working Group and is thought representative of the methods used up to this time and the verdicts expressed. The consensus of the forecasts clearly falls within the framework outlined in the paragraphs above and attempts some further detail (with some apparent success in the years elapsed up to 1975). The position encourages belief in the possibility of developing a real forecasting skill in relation to such periods ahead, and suggests that some useful guidance has already been obtained, from this scientific research.[1]

1. Since this chapter was written, however, an official statement issued by the World Meteorological Organization in June 1976 places most emphasis on the prospect of Man's impact on the global climate, through the increasing production of carbon dioxide and waste heat, both producing a warming effect expected to become dominant over the natural climate fluctuations by about A.D. 2000. The statement warned of dire consequences to be expected within the next 50 to 100 years through displacement of the natural vegetation and crop belts and melting of ice caps.

Appendix to Part IV

VI Scientifically based climate forecasts known to have been issued

(1) *Author:* C. EASTON (in 'Klimaatschommelingen en Weervoorspelling', *Tds. Kon. Ned. Aard. Genootsch.*, **34**, No. 5; Amsterdam 1917).
Basis: 89-year cycle of solar activity reflected in the history of the severity of the winters in Europe from A.D. 760 to 1916.
Forecast: The general mean winter temperature over the 22 years after 1916 expected to be lower than in the preceding 22 years in western Europe 'with one or two severe winters, one of them possibly very severe'.
Comment: The two 22-year periods shared almost equally the highest level of winter temperatures for several centuries past, though the later of the two periods had marginally the lower mean winter temperature and there was literally just one severe winter (or two including 1917 itself). This pioneer attempt seems to have had little success, perhaps because the variation used as its basis was at the time overlaid by bigger amplitude and apparently longer-term variations.

(2) *Author:* R. SCHERHAG (in an article on the warming of the Arctic in *Ann. Hydr. und maritimen Met.*, **67**, 57–67; Berlin 1939).
Basis: Periodicities of unascertained origin observed in the circulation of the atmosphere.
Forecast: About the cause of the 110- and 220-year periods little can be said. In any case one can conclude that the [European winter] temperature maximum must be attained in this decade [the 1930s] and that then a gradual increase in the frequency of severe winters is to be expected.

Comment: Became famous because the next three winters, 1940–2, were the coldest for 50 or more years previously. The forecast appears, however, to be literally verified by events up to at least 1970.

(3) *Author:* H. C. WILLETT (in 'Extrapolation of sunspot-climate relationships', *J. Met.*, **8** (1), 1–6; Lancaster, Pa., 1951).

Basis: Extrapolation of cyclic trends tentatively associated with sunspot activity.

Forecast: 'The next sunspot maximum will be delayed until . . . 15 years or more from the 1947 maximum. The maximum annual sunspot number will be one of the smallest on record, probably in the vicinity of 50. . . . The temperature level over much of the world will fall significantly during the next 15 years, probably reaching a first minimum level during . . . 1960–5. This temperature fall will be sharpest where the anomalous warmth of the past 25 years has been most extreme. The greatest cooling, then, should occur in the Spitsbergen-Greenland-Iceland area, while significant cooling will occur in northern Europe, the eastern United States and in middle and lower latitudes of both hemispheres. Exceptions to this downward trend of temperature may be noted in the Antarctic, the interior of western Canada and the northwestern United States, and particularly from the eastern Mediterranean northeastward into Russia. . . . The rainfall in lower middle latitudes, south of 50°N, will be substantially higher during the next 20 years. . . . The general recent recession of glaciers in all regions is due to be reversed in the very near future. The change of trend should be most pronounced during the next 20 years. . . .'

Comment: The change to weaker and longer sunspot cycles foreseen took place one cycle later than expected, i.e. in the 1960s–70s, not in the 1950s, so the forecast of the sunspot maximum which actually fell in 1957 was utterly wrong, it being the highest, not the lowest, ever observed. Nevertheless, because the spot areas exceeded the other types of solar disturbance such as faculae and flares in an exceptional manner, it may have reproduced some of the characteristics of a weak activity cycle, particularly as regards total energy output of the sun. The lengthening of the sunspot cycles also was less than forecast – i.e. the 1969 sunspot maximum followed about $11\frac{1}{2}$ years after its predecessor and the highest yearly sunspot number was about 107 against 190 in 1957 and 152 in 1947. The statements about temperature and rainfall were, however, all completely right. Glaciers generally have been slower to respond to the change of temperature trend than WILLETT expected (see Chapter 13, pp. 149–55).

(4) *Author:* A. A. GIRS (in a number of papers, e.g. 'Modification of forms of atmospheric circulation and their analysis', *Met. i Gidr.* No. 3; Leningrad, 1956).

Basis: Succession of predominance of various large-scale (half-hemisphere) circulation pattern types over periods ranging in length from 9 to 29 years.

Forecast: After predominance of the WANGENHEIM/GIRS types W + C 1891–9, W 1900–28, E 1929–39, C 1940–8, and E + C 1949–64, symptoms of increasing frequency of type W have been detected from 1950 onwards and type W is expected to dominate the next 20 years to the mid 1980s or 1990.

Comment: Classification of circulation patterns over the British Isles from 1861 shows the frequency of zonal W forms in this sector in decline from about 1950 and reaching in 1968–76 a level lower than any seen in the previous 116 years. Whether because of a pattern of solar behaviour which has not recurred since the great sunspot maxima in the 1770s and 1780s or for other reasons unknown, it appears that a much longer span of years of circulation pattern records is needed to show situations analogous to those occurring in the 1960s and 1970s or to draw firm conclusions. In fact, the only analogous case so far identified of the frequency of the W type in the British Isles as low as in 1968–76 is provided by the years 1781–5.

(5) *Author:* A. L. KATZ (in *Seasonal Changes in the General Circulation of the Atmosphere and Long-Range Forecasts*, Leningrad (Gidrometeoizdat) 1960, 270 pp.: in Russian, translation in U.K. Met. Office Library). Summarizes also much earlier, similar Soviet work and forecasts made in the 1950s.

Basis: Classification of hemispheric circulation pattern types, meridional and zonal indices and the ratio of the one to the other, and their associations with the prevailing level of solar disturbance and its 11-year and longer cycles.

Forecast: Weaker solar activity expected after the '11-year' cycle minimum in the mid 1960s and therewith reduced intensity of the general atmospheric circulation, both meridional and zonal components, for the rest of this century, though more days of zonal type. This should lead to wetter years and snowier winters in the Volga basin, and with decreased evaporation in cloudier summers the natural trend should be towards renewed rise of the level of the Caspian Sea after the decline over recent decades.

Comment: The predictions as regards solar activity and general atmospheric circulation intensity so far appear sound. The continued, alarming fall in level of the Caspian Sea may be largely due to increasing diversion of the Volga River water for irrigation and industrial purposes.

(6) *Author:* L. A. VITELS (reported by A. P. GALTZOV in the proceedings of the Conference on the Modification of Climate, *Izvestia, Ser.Geogr.*, **5**, 128–133; Moscow, Akad. Nauk., 1961: in Russian, translation in U.K. Met. Office Library).

Basis: 'A 100-year cycle of solar activity, observed in the variations of WOLF (relative sunspot) number over the last 200–250 years and confirmed by evidence of polar aurorae, tree rings, Nile flood levels and other phenomena associated with solar activity over periods up to 3200 years past'.

Forecast: A particularly low level of solar activity is expected in the last 30 years of the present century and in the beginning of the twenty-first century, accompanied by a weakening of the atmospheric circulation, a change in the paths of depressions from our northern seas to the south, and an intensification of the continental character of the climate of European Russia and western Siberia with an increase in the natural drainage (i.e. flow) of the River Volga.

Comment: Circulation pattern changes so far well described (see comment on (5) regarding Caspian Sea).

(7) *Author:* G. SIRÉN (from 'Skogsgränstallen som indikator för klimatfluktuationerna i norra Fennoskandien under historisk tid', *Communicationes Instituti Forestalis Fenniæ,* **54** (2); Helsingfors 1961).

Basis: Periodogram analysis of the growth ring series of birch and pine trees near the northern forest limit in Finland 1181–1960. Prediction on basis of extrapolating six, and especially the three largest amplitude, cyclic oscillations of the tree ring series (73, 93 and 204 years), which were also shown to be correlated with the observed fluctuations of average summer temperature in northern Finland (which was 2·7°C higher in 1930–40 than in 1900–10).

Forecast: 'A marked deterioration of climate (after 1960) right up to *c.* 1975, whereafter a weak oscillation around the mean level (of that time) will follow'. The forecast covered the period 1960 to 2060.

Comment: Believed so far right, at least up to 1971, as regards general direction, and supported as regards magnitude by the changes in the extent of ice and prevailing temperatures in northeast Greenland and Franz Josefs Land, but complicated in 1972–5 by the run of mild winters in those years and two very warm summers, in 1972 and 1974.

This forecast is of special interest because meteorological data played a very small part and no hypothesis about solar (or any other) causation was used. The data were thought to confirm a conclusion reached by palaeobotanists examining data from much longer periods past.

(8) *Author:* B. L. DZERDZEEVSKI (in 'Long-term variability of the general circulation of the atmosphere as a basis for forecasting climate', *Proceedings of a Conference on the General Circulation of the Atmosphere,* Leningrad (Gidrometeoizdat) 1962: in Russian).

Basis: Classification of daily hemispheric circulation pattern types and study of changes in the predominance of this or that group of types with solar activity, particularly the solar cycle of supposedly about 80 years in length.

Forecast: A 20- to 25-year epoch of predominantly zonal circulation character according to DZERDZEEVSKI's classification is considered to have ended about 1952. Increasing meridionality was expected to continue and produce a maximum of meridional character in the 10-year average 1959–68, or possibly 1961–70, followed by a decline. In 1968 DZERDZEEVSKI suggested that the end of this predominantly meridional phase might come as late as 1975, since there was up to that time no sign of downturn: the ensuing maximum of zonal character should be expected about 20 years after the meridional maximum.

Comment: Forecast evidently right as regards the meridional character of the 1960s which is believed to have affected most sectors of both hemispheres. Classification of circulation types over the British Isles shows, however, the meridional character being maintained at least up to 1968–76 and possibly becoming still more extreme.

(9) *Author:* J. Cosimo (in 'On the change of temperature in midsummer in Milan', *Atti 13 Conv. Ann. Roma, Assoc. Geofis. Ital.,* 243–245; Rome 1963).
Basis: Statistical analysis of the very long series of temperature observations at Milan, and extrapolation.
Forecast: A lower level of summer temperatures expected to prevail in the remainder of the twentieth century.

(10) *Author:* M. I. Martinov (in 'Climate in the Earth's evolution, its present state and possible future trend', *Priroda,* **12** (6), 39–42; Sofia 1963: in Bulgarian).
Basis: Sun–weather relationships.
Forecast: More severe climatic conditions expected till 1980.
Comment: No hint here of the run of mild winters in Europe 1971–5, but criticism may be premature – the forecast was well verified up to 1970.

(11) *Author:* P. G. Rackliff (in 'Summer and winter indices at Armagh', *Weather,* **20**, 38–44; London 1965).
Basis: Statistical analysis of the temperature, rainfall and sunshine observations and number of days with snow or sleet, at Armagh Observatory, Northern Ireland from 1880 to 1962, combined in the form of summer and winter indexes. Cycles of 30–35 years length in the summers, and 80–90 years length in the winter conditions prevailing, seemed indicated by the patterns of occurrence of deviations exceeding one standard deviation from the mean.
Forecast: 'Extrapolation can be presumptuous and misleading; nevertheless the winter curve clearly shows the period of warming that occurred during the first quarter of the present century together with the more recent decline and there is at least a hint of harsher conditions to come, and a possible excess of cold winters during the decade centred about 1975 or 1980.'

(12) *Author:* C. G. Abbot (in 'Solar variation and weather', written over 91 years of age; *Smithsonian Misc. Collections,* **146** (No. 3), 67 pp.; Washington 1963).
Basis: One principal cycle of 273 months ($22\frac{3}{4}$ years) in solar variation and in weather data, together with consideration of 26 harmonics (or submultiples of 273 months), among which a periodicity of $30\frac{1}{3}$ months was stressed.
Forecast: The technique was first used to 'predict' the monthly rainfall values at St Louis, Missouri, from 1875 to 1879 from the analysis of the whole record from 1854 to 1939. A correlation coefficient of $+0.80$ was claimed between the predicted and the actual sixty monthly values. Other 'predictions' of precipitation at other places, including Buenos Aires, for periods already past were tried by this method, with more or less similar results claimed. Finally, a forecast was made of precipitation at Nashville, Tennessee, month by month from 1965 to 1970.
Comment: No physical reasoning advanced for the supposed association which accords a dominant role to very precise periodicities in weather and climate.

(13) *Authors:* I. V. MAKSIMOV and co-workers, by I. V. MAKSIMOV, N. P. SMIRNOV and V. N. VOROBIEV (in 'Long-range forecasting of secular changes of the general ice formation of the Barents Sea by the component-harmonic method', *Publ. Res. Fish. Ind. Inv. North Basin*, **4**, 1964; Murmansk (Polar Sci. Res. Inst.)).

Basis: 'Assuming that the climatic observation series handled has a complex structure, partly attributable to errors and random elements but also containing a many-component pseudo-periodic or harmonic part, the nature of the more regular undulatory elements is largely explicable and forms a cosmic-cum-geophysical background to the long-period changes of the atmospheric and oceanic circulations'. The main elements of this cosmic-geophysical background for the ocean are taken to be waves produced by:

  (i) Free oscillations of the Earth's rotation axis, the 'polar tide', mean period 6 years.
 (ii) Solar activity variations, period averaging 11 years but ranging more widely.
(iii) The lunar declination variations producing an ocean tide with period 17–22 years.
(iv) The longer, so-called 'secular' cycle of solar activity, period 70–90 years.

Forecast: On average ice formation in the Barents Sea from 1965 to 1979 will remain comparatively low and will not go beyond the limits of variation observed between 1930 and 1960, though as much as 40–41 % of the whole sea area may be covered in the summers of 1968 and 1972–5, decreasing to 26 % in the summer of 1979. A rapid weakening of the meridional circulation of the northern hemisphere atmosphere will set in in 1980, increasing the continentality of the climate of western Europe, this being associated with a great secular fall in the level of solar activity to a minimum in 1990–2000. Ice formation on the polar seas is expected to increase sharply at that time, with 52 % of the Barents Sea expected to be ice-covered in the summers of 1982–8 and 1991–2 and as much as 62 % in the summers of 1997–2000, equalling the extent of ice in 1810–20, 'when navigation of the Barents Sea to the straits of Novaya Zemlya was only possible in isolated months of a few years'.

A very low incidence of westerly type atmospheric circulation in the Atlantic sector of the northern hemisphere at first was predicted to be followed by above-normal incidence 1968–73, after which it would fall even farther below normal around 1975. Zonality should also be very low in the Pacific sector around 1975.

Comment: The expectation of years of low incidence of westerly type atmospheric circulation in the Atlantic sector in the early part of the forecast period seems to have been right, but instead of recovering between 1968 and 1973 the frequency then reached a low extreme while the ice extent on the Barents Sea declined, in association with warm SE'ly winds in the summers and S'ly and SW'ly winds in the winters. It is also reported that the forecast water temperatures on the Kola meridian in the extreme ice year on the Barents Sea in 1966 were too high by 1°C. An impression remains (1975) that the forecast has so far indicated the right sequence but has been very rough in the suggested timing (cf. fig. 18.18, p. 502).

(14) *Author:* C. J. E. SCHUURMANS (in 'The influence of solar flares on the tropospheric circulation', *Med. en Verhandl*, KNMI, No. **92**, 1969).

Basis: The particle radiation from solar flares causes the circulation in the troposphere to become more meridional in middle latitudes in the northern and southern hemispheres. A periodicity of 80–90 years is recognized in this flare-troposphere effect, and it leads to predictable changes of known character in the atmospheric circulation and climatic fluctuations with the same period length.

Forecast: In the quarter of a century which started in 1962, the atmospheric circulation will behave in a way which will produce monthly mean temperatures in western Europe north of about 50°N that are generally lower throughout the year than in the quarter of a century preceeding.

Comment: Over the first 13 years, 1962–74, the forecast must be accounted right; though from 1971 to 1976 the prevailing temperatures in the winter 3 months constituted a marked exception, and the historic warm summers of 1975 and 1976 were not hinted at.

(15) *Authors:* E. I. Saruhanjan and N. P. Smirnov (in 'Genetic method for prediction of many-year temperature fluctuations: the Barents Sea' *Okeanologija*, **10**, 614–622; Moscow (Akad. Nauk, Okeanograf. Kom.) 1970).

Basis: Fluctuations of water temperature in the top 200 m of the Barents Sea on the Kola meridian from 1943 to 1965 were correlated with the 18-year variations of the tidal force, the 4·8-year circulation of the ocean, the 11-year and 80- to 90-year variations of solar activity and the Earth's polar wobble. The quasibiennial oscillation in the atmosphere was also used. Components of the temperature variation could then be computed for each of these variables which were considered jointly responsible for the genesis of the variations of surface water temperature; the values of these components were added up for each particular year for which a forecast was required (this is called the 'genetic method').

Forecast: The paper reports the results of forecasting the water temperature on the Kola meridian in the Barents Sea for each year and each July 1966 to 1968, the errors ranging from −0·6° to +0·3°C for the annual mean and −0·4° to +0·6°C for the Julys.

(16) *Author:* V. A. Bugaev (source unknown, about 1968; forecast presumably applicable to Russia, probably summer).

Basis: Behaviour of the stratospheric vortex. Reasoning believed similar to many unpublished expectations based on the predominant behaviour of the quasibiennial oscillation over the period 1900–63.

Forecast: 1972 expected to be a 'poor year'.

Comment: This forecast is included as a sample probably typical of many about which no details are known. 1972 was a great drought year in Russia and central Asia, with a severe shortfall of the harvest.

(17) *Author:* H. H. Lamb (in 'Die Zunahme der Wellenhöhen in jüngster Zeit in den Operationsgebieten der Bundesmarine, ihre vermutlichen Ursachen und ihre voraussichtliche weitere Entwicklung' *Fachl. Mitteilungen*, No. **160**; Porz-Wahn (Geophys. BDBw, Luftwaffenamt) 1970).

Basis: 600 years history of the frequency of various wind directions over Britain and other data bearing on the wind situation over the North Sea. Sunspot activity and radiocarbon data indicating that the pronounced declines of west wind frequency at about 200-year intervals, specifically in the 1960s, 1760s, 1560s and around 1350–1400, may be related to a fairly regular solar energy fluctuation traceable over the last several thousand years in radiocarbon data.

Forecast: The reduced frequency of W and SW winds over the North Sea noted in the last 20 years as compared with 1900–50 should be expected to continue for 80 to 100 years, or possibly rather longer, and the enhanced frequency of winds from various northerly points likely to continue during that time, with correspondingly rougher waters in the North Sea which is open to seas from that quarter, except that for 10 to 20 years around the 1990s SE'ly, with also some enhanced E'ly and NE'ly winds may prevail. These conditions were thought to imply an increased frequency of slow-moving cyclonic storms as compared with the period 1850–1950, probably bringing enhanced frequency of rain floods in the English, Dutch and German lowlands and possibly of sea-storm floods on the low-lying North Sea coasts, particularly while sea level remains high. In several decades, and most likely in the 1970s, the frequency of NW'ly and N'ly winds could greatly exceed even the recently enhanced figures. A somewhat colder climate than in 1900–50 is probably implied.

(18) *Authors:* S. J. JOHNSEN, W. DANSGAARD and H. B. CLAUSSEN (in 'Climatic oscillations 1200–2000 A.D.', *Nature,* **227**, 482–3; London, 1 August 1970).

Basis: Oxygen and hydrogen isotope analysis of the snow and ice at different levels in a core bored from the ice cap in northwest Greenland, increases of O^{18} and deuterium corresponding to increase of the temperatures prevailing in the atmosphere at the point where the snow is formed by condensation. Fourier analysis of the oxygen-18 variations indicated systematic long-term oscillations of climate with periods 2400 and 400 years and closer analysis of the ice of the last eight centuries showed further dominant quasiperiodicities of about 78 and 181 years. Similarity of several of these oscillation periods to periodicities reported in solar activity and its associations, including radiocarbon, suggested that solar variation was the underlying basis.

Forecast: 'Extrapolation suggests the probable future climatic development as a continued cooling through the next one or two decades, followed by a warming trend towards a new climatic optimum about A.D. 2015'.

The authors' graphical presentation of this suggests that the warm peak around 2015, or between 2010 and 2020, will be a very modest one, bringing no better than a return to the conditions of the 1960s and less warm than any of the main or even the subsidiary maxima of the last eight centuries.

(19) *Author:* J. R. BRAY (in 'Solar-climatic relationships in the post-Pleistocene', *Science,* **171**, pp. 1242–3, 26 March 1971).

Basis: Correlations between length of the '11-year' sunspot cycle and activity of the cycle between A.D. 1699 and 1964, as well as a solar index based on Chinese observations of sunspots and aurorae from 522 B.C. to the present, and between both these constructions of solar data and indices of climate such as glacier advances and retreats and forest growth, indicate that a general high level of solar disturbance over a number of cycles, especially if it characterizes a century or two, is associated with warmth in most parts of the world and weak solar activity is associated with cold-climate phases and glacier advances. Suggestions of a 2400- to 2600-year periodicity in glacier advances, which is found also in the stratigraphy of the Greenland ice cap, appear likely to reflect a further periodicity of this length in solar activity.

Forecast: If one superposes a 2500-year cycle on the radiation changes expected from extrapolation of the Earth's orbital variables (MILANKOVITCH), the next severe low level of the former around A.D. 4300, falling at a generally lower level of radiation associated with the latter, may be sufficient to develop a new glaciation on the continents in middle latitudes about that time.

(20) *Author:* H. GRÜNHAGEN of Lemgo, Niedersachsen, Germany (in 'Zur vergangenen und zukünftigen Klimageschichte Niedersachsens', *Niedersachsen*, Hildesheim (Verlag August Lax) 1972; the same author's work *Wie sind Klimaprognosen möglich?*, published in Detmold (Merkur Verlag) in 1965, gives further details of his data and analysis).

Basis: Statistical analysis of the longest available series of weather data and of very long series of proxy data (glaciology, tree rings, archaeology, etc.) indicative of the climatic record have led the author to stress a period length of about 2520 years and its submultiples (1260, 630 and 315 years) as most important (the proposed period length is virtually the same as favoured by J. R. BRAY). Within the times of meteorological instrument observations GRÜNHAGEN finds as the chief element a periodicity of about 170–180 years.

Forecast: 5-year means of Berlin winter temperatures, arranged as '*Kältesummen*' (totals of negative degree days), since 1943 show a strong parallelism with those observed 179 years earlier. This indicates a concentration of cold winters in a period of ten or more years duration centred about the early 1980s, colder than the averages for the 1960s from which the episode is expected to be separated by some milder winters about 1971 and 1975. On the basis of the longer periodicities mentioned, apart from a brief phase of warmer summers around the late 1980s, a colder climate is indicated as lasting from about the 1990s to nearly 2070 and then a warm phase culminating around A.D. 2100.

(21) *Author:* R. R. DICKSON (in 'A recurrent and persistent pressure-anomaly pattern as the principal cause of intermediate-scale hydrographic variation in the European shelf seas', *Deutsche Hydrographische Zeitschrift*, **24** (3), 97–119; Hamburg 1971; and 'The beginnings of a new Baltic inflow?', I.C.E.S. Hydr. Comm., C.M. 1972/C:10, Copenhagen).

Basis: The occurrence over the previous 65 years of a continual cyclic fluctuation of salinity values, with typical amplitudes of 0·5 to 1 part per 1000, affecting all the sea areas of the European continental shelf around the British Isles, the period being variable though spectral analysis indicated a fairly conspicuous peak of power at around a 5-year periodicity. A major inflow of salt water into the Baltic followed as each salinity maximum reached the deep water in the Skagerrak. The build-up to each salinity maximum was associated with an anomalous tendency in the large-scale atmospheric circulation with a deep meridional upper cold trough (500 mb level charts were used) over the western North Atlantic and an intense upper warm ridge over northwest Europe. The associated surface pressure features show an excess of S'ly winds and anticyclonic influence along the European seabord.

Forecast: By watching the progress of each cycle, and studying the variations of position and intensity of the controlling features in the atmospheric circulation that distinguish one cycle from another, DICKSON was able to point out that in 1971–2 the European anticyclone was near the westernmost position ever observed during the phases of rising salinity, and that minor or short-term variations of its position would introduce N'ly instead of S'ly winds over the North Sea, so reversing the water transport. On this account only a weak inflow into the Baltic was to be expected in that cycle.

(22) *Author:* B. M. GRAY (in 'Medium term fluctuations of rainfall and temperature in southeastern England', *Final Report on a Research Contract to the Central Water Planning Unit*, Department of the Environment, Reading, carried out in the Climatic Research Unit, University of East Anglia, Norwich, 1975).

Basis: Various time series models were constructed and applied to the longest available series of monthly rainfall data in the southeastern half of England; the Kew (London) observation series from 1697 to 1896 was used to set up the models and the results of the simulation were compared with the observations for 1897–1946. The most satisfactory model used an incomplete gamma distribution. Comparisons with rainfall in the Low Countries were encouraging. Periodicities of about 245, 100, 50 and 40 years, indicated by spectral analysis, were used. Finally, the residual variance of the decade means could be reduced to 8%. No equally satisfactory model was found for simulating the temperature series for central England from 1698, although spectral peaks indicated quasiperiodicities of about the same lengths. There was a marked spectral peak in the series of an index of cyclonic activity over the British Isles at about 50 years.

Forecast: The observed decade mean values of average rainfall over southeastern England from 1840–9 in inches per average month, and of the cyclonic activity index from 1870–9, and those produced by the model for each decade to 2010–19 are shown in fig. VI.1.

(23) *Author:* H. C. WILLETT (in unpublished memorandum prepared for a Symposium on Atmospheric Quality and Climatic Change, University of North Carolina, 20 and 21 March 1975).

Basis: After deducing that 'the pollution hypothesis cannot properly be made to account for recent climatic fluctuations', WILLETT states that the solar-climatic hypothesis 'has performed remarkably in a number of long-term forecasts of climatic trend, without one serious error' and notes that 'the long secular solar climatic cycle, formerly treated as an 80–

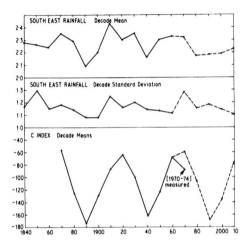

Fig. App. VI.1   Mean values:
(*a*) Of the rainfall for the southeastern half of England (expressed as one twelfth part of the year's total in inches, i.e. as the rainfall in an average month) by decades including (broken line) forecast values up to 2010–19.
(*b*) Standard deviation of the year by year values within each decade.
(*c*) Decade averages of a cyclonicity index (high points on the curve indicate much cyclonc activity over the British Isles).
Decades labelled by the first year of the decade, i.e. 1840 stands for 1840–9, etc.
(*From* B. M. GRAY, *work cited in text of forecast; reproduced by kind permission.*)

90 year cycle . . . has been rather rigorously established by H. SLEEPER [in a number of papers 1970–3] on the basis of planetary configurations and sunspot and solar magnetic activity, as alternately of approximately 100 and 80 years duration'. This is used as the basis of 'predictions for the next double secular cycle of 180 years, 1975–2155'.

Forecast (*excerpts*): 'The cooling that started after the peak warmth in each [of the last two] 100-year cycles began . . . 15 years sooner in higher than in lower middle latitudes . . . and in both cycles bottomed out in a minor minimum just prior to the end of the cycle. This was followed in the earlier cycle, and probably will be in the current cycle just ending, by a substantially colder minimum some 20–25 years later (1988–90?). The intermediate 80-year cycle . . . presumably . . . of most predictive significance at the present time . . . [showed only a modest] maximum rise of temperature and peak warmth . . . occurred by mid cycle. This was followed by a return of temperature to severe cold with the falling sunspot activity and only a modest upturn [of temperature and sunspot activity] toward the end of the cycle.' WILLETT adds a note of reservations about the timing of a renewed fall of temperature in the later 1970s or 1980s because of ignorance about the manifestations of the double sunspot cycle between 1795 and 1875 (the last 80-year cycle), 'uncertainty about its imminent reversal of phase' and 'its tendency to be less dominant during the first relatively quiet half of the long secular cycle'.

In a further outlook, referring to much longer cycles, WILLETT dismisses any early prospect of a major ice age but expects a further Little Ice Age and another peak of interglacial warmth.

(24) *Author:* JAPAN METEOROLOGICAL AGENCY, Tokyo.
Report: In response to public demand for information about the apparent increase in incidence of exceptional weather in recent years, and anxiety about food supply problems in

many parts of the world, the Japan Meteorological Agency in 1973 established a Study Group to survey the question, as well as establishing a joint programme of the J.M.A. and the Japanese Ministry of Agriculture and Forestry. A summary of the Study Group's report was issued in English in June 1974. The results of sending a questionnaire to the meteorological services of about thirty countries produced different opinions: one country's meteorological service gave the view that 'unusual weather events are not considered to be on the increase' and two others that 'the development of the mass media is responsible for leading the public into a misconception by bringing them more news than before'; most considered that too little was known to establish the current climatic trend, although several indicated a recent cooling trend in large parts of the world. The Study Group's own examination of monthly mean temperature and precipitation totals from about 150 stations all over the world showed that the frequency of values showing statistically significant departures from the previous 30-year average

(i) in the case of monthly mean temperature, had increased with each successive decade since 1940, with a predominance of departures towards the low temperature side. From 1961 onwards (up to 1972) the frequency of significantly cold months was more than double that of significantly warm months; whereas

(ii) There was no noteworthy variation in the occurrence of precipitation anomalies until 1960, but in the later years their frequency was very significantly increased, with cases of deficient rainfall about 1·6 times as frequent as cases of excessive precipitation.

Summary comment

These forecasts have all been attempts at prognosis of the natural climate. They should all be subject to the provisos that the tendency of the climate is not affected by (a) any great outburst of volcanic activity, or (b) any new impact of human activity during the period covered. Survey of the forecasts quoted shows that about half were based, at least in part, on an analysis of the history of atmospheric circulation patterns, the variations being commonly thought by the authors to be associated with long-term variations of solar disturbance. Thirteen of the forecasts mentioned probable associations with variations of solar activity, though these may be considered to have been adequately demonstrated in only six cases. Two forecasts attributed the climatic variations concerned to variations of tidal force. As regards the quasiperiodicities principally used, eleven forecasts were based on a cyclic variation of about 90 years (figures mentioned ranged from 73 to 110 years), six (or possibly eight) forecasts used a cyclic variation of about 200 years (figures mentioned ranged from 170 to 245 years, most being close to 200 years), two forecasts stressed cycles of about 22–23 years, two about 19 years, three others about 4–6 years; one forecast made use of the quasibiennial oscillation in the stratosphere.

Despite the use in most cases of a single line of reasoning, i.e. postulation of the prime importance of just one controlling variable, this variable differing in different cases, there is a considerable measure of agreement between the forecasts listed. Expectation of a trend towards colder climates with weakened general atmospheric circulation from 1950 or 1960 onwards seems to have been well verified by the actual weather to date. Most forecasts expect this regime to continue into the twenty-first century, possibly into the second half of that century, in some cases with a sharp further cooling about 1980 and somewhat easier conditions for a time in the first half of the next century. Several forecasts indicated a temporary slackening of the cooling trend and diminution of the Barents Sea ice in the 1970s. The rainfall forecast (No. 22), indicating a run of drier and relatively anticyclonic decades in England from the 1980s onwards seems consistent with the general implication of the temperature forecasts. The Japanese global survey and forecast (No. 24) usefully stresses the increased variability and incidence of extremes of temperature and rainfall in recent years.

References

References

Chapter 12 Man's awareness of climatic changes

BELL, B. (1971) The dark ages in ancient history, I: The first Dark Age in Egypt, *Amer. J. Archæol.*, **75**, 1–26. New York.

BERGERON, T., FRIES, M., MOBERG, C.-A. and STRÖM, F. (1956) Fimbulvinter, *Fornvännen*, **1**, 1–18. Stockholm.

BUCHINSKY, I. E. (1957) *The Past Climate of the Russian Plain*. Leningrad (Gidrometeoizdat, 2nd ed.) (in Russian).

CHU KO-CHEN (COCHING CHU) (1961) The pulsation of world climate during historical times, *New China Monthly*, **6**, June. Peking. (In Chinese: translation available in Meteorological Office Library, Bracknell, England.)

CHU KO-CHEN (COCHING CHU) (1973) A preliminary study on the climatic fluctuations during the last 5000 years in China, *Scientia Sinica*, **16** (2), 226–256. Peking (May 1973).

DALBY, D. and HARRISON CHURCH, R. J. (eds) (1973) *Drought in Africa*. Report of the symposium held on 19 July 1973 in the Centre for African Studies (School of Oriental and African Studies), University of London. 124 pp.

EYTHORSSON, J. (1952) History of Breiða. *Jökull*, **2**, 17–20. Reykjavik. (In Icelandic.)

FLOHN, H. (1970) Climatology – descriptive or physical science? *W.M.O. Bull.*, **19** (4), 223–9. Geneva.

GRAHAM, H. G. (1899) *The Social Life of Scotland in the Eighteenth Century*. London (Black).

HOEL, A. and WERENSKIØLD, W. (1962) Glaciers and snowfields in Norway, *Norsk Polarinstituttets Skrifter*, Nr. 114. Oslo (Univ. Press).

HOINKES, H. (1964) Glaciers and avalanches in the Alps, pp. 9–19 in *The Alps in Colour*. Innsbruck (Pinguin Verlag).

HOINKES, H. and RUDOLPH, R. (1962) Mass balance studies on the Hintereisferner, Ötztal Alps, 1952–61, *J. Glaciol.*, **4** (33), 266–80. Cambridge.

KINZL, H. (1932) Die grössten nacheiszeithlichen Gletschervorstösse in den schweizer Alpen und in der Mont-Blanc Gruppe, *Z. f. Gletscherkunde*, **20**, 269–397. Berlin.

KLEIN, H. (1914) *Das Klima Palästinas, auf Grund der alten hebräischen Quellen* (Inaugural dissertation, Universität Würzburg). Halle (Ehrhardt Karras G.m.b.H.). 77 pp.

KOCH, L. (1945) The East Greenland ice, *Meddelelser om Grønland*, **130** (3). Copenhagen.

KRISTJANSSON, L. (1969) The ice drifts back to Iceland, *New Scientist*, **41**, 508–9. London. (See also the special volume of *Jökull*, **19** (Reykjavik, 1969), presenting the papers given at a specially convened conference on the Arctic sea ice.)

KUKLA, G. J. and MATTHEWS, R. K. (1972) When will the present interglacial end? *Science*, **178**, 190–1. Washington (13 October 1972).

LADURIE, E. LE R. (1967) *Histoire du climat depuis l'an mil*. Paris (Flammarion). 377 pp.

LADURIE, E. LE R. (1971) *Times of Feast, Times of Famine*. New York (Doubleday). 426 pp.

MAUNDER, W. J. (1970) *The Value of the Weather*. London (Methuen). 388 pp.

OESCHGER, H. and RÖTHLISBERGER, H. (1961) Datierung eines ehemaligen Standes des Aletschgletschers, *Z.f. Gletscherkunde und Glazialgeol.*, **4** (3), 191–205. Berlin.

PETTERSSON, O. (1914) Climatic variations in historic and prehistoric time. *Svenska Hydrografiska-Biologiska Kommissionens Skrifter*, **5**. Göteborg.

RÖTHLISBERGER, H. (1974) Möglichkeiten und Grenzen der Gletscherüberwächung. *Neue Zürcher Zeitung*, No. 196, 29 April 1974. Zürich.

ROTHMAN, R. W. (1848) *Observations on the Climate of Italy and Other Countries in Ancient Times*. London (R. & J. E. Taylor). 40 pp.

RUDLOFF, H. VON (1967) *Die Schwankungen und Pendelungen des Klimas in Europa seit dem Beginn der regelmässigen Instrumenten-Beobachtungen (1670)*. Braunschweig (Vieweg: *Die Wissenschaft* Band 122).

SHAW, N. (1926) *Manual of Meteorology, Vol. I: Meteorology in History*, p. 84. Cambridge (Univ. Press).

STEFANSSON, V. (1943) *Greenland*. London (Harrap). 240 pp.

TREVELYAN, G. M. (1942) *English Social History*. London (Longmans). 628 pp.

WALTON, K. (1952) Climate and famines in northeast Scotland, *Scot. Geogr. Mag.*, **68** (1), 13–21. Edinburgh.

WILLIAMS FREEMAN, J. P. (1928) Note in *Antiquity*, p. 208. Gloucester.

WINSTANLEY, D. (1973a) Recent rainfall trends in Africa, the Middle East and India, *Nature*, **243**, 464–5. London (22 June 1973).

Winstanley, D. (1973*b*) Rainfall patterns and general atmospheric circulation, *Nature*, **245**, 190–4. London (28 September 1973).

Ziegler, P. (1969) *The Black Death*. London (Collins). 319 pp.

Chapter 13 Evidence of past weather and climate

13.1 Meteorological evidence

Ahlmann, H. W. (1953) *Glacier variations and climatic fluctuations*. Bowman Memorial Lecture. New York (American Geographical Soc.).

Ångström, A. (1935) Teleconnections of climatic changes in present time, *Geogr. Annaler*, **17**, 242–58. Stockholm.

Arago, F. (1858) Sur l'état thermométrique du globe terrestre, pp. 184–487 in *Œuvres complètes*, Vol. VIII. Paris (Gide, for Acad. des Sciences) and Leipzig (Weigel).

Arakawa, H. (1954) Fujiwhara on five centuries of freezing dates of Lake Suwa in central Japan, *Archiv f. Met. Geophys. Biokl.*, **B, 6**, 152–66. Vienna.

Arakawa, H. (1955) Twelve centuries of blooming dates of the cherry blossoms at the city of Kyoto and its own vicinity, *Geofisica pura e applicata*, **30**, 147–50. Milan.

Arakawa, H. (1956) Dates of first or earliest snow covering for Tokyo since 1632, *Quart. J. Roy. Met. Soc.*, **82**, 222–6. London.

Baker, J. N. L. (1932) Climate of England in the seventeenth century, *Quart. J. Roy. Met. Soc.*, **58**, 421–38. London.

Baker, J. N. L. (unpublished) Extensive MSS collection of reports on English weather through the centuries. (Available in Met. Office Library, Bracknell.)

Brimblecombe, P. and Ogden, C. (1977) Air pollution in art and literature, *Weather*, **32**, in press.

Britton, C. E. (1937) A meteorological chronology to A.D. 1450, *Geophys. Mem.*, **70**. London (H.M.S.O. for Met. Office).

Bryson, R. A., Baerreis, D. A. and Wendland, W. M. (1970) The character of Late Glacial and postglacial climatic changes, pp. 53–74 in *Pleistocene and Recent Environments of the Central Great Plains*. Dept. Geol. Special Pubn No. 3. Lawrence (Univ. Kansas Press).

Buchinsky, I. E. (1957) *The Past Climate of the Russian Plain*. Leningrad (Gidrometeoizdat), 2nd edn. (In Russian.)

Budyko, M. I. (1966) Polar ice and climate, *Proc. Symp. Arctic Heat Budget and Atmospheric Circulation*. RM-5233-NSF. Santa Monica, Calif. (Rand Corp.).

Bushby, F. H. and Timpson, M. S. (1967) A 10-level atmospheric model and frontal rain, *Tellus*, **21**, 289–307. Stockholm.

Central Met. Office, Republic of Korea (1961) *Seoul Monthly Precipitation Records 1770–1960*. Seoul.

CHU KO-CHEN (COCHING CHU) (1961) The pulsation of world climate during historical times, *New China Monthly*, **6** (June). Peking. (In Chinese: English translation available in Meteorological Office Library, Bracknell, England.)

DEFANT, A. (1921) Die Zirkulation der Atmosphäre in dem gemässigten Breiten der Erde, *Geogr. Annaler*, **3**, 209-66. Stockholm.

EASTON, C. (1928) *Les hivers dans l'Europe occidentale*. Leyden (Brill). 208pp.

FLOHN, H. (1949) Klima und Witterungs-Ablauf in Zürich im 16. Jahrhundert, *Viertel-jahresschrift der Naturf. Gesell. in Zürich*, **95**, 28-41. Zürich.

FRANKLIN, B. (1786) A letter from Dr. BENJAMIN FRANKLIN to Mr. Alphonsus le Roy, Member of several Academies, at Paris. Containing sundry Maritime Observations, *Trans. Amer. Phil. Soc.*, **2**, 294-329. Philadelphia.

GIOVINETTO, M. B. and SCHWERDTFEGER, W. (1966) Analysis of a 200-year snow accumulation series from the South Pole. *Archiv f. Met., Geophys., Biokl.*, A **15** (2), 227-250. Vienna.

GROVE, J. M. (1972) The incidence of landslides, avalanches, and floods in western Norway during the Little Ice Age. *Arctic & Alpine Research*, **4** (2), 131-8. Boulder, Colorado.

HAVENS, J. M. (1958) *An Annotated Bibliography of Meteorological Observations in the United States, 1715-1818*. Washington (U.S. Dept. of Commerce, Weather Bureau).

HELLMANN, G. (1908) The dawn of meteorology, *Quart. J. Roy. Met. Soc.*, **34**, 221-32. London.

HELLMANN, G. (1916) Über die Ägyptischen Witterungsangaben im Kalender von Claudius Ptolemaeus, *Sitz.-Ber. der Kgl. Preuss. Adad. Wiss.*, **XIII**, 332-41. Berlin.

HELLMANN, G. (1926) Die Entwicklung der meteorologischen Beobachtungen in Deutschland von den ersten Anfangen bis zur Einrichtung staatlicher Beobachtungs-netze, *Abhandl. der Preuss. Akad. Wiss. Phys.-math Klass*, Nr. 1. Berlin.

HELLMANN, G. (1927) Die Entwicklung der meteorologischen Beobachtungen bis zum Ende des XVIII Jahrhunderts, *Abhandl. der Preuss. Akad. Wiss. Phys.-math Klasse*, Nr. 1. Berlin.

HENNIG, R. (1904) Katalog bemerkenswerter Witterungsereignisse von den ältesten Zeiten bis zum Jahre 1800, *Abhandl. der Preuss. Met. Inst.*, **II**, No. 4. Berlin.

HOOKE, R. (1663) *A Scheme, at one View Representing to the Eye the Observation of the Weather for a Month*. MSS. in Royal Society Archives, Register Book No. 2. London.

KINGTON, J. A. (1970) A late eighteenth century source of meteorological data, *Weather*, **25**, 169-75. London.

KINGTON, J. A. (1975) Daily synoptic weather maps from the 1780s: a research project of synoptic climatology, *Met. Mag.*, **104**, 33-52. London.

LABRIJN, A. (1945) Het Klimaat van Nederland gedurende de laatste twee en een halve eeuw, *Mededelingen en Verhandlingen*, **49**, 11-105. De Bilt (Koninklijk Nederlands Met. Inst. No. 102).

LA COUR, P. (1876) Tyge Brahes Meteorologiske Dagbog, holdt paa Uranienborg for Aarene 1582-1597 (Appendix to) *Collectanea Meteorologica*. Copenhagen (Kgl. Danske Videnskabernes Selskab).

LADURIE, E. LE R. (1967) *Histoire du climat depuis l'an mil*. Paris (Flammarion).

LAMB, H. H. (1963) On the nature of certain climatic epochs which differed from the modern (1900-39) normal, pp. 125-50 in *Changes of Climate: Proc. UNESCO/WMO, Rome 1961 Symp.* (UNESCO Arid Zone Res. Series, XX). Paris.

LAMB, H. H. (1965) The early medieval warm epoch and its sequel. *Palaeogeogr., Palaeoclim, Palaeoecol.*, **1**, 13-37. Amsterdam (Elsevier).

LAMB, H. H. and JOHNSON, A. I. (1966) Secular variations of the atmospheric circulation since 1750, *Geophys. Mem.*, **110**. London (H.M.S.O., for Met. Office).

LAMB, H. H., LEWIS, R. P. W. and WOODROFFE, A. (1966) Atmospheric circulation and the main climatic variables between 8000 and 0 B.C.: meteorological evidence, pp. 174-217 in *World Climate 8000-0 B.C.: Proc. Internat. Conf.* (ed. J. S. SAWYER). London (Roy. Met. Soc.).

LAMB, H. H. and WOODROFFE, A. (1970) Atmospheric circulation during the last ice age, *Quaternary Res.*, **1** (1), 29-58. Seattle.

LANDSBERG, H. E., YU, C. S. and HUANG, L. (1968) *Preliminary Reconstruction of a Long Time Series of Climatic Data for the Eastern United States*. Tech Note BN-571. Univ. of Maryland (Inst. Fluid Dynamics).

LAWRENCE, E. N. (1972) The earliest known journal of the weather, *Weather*, **27**, 494-501. London.

LENKE, W. (1960) Klimadaten von 1621-1650 nach Beobachtungen des Landgrafen Hermann IV von Hessen, *Ber. des dt. Wetterdienstes*, **9**, No. 63, 1-52. Offenbach.

LENKE, W. 1961) Neuberechnung der Temperaturwerte von Berlin für die Jahre 1730-1750, *Met. Rundschau*, **14**, 162-70. Berlin, Göttingen, Heidelberg.

LERMAN, J. C., MOOK, W. G. and VOGEL, J. C. (1970) C14 in tree rings from different localities, pp. 275-301 in *Radiocarbon Variations and Absolute Chronology: Proc. Twelfth Nobel Symp., Uppsala, 1969* (ed. I. U. OLSSON). Stockholm (Almqvist & Wiksell) and New York (Wiley). 657 pp.

LORENZ, E. N. (1968) Climatic determinism, *Met. Monogs*, **5**, 1-3. Boston, Mass. (Amer. Met. Soc.).

LORENZ, E. N. (1969) The predictability of a flow which possesses many scales of motion, *Tellus*, **21**, 289-307. Stockholm.

LORENZ, E. N. (1970) Climatic change as a mathematical problem. *J. Appl. Met.*, **9**, 325-9. Lancaster, Pa.

LOWE, E. J. (1870) *Natural Phenomena and a Chronology of the Seasons*. London (Bell & Dalby).

MANABE, S. and BRIAN, K. (1969) Climate calculations with a combined ocean-atmosphere model. *J. Atmos. Sci.*, **26**, 786-9.

MANLEY, G. (1953) The mean temperature of central England, 1968–1952, *Quart. J. Roy. Met. Soc.*, **79**, 242–61. London.

MANLEY, G. (1959) Temperature trends in England 1698–1957, *Archiv f. Met. Geophys. Biokl.*, **B, 9**, 413–33. Vienna.

MANLEY, G. (1961) A preliminary note on early meteorological observations in the London region with estimates of monthly mean temperatures 1680–1706, *Met. Mag.*, **90**, 303–10. London.

MANLEY, G. (1962) Early meteorological observations and the study of climatic fluctuation, *Endeavour*, **21**, 43–50. London (Imp. Chem. Ind.)

MANLEY, G. (1974) Central England temperatures: monthly means 1659–1973, *Quart. J. Roy. Met. Soc.*, **100**, 389–405. London.

MÜLLER, K. (1953) *Geschichte des badischen Weinbaus*. Laar in Baden (von Moritz Schauenburg).

MYAKODA, K., SMAGORINSKY, J., STRICKLER, R. F. and HEMBREE, G. D. (1969) Experimental extended predictions with a 9-level atmospheric model, *Mon. Weather Rev.*, **97**, 1–76. Washington.

NICHOLAS, F. J. and GLASSPOOLE, J. (1931) General monthly rainfall over England and Wales 1727–1931, *British Rainfall* (1931), 299–306. London.

NORLIND, A. (1914) Einige Bemerkungen über das Klima der historischen Zeit, nebst einem Verzeichnis mittelalterlicher Witterungserscheinungen, *Lund Universitets Årsskrift*, S. I Afd. I, Bd. 10, Nr. I, pp. 1–53. (Includes 19 pp. chronicle of weather A.D. 709–1499 in Europe and a supplement on freezing of the Baltic.)

OLIVER, J. and KINGTON, J. A. (1970) The usefulness of ships' log books in the synoptic analysis of past climates, *Weather*, **25**, 520–8. London.

ROBINSON, G. D. (1970) *Long-Term Effects of Air Pollution: A Survey*. CEM 4029–400. Hartford, Conn. (Center for the Environment and Man).

RUDLOFF, H. von (1967) *Die Schwankungen und Pendelungen des Klimas in Europa seit dem Beginn der regelmässigen Instrumenten-Beobachtungen (1670)*. Braunschweig (Vieweg-Die Wissenschaft, **122**).

SANSON, J. (1955) La prévision du temps a cent ans d'existence, *La Météorologie*, **37**, 167–72. Paris.

SELLERS, W. D. (1969) A global climatic model based on the energy balance of the Earth-atmosphere system, *J. Appl. Met.*, **8**, 392–400. Lancaster, Pa.

SHORT, THOS. (1749) *A General Chronological History of the Air Weather, Seasons, Meteors, etc.* London, 2 vols.

SHORT, THOS. (1750) *New Observations, Natural, Moral, Civil, Political and Medical ... With an Appendix on the Weather and Meteors*. London.

SHORT, THOS. (1767) *A Comparative History of the Increase and Decrease of Mankind, etc.* London. (Includes a chronology 1468–1765 and list of wet and dry years from 1400.)

SMAGORINSKY, J. (1963) General circulation experiments with primitive equations, I: The basic experiment, *Mon. Weather Rev.*, **91**, 99–165. Washington.

SPEERSCHNEIDER, C. I. H. (1915) Om Isforholdene i danske Farvande i aeldre og nyere Tid: Aarene 690–1860, *Meddelelser*, No. 2. Copenhagen (Danske Met. Inst.).

SPEERSCHNEIDER, C. I. H. (1927) Om Isforholdene i danske Farvande, *Meddelelser*, Nr. 6. Copenhagen (Danske Met. Inst.).

SUTCLIFFE, R. C. (1939) Cyclonic and anticyclonic development, *Quart. J. Roy. Met. Soc.*, **65**, 519–24. London.

SUTCLIFFE, R. C. (1947) A contribution to the problem of development, *Quart. J. Roy. Met. Soc.*, **73**, 370–83. London.

SYMONS, G. J. (1891) *Considerationes Temperici pro 7 Annis: The Earliest Known Journal of the Weather, Kept by the Rev. W. Merle, Rector of Driby, Lincolnshire, 1337–44*. Reproduced and translated. London.

THORODDSEN, Th. (1916–17) *Árferði á Islandi i thusund ár*. Copenhagen.

TITOW, J. (1970) Le climat a travers les rôles de comptabilité de l'évêche de Winchester (1350–1450), *Annales: Économies, Sociétés, Civilisations*, No. 2. Paris (Armand Colin). (Fold-out diagram gives data over a longer period.)

VANDERLINDEN, E. (1924) Chronique des événements météorologiques en Belgique jusqu'en 1834, *Mém. Acad. Roy. Belge*, 2nd Series, Vol. 5. Brussels.

WALES-SMITH, B. G. (1971) Monthly and annual totals of rainfall representative of Kew, Surrey, from 1697–1970, *Met. Mag.*, **100**, 345–62. London.

WEIKINN, C. (1958) *Quellentexte zur Witterungsgeschichte Europas von der Zeitwende bis zum Jahre 1850, I: Hydrographie*, (Zeitwende-1500). Berlin (Akad. Verlag).

WEIKINN, C. (1960) Ibid., **I**, 2 (1501–1600).

WEIKINN, C. (1961) Ibid., **I**, 3 (1601–1700).

WEIKINN, C. (1963) Ibid., **I**, 4 (1701–1750).

WOLF, R. (1865, 1872, 1873, 1885) Meteorologische Beobachtungen, *Schweizerische Meteorologische Beobachtungen*, **II**: (1865) 147. (Regenmengen in Zürich, 1708–53; 1830–63) **IX** (1872) 411–13, 462–5, 513–17. 566–9; **X** (1873), 44–8, 96–101; **Suppl.-Band.** (1885) 180–1 (HALLER's observations). Zürich (Met. Anstalt).

World Weather Records (1927, 1934, 1947) *Smithsonian Misc. Collections:* **79** (1927), 1119 pp.; **90** (1934), 618 pp.; **105** (1947), 647 pp. Washington.

World Weather Records, 1941–50 (1959) U.S. Dept. of Commerce, Weather Bureau. Washington.

World Weather Records, 1951–60 (1965–8) U.S. Dept. of Commerce, Environmental Data Service (in 6 volumes). Washington.

13.2 Dating methods

ABELSON, P. H. (1956) Paleobiochemistry, *Scientific American*, **195**, 83–92. New York.

ADAMENKO, G. (1963) On the similarity in the growth of trees in northern Scandinavia and the polar Urals, *J. Glaciol.*, **4**, 449–51. Cambridge.

AITKEN, M. J. (1970) Dating by archaeomagnetic and thermoluminescent methods, *Phil. Trans. Roy. Soc.*, **A, 269**, 77–88. London.

ALLIBONE, T. E. (1970) *The Impact of the Natural Sciences on Archaeology*. London (British Academy and Oxford Univ. Press). 185 pp. Also as *Phil. Trans. Roy. Soc.*, **A, 269**. (No. 1193).

ANTEVS, E. (1947) Dating the past (review), *J. Geol.*, **55**, 527–30. Chicago.

ANTEVS, E. (1953) Geochronology of the Deglacial and Neothermal ages, *J. Geol.*, **61**, 195–230. Chicago.

AUER, V. (1956) The Pleistocene of Fuego-Patagonia, Part I: The ice age and interglacial ages, *Suomal. Tiedeakat. Toim. Helsinki*, **A, III**. *Geologica-Geographica*, **45**. 226 pp.

AUER, V. (1958) The Pleistocene of Fuego-Patagonia, Part II: The history of the flora and vegetation, *Ibid.*, **50**. 239 pp.

AUER, V. (1959) The Pleistocene of Fuego-Patagonia, Part III: Shoreline displacements, *Ibid.*, **60**. 247 pp.

AUER, V. (1965) The Pleistocene of Fuego-Patagonia, Part IV: Bog profiles, *Ibid.*, **80**. 160 pp.

BADA, J. L., PROTSCH, R. and SCHROEDER, R. A. (1973) The racemization reaction of isoleucine used as a palaeotemperature indicator, *Nature*, **241**, 394–5. London (9 February).

BADA, J. L. and DEEMS, L. (1975) Accuracy of dates beyond the ^{14}C dating limit using the aspartic acid racemization reaction, *Nature*, **255**, 218–19. London (15 May).

BANNISTER, B. (1963) Dendrochronology, Ch. 17, pp. 162–76 in *Science in Archaeology* (ed. D. BROTHWELL and E. HIGGS). Bristol (Thames & Hudson).

BAXTER, M. S. and WALTON, A. (1971) Fluctuations of atmospheric carbon-14 concentration during the past century, *Proc. Roy. Soc.*, **A, 321**, 105–27. London.

BECKER, B. and GIERTZ-SIEBENLIST, V. (1970) Eine über-1100-jährige mitteleuropäische Tannenchronologie, *Flora*, **159**, 310–46. Berlin.

BENSON, C. (1967) Polar regions snow cover, pp. 1039–63 in *Physics of Ice and Snow: Proc. Internat. Conf. on Low Temp. Science, 1966*, Vol. I, Part 2 (ed. H. OURA). Sapporo (Hokkaido Univ. Inst. Low Temp. Science).

BERGER, R. (1970) Ancient Egyptian radiocarbon chronology, *Phil. Trans. Roy. Soc.*, **A, 269**, 23–36. London.

BERNARD, E. A. (1962) Théorie astronomique des pluviaux et interpluviaux du Quaternaire africain, *Acad. Royale des Sciences d'Outre-Mer, Classe des Sciences Nat. et Méd.* (Nouvelle Serie), XII, 1. Brussels.

BLYTT, A. (1876a) Forsøg til en Theorie om Indvandringen af Norges Flora under vekslende regnfulde og tørre Tider, *Nyt Magazin f. Naturv.* Christiania.

BLYTT, A. (1876b) *Essay on the Immigration of the Norwegian Flora*. Christiania.

BLYTT, A. (1909) Theorien om den norske floras invandring under vekslende tørre og fugtige perioder, *Bergens Museums Aarbok*, **8**. Bergen.

BORELL, R. and OFFERBERG, J. (1955) Geokronologiska undersökningar inom Indalsälvensdalgang mellan Bergeforsen och Ragunda, *Sveriges Geol. Undersökning*, Ser. Ca., No. 31. Stockholm. 22 pp.

BRAY, J. R. and STRUIK, G. J. (1963) Forest growth and glacial chronology in eastern British Columbia and their relation to recent climatic trends, *Canad. J. Bot.*, **41**, 1245–71. Ottawa.

BREHME, K. (1951) Jahrringchronologische und klimatologische Untersuchungen an Hochgebirgslärchen des Berchtesgadener Landes, *Z. f. Weltforst-wirtschaft*, **14**, 65–80. Berlin.

BROECKER, W. S., EWING, M. and HEEZEN, B. C. (1960) Evidence for an abrupt change in climate close to 11 000 years ago, *Amer. J Sci.*, **258**, 429–48. New Haven, Conn.

BRYSON, R. A., BAERREIS, D. A. and WENDLAND, W. M. (1970) The character of Late-Glacial and postglacial climatic changes, pp. 53–74 in *Pleistocene and Recent Environments of the Central Great Plains*. (Dept. Geol., Special Pubn 3). Lawrence, Kansas (Univ. Kansas Press).

BUCHA, V. (1970) Evidence for changes in the Earth's magnetic field intensity, *Phil. Trans. Roy. Soc.*, **A, 269**, 47–55. London.

CHAPMAN, S. (1951) *The Earth's Magnetism*. London (Methuen) and New York (Wiley), 2nd edn.

CHILDE, V. G. (1954) *What Happened in History*. Harmondsworth, Middlesex (Penguin Books), 2nd rev. edn.

CHILDE, V. G. (1958) *The Prehistory of European Society*. Harmondsworth, Middlesex (Penguin Books).

CHU KO-CHEN (1961) The pulsation of world climate during historical times, *New China Monthly*, **6**, June. Peking. (In Chinese: English translation available in Met. Office Library, Bracknell, England.)

CLARK, G. and PIGGOTT, S. (1965) *Prehistoric Societies*. Harmondsworth, Middlesex (Penguin Books).

COX, A. (1969) Geomagnetic reversals, *Science*, **163** (3864), 237–45. New York (17 January).

DAMON, P. E. (1968) Radiocarbon and climate, *Met. Monogs*, **8** (30), 151–4. Boston, Mass. (Amer. Met. Soc.).

DANSGAARD. W. and JOHNSEN, S. J. (1969a) A flow model and a time scale for the ice core from Camp Century, Greenland, *J. Glaciol.*, **8** (53), 215–23. Cambridge.

DANSGAARD, W., JOHNSEN, S. J., MØLLER, J. and LANGWAY, C. C. (1969b) One thousand centuries of climatic record from Camp Century on the Greenland ice sheet. *Science*, **166**, 377–81. New York (17 October).

DANSGAARD, W., JOHNSEN, S. J. CLAUSEN, H. B. and LANGWAY, C. C. (1971) Climatic record revealed by the Camp Century ice core, in *The Late Cenozoic Glacial Ages: Proc. Yale Univ. Symp., December 1969* (ed. K. K. TUREKIAN). New Haven, Conn. (Yale Univ. Press). 606 pp.

DE GEER, E. H. (1954) Skandinaviens geokronologi, *Geol. Fören. Förhandl.*, **76**, 299–329. Stockholm.

DE GEER, E. H. (1957) Old and new datings of Swedish ice lakes, etc., *Geol. Fören, Förhandl.*, **79**, 93–100. Stockholm.

DE GEER, G. (1912) A geochronology of the last 12 000 years, *Congrès géol. internat. Stockholm 1910, Comptes rendues*, **XI**, 241–58.

DE GEER, G. (1940) Geochronologia Suecica Principles, *Kgl. Vet. Akad. Handlingar* (III serien), **18** (6). Stockholm.

DE VRIES, H. (1958) Variation in concentration of radiocarbon with time and location on Earth, *Koninklijk Nederlandse Akad. Wet., Proc.*, **B, 61** (2), 1–9. Amsterdam.

EDWARDS, I. E. S. (1970) Absolute dating from Egyptian records and comparison with carbon-14 dating, pp. 11–18 in *The Impact of the Natural Sciences on Archaeology* (ed. T. E. ALLIBONE). London (Oxford Univ. Press for The British Academy). Also *Phil. Trans. Roy. Soc. Lond.*, **A, 269**, 11–18.

EMILIANI, C. (1955) Pleistocene temperatures, *J. Geol.*, **63** (6), 538–78. Chicago.

EMILIANI, C. (1958) Palaeotemperature analysis of core 280 and Pleistocene correlations, *J. Geol.*, **66** (3), 264–75. Chicago.

EMILIANI, C. (1961) Cenozoic climatic changes, as indicated in the stratigraphy and chronology of deep sea cores, *New York Acad. Sci.*, **95**, Art. 1, 521–36.

EMILIANI, C. (1966) Palaeotemperature analysis of Caribbean cores P 6304-8 and P 6304-9 and a generalized temperature curve for the past 425 000 years, *J. Geol.*, **74** (2), 109–26. Chicago.

EPSTEIN, S., SHARP, R. P. and GOW, A. J. (1970) Antarctic ice sheet: stable isotope analyses of Byrd station cores and interhemispheric climatic implications, *Science*, **168**, 1570–2. New York (13 April).

ERICSON, D. B., EWING, M. and WOLLIN, G. (1964) The Pleistocene epoch in deep-sea sediments, *Science*, **146** (3645), 723–32. New York (6 November).

ERICSON, D. B. and WOLLIN, G. (1966) *The Deep and the Past.* London (Jonathan Cape).

ERICSON, D. B. and WOLLIN, G. (1968) Pleistocene climates and chronology in deep-sea sediments, *Science*, **162** (3859), 1227–34. New York (13 December).

EVANS, P. (1972) Towards a Pleistocene time scale, Part 2, in *The Phanerozoic Time Scale* (Special pubn No. 5). London (Geol. Soc).

FERGUSON, C. W. (1968) Bristlecone pine: science and aesthetics, *Science*, **159** (3817), 839–46. New York.

FERGUSON, C. W. (1969) A 7104-year annual tree-ring chronology for bristlecone pine, *Pinus aristata*, from the White Mountains, California, *Tree-ring Bull.*, **29** (3–4), 3–29. Tucson (Univ. Arizona).

FIRBAS, F. (1949) *Spät- und nacheiszeitliche Waldgeschichte Mitteleuropas nördlich der Alpen* (I: Allgemeine Waldgeschichte). Jena (Fischer).

FLECK, R. J., MERCER, J. H., NAIRN, A. E. M. and PETERSON, D. N. (1972) Chronology of late Pliocene and early Pleistocene glacial and magnetic events in southern Argentina, *Earth & Planetary Sci. Letters*, 16, 15–22. Amsterdam (North Holland).
(See also MERCER, J. H., FLECK, R. J., MANKINEN, E. A. and SANDER, W. (1972) Glaciation in southern Argentina before 3·6 million years ago and origin of the Patagonian gravels (rodados Patagonicos). Minneapolis 1972 meeting of the Geological Society of America.)

FLETCHER, J. M. and HUGHES, J. R. (1970) Uses of X-rays for density determinations and dendrochronology, *Bull. Faculty of Forestry*, 7, 41–54. Vancouver (Univ. Brit. Columbia).

FLINT, R. F. (1957) *Glacial and Pleistocene Geology*. New York (Wiley).

FRITTS, H. C. (1965) Dendrochronology, in *The Quaternary of the United States* (ed. H. E. WRIGHT and D. G. FREY). Princeton (Univ. Press).

FROMM, E. (1970) An estimation of the errors in the Swedish varve chronology, pp. 163–72 in *Radiocarbon Variations and Absolute Chronology: Proc. Twelfth Nobel Symp., Uppsala, 1969*, (ed. I. V. OLSSON). Stockholm (Almqvist & Wiksell) and New York (Wiley).

FUCHS, V. E. (1947) The volcanics of East Africa and pluvial periods, in V. E. FUCHS and T. T. PATERSON, The relation of volcanicity and orogeny to climatic change, *Geol. Mag.*, 84, 321–33. London.

GAMS, H. and NORDHAGEN, R. (1923) Postglaziale Klimaänderungen und Erdkrustenbewegungen in Mitteleuropa, *Landeskundl. Forschung*, 25, 283–36. Munich (*Mitteil.d. Geogr. Gesell.*, 16 (2)).

GEYH, M. A. and ROHDE, P. (1972) Weichselian chronostratigraphy, ^{14}C dating and statistics, pp. 27–36 in Section 12, *Proc. 24th Internat. Geol. Congress*, Montreal.

GIDDINGS, J. L. (1941) Drendrochronology in northern Alaska, *Univ. Arizona Bull.*, 12 (4). Tucson.

GIOVINETTO, M. B. and SCHWERDTFEGER, W. (1966) Analysis of a 200-year snow accumulation series from the South Pole, *Archiv. f. Met. Geophys. Biokl.*, A, 15, 227–50. Vienna.

GLASS, B., ERICSON, D. B., HEEZEN, B. C., OPDYKE, N. D. and GLASS, J. A. (1967) Geomagnetic reversals and Pleistocene chronology, *Nature*, 216, 437–42. London (4 November).

GLOCK, W. S. (1937) *Principles and Methods of Tree Ring Analysis*. Washington (Carnegie Inst. Pubn 486).

GODWIN, H. (1954) Recurrence surfaces, *Danmarks Geol. Undersøgelse*, II Raekke, 80. Copenhagen.

GODWIN, H. (1956) *The History of the British Flora*. Cambridge (Univ. Press).

GODWIN, H. (1962) Radiocarbon dating: Fifth International Conference, *Nature*, **195**, 943–5. London.

GRANLUND, E. (1932) *De svenska högmossarnas geologi.* Sveriges Geologiska Undersökning, Ser. C, No. 373. Stockholm.

GRIGGS, G. B., KULM, L. D., DUNCAN J. R. and FOWLER, G. A. (1970) Holocene faunal stratigraphy and palaeoclimatic implications of deep sea sediments in Cascadia Basin, *Palaeogeogr., Palaeoclim., Palaeoecol.* **7**, 5–12. Amsterdam.

HAFSTEN, U. (1960) Pollen-analytic investigations in South Norway, *Norges Geol. Undersøkelse*, **208**, 434. Oslo.

HAGEN, A. (1961) *Steinalder, Bronsealder.* Oslo (Universitetets Oldsaksamling).

HARE, P. E. (1969) Geochemistry of proteins, peptides and amino acids, Ch. 18, pp. 438–63, in *Organic Geochemistry* (ed. G. EGLINTON and M. T. J. MURPHY). Berlin, Heidelberg, New York (Longman/Springer Verlag).

HARRISON, C. G. A. and FUNNELL, B. M. (1964) Relationship of palaeomagnetic reversals and micropalaeontology in two late Cenozoic cores from the Pacific Ocean, *Nature*, **204**, 566. London.

HEUSSER, C. J. (1966) Polar hemispheric correlation: palynological evidence from Chile and the Pacific north-west of America, pp. 124–41 in *World Climate* 8000–0 *B.C.: Proc. Internat. Conf.* (ed. J. S. SAWYER). London (Roy. Met. Soc.).

HOUTERMANS, J., SUESS, H. E. and MUNK, W. (1967) Effect of industrial fuel combustion on the carbon-14 level of atmospheric CO_2, pp. 57–68 in *Radioactive Dating and Methods of Low-Level Counting.* Vienna (Internat. Atomic Energy Agency).

HUBER, B. (1960) Dendrochronologie, *Geol, Rundsch.*, **49**, 120–31. Stuttgart.

HUBER, B. and HOLDHEIDE, W. (1942) Jahrringchronologische Untersuchungen an Hölzern der bronzezeitlichen Wasserburg Buchau am Federsee, *Ber. der dt. Bot. Gesell.*, **60**, 261–83. Berlin.

HUBER, B. and SIEBENLIST, V. (1963) Das Watterbacher Haus im Odenwald, ein wichtiges Brückenstück unserer tausend-jährigen Eichenchronologie, *Mitt. der Floristisch-soziologischen Arbeitsgemeinschaft* (Neue Folge), Heft 10, 256. Stolzenau.

IVERSEN, J. (1954) The late glacial flora of Denmark and its relation to climate and soil, *Danmarks Geol. Undersøgelse*, II Raekke, **80**, 87. Copenhagen.

IVERSEN, J. (1958) The bearing of glacial and interglacial epochs on the formation and extinction of plant taxa, pp. 210–15 in *Systematics of Today* (ed. O. HEDBERG). *Uppsala Univs. Årsskrift.*

JAPANESE COMMITTEE ON TEPHROCHRONOLOGY (K. KOBAYASHI, Chairman) (1969) Short review of recent studies on teprochronology in Japan. Tokyo (for INQUA, Paris Meeting).

JÄRNEFORS, B. (1963) Lervarvskronologien och isrecessionen i östra Mellansverige (Varve chronology and ice recession in eastern central Sweden), *Sveriges Geol. Undersökning*, Ser. C, No. 594, Årsbok **57** (1963) No. 5. Stockholm. 67 pp., incl. 4 pp. English summary.

JESSEN, K. (1949) Studies in Late Quaternary deposits and flora history of Ireland, *Proc. Roy. Irish Acad.*, **52** (B6), 85. Dublin.

KENNETT, J. P. and WATKINS, N. D. (1970) Geomagnetic polarity change, volcanic maxima and faunal extinction in the South Pacific, *Nature*, **227**, 930–4. London (29 August).

KOBAYASHI, K. (1965) Problems of Late Pleistocene history of central Japan, *Geol. Soc. Amer. Spec. Paper*, **84**, 367–91. Boulder, Colorado.

KUTZBACH, J. E., BRYSON, R. A. and SHEN, W. C. (1968) An evaluation of the thermal Rossby number in the Pleistocene, *Met. Monogs*, **8** (30), 134–8. Boston (Amer. Met. Soc.).

LaMARCHE, V. C. and HARLAN, T. P. (1973) Accuracy of tree-ring dating of bristlecone pine for calibration of the radiocarbon time scale, *J. Geophys. Res.*, **78** (36), 8849–58. Baltimore.

LAMB, H. H. (1970) Volcanic dust in the atmosphere, with a chronology and assessment of its meteorological significance, *Phil. Trans. Roy. Soc.*, **A, 266**, 425–533. London.

LAMB, H. H., LEWIS, R. P. W. and WOODROFFE, A. (1966) Atmospheric circulation and the main climatic variables, pp. 174–217 in *World Climate 8000-0 B.C.: Proc. Internat. Conf.* (ed. J. S. SAWYER). London (Roy. Met. Soc.).

LANGWAY, C. C. (1967) Stratigraphic analysis of a deep ice core from Greenland, *Res. Rep.*, **77**, Hanover, New Hampshire (U.S. Army Cold Regions Res. & Eng. Lab.).

LERMAN, J. C., MOOK, W. G. and VOGEL, J. C. (1970) C14 in tree rings from different localities, pp. 275–301 in *Radiocarbon Variations and Absolute Chronology: Proc. Twelfth Nobel Symp., Uppsala, 1969* (ed. I. U. OLSSON). Stockholm (Almqvist & Wiksell) and New York (Wiley). 657 pp.

LIBBY, W. F. (1946) Atmospheric helium-3 and radiocarbon from cosmic radiation, *Phys. Rev.*, **69**, 671–2. Lancaster, Pa.

LIBBY, W. F. (1970) Radiocarbon dating, *Phil. Trans. Roy. Soc.*, **A, 269**, 1–10. London.

LIDÉN, R. (1913) Geokronologiska studier öfver det Finiglaciala skedet, *Sveriges Geol. Undersökning*, **9**. Stockholm.

LIDÉN, R. (1938) Den senkvartära strandförskjutningens forlöpp och kronologi i Ångermanland, *Geologiska Fören. Förhandl.*, **60**, 397–404. Stockholm.

MACHIDA, H. (1964) Tephrochronological study of volcano Fuji and adjacent areas, *Chigaku Zasshi*, **73**, 293–308, 337–50. Tokyo.

MICHAEL, H. N. and RALPH, E. K. (1971) *Dating Techniques for the Archaeologist.* MASCA Handbook. Boston (M.I.T. Press).

MILANKOVITCH, M. (1930) Mathematische Klimalehre, *Handbuch der Klimatologie* (ed. KÖPPEN and GEIGER), Band I, Teil A. Berlin (Borntraeger).

MIYADI, D. (1939) Limnological survey of Taiwan (Formosa), *Archiv für Hydrobiologie*, **35**, 1–27. Stuttgart.

MOAR, N. T. (1971) Contributions to the Quaternary history of the New Zealand flora, 6: Araunian pollen diagrams from Canterbury, Nelson and North Westland, South Island, *N.Z. J. Bot.*, **9** (1), 80–145. Wellington, N.Z.

MÖRNER, N.-A. (1974) The Greenland o^{18} curve: a time-scale problem, pp. 39–42 in *Les méthodes quantitatives d'étude des variations du climat au cours du Pléistocène*. Gif-sur-Yvette (C.N.R.S.). 317 pp.

NICHOLS, H. (1967) The postglacial history of vegetation and climate at Ennadai Lake, Keewatin and Lynn Lake, Manitoba (Canada), *Eiszeitalter und Gegenwart*, **18**, 176–97. Öhringen.

NILSSON, E. (1964) Geochronological investigations in southern Sweden, *Geol. Fören. Förhandl.*, **86**, 162–70. Stockholm.

NILSSON, E. (1968a) The late Quaternary history of southern Sweden, *Kungl. Svensk. Vetensk. Akad. Handl.*, Ser. 4, **12**, No. 1. Stockholm. 117 pp.

NILSSON, E. (1968b) Skånes senkvartära historia med ny datering, *Geol. Föreningens Förhandl.*, **90**, 468. Stockholm.

NOEL, M. and TARLING, D. H. (1975) The Laschamp geomagnetic 'event', *Nature*, **253**, 705–7. London (27 February).

NYDAL, R. (1966) Variation in C^{14} concentration in the atmosphere during the last several years, *Tellus*, **18** (2), 271–9. Stockholm.

NYE, J. F. (1959) The motion of ice sheets and glaciers, *J. Glaciol.*, **3**, 493–507. Cambridge.

NYE, J. F. (1963) Correction factor for accumulation measured by the thickness of annual layers in an ice sheet, *J. Glaciol.*, **4**, 785–8. Cambridge.

OGDEN, J. G. (1967) Radiocarbon and pollen evidence for a sudden change in climate in the Great Lakes region approximately 10 000 years ago, pp. 117–27 in *Quaternary Palaeoecology* (ed. E. J. CUSHING and H. E. WRIGHT). New Haven and London (Yale Univ. Press).

OLAUSSON, E. (1971) Oceanographic aspects of the Pleistocene of Scandinavia, *G.F.F.*, **93** (3), 459–75. Stockholm (Geol. Fören.)

OLAUSSON, E. and OLSSON. I. U. (1969) Varve stratigraphy in a core from the Gulf of Aden, *Palaeogeogr., Palaeoclim., Palaeoecol.*, **6**, 87–103. Amsterdam.

OLAUSSON, E., BILAL UL HAQ, U. Z., KARLSSON, G. B. and OLSSON, I. U. (1971) Evidence in Indian Ocean cores of later Pleistocene changes in oceanic and atmospheric circulation, *G.F.F.*, **93** (1), 51–84. Stockholm (Geol. Fören.).

OLSSON, I. U. (1968) Modern aspects of radiocarbon datings, *Earth Sci. Rev.*, **4**, 203–18. Amsterdam.

OLSSON, I. U. (ed.) (1970) *Radiocarbon Variations and Absolute Chronology: Twelfth Nobel Symp., Uppsala, 1969.* Stockholm (Almqvist & Wiksell) and New York (Wiley), 657 pp.

OUTI, M. (1964) The relation between the growth rate of trees and climatic changes in the northeastern section of Japan, *Bull. Kyoto Gakugei Univ.*, **B**, **25**, 89–107. Kyoto.

OVERBECK, F., MÜNNICH, K. O., ALETSEE, L. and AVERDIECK, F. R. (1957) Das Alter des 'Grezhorizonts' norddeutscher Hochmoore nach Radio-Carbon Datierungen, *Flora*, **145**, 35–71. Jena.

PITMAN, G. T. K. (1973) A lichenometrical study of snow patch variation in the Frederikshåb district, southwest Greenland, *Grønlands Geol. Undersøgelse*, Bull. No. 104. Copenhagen. 31 pp.

Radiocarbon (1959 ff.) Bulletin giving comprehensive reporting of the radiocarbon dating results of all the world's laboratories concerned. First published 1959 as the *Radiocarbon Supplement* to *Amer. J. Sci.*, but from 1961 (Vol. 3) onwards published independently under the sole title *Radiocarbon*. New Haven, Conn.

RALPH, E. K. and MICHAEL, H. N. (1967) Problems of the radiocarbon calendar. *Archaeometry*, **10**, 3–11, Oxford.

RONA, E. and EMILIANI, C. (1969) Absolute dating of Caribbean cores P 6304–8 and P 6304–9, *Science*, **163**, 66–8. New York.

RUDAKOV, V. E. (1968) The tree as a chronicler of climate, *Priroda*, **9**, 60–3. Moscow (Akad. Nauk). (In Russian.)

SACHS, A. (1970) Absolute dating from Mesopotamian records, pp. 19–22 in *The Impact of the Natural Sciences on Archaeology* (ed. T. E. ALLIBONE). London (Oxford Univ. Press for the British Academy). Also *Phil. Trans. Roy. Soc. Lond.*, A, **269**, 11–18.

SAURAMO, M. (1923) Studies on the Quaternary varve sediments in southern Finland, *Bull. Comm. geol. Finlande*, **60**. Helsinki. 164 pp.

SCHOFIELD, J. C. (1965) The Hinuera Formation and associated Quarternary events, *N.Z. J. Geol. & Geophys.*, **8** (5), 772–91. Wellington, N.Z.

SCHOSTAKOWITSCH, W. B. (1934) Bodenablagerungen der Seen und periodische Schwankungen der Naturerscheinungen. Reprinted by C. E. P. BROOKS in *Met. Mag.*, **70**, 134–8 (1935) from *Memoirs of Leningrad Hydr. Inst.*

SCHOVE, D. J. and LOWTHER, A. W. G. (1957) Tree rings and medieval archaeology, *Medieval Archaeology*, **1**, 78–95. London.

SCHWARZBACH, M. (1961) *Das Klima der Vorzeit*. Stuttgart (Enke).

SEELEY, M.-A. (1975) Thermoluminescent dating in its application to archaeology, *J. Archaeol. Sci.*, **2**, 17–43. London (Academic Press).

SERNANDER, R. (1908) On the evidence of postglacial changes of climate furnished by the peat mosses of northern Europe, *Geol. Fören*, **30**, 465–78. Stockholm.

SERNANDER, R. (1910) *Die schwedischen Torfmoore als Zeugen postglazialer Klimaschwankungen*. Stockholm (Geol. Fören.).

SIRÉN, G. (1961) Skogsgränstallen som indikator för klimatfluktuationerna i norra Fennoskandien under historisk tid. Helsingfors, *Communicationes Instituti Forestalis Fenniae*, **54** (2).

SMITH, A. G. (1965) Problems of inertia and threshold related to postglacial habitat changes, *Proc. Roy. Soc.*, B, **161**, 331–45. London.

STEVEN, H. M. and CARLISLE, A. (1959) *The Native Pinewoods of Scotland.* Edinburgh (Oliver & Boyd).

STUIVER, M. (1961) Variations in radiocarbon concentration and sunspot activity, *J. Geophys. Res.*, **66** (1), 273–6. Baltimore, Md.

STUIVER, M. (1970) Long-term C14 variations, pp. 197–213 in *Radiocarbon Variations and Absolute Chronology: Proc. Twelfth Nobel Symp., Uppsala 1969* (ed. I. U. OLSSON). Stockholm (Almqvist & Wiksell) and New York (Wiley).

SUESS, H. E. (1968) Climatic changes, solar activity, and the cosmic-ray production of natural radiocarbon, *Met. Monogs*, **8** (30), 146–50. Boston, Mass.

SUESS, H. E. (1970) The three causes of the secular C14 fluctuations and their time constants, pp. 595–605 in *Radiocarbon Variations and Absolute Chronology: Proc. Twelfth Nobel Symp., Uppsala 1969* (ed. I. U. OLSSON). Stockholm (Almqvist & Wiksell) and New York (Wiley).

TARLING, D. H. (1975) Archeomagnetism: the dating of archaeological materials by their magnetic properties, *World Archaeology*, **7** (2), 185–97. Henley-on-Thames (Routledge & Kegan Paul).

TAUBER, H. (1970) The Scandinavian varve chronology and C14 dating, pp. 173–96 in *Radiocarbon variations and Absolute Chronology: Proc. Twelfth Nobel Symposium, Uppsala 1969* (ed. I. U. OLSSON). Stockholm (Almqvist & Wiksell) and New York (Wiley).

THOMPSON, M. W. (1967) *Novgorod the Great.* London (Evelyn, Adams & Mackay).

THOMPSON, R. (1973) Palaeolimnology and palaeomagnetism, *Nature*, **242**, 182–4. London (16 March).

THORARINSSON, S. (1944) Tefrokonologiska studier på Island, *Geografiska Annaler*, **26**, 1–217. Stockholm.

THORARINSSON, S. (1958) The Öraefajökull eruption of 1362, *Acta nat. Islandica*, Vol. II, (2). Reykjavik.

THORARINSSON, S., EINARSSON, T. and KJARTANSSON, G. (1959) On the geology and geomorphology of Iceland, *Geografiska Annaler*, **41**, 138–69. Stockholm.

VERNEKAR, A. D. (1968) *Research on the Theory of Climate, Vol. 2: Long-Period Global Variations of Incoming Solar Radiation.* Hartford, Conn. (Travelers Research Center, Contract No. E22-137-67 (N)).

WARDLE, P. (1963) Vegetation studies on Secretary Island, Fjordland, Part 5, *N.Z. J. Bot.*, **1** (2), 208–14. Wellington.

WATERBOLK, H. T. (1968) Food production in prehistoric Europe, *Science*, **162**, 1093–1102. New York.

WEST, R. G. (1961) Late- and postglacial vegetational history in Wisconsin, particularly changes associated with the Valders readvance, *Amer. J. Sci.*, **259**, 766–83. New Haven, Conn.

WEST, R. G. (1968) *Pleistocene Geology and Biology.* London (Longmans). 377 pp.

WILLIS, E. H. (1961) The validity of radiocarbon dating, *Times Sci. Rev.*, 6–9, London (Winter issue).

WOOD, E. S. (1972) *Collins Field Guide to Archaeology*. London (Collins), 3rd edn.

ZEUNER, F. E. (1958) *Dating the Past*. London (Methuen), 4th edn.

13.3 Physical evidence

AARIO, R. (1969) The northern discharge channel of ancient Päijänne and the palaeohydrology of the Atlantic period. *Bull. Geol. Soc. Finland*, **41**, 3–20. Helsinki.

AHLMANN, H. W. (1953) *Glacier Variations and Climatic Fluctuations*. Bowman Memorial Lecture. New York (American Geog. Soc.).

ANDERSEN, B. G. (1960) Sørlandet i sen og postglacial tid, *Norges Geol. Undersøgelse*, **210**, 1–142. Oslo.

ARAKAWA, H. (1954) Five centuries of freezing dates of Lake Suwa in central Japan, *Arch. Met. Geophys. Biokl.*, **B, 6**, 152–66. Vienna.

ARRHENIUS, G. (1952) Sediment cores from the east Pacific, *Swedish Deep-Sea Expedition, 1947–8, Reports*, **5** (1). Göteborg (Elanders Boktryckeri A/B). 227 pp.

BAKKER, E. VAN ZINDEREN (1969) Intimations on Quaternary ecology of Africa, *Acta Bot. Neerland.*, **18** (1), 230–9. Amsterdam.

BELL, B. (1970) The oldest records of the Nile floods, *Geogr. J.*, **136** (4), 569–73. London.

BELL, B. (1971) The Dark Ages in ancient history, I: The first Dark Age in Egypt, *Amer. J. Archaeol.*, **75**, 1–26. New York.

BERG, L. S. (1934) The level of the Caspian Sea in historical times, *Problemy fisic. geograf.*, **1** (1). Moscow. (In Russian.)

BETIN, V. V. (1957) Ice conditions in the Baltic and its approaches and their long-term variations, *Trudy*, **41**, 54–125. Leningrad (Grosudarst-Okeanogr. Inst.). (In Russian.)

BLACK, R. F. (1954) Permafrost: a review, *Geol. Soc. Amer. Bull.*, **65**, 839–56. Rochester, N.Y.

BLAKE, I. (1969) Climate, survival and the second-class societies in Palestine before 3000 B.C., *Advancement of Sci.*, **25** (126), 409–21. London.

BLOOM, A. L. (1967) Pleistocene shorelines: a new test of isostasy, *Geol. Soc. America Bull.*, **78**, 1477–94. Rochester, N.Y.

BRADLEY, R. S. and MILLER, G. H. (1972) Recent climatic change and increased glacierization in the eastern Canadian Arctic, *Nature*, **237**, 385–7. London (16 June).

BRAY, J. R. (1971) Solar-climate relationships in the post-Pleistocene, *Science*, **171**, 1242–3. Washington (26 March). (Originally reported by J. R. BRAY and G. J. STRUIK in *Canad. J. Bot.*, **41**, 1245 (1963). Ottawa.)

BROECKER, W. S. and ORR, P. C. (1958) Radiocarbon chronology of Lake Lahontan and Lake Bonneville. *Geol. Soc. Amer. Bull.*, **69**, 1009–32. Rochester, N.Y.

BROECKER, W. S., EWING, M. and HEEZEN, B. C. (1960) Evidence for an abrupt change in climate close to 11 000 years ago, *Amer. J. Sci.*, **258**, 429–48. New Haven, Conn.

BROOKS, C. E. P. (1949) *Climate Through the Ages*. London (Benn), 2nd edn.

BRÜCKNER, E. A. (1890) *Klima-Schwankungen seit 1700, nebst Bemerkungen über die Klimaschwankungen der Diluvialzeit* (Geogr. Abhandl. von A. Penck, **4** (2)). Vienna.

BUCHINSKY, I. E. (1957) *The Past Climate of the Russian Plain*. Leningrad (Gidrometeoizdat), 2nd edn. (In Russian: English trans. in Met. Office Library.)

BUTZER, K. W. (1959) Some recent geological deposits of the Egyptian Nile valley, *Geogr. J.*, **125**, 75–9. London.

BUTZER, K. W., ISAAC, G. L., RICHARDSON, J. L. and WASHBOURN-KAMAU, C. (1972) Radiocarbon dating of East African lake levels, *Science*, **175** (4027), 1069–76. New York.

CERMAK, V. (1971) Underground temperature and inferred climatic temperature of the past millennium, *Palaeogeogr., Palaeoclim., Palaeoecol.*, **10**, 1–19. Amsterdam.

CHAPPEL, J. E. (1970) Another look at the 'Pulse of Asia', *Geogr. Rev.*, **60** (3), 347–73. New York.

CHU KO-CHEN (COCHING CHU) (1926) Climatic pulsations during historical time in China, *Science Essays*, **3**, 1–12. Shanghai (China Science Publishing House).

CHU KO-CHEN (1961) The pulsation of world climate during historical times, *New China Monthly* **6**, June. Peking. (In Chinese: English translation in Meteorological Office Library, Bracknell.)

CHURCHILL, D. M. (1965) The displacement of deposits formed at sea level 6500 years ago in southern Britain, *Quaternaria*, **7**, 239–49. Rome.

COLINVAUX, P. A. (1972) Climate in the Galapagos, *Nature*, **240**, 17–20. London (3 November).

CORNWALL, I. (1970) *Ice Ages: Their Nature and Effects*. London (John Baker).

DANSGAARD, W. (1964) Stable isotopes in precipitation, *Tellus*, **16** (4), 436–68. Stockholm.

DANSGAARD, W. and TAUBER, H. (1969) Glacier oxygen-18 content and Pleistocene ocean temperatures, *Science*, **166**, 499–502. New York (24 October).

DANSGAARD, W., JOHNSEN, S. J., CLAUSEN, H. B. and LANGWAY, C. C. (1971) Climatic record revealed by the Camp Century ice core, pp. 37–56 in *The Late Cenozoic Glacial Ages* (ed. K. K. TUREKIAN). New Haven, Conn. (Yale Univ. Press). 606 pp. *Note:* A rather full account of the potentialities of stable isotope work on ice sheets and sample results explained from Greenland and Antarctica is given in the next reference entry.

DANSGAARD, W., JOHNSEN, S. J., CLAUSEN, H. B. and GUNDESTRUP, N. (1973) Stable isotope glaciology, *Medd. øm Gronland*, **197** (2). Copenhagen (Reitzel). 53 pp.

DE MARCHI, L. (1895) *Reale istituto lombardo di scienze e lettere, Rendiconti*, ser. 2, **28**, 1018–31. Milan.

DERBYSHIRE, E. (1971) A synoptic approach to the atmospheric circulation of the last glacial maximum in southeastern Australia, *Palaeogeogr. Palaeoclim. Palaeoecol.*, **10**, 103–24. Amsterdam.

DIMBLEBY, G. W. (1965) Postglacial changes in soil profiles, *Proc. Roy. Soc.*, **B, 161** (984), 355–61. London.

EASTON, A. K. (1970) *The Tides of the Continent of Australia*. Adelaide (Horace Lamb Centre for Oceanographic Research, Flinders Univ.).

EMILIANI, C. (1955) Pleistocene temperatures, *J. Geol.*, **63** (6), 538–78. Chicago.

EMILIANI, C. (1966a) Isotopic paleotemperatures, *Science*, **154**, 851–7. New York (18 November).

EMILIANI, C. (1966b) Palaeotemperature analysis of Caribbean cores P6304–8 and P6304–9 and a generalized temperature curve for the past 425 000 years, *J. Geol.*, **74** (2), 109–26. Chicago.

EMILIANI, C. and SHACKLETON, N. J. (1974) The Brunhes epoch: isotopic palaeotemperatures and geochronology, *Science*, **183**, 511–14. New York (8 February).

EPSTEIN, S., BUCHSBAUM, R., LOWESTAM, H. and UREY, H. C. (1953) Revised carbonate-water isotopic temperature scale, *Geol. Soc. Amer. Bull.*, **64**, 1315–25. Rochester, N.Y.

EPSTEIN, S., SHARP, R. P. and GOW, A. J. (1970) Antarctic ice sheet: stable isotope analyses of Byrd station cores and interhemispheric climatic implications, *Science*, **168**, 1570–2. New York. (26 June).

FAIRBRIDGE, R. W. (1961) Eustatic changes in sea level, Ch. 3, pp. 99–185, in *Physics and Chemistry of the Earth*. New York (Pergamon).

FLINT, R. F. (1971) *Glacial and Quaternary Geology*. New York (Wiley). 892 pp. (New, enlarged edition of *Glacial and Pleistocene Geology*, 1957.)

FLINT, R. F. and GALE, W. A. (1958) Stratigraphy and radiocarbon dates at Searles Lake, California, *Amer. J. Sci.*, **256**, 689–714. New Haven, Conn.

FRASER, LADY A. (1970) *Mary Queen of Scots*. London.

FRENZEL, B. (1959) Die Vegetations- und Landschaftszonen Nord-Eurasiens während der letzten Eiszeit und während der postglazialen Wärmezeit. Mainz (Akad. der Wiss. und Lit., *Abhandl. math.-naturw. Klasse*, I (13)).

FRENZEL, B. (1966) Climatic change in the Atlantic/Sub-Boreal transition on the northern hemisphere: botanical evidence, pp. 99–123 in *World Climate 8000–0 B.C.: Proc. Internat. Conf.* (ed. J. S. SAWYER). London (Roy. Met. Soc.).

FRENZEL, B. (1967) *Die Klimaschwankungen des Eiszeitalters*. Braunschweig (Vieweg—*Die Wissenschaft*, Band 129).

GIOVINETTO, M. B. and SCHWERTFEGER, W. (1966) Analysis of a 200-year snow accumulation series from the South Pole, *Archiv. f. Met. Geophys. Biokl.*, A, 15 (2), 227–50. Vienna.

GLEN, J. W. (1952) Experiments on the deformation of ice, *J. Glaciol.*, **2**, 111–14. Cambridge.

GLEN, J. W. (1953) The rate of flow of polycrystalline ice, *Nature*, **172** (4381), 721–2. London.

GLEN, J. W. (1956) Measurements of the deformation of ice in a tunnel at the foot of an ice fall, *J. Glaciol.* **2** (20), 735–45. Cambridge.

GLEN, J. W. (1967) The physics of the flow of glaciers, *Bull. Inst. Phys. & Phys. Soc.*, **18**, 135–40. London.

GODWIN, H., SUGGATE, R. P. and WILLIS, E. H. (1958) Radiocarbon dating of the eustatic rise in ocean level, *Nature*, **181**, 1518–19. London (31 May).

GOW, A. J. (1970) *International Symposium on Antarctic Glaciological Exploration*. Internat. Assn. of Sci. Hydrol. Pubn No. 86. Gentbrugge, Belgium (U.G.G.I.).

GRANT, C. P. (1938) *The Syrian Desert*. London.

GRIGORIEV, A. A. (1930) Die ewige Gefrornis und die ehemalige Vergletscherung. *Kom. po izuč. estestven. proizvod. sil. SSSR, Materialy* **80**, *Sbornik Večnaja Merzlota*, 43–104. Moscow (Akad. Nauk).

GROVE, A. T. and WARREN, A. (1968) Quaternary landforms and climate on the south side of the Sahara, *Geogr. J.*, **134** (2), 194–208. London.

GROVE, J. M. (1972) The incidence of landslides, avalanches, and floods in western Norway during the Little Ice Age, *Arctic and Alpine Res.*, **4** (2), 131–8. Boulder, Colorado.

GUMILYOV (GUMILEV), L. N. (1964) Khazaria and the Caspian (landscape and ethos, Part I). *Soviet Geography: Review and Translation*, **5** (6), 54–68. (Originally published in Russian in *Vestnik Leningrad Univ., Ser. Geogr.*, **19** (1:6), 83–95.)

HAMILTON, W. L. and SELIGA, T. A. (1972) Atmospheric turbidity and surface temperature on polar ice sheets, *Nature*, **235**, 320–2. London (11 February).

HASTENRATH, S. (1971) On snowline depression and atmospheric circulation in the tropical Americas during the Pleistocene, *S. African Geogr. J.*, **53**, 53–68. Johannesburg.

HENDY, C. H. and WILSON, A. T. (1968) Palaeoclimatic data from speleothems, *Nature*, **219**, 48–51. London (6 July).

HENNIG, R. (1904) Katalog bemerkenswerter Witterungsereignisse von den ältesten Zeiten bis zum Jahre 1800. *Abhandl. preuss. met. Inst.*, Band II (4). Berlin.

HIGGS, E. S., FAGG, A. E., VITA FINZI, C., HARRIS, D. R. and BOTTEMA, S. (1967) The climate, environment and industries of Stone Age Greece, Part III, *Proc. Prehist. Soc.*, **33**, 1–29. Cambridge.

HOINKES, H. C. (1964) Glacial meteorology, Ch. 15, pp. 391–424, in *Research in Geophysics, Vol. 2: Solid Earth and Interface Phenomena*. Cambridge, Mass. (M.I.T. Press).

HOINKES, H. C. and RUDOLPH, R. (1962) Variations in the mass-balance of Hintereisferner (Oetztal Alps) 1952–1961 and their relation to variations of climatic elements, pp. 16–28 in *Publication No. 58, Internat. Comm. Snow and Ice, I.A.S.H.* Gentbrugge, Belgium (U.G.G.I.)

HOLMES, J. W. (1969) On the absolute fall of sea level during the Quaternary, *Palaeogeogr., Palaeoclim., Palaeoecol.*, **6**, 237–9. Amsterdam.

IMBRIE, J. and KIPP, N. G. (1971) A new micropalaeontological method for quantitative palaeoclimatology: application to a Late Pleistocene Caribbean core, pp. 71–181 in *The Late Cenozoic Glacial Ages: Proc. Yale Univ. Symp., December 1969* (ed. K. K. TUREKIAN). New Haven, Conn. (Yale Univ. Press). 606 pp.

JAKOVLEV, S. A. (1956) *Osnovy geologii četvertičnych otloženii Russkoi ravniny.* Moscow.

JELGERSMA, S. (1966) Sea level changes during the last 10 000 years, pp. 54–71 in *World Climate 8000–0 B.C.: Proc. Internat. Conf.* (ed. J. S. SAWYER). London (Roy. Met. Soc.).

JELGERSMA, S., DE JONG, J., ZAGWIJN, W. H. and VAN REGTEREN ALTENA, J. F. (1970) The coastal dunes of the western Netherlands; geology, vegetational history and archaeology, *Medd. Rijks Geol. Dienst*, nieuwe serie, No. 21. Maastricht.

JOHNSEN, S. J., DANSGAARD, W., CLAUSEN, H. B. and LANGWAY, C. C. (1972) Oxygen isotope profiles through the Antarctic and Greenland ice sheets, *Nature*, **235**, 429–34. London (25 February).

KENNETT, J. P. and WATKINS, N. D. (1970) Geomagnetic change, volcanic maxima and faunal extinction in the South Pacific, *Nature*, **227** (5261), 930–4. London (29 August).

KESSLER, A. (to be published) Atmospheric circulation anomalies and level fluctuations of Lake Titicaca, South America, *Proc. Internat. Trop. Met. Meeting, 1974, Nairobi, Kenya.* Boston, Mass. (Amer. Met. Soc.).

KLEIN, C. (1961) *On the Fluctuations of the Level of the Dead Sea Since the Beginning of the Nineteenth Century.* Hydrol. Paper No. 7. Jerusalem (Israel Hydrol. Service).

KOCH, L. (1945) The east Greenland ice, *Medd. om Grønland*, **130** (3). Copenhagen.

LAMB, H. H. (1959) The southern westerlies. *Quart. J. Roy. Met. Soc.*, **85**, 1–23. London.

LAMB, H. H. (1966) Climate in the 1960s: changes in the world's wind circulation . . . and the levels of the African lakes, *Geogr. J.*, **132** (2), 183–212.

LAMB, H. H. (1967) On climatic variations affecting the Far South, pp. 428–53 in *Polar Meteorology: W.M.O. Tech. Note No. 87.* Geneva (World Met. Org., No. 211, T.P. 111).

LAMB, H. H. and JOHNSON, A. I. (1961) Climatic variation and observed changes in the general circulation, III, *Geogr. Annaler*, **43**, 393–400. Stockholm.

LAMBERT, G., LE ROULLEY, J.-C. and SANAK, J. (1971) Corrélations entre le dépôt de plomb 210 dans l'Antarctique et les fluctuations du champ magnétique terrestre, *Comptes rendues*, **272**, 177–80. Paris (Acad. Sci.).

LAMBERT, G., SANAK, J. and ARDOUIN, B. (1971) Corrélations entre la concentration du plomb 210 en Terre Adélie et l'activité solaire, *Comptes rendues*, **272**, 497–500. Paris (Acad. Sci.).

LANGWAY, C. C. (1967) Stratigraphic analysis of a deep ice core from Greenland, *Res. Rep.* **77**. Hanover, New Hampshire (U.S. Army Cold Regions Research and Engineering Laboratories).

LATTER, J. H. (1969) Natural disasters, *Advancement of Science*, **25** (126), 362–80. London (British Assoc. Adv. Sci.).

LEONTEV, O. K. and FEDEROV, P. V. (1953) History of the Caspian Sea in late- and post-Hvalynsk time, *Izvestia. Ser. Geogr.* No. 4, 64–74. Moscow (Akad. Nauk).

LIBBY, L. M. (1972) Multiple thermometry in palaeoclimate and historic climate. *J. Geophys. Res.*, **77** (23), 4310–7. Baltimore, Md.

MACHATSCHEK, F. (1969) *Geomorphology* (9th edn of *Geomorphology*, translated into English by D. J. DAVIS). *Geomorphology Texts No. 1* (ed. K. M. CLAYTON). Edinburgh (Oliver & Boyd).

MACKERETH, F. J. H. (1965) Chemical investigation of lake sediments and their interpretation, *Proc. Roy. Soc.*, **B, 161** (984), 295–309. London.

MAKSIMOV, I. V. (1971) Causes of the rise of sea level in the present century, *Okeanologija*, **11**, 530–41. Moscow (Akad. Nauk, Okean. Kom.).

MALEY, J. (1973) Paléoclimatologie, *Comptes rendus*, **276** (series D), 1673–5. Paris (Acad. Sci.).

MARGARITZ, M. and KAUFMAN, A. (1973) Changes in isotopic composition of the eastern Mediterranean sea water during the Holocene, *Nature*, **243**, 462–4. London (22 June).

MARGOLIS, S. V. and KENNETT, J. P. (1971) Cenozoic palaeoglacial history of Antarctica recorded in deep sea cores, *Amer. J. Sci.*, **271**, 1–36. New Haven, Conn.

MERCER, J. H. (1969) The Allerød oscillation: a European climatic anomaly?, *Arctic and Alpine Res.*, **1** (4), 227–34. Boulder, Colorado.

MESOLELLA, K. J., MATTHEWS, R. K., BROECKER, W. S. and THURBER, D. L. (1969) The astronomical theory of climatic change: Barbados data. *J. Geol.*, **77**, 250–74. Chicago.

MONGAIT, A. L. (1961) *Archaeology in the U.S.S.R.* London (Penguin Books). (Originally published in Russian as *Arkheologiya v SSSR*, 1955.)

MÖRNER, N.-A. (1969) The late Quaternary history of the Kattegatt Sea and the Swedish west coast, *Sveriges Geol. Undersökning*, ser. C, Nr. 640, Årsbok 63, Nr. 3 (esp. pp. 414–47). Stockholm.

MÖRNER, N.-A. (1973) Eustatic changes during the last 300 years, *Palaeogeogr., Palaeoclim., Palaeoecol.*, **13**, 1–14. Amsterdam.

NIKOLAEVA, R. V. and HAN-MAGOMEDOV, S. O. (1962) New data on the level of the Caspian Sea during the historical period, *Trudy*, **60**, 178–88. Moscow (Akad. Nauk, Inst. Okeanologii).

NYE, J. F. (1953) The flow law of ice from measurements in glacier tunnels, laboratory experiments and the Jungfraufirn borehole experiment, *Proc. Roy. Soc.*, **A, 219**, 477–89. London.

NYE, J. F. (1960) The response of glaciers and ice sheets to seasonal and climatic changes, *Proc. Roy. Soc.*, **A, 256**, 559–84. London.

NYE, J. F. (1961) The influence of climatic variations on glaciers, pp. 397–404 in *Publication No. 54, Internat. Comm. Snow and Ice, I.A.S.H.* Gentbrugge, Belgium (U.G.G.I.).

NYE, J. F. (1963a) The response of a glacier to changes in the rate of nourishment and wastage, *Proc. Roy. Soc.*, **A, 275**, 87–112. London.

NYE, J. F. (1963b) On the theory of the advance and retreat of glaciers. *Geophys. J.*, **7** (4), 431–56. London (R. Astronom. Soc.).

OLAUSSON, E. (1967) *Marine Sediments.* Unpublished memorandum, Maringeologiska Laboratoriet, Göteborg. 21 pp.

OLAUSSON, E. (1971) Quaternary correlations and the geochemistry of oozes, pp. 375–98 in *The Micropalaeontology of Oceans* (ed. B. M. FUNNELL). Cambridge (Univ. Press).

OLAUSSON, E. and OLSSON, I. U. (1969) Varve stratigraphy in a core from the Gulf of Aden, *Palaeogeogr., Palaeoclim., Palaeoecol.,* 6, 87–103. Amsterdam.

OLLIER, C. (1969) *Weathering. Geomorphology Texts No. 2* (ed. K. M. CLAYTON). Edinburgh (Oliver & Boyd).

ORHEIM, O. (1972) *A 200-year Record of Glacier Mass Balance at Deception Island, Southwest Atlantic Ocean, and its Bearing on Models of Global Climatic Change.* Report No. 42 (RF 3231-A1). Columbus, Ohio (Inst. Polar Studies, Ohio State Univ. Research Foundation).

PARKIN, D. W. (1974) Trade winds during the glacial cycles, *Proc. Roy. Soc. London,* A, 337, 73–100.

PARKIN, D. W., PHILLIPS, D. R., SULLIVAN, R. A. L. and JOHNSON, L. R. (1972) Airborne dust collections down the Atlantic, *Quart. J. R. Met. Soc.,* 98, 798–808. London.

PENNINGTON, W. and LISHMAN, J. P. (1971) Iodine in lake sediments in northern England and Scotland, *Biol. Rev.,* 46, 279–313. Cambridge (Phil. Soc.).

PETERSON, J. A. and HOPE, G. S. (1972) Lower limit and maximum age for the last major advance of the Carstensz glaciers, West Irian, *Nature,* 240, 36–7. London (3 November).

PÉWÉ, T. L. (1966) Palaeoclimatic significance of fossil ice wedges, *Biulletyn Peryglacjalny,* 15, 65–73. Lodz, Poland, Seventh INQUA Report.

PIKE, J. G. (1965) The sunspot/lake level relationship and the control of Lake Nyasa, *J. Inst. Water Engineers,* 19, 221–6. London.

RAYNAUD, D. and LORIUS, C. (1974) Isotopic implications of total gas content in ice at Camp Century, Greenland, pp. 289–93 in *Colloques Internationaux du C.R.N.S., No. 219, Les méthodes quantitatives d'étude des variations du climat au cours du Pléistocène.* Paris (Éditions du Centre National de la Recherche Scientifique).

ROBIN, G. DE Q. (1955) Ice movement and temperature distribution in glaciers and ice sheets, *J. Glaciol.,* 2, 523–32. Cambridge.

ROSSITER, J. R. (1972) Sea level observations and their secular variation, *Phil. Trans.,* A, 272, 131–9. London (Roy. Soc.).

RUDDIMAN, W. F. (1973) Counterclockwise Circulation in the North Atlantic subpolar gyre during the Quaternary. In *CLIMAP Conference on Reconstruction of the Climate and Circulation Regime at the Time of the last Glacial Maximum 17 000 years ago.* Research Pubn No. 2. Norwich (Climatic Research Unit, Univ. of East Anglia).

RUDDIMAN, W. F. and GLOVER, L. K. (1975) Subpolar North Atlantic circulation at 9300 year B.P.: faunal evidence, *Quaternary Res.,* 5, 361–89. Seattle.

RUSSELL, E. J. (1957) *The World of the Soil.* London (Collins New Naturalist Series).

SCHELL, I. I. (1970) On the absolute fall of sea level during the Quarternary: a discussion, *Palaeogeogr., Palaeoclim., Palaeoecol.* 8, 67–8. Amsterdam.

SCHOFIELD, J. C. and THOMPSON, H. R. (1964) Postglacial sea levels and isostatic uplift, *N.Z. J. Geol. & Geophys.*, **7** (2), 359–70. Wellington, New Zealand.

SCHOVE, D. J. (1949) Chinese 'raininess' through the centuries. *Met. Mag.*, **78**, 11–16. London.

SCHWARZBACH, M. (1961) *Das Klima der Vorzeit*. Stuttgart (Enke), 2nd edn.

SCHYTT, V., HOPPE, G., BLAKE, W. and GROSSWALD, M. G. (1968) The extent of the Würm glaciation in the European Arctic, *Internat. Assn of Sci. Hydrol. U.G.G.I., Bern Assembly, 1967, Comm. Snow & Ice. Publication No. 79*. Gentbrugge, Belgium. (Reprinted as *Medd. Naturgeograf. Inst. Stockholms Univ., No. A18*. Stockholm.)

SHACKLETON, N. J. (1967) Oxygen isotope analyses and Pleistocene temperatures reassessed, *Nature*, **215**, 15–17. London (1 July).

SHACKLETON, N. J. (1969) Radiometric dating of periods of high sea level, *Proc. Roy. Soc.*, **B, 174**, 134–54. London.

SHACKLETON, N. J. (1970) Stable isotope study of the palaeoenvironment of the Neolithic site of Nea Nikomedeia, Greece, *Nature*, **227**, 943–4. London (29 August).

SHACKLETON, N. J., WISEMAN, J. D. H. and BUCKLEY, H. A. (1973) Non-equilibrium isotopic fractionation between sea water and planktonic foraminiferal tests, *Nature*, **242**, 177–9. London (16 March).

SHEPARD, F. P. (1963) Thirty-five thousand years of sea level, *Essays in Marine Geology*. Los Angeles (Univ. California Press). 201 pp.

SMITH, S. M. (1971) Palaeoecology of postglacial beaches in East Lothian, *Scot. J. Geol.*, **8** (1), 31–49. Edinburgh.

ŠNITNIKOV, A. V. (1957) Variability in the General Moistness of the Continents of the Northern Hemisphere. Moscow (*Zap. Geogr. Obs. S.S.S.R.*, **16**, new series, Akad. Nauk). (In Russian.)

ŠNITNIKOV, A. V. (1961) The present phase of intrasecular variability of mountain glaciation in the northern hemisphere, pp. 449–50 in *Internat. Comm. Snow and Ice, I.A.S.H.*, Pubn No. 54. Gentbrugge, Belgium (U.G.G.I.).

ŠNITNIKOV, A. V. (1969) Some Material on Intrasecular Fluctuations of the Climate of Northwestern Europe and the North Atlantic in the 18th–20th Centuries, *Geogr. Obs. S.S.S.R.*, pp. 5–29. Moscow (Akad. Nauk). (In Russian.)

SPARKS, B. W. and WEST, R. G. (1972) *The Ice Age in Britain*. London (Methuen).

SPEERSCHNEIDER, C. I. H. (1915) Om Isforholdene i danske Farvande i aeldre og nyere Tid: Aarene 690–1860. *Meddelelser*, Nr. 2. Copenhagen (Dansk. Met. Inst.).

STODDART, D. R. (1969) Climatic geomorphology, pp. 473–85 in *Water, Earth and Man* (ed. R. J. CHORLEY). London (Methuen).

STUIVER, M. (1970) Oxygen and carbon isotope ratios of freshwater carbonates as climatic indicators, *J. Geophys. Res.*, **75** (27), 5247–57. Baltimore, Md.

TANNER, W. F. (1968) Tertiary sea level symposium, *Palaeogeogr, Palaeoclim., Palaeoecol.*, **5**, 7–14. Amsterdam.

TEH-LUNG KU, KIMMEL, M. A., EASTON, W. H. and O'NEIL, T. J. (1974) Eustatic sea level 120 000 years ago on Oahu, Hawaii, *Science*, **183**, 959–62. New York.

THORÉN, R. (1969) *Picture Atlas of the Arctic*. Amsterdam (Elsevier).

THORODDSEN, Th. (1884) Den Grönlandska drifisen vid Island, *Ymer*, **4**, 145–60. Stockholm.

THORODDSEN, Th. (1916) *Árferði á Íslandi i þusund ár*. Hið. íslenzka fraeðafélag. Kaupmannahöfn. (Copenhagen).

TOPOR, N. (1963) Rainy and drought years in the Rumanian People's Republic, pp. 9–21 in *Ani ploiosi si secetosi in Republica Populara Romina*. Bucharest (Institutul Meteorologic).

TOUSSOUN, Prince OMAR (1925) Mémoire sur l'histoire du Nil, *Mem. Inst. Egypt*, **9**. Cairo.

UREY, H. C. (1947) The thermodynamic properties of isotopic substances, *J. Chem. Soc.*, **152**, 190–219. London.

VALLENTYNE, J. R. and SWABEY, Y. S. (1955) A reinvestigation of the history of Lower Linsley Pond, Connecticut. *Amer. J. Sci.*, **253**, 313–40. New Haven, Conn.

VEEH, H. H. and CHAPPELL, J. (1970) Astronomical theory of climatic change: support from New Guinea, *Science*, **167**, 862–5. New York (6 February).

VEEN, J. VAN (1954) Tide gauges, subsidence gauges and flood-stones in the Netherlands, *Geol. en Mijnbouw*, **16** (New Series), 214–19. Den Haag.

VITA-FINZI, C. (1960) Post-Roman erosion and deposition in the *wadis* of Tripolitania, *I.A.S.H.*, **53**, 61–4. Helsinki.

VITA-FINZI, C. (1963) Carbon-14 dating of medieval alluvium in Libya, *Nature*, **198**, 880. London.

VITA-FINZI, C. (1964) Synchronous stream deposition throughout the Mediterranean area in historical times, *Nature*, **202**, 1324. London.

VITA-FINZI, C. (1966) The new Elysian fields, *Amer. J. Archaeol.*, **70**, 175–8. New York.

WARD, W. T., ROSS, P. J. and COLQUHOUN, D. J. (1971) Interglacial high sea levels – an absolute chronology derived from shoreline elevations, *Palaeogeogr., Palaeoclim., Palaeoecol.*, **9**, 77–99. Amsterdam.

WARDLE, P. (1973) Variations of the glaciers of Westland National Park and the Hooker Range, New Zealand, *N.Z. J. Bot.*, **11**, 349–88. Wellington, New Zealand (D.S.I.R.).

WEST, R. G. (1968) *Pleistocene Geology and Biology*. London (Longmans).

WILSON, A. T. (1964) Origin of ice ages: an ice-shelf theory for Pleistocene glaciation. *Nature*, **201** (4915), 147–9. London (11 January).

WILSON, A. T. and HENDY, C. H. (1971) Past wind strength from isotope studies, *Nature*, **234**, 344–5. London (10 December).

WILSON, J. T. (1957) The crust, Ch. 4, pp. 48–73, in *The Planet Earth* (ed. D. R. BATES). London (Pergamon).

WOLDSTEDT, P. (1958) *Das Eiszeitalter, II: Europa, Vorderasien und Nordafrika*. Stuttgart (Enke).

WOLDSTEDT, P. (1965) *Das Eiszeitalter, III: Afrika, Asien, Australien und Amerika.* Stuttgart (Enke).

YAMAMOTO, T. (1972) Long-term variation of rainfall in the Far East, *Japanese Physiographical Mag.*, **81** (4), 199–222.

YAO, SHAN-YU (1943) The geographical distribution of floods and droughts in Chinese history 206 B.C.–A.D. 1911, *Far East Quarterly*, p. 357. New York.

YAO, SHAN-YU (1944) Flood and drought data in the T'u-shu Chi Chung and Ch'ing Shi Kao, *Harvard J. Asiatic Studies*, **8**, 214. Cambridge, Mass.

YOSHINO, M. M. and TABUCHI, H. (1975) Wind variation in the lower troposphere over Japan in the late Quaternary period estimated by the distribution of volcanic ashes, pp. 31–8 in *Proc. WMO/IAMAP Symposium on Long-Term Climatic Fluctuations, Norwich, 18–23 August 1975.* W.M.O. No. 421. Geneva (World Met. Org.).

YOUNG, A. (1972) *Slopes.* Geomorphology texts No. 3 (ed. K. M. CLAYTON). Edinburgh (Oliver & Boyd).

ZEUNER, F. E. (1959) *The Pleistocene Period.* London (Hutchinson).

13.4 Biological evidence

ADAMENKO, V. N. (1963) On the similarity in the growth of trees in northern Scandinavia and in the polar Ural Mountains, *J. Glaciol.*, **4**, 449–51. Cambridge.

ADDICOTT, W. O. (1970) Latitudinal gradients in Tertiary molluscan faunas of the Pacific coast, *Palaeogeogr., Palaeoclim., Palaeoecol.*, **8**, 287–312. Amsterdam.

ALISSOW, B. P., DROSDOW, O. and RUBINSTEIN, E. S. (1956) *Lehrbuch der Klimatologie.* Berlin (Deutscher Verlag der Wiss.). 536 pp.

ANDERSEN, SV. TH. (1970) The relative pollen productivity and pollen representation of north European trees, and correction factors for tree pollen spectra, *Danmarks Geol. Undersøgelse*, II Række, **96**. Copenhagen (Reitzel). 99 pp.

ANDERSEN, SV. TH. (1973) The differential pollen productivity of trees and its significance for the interpretation of a pollen diagram from a forested region, pp. 108–14 in *Quaternary Plant Ecology* (ed. H. J. B. BIRKS and R. G. WEST). Oxford (Blackwell Scientific Publications).

ANTEVS, E. (1925) *The Big Tree as a Climate Measure.* Publication No. 352, pp. 115–53. Washington (Carnegie Institution).

ARRHENIUS, G. (1952) Sediment cores from the east Pacific, *Swedish Deep-Sea Expedition, 1947–8, Reports*, **5** (1). Göteborg (Elanders Boktryckeri A/B).

AUER, V. (1958) The Pleistocene of Fuego-Patagonia, II: the history of the flora and vegetation, *Annales Acad. Sci. Fennicae, Series A III Geologica-Geographica*, **50**. Helsinki.

AUER, V. (1965) The Pleistocene of Fuego-Patagonia, IV: Bog profiles, *Ibid.*, **80**. Helsinki.

BAKKER, R. T. (1972) Anatomical and ecological evidence of endothermy in dinosaurs, *Nature*, **238**, 81–5. London (14 July).

BAKKER, E. M. VAN ZINDEREN (1969) Intimations on Quaternary palaeoecology of Africa, *Acta Bot. Neerland.*, ₁8 (1), 230–9. Amsterdam.

BELL, B. (1970) The oldest records of the Nile Floods, *Geogr. J.*, **136**, 569–73. London.

BELL, B. (1971) The Dark Ages in ancient history, I: The first Dark Age in Egypt, *Amer. J. Archaeol.*, **75**, 1–26. New York.

BELL, F. G. (1970) Late Pleistocene floras from Earith, Huntingdonshire, *Phil. Trans. Roy. Soc. Lond.*, **B, 258** (826), 347–78.

BERNABO, J. C. and WEBB, T. (1977) Postglacial isopollen maps of major forest genera and herbs in northeastern eastern North America. (To be published in *Quaternary Res.*, **7**.)

BEUG, H.-J. (1967) Contributions to the postglacial vegetational history of northern Turkey, pp. 349–56 in *Quaternary Palaeoecology* (ed. E. J. CUSHING). New Haven and London (Yale Univ. Press).

BEVERTON, R. J. H. and LEE, A. J. (1965) Hydrographic fluctuations in the North Atlantic Ocean and some biological consequences, pp. 79–107 in *The Biological Significance of Climatic Changes in Britain* (ed. C. G. JOHNSON and L. P. SMITH). London (Institute of Biology and Academic Press). 222 pp.

BIRKS, H. H. (1970) Studies in the vegetational history of Scotland, I: A pollen diagram from Abernethy Forest, Inverness-shire, *J. Ecol.*, **58**, 827–46. Oxford and Edinburgh (Blackwell Sci. Pubns).

BIRKS, H. J. B., DEACON, J. and PEGLAR, S. (1975) Pollen maps for the British Isles 5000 years ago, *Proc. Roy. Soc. London*, **B, 189**, 87–105.

BONNY, A. P. (1972) A method of determining absolute pollen frequencies in lake sediments, *New Phytol.*, **71**, 393–405. Oxford and Edinburgh (Blackwell Sci. Pubns).

BRICE, W. C. (1955) The history of forestry in Turkey, *Orman fakultesi dergisi*, **5** (1–2), 29–38. Istambul (Univ. Faculty of Forest Science).

BRIDEN, J. C. and IRVING, E. (1964) Palaeolatitude spectra of sedimentary palaeoclimatic indicators, pp. 199–224 in *Problems in Palaeoclimatology* (ed. A. E. M. NAIRN). London, New York and Sydney (Interscience Publishers). 705 pp.

BROOKS, C. E. P. (1949) *Climate Through the Ages*. London (Benn, 2nd edn). 395 pp.

BRYSON, R. A. (1966) Air masses, streamlines, and the boreal forest, *Geogr. Bull.*, **8** (3), 228–69. Ottawa.

BRYSON, R. A., BAERREIS, D. A. and WENDLAND, W. M. (1970) *The Character of Late-Glacial and Post-Glacial Climatic Changes*. Dept. Geol., Univ. of Kansas, Special Pubn No. 3. Lawrence/Manhatten/Wichita (Univ. Kansas Press).

BUTZER, K. W. (1958) Studien zur vor- und frühgeschichtlichen Landschaftswandel der Sahara, *Abhandl. math.-naturwiss. Klasse*, No. 1. Mainz (Akad. der Wiss. und Lit.). 49 pp.

CARPENTER, R. (1966) *Discontinuity in Greek civilization*. Cambridge (Univ. Press).

CHU KO-CHEN (COCHING CHU) (1973) A preliminary study on the climatic fluctuations during the last 5000 years in China, *Scientia Sinica*, **16** (2), 226–56. Peking.

CHURCHILL, D. M. (1968) The distribution and prehistory of *Eucalyptus diversicolor, E. marginata* and *E. calophylla* in relation to rainfall, *Australian J. Bot.*, **16**, 125–51. Melbourne.

COETZEE, J. A. and BAKKER, E. M. VAN ZINDEREN (1970) Palaeoecological problems of the Quaternary of Africa, *Suid-Afrikaanse Tydskrif vir Wetenskap*, Paper No. 4, pp. 78–84. Johannesburg.

CONOLLY, A. P. and DAHL, E. (1970) Maximum summer temperature in relation to the modern and Quaternary distributions of certain Arctic-montane species in the British Isles, pp. 159–223 in *Studies in the Vegetational History of the British Isles: Essays in Honour of Harry Godwin* (ed. D. WALKER and R. G. WEST). Cambridge (Univ. Press). 266 pp.

COOPE, G. R. (1969a) The response of coleoptera to gross thermal changes during the mid Weichselian interstadial, *Mitt. Internat. Verein. Limnol.*, **17**, 173–83. Stuttgart.

COOPE, G. R. (1969b) Insect remains from mid Weichselian deposits at Peelo, Netherlands, *Med. Rijks Geol. Stichting*, **20**, new series, 79–83. 's Gravenhage.

COOPE, G. R. (1970) Interpretations of Quaternary insect fossils, *Annual Rev. Entomology*, **15**, 97–120. Stanford, Palo Alto, Calif.

COOPE, G. R., MORGAN, A. and OSBORNE, P. J. (1971) Fossil coleoptera as indicators of climatic fluctuations during the last glaciation in Britain, *Palaeogeogr., Palaeoclim., Palaeoecol.*, **10**, 87–101. Amsterdam (Elsevier).

CRAIG, R. E. (1960) A note on the dependence of catches on temperature and wind in the Buchan pre-spawning herring fishery, *J. Cons. Internat. de l'Explor. de la Mer*, **25**, 185–90. Copenhagen.

CRANWELL, L. M. and VON POST, L. (1936) Pollen diagrams from New Zealand, *Geogr. Annaler*, **18**, 308–47. Stockholm.

CRISP, D. J. (1959) The influence of climatic changes on animals and plants, *Geogr. J.*, **125**, 1–19, London.

CRISP, D. J. (1965) Observations on the effects of climate and weather in marine communities, pp. 63–77 in *The Biological Significance of Climatic Changes in Britain* (ed. C. G. JOHNSON and L. P. SMITH). London (Inst. Biol. and Acad. Press). 222 pp.

DANSGAARD, W., JOHNSEN, S. J., CLAUSEN, H. B. and LANGWAY, C. C. (1971) Climatic record revealed by the Camp Century ice core, in *The Late Cenozoic Ice Ages: Proc. Yale Univ. Symp. December 1969* (ed. K. K. TUREKIAN). New Haven, Conn. (Yale Univ. Press). 606 pp.

DAVIS, M. B. (1961) Pollen diagrams as evidence of late glacial climatic change in southern New England, *Ann. New York Acad. Sci.*, **95** (1), 623–31.

DAVIS, M. B. (1963) On the theory of pollen analysis, *Amer. J. Sci.*, **261**, 897–912. New Haven, Conn.

DE BOER, H. J. (1951) Tree-ring measurements and weather fluctuations in Java from A.D. 1514, *Proc. Kon. Ned. Akad. Wetenschappen Amsterdam*, Serie B, 194–209. Amsterdam.

DEGERBØL, M. and KROG, H. (1951) Den europeiske Sumpskildpadde (*Emys orbicularis* L.) i Danmark, *Danmarks Geologiske Undersøgelse*, II Række, No. 78. Copenhagen.

DENTON, G. H. (1972) Holocene glacier fluctuations and their possible cause, pp. 267–306 in *The Late Cenozoic Glacial Ages: Proc. Yale Univ. Symp., December 1969* (ed. K. K. TUREKIAN). New Haven, Conn. (Yale Univ. Press). 606 pp.

DE WORMS, C. G. M. (1958) The northward movement of certain species of European lepidoptera and its relationship to climatic changes and other factors with special reference to the British Isles, *Proc. Tenth Internat. Congress of Entomol., 1956*, **1**, 737–40.

DICKSON, J. H. (1967) The British moss flora of the Weichselian glacial, *Rev. Palaeobot. Palynol.*, **2**, 245–53. Amsterdam (Elsevier).

DICKSON, R. R. and LEE, A. J. (1972) Recent hydrometeorological trends on the North Atlantic fishing grounds, *Fish Industry Review*, **2** (2), 1–8 London (White Fish Authority).

DRAKE, R. J. (1960–2) *Molluscs in Archaeology and the Recent.* An unpublished bibliography in 6 parts, available from Department of Zoology, University of British Columbia, Vancouver.

EMILIANI, C. (1955) Pleistocene temperatures, *J. Geol.*, **63** (6), 538–78. Chicago.

EMILIANI, C. (1966) Palaeotemperature analysis of Caribbean cores P.6304-8 and P.6304-9 and a generalized temperature curve for the past 425 000 years, *J. Geol.*, **74** (2), 109–26. Chicago.

ERDTMAN, O. G. E. (1921) Pollenanalytische Untersuchungen von Torfmooren und marinen Sedimenten in Südwest-Schweden, *Arkiv f. Botanik*, **17** (10), 1–173. Uppsala/Stockholm.

ERDTMAN, O. G. E. (1937) Pollen grains recorded from the atmosphere over the Atlantic. *Meddel. från Göteborgs bot. trädgård*, **12**, 185–96. Göteborg.

ERICSON, D. B., WOLLIN, G. and WOLLIN, J. (1954) Coiling direction of *Globorotalia truncatulinoides* in deep-sea cores, *Deep Sea Res.*, **2**, 152–8. London (Pergamon).

ERICSON, D. B., EWING, M. and WOLLIN, G. (1964) The Pleistocene epoch in deep-sea sediments, *Science*, **146** (3645), 723–732. New York.

ERICSON, D. B. and WOLLIN, G. (1966) *The Deep and the Past.* London (Jonathan Cape). 292 pp.

ERICSON, D. B. and WOLLIN, G. (1970) Pleistocene climates in the Atlantic and Pacific Oceans: a comparison based on deep-sea sediments, *Science*, **167**, 1483–5, New York (13 March).

FÆGRI, K. and IVERSEN, J. (1964) *Textbook of Pollen Analysis.* Copenhagen (Munksgaard), Oxford (Blackwell Scientific Publications) and New York (Hafner). 237 pp.

FIRBAS, F. (1949) *Spät- und nacheiszeitliche Waldgeschichte Mitteleuropas nördlich der Alpen*, Vol. I. Jena (Fischer Verlag). 480 pp.

FIRBAS, F. and LOSERT, H. (1949) Untersuchungen über die Entstehung der heutigen Waldstufen in den Sudeten, *Planta*, **36**, 478–506. Berlin.

FLETCHER, J. M. (1974) Annual rings in modern and medieval times, pp. 80–97 in *The British Oak* (ed. M. G. MORRIS and F. H. PERRING). Farringdon, Berkshire (E. W. Classey for Bot. Soc. Brit. Is.).

FRASER DARLING, F. and BOYD, J. M. (1964) *The Highlands and Islands*. London and Glasgow (Collins, New Naturalist Series). (Fontana paperback edn 1969). 405 pp.

FREDSKILD, B. (1973) Studies in the vegetational history of Greenland, *Medd. om Grønland*, **198** (4), Copenhagen. 245 pp.

FRENZEL, B. (1966) Climatic change in the Atlantic/Sub-Boreal transition on the northern hemisphere: botanical evidence, pp. 99–123 in *World Climate 8000–0 B.C.: Proc. Internat. Conf.* (ed. J. S. SAWYER). London (Roy. Met. Soc.).

FRENZEL, B. (1967) *Die Klimaschwankungen des Eiszeitalters*. Braunschweig (Vieweg). 296 pp.

FRITTS, H. C. (1962) The relation of growth ring widths in American beech and white oak to variations in climate, *Tree-Ring Bull.*, **25** (1), 2–10. Tucson (Univ. Arizona).

FRITTS, H. C. (1965) Tree-ring evidence for climatic changes in western North America, *Mon. Weather Rev.*, **93**, 421–43. Washington.

FRITTS, H. C. (1966) Growth rings of trees: their correlation with climate, *Science*, **154** (3752), 973–9. New York.

FRITTS, H. C. (1971) Dendroclimatology and dendroecology, *Quaternary Res.*, **1** (4), 419–49. Seattle.

FRITTS, H. C. (1976) *Tree Rings and Climate*. London, New York, San Francisco (Acad. Press). 567 pp.

FRITTS, H. C., SMITH, D. G. and STOKES, M. A. (1965) The biological model for palaeoclimatic interpretation of Mesa Verde tree-ring series, *Amer. Antiquity*, **31** (2), Part 2, 101–21. Menasha.

FRITTS, H. C., MOSSIMAN, J. E. and BOTTORFF, C. P. (1969) A revised computer program for standardizing tree-ring series, *Tree-Ring Bull.*, **29** (1–2), 15–20. Tucson (Univ. of Arizona).

FRITTS, H. C., BLASING, T. J., HAYDEN, B. P. and KUTZBACH, J. E. (1971) Multivariate techniques for specifying tree growth and climate relationships and for reconstructing anomalies in palaeoclimate, *J. Appl. Met.*, **10** (5), 845–64. Boston, Mass. (Amer. Meteorol. Soc.).

FRY, C. H. (1973) *Avian Indicators of Increasing Aridity in Northern Nigeria*. Dept of Zoology, Univ. Aberdeen, unpublished report.

GAMS, H. (1937) Aus der Geschichte der Alpenwälder, *Zeit. des deutschen und österreichischen Alpenvereins*, **68**, 157–170.

GAUNT, G. D., COOPE, G. R. and FRANKS, J. W. (1970) Quaternary deposits at Oxbow opencast coal site in the Aire valley, Yorkshire, *Proc. Yorks. Geol. Soc.*, **38** (2:9), 175–200. Leeds.

GLAHN, H. (1968) Canonical correlation and its relationship to discriminant analysis and multiple regression, *J. Atmos. Sci.*, **25**, 23–31. Lancaster, Pa.

GLASS, B., ERICSON, D. B., HEEZEN, B. C., OPDYKE, N. D. and GLASS, J. A. (1967) Geomagnetic reversals and Pleistocene chronology, *Nature*, **216**, 437–442. London (4 November).

GODWIN, H. (1935) Botanical evidence for postglacial climates, *Quart. J. Roy. Met. Soc.*, **61**, 95–8, London.

GODWIN, H. (1954) Recurrence surfaces, *Danmarks Geol. Undersøgelse*, II. Række, No. 80. Copenhagen.

GODWIN, H. (1956) *The History of the British Flora*. Cambridge (Univ. Press). 384 pp.

GODWIN, H. (1975) History of the natural forests of Britain: establishment, dominance and destruction, *Phil. Trans. Roy. Soc.*, **B, 271**, 45–67. London.

GODWIN, H. and WILLIS, E. H. (1959) Radiocarbon dating of prehistoric wooden trackways, *Nature*, **184**, 490–1. London.

GORDON, A. D. and BIRKS, H. J. B. (1972) Numerical methods of Quaternary palaeoecology, I: Zonation of pollen diagrams, *New Phytol.*, **71**, 961–79. Oxford and Edinburgh (Blackwell Sci. Pubns).

GRANLUND, E. (1932) De svenska högmossernas geologi, *Sveriges geol. Undersöknings Afhandlingar, Årsbok* **26** (1), Serie C **373**, 1–193. Stockholm.

GREGORY, P. H. (1945) The dispersion of airborne spores, *Trans. Brit. Mycol. Soc.*, **28**, 26–72.

GRIGGS, G. B., KULM, L. D., DUNCAN, J. R. and FOWLER, G. A. (1970) Holocene faunal stratigraphy and palaeoclimatic implications of deep-sea sediments in Cascadia Basin. *Palaeogeogr., Palaeoclim., Palaeoecol.*, **7**, 5–12. Amsterdam (Elsevier).

HAMMEN, T. VAN DER (1963) A palynological study on the Quaternary of British Guiana, *Leidse Geol. Med.*, **29**, 125–80. Leiden.

HAMMEN, T. VAN DER (1968) Climatic and vegetational succession in the equatorial Andes of Colombia. *Colloquium Geographicum*, **9**, 187–94. Bonn.

HAMMEN, T. VAN DER, WIJMSTRA, T. A. and ZAGWIJN, W. H. (1971) The floral record of the late cenozoic of Europe, pp. 391–424 in *The Late Cenozoic Glacial Ages*, ed. K. K. TUREKIAN. New Haven, Conn. (Yale Univ. Press).

HARRIS, G. (1964) Climatic changes since 1860 affecting European birds, *Weather*, **19** (3), 70–9. London.

HARRISON, C. G. A. and FUNNELL, B. M. (1964) Relationship of palaeomagnetic reversals and micropalaeontology in two late Cainozoic cores from the Pacific ocean, *Nature*, **204** (4958), 566. London (7 November).

HARRISON, J. M. (1968) *Bristow and the Hastings Rarities Affair*. St Leonards on Sea (A. H. Butler). 160 pp.

HASTINGS, J. R. and TURNER, R. M. (1965) *The Changing Mile: An Ecological Study of Vegetation Change with Time in the Lower Mile of an Arid and Semi-Arid Region*. Tucson (Univ. of Arizona Press). 317 pp.

HAYNES, C. V. (1966) Elephant-hunting in North America, *Scientific American*, **214** (6), 104–12. New York.

HAYS, J. D. (1969) Climatic record of late Cenozoic Antarctic ocean sediments related to the record of world climate, Ch. 8, pp. 139–63, in *Palaeoecology of Africa and of the Surrounding Islands and Antarctica*, Vol. 5 (Ed. E. M. VAN ZINDEREN BAKKER). Cape Town (Balkema).

HAYS, J. D. and OPDYKE, N. D. (1967) Antarctic radiolaria, magnetic reversals and climatic change, *Science*, **158**, 1001. New York.

HAYS, J. D., SAITO, T., OPDYKE, N. D. and BURCKLE, L. H. (1969) Plio-Pleistocene sediments of the equatorial Pacific: their palaeomagnetic, biostratigraphic and climatic record, *Geol. Soc. of America Bull.*, **80**, 1481–514. Rochester, N.Y.

HEINZELIN, J. DE (1967) Geological history of the Nile valley in Nubia, pp. 1–55 in *The Prehistory of Nubia* (ed. F. WENDORF). Dallas, Texas (South Methodist Univ. Press).

HEPTING, G. H. (1971) Climate and forest diseases, pp. 203–26 in *Man's Impact on Terrestrial and Oceanic Ecosystems* (ed. W. H. MATTHEWS, F. E. SMITH and E. D. GOLDBERG). Cambridge, Mass. (M.I.T. Press). 540 pp.

HEUSSER, C. J. (1966) Polar hemispheric correlation: palynological evidence from Chile and the Pacific north-west of America, pp. 124–41 in *World Climate 8000–0 B.C.: Proc. Internat. Conf.* (ed. J. S. SAWYER). London. (Roy. Met. Soc.). 229 pp.

HOLLOWAY, J. T. (1954) Forests and climates in the South Island of New Zealand, *Proc. Roy. Soc. N.Z.*, **82**, 329–410. Wellington.

HOTELLING, H. (1936) Relations between two sets of variates, *Biometrika*, **28**, 321–77. Cambridge.

HUBER, B. (1964) Durchschnittliche Schwankung und Periodenlänge von Jahresring-Breitenkurven als Klima-Indikatoren, *Geol. Rundschau*, **54** (1) 441–8. Stuttgart (Enke).

HUBER, B. and VON JAZEWITSCH, W. (1958) Jahrringuntersuchungen an Pfahlbauhölzern, *Flora*, **146**, 445–71. Jena.

HUNKINS, K., BÉ, A. W. H., OPDYKE, N. D. and MATHIEU, G. (1971) The late Cenozoic history of the Arctic Ocean, pp. 215–37 in *Late Cenozoic Glacial Ages: Proc. Yale Univ. Symp., December 1969* (ed. K. K. TUREKIAN). New Haven, Conn. (Yale Univ. Press). 606 pp.

HUNTINGTON, E. (1913) *The Secret of the Big Trees*. Washington (U.S. Dept of the Interior). 24 pp. (Reprinted from *Harper's Magazine*, July 1912.)

HURST, G. W. (1964) Meteorological aspects of the migration to Britain of *Laphygma exigua* and certain other moths on specific occasions, *Agric. Meteorol.*, **1**, 271–81. Amsterdam (Elsevier).

HUSTICH, I. (1945) The radial growth of the pine at the forest limit and its dependence on climate, *Comm. Biol. Soc. Scient. Fenn.*, **9** (11). Helsinki.

HUSTICH, I. (1956) Correlation of tree-ring chronologies of Alaska, Labrador and northern Europe, *Acta geographica*, **15** (3). Helsinki. 26 pp.

IMBRIE, J. and KIPP, N. G. (1971) A new micropalaeontological method for quantitative palaeoclimatology: application to a late Pleistocene Caribbean core, Ch. 5, pp. 71–181, in *The Late Cenozoic Glacial Ages: Proc. Yale Univ. Symp., December 1969* (ed. K. K. TUREKIAN). New Haven, Conn. (Yale Univ. Press). 606 pp.

IVERSEN, J. (1944) *Viscum, Hedera* and *Ilex* as climate indicators, *Geol. Fören. Stockholm Förhandl.,* **66,** 463–83.

IVERSEN, J. (1947) Pp. 205–6 in the discussion in the *Nordiskt kvartærgeologiskt möte* (Nordic Meeting for Quaternary Geology), 1945. *Geol. Forening. Stockholm. Förhandl.,* **69.**

JAKOBSSON, J. (1969) On herring migrations in relation to changes in sea temperature, *Jökull,* **19,** 134–45. Reykjavik.

JESSEN, K. (1949) Studies in Late Quaternary deposits and flora history of Ireland, *Proc. Roy. Irish Acad.,* **52, B,** 85–290. Dublin.

KALELA, O. (1949) Changes in geographic ranges in the avifauna of northern and central Europe in relation to recent changes in climate, *Bird Banding,* **20** (2), 77–103. Boston, Mass.

KEANY, J. and KENNETT, J. P. (1972) Pliocene-early Pleistocene palaeoclimatic history recorded in Antarctic-Subantarctic deep-sea cores, *Deep-Sea Res.,* **19,** 529–48. Oxford and New York (Pergamon).

KENNETT, J. P. (1968a) *Globerotalia truncatulinoides* as a paleo-oceanographic index, *Science,* **159,** 1461–3. New York.

KENNETT, J. P. (1968b) Paleo-oceanographic aspects of the foraminiferal zonation in the upper Miocene–lower Pliocene of New Zealand, *Giornale di Geologia,* (2) **XXXV,** fasc. III, 143–56. Bologna.

KENNETT, J. P. (1970) Pleistocene paleoclimates and foraminiferal biostratigraphy in sub-Antarctic deep-sea cores, *Deep-Sea Res.,* **17,** 125–40. Oxford and New York (Pergamon).

KENNETT, J. P. (1972) The climatic and glacial record in Cenozoic sediment of the Southern Ocean, pp. 59–78 in *Palaeoecology of Africa: The Surrounding Islands and Antarctica,* Vol. 6 (ed. E. M. VAN ZINDEREN BAKKER). Cape Town (Balkema).

KENNETT, J. P. and GEITZENAUER, K. R. (1969) Pliocene-Pleistocene boundary in a South Pacific deep-sea core, *Nature,* **224,** 899–901. London (29 November).

KENNETT, J. P. and HUDDLESTUN, P. (1972a) Late Pleistocene paleoclimatology, foraminiferal biostratigraphy and tephrochronology, western Gulf of Mexico, *Quaternary Res.,* **2** (1), 38–69. Seattle.

KENNETT, J. P. and HUDDLESTUN, P. (1972b) Abrupt climatic change at 90 000 years B.P.: faunal evidence from Gulf of Mexico cores, *Quaternary Res.,* **2** (3), 384–95. Seattle.

KENNETT, J. P. and WATKINS, N. D. (1970) Geomagnetic polarity change, volcanic maxima and faunal extinction in the South Pacific, *Nature,* **227** (5261), 930–4. London (29 August).

748   References

KERNEY, M. P., BROWN, E. H. and CHANDLER, T. J. (1964) The Late-Glacial and post-glacial history of the chalk escarpment near Brook, Kent, *Phil. Trans. Roy. Soc. Lond.*, **B, 248** (745), 135–204.

KHOTINSKY, N. A. (1971) The problem of the boundary horizon with special reference to the Shuvaloff peat bog, III. *International Palynological Conference, Appendix to Guide for Field Route No. 1 B.* Novosibirsk.

KITCHING, J. A. (1950) The distribution of the littoral barnacle *Chthamalus stellatus* around the British Isles. *Nature*, **165**, 820. London.

KREBS, J. S. and BARRY, R. G. (1970) The Arctic front and the tundra-taiga boundary in Eurasia, *Geogr. Rev.*, **60** (4), 548–54. New York.

KROGH, K. J. (1967) *Viking Greenland.* Copenhagen (National Museum). 187 pp.

LA MARCHE, V. C. (1972) Climatic history since 5100 B.C. from treeline fluctuations, White Mountains, east-central California, *Geol. Soc. Amer. Abstracts with Programs*, **4** (3), 189.

LA MARCHE, V. C. and FRITTS, H. C. (1971) Anomaly pattern of climate over the western United States, 1700–1930, derived from principal component analysis of tree-ring data, *Mon. Weather Rev.*, **99** (2), 138–42. Washington.

LA MARCHE, V. C. and MOONEY, H. A. (1967) Altithermal timberline advance in western United States, *Nature*, **213**, 980–2. London (11 March).

LAMB, H. H. (1968) The climatic background to the birth of civilization, *Adv. Sci.*, **25**, 103–20. London.

LANDSBOROUGH THOMSON, A. (ed.) (1964) *A New Dictionary of Birds.* London and Edinburgh (Nelson). 928 pp.

LERMAN, J. C. (1974) Isotope 'palaeothermometers' on continental matter: assessment, pp. 163–81 in *Les méthodes quantitatives d'étude des variations du climat au cours du Pléistocene* (ed. J. LABEYRIE). Colloques internat. du C.N.R.S. No. 219. Paris (C.N.R.S.). 317 pp.

LIBBY, L. M. and PANDOLFI, L. T. (1974) Calibration of two isotope thermometers C^{13}/C^{12} and O^{18}/O^{16} in a European oak using official weather records, in *Proc. Internat. Conf. on Mapping the Atmospheric and Oceanic Circulations and other Climatic Parameters at the Time of the last Glacial Maximum 17 000 Years Ago.* Norwich (Univ. of East Anglia, Climatic Research Unit, CRU RP2).

LINDROTH, C. H. (1943) *Oodes gracilis* Villa: Eine thermophile Carabide Schwedens, *Notulae Entomol.*, **22**, 109–57. Helsingfors.

LINDROTH, C. H. (1945) Die fennoskandischen Carabidae: Eine tiergeographische Studie, 1, 2. *Göteborgs Kgl. Vetensk. och Vittern Samh. Handl.*, **B, 6**. Göteborg.

LOZEK, V. (1965) Problems of analysis of the Quaternary non-marine molluscan fauna in Europe, *Geol. Soc. Amer. Special Paper*, **84**, 201–18.

McINTYRE, A. (1967) Coccoliths as paleoclimatic indicators of Pleistocene glaciation, *Science*, **158**, 1314. New York.

McINTYRE, A., RUDDIMAN, W. F. and JANTZEN, R. (1972) Southward penetrations of the North Atlantic polar front: faunal and floral evidence of large-scale water-mass movements over the last 225 000 years, *Deep-Sea Res.*, **19**, 61–77. Oxford and New York (Pergamon).

MARTIN, P. S. (1966) Africa and Pleistocene overkill, *Nature*, **212**, 339–42. London (22 October).

MARTINI, E. (1952) Klima und Krankheitserreger, pp. 190–225 in *Klima, Wetter, Mensch* (ed. A. SEYBOLD and H. WOLTERECK). Heidelberg (Quelle & Meyer). 291 pp.

MERE, R. M. (1961) The recent colonization of England by new species of Macro-lepidoptera, pp. 63–73 in *Proc. South London Entomol. and Nat. Hist. Soc.* (Presidential address for 1960).

MESOLELLA, K. J., MATTHEWS, R. K., BROECKER, W. S. and THURBER, D. L. (1969) The astronomical theory of climatic change: Barbados data, *J. Geol.*, **77** (3), 250–74. Chicago.

MITCHELL, G. F. (1956) Post-Boreal pollen diagrams from Irish raised-bogs, *Proc. Royal Irish Acad.*, **57**, **B** (14), 185–251. Dublin.

MOAR, N. T. (1969a) Possible long-distance transport of pollen to New Zealand, *N.Z. J. Bot.*, **7** (4), 424–6. (D.S.I.R.). Wellington.

MOAR, N. T. (1969b) Two pollen diagrams from the Mainland, Orkney Islands, *New Phytol.*, **68**, 201–8. Oxford and Edinburgh (Blackwell Sci. Pubns).

MOAR, N. T. (1973) Late Pleistocene vegetation and environment in southern New Zealand, *Palaeoecology of Africa and of the Surrounding Islands and Antarctica*, **8**, 179–98. Cape Town (Balkema).

MONOD, T. (1964) The late Tertiary and Pleistocene in the Sahara, pp. 117–229 in *African Ecology and Human Evolution* (ed. F. C. HOWELL and F. BOURLIÈRE). London (Methuen).

MOORE, P. D. (1972) The initiation of peat formation and the development of peat deposits in mid Wales, pp. 89–100 in *Proc. 4th Internat. Peat Congress I–IV*. Helsinki.

MOORE, P. D. (1973) The influence of prehistoric cultures upon the initiation and spread of blanket bog in upland Wales, *Nature*, **241**, 350–3. London (2 February 1973).

MOORE, T. C. (1974) Late Pleistocene-Holocene oceanographic changes in the northeast Pacific, in *Proc. Internat. Conf. on Mapping the Atmospheric and Oceanic Circulations and Other Climatic Parameters at the Time of the last Glacial Maximum 17 000 Years Ago.* Norwich (Univ. of East Anglia, Climatic Research Unit, CRU RP2).

MOREAU, R. E. (1963) Vicissitudes of the African biomes in the late Pleistocene, *Proc. Zool. Soc. Lond.*, **141** (2), 395–421.

MOREAU, R. E. (1969) Climatic changes and the distribution of forest vertebrates in West Africa, *J. Zool.*, **158**, 39–61. London.

MORRISON, D. F. (1968) *Multivariate Statistical Techniques.* New York (McGraw-Hill).

MÜLLER, K. (1953) *Geschichte des Badischen Weinbaus*. Lahr in Baden (Moritz Schauenburg). 283 pp.

NEJSTADT, M. I. (1957) *History of the Forest and Palaeogeography of the U.S.S.R. in the Holocene*. Moscow. (In Russian.)

NELDER, J. A. (1962) A statistical examination of the Hastings rarities, *British Birds*, **55** (8), 283–98. London.

NICHOLS, H. (1967a) Central Canadian palynology and its relevance to northwestern Europe in the Late-Quaternary period, *Rev. Palaeobot. and Palynol.*, **2**, 231–43. Amsterdam (Elsevier).

NICHOLS, H. (1967b) The suitability of certain categories of lake sediments for pollen analysis, *Pollen et spores*, **9** (3), 615–20. Paris.

NICHOLS, H. (1969) Chronology of peat growth in Canada, *Palaeogeogr., Palaeoclim., Palaeoecol.*, **6**, 61–5. Amsterdam (Elsevier).

NICHOLSON, E. M. and FERGUSON-LEES, I. J. (1962) The Hastings rarities, *British Birds*, **55** (8), 299–345. London.

OLAUSSON, E., BILAL, UL HAQ, U. Z., KARLSSON, G. B. and OLSSON, I. U. (1971) Evidence of Indian Ocean cores of late Pleistocene changes in oceanic and atmospheric circulation, *Geol. Fören. i Stockholm Förhandlingar (GFF)*, **93** (1), 51–84.

OLDFIELD, F. (1970) Some aspects of scale and complexity in pollen-analytically based paleaoecology, *Pollen et spores*, **12** (2), 163–71. Paris.

OSBORNE, P. J. (1969) An insect fauna of late Bronze Age date from Wilsford, Wiltshire, *J. Anim. Ecol.*, **38**, 555–66. Oxford and Edinburgh (Blackwell Sci. Pubns).

OUTI, M. (1961) Climatic variations in the North Pacific subtropical zone and solar activity during the past ten centuries, 1, *Bull. Kyoto Gakugei Univ.*, **B, 19**, 41–61. Kyoto.

OUTI, M. (1962) Ibid., 2, *Bull. Kyoto Gakugei Univ.*, **B, 20**, 25–48. Kyoto.

OUTI, M. (1964) The relation between the growth rate of trees and climate changes in the northeastern section of Japan, *Bull. Kyoto Gakugei Univ.*, **B, 25**, 89–107. Kyoto. (In Japanese, with English summary and titles).

OVERBECK, F., MÜNNICH, K. O., ALETSEE, L. and AVERDIECK, F. R. (1957) Das alter des 'Grenzhorizonts' norddeutscher Hochmoore nach Radiocarbon-Datierungen, *Flora*, **145**, 37–71. Jena.

PARRISH, B. B. and CRAIG, R. E. (1963) The herring of the northwestern North Sea: postwar changes in the stock fished by Scottish drifters, *Rapports et Proc. Verb.*, **154**, 139–58. Copenhagen (Cons. Internat. de l.Explor. de la Mer.).

PASQUILL, F. (1961) *Atmospheric Diffusion*. London (Van Nostrand). 286 pp.

PENNY, L. F., COOPE, G. R. and CATT, J. A. (1969) Age and insect fauna of the Dimlington silts, east Yorkshire, *Nature*, **224**, 65–67. London.

PETTERSSON, O. (1914) Climatic variations in historic and prehistoric time, *Svenska Hydrografisk-Biologiska Kommissionens Skrifter*, Häft 5. Göteborg.

PHILLIPI, E. (1910) Die Grundproben der Deutschen Südpolarexpedition 1901–3, *Dt. Südpolarexpedition*, **2** (6) (Geogr. Geol.), 411–616.

PICKETT, R. L. and ATHEY, G. L. (1968) Iceland sea surface temperature survey, April 1968, *I.R.*, 68-46. Washington (Naval Oceanographic Office).

PISIAS, N. G., DAUPHIN, J. P. and SANCETTA, C. (1973) Spectral analysis of Late Pleistocene-Holocene sediments, *Quaternary Res.*, **3** (1), 3–9. Seattle.

POLGE, H. (1970) The use of X-ray densitometric methods in dendrochronology, *Tree-Ring Bull.*, **30** (1–4), 1–10. Tucson (Univ. of Arizona).

POST, L. VON (1918) Skogsträdpollen i sydsvenska torvmosslagerföljder, *Forhandl. 16. Skandinaviske Naturforskermöte 1916*, pp. 433–65.

RAINEY, R. C. (1973) Airborne pests and the atmospheric environment, *Weather*, **28** (6), 224–39. London.

RALPH, E. K. and MICHAEL, H. N. (1967) Problems of the radiocarbon calendar, *Archaeometry*, **10**, 3–11. Oxford.

RITCHIE, J. C. and HARE, F. K. (1971) Late-Quaternary vegetation and climate near the Arctic tree line of northwestern North America, *Quaternary Res.*, **1** (3), 331–42. New York and London (Acad. Press).

ROBINSON, W. J. and DEAN, J. S. (1969) *Tree-Ring Evidence for Climatic Changes in the Prehistoric Southwest from A.D. 1000 to 1200*. Annual Report (1967–8) to the National Park Service, Project Southwestern Paleoclimate, Contract No. 14-10-7:931–8. Washington (Dept of the Interior). 9 pp.

RUDDIMAN, W. F., TOLDERLUND, D. S. and BÉ, A. W. H. (1970) Foraminiferal evidence of a modern warming of the North Atlantic Ocean, *Deep-Sea Res.*, **17**, 141–55. Oxford and New York (Pergamon).

RUDY, H. (1925) Die Wanderheuschrecke (*Locusta migratoria* L. *Phasa migratoria* L. et *Phasa danica* L.). Beiträge zu einer Monographie. *Sonderbeilage zur Badischen Blätter für Schädlings-bekämpfung*. Freiburg.

RUSSELL, D. and TUCKER, W. (1971) Supernovae and the extinction of the dinosaurs, *Nature*, **229**, 553–4. London (19 February 1971).

SCHOTT, W. (1935) Die Foraminiferen in den Äquatorialen Teil des Atlantischen Ozeans, *Deutsche Atlantische Expedition 1925–7, Wissenschaftliche Ergebnisse*, **3** (3), 42–134; **11** (6), 411. Berlin, Leipzig.

SCHULTZ, C. B. (1972) Holocene interglacial migrations of mammals and other vertebrates, *Quaternary Res.*, **2**, 337–40. New York and London (Acad. Press).

SCHWARZBACH, M. (1949) Fossile Korallenriffe und Wegeners Drifthypothese, *Naturwiss.*, **36**. Berlin.

SCHWARZBACH, M. (1961) *Das Klima der Vorzeit*. Stuttgart (Enke). 275 pp.

SEARS, P. B. (1937) Pollen analysis as an aid in dating cultural deposits in the United States, pp. 61–6 in *Early Man* (ed. G. G. MACCURDY). New York (Lippincott).

SHOTTON, F. W. (1962) The physical background of Britain in the Pleistocene, *Adv. Sci.*, **19**, 193–206. London.

SIMMS, E. (1971) *Woodland Birds.* London (Collins New Natural Naturalist Series). 391 pp.

SINGH, G. (1963) A preliminary survey of the postglacial vegetational history of the Kashmir valley, *Palaeobotanist*, **12**, 73–108. Lucknow.

SIRÉN, G. (1961) Skogsgränstallen som indikator för klimafluktuationerna i norra Fennoskandien under historisk tid, *Communicationes Instituti Forestalis Fenniae*, **54** (2). Helsinki. 66 pp.

SMITH, A. G., PEARSON, G. W. and PILCHER, J. R. (1971) Belfast radiocarbon dates, III, *Radiocarbon*, **13**, 103–25. New Haven, Conn.

SMITH, L. P. (1956) Potato blight forecasting by 90% humidity criteria, *Plant Pathol.*, **5**, 83–7. London.

SMITH, L. P. and WALKER, J. (1966) Simplified weather criteria for potato blight infection periods, *Plant Pathol.*, **15**, 113–16. London.

SOUTHWARD, A. J. and CRISP, D. J. (1954) Recent changes in the distribution of the intertidal barnacles *Chthamalus stellatus* and *Balanus Balanoides* in the British Isles, *J. Animal Ecol.*, **23** (1), 163–77. Oxford and Edinburgh (Blackwell Sci. Pubns).

SPARKS, B. W. and WEST, R. G. (1972) *The Ice Age in Britain.* London (Methuen). 302 pp.

STEUERWALD, B. A. and CLARK, D. L. (1972) Globerigina pachyderma in Pleistocene and Recent Arctic Ocean sediment, *J. Palaeontol.*, **46** (4), 573–80. Tulsa, Oklahoma.

STOCKTON, C. W. (1971) *The Feasibility of Augmenting Hydrologic Records Using Tree-Ring Data.* Doctoral dissertation, Univ. of Arizona. 172 pp. (Available in University Microfilms, Ann Arbor.)

STOCKTON, C. W. and FRITTS, H. C. (1971) Augmenting annual run-off records using tree-ring data, *Hydrology and Water Resources in Arizona and the Southwest, Vol. 1: Proceedings of the 1971 Meetings of the Arizona Section Acad. Sci., 22–3 April 1971.* Tempe, Arizona.

STRAUCH, F. (1968) Determination of Cenozoic sea temperatures using *Hiatella arctica* (Linne), *Palaeogeogr., Palaeoclim., Palaeoecol.*, **5**, 213–33. Amsterdam (Elsevier).

SUESS, H. E. (1970) Bristlecone pine calibration of the radiocarbon time-scale 5200 B.C. to the present, pp. 595–605 in *Radiocarbon Variations and Absolute Chronology: Proc. Twelfth Nobel Symp., Uppsala, 1969* (ed. I. U. OLSSON). New York (Wiley) and Stockholm (Almqvist & Wiksell).

SUTTON, O. G. (1932) A theory of eddy diffusion in the atmosphere, *Proc. Roy. Soc.*, A, **135**, 143–65. London.

SUTTON, O. G. (1947) The problem of diffusion in the lower atmosphere, *Quart. J. Roy. Met. Soc.*, **73**, 257–76. London.

SUTTON, O. G. (1953) *Micrometeorology.* New York.

SWAIN, A. M. (1973) A history of fire and vegetation in northeastern Minnesota as recorded in lake sediments, *Quaternary Res.*, **3** (3), 383–96. Seattle.

TAUBER, H. (1965) Differential pollen dispersion and the interpretation of pollen diagrams, *Danmarks Geol. Undersogelse*, II Raekke, Nr. 89. Copenhagen (Reitzel). 70 pp.

TEICHERT, C. (1964) Some biological and palaeogeographical factors in the evaluation of ancient climates, pp. 597–600 in *Problems in Palaeoclimatology* (ed. A. E. M. NAIRN). New York and Sydney (Interscience Publishers). 705 pp.

TERASMAE, J. (1958) Contributions to Canadian palynology, *Geol. Survey Can. Bull.*, **46**. Ottawa.

THIEDE, J. (1971) Planktonische Foraminiferen in Sedimenten vom ibero-marokkanischen Kontinentalrand, '*Meteor*' *Forschungs-Ergebnisse*, C, No. 7, 15–102. Berlin and Stuttgart.

THOMPSON, M. W. (1967) *Novgorod the Great*. London (Evelyn Adams). 104 pp.

TRAILL, W. (1868) On submarine forests and other remains of indigenous wood in Orkney, *Trans. Bot. Soc. Edinburgh*, **9**, 146–54'

TSUKADA, M. (1967) Pollen succession, absolute pollen frequency and recurrence surfaces in central Japan, *Amer. J. Bot.*, **54** (7), 821–31. Columbus, Ohio (Bot. Soc. Amer.).

TURNER, C. (1970) The Middle Pleistocene deposits at Marks Tey, Essex, *Phil. Trans. Roy. Soc.*, **B, 257** (817), 373–440. London.

TURNER, J. (1965a) A recent study of Tregaron Bog, Cardiganshire, pp. 33–40 in *Climatic Change with Special Reference to Wales and its Agriculture* (ed. J. A. TAYLOR). Univ. Coll. of Wales, Aberystwyth, Symposia in Agricultural Meteorology, Memorandum No. 8.

TURNER, J. (1965b) A contribution to the history of forest clearance, pp. 343–53 in *A Discussion on the Development of Habitats in the Postglacial* (ed. W. H. PEARSALL), *Proc. Roy. Soc.*, **B, 161**. London.

TYLDESLEY, J. B. (1973) Long-range transmission of tree pollen to Shetland, *New Phytol.*, **72**, 175–81, 183–90. Oxford and Edinburgh (Blackwell Sci. Pubns).

UFFEN, R. J. (1963) Influence of the Earth's core on the origin and evolution of life, *Nature*, **198**, 143. London.

WALKER, D. (1966) A commentary on botanical data from New Guinea, Australia and New Zealand, pp. 149–53 in *World Climate 8000–0 B.C.: Proc. Internat. Conf.* (ed. J. S. SAWYER). London (Roy. Met. Soc.). 229 pp.

WARDLE, P. (1963a) Vegetation studies on Secretary Island, Fiordland, part 5: Population structure and growth of rimu, *N.Z.J. Bot.*, **1**, 208–14. Wellington (D.S.I.R.).

WARDLE, P. (1963b) The regeneration gap of New Zealand Gymnosperms, *N.Z.J. Bot.*, **1**, 301–15. Wellington (D.S.I.R.).

WEBB, T. and BRYSON, R. A. (1972) Late- and postglacial climatic change in the northern midwest U.S.A.: quantitative estimates derived from fossil pollen spectra by multivariate statistical analysis, *Quaternary Res.*, **2**, 70–115. New York and London (Acad. Press).

WEBB, T. and MCANDREWS, J. H. (1976) Corresponding patterns of contemporary pollen and vegetation in central North America. *Geol. Soc. America Mem.* **145**, 267–99. Boulder, Colorado.

WEBER, C. A. (1900) Über die Moore, mit besonderen Berücksichtigung der zwischen Unterweser und Unterelbe liegenden, *Jahresbericht Männer vom Morgenstern*, **3**. Bremerhaven.

WEST, R. G. (1968) *Pleistocene Geology and Biology*. London (Longman). 378 pp.

WEST, R. G. (1970) Pleistocene history of the British flora, pp. 1–11 in *Studies in the Vegetational History of the British Isles: Essays in Honour of Harry Godwin* (ed. D. WALKER and R. G. WEST). Cambridge (Univ. Press). 266 pp.

WHITE, M. G. (1954) The house long-horn beetle *Hylotrupes bajulus* L. (*Col. Cerambycidae*) in Great Britain, *Forestry*, **27** (1), 31–40. Oxford.

WILLIAMSON, K. (1973) New bird species admitted to the British and Irish lists since 1800, in Systematics Association Symposium on *The Changing Flora and Fauna of Britain* (ed. D. L. HAWKSWORTH). London and New York (Acad. Press).

WISEMAN, J. D. H. (1966) Evidence for recent climatic changes in cores from the ocean bed, pp. 84–98 in *World Climate 8000–0 B.C.: Proc. Internat. Conf.* (ed. J. S. SAWYER). London (Roy. Met. Soc.). 229 pp.

WISEMAN, J. D. H. and OVEY, C. D. (1950) Recent investigations on the deep-sea floor, *Proc. Geol. Ass.*, **61** (1), 28–84. London.

WOILLARD, G. (1975) Recherches palynologiques sur le Pléistocene dans l'Est de la Belgique et dans les Voges Lorraines. *Acta Geographica Lovaniensia*, **14**. Louvain (Université Catholique).

WOLLIN, G., ERICSON, D. B. and EWING, M. (1970) Late Pleistocene climate record in Atlantic and Pacific deep-sea sediments, in *Sillman Lecture Volume Honouring Richard Foster Flint*. New Haven, Conn. (Yale Univ. Press).

WRIGHT, H. E. (1966) Stratigraphy of lake sediments and the precision of the palaeoclimatic record, pp. 157–73 in *World climate 8000–0 B.C.: Proc. Internat. Conf.* (ed. J. S. SAWYER). London (Roy. Met. Soc.). 229 pp.

ZAGWIJN, W. H. (1959) Zur stratigraphischen und pollenanalytischen Gliederung der Pliozänen-Ablagerungen im Roertal-Graben und Venloer Graben der Niederlande, *Fortschr. Geol. Rheinld. und Westfalen*, **4**, 5–26. Krefeld.

ZEUNER, F. E. (1958) *Dating the Past*. London (Methuen), 4th edn. 516 pp.

ZEUNER, F. E. (1959) *The Pleistocene Period*. London (Hutchinson). 447 pp.

13.5 Evidence from history and archaeology

ADAM OF BREMEN (1067–85) *History of the Archbishops of Hamburg-Bremen*. Edition translated from the four books of Adam's *Gesta* by F. J. TSCHAN. New York (Columbia Univ. Press, 1959).

ARUTYUNOV, S. A., LEVIN, M. G. and SERGEYEV, D. A. (1964) Ancient burials on the Chuckchi peninsula, pp. 333–46 in *The Archaeology and Geomorphology of Northern Asia.* (ed. H. N. MICHAEL). Arctic Inst. of N. America, Anthropology of the North: Translations from Russian Sources No. 5. Toronto (Univ. Press). 512 pp.

ASHE, G. (1962) *Land to the West.* London (Collins). 352 pp.

BERGTHORSSON, P. (1962) Unpublished report to the History Section of the Conference on the Climate of the Eleventh and Sixteenth Centuries, Aspen, Colorado, 16–24 June 1962. (See *N.C.A.R. Tech Notes 63–1.* Boulder, Colorado.)

BEVERIDGE, W. H. (1921) Weather and harvest cycles, *Econ. J.,* **31**, 421–53. London.

BEVERIDGE, W. H. (1922) Wheat prices and rainfall, *J. Roy. Stat. Soc.,* **85**, 418–54. London.

BEVERIDGE, W. H. (1929) A statistical crime of the seventeenth century, *J. Econ. & Bus. Hist.,* **1**, 503–33. Cambridge, Mass.

BLOCH, M. R. (1965) A hypothesis for the change of ocean levels depending on the albedo of the polar ice caps, *Palaeogeogr., Palaeoclim., Palaeoecol.,* **1**, 127–42. Amsterdam (Elsevier).

BLOCH, M. R. (1970) Zur Entwicklung der vom Salz abhängigen Technologien: Auswirkung von postglazialen Veränderungen der Ozeanküsten, *Saeculum,* **21** (1), 1–33. Munich.

BONSER, W. (1963) *The Medical Background of Anglo-Saxon England.* London (Wellcome Historical Medical Library). 448 pp.

BRAUDEL, F. (1949) *La Mediterranée et le monde mediterranéen a l'époque de Philippe II.* Paris.

BRIMBLECOMBE, P. and OGDEN, C. (1977) Air pollution in art and literature. *Weather,* **32**, in press.

BROOKS, C. E. P. (1949) *Climate Through the Ages.* London (Benn), 2nd edn. 395 pp.

BROOKS, C. E. P. (1951) *Climate in Everyday Life.* New York (Philosophical Library). 314 pp.

BULL, E. (1916) Islands klima i oldtiden, *Geografisk Tidsskrift,* **23**, 1–5. Copenhagen.

BUTZER, K. W. (1958) Studien zum vor- und frühgeschichtlichen Landschaftswandel der Sahara, *Abhandl. math.- naturw. Klasse,* No. 1. Mainz (Akad. der Wiss. und Litt.).

CAMDEN, W. (1586) *Britannia – or a Chorographical Description of Great Britain and Ireland.* (Translated into English, revised and digested by E. GIBSON, London 1722).

CARY, M. and WARMINGTON, E. H. (1929) *The Ancient Explorers.* London (Methuen). (Republished in Pelican Books, 1963.) 319 pp.

CHAPPELL, J. E. (1970) Climatic change reconsidered: another look at the 'Pulse of Asia'. *Geogr. Rev.,* **60** (3), 347–73. New York.

CHARD, C. S. and GRIFFIN, J. B. (1962) Summary report of the Anthropology section, pp. 17–19 in *Proc. Conf. on the Climate of the Eleventh and Sixteenth Centuries, Aspen, Colorado 16–24 June 1962* (AFCRL-63-660). Boulder, Colorado (N.C.A.R. Tech. Note 63–1).

CHILDE, G. (1942) *What Happened in History.* London (Pelican Books). 288 pp.

CHU KO-CHEN (COCHING CHU) (1973) A preliminary study on the climatic fluctuations during the last 5000 years in China, *Scientia Sinica*, **16** (2), 226–56. Peking.

CLARK, G. and PIGGOTT, S. (1965) *Prehistoric Societies.* London (Hutchinson and Penguin Books). 352 pp.

COMFORT, A. (1969) *Nature and Human Nature.* Harmondsworth (Pelican Books). 221 pp.

DARBY, H. C. (ed.) (1952) *The Domesday Geography of England.* Cambridge (Univ. Press).

EASTON, C. (1928) *Les hivers dans l'Europe occidentale.* Leyden (E. J. Brill).

ELLIS, H. K. (1833) *A General Introduction to Domesday Book.* London (Commissioners of Public Records).

FLINT, R. F. (1971) *Glacial and Quaternary Geology.* New York and London (Wiley). 892 pp.

FLOHN, H. (1950) Klimaschwankung im Mittelalter und ihre historisch-geographische Bedeutung, *Berichte zur deutschen Landeskunde*, **7**, 347–57. Remagen, Stuttgart.

FOX, C. (1932) *The Personality of Britain: Its Influence on Inhabitant and Invader in Prehistoric and Early Historic Times.* Cardiff (National Museum of Wales). 84 pp.

GAMS, H. (1937) Aus der Geschichte der Alpenwälder, *Zeit. des Deutschen und Öster-reichischen Alpenvereins*, **68**, Jahrb. 1937, 157–70. Innsbruck.

GIDDINGS, J. L. (1960) The archaeology of Bering Strait, *Current Anthropology*, **1** (2).

GIRALT, E. (1962) *Number of Days of Rogation for Rain at Barcelona, and the Price of one Quartera of Wheat in 'sous' and 'diners' of Barcelona.* Unpublished report to the History Section of the Conference on the Climate of the Eleventh and Sixteenth Centuries, Aspen, Colorado, 16–24 June 1962. (See N.C.A.R Tech. Notes. 63–1. Boulder, Colorado).

GRIFFIN, J. B. (1961) Some correlations of climatic and cultural change in eastern North American prehistory, *Ann. New York Acad. Sci.*, **95** (1), 710–17.

GUMILYOV, L. N. (GUMILÉV) (1968) Heterochronism in the moisture supply of Eurasia in the Middle Ages, *Soviet Geogr.: Review and Translation*, **9** (1), 23–35. New York (Amer. Geogr. Soc.). See also GUMILYOV (1966), Heterochronism in the moisture supply of Eurasia in Antiquity, *Ibid.*, **7** (10), 34–45.

HACKETT, C. J. (1963) On the origin of the human treponematoses (pinta, yaws, endemic syphilis and veneral syphilis), *W.H.O. Bull.*, **29**, 7–41 Geneva (World Health Org.).

HAGEN, A. (1961) *Steinalder, Bronsealder.* Oslo (Universitetets Oldsaksamling). 45 pp.

HALLAM, H. E. (1959–60) Salt making in the Lincolnshire Fenland during the Middle Ages, *Lincs. Architect. Archaeol. Soc. Reports & Papers*, New Series, **8**, 85–112. Spalding.

HAMPDEN, J. (1958) *Richard Hakluyt: Voyages and Documents.* Oxford (Univ. Press World's Classics Series).

HAWKINS, G. S. (1966) *Stonehenge Decoded.* London (Souvenir Press). (Published in paperback by Fontana, 1970. 253 pp.)

HIGGS, E. S. and VITA FINZI, C. (1966) The climate, environment and industries of Stone Age Greece, II, *Proc. Prehist. Soc.*, **32**, 1–29. Cambridge.

HOLLOWAY, J. T. (1954) Forests and climates in the South Island of New Zealand, *Proc. Roy. Soc. N.Z.*, **82**, 329–410. Wellington.

HOLLSTEIN, E. (1965) Jahrringchronologische Datierung von Eichenhölzern ohne Waldkante (Westdeutsche Eichenchronologie), *Bonner Jahrbücher*, **165**, 1–27. Bonn.

HOLMSEN, A. (1961) *Norges Historie*. Oslo (Universitets forlaget), 3rd edn. 480 pp.

HOPKINS, D. M. (ed.) (1967) *The Bering Land Bridge*. Stanford (Univ. Press). 495 pp.

HOSKINS, W. G. (1964) Harvest fluctuations and English economic history, 1480–1619, *Agric. Hist. Rev.*, **12** (1), 28–46. London and Oxford (British Agric. History Soc.).

HOSKINS, W. G. (1968) Harvest fluctuations and English economic history, 1620–1759, *Agric. Hist. Rev.*, **16** (1), 15–31.

HUNTINGTON, E. (1907) *The Pulse of Asia*. Boston, Mass. (Houghton Mifflin). 415 pp.

HUNTINGTON, E. (1924) *Civilization and Climate*. New Haven, Conn., 3rd edn. 453 pp.

HYAMS, E. (1953) *Vineyards in England*. London.

IVES, R. L. (1953) Climatic studies in western North America. *Proc. Toronto Met. Conf.*, 218–22. London (Roy. Met. Soc.).

JONES, G. (1968) *The Vikings*. Oxford (Univ. Press). 504 pp.

KIRWAN, L. P. (1959) *The White Road*. London (Hollis & Carter). (Republished as *A History of Polar Exploration*, Penguin Books, 1962.) 408 pp.

KROGH, K. J. (1967) *Viking Greenland*. Copenhagen (National Museum). 182 pp.

LADURIE, E. LE R. (1971) *Times of Feast, Times of Famine*. New York (Doubleday). 426 pp.

LAHR, E. (1950) *Un siècle d'observations météorologiques appliqueés á l'étude du climat luxembourgeois*. Luxemburg (Bourg-Bourger Verlag).

LAMB, H. H. (1965) The early medieval warm epoch and its sequel, *Palaeogeogr., Palaeoclim., Palaeoecol.*, **1**, 13–37. Amsterdam (Elsevier).

LAMB, H. H. (1967) Britain's changing climate, *Geogr. J.*, **133** (4), 445–68. London.

LAMB, H. H. (1974) Climate, vegetation and forest limits in early civilized times, *Phil. Trans. Roy. Soc.*, **A, 276** (1257), 193–230. London.

LANDSBERG, H. E. (1962) *Physical Climatology*. Du Bois, Pa. (Gray). 446 pp.

LEIDLMAIR, A. (1962) Klimamorphologische Probleme in Hadramaut, pp. 162–80 in *Hermann von Wissmann Festschrift*. Tübingen (Geogr. Inst. der Universität).

LEVIN, M. G. and SERGEYEV, D. A. (1964) The penetration of iron into the Arctic, pp. 319–26 in *The Archaeology and Geomorphology of Northern Asia* (ed. H. N. MICHAEL). Arctic Inst. of N. America, Anthropology of the North: Translations from Russian sources No. 5. Toronto (Univ. Press). 512 pp.

LEVISON, M., WARD, R. G. and WEBB, J. W. (1972) The settlement of Polynesia: report on a computer simulation, *Archaeol. and Physical Anthropol. in Oceania*, **7** (3), 234–45. Glebe, N.S.W. (Australian Medical Publishing Co.).

MANLEY, G. (1958) The revival of climatic determinism, *Geogr. Rev.*, **48** (1), 98–105. New York.

MARKHAM, S. F. (1942) *Climate and the Energy of Nations*. Oxford (Univ. Press) and London (Milford). 144 pp.

MARTINI, E. (1954) Klima und Krankheitserreger, pp. 190–225 in *Klima, Wetter, Mensch* (ed. A. SEYBOLD and H. WOLTERECK). Heidelberg (Quelle & Meyer). 293 pp.

MEYER, G. M. (1927) Early water mills in relation to changes in the rainfall of east Kent, *Quart. J. Roy. Met. Soc.*, **53**, 407. London.

MÜLLER, K. (1953) *Geschichte des Badischen Weinbaus*. Lahr in Baden (von Moritz Schauenburg). 283 pp.

NEUBERGER, H. (1970) Climate in art, *Weather*, **25** (2), 46–56. London.

NORDHAGEN, R. (1933) *De senkvartære klimavekslinger i Nordeuropa og deres betydning for kulturforskningen*. Oslo (Aschehoug).

ORDISH, G. (1953) *Wine Growing in England*. London (Hart-Davis).

RAESIDE, J. D. (1948) Some postglacial climatic changes in Canterbury and their effect on soil formation, *Trans. Roy. Soc. N.Z.*, **77** (1), 153–71. Wellington.

RIEHM, K. (1960) Die Formalsalzproduktion der vorgeschichtlichen Salzsiedestätten Europas, *Jahresschrift für mitteldeutscher Vorgeschichte*, **44**, 180–218. Halle/Saale.

RUSSELL, J. C. (1958) Late Ancient and medieval population, *Trans. Amer. Phil. Soc.*, New Series, **48** (3), 1–152. Philadelphia.

SALTMARSH, J. (1941) Plague and economic decline in England in the later Middle Ages, *Cambridge Hist. J.*, **7**, 23–41.

SETH, S. K. (1963) A review of evidence concerning changes of climate in India during the protohistorical and historical periods, pp. 443–54 in *Changes of Climate: Proc. UNESCO/WMO Rome 1961 Symp.* (UNESCO Arid Zone Res. Series, XX). Paris.

SIMON, A. L. (1946) *English Wines and Cordials*. London.

SINGH, G. (1971) The Indus Valley culture, *Archaeol. & Phys. Anthropol. in Oceania*, **6** (2), 177–89. Sydney, N.S.W.

SLICHER VAN BATH, B. H. (1963) *The Agrarian History of Western Europe A.D. 500–1850* (translation from the Dutch by O. ORDISH). London (Arnold).

STEENSBERG, A. (1951) Archaeological dating of the climatic change in north Europe about A.D. 1300, *Nature*, **168**, 672–4. London (22 October 1951).

TANNER, J. M. (1962) *Growth at Adolescence*. Oxford (Blackwell), 2nd edn.

TAYLOR, E. G. R. (1956) A letter dated 1577 from Mercator to John Dee, *Imago Mundi*, **13**, 56–68. The Hague (Mouton & Co.).

THOM, A. (1964) Megalithic geometry in standing stones, *New Scientist*, **21** (382), 690–2. London (12 March).

THORARINSSON, S. (1956) *The Thousand Years Struggle Against Ice and Fire*. Dept. Geol. and Geogr. Reykjavik Misc. Papers No. 14. Reykjavik (Museum of Natural History).

THORARINSSON, S. (1961) Population changes in Iceland, *Geogr. Rev.*, **51**, 519–33. New York.

TITOW, J. Z. (1960) Evidence of weather in the account rolls of the bishopric of Winchester, 1209–1350, *Econ. Hist. Rev.*, 2nd Series, **12**(3), 360–407. London and Utrecht.

TOYNBEE, A. J. (1946) *A Study of History*. (Abridged edn by D. C. SOMERVELL). Oxford (Univ. Press). 617 pp.

UTTERSTRÖM, G. (1955) Climatic fluctuations and population problems in early modern history, *Scandinavian Econ. Hist. Rev.*, **3** (1), 3–47. Copenhagen.

VITA-FINZI, C. (1964) Synchronous stream deposition throughout the Mediterranean area in historical times, *Nature*, **202**, 1324. London (27 June).

VITA-FINZI, C. (1966) Archaeological notes, *Amer. J. Archaeol.*, **70**, 175–8. New Haven, Conn.

WALTON, K. (1961) Population changes in North-East Scotland 1696–1951, *Scottish Studies*, **5**, (2), 149–80. Edinburgh (Univ. School of Scottish Studies).

WARREN, A. (1970) Dune trends and their implications in central Sudan, *Zeit. Geomorph.*, Suppl., **10**, 154–80. Berlin, Leipzig.

WELLS, C. (1964) *Bones, Bodies and Disease.* London (Thames & Hudson). (See also: Disease before history, *The Listener* (26 March 1964), 520–1.)

WICKENS, G. E. (1975) Changes in the climate and vegetation of the Sudan since 20 000 B.P., *Boissiera*, **24**, 43–65. Geneva.

WILLIAMS, D. P. (1973) Environmental archaeology in the western Negev, *Nature*, **242**, 501–3. London (20 April).

WINSTANLEY, D. (1973) Rainfall patterns and general atmospheric circulation, *Nature*, **245**, 190–4. London (28 September).

WRIGLEY, E. A. (1966) Marriage and fertility in pre-industrial England, *The Listener* (10 February), 199–201. London. (See also *An Introduction to English Historical Demography* (ed. E. A. WRIGLEY) for the Cambridge Group for the History of Population and Social Structure. London (Weidenfeld & Nicolson, 1965).)

YULE, G. U. and KENDALL, M. G. (1950) *An Introduction to the Theory of Statistics.* London (Charles Griffin). 701 pp.

ZIEGLER, P. (1969) *The Black Death.* London (Collins), 319 pp.

Chapter 14   Climate and the long history of the Earth

ADDICOTT, W. O. (1970) Latitudinal gradients in Tertiary molluscan faunas of the Pacific coast, *Palaeogeogr., Palaeoclim., Palaeoecol.*, **8**, 287–312. Amsterdam (Elsevier).

ARRHENIUS, S. (1896) On the influence of carbonic acid in the air upon the temperature of the ground, *Phil. Mag.*, **41**, 237. London.

BLACKETT, P. M. S. (1956) *Lectures in Rock Magnetism.* Jerusalem (Weiszmann Science Press of Israel).

BRIDEN, J. C. and IRVING, E. (1964) Palaeolatitude spectra of sedimentary palaeoclimatic indicators, pp. 199–224, 249–50, in *Problems in Palaeoclimatology*, (ed. A. E. M. NAIRN). New York (Wiley). 705 pp.

BROOKS, C. E. P. (1949) *Climate Through the Ages.* London (Benn), 2nd edn.

BULLARD, E. C., EVERETT, J. E. and SMITH, A. G. (1965) The fit of the continents around the Atlantic, in a Symposium on Continental Drift, *Phil. Trans. Roy. Soc.*, A, **258**, 41–51. London.

CHAMBERLIN, T. C. (1899) An attempt to frame a working hypothesis of the cause of glacial periods on an atmospheric basis, *J. Geol.*, **7**, 545, 667, 752. Chicago.

COX, A. (1968) Polar wandering, continental drift, and the onset of Quaternary glaciation, *Met. Monogs.*, **8**, (30), 112–25. Boston (Amer. Met. Soc.).

DAMON, P. E. (1968) The relationship between terrestrial factors and climate, *Met. Monogs*, **8** (30), 106–11. Boston, Mass. (Amer. Met. Soc.).

DENTON, G. H., ARMSTRONG, R. K. and STUIVER, M. (1971) The late Cenozoic glacial history of Antarctica, pp. 267–306 in *The Late Cenozoic Glacial Ages: Proc. Yale Univ. Symp., December 1969* (ed. K. K. TUREKIAN). New Haven, Conn. (Yale Univ. Press). 606 pp.

DILKE, F. W. W. and GOUGH, D. O. (1972) The solar spoon, *Nature*, **240**, 262–4, 293–4. London (1 December).

EMILIANI, C. (1961) The temperature decrease of surface sea water in high latitudes and of abyssal-hadal water in open ocean basins during the past 75 million years, *Deep-Sea Res.*, **8**, 144–7. London (Pergamon).

FORBES, J. D. (1859) Inquiries about terrestrial temperature, *Trans. Roy. Soc. Edin.*, **22**, 75. Edinburgh.

FRAKES, L. A. and KEMP, E. M. (1972) Influence of continental positions on early Tertiary climates, *Nature*, **240**, 97–100. London (10 November).

FULTZ, D. (1961) Developments in controlled experiments on large-scale geophysical problems, *Adv. in Geophysics*, **7**, 1–103. New York and London (Acad. Press).

GLENIE, R. C., SCHOFIELD, J. C. and WARD, W. T. (1968) Tertiary sea levels in Australia and New Zealand, *Palaeogeogr., Palaeoclim., Palaeoecol.*, **5**, 141–63. Amsterdam (Elsevier).

HALLAM, A. (1973) *A Revolution in the Earth Sciences*. Oxford (Clarendon Press). 127 pp.

HARLAND, W. B. (1964) Evidence of Late Pre-Cambrian glaciation and its significance, pp. 119–49, 179–84, in *Problems in Palaeoclimatology* (ed A. E. M. NAIRN). London, New York, Sydney (Interscience Publishers/Wiley). 705 pp.

HARLAND, W. B. and HEROD, K. N. (1975) Glaciations through time, pp. 189–216 in *Ice Ages: Ancient and Modern*. Liverpool (Seel House Press), *Geol. J. Special Issue* No. 6. 322 pp.

HESS, H. H. (1962) History of ocean basins, pp. 599–620 in *Petrologic Studies: A Volume to Honour A. F. Buddington* (ed. A. E. J. ENGEL, H. L. JAMES and B. F. LEONARD). (Geol. Soc. of America: Buddington volume).

HIDE, R. (1970) Some laboratory experiments on free thermal convection in a rotating fluid subject to a horizontal temperature gradient and their relation to the theory of the global atmospheric circulation, pp. 196–221 in *The Global Atmospheric Circulation* (ed. G. A. CORBY). London (Roy. Met. Soc.). 255 pp.

HIDE, R. (1971) Motions in planetary atmospheres: a review, *Met. Mag.*, **100**, 268–76. London.

HOPKINS, D. M. (1972) Changes in oceanic circulation and Late Cenozoic cold climates, *24th Internat. Geol. Congress*, Montreal 1972 (P.370 in Abstracts volume).

HOPKINS, D. M., MATTHEWS, J. V., WOLFE, J. A. and SILBERMAN, M. L. (1971) A Pliocene flora and insect fauna from the Bering Strait region, *Palaeogeogr., Palaeoclim., Palaeoecol.*, **9**, 211–31. Amsterdam (Elsevier).

KENNETT, J. P. and THUNELL, R. C. (1975) Global increase in Quaternary explosive volcanism, *Science*, **187** (4176), 497–503. New York (14 February).

KENNETT, J. P. and BRUNNER, C. A. (1973) Antarctic Late Cenozoic glaciation: evidence for the initiation of ice-rafting and inferred increase bottom-water activity, *Geol. Soc. of Amer. Bull.*, **84**, 2043–52. Rochester, N.Y.

KENNETT, J. P., BURNS, R. E., ANDREWS, J. E., CHURKIN, M., DAVIES, T. A., DUMITRICA, P., EDWARDS, A. R., GALEHOUSE, J, S., PACKHAM, G. H. and VANDER LINGEN, G. J. (1972) Australian-Antarctic continental drift, palaeocirculation changes and Oligocene deep-sea erosion, *Nature*, **239**, 51–5. London (25 September).

KOMAROV, A. G. (1960) Polar wandering in the Earth's history, *Priroda*. Moscow. (In Russian).

KÖPPEN, W. and WEGENER, A. (1924) *Die Klimate der geologischen Vorzeit*. Berlin (Borntraeger).

KUIPER, G. P. (1957) Origin, age and possible ultimate fate of the Earth, Ch. 2, pp. 12–30, in *The Planet Earth* (ed. D. R. BATES). London (Pergamon). 312 pp.

MCKEE, E. D. (1964) Problems on the recognition of arid and of hot climates of the past, pp. 367–77, 421–2 in *Problems in Palaeoclimatology* (ed. A. E. M. NARIN). London, New York, Sydney (Interscience Publishers/Wiley). 705 pp.

NAGATA, T. (1953) *Rock Magnetism*. Tokyo (Maruzen).

NAIRN, A. E. M. (ed.) (1961) *Descriptive Palaeoclimatology* (esp. Ch. 7, pp. 156–182: The application of geophysics to palaeoclimatology, by A. E. M. NAIRN and N. THORLEY). New York (Interscience Publishers). 380 pp.

ÖPIK, E. J. (1958) Solar variability and palaeoclimatic changes, *Irish Astronom. J.*, **5** (4), 97–109. Armagh Observatory.

ÖPIK, E. J. (1965) Climatic change in cosmic perspective, *Icarus*, **4**, 289–307. Dublin (Trinity College).

ROBINSON, P. L. (1972) *Palaeoclimatology and Continental Drift*. Unpublished paper given before the British Association for the Advancement of Science (Geology Section), 1972 Leicester Meeting.

RUNCORN, S. K. (1962) Continental drift, *Research*, **15**, 103–8.

RUNCORN, S. K. (1964) Geophysical techniques and ancient climates, pp. 189–99, 248–9 in *Problems in Palaeoclimatology* (ed. A. E. M. NAIRN). New York (Wiley). 705 pp.

SCHWARZBACH, M. (1961) *Das Klima der Vorzeit*. Stuttgart (Enke), 2nd edn. 275 pp.

SMITH, A. G., BRIDEN, J. C. and DREWRY, G. E. (1973) Phanerozoic world maps, pp. 1–42 in *Organisms and Continents Through Time*. Special Papers in Palaeontology No. 12. London (Palaeontological Association).

SPITALER, R. (1886) Die Wärmevertheilung der Erdoberfläche, *Denkschrift Kgl. Akad. Wiss.*, Abt. **3** (1). Vienna. (The investigation referred to was later carried further, including into monthly detail, by SPITALER in a lithographed monograph *Das Klima des Eiszeitalters*, Prague, 1921).

STEINER, J. (1967) The sequence of geological events and the dynamics of the Milky Way galaxy, *J. Geol. Soc. Australia*, **14** (1), 99–132. Adelaide.

UMBGROVE, J. H. F. (1947) *The Pulse of the Earth*. Den Haag (Martinus Nijhoff.)

VINE, F. J. (1966) Spreading of the ocean floor, *Science*, **154**, 1405–15. New York.

VINE, F. J. and MATTHEWS, D. H. (1963) Magnetic anomalies over oceanic ridges, *Nature*, **199**, 947–9. London.

WEGENER, A. (1912) *Die Entstehung der Kontinente und Ozeane*. Braunschweig (Vieweg). (English language edn: *The Origin of the Continents and Oceans*, 1941.)

WOLDSTEDT, P. (1954) Die Klimakurve des Tertiärs und Quartärs in Mitteleuropa, *Eiszeitalter und Gegenwart*, **4**, 5–9. Ohringen/Wurtt.

WOLFE, J. A. (1971) Tertiary climatic fluctuations and methods of analysis of Tertiary floras, *Palaeogeogr., Palaeoclim., Palaeoecol.*, **9**, 25–57. Amsterdam (Elsevier).

Interesting material is also contained in a chapter on 'Tertiary climate' by N. DE B. HORNIBROOK, in *The Geology of New Zealand* (ed. R. P. SUGGATE). Wellington, N.Z. (D.S.I.R., publication pending).

Chapter 15   The Quaternary ice ages and interglacial periods

ABBOT, C. G. and FOWLE, F. E. (1913) Volcanoes and climate, *Annals of the Astrophys. Observatory*, **3**. Washington (Smithsonian Institution). (Also *Smithsonian Misc. Coll.*, **60**, (29), 1913.)

ADIE, R. J. (1964) Geological history, Ch. 9, pp. 118–62, in *Antarctic Research* (eds. R. PRIESTLEY, R. J. ADIE and G. de Q. ROBIN). London (Butterworth). 360 pp.

ANDEL, T. VAN., ROSS HEATH, G., MOORE, T. C. and McGEARY, D. F. R. (1967) Late Quaternary history, climate and oceanography of the Timor Sea, northwestern Australia, *Amer. J. Sci.*, **265**, 737–58. New Haven, Conn.

ANDERSON, SV. TH. (1961) Vegetation and its environment in Denmark in the Weichselian glacial, *Danmarks geologiske Undersøgelse*, II Raekke, No. 75. Copenhagen.

ARCTOWSKI, H. (1915) Volcanic dust veils and climatic variations. *Ann. New York Acad. Sci.*, **26**, 149.

BLOOM, A. L. (1971) Glacial-eustatic and isostatic controls of sea level since the last glaciation, pp. 355–80 in *The Late Cenozoic Glacial Ages: Proc. Yale Univ. Symp., December 1969* (ed. K. K. TUREKIAN). New Haven, Conn. (Yale Univ. Press). 606 pp.

BONATTI, E. (1966) North Mediterranean climate during the last Würm glaciation, *Nature*, **209**, 184–5. London (5 March).

BOWLER, J. M., HOPE, G. S., JENNINGS, J. N., SINGH, G. and WALKER, D. (1975) Late Quaternary climates: evidence from Australia and Papua/New Guinea, *Abstracts in Advance: Australian Conf. on Climate and Climatic Change, Monash Univ., 7–12 December 1975*. Clayton, Victoria (Roy. Met. Soc., Australian Branch).

BRAIDWOOD, R. J. (1963) Summary of prehistoric investigations in Kurdistan in relation to climatic change, pp. 251–4 in *Changes of Climate: Proc. UNESCO/WMO Rome 1961 Symp.* Paris (UNESCO Arid Zone Research Series, XX).

BROECKER, W. S., EWING, M. and HEEZEN, B. C. (1960) Evidence for an abrupt change in climate close to 11 000 years ago, *Amer. J. Sci.*, **258**, 429–48. New Haven, Conn.

BROOKS, C. E. P. (1925) The problem of mild polar climates, *Quart. J. Roy. Met. Soc.*, **51**, 83–94. London.

BROOKS, C. E. P. (1949) *Climate through the Ages*. London (Benn), 2nd edn. 395 pp.

BÜDEL, J. (1951) Die Klimazonen des Eiszeitalters, *Das Eiszeitalter* (ed. P. WOLDSTEDT), **1**, 16–26. Stuttgart (Enke).

BUDYKO, M. I. (1968) On the causes of climatic variations, *Sveriges Met. och Hydr. Institut: Meddelanden*, Serie B, No. 28 (*Scientific Papers Dedicated to Dr Anders Ångström*), 6–13. Stockholm.

BUDYKO, M. I. (1969) The effect of solar radiation variations on the climate of the Earth, *Tellus*, 21, 611–19. Stockholm.

BUTZER, K. W. (1974) Geological and ecological perspectives on the Middle Pleistocene, *Quaternary Res.*, 4, 136–48. Seattle.

CALDER, N. (1974) Arithmetic of ice ages, *Nature*, **252**, 216–18. London (15 November).

CHAPPELL, J. (1973) Astronomical theory of climatic change: status and problem, *Quaternary Res.*, 3, 221–36. Seattle.

COOPE, G. R., SHOTTON, F. W. and STRACHAN, I. (1961) A late Pleistocene fauna and flora from Upton Warren, Worcestershire, *Phil. Trans. Roy. Soc. B*, **244**, 379–418. London.

COOPE, G. R. and BROPHY, J. A. (1972) Late-Glacial environmental changes indicated by a Coleopteran succession from North Wales, *Boreas*, **1** (2), 97–142. Oslo (Universitetsforlaget).

COOPE, G. R. and ANGUS, R. B. (1975) An ecological study of a temperate interlude in the middle of the last glaciation, based on fossil coleoptera from Isleworth, Middlesex. *J. Animal Ecol.*, **44**, 365–91. Oxford (Blackwell).

CRANWELL, L. M., HARRINGTON, H. J. and SPEDEN, I. G. (1960) Lower Tertiary microfossils from McMurdo Sound, Antarctica, *Nature*, **186**, 700. London.

CROLL, (1875) *Climate and Time in their Geological Relations: A Theory of Secular Changes of the Earth's Climate.* London (Stanford). 577 pp.

DANSGAARD, W., JOHNSEN, S. J., CLAUSEN, H. B. and LANGWAY, C. C. (1971) Climatic record revealed by the Camp Century ice core, pp. 37–56 in *The Late Cenozoic Glacial Ages: Proc. Yale Univ. Symp., December 1969* (ed. K. K. TUREKIAN). New Haven, Conn. (Yale Univ. Press). 606 pp.

DENTON, G. H. and ARMSTRONG, R. L. (1969) Miocene-Pliocene glaciations in southern Alaska, *Amer. J. Sci.*, **267**, 1121–42. New Haven, Conn.

DENTON, G. H., ARMSTRONG, R. L. and STUIVER, M. (1971) The Late Cenozoic glacial history of Antarctica, pp. 267–306 in *The Late Cenozoic Glacial Ages: Proc. Yale Univ. Symp., December 1969* (ed. K. K. TUREKIAN) New Haven, Conn. (Yale Univ. Press). 606 pp.

DERBYSHIRE, E. (1971) A synoptic approach to the atmospheric circulation of the last glacial maximum in southeastern Australia, *Palaeogeogr., Palaeoclim., Palaeoecol.*, **10**, 103–24. Amsterdam (Elsevier).

DESTOMBES, J.-P., SHEPHARD-THORN, E. R. and REDDING, J. H. (1975) A buried valley system in the Strait of Dover, *Phil. Trans. Roy. Soc.*, A, **279**, 243–56. London.

DICKSON, J. (1967) The British moss flora of the Weichselian glacial, *Rev. Palaeobot. & Palynol.*, **2**, 245–53. Amsterdam (Elsevier).

DONN, W. L. and EWING, W. (1968) The theory of an ice-free Arctic Ocean, *Met. Monogs*, **8** (30), 100–5. Boston, Mass. (Amer. Met. Soc.).

DUPLESSY, J. C., LABEYRIE, J., LALOU, C. and NGUYEN, H. V. (1970) Continental climatic variations between 130 000 and 90 000 years B.P., *Nature*, **226**, 631–3. London.

EMILIANI, C. (1955) Pleistocene temperatures, *J. Geol.*, **63** (6), 538–78. Chicago.

EMILIANI, C. (1961) Cenozoic climatic changes as indicated by the stratigraphy and chronology of deep-sea cores of *Globerigina* ooze facies, *Ann. N.Y. Acad. Sci.*, **95** (1) 521–36. New York.

EMILIANI, C. (1968) Paleotemperature analysis of Caribbean cores P. 6304–8 and P. 6304–9 and a generalized temperature curve for the past 425 000 years. *J. Geol.*, **74**, 109–126. Chicago.

EMILIANI, C. (1969) Interglacial high sea levels and the control of the Greenland ice by the precession of the equinoxes, *Science*, **166**, 1503–4. New York (19 December).

EMILIANI, C. and GEISS, J. (1957) On glaciations and their causes, *Geol. Rundschau*, **46** (2), 576–601. Stuttgart (Enke).

EMILIANI, C. and RONA, F. (1969) Caribbean cores P. 6304–8 and P. 6304–9: new analysis of absolute chronology, *Science*, **166**, 1551–2. New York (19 December).

EWING, M. and DONN, W. L. (1956) A theory of ice ages, *Science*, **123**, 1061–6. New York.

FAIRBRIDGE, R. W. (1961) Eustatic changes in sea level, pp. 99–185 in *Physics and Chemistry of the Earth*, Vol. 4. New York (Pergamon Press).

FALCONER, G., IVES, J. D., LØKEN, O. H. and ANDREWS, J. T. (1965) Major end-moraines in eastern and central Arctic Canada, *Geogr. Bull.* **7**, 137–53. Ottawa.

FIRBAS, F. (1949) *Spät- und nacheiszeitliche Waldgeschichte Mitteleuropas nördlich der Alpen.* Jena (Fischer). 480 pp.

FIRBAS, F. (1950) The Late-Glacial vegetation of central Europe, *New Phytol.*, **49**, 163–73. Oxford and Edinburgh (Blackwell).

FLECK, R. J., MERCER, J. H., NAIRN, A. E. M. and PETERSON, D. N. (1972) Chronology of late Pliocene and early Pleistocene glacial and magnetic events in southern Argentina, *Earth and Planetary Sci. Letters*, **16**, 15–22. Amsterdam (North Holland Publ. Co.)

FLINT, R. F. (1957) *Glacial and Pleistocene Geology.* Wiley, New York. 553 pp.

FLINT, R. F. (1971) *Glacial and Quaternary Geology.* New York, London, Sydney and Toronto (Wiley) 892 pp.

FLOHN, H. (1952) Allgemeine atmosphärische Zirkulation und Paläoklimatologie, *Geol. Rundschau*, **40** (1), 153–78. Stuttgart (Enke).

FLOHN, H. (1969) Ein geophysikalisches Eiszeit-Modell, *Eiszeitalter und Gegenwart*, **20**, 204–31. Öhringen/Württ.

FRENZEL, B. (1960) Die Vegetations- und Landschaftszonen Nord-Eurasiens während der postglazialen Wärmezeit, II Teil: Rekonstruktions-versuch der letzteiszeitlichen und wärmezeitlichen Vegetation, *Akademie der Wissenschaften und der Literatur (Mainz); Abhandlungen der mathematisch-naturwissenschaftlichen Klasse*, **6**. Mainz.

FRENZEL, B. (1967) *Die Klimaschwankungen des Eiszeitalters.* Braunschweig (Vieweg).

GEEL, B. VAN and HAMMEN, T. VAN DER (1973) Upper Quaternary vegetational and climatic sequence of the Fuquene area (eastern Cordillera, Colombia), *Palaeogeogr., Palaeoclim., Palaeoecol.*, **14**, 9–92. Amsterdam (Elsevier).

GERASIMOV, I. P. (1965) *The Last European Glaciation.* International Association for Quaternary Research (INQUA) VIIth Congress, Academy of Science of the U.S.S.R., Polish Academy of Science and German Academy of Science.

GERASIMOV, I. P. (1969) Degradation of the last European ice sheet, pp. 72–8 in *Quaternary Geology and Climate* (Proc. VII INQUA, Vol. 16). Washington (Nat. Acad. Sci.).

GEYH, M. A. and ROHDE, P. (1972) Weichselian chronostratigraphy C^{14} dating and statistics, *Proc. 24th Internat. Geol. Congress* (Montreal 1972), Section 12, pp. 27–36.

GONZALES, E., HAMMEN, TH. VAN DER and FLINT, R. F. (1966) Late Quaternary glacial and vegetational sequence in Valle de Lagunillas, Sierra Nevada del Cocuy, Columbia, *Leidse Geologische Mededelingen*, **32**, 157–82. Leiden, Netherlands.

GORDON, H. B. and DAVIES, D. R. (1975) Mathematical prediction of climatic change, *Nature*, **253**, 419–20. London (6 February).

GROSS, H. (1958) Die bisherige Ergebnisse von C^{14}-Messungen und paläolithischen Untersuchungen für die Gliederung und Chronologie des Jungpleistozäns in Mitteleuropa und den Nachbargebieten, *Eiszeit und Gegenwart*, **9**, 155–87. Ohringen/Württ.

HAMILTON, W. L. and SELIGA, T. A. (1972) Atmospheric turbidity and surface temperature on the polar ice sheets, *Nature*, **235**, 320–2. London (11 February).

HAMMEN, T. VAN DER, MAARLEVELD, G. C., VOGEL, J. C. and ZAGWIJN, W. H. (1967) Stratigraphy, climatic succession and radiocarbon dating of the last glacial in the Netherlands, *Geologie en Mijnbouw*, **46**, 79–95. Den Haag.

HAMMEN, T. VAN DER, WIJMSTRA, T. A. and ZAGWIJN, W. H. (1971) The floral record of the late Cenozoic of Europe, pp. 391–424 in *The Late Cenozoic Glacial Ages: Proc. Yale Univ. Symp., December 1969* (ed. K. K. TUREKIAN). New Haven, Conn. (Yale Univ. Press). 606 pp.

HAYS, J. D., IMBRIE, J. and SHACKLETON, N. J. (1976) Variations in the Earth's orbit: pacemaker of the ice ages. *Science*, **194**, 1121–32. New York.

HEATH, G. R. (1969a) Carbonate sedimentation in the abyssal equatorial Pacific during the past 50 million years, *Geol. Soc. Amer. Bull.*, **80**, 689. Rochester, N.Y.

HEATH, G. R. (1969b) Mineralogy of Cenozoic deep-sea sediments from the equatorial Pacific Ocean, *Geol. Soc. Amer. Bull.*, **80**, 1997. Rochester, N.Y.

HERMAN, Y. and O'NEIL, J. R. (1975) Arctic palaeosalinities during Late Cenozoic time. *Nature*, **258**, 591–5. London (Dec. 18, 1975).

HEUSSER, C. J. (1974) Vegetation and climate of the southern Chilean lake district during and since the last interglaciation, *Quaternary Res.*, **4**, 290–315. Seattle.

HEUVEL, E. P. J. VAN DEN (1966) Ice shelf theory of Pleistocene glaciations, *Nature*, **210**, 363–5. London (23 April).

HOLLIN, J. T. (1962) On the glacial history of Antarctica, *J. Glaciol.*, **4**, 173–95. Cambridge.

HOLLIN, J. T. (1965) Wilson's theory of ice ages, *Nature*, **208**, 12–16. London (2 October).

HOPPE, G. (1968) Grimsey and the maximum extent of the last glaciation of Iceland, *Geogr. Annaler*, **A, 50** (1), 16–24. Stockholm.

HOPPE, G. (1971) Nordvästeuropas inlandsisar under den sista istiden, *Svensk Naturvetenskap* (1971), 31–40. Stockholm.

HORIE, S. (1974) *Palaeolimnology of Lake Biwa and the Japanese Pleistocene*, 2nd issue. Otsu, Japan (Kyoto Univ.). 288 pp.

HUGHES, T. (1970) Convection in the Antarctic ice sheet leading to a surge of the ice sheet and possibly to a new ice age, *Science*, **170**, 630–3. New York.

HUMPHREYS, W. J. (1913) Volcanic dust and other factors in the production of climatic changes and their possible relation to ice ages, *J. Franklin Inst.*, **176**, 131. Philadelphia.

HUNKINS, K. and KUTSCHALE, H. (1965) Quaternary sedimentation in the Arctic Ocean. *Progress in Oceanography*, **4**, 89–94. London (Pergamon).

IVERSEN, J. (1973) *The development of Denmark's nature since the last Glacial*. Danmarks Geol. Undersøgelese, 5 Raekke, No. 7-C. Copenhagen (Reitzel). 126 pp.

IVES, J. D. and ANDREWS, J. T. (1963) Studies in the physical geography of north-central Baffin Island, N.W.T., *Geogr. Bull.*, **19**, 5–48. Ottawa.

JELGERSMA, S. (1966) Sea level changes during the last 10 000 years, pp. 54–71 in *World Climate 8000–0 B.C.: Proc. Internat. Conf.* (ed. J. S. SAWYER). London (Roy. Met. Soc.).

JOHN, B. S. (1965) A possible main Würm glaciation in west Pembrokeshire, *Nature*, **207**, 622–3. London (7 August).

JOHNSEN, S. J., DANSGAARD, W., CLAUSEN, H. B. and LANGWAY, C. C. (1972) Oxygen isotope profiles through the Antarctic and Greenland ice sheets, *Nature*, **235**, 429–34. London (25 February).

KELLAWAY, G. A., REDDING, J. H., SHEPHARD-THORN, E. R. and DESTOMBES, J.-P. (1975) The Quaternary history of the English Channel, *Phil. Trans. Roy. Soc.*, A, **279**, 187–218. London.

KENNETT, J. P. and HUDDLESTUN, P. (1972) Abrupt climatic change at 90 000 years B.P.: faunal evidence from a Gulf of Mexico core, *Quaternary Res.*, **2** (3), 384–95. Seattle.

KERNEY, M. P., BROWN, E. H. and CHANDLER, T. J. (1964) The Late Glacial and post-glacial history of the chalk escarpment near Brook, Kent, *Phil. Trans. Roy. Soc.*, B, **248**, 135–204. London.

KLEIN, A. (1953) Die Niederschläge in Europa im Maximum der letzten Eiszeit, *Petermanns Geogr. Mitteilungen*, **97**, 98–104. Gotha.

KNETSCH, G., SHATA, A., DEGENS, E., MÜNNICH, K. O., VOGEL, J. C. and SHAZLY, M. M. (1962) Untersuchungen an Grundwassern der Ost-Sahara, *Geol. Rundschau*, **52**, 587–610. Stuttgart.

KU, T. L. and BROECKER, W. S. (1965) Rates of sedimentation in the Arctic Ocean, *Progress in Oceanography*, **4**, 95–104. London (Pergamon).

KUKLA, J. (1970) Correlations between loesses and deep-sea sediments, *Geol. Föreningen i Stockholms Förhandlingar*, **92**, 148–80.

LA MARCHE, V. C. (1974) Palaeoclimatic inferences from long tree-ring records, *Science*, **183**, 1043–8. New York (15 March).

LAMB, H. H. (1959) The southern westerlies, *Quart. J. Roy. Met. Soc.*, **85**, 1–23. London.

LAMB, H. H. (1963) On the nature of certain climatic epochs which differed from the modern (1900–39) normal, pp. 125–50 in *Changes of Climate: Proc. UNESCO/WMO Rome 1961 Symp.* (UNESCO Arid Zone Research Series, XX). Paris.

LAMB, H. H. (1969) Climatic fluctuations, pp. 173–249 in *World Survey of Climatology*, (general ed. H. E. LANDSBERG), Vol. 2: *General Climatology* (ed. H. FLOHN). Amsterdam (Elsevier).

LAMB, H. H. (1970) Volcanic dust in the atmosphere; with a chronology and assessment of its meteorological significance, *Phil. Trans. Roy. Soc.*, A, **266** (1178), 425–533. London.

LAMB, H. H. and WOODROFFE, A. (1970) Atmospheric circulation during the last ice age, *Quaternary Res.*, **1**, 29–58. Seattle.

LEONTEV, O. K. and FEDEROV, P. V. (1953) History of the Caspian Sea in late- and post-Hvalynsk time, *Akad, Nauk., Izvestia, Ser. Geograf.* **4**, 64–74. Moscow.

768   References

MCINTYRE, A. (1974) The CLIMAP 17 000 years B.P. North Atlantic map, pp. 41–7 in *Mapping the Atmospheric and Oceanic Circulations and Other Climatic Parameters at the Time of the Last Glacial Maximum about 17 000 Years Ago.* Research Publication CRU RP2. Norwich (Univ. of East Anglia, Climatic Research Unit).

MCINTYRE, A., RUDDIMAN, W. F. and JANTZEN, R. (1972) Southward penetrations of the North Atlantic polar front: faunal and floral evidence of large-scale surface water-mass movements over the last 225 000 years, *Deep-Sea Res.,* **19,** 61–77. London (Pergamon).

MCINTYRE, O. J. and WILSON, G. J. (1966) Preliminary palynology of some Antarctic Tertiary erratics, *N.Z. J. Bot.,* **4,** 315. Wellington, N.Z.

MANLEY, G. (1959) The Late-Glacial climate of northwest England, *Liverpool and Manchester Geol. J.,* **2,** 188–215. (Also *Weather,* **17** (1962), 303–10. London.)

MARGOLIS, S. V. and KENNETT, J. P. (1971) Cenozoic paleoglacial history of Antarctica recorded in deep-sea cores, *Amer. J. Sci.,* **271,** 1–36. New Haven, Conn.

MERCER, J. H. (1962) Glacier variations in the Antarctic, *Glaciol. Notes,* No. 11, 5. New York (Amer. Geogr. Soc.).

MERCER, J. H. (1968) The discontinuous glacio-eustatic fall in Tertiary sea level, *Palaeogeogr., Palaeoclim., Palaeoecol.,* **5** (1), 77–85. Amsterdam (Elsevier).

MERCER, J. H. (1969) The Allerød oscillation: a European climatic anomaly? *Arctic & Alpine Res.,* **1** (4), 227–34. Boulder, Colorado.

MERCER, J. H., FLECK, R. J., MANKINEN, E. A. and SANDER, W. (1973) *Southern Argentina: Glacial Events Between 4 Million and 1 Million Years Ago During the Gilbert, Gauss and Matuyama Polarity Epochs.* Abstract prepared for INQUA Meeting in New Zealand, December 1973. Christchurch (Univ. of Canterbury, Geogr. Dept).

MILANKOVITCH, M. (1930) Mathematische Klimalehre und astronomische Theorie der Klimaschwankungen, in *Handbuch der Klimatologie,* **I,** Teil A. (ed. W. KÖPPEN and R. GEIGER). Berlin (Borntraeger).

MITCHELL, G. F., PENNY, L. F., SHOTTON, F. W. and WEST, R. G. (1973) *A Correlation of Quaternary Deposits in the British Isles.* Special Report No. 4. London (Geol. Soc.).

MITCHELL, J. M. (1963) On the world-wide pattern of secular temperature change, pp. 161–81 in *Changes of Climate: Proc. UNESCO/WMO Rome 1961 Symp.* (UNESCO Arid Zone Research Series, XX) Paris.

MOAR, N. T. and SUGGATE, R. P. (1973) Pollen analysis of late Otiran and Aranuian sediments at Blue Spur road (S51), North Westland, *N.Z. J. Geol. & Geophys.,* **16** (3), 333–44.

MOORE, P. D. (1974) When the ice age seemed to end, *Nature,* **251,** 185–6. London (20 September).

MORAN, J. M. (1972) *An Analysis of Periglacial Climatic Indicators of Late Glacial Time in North America.* Ph.D. thesis, University of Winconsin.

MORGAN, A. (1973) Late Pleistocene environmental changes indicated by fossil insect faunas of the English Midlands, *Boreas,* **2** (4), 173–212. Oslo (Universitetsforlaget).

MULLER, E. H. (1964) Quaternary sections at Otto, New York. *Amer. J. Sci.*, **262**, 461. New Haven, Conn.

NATIONAL ACADEMY OF SCIENCES (1975) *Understanding Climatic Change: A Program for Action.* By the Panel on Climatic Variation of the U.S. Committee for the Global Atmospheric Research Program, National Research Council. Washington. 239 pp.

NEWELL, R. E. (1974) Changes in the poleward energy flux by the atmosphere and ocean as a possible cause for ice ages, *Quaternary Res.*, **4**, 117–27. Seattle.

NEWELL, R. E., HERMAN, G. F., GOULD-STEWART, S. and TANAKA, M. (1975) Decreased global rainfall during the past ice age, *Nature*, **253**, 33–4. London (3 January).

OLAUSSON, E. (1971) Oceanographic aspects of the Pleistocene of Scandinavia, *Geol. Föreningen i Stockholm Förhandl. (GFF)*, **93** (3), 459–75.

OLDFIELD, F. (1965) Late Quaternary environments and early Man on the Southern High Plains, *Antiquity*, **39**, 226–9. Cambridge.

PARKIN, D. W. (1974) Trade winds during glacial cycles, *Proc. Roy. Soc.*, A, **337**, 73–100. London. (See also PARKIN, D. W. and PADGHAM, R. C. (1975) Further studies on trade winds during glacial cycles, *Proc. Roy. Soc.*, A, **346**, 245–60. London.)

PARKIN, D. W., PHILLIPS, D. R., SULLIVAN, R. A. L. and JOHNSON, L. R. (1972) Airborne dust collections down the Atlantic, *Quart. J. Roy. Met. Soc.*, **98**, 798–808. London.

PENCK, A. and BRÜCKNER, E. (1901–9) *Die Alpen im Eiszeitalter.* Leipzig (Tauchnitz), 3 vols. 1189 pp.

PENNINGTON, W. (1973) Absolute pollen frequencies in the sediments of lakes of different morphometry, pp. 79–104 in *Quaternary Plant Ecol.* (ed. H. J. B. BIRKS and R. G. WEST). Oxford (Blackwell Sci. Pubns).

POSER, H. (1954) Die nördliche Lössgrenze in Mitteleuropa und das spätglaziale Klima, pp. 27–53 in *Das Eiszeitalter, 1: Allgemeine Erscheinungen des Eiszeitalters* (ed. P. WOLDSTEDT). Stuttgart (Enke).

QUINN, W. H. (1971) Late Quaternary meteorological and oceanographic developments in the equatorial Pacific, *Nature*, **229**, 330–1. London (29 January).

RAMPTON, V. (1971) Late Quaternary vegetational and climatic history of the Snag-Klutlan area, southwestern Yukon Territory, Canada, *Geol. Soc. Amer. Bull.*, **82**, 959–78. Rochester, N.Y.

ROBIN, G. de Q. (1955) Ice movement and temperature distribution in glaciers and ice sheets, *J. Glaciol.*, **2** (18), 523–32. Cambridge.

ROBIN, G. de Q. (1962) The ice of the Antarctic, *Sci. Amer.*, **207** (September), 132–46. New York.

SAARNISTO, M. (1974) The deglaciation history of the Lake Superior region and its climatic implications, *Quaternary Res.*, **4**, 316–39. Seattle.

SANCETTA, C., IMBRIE, J. and KIPP, N. G. (1973) Climatic record of the past 130 000 years in North Atlantic deep-sea core V23-82: correlation with the terrestrial record, *Quaternary Res.*, **3**, 110–16. Seattle.

SCHNEIDER, J. L. (1967) Évolution du dernier lacustre et peuplements préhistoriques aux pays bas du Tchad, *Assoc. Sénégalaise pour l'etude du Quaternaire de l'Ouest Africain, Bull. de Liaison*, 14–15. Dakar (June).

SCHOFIELD, J. C. (1964) Postglacial sea levels and isostatic uplift, *N.Z. J. Geol. & Geophy.*, **7**, 359–70. Wellington, N.Z.

SCHWARZBACH, M. (1961) *Das Klima der Vorzeit*. Stuttgart (Enke). 275 pp.

SCHYTT, V., HOPPE, G., BLAKE, W. and GROSSWALD, M. G. (1968) The extent of the Würm glaciation in the European Arctic. *Publication No. 79, Internat. Ass. of Sci. Hydrol., I.U.G.G., General Assembly Bern 1967, Comm. Snow and Ice*, pp. 207–216. Gentbrugged, Belgium. (Also available as *Meddelanden*, Nr. *A.18:* Stockholm (Naturgeografiska Inst. vid Stockholms Universitet).)

SHACKLETON, N. J. (1969) The last interglacial in the marine and terrestrial records, *Proc. Roy. Soc.*, **B, 174**, 135–54. London.

SHACKLETON, N. J. and OPDYKE, N. D. (1973) Oxygen isotope and palaeomagnetic stratigraphy of equatorial Pacific core V28-238: oxygen isotope temperatures and ice volumes on a 10^5-year and 10^6-year scale, *Quaternary Res.*, **3**, 39–55. Seattle.

SHACKLETON, N. J. and OPDYKE, N. D. (1976) Oxygen isotope and palaeomagnetic stratigraphy of Pacific core V28-239, Late Pliocene to Latest Pleistocene, *Geol. Soc. Amer. Mem.* 145 (ed. J. D. HAYS and R. M. CLINE), pp. 449–464. Boulder. Colorado.

SHACKLETON, N. J. and TURNER, C. (1967) Correlation between marine and terrestrial Pleistocene successions, *Nature*, **216**, 1079–82. London (16 December).

SHAW, D. M. and DONN, W. L. (1971) A thermodynamic study of Arctic palaeoclimatology, *Quaternary Res.*, **1** (2), 175–87. Seattle.

SHEPARD, F. P. (1963) Thirty-five thousand years of sea level, *Essays in Marine Geology*. Los Angeles (Univ. of California Press).

SHOTTON, F. W. (1962) The physical background of Britain in the Pleistocene, *Adv. Sci.*, **19**, 193–206. London (Brit. Assn for the Adv. of Sci.).

SHOTTON, F. W. (1967) Age of the Irish Sea glaciation of the Midlands, *Nature*, **215**, 1366. London (23 September).

SIMPSON, B. B. (1975) Glacial climates in the eastern tropical South Pacific, *Nature*, **253**, 34–5. London (3 January).

SIMPSON, G. C. (1934) World climate during the Quaternary period, *Quart. J. Roy. Met. Soc.*, **59**, 425–71. London.

SIMPSON, G. C. (1957) World temperature during the Pleistocene, *Quart. J. Roy. Met. Soc.*, **85**, 332–49. London.

SIMPSON, I. M. and WEST, R. G. (1958) On the stratigraphy and palaeobotany of a late Pleistocene deposit at Chelford, Cheshire, *New Phytol.*, **57**, 239–50. Oxford and Edinburgh (Blackwell).

SISSONS, J. B. (1974) Glacial readvances in Scotland, pp. 5–15 in *Problems of the Deglaciation of Scotland* (ed. C. J. CASSELDINE and W. A. MITCHELL). STAG, Special Publication No. 1. St Andrews (Univ. Dept Geogr.).

SPARKS, B. W. and WEST, R. G. (1972) *The Ice Age in Britain*. London (Methuen). 302 pp.

VALERY, N. (1972) Water mining to make the desert bloom, *New Scientist*, **56** (819), 322–4. London (9 November).

VEEH, H. H. and CHAPPELL, J. (1970) Astronomical theory of climatic change: support from New Guinea, *Science*, **167**, 862–5. New York (6 February).

VERNEKAR, A. D. (1968) *Research on the Theory of Climate, Vol. 2: Long-Period Global Variations of Incoming Solar Radiation*. Hartford, Conn. (Travelers Research Center, Final Report of the U.S. Department of Commerce, Environmental Science Services Administration, Contract No. E22-137-67(N).)

VERNEKAR, A. D. (1972) Long-period global variations of incoming solar radiation, *Met. Monogs*, **12**, (34). Boston, Mass. (Amer. Met. Soc.).

WEBB, T. and BRYSON, R. A. (1972) Late- and postglacial climatic change in the northern Midwest USA: quantitative estimates derived from fossil pollen spectra by multivariate statistical analysis, *Quaternary Res.*, **2**, 70–115. Seattle.

WEBSTER, P. J. and STRETEN, N. A. (1972) *Aspects of Late Quaternary Climate in Tropical Australasia*. Unpublished internal scientific report of the Commonwealth Met. Res. Centre, Melbourne.

WEERTMAN, J. (1964) Rate of growth or shrinkage of non-equilibrium ice sheets, *U.S. Army Cold Regions Research and Engineering Laboratory (CRREL Report)*, **145**. Hanover, New Hampshire.

WEST, R. G. (1968) *Pleistocene Geology and Biology*. London (Longmans). 377 pp.

WEST, R. G., DICKSON, C. A., CATT, J. A., WEIR, A. H. and SPARKS, B. W. (1974) Late Pleistocene deposits at Wretton, Norfolk, II: Devensian deposits, *Phil. Trans. Roy. Soc.*, **B, 267** (887), 337–420. London.

WEXLER, H. (1956) Variations in insolation, atmospheric circulation and climate, *Tellus*, **8**, 480–94. Stockholm.

WEYL, P. K. (1968) The role of the oceans in climatic change: a theory of the ice ages, *Met. Monogs*, **8** (30), 37–62. Boston, Mass. (Amer. Met. Soc.).

WILLIAMS, J., BARRY, R. G. and WASHINGTON, W. M. (1974) Simulation of the atmospheric circulation using the NCAR global circulation model with ice-age boundary conditions, *J. Appl. Met.*, **13** (3), 305–17. Boston, Mass.

WILSON, A. T. (1964) Origin of ice ages: an ice-shelf theory for Pleistocene glaciation, *Nature*, **201**, 147–9. London (11 January).

WOLDSTEDT, P. (1958) *Das Eiszeitalter: Grundlinien einer Geologie des Quartärs*, Bd. **II**. Stuttgart (Enke). 438 pp.

WOLDSTEDT, P. (1965) *Das Eiszeitalter: Grundlinien einer Geologie des Quartärs*, Bd. **III**. Stuttgart (Enke). 328 pp.

WRIGHT, H. E. (1961) Late Pleistocene climate of Europe: a review, *Geol. Soc. Amer. Bull.*, **72**, 933–984. Rochester, N.Y.

WRIGHT, H. E., BENT, A. M., SPROSS HANSEN, B. and MAHER, L. J. (1973) Present and past vegetation of the Chuska mountains, northwestern New Mexico, *Geol. Soc. Amer. Bull.*, **84**, 1155–80. Rochester, N.Y.

WRIGHT, P. B. and LAMB, H. H. (1974) A second approximation to the circulation patterns prevailing at the time of the last glacial maximum, pp. 104–7 in *Mapping the Atmospheric and Oceanic Circulations and Other Climatic Parameters at the Time of the Last Glacial Maximum about 17 000 Years Ago. Research Publication CRU RP2.* Norwich (Univ. of East Anglia, Climatic Research Unit).

ZAGWIJN, W. H. (1961) Vegetation, climate and radiocarbon datings in the Late Pleistocene of the Netherlands, *Mededelingen van de Geologische Stichting*, **14**, 15–45.

ZEUNER, F. E. (1958) *Dating the Past.* London (Methuen, 4th edn). 516 pp.

VAN ZINDEREN BAKKER, E. (1969) Intimations of Quaternary ecology in Africa, *Acta Bot. Neerl.*, **18** (1), 230–9. Amsterdam.

Chapter 16 Postglacial times

ADAM, D. P. (1967) Late Pleistocene and Recent palynology in the Central Sierra Nevada, California, pp. 275–301 in *Quaternary Palaeoecology* (ed. E. J. CUSHING and H. E. WRIGHT). New Haven and London (Yale Univ. Press).

ANDEL, T. H. VAN, ROSS HEATH, G., MOORE, T. C. and McGEARY, D. F. R. (1967) Late Quaternary history, climate and oceanography of the Timor Sea, northwestern Australia, *Amer. J. Sci.*, **265**, 737–58. New Haven, Conn.

ANDERSON, SV. TH. (1973) The differential pollen productivity of trees and its significance for the interpretation of pollen diagrams, pp. 108–14 in *Quaternary Plant Ecology* (ed. H. J. B. BIRKS and R. G. WEST). Oxford (Blackwell Sci. Pubns).

ANDERSSON, G. (1902) Hasseln i Sverige fordom och nu, *Sveriges Geol. Undersökning, Afhand.*, serie C2, No. 3. Stockholm.

ANDREWS, J. T. and BARRY, R. G. (1972) *Present and Paleoclimatic Influences on the Glacierization and Deglacierization of Cumberland Peninsula, Baffin Island, N.W.T., Canada.* Occasional Paper No. 2. Boulder, Colorado (Inst. Arct. and Alpine Res.).

ANDREWS, J. T. and IVES, J. D. (1972) Late- and postglacial events (< 10 000 B.P.) in the eastern Canadian Arctic with particular reference to the Cockburn moraines and break-up of the Laurentide ice sheet, pp. 149–74 in *Climatic Changes in Arctic Areas During the Last Ten Thousand Years* (ed. Y. VASARI, H. HYVÄRINEN and S. HICKS). Series A, Scientiae Rerum Nat. No. 3, Geologica No. 1. Oulu (Univ. Press).

ASHBEE, P. (1963) The Wilsford shaft, *Antiquity*, **37**, 116–20. Cambridge (Heffer).

ASHBEE, P. (1966) The dating of the Wilsford shaft, *Antiquity*, **40**, 227–8. Cambridge.

AUER, V. (1960) The Quaternary history of Fuego-Patagonia, *Proc. Roy. Soc.*, **B, 152**, 507–16 (and discussion, 533–8). London.

AVERDIECK, F. -R. and DÖBLING, H. (1959) Das Spätglazial am Niederrhein, *Fortschr. Geol. Rheinld. Westfalen*, **4**, 341–62. Krefeld.

BAKKER, E. M. VAN ZINDEREN (1973) The glaciation of Marion Island, *Palaeoecol. of Africa and Surrounding Islands and Antarctica*, **8**, 161–78. Cape Town (Balkema).

BEUG, H.-J. (1964) Untersuchungen zur spät- und postglazialen Vegetationsgeschichte im Gardaseegebiet unter besonderer Berücksichtigung der mediterranen Arten, *Flora*, **154** (Neue Folge), 401–44. Jena.

BEUG, H. -J. (1967) Contributions to the postglacial vegetational history of northern Turkey, pp. 349–56 in *Quaternary Palaeoecology* (ed. H. J. CUSHING and H. E. WRIGHT). New Haven and London (Yale Univ. Press).

BIRKS, H. H. (1972) Studies in the vegetational history of Scotland, III: a radiocarbon-dated pollen diagram from Loch Maree, *New Phytol.*, **71**, 731–54. Oxford and Edinburgh (Blackwell).

BIRKS, H. H. (1975) Studies in the vegetational history of Scotland, IV: Pine stumps in Scottish blanket peats, *Phil. Trans. Roy. Soc.*, **B, 270** (905), 181–226. London.

BIRKS, H. J. B., DEACON, J. and PEGLAR, S. (1975) Pollen maps for the British Isles 5000 years ago, *Proc. Roy. Soc.*, **B, 189**, 87–105. London.

BLAKE, I. (1969) Climate, survival and the second class societies in Palestine before 3000 B.C., *Adv. Sci.*, **25**, 409–21. London.

BLAKE, J. W. (1970) Studies of glacial history in Arctic Canada, I: Pumice, radiocarbon dates, and differential, postglacial uplift in the eastern Queen Elizabeth Islands, *Canad. J. Earth Sci.*, **7**, 634–64. Ottawa.

BRAY, J. R. (1970) Temporal patterning of post-Pleistocene glaciation, *Nature*, **228** (5269), 353. London (24 October).

BRAY, J. R. (1971) Solar-climate relationships in the post-Pleistocene, *Science*, **171**, 1242–3. New York (26 March).

BRAY, J. R. (1972) Cyclic temperature oscillations from 0–20 300 years B.P., *Nature*, **237** (5353). 277–9, London (2 June).

BROOKS, C. E. P. (1949) *Climate Through the Ages*. London (Benn), 2nd edn. 395 pp.

BRYSON, R. A. (1966) Airmasses, streamlines and the boreal forest, *Geogr. Bull.*, **8**, 228–69. Ottawa.

BRYSON, R. A., BAERREIS, D. A. and WENDLAND, W. M. (1970) The character of Late-Glacial and postglacial climatic changes, pp. 53–74 in *Pleistocene and Recent Environments of the Central Great Plains*. Dept Geol. Special Pbn 3. Lawrence, Kansas (Univ. Kansas Press).

BRYSON, R. A., LAMB, H. H. and DONLEY, D. L. (1974) Drought and the decline of Mycenae, *Antiquity*, **48**, 46–50. Cambridge (Heffer).

BRYSON, R. A. and WENDLAND, W. M. (1967) Tentative climatic patterns for some Late-Glacial and postglacial episodes in central North America. *Tech. Rep.* 34, Nonr. 1202 (07). Madison (Univ. Wisconsin, Dept Meteorol.). 32 pp.

BRYSON, R. A., WENDLAND, W. A., IVES, J. D. and ANDREWS, J. T. (1969) Radiocarbon isochrones on the disintegration of the Laurentide ice sheet, *Arctic and Alpine Res.*, **1** (1), 1–14. Boulder, Colorado.

BURZER, K. W. (1958) Studien zum vor- und frühgeschichtlichen Landschaftswandel in Sahara, *Afhandl. math.-naturw. Klasse, No. 1.* Mainz (Akad. der Wiss, und Litt). 49 pp.

BUTZER, K. W. and HANSEN, C. L. (1968) *Desert and Rain in Nubia: Geomorphology and Prehistoric Environments at the Aswan Reservoir.* Madison and London (Univ. Wisconsin Press). 562 pp.

CARPENTER, R. (1966) *Discontinuity in Greek Civilization.* Cambridge (Univ. Press). 80 pp.

CASSELDINE, C. J. and MITCHELL, W. A. (eds) (1974) *Problems of the Deglaciation of Scotland.* STAG, Special Publication No. 1. St Andrews (Univ. Dept Geogr.).

CHAPPELL, J. E. (1970) Climatic change reconsidered: a new look at 'The Pulse of Asia'. *Geogr. Rev.*, **60**, 347–73. New York.

CHAPPELL, J. E. (1971) Climatic pulsations in inner Asia and correlations between sunspots and weather. *Palaeogeogr., Palaeoclim., Palaeoecol.*, **10**, 177–97. Amsterdam (Elsevier).

CHILDE, G. (1954) *What Happened in History.* Harmondsworth, Middlesex (Penguin Books, rev. edn). 288 pp.

CHILDE, G. (1958) *The Prehistory of European Society.* Harmondsworth, Middlesex (Penguin Books). 185 pp.

CHU KO-CHEN (1973) A preliminary study of the climatic fluctuations during the last 5000 years in China, *Scientia Sinica*, **16** (2), 226–56. Peking.

CHURCHILL, D. M. (1968) The distribution and prehistory of *Eucalyptus diversicolor, E. marginata* and *E. calophylla* in relation to rainfall. *Australian J. Bot.*, **16**, 125–51. Melbourne.

CLARK, G. and PIGGOTT, S. (1965) *Prehistoric Societies.* Harmondsworth, Middlesex (Penguin Books). 352 pp.

CURRAY, J. R. (1961) Late Quaternary sea level: a discussion, *Geol. Soc. Amer. Bull.*, **72**, 1707–12. Rochester, N.Y.

DE GEER, E. H. (1954) Skandinaviens geokronologi, *Geol. Fören. Förhandl.*, (*GFF*), **76**, 299–329. Stockholm.

DEGENS, E. T. and HECKY, R. E. (1974) Palaeoclimatic reconstruction of Late Pleistocene and Holocene based on biogenic sediments from the Black Sea and a tropical African lake, pp. 13–24 in *Les méthodes quantitatives d'étude des variations du climat au cours du Pléistocène.* (ed. J. LABEYRIE). *Colloques internat. du C.N.R.S., No. 219.* Paris (C.N.R.S.). 317 pp.

FALCONER, G., IVES, J. D., LØKEN, O. H. and ANDREWS, J. T. (1965) Major end-moraines in eastern and central Arctic Canada, *Geogr. Bull.*, **7** (2), 137–53. Ottawa.

FIRBAS, F. (1949) *Spät- und nacheiszeitliche Waldgeschichte Mitteleuropas nördlich der Alpen*. Jena (Fischer). 480 pp.

FLINT, R. F. (1971) *Glacial and Quaternary Geology*. New York and London (Wiley). 892 pp.

FREDSKILD, B. (1967) Palaeobotanical investigations at Sermermiut, Jakobshavn, west Greenland, *Medd. om Grønland*, **178** (4). Copenhagen (Reitzel). 54 pp.

FREDSKILD, B. (1969) A postglacial standard pollen diagram from Peary Land, north Greenland (1), *Pollen et spores*, **11** (3), 573–83. Paris.

FREDSKILD, B. (1972) Palynological evidence for Holocene climatic changes in Greenland, pp. 277–306 in *Climatic Changes in Arctic Areas During the Last Ten Thousand Years* (ed. Y. VASARI, H. HYVÄRINEN and S. HICKS). *Acta Univ, Oulu*, **A, 3**, *Geologica* 1. Oulu. 511 pp.

FREDSKILD, B. (1973a) Studies in the vegetational history of Greenland, *Medd. om* **198** (4). Copenhagen (Reitzel). 245 pp.

FREDSKILD, B. (1973b) *Report about the Holocene Investigations in Greenland*. Un-Grønland, **198** (4). Copenhagen (Reitzel). 245 pp.   for the Holocene.

FRENZEL, B. (1966) Climatic change in the Atlantic/Sub-Boreal transition, pp. 99–123 in *World Climate 8000–0 B.C.: Proc. Internat. Conf.* (ed. J. S. SAWYER). London (Roy. Met. Soc.). 229 pp.

FRENZEL, B. (1967) *Die Klimaschwankungen des Eiszeitalters*. Braunschweig (Vieweg – *Die Wissenschaft* **129**). 296 pp.

FROMM, E. (1953) *Atlas över Sverige*. Stockholm.

FÜRST, O. (1963) Vergleichende Untersuchungen über räumliche und zeitliche Unterschiede interannueller Jahrringbreitenschwankungen und ihre klimatologische Auswertung, *Flora*, **153**, 469–508. Jena.

GAMS, H. (1937) Aus der Geschichte der Alpenwälder, *Zeitschrift des deutschen und österreichischen Alpenvereins*, **68**, 157–70. Stuttgart.

GAMS, and NORDHAGEN, R. (1923) Postglaziale Klimaänderungen und Erdkrustenbewegungen in Mitteleuropa, *Forschungen der geogr. Gesell. f. Landeskunde*, Heft 25. Munich.

GEYH, M. A. and JÄKEL, D. (1974) Late glacial and Holocene climatic history of the Sahara desert derived from a statistical assay of ^{14}C dates, *Palaeogeogr., Palaeoclim., Palaeoecol.*, **15**, 205–8. Amsterdam (Elsevier).

GEYH, M. A. and ROHDE, P. (1972) Weichselian chronostratigraphy, C^{14} dating and statistics, *Proc. 24th International Geological Congress* (*Montreal 1972*), Section 12, pp. 27–36.

GJESSING, J. (1960) *Isavsmeltningstidens drenering, dens forløp og formdannende virkning*. Oslo (Universitetsforlaget). 492 pp.

GJESSING, J. (1966) Deglaciation of south-east and east-central south Norway, *Norsk geografisk tidsskrift*, **20** (5), 133–49. Oslo.

GODWIN, H. (1956) *The History of the British Flora*. Cambridge (Univ. Press). 384 pp.

GODWIN, H. and TANSLEY, E. G. (1941) Prehistoric charcoals as evidence of former vegetation, soil and climate, *J. Ecol.*, **29**, 117–26. London.

GODWIN, H., WALKER, D. and WILLIS, E. H. (1957) Radiocarbon dating and postglacial vegetational history: Scaleby Moss, *Proc. Roy. Soc.*, **B, 147**, 352–66. London.

GODWIN, H. and WILLIS, E. H. (1959) Radiocarbon dating of prehistoric wooden trackways, *Nature*, **184**, 490–1. London.

GRIFFEY, N. J. (1976) Stratigraphical evidence for an early Neoglacial glacier maximum of Steikvassbreen, Okstindan, North Norway. *Norsk Geol. Tidsskrift*, **56**, 187–94. Oslo. (GRIFFEY adds information about south Norway and the first millennium A.D. in a personal communication, 26 January 1977).

GROEBEN, F. (1694) *Guineische Reise*. Leipzig (Insel Verlag).

GROVE, A. T. (1974) *The Late Quaternary in Africa*. Unpublished MS. Cambridge, Univ. Dept. Geogr. 45 pp.

HAGEN, A. (1961) *Steinalder, bronsealder*. Oslo (Universitetets oldsaksamling). 47 pp.

HEUSSER, C. J. (1966a) Polar hemispheric correlation: palynological evidence from Chile and the Pacific north-west of America, pp. 124–41 in *World Climate 8000–0 B.C.: Proc. Internat. Conf.* (ed. J. S. SAWYER). London (Roy. Met. Soc.). 229 pp.

HEUSSER, C. J. (1966b) Late Pleistocene pollen diagrams from the province of Llanquihue, southern Chile, *Proc. Amer. Phil. Soc.*, **110**, 269–305. Philadelphia.

HICKS, S. (1975) Variations in pollen frequency in a bog at Kangerjoki, northeast Finland during the Flandrian, *Commentationes Biol.*, **80**, 4–28. Helsinki (Soc. Sci. Fenn.).

HILLEFORS, Å. (1969) Västsveriges glaciala historia och morfologi, *Meddelanden från Lunds Universitets Geografiska Institutionen*, Avhandl. **60**. Lund.

HOPPE, G. and FRIES, M. (1965) Submarine peat in the Shetland islands, *Geografiska Annaler*, **47, A**, 195–203. Stockholm.

HULTEN, E. (1971) *Atlas över växternas utbredning i norden*. Stockholm (Generalstabens Litografiska Anstalt). 531 pp.

IVERSEN, J. (1973) *The Development of Denmark's Nature Since the Last Glacial*. Danmarks Geol. Undersøgelse, V. Række, No. 7C. Copenhagen (Reitzel). 126 pp.

IVES, J. D. (1960) The deglaciation of Labrador-Ungava: an outline, *Cahiers de géographie de Quebec*, **4** (8), 323–43. Quebec.

KING, J. W. (1974) Weather and the Earth's magnetic field, *Nature*, **247**, 131–4. London (18 January).

KIRK, W. (1975) The role of India in the diffusion of early cultures, *Geogr. J.*, **141** (1), 19–34. London (Roy. Geogr. Soc.).

KLEIN, H. (1914) *Das Klima Palästinas auf Grund der alten hebräischen Quellen*. Inaugural Dissertation, Universität Würzburg. Halle (Ehrhardt Karras GmbH.). 77 pp.

KNOX, J. C. (1973) Alluvial response to climatic fluctuations since 12 000 radiocarbon years B.P., *International Union for Quaternary Research (INQUA), Abstracts for IX Congress*, 189–90. Christchurch, New Zealand.

KNOX, J. C. (1976) Concept of the graded stream. Chapter 10, pp. 169–98, in *Theories of landform development: Proc. Sixth Annual Geomorphol. Symposia Series, Binghamton, New York 1975.* New York (State Univ.). (W. N. MELHORN and R. C. FLEMAL eds.).

KRINSLEY, D. R. (1970) *A Geomorphological and Palaeoclimatological Study of the Playas of Iran.* Final Sci. Rep., Contract No. PRO CP 70–800 Washington (U.S. Geol. Survey for Air Force Cambridge Research Laboratories), 2 vols.

KUTZBACH, J. E. and BRYSON, R. A. (1974) *Variance Spectrum of Holocene Climatic Fluctuations in the North Atlantic Sector.* Unpublished memo. Madison, Wisconsin (Dept. Meteorol., Univ. Wisconsin). 15 pp.

LAMARCHE, V. C. (1973) Holocene climatic variations inferred from treeline fluctuations in the White Mountains California, *Quaternary Res.*, **3**, 632–60. Seattle.

LAMB, H. H. (1964) Trees and climatic history in Scotland, *Quart. J. Roy. Met. Soc.*, **90**, 382–94. London.

LAMB, H. H. (1965) Trees and climatic history in Scotland: discussion, *Quart. J. Roy. Met. Soc.*, **91**, 542–50. London.

LAMB, H. H. (1974) Climate, vegetation and forest limits in early civilized times, *Phil. Trans. Roy. Soc.*, **A, 276**, 195–230. London.

LAMB, H. H., LEWIS, R. P. W. and WOODROFFE, A. (1966) Atmospheric circulation and the main climatic variables between 8000 and 0 B.C.: meteorological evidence, pp. 174–217 in *World Climate 8000–0 B.C.: Proc. Internat. Conf.* (ed. J. S. SAWYER). London (Roy. Met. Soc.). 229 pp.

LANGWAY, C. C. (1967) *Stratigraphic Analysis of a Deep Ice Core from Greenland.* Res. Rep. 77. Hanover, New Hampshire (U.S. Army Cold Regions Research and Engineering Laboratory).

LINK, F. (1958) Kometen, Sonnentätigkeit und Klimaschwankungen, *Die Sterne*, **34**, 129–40. Leipzig (A. Barth).

MANLEY, G. (1955) A climatological survey of the retreat of the Laurentide ice sheet, *Amer. J. Sci.*, **253**, 256–73. New Haven, Conn.

MARKGRAF, V. (1970) Palaeohistory of the spruce in Switzerland, *Nature*, **228**, 249–51. London (17 October).

MARKGRAF, V. (1972) Die Ausbreitungsgeschichte der Fichte (*Picea abies* H. Karst) in der Schweiz, *Ber. der deutscher Bot. Gesellschaft*, **85**, 165–72. Berlin.

MARKGRAF, V. (1974) Palaeoclimatic evidence derived from timberline fluctuations, pp. 67–83 in *Les Méthodes quantitatives d'étude des variations du climat au cours du Pléistocène* (ed. J. LABEYRIE). Colloques internat. du C.N.R.S. No. 219. Paris (C.N.R.S.). 317 pp.

MATTHES, F. W. (1939) Report of Committee on Glaciers, *Trans, Amer. Geophys. Union*, **I**, 518–20. Washington.

MITCHELL, G. F. and STEPHENS, N. (1974) Is there evidence for a Holocene sea level higher than that of today on the coasts of Ireland?, pp. 115–25 in *Les méthodes quantitatives d'étude des variations du climat au cours du Pléistocène* (ed. J. LABEYRIE). *Colloques internat, du C.N.R.S.* Paris (C.N.R.S.). 317 pp.

MOAR, N. T. (1965) Contribution to the discussion in LAMB (1965).

MOAR, N. T. (1969a) Two pollen diagrams from the mainland, Orkney Islands, *New Phytol.*, **68**, 201–8. Oxford and Edinburgh (Blackwell).

MOAR, N. T. (1969b) A radiocarbon-dated pollen diagram from northwest Scotland, *New Phytol.*, **68**, 209–14.

MONOD, T. (1963) The late Tertiary and Pleistocene in the Sahara and adjacent southerly regions, pp. 117–229 in *African Ecology and Human Evolution* (ed. F. C. HOWELL and F. BOURLIÈRE). Chicago (Aldine).

NEJSTADT, M. I. (1957) *History of the Forests of the Soviet Union in the Holocene.* Moscow (Izdat. Akad. Nauk). (In Russian.)

NICHOLS, H. (1967a) Central Canadian palynology and its relevance to northwestern Europe in the Late Quaternary period, *Review of Palaeobot. and Palynol.*, **3**, 231–43. Amsterdam (Elsevier).

NICHOLS, H. (1967b) The postglacial history of vegetation and climate at Ennadai Lake, Keewatin and Lynn Lake, Manitoba (Canada), *Eiszeitalter und Gegenwart*, **18**, 176–97. Öhringen/Württ.

NICHOLS, H. (1967c) Vegetational change, shoreline displacement and the human factor in the Late Quaternary history of southwest Scotland, *Trans. Roy. Soc. Edin.*, **67**, 145–87.

NICHOLS, H. (1970) Late Quaternary pollen diagrams from the Canadian Arctic barren grounds at Pelly Lake, Keewatin, N.W.T., *Arctic & Alpine Res.*, **2**, 43–61. Boulder, Colorado.

NORDHAGEN, R. (1933) *De senkvartære klimavekslinger i Nordeuropa og deres betydning for kulturforskningen.* Oslo (Aschehoug). 246 pp.

OLAUSSON, E. and OLSSON, I. U. (1969) Varve stratigraphy in the Gulf of Aden, *Palaeogeogr., Palaeoclim., Palaeoecol.*, **6**, 87–103. Amsterdam (Elsevier).

d'OLIER (1972) Subsidence and sea-level rise in the Thames estuary, *Phil. Trans. Roy. Soc.*, **A, 272** (1221), 121–30. London.

PEJML, K. (1962) A contribution to the historical climatology of Morocco and Mauritania, *Studia geoph. et geod.*, **6**, 257–79. Prague.

PENMAN, H. L. (1956) Estimating evaporation, *Trans. Amer. Geophys. Union*, **37**, 43–6. Washington.

PENNINGTON, W. (1970) Vegetation history in the northwest of England: a regional synthesis, pp. 41–79 in *Studies in the Vegetational History of the British Isles: Essays in Honour of Harry Godwin* (ed. D. WALKER and R. G. WEST) Cambridge (Univ. Press). 266 pp.

PENNINGTON, W. (1973) Absolute pollen frequencies in the sediments of lakes of different morphometry, pp. 79–104 in *Quaternary Plant Ecology* (ed. H. J. B. BIRKS and R. G. WEST). Oxford (Blackwell Sci. Pubns).

PENNINGTON, W., HAWORTH, E. Y., BONNY, A. P. and LISHMAN, J. P. (1972) Lake sediments in northern Scotland, *Phil. Trans. Roy. Soc.*, **B, 264**, 191–294. London.

PISIAS, N. G., DAUPHIN, J. P. and SANCETTA, C. (1973) Spectral analysis of Late Pleistocene-Holocene sediments, *Quaternary Res.* 3, 3–9. Seattle.

RAMASWAMY, C. (1968) Monsoon over the Indus valley during the Harappan period, *Nature*, **217**, 628–9. London (17 February).

REYNAUD, C. (1976) Palaeoecological significance of *Hypophaë rhamnoides*, with an example of the protocratic vegetational stage in NE Fennoscandia. *Boreas*, **5**, 9–24. Oslo.

SAURAMO, M. (1939) The mode of land upheaval in Fennoscandia during Late Quaternary time, *Comptes rendus*, **13**. Helsinki (Geol. Soc. of Finland). 26 pp.

SCHIRMER, W. (1972) *Zur Hang- und Flusstalentwicklung im Holozän Süddeutschlands.* Unpublished paper given to the INQUA Commission for Study of the Holocene at its meeting in Poland, 10–20 September 1972.

SCHOSTAKOWITSCH, W. B. (1934) Bodenablagerungen der Seen und periodische Schwankungen der Naturerscheinungen. (Reprinted in English by C. E. P. BROOKS in *Met. Mag.*, **70**, 134–8 (1935) from *Memoirs of the Leningrad Hydr. Inst.*)

SETH, S. K. (1963) A review of evidence concerning changes of climate in India during the protohistorical and historical periods, pp. 443–52 in *Changes of Climate: Proc. UNESCO/WMO Rome 1961 Symp.* Paris (UNESCO, Arid Zone Research Series XX).

SINGH, G. (1963) A preliminary survey of the postglacial vegetational history of the Kashmir Valley, *Palaeobotanist*, **12** (1), 73–108. Lucknow.

SINGH, G. (1971) The Indus Valley culture, *Archaeol. & Phys. Anthropol. in Oceania*, **6** (2), 177–89. Sydney.

SINGH, G., JOSHI, R. D. and SINGH, A. B. (1972) Stratigraphic and radiocarbon evidence for the age and development of three salt lake deposits in Rajasthan, India, *Quaternary Res.*, **2** (4), 496–505. Seattle.

SINGH, G., JOSHI, R. D., CHOPRA, S. K. and SINGH, A. B. (1974) Late Quaternary history of vegetation and climate of the Rajasthan desert, India, *Phil. Trans. Roy. Soc.*, **B, 267** (889), 467–501. London.

SUGGATE, R. P. (1965) Late Pleistocene geology of the northern part of South Island, New Zealand, *N.Z. Geol. Survey Bull.*, N.S., **77**. Wellington, N.Z.

SUZUKI, H. (1975) World precipitation, present and hypsithermal, *Bull. Dept. Geogr. Univ. Tokyo*, No. 7 (March), 4–17.

TAIRA, K. (1975) Temperature variation of the 'Kuroshio' and crustal movements in eastern and southeastern Asia 7000 years B.P., *Palaeogeogr., Palaeoclim., Palaeoecol.*, **17**, 333–8. Amsterdam (Elsevier).

TALLANTIRE, P. A. (1972a) The regional spread of spruce (*Picea abies* (L.) Karst.) within Fennoscandia: a reassessment, *Norwegian J. Bot.*, **19**, 1–16. Oslo (Universitetsforlaget).

TALLANTIRE, P. A. (1972b) Spread of spruce (*Picea abies* (L.) Karst) in Fennoscandia and possible climatic implications, *Nature*, **236**, 64–5. London (10 March).

TALLANTIRE, P. A. (1973) Apropos the postglacial spread of spruce in Switzerland, *Grana*, **13**, 79–84. Stockholm.

TAUBER, H. (1956) Differential pollen dispersion and the interpretation of pollen diagrams: with a contribution to the interpretation of the elm fall, *Danmarks Geol. Undersøgelse*, II Raekke, Nr. 89. Copenhagen (Reitzel). 70 pp.

TEN BRINK, N. W. and WEIDICK, A. (1974) Greenland ice sheet history since the last glaciation, *Quaternary Res.*, **4**, 429–40. Seattle.

TRAILL, W. (1868) On submarine forests and other remains of indigenous woods in Orkney, *Trans. Bot. Soc. Edin.*, **9**, 146–54.

TROLL, C. (1972) Absolute Chronologie der Nacheiszeit (Holozän) mit Hilfe der Dendrochronologie, pp. 143–5 in *Jahrbuch der Akad. der Wiss. und Lit.*, (*Kommission für Erdwiss. Forschung*). Mainz.

TSUKADA, M. (1967) Pollen succession, absolute pollen frequency and recurrence surfaces in central Japan, *Amer. J. Bot.*, **54**, 821–31. Lancaster, Pa.

TURC, L. (1958) Le bilan d'eau des sols: relations entre les précipitations, l'évaporation et l'écoulement, *Soc. hydrotech. de France: Compte rendu des troisièmes journées de l'hydraulique*, Algiers, 36–44. Paris.

TURNER, J. (1965) A contribution to the history of forest clearance, *Proc. Roy. Soc.*, **B, 161**, 343–54. London.

VISHNU-MITTRE (1966) Some aspects concerning pollen-analytical investigations in the Kashmir valley, *Palaeobotanist*, **15**, 157–75. Lucknow.

VISHNU-MITTRE and SHARMA, B. D. (1966) Studies of postglacial vegetational history from the Kashmir valley, I: Haigan Lake. *Palaeobotanist*, **15**, 185–212. Lucknow.

WADIA, D. N. (1960) The postglacial desiccation of central Asia: evolution of the arid zone of Asia. *Nat. Inst. Sci. India Monograph*. Delhi.

WATERBOLK, H. T. (1968) Food production in prehistoric Europe, *Science*, **162**, 1093–1102. New York (6 December).

WEIDICK, A. (1972) Notes on Holocene glacial events in Greenland, pp. 177–204 in *Climatic Changes in Arctic Areas During the Last Ten Thousand Years*. (ed. Y. VASARI, H. HYVÄRINEN and S. HICKS). *Acta Univ. Oulu.*, **A, 3**, *Geologica*, 1. Oulu. 511 pp.

WENDLAND, W. M. and BRYSON, R. A. (1974) Dating climatic episodes of the Holocene, *Quaternary Res.*, **4**, 9–24. Seattle.

WOLDSTEDT, P. (1958) *Das Eiszeitalter*, II: *Europa, Vorderasien und Nordafrika im Eiszeitalter*. Stuttgart (Enke), 2 Auflage. 438 pp.

WOOD, E. S. (1963) *Collins Field Guide to Archaeology*. London (Collins). 384 pp.

WOODBURY, R. G. (1961) Climatic changes and prehistoric agriculture in the southwestern United States, *Ann. New York Acad. Sci.*, **95** (1), 705–9.

WRIGHT, H. E. (1968) History of the Prairie Peninsula, pp. 78–88 in *The Quaternary of Illinois: Special Publications of Univ. of Illinois Coll. Agric.*, **14**. 179 pp.

WRIGHT, H. E. (1971a) Retreat of the Laurentide ice sheet from 14 000 to 9000 years ago, *Quaternary Res.*, **1**, 316–330. Seattle.

WRIGHT, H. E. (1971b) Late Quaternary vegetational history of North America, pp. 425–64 in *The Late Cenozoic Glacial Ages: Proc. Yale Univ. Symp., December 1969* (ed. K. K. TUREKIAN). New Haven, Conn. (Yale Univ. Press). 606 pp.

ZEUNER, F. E. (1958) *Dating the Past.* London (Methuen), 4th edn. 516 pp.

Chapter 17   Climate in historical times

ABEL, W. (1943) *Die Wüstungen des ausgehenden Mittelalters.* Jena. 165 pp.

ALLISON, K. J. (1955) The lost villages of Norfolk, *Norfolk Archaeology*, **31**, 131. Norwich (Norfolk Archaeol. Soc.).

ALLISON, K. J., BERESFORD, M. W. and HURST, J. G. (1965) *The Deserted Villages of Oxfordshire.* Dept of English Local History Occas. Pap. 17. Leicester (Univ. Press).

ALLISON, K. J., BERESFORD, M. W. and HURST, J. G. (1966) *The Deserted Villages of Northamptonshire.* Dept of English Local History Occas. Pap. 18. Leicester (Univ. Press).

ARAGO, F. (1858) Sur l'état thermométrique du globe terrestre, *Œuvres de François Arago* (Barral), Vol. V. Paris and Leipzig.

AUER, V. (1960) The Quaternary history of Fuego-Patagonia, *Proc. Roy. Soc.*, **B**, **152**, 507–16, 533–8. London.

BACON, SIR FRANCIS (1653) *The Naturall and Experimentall History of Winds.* London.

BAHNSON, H. (1973) Spor af muldflugt i keltisk jernalder påvist i højmoseprofiler, *Danmarks Geol. Undersøgelse, Årbog 1972*, 7–12. Copenhagen.

BANG, N. E. (ed.) (1906, 1922) *Tabeller over Skibsfart og Varetransport gennem Øresund 1497–1660.* Copenhagen.

BENNEMA, J. (1954) Holocene movements of land and sea level in the coastal area of the Netherlands, *Geologie en Mijnbouw*, **16** (new Series), 254–64. Den Haag.

BERESFORD, M. (1951) The lost villages of medieval England, *Geogr. J.*, **117** (2), 129–49. London.

BERESFORD, M. (1954) *The Lost Villages of England.* London (Lutterworth). 445 pp.

BERGTHORSSON, P. (1962) Julihiimn á Islandi (July temperature in Iceland), *Veðrið*, **7**, 22–5. Reykjavik.

BLANTON, R. E. (1972) Prehispanic adaptation in the Ixtapalapa region, Mexico, *Science*, **175** (4028), 1317–26. Washington (24 March).

BOECE, H. (HECTOR BOETHIUS) (c. 1536) *The History and Chronicles of Scotland.* (Edition translated from the Latin into English by JOHN BELLENDEN. Edinburgh (Tait) 1821.)

BRITTON, C. E. (1937) A meteorological chronology to A.D. 1450, *Geophys. Mem.*, **70**. London (H.M.S.O. for Met. Office).

BROOKS, C. E. P. (1949) *Climate Through the Ages.* London (Benn), 2nd edn. 395 pp.

BRYSON, R. A., BAERREIS, D. A. and WENDLAND, W. M. (1970) The character of Late-Glacial and postglacial climatic changes, pp. 53–74 in *Pleistocene and Recent Environments of the Central Great Plains.* Lawrence (Univ. Kansas Press; Dept Geol. Special Pubn 3).

BRYSON, R. A., IRVING, W. and LARSEN, J. (1965) Radiocarbon and soil evidence of a former forest in the southern Canadian tundra, *Science*, **147** (3653), 46–8. New York (1 January).

BUCHINSKY, I. E. (1957) *The Past Climate of the Russian Plain.* Leningrad (Gidrometeoizdat), 2nd edn. (In Russian: English translation available in Meteorological Office Library.)

BULL, E. (1925) Klimaskifte og nedgang i Noreg i seinomillomalderen, *Syn og segn*, **31** (1), 12–19. Oslo.

CARPENTER, R. (1966) *Discontinuity in Greek Civilization.* Cambridge (Univ. Press). 80 pp.

CHAPPELL, J. E. (1971) Climatic pulsations in inner Asia and correlations between sunspots and weather, *Palaeogeogr., Palaeoclim., Palaeoecol.*, **10**, 177–97. Amsterdam (Elsevier).

CHRISTENSEN, A. E. (1938) Danmarks befolkning og bebyggelse i Middelalderen, *Nordisk Kultur*, **2**, 1–57. Copenhagen, Oslo, Stockholm.

CHU KO-CHEN (COCHING CHU) (1961) The pulsation of world climate during historical times, *New China Monthly*, **6**. Peking.

CHU KO-CHEN (COCHING CHU) (1973) A preliminary study of the climatic fluctuations during the last 5000 years in China, *Scientia Sinica*, **16** (2), 226–56. Peking.

CRANWELL, L. M. and VON POST, L. (1936) Post-Pleistocene pollen diagrams from the southern hemisphere, I: New Zealand, *Geogr. Annaler*, **18**, 308–47. Stockholm.

CROMERTIE, GEORGE EARL OF (1710) An account of the mosses of Scotland, *Phil. Trans. Roy. Soc.*, **27**, 296–301. London.

CROSBY, A. W. (1972) *The Columbian Exchange: Biological and Cultural Consequences of 1492.* Westport, Conn. (Greenwood Publishing Co. – Contributions in American Studies No. 2), 268 pp.

CROWN, A. D. (1973) Towards a reconstruction of the climate of Palestine 8000–0 B.C., *J. Near Eastern Studies*, 312–30.

DANSGAARD, W., JOHNSEN, S. J., REEH, N., GUNDESTRUP, N., CLAUSEN, H. B. and HAMMER, C. U. (1975) Climatic changes, Norsemen and modern man, *Nature*, **255**, 24–8. London (1 May).

DAYTON, J. (1975) The problem of climatic change in the Arabian peninsula, *Proc. Seminar for Arabian Studies*, **5**, 33–59. London (Inst. Archaeol.).

DELIBRIAS, G., LADURIE, M. and LADURIE, E. L. (1975) Le fossil forêt de Grindelwald: nouvelles datations, *Annales: économies, sociétés, civilisations*, No. 1, 137–47. Paris (Armand Colin).

EASTON, C. (1928) *Les hivers dans l'Europe occidentale*. Leyden (Brill). 208 pp.

FIRBAS, F. and LOSERT, H. (1949) Untersuchungen über die Entstehung der heutigen Waldstufen in den Sudeten, *Planta*, **36**, 478–506. Berlin.

FISHER, H. A. L. (1936) *A History of Europe*. London (Arnold). 1301 pp.

FLETCHER, J. M. (1975) Relation of abnormal earlywood in oaks to dendrochronology and climatology, *Nature*, **254**, 506–7. London (10 April).

FLOHN, H. (1949) Klima und Witterungsablauf in Zürich im 16. Jahrhundert, *Vierteljahresheft der Naturf. Gesell. Zürich*, **94**, 28–41.

FLOHN, H. (1950) Klimaschwankungen im Mittelalter und ihre historisch-geographische Bedeutung, *Berichte zur dt. Landeskunde*, **7**, 347–57. Remagen and Stuttgart.

FREDSKILD, B. (1967) Palaeobotanical investigations at Sermermiut, Jakobshavn, west Greenland, *Medd. om Grønland*, **178** (4). Copenhagen (Reitzel), 54 pp.

FRIEDMAN, I. and SMITH, G. I. (1972) Deuterium content of snow as an index to winter climate in the Sierra Nevada area, *Science*, **176**, 790–3. New York (19 May).

FRIIS, A. (1925) Øresundstoldregnskaberne, *Dansk Historisk Tiddsskrift*.

GAMS, H. (1937) Aus der Geschichte der Alpenwälder, *Zeit. des. dt. und österr. Alpenvereins*, **68**, 157–70. Stuttgart.

GEIKIE, A. (1901) *The Scenery of Scotland*. London (Macmillan). 540 pp.

GISSEL, S. (1972) Forskningsrapport for Danmark, pp. 1–71 in *Nasjonale forskningsoversikter*, Det Nordiske Ødegårdsprojekt, Pubn No. 1 Copenhagen (Landbohistorisk Selskab). 223 pp.

GREIG, J. R. A. and TURNER, J. (1974) Some pollen diagrams from Greece and their archaeological significance, *J. Archaeol. Sci.*, **1** (2), 177–94. London and New York (Acad. Press).

GRIFFIN, J. B. (1961) Some correlations of climatic and cultural change in eastern North American prehistory, *Ann. New York Acad. Sci.*, **95** (1), 710–17.

GROEBEN, F. (1694) *Guineische Reise*. Leipzig (Insel Verlag).

GROVE, A. T., STREET, F. A. and GOUDIE, A. S. (1975) Former lake levels and climatic change in the Rift Valley of southern Ethiopia, *Geogr. J.*, **141** (2), 177–202. London.

HARRINGTON, H. J. and McKELLAR, I. C. (1958) A radiocarbon date for penguin colonization of Cape Hallett, Antarctica, *N.Z. J. Geol. Geophys.*, **1**, 571–6. Wellington.

HARTING, J. E. (1880) *British Animals Extinct Within Historical Times*. London (Trübner & Co.).

HENDERSON, E. (1879) *The Annals of Dunfermline*. Glasgow (J. Tweed).

HENNIG, R. (1904) Katalog bemerkenswerter Witterungsereignisse von den ältesten Zeiten bis zum Jahre 1800, *Abhandl. des kgl. Preuss. Met. Inst.*, **2**, (4). Berlin (Ascher).

HEUBERGER, H. (1968) Die Alpengletscher im Spät- und Postglazial: eine chronologische Übersicht, *Eiszeitalter und Gegenwart*, **19**, 270–5. Öhringen/Württ.

HOLLOWAY, J. T. (1954) Forests and climates in the South Island of New Zealand, *Proc. Roy. Soc. N.Z.*, **82**, 329–410. Wellington.

HOLMSEN, A. (1961) *Norges historie*. Oslo and Bergen (Universitetsforlaget). 480 pp.

HOSKINS, W. G. (1964) Harvest fluctuations and English economic history, 1480–1619, *Agric. Hist. Rev.*, **12** (1), 28–46. London.

HOSKINS, W. G. (1968) Harvest fluctuations and English economic history, 1620–1759, *Ibid.*, **16** (1), 15–31.

HUBER, B., COURTOIS, H. and ELLING, W. (1964) Die Periodenlänge von Jahrringbreitenkurven, *Meteorol. Rundschau*, **17**, 122–5. Heidelberg.

HUBER, B. and GIERTZ-SIEBENLIST, V. (1969) Unserer tausendjährige Eichen-Jahrringchronologie durchschnittlich 57(10–150)-fach belegt, *Sitzungsberichte, Abt. I: Biol., Mineral., Erdkunde und verwandte Wiss*, **178** (1–4), 37–42. Vienna (Österr. Akad. Wiss.).

HUNTINGTON, E. (1907) *The Pulse of Asia*. Boston, Mass. (Houghton Mifflin). 415 pp.

HUTCHINS, D. E. (1889) *Cycles of Drought and Good Seasons in South Africa*.

JONES, GWYN (1968) *A History of the Vikings*. Oxford (Univ. Press). 504 pp.

KALELA, O. (1949) Changes in geographical ranges in the avifauna of northern and central Europe in relation to recent changes in climate, *Bird Banding*, **20**, 77–103. Boston, Mass. (April).

KOCH, L. (1945) The East Greenland ice, *Medd. om Grønland*, **130**, Nr. 3. Copenhagen.

LA COUR, P. (1876) *Tyge Brahes Meteorologiske Dagbog, holdt paa Uranienborg for Aarene 1582–1597*. Copenhagen (Kgl. Danske Videnskabernes Selskab: *Collectanea Danica*).

LADURIE, E. L. (1956–7) Fluctuations météorologiques et bans de vendages au XVIIIe siècle, *Féd. hist. du Lang. médit. et du Rouss.*, 30th and 31st Congresses. Sète-Beaucaire, Montpelier.

LADURIE, E. L. (1966) *Les paysans de Languedoc*. Paris.

LADURIE, E. L. (1971) *Times of Feast, Times of Famine*. New York (Doubleday). 426 pp.

LA MARCHE, V.C. (1974) Frequency-dependent relationships between tree-ring series along an ecological gradient and some dendroclimatic implications, *Tree-Ring Bull.*, **34**, 1–20. Tucson, Arizona.

LAMB, H. H. (1967) Britain's changing climate, *Geogr. J.*, **133** (4), 445–68. London.

LARSSON, L.-O. (1972) Översikt över det svenska forskningsläget inom projektets arbetsfält, pp. 193–223 in *Nasjonale forskningsoversikter, Det Nordiske Ødegårdsprojekt*, Pubn No. 1. Copenhagen (Landbohistorisk Selskab). 223 pp.

LENKE, W. (1960) Klimadaten von 1621–1650 (Landgraf Hermann von Hessen). *Berichte des dt. Wetterdienstes*, **9**, (63). Offenbach.

LIBBY, L. M. and PANDOLFI, L. J. (1974) Temperature dependence of isotope ratios in tree rings, *Proc. Nat. Acad. Sci. U.S.A.*, **71** (6) 2482–6. Washington (June).

LYTHE, S. G. E. (1955) Scottish trade with the Baltic, *Econ. Essays in Commemoration of the Dundee School of Economics*, **193** (5), 63–81. (ed. J. K. EASTHAM). Coupar Angus (Wm Culross).

MALEY, J. (1973) Mécanisme des changements climatiques aux basses latitudes, *Palaeogeogr., Palaeoclim., Palaeoecol.*, **14**, 193–227. Amsterdam (Elsevier).

MANLEY, G. (1965) Possible climatic agencies in the development of postglacial habitats, *Proc. Roy. Soc.*, **B, 161**, 363–75. London.

MANLEY, G. (1974) Central England temperatures: monthly means from 1659 to 1973. *Quart. J. Roy. Met. Soc.*, **100**, 389–405. London.

MÜLLER, K. (1953) *Geschichte des badischen Weinbaus*. Lahr in Baden (Moritz Schauenburg). 283 pp.

NICHOLS, H. (1970) Late Quaternary pollen diagrams from the Canadian Arctic Barren Grounds at Pelly Lake, northern Keewatin, N.W.T., *Arctic & Alpine Res.*, **2** (1), 43–61. Boulder, Colorado.

NICHOLSON, S. E. (1976) *A climatic chronology for Africa: synthesis of geological, historical and meteorological information and data*. Unpublished Ph.D. Thesis, Univ. of Wisconsin, Madison (Dept. Meteorology). 324 pp.

NØRLUND, P. (1924) Buried Norsemen at Herjolfsnes: an archaeological study. *Medd. om Grønland*, **67** (1). 270 pp.

PARRY, M. L. (1975) Secular climatic change and marginal agriculture, *Trans. Inst. British Geographers*, Pubn No. 64, 1–13. London.

PEJML, K. (1962) A contribution to the historical climatology of Morocco and Mauretania, *Studia geoph. et geodet.*, **6**, 257–79. Prague.

POST, J. D. (1973) Meteorological historiography (a review), *J. Interdisciplinary History*, **3** (4), 721–32. Boston, Mass. (M.I.T.).

RÖTHLISBERGER, F. (1976) Gletscher-und Klimaschwankungen im Raum Zermatt, Ferpècle und Arolla. *Die Alpen*, **52** (3/4), 59–152. Luzern (Swiss Alpine Club).

SANCHEZ, W. A. and KUTZBACH, J. E. (1974) Climate of the American tropics and subtropics in the 1960s and possible comparisons with climatic variations of the last millennium. *Quaternary Research*, **4**, 128–35.

SCHNEEBELI, W. (1976) Untersuchungen von Gletscherschwankungen in Val des Bagnes. *Die Alpen*, **52** (3/4), 5–57. Luzern (Swiss Alpine Club).

SANDNES, J. (1971) *Ødetid og gjenreisning*. Oslo (Universitetsforlaget). 385 pp.

SKELTON, R. A., MARSTON, T. E. and PAINTER, G. D. (1965) *The Vinland Map and the Tartar relation*. New Haven and London (Yale Univ. Press). 291 pp.

SMILEY, T. L. (1961) Evidences of climatic fluctuations in southwestern prehistory, *Ann. N.Y. Acad. Sci.*, **95** (1), 697–704.

SMOUT, T. C. S. (1963) *Scottish Trade on the Eve of Union, 1660–1707*. Edinburgh (Oliver & Boyd). 320 pp.

SOLVANG, J. (1942) Korndyrking nordom Malangen i eldre tid, *Haløygminne*, **6** (3), 249–55. Harstad, Norway.

SOUTAR, A. and ISAACS, J. D. (1969) *History of Fish Populations Inferred from Fish Scales in Anaerobic Sediments off California.* La Jolla (Scipps Institution of Oceanography).

STEENSBERG, A. (1951) Archaeological dating of the climatic change in north Europe about A.D. 1300, *Nature*, **168**, 672–4. London (20 October).

STEENSTRUP, J. (1907) Danmarks Tab til Havet i den historiske Tid, *Historisk Tidsskrift* (8. Række) **1**, 153–66. Oslo (Universitetsforlaget).

STEFANSSON, V. (1943) *Greenland.* London (Harrap). 240 pp.

THOMPSON, M. W. (1967) *Novgorod the Great.* London (Evelyn, Adams & Mackay). 104 pp.

TREVELYAN, G. M. (1928) *History of England.* London (Longmans, Green & Co.). 723 pp.

WALLIS, H., MADDISON, F. R., PAINTER, G. D., QUINN, D. B., PERKINS, R. M., CRONE, G. R., BAYNES-COPE, A. D., MCCRONE, W. C. and MCCRONE, L. B. (1974) The strange case of the Vinland map, *Geogr. J.*, **140** (2), 183–214. London.

WARNER ALLEN, H. (1961) *A History of Wine.* London (Faber & Faber). 304 pp.

WOLF, R. (1885) *Schweizerische Meteorologische Beobachtungen: Supplement-Band*, 181–2. Zurich (Met. Anstalt).

WOODBURY, R. B. (1961) Climatic changes and prehistoric agriculture in the southwestern United States, *Ann. N.Y. Acad. Sci.*, **95** (1), 705–9.

ZIEGLER, P. (1969) *The Black Death.* London (Collins). 319 pp.

Chapter 18   Climate since instrument records began

ANGELL, J. K. and KORSHOVER, J. (1976) *Estimate of the Global Change in Temperature Surface to 100 mb Between 1958 and 1975.* Unpublished memorandum, Air Resources Laboratories, NOAA, Washington.

ARAKAWA, H. (1956) Dates of first or earliest snow covering for Tokyo since 1632, *Quart. J. Roy. Met. Soc.*, **82**, 222–6. London.

ARMSTRONG, T. E. (1958) *Sea Ice North of the U.S.S.R.*, Part II. London (Admiralty).

AURROUSSEAU, M. (1958) Surface temperatures of the Australian seas, *J. & Proc. Roy. Soc. N.S.W.*, **92**, Part (IV), p. 104. Sydney.

BECHER, A. B. (1843, 1852) Bottle chart of the Atlantic Ocean, *Nautical Mag.* (1843) pp. 181–4, (1852) pp. 568–72. Glasgow.

BEVERIDGE, W. H. (1921) Weather and harvest cycles, *Econ. J.*, **31**, 421–53. London.

BLAGDEN, C. (1781) On the heat of the water in the Gulf Stream, *Phil. Trans. Roy. Soc.*, **71**, 334–44. London.

BRADLEY, R. S. (1973) Recent freezing level changes and climatic deterioration in the Canadian Arctic archipelago, *Nature*, **243**, 398–9. London (15 June).

BRADLEY, R. S. and MILLER, G. H. (1972) Recent climatic change and increased glacierization in the eastern Canadian Arctic, *Nature*, **237**, 385–7. London (16 June).

BROWN, C. W. and ROBERTS, W. O. J. (1935) The distribution and frequency of tornadoes in the United States from 1880 to 1931, *Trans. Amer. Geophys. U.*, **16**, 151. Washington.

BROWN, P. R. (1963) Climatic fluctuation over the oceans and in the tropical Atlantic, pp. 109–23 in *Changes of Climate: Proc. UNESCO/WMO Rome 1961 Symp.* Paris (UNESCO, Arid Zone Research Series XX).

BRYSON, R. A. (1973) *World Food Prospects and Climatic Change*. Unpublished memorandum, Center for Climatic Research, Univ. of Wisconsin, Madison. 26 pp.

BRYSON, R. A. (1974) A perspective on climatic change, *Science*, **184**, 753–60. Washington (17 May).

COLEBROOK, J. M. (1975) *Sea Surface Temperatures in the Celtic Sea – Bay of Biscay Area 1854–1968*. Unpublished memorandum, Marine Laboratory, Plymouth.

COLEMAN, F. (1972) *Frequencies, Tracks and Intensities of Tropical Cyclones in the Australian Region, 1909–1969*. Melbourne (Bureau of Meteorology, Dept of the Interior).

COURT, A. (1970) *Tornado Incidence Maps*. ESSA Tech Memo.: ERLTM-NSSL 49. Norman, Oklahoma (National Severe Storms Lab.). 76 pp.

CRADDOCK, J. M. (1976) Annual rainfall in England since 1725. *Quart. J. Roy. Met. Soc.*, **102**, 823–40. London.

CRY, G. W., HAGGARD, W. H. and WHITE, H. S. (1959) *North Atlantic Tropical Cyclones*. Tech. Paper 36. Washington (U.S. Weather Bureau).

DE LISLE, J. F. (1956) *Secular Variations of West Coast Rainfall in New Zealand and Their Relation to Circulation Changes*. N.Z. Met. Office Note No. 42. Wellington, N.Z.

DICKSON, R. R. (1971) A recurrent and persistent pressure-anomaly pattern as the principal cause of intermediate-scale hydrographic variation in the European Shelf seas, *Deutsche Hydrographische Zeit.*, **24** (3), 97–119. Hamburg.

DICKSON, R. R. and LAMB, H. H. (1972) A review of recent hydrometeorological events in the North Atlantic sector, pp. 35–62 in *Special Pubn No. 8*. Dartmouth, Nova Scotia (Internat. Comm. for the Northwest Atlantic Fisheries).

DIETRICH, G. and KALLE, K. (1957) *Allgemeine Meereskunde*. Berlin (Gebrüder Borntraeger). 492 pp. (Republished in English language 1963 as *General Oceanography*.)

DOUGLAS, A. V. (1974) Past air–sea interaction off southern California as revealed by coastal tree-ring chronologies, *Amer. Quaternary Ass. Third Biennial Meeting, Abstracts*, p. 7. Madison (Univ. Wisconsin).

DRONIA, H. (1967) Der Stadteinfluss auf den weltweiten Termperaturtrend, *Met. Abhandlungen*, **74** (4). Berlin (Inst. f. Met. Geophys. der Freien Universität).

DRONIA, H. (1974) Klimaverschlechterung an den deutschen Küsten? *Beilage zur Berliner Wetterkarte*, 106/74 SO 24/74. Berlin (Inst. f. Met. Fr. Univ. Berlin, 17 October).

FARMER, S. A. (1973) A note on the long-term effects on the atmosphere of sea surface temperature anomalies in the North Pacific Ocean, *Weather*, **28**, 102–5. London.

FINDLAY, A. G. (1884) *Directory for the South Pacific Ocean*. London, 5th edn, p. 1144.

FINDLAY, A. G. (1920) *Directory for the South Atlantic Ocean*. London, 10th edn, p. 172.

FLOHN, H. and RODEWALD, M. (1975) Beiträge zum Problem der aktuellen Klimavariationen, *Beilage zur Berliner Wetterkarte*, 89/75 SO 25/75. Berlin (Inst. f. Met. Fr. Univ., 17 July).

FRANKLIN, B. (1786) A Letter from Dr Benjamin Franklin, to Mr Alphonsus le Roy, Member of several Academies, at Paris, containing sundry maritime observations, *Trans. Amer. Phil. Soc.*, **2**, 330.

FRITTS, H. C., BLASING, T. J., HAYDEN, B. P. and KUTZBACH, J. E. (1971) Multivariate techniques for specifying tree-growth and climate relationships and for reconstructing anomalies in paleoclimate, *J. Appl. Met.*, **10** (5), 845–64. Boston, Mass.

GENTRY, R. C. (1963) Historical survey and climatology of hurricanes and tropical storms, pp. 17–30 in *Proc. Interregional W.M.O. Seminar on Tropical Cyclones, Tokyo, January 1962*. Tokyo (Jap. Met. Agency).

GLASSPOOLE, J. (1933) The rainfall over the British Isles of each of the eleven decades during the period 1820 to 1929, *Quart. J. Roy. Met. Soc.*, **59**, 253–60. London.

GRAY, B. M. (1974) Early Japanese winter temperatures, *Weather*, **29** (3), 103–7. London.

GRAY, B. M. (1975) Japanese and European winter temperatures, *Weather*, **30**, 359–68. London.

GRAY, B. M. (1976) Medium term fluctuations of rainfall in southeastern England. *Quart. J. Roy. Met. Soc.*, **102**, 627–38. London.

GROISSMAYR, F. B. (1939) Schwere und leichte Eisjahre bei Neufundland und das Vorwetter, *Ann. Hydr. u. maritimen Met.*, **67**, 26–30. Berlin.

HARVEY, J. (1965) Factors affecting water temperature in the seas north of Norway, pp. 869–79 in *Special Pubn* No. 6. Dartmouth, Nova Scotia (Internat. Comm. Northwest Atlantic Fisheries).

HASTENRATH, S. (1975) Large-scale circulation and weather anomalies in the Caribbean and Central America, pp. 127–33 in *Proc. WMO/IAMAP Symp. on Long-term Climatic Fluctuations, Norwich, 18–23 August 1975*. W.M.O. No. 421. Geneva (World Met. Org.). 503 pp.

HENRY, P. W. T. (1974) Forestry on St Helena, *Land Resources Rep. 2*. London (Min. of Overseas Development, Land Resources Div.), 2 vols.

HERMANN, F., LENZ, W. and BLACKER, R. W. (1973) Hydrographic sections off west Greenland in 1972, *Red Book*, Part III, pp. 27–32. Dartmouth, Nova Scotia (Internat. Comm. Northwest Atlantic Fisheries).

I.C.N.A.F. (1967) Selected papers on fluctuations in sea and air temperatures in the ICNAF area since 1950, *Red Book*, Part IV. Dartmouth, Nova Scotia (Internat. Comm. Northwest Atlantic Fisheries).

IVES, J. D. (1962) Indications of recent extensive glacierization in north-central Baffin Island, N.W.T., *J. Glaciol.*, **4**, 197–205. Cambridge.

KING, J. W. (1974) Weather and the Earth's magnetic field, *Nature*, **247**, 131–4. London (18 January).

KINGTON, J. A. (1975) An analysis of the seasonal characteristics of 1781–4 and 1968–71 using the PSCM Indices, *Weather*, **30** (4), 109–14. London.

KIRCH, R. (1966) Temperaturverhältnisse in der Arktis während der letzten 50 Jahre, *Met. Abhandlungen*, **69** (3). Berlin (Inst. f. Met. Geophys. der Freien Universität).

KOCH, L. (1945) The East Greenland ice, *Medd. om Grønland*, **130**, (3). Copenhagen.

KOERNER, R. M. and PATERSON, W. S. B. (1974) Analysis of a core through the Meighen Ice Cap, Arctic Canada, and its paleoclimatic implications, *Quaternary Res.*, **4**, 253–63. Seattle.

KOHL, J. G. (1868) *Geschichte des Golfstroms und seiner Erforschung*. Bremen (Meridian Publishing Co., Amsterdam reprint 1966).

KONDRATIEV, K. YA. (1970) Solar radiation and solar activity, *Quart. J. Roy. Met. Soc.*, **96**, 509–22. London.

KRAUS, E. B. (1954) Secular changes in the rainfall regime of southeast Australia, *Quart. J. Roy. Met. Soc.*, **80**, 591–601. London.

KRAUS, E. B. (1955a) Secular changes in tropical rainfall regimes, *Quart. J. Roy. Met. Soc.*, **81**, 198–210. London.

KRAUS, E. B. (1955b) Secular changes of east coast rainfall regimes, *Quart. J. Roy. Met. Soc.*, **81**, 430–9. London.

KRAUS, E. B. (1956) Secular changes of the standing circulation, *Quart. J. Roy. Met. Soc.*, **82**, 289–300. London.

KRAUS, E. B. (1958) Recent climatic changes, *Nature*, **181**, 666–8. London (8 March).

KRAUS, E. B. (1959) The evaporation-precipitation cycle of the Trades, *Tellus*, **11**, 147–58. Stockholm.

KRISTJANSSON, L. (1971) Úr heimildahandraða seytjándu og atjándu aldar. In *Saga*. Reykjavik.

KUKLA, G. J. and KUKLA, H. J. (1974) Increased surface albedo in the northern hemisphere, *Science*, **183**, 709–14. New York.

LAMB, H. H. (1957) Tornadoes in England, 21 May 1950, *Geophys. Mem.*, **99**. London (H.M.S.O. for Met. Off.). 38 pp.

LAMB, H. H. (1963) On the nature of certain climatic epochs which differed from the modern (1900–39) normal, pp. 125–50 in *Changes of Climate: Proc UNESCO/WMO Rome 1961 Symp.* Paris (UNESCO, Arid Zone Research Series XX).

LAMB, H. H. (1965) Britain's changing climate, pp. 3–31 in *The Biological Significance of Climatic Changes in Britain* (ed. C. G. JOHNSON and L. P. SMITH). London and New York (Academic Press for Inst. Biol.).

LAMB, H. H. (1966) Climate in the 1960s, *Geogr. J.*, **132** (2), 183–212. London.

LAMB, H. H. (1967a) Britain's climate, its variability and some of its extremes, pp. 1–27 in *Symposium on Heating and Ventilation for a Human Environment*. London (Inst. Mech. Eng.).

LAMB, H. H. (1967b) On climatic variations affecting the Far South, pp. 428–53 in *Polar Meteorology, Tech. Note No. 87*. W.M.O. No. 211, TP 111. Geneva (World Met. Org.).

LAMB, H. H. (1970) Volcanic dust in the atmosphere; with a chronology and assessment of its meteorological significance. *Phil. Trans. Roy. Soc.*, A, **266** (1178), 425–533. London.

LAMB, H. H. (1974) Contributions to historical climatology: the Middle Ages and after; Christmas weather and other aspects, pp. 549–67 in *Klimatologische Forschung/Climatological Research* (*The Hermann Flohn 60th Anniversary Volume*). Bonn (Dümmler, Bonner Met. Abhandl., Heft 17).

LAMB, H. H. and JOHNSON, A. I. (1959) Climatic variation and observed changes in the general wind circulation, Parts I and II, *Geografiska Annaler*, **41**, 94–134. Stockholm.

LAMB, H. H. and JOHNSON, A. I. (1961) Climatic variation and observed changes in the general wind circulation, Part III, *Ibid.*, **43**, 363–400.

LAMB, H. H. and JOHNSON, A. I. (1966) Secular variations of the atmospheric circulation since 1750, *Geophys. Mem.*, **110**. London (Met. Office). 125 pp.

LANDSBERG, H. E. (1975) *Effective Use of Weather Data and Long-Range Weather Forecasting*. Unpublished paper delivered to the Soil and Water Management Workshop, Agency for International Development, Office of Agriculture (19 February).

LIMBERT, D. W. S. (1974) Variations in the mean annual temperature for the Antarctic Peninsula, 1904–72, *Polar Record*, **17** (108), 303–6. Cambridge.

LOEWE, F. (1937) A period of warm winters in western Greenland and the temperature seesaw between western Greenland and Europe, *Quart. J. Roy. Met. Soc.*, **63**, 365–71. London.

LYSGAARD, L. (1949) Recent climatic fluctuations, *Folia Geographica Danica*, **5**. Copenhagen. 86 pp.

MALMBERG, S.-A. (1969) Hydrographic changes in the waters between Iceland and Jan Mayen in the last decade, *Jökull*, **19**, 30–43. Reykjavik.

MANLEY, G. (1949) The snowline in Britain, *Geografiska Annaler*, **36**, 179–93. Stockholm.

MANLEY, G. (1953) Mean temperature of central England, 1698–1952, *Quart. J. Roy. Met. Soc.*, **79**, 242–61. London.

MANLEY, G. (1957) Climatic fluctuations and fuel requirements, *Scot. Geogr. Mag.*, **73** (1), 19–28. Edinburgh.

MANLEY, G. (1959) Temperature trends in England 1698–1957, *Archiv f. Met. Geophys. Biokl.*, B, **9**, 413–33. Vienna.

MANLEY, G. (1969) Snowfall in Britain over the past 300 years, *Weather*, **24**, 428–37. London.

MANLEY, G. (1971) The mountain snows of Britain, *Weather*, **26**, 192–200. London.

MANLEY, G. (1974) Central England temperatures: monthly means 1659 to 1973, *Quart. J. Roy. Met. Soc.*, **100**, 389–405. London.

MAURITIUS, ROYAL ALFRED OBSERVATORY (1856–1959) *Annual Reports*.

MILTON, D. (1974) Some observations on global trends in tropical cyclone frequencies, *Weather*, **29**, 267–70. London.

MITCHELL, J. M. (1961) Recent secular changes of global temperature, *Ann. N.Y. Acad. Sci.*, **95** (1), 235–50.

MITCHELL, J. M. (1963) On the world-wide pattern of secular temperature change, pp. 161–81 in *Changes of Climate: Proc. UNESCO/WMO Rome 1961 Symp.* Paris (UNESCO Arid Zone Research Series, XX).

MITCHELL, J. M., DZERDZEEVSKII, B., FLOHN, H., HOFMEYR, W. L., LAMB, H. H., RAO, K. N. and WALLÉN, C. C. (1966) *Climatic Change.* Tech. Note. 79, W.M.O. 195, T.P. 100. Geneva (World Met. Org.). 79 pp.

MOODIE, D. W. and CATCHPOLE, A. J. W. (1975) Environmental data from historical documents by content analysis: freeze-up and break-up of estuaries on Hudson Bay 1714–1871, *Manitoba Geogr. Studies*, **5**. Winnipeg. 119 pp.

NAMIAS, J. (1963) Large-scale air–sea interactions over the North Pacific from summer 1962 through the following winter, *J. Geophys. Res.*, **68** (22), 6171–86. Washington.

NEWELL, R. E., KIDSON, J. W., VINCENT, D. G. and BOER, G. J. (1974) *The General Circulation of the Tropical Atmosphere and Interactions with Extratropical Latitudes*, Vol. 2. Cambridge, Mass. (M.I.T. Press). 371 pp.

OTTESTAD, P. (1942) On periodical variations in the yield of the great sea fisheries and the possibility of establishing yield prognoses, *Fiskeridirektoratets Skrifter, Serie Havundersøkelser:* **7** (5). Bergen (A/S John Grieg).

PENNANT, T. (1771) *A Tour of Scotland, 1769.* Chester.

PETTERSSEN, S. (1949) Changes in the general circulation associated with the recent climatic variation, *Geografiska Annaler*, **31**, 212–21. Stockholm.

PFISTER, CHR. (1974) Historische Quellen und ihre klimageschichtliche Aussage, *Informationen und Beiträge zur Klimaforschung*, No. 12, p. 6. Bern (Geogr. Inst. der Univ.).

PICKETT, R. L. and ATHEY, G. L. (1968) *Iceland Sea Surface Temperature Survey, April 1968.* U.S. Naval Oceanographic Office IR 68–46. Washington.

RAMPTON, V. (1971) Late Quaternary vegetational and climatic history of the Snag-Klutlan area, southwestern Yukon Territory, Canada, *Geol. Soc. Amer. Bull.*, **82**, 959–78. Rochester, N.Y.

ROBIN, G. de Q. (1955) Ice movement and temperature distribution in glaciers and ice sheets. *J. Glaciol.*, **2**, 523–32. Cambridge.

RODEWALD, M. (1972) Einige hydroklimatische Besonderheiten des Jahrzehnts 1961–70 im Nordatlantik und im Nordpolarmeere, *Deutsche Hydrogr. Zeit.*, **25** (3), 97–117. Hamburg.

RODEWALD, M. (1973) Der Trend der Meerestemperatur im Nordatlantik, *Beilagen zur Berliner Wetterkarte*, SO 27/73 and SO 29/73. Berlin (Inst. f. Met. der Freien Univ.).

RODEWALD, M. (1974) (a) Vom Wechselschritt der Witterung, *Wetterkarte des dt. Wetterdienstes*, No. 145–7 (31 July – 2 August). (b) Die Ursachen des Wechselschritt der Witterung, *Ibid.*, No. 159 (20 August). (c) Das Nordseewasser im September, *Ibid.*, No. 209 (29 October). (d) Das Ostseewasser im September, *Ibid.*, No. 216 (7 November) Hamburg (Seewetterampt).

RODEWALD, M. (1975) Der Trend der Frontalzonen-Verschärfung im westlichen Nordatlantik, *Beilage zur Berliner Wetterkarte* SO 13/75. Berlin (Inst. f. Met. der Freien Univ.).

RUDLOFF, H. VON (1967) *Die Schwankungen und Pendelungen des Klimas in Europa seit dem Beginn der regelmässigen Instrumenten-Boebachtungen*. Braunschweig (Vieweg, *Die Wissenschaft* 122). 370 pp.

SALINGER, M. J. and GUNN, J. M. (1975) Recent climatic warming in the New Zealand area, *Nature*, **256**, 396–8. London.

SANCHEZ, W. A. and KUTZBACH, J. E. (1974) Climate of the American tropics and subtropics in the 1960s and possible comparisons with climatic variations of the last millenium, *Quaternary Res.*, **4**, 128–35. Seattle.

SAUR, J. F. T. (1963) A study of the quality of sea water temperatures reported in logs of ships' weather observations, *J. Appl. Met.*, **2**, 417–25. Boston, Mass.

SCHELL, I. I. (1974) The lag in the response of the ocean during a climatic change, pp. 85–93 in *Mapping the Atmosphere and Oceanic Circulations and Other Climatic Parameters at the Time of the Last Glacial Maximum about 17 000 Years Ago* (CLIMAP Conference). Research Publication CRU RP2. Norwich (Univ. of East Anglia, Climatic Res. Unit).

SCHNEIDER, S. H. and MASS, C. (1975) Volcanic dust, sunspots and long-term climatic trends: theories in search of verification, pp. 365–72 in *Proc. WMO/IAMAP Symp. on Long-term Climatic Fluctuations, Norwich, 18–23 August 1975*. W.M.O. No. 421. Geneva (World Met. Org.) 503 pp.

SCHWERDTFEGER, and KACHELHOFFER, St J. (1974) The frequency of cyclonic vortices over the Southern Ocean in relation to the extension of the pack-ice belt, *Antarctic J.*, p. 234. Washington, D.C.

STARR, V. P. and OORT, A. H. (1973) Five-year climatic trend for the northern hemisphere, *Nature*, **242**, 310–13. London (30 March).

STOLLE, H. J. (1975) *Climatic Change and the Gulf Stream*. Unpublished memorandum. Madison (Univ. Wisconsin, Center for Climatic Res.).

STRICKLANDER, W. (1802) On the use of the thermometer in navigation, *Trans. Amer. Phil. Soc.*, **5**, 90–103. Philadelphia.

SUGDEN, D. E. (1974) Deglaciation of the Cairngorms and its wider implications, pp. 17–28 in *Problems of the Deglaciation of Scotland* (ed. C. J. CASSELDINE and W. A. MITCHELL). STAG Special Publication No. 1. St Andrews (Univ. Geogr. Dept).

SVABO, J. C. (1959) *Indberetninger fra en Reise i Færøe 1781 og 1782*. Copenhagen (Selskabet til Udgivelse af Færøske Kildeskrifter og Studier).

TAIT, J. B. and MARTIN, J. H. A. (1964) Inferential biological effects of long-term hydrographical trends deduced from investigations in the Faroe-Shetland Channel. *Special Publication No.* 6, pp. 855–8. Dartmouth, Nova Scotia (Internat. Comm. Northwest Atlantic Fisheries).

TAYLOR, J. (1618) *The Penniless Pilgrimage.*

THOMAS, M. K. (1975) *Recent Climatic Fluctuations in Canada.* Climatological Studies No. 28. Toronto (Environment Canada). 92 pp.

THOMSON, L. M. (1969) Weather and technology in the production of wheat in the United States. *J. Soil and Water Conservation,* **24** (6), 219–24.

TOKYO CENTRAL MET. OBSERVATORY (1912–) Japanese typhoon records. (See also *Geophys. Mag.,* **30**, 131–48. Tokyo, 1960.)

TOWSON, J. T. (1859) *Icebergs in the Southern Ocean.* Liverpool.

TRENBERTH, K. E. (1975) Fluctuations and trends in indices of the southern hemisphere circulation, pp. 243–53 in *Proc. WMO/IAMAP Symp. on Long-Term Climatic Fluctuations, Norwich, 18–23 August 1975.* W.M.O. No. 421. Geneva (World Met. Org.). 503 pp.

TRUXTON, T. (1794) *Remarks, Instructions and Examples Relating to the Latitude and Longitude.* Philadelphia.

WAHL, E. W. (1968) A comparison of the climate of the eastern United States during the 1830s with the current normals, *Mon. Weather Rev.,* **96**, (2) 73–82. Washington.

WAHL, E. W. and BRYSON, R. A. (1975) Recent changes in Atlantic surface temperatures, *Nature,* **254**, 45–6. London (6 March).

WAHL, E. W. and LAWSON, T. L. (1970) The climate of the mid nineteenth century United States compared to the current normals, *Mon. Weather Rev.,* **98** (4), 259–65. Washington.

WALLÉN, C. C. (1953) The variability of summer temperature in Sweden and its connection with changes in the general circulation, *Tellus,* **5** (2), 157–78. Stockholm.

WEST, J. F. (1970) *The Journals of the Stanley Expedition to the Faroe Islands and Iceland in 1789, Vol. 1: Introduction and Diary of James Wright* (edited and annotated by J. F. West). Torshavn (Føroya Fróðskaparfelag). 216 pp.

WEXLER, H. (1956) Variations in insolation, atmospheric circulation and climate, *Tellus,* **8**, 480–94. Stockholm.

WEXLER, H. (1959) A warming trend at Little America, Antarctica, *Weather,* **14**, 191–7. London.

WHITE, W. B. (1975) Low frequency development in the westerly wind system of the North Pacific, *Norpax Highlights,* **3** (4), 1–3. La Jolla, California.

WILLETT, H. C. (1950) Temperature trends of the past century, pp. 195–206 in *Centenary Proceedings.* London (Roy. Met. Soc.).

WINSTANLEY, D. (1973) Rainfall patterns and general atmospheric circulation, *Nature,* **245**, 190–4. London (28 September).

794   References

WOOD, C. A. and LOVETT, R. R. (1974) Rainfall, drought and the solar cycle, *Nature*, **251**, 594–6. London (18 October).

WRIGHT, P. B. (1965) Four mild winters – In Europe. *Weather*, **30**, 125–6. London.

YAMAMOTO, T. (1971) On the nature of the climatic change in Japan since the 'Little Ice Age' around A.D. 1800, *J. Met. Soc. Japan*, **49** (Special Issue), 75–89. Tokyo.

Appendix V   Supplementary tables and figures

ARAKAWA, H. (1954) Fujiwhara on five centuries of freezing dates of Lake Suwa in the central Japan, *Archiv f. Met. Geophys. Biokl.*, **B, 6**, 152–66. Vienna.

ARAKAWA, H. (1955a) Remarkable winters in Japan from the seventh century, *Geofis. pura e applicata*, **30**, 144–6. Milan.

ARAKAWA, H. (1955b) Twelve centuries of blooming dates of the cherry blossoms at the city of Kyoto and its own vicinity, *Geofis. pura e applicata*, **30**, 147–50. Milan.

ARAKAWA, H. (1956) Dates of first or earliest snow covering for Tokyo since 1632, *Quart. J. Roy. Met. Soc.*, **82**, 222–6. London.

ARAKAWA, H. (1957) Climatic change as revealed by the data from the Far East, *Weather*, **12** (2), 46–51. London.

ARAKAWA, H. (1959) Relation between beginning of dry season in Java, establishment of southwest monsoon in India and onset of Bai-U in the Far East, *Proc. I.G.U. Regional Conf. in Japan, 1957*, pp. 589–92.

BELL, B. (1975) Climate and the history of Egypt: the Middle Kingdom, *Amer. J. Archaeol.*, **79**, 223–69. New York.

BERGTHORSSON, P. (1969) Forecasting drift ice at Iceland by means of Jan Mayen air temperature, *Jökull*, **19**, 44–52. Reykjavik.

BERLAGE, H. P. (1931) *Tectona*, **24**, 939. Semarang (Java).

BETIN, V. V. and PREOBAZENSKY, Ju. V. (1959) Variations in the state of the ice on the Baltic Sea and in the Danish Sound, *Trudy*, **37**, 3–13. Moscow (Gosudarst. Okean. Inst.). (In Russian.)

BHULLAR, G. S. (1952) Onset of monsoon in Delhi, *Indian J. Met. & Geophys.*, **3** (1), 25–30. New Delhi.

BROOKS, C. E. P. (1935) Lake deposits in the Crimea and the rainfall of Europe since 2000 B.C., *Met. Mag.*, **70**, 134–8. London.

CENTRAL METEOROLOGICAL OFFICE, REPUBLIC OF KOREA (1961) *Seoul Monthly Precipitation Records (1770–1960)*. Seoul. 15 pp.

CHU KO-CHEN (COCHING CHU) (1973) A preliminary study on the climatic fluctuations during the last 5000 years in China, *Scientia Sinica*, **16** (2), 226–56. Peking (Academia Sinica).

CRY, G. W., HAGGARD, W. H. and WHITE, H. S. (1959) *North Atlantic Tropical Cyclones*. Tech. Paper No. 36. Washington (U.S. Weather Bureau).

DE BOER, H. J. (1951) Tree ring measurements and weather fluctuations in Java from A.D. 1514, *Proc. Kon. Ned. Akad. Wetenschappen*, **B**, 194–209. Amsterdam.

DEWEY, E. R. (1964) The 17⅓-year cycle in Lake Saki varves, 2295 B.C.–A.D. 1894, *Res. Bull. 1964–1*, pp. 7–70. Pittsburgh, Pa. (Foundation for the Study of Cycles).

EASTON, C. (1928) *Les hivers dans l'Europe occidentale.* Leyden (Brill). 208 pp.

FLOHN, H. (1973) Globale Energiebilanz und Klimaschwankungen. *Bonn. Met. Abhandl.* (Sonderheft). (Also in *Rheinisch-Westfälische Akad. Wiss.*, No. 234.)

GRAY, B. M. (1974) Early Japanese winter temperatures, *Weather*, **29** (3), 103–7. London.

GROISSMAYR, F. B. (1939) Schwere und leichte Eisjahre bei Neufundland und das Vorwetter, *Ann. der Hydr. und maritimen Met.*, **67**, 26–30. Berlin.

HANN-SÜRING (1940, 1951) *Lehrbuch der Meteorologie.* Leipzig, 2 vols. 1092 pp. (Original edition was by J. von HANN, 1901.)

HOLLSTEIN, E. (1965) Jahrringchronologische Datierung von Eichenhölzern ohne Waldkante (Westdeutsche Eichenchronologie), *Bonner Jahrbücher*, **165**, 1–27. Bonn.

HUBER, B. and GIERTZ-SIEBENLIST, V. (1969) Unsere tausendjährige Eichen-Jahrringchronologie durchschnittlich 57 (10–150)-fach belegt, *Sitz.- berichte I: Biol., Mineral., Erdkunde und verwandte Wiss.*, **178** (1–4), 37–42. Vienna (Österreichische Akad. der Wiss.).

KOCH, L. (1945) The East Greenland ice, *Medd. om Grønland*, **130** (3). Copenhagen.

KUIPER, G. P. (1957) Origin, age and possible ultimate fate of the Earth, Chapter 2, pp. 12–30, in *The Planet Earth* (ed. D. R. BATES). London (Pergamon). 312 pp.

LANDSBERG, H. E., YU, C. S. and HUANG, L. (1968) *Preliminary Reconstruction of a Long Time Series of Climatic Data for the Eastern United States.* Tech. Note BN-571. Univ. Maryland (Inst. Fluid Dynamics).

MANLEY, G. (1959) Temperature trends in England, 1698–1957, *Archiv f. Met. Geophys. Biokl.*, **B, 9**, 413–33. Vienna.

MANLEY, G. (1961) A preliminary note on early meteorological observations in the London region with estimates of monthly mean temperature 1680–1706, *Met. Mag.*, **90**, 303–10. London.

MANLEY, G. (1974) Central England temperatures: monthly means 1659 to 1973, *Quart. J. Roy. Met. Soc.*, **100**, 389–405. London.

MILES, M. K. (1974) An index of pack-ice severity off Newfoundland and its secular variation, *Met. Mag.*, **103**, 121–5. London.

MILTON, D. (1974) Some observations of global trends in tropical cyclone frequencies, *Weather*, **29** (7), 267–70. London.

MOTHE, P. D. DE LA and WRIGHT, P. B. (1969) The onset of the Indian southwest monsoon and extratropical 500-mb trough and ridge patterns over Europe and Asia., *Met. Mag.*, **98**, 145–55. London.

NICHOLAS, F. J. and GLASSPOOLE, J. (1931) General monthly rainfall over England and Wales 1727 to 1931, *British Rainfall 1931*, pp. 299–306. London.

RAMDAS, L. A., JAGANATHAN, P. and GOPAL RAO, S. (1954) Prediction of the date of establishment of southwest monsoon along the west coast of India, *Indian J. Met. & Geophys.*, **5** (4), 305–14.

RODEWALD, M. (1971) Das grosse Erlahmen der Eisberg-Drift bei Neufundland, *Hansa (Nautik)*, **108** (23), 2341–4. Hamburg.

RODEWALD, M. (1974) Die jungste Entwiddung der Eisberg-Drift bei Neufundland und ihre Vorhersage, *Beilage zur Berliner Wetterkarte*, 37/74 SO 7/74. Berlin (Inst. f. Met., Fr. Univ.).

SCHELL, I. I. (1962) On the iceberg severity off Newfoundland and its prediction, *J. Glaciol.*, **4** (32), 161–72. Cambridge.

SCHELL, I. I. (1974) On the lag in the response of the ocean during a climatic change, pp. 85–93 in *Mapping the Atmospheric and Oceanic Circulations and Other Climatic Parameters at the Time of the Last Glacial Maximum about 17 000 Years Ago*. Research Publication CRU RP2 Norwich (Univ. of East Anglia, Climatic Research Unit).

SCHOSTAKOWITSCH, W. B. (1934) Bodenablagerungen der Seen und periodische Schwankungen der Naturerscheinungen, *Mem. de l'Inst. Hydrol.*, **13**, 95–140. Leningrad. (In Russian with a German summary.)

SCHWERDTFEGER, W. (1970) The climate of the Antarctic, pp. 253–355 in *Climates of the Polar Regions* (ed. S. ORVIG), Vol. 14 of *World Survey of Climatology* (ed. H. E. LANDSBERG). Amsterdam (Elsevier).

SIRÉN, G. (1961) Skogsgränstallen som indikator för klimatfluktuationerna i norra Fennoskandien under historisk tid, *Communicationes Instituti Forestalis Fenniae*, **54**. 2. Helsingfors.

SMITH, E. H. (1931) The Marion Expedition to Davis Strait and Baffin Bay, 1928, *U.S. Coast Guard Bulletin*, No. 19 (Part 3), 1–221. Washington, D.C.

SPEERSCHNEIDER, C. I. H. (1915) Om Isforholdene i danske Farvande i ældre og nyere Tid: aarene 690–1860, *Meddelelser*, Nr. 2. Copenhagen (Dansk. Meteorol. Inst.). 123 pp.

SPEERSCHNEIDER, C. I. H. (1927) Om Isforholdene i danske Farvande: Aarene 1861–1906. *Meddelelser*, Nr. 6. Copenhagen (Danske. Meteorol. Inst.). 84 pp.

TAGUCHI, T. (1939) *Japanese Meteorological Chronology*. Tokyo (Central Meteorological Observatory). 770 pp. and two supp. vols. (In Japanese.)

TAULIS, E. (1934) De la distribution des pluies au Chili, *Matériaux pour l'étude des calamités*, No. 33, pp. 3–20. Geneva (Soc. de Géographie).

THORARINSSON, S. (1956) The thousand years struggle against ice and fire, *Misc. Papers* No. 14. Reykjavik (Museum of Natural History, Dept Geol. and Geogr.).

TOUSSOUN, PRINCE OMAR (1925) Mémoire sur l'histoire du Nil, vol II, pp. 361–404, *Mémoires presentés a l'Institut d'Égypte*, **IX**. Cairo (Imprimerie de l'Inst. Français d'Archéol. Orientale).

TRENKLE, H. (1956) Näherungswerte der Zonalgeschwindigheit im atlantisch-europäischen Sektor für den Zeitraum 1881–1955, *Met. Rundsch.*, **9**, 153–6. Berlin, Göttingen, Heidelberg.

VOWINCKEL, E. and ORVIG, S. (1970) The climate of the north polar basin, pp. 129–252 in *Climates of the Polar Regions* (ed. S. ORVIG), Vol. 14 of World Survey of Climatology (ed. H. E. LANDSBERG). Amsterdam (Elsevier).

WALES-SMITH, B. G. (1971) Monthly and annual totals of rainfall representative of Kew, Surrey, from 1697 to 1970, *Met. Mag.*, **100**, 345–62. London.

WASHINGTON, W. W. (1972) Numerical climatic-change experiments: the effect of man's production of thermal energy, *J. Appl. Met.*, **11**, 768–72.

WRIGHT, P. B. (1975) *An Index of the Southern Oscillation*. Research Publication CRU RP4. Norwich (Univ. of East Anglia, Climatic Research Unit).

Chapter 19   Man-made climatic changes

ADABASHEV, I. (1966) *Global Engineering*. Moscow (Progress Publishers). 237 pp.

BERGERON, T. (1960) Possible Man-made great-scale modifications of precipitation climate in Sudan, *Geophys. Monogr.*, No. 5, 399–401. Washington (Amer. Geophys. Union).

BERGERON, T. (1965) Opening address at the International Conference on Cloud Physics, Tokyo and Sapporo, 24 May–1 June 1965, *Meddelande 91*. Uppsala (Met. Inst. vid Kungl. Universitetet).

BORISOV, P. M. (1962) The problem of the fundamental amelioration of climate, *Izvestia Vses. Geogr. Obshchestva*, **94**, 304–318. Voronezh. MGA 14:249.

BRYSON, R. A. (1968) 'All other factors being constant...': a reconciliation of several theories of climatic change, *Weatherwise*, **21** (2), 56–60 and 94. Boston, Mass. (Amer. Met. Soc.).

BRYSON, R. A. (1974) A perspective on climatic change, *Science*, **184**, 753–60. New York (17 May).

BUDYKO, M. I. (1962) Some ways of influencing the climate, *Met. i. Gidr.* No. 2, 3–8. Leningrad. MGA 14A–43.

BUDYKO, M. I. (1966) The possibility of changing the climate by action on the polar ice, pp. 347–57 in *Sovremennye Problemy Klimatologii (Contemporary Problems of Climatology)* (ed. M. I. BUDYKO). Leningrad (Gidrometeoizdat). (In Russian: English translation by U.S. Joint Pub. Res. Service, JPRS 43, 482, MGA 20.5-182 (1967).)

BUDYKO, M. I. (1969) *Poliarnye l'dy i Klimat (The polar ice and climate)*. Leningrad (Gidrometeoizdat). 36 pp. (In Russian: English translation available in U.K. Met. Office Library.)

BUDYKO, M. I. (1974) *Climate and Life* (English edition, ed. by D. H. MILLER). New York and London. 508 pp.

BUDYKO, M. I., DROZDOV, O. A. and JUDIN, M. L. (1966) Vliyani gozyaistvenoi deyatelnosti no klimat, *Sovramenii problemi klimatologii*. Leningrad (Gidrometeoizdat). (In Russian: available in English in *Soviet Geogr. Rev. Translations*, **12** (10), 666–679 (1971).)

798 References

CHARNEY, J. G. (1975) Dynamics of deserts and drought in the Sahel, *Quart. J. Roy. Met. Soc.*, **101**, 193–202. London.

DAVITAYA, F. F. (1969) Atmospheric dust content as a factor affecting glaciation and climatic change, *Ann. Ass. Amer. Geogr.*, **59**, 552–60. Washington, D.C.

DORONIN, JU P. (1968) On the problem of eradicating the Arctic ice, *Probl. Arktiki Antarktiki*, **28**, 21–8. (In Russian: English translation available in U.K. Met. Office Library.)

DROZDOV, O. A. (1966) On the variation of precipitation over the northern hemisphere with variation of the temperature of the polar basin, *Trudy*, **138**, 3–16. Leningrad (Glav. Geofiz. Obs.). (In Russian.)

DROZDOV, O. A. and GRIGORIEVA, A. S. (1963) *Vlagooborot v Atmosfere.* Leningrad (Gidrometeoizdat). (Published as *Moisture Exchange in the Atmosphere*, Jerusalem, 1965 (Israel Program of Sci. Translations).)

FEDEROV, E. K. (1962) Modifications of meteorological processes. *Izvestia*, **9**. Moscow (Akad. Nauk SSSR).

FLETCHER, J. O. (1969) *Managing Climatic Resources.* Santa Monica, California (Rand Corporation P-4000-1).

FLOHN, H. (1963) *Klimaschwankungen und grossräumige Klimabeeinflussung* (Arbeitsgemeinschaft für Forschung des Landes Nordrhein-Westfalen, Heft 115). Köln & Opladen (Westdeutscher Verlag). 61 pp.

FLOHN, H. (1971) Klimaschwankung oder Klimamodifikation?, pp. 291–309 in *Arbeiten zur allgemeinen Klimatologie.* Darmstadt (Wissenschaftliche Buchgesellschaft). 319 pp.

GALTSOV, A. P. (1961) Conference on the modification of climate, *Izvestia, Seria Geogr.*, **5**, 128–33. Moscow (Akad. Nauk). (In Russian: English translation available in U.K. Met. Office Library.)

HAEFELE, W. (1974a) *Energy Systems.* Laxenburg/Baden, Austria (Internat. Inst. for Applied Systems Analysis). 39 pp.

HAEFELE, W. (1974b) A systems approach to energy, *Amer. Scientist*, **62** (4), 438–47. New Haven, Conn.

IDSO, S. B. (1973) Thermal radiation from a tropospheric dust suspension, *Nature*, **241**, 448–9. London (16 February).

KOMHYR, W. D., BARRETT, E. W., SLOCUM, G. and WEICKMANN, H. K. (1971) Atmospheric total ozone increases during the 1960s, *Nature*, **232**, 390–1 London.

KONDRATIEV, K. IA. and NIILISK, KH. IU. (1963) Thermal radiation of CO_2 in the atmosphere, *Probl. Fiz. Atmos.*, **2**, 28–47. Leningrad (Gos. Univ.). (In Russian: English translation NASA Tech. Transl. TTF-208 (1964). MGA 17.3–281.)

LAMB, H. H. (1971) Climate-engineering schemes to meet a climatic emergency, *Earth-Sci. Reviews*, **7**, 87–95. Amsterdam (Elsevier).

LANDSBERG, H. E. (1970) Man-made climatic changes, *Science*, **170**, 1265–74. New York (18 December).

LANDSBERG, H. E. (1971) *Man-Made Climatic Changes.* Tech. Note BN-705. College Park, Md. (Univ. Maryland Inst. for Fluid Dynamics and Appl. Maths.).

MACHTA, L. (1971) Stratospheric water vapor, pp. 429–33 in *Man's Impact on the Climate* (ed. W. H. MATTHEWS, W. W. KELLOGG and C. D. ROBINSON). Cambridge, Mass. (M.I.T. Press). 594 pp.

MACHTA, L. and CARPENTER, T. (1971) Trends in high cloudiness at Denver and Salt Lake City, pp. 410–15 in *Man's Impact on the Climate* (ed. W. H. MATTHEWS, W. W. KELLOGG and G. D. ROBINSON). Cambridge, Mass. (M.I.T. Press). 594 pp.

MANABE, S. and WETHERALD, R. T. (1967) Thermal equilibrium of the atmosphere with a given relative humidity, *J. Atmos. Sci.*, **24**, 241–59. Boston, Mass.

MASTENBROOK, H. J. (1971) The variability of water vapor in the stratosphere, *J. Atmos. Sci.*, **28**, 1495.

MATTHEWS, W. H., KELLOGG, W. W. and ROBINSON, G. D. (eds.) (1971) *Man's Impact on the Climate.* Cambridge, Mass. (M.I.T. Press). 594 pp.

MAUNDER, W. J. (1970) *The Value of the Weather.* London (Methuen). 388 pp.

MITCHELL, J. M. (1971) The effect of atmospheric aerosols on climate with special reference to temperature near the Earth's surface, *J. Appl. Met.*, **10**, 703–14. Boston, Mass.

MITCHELL, J. M. (1973) A reassessment of atmospheric pollution as a cause of long-term changes of global temperature, in *Global Effects of Environmental Pollution* (ed. S. F. SINGER). Dordrecht (Reidel).

NATIONAL ACADEMY OF SCIENCES, WASHINGTON (1966) *Weather and Climate Modification.* Nat. Res. Council Pubn No. 1350.

NATIONAL ACADEMY OF SCIENCES, WASHINGTON (1975) *Understanding Climatic Change: A Program for Action.* Prepared by the Panel on Climatic Variation (Co-Chairmen W. L. GATES and Y. MINTZ) of the U.S. Committee for the Global Atmospheric Research Program. Washington (National Research Council). 239 pp.

NEWSON, R. L. (1973) Response of a general circulation model of the atmosphere to removal of the Arctic ice cap, *Nature*, **241**, 39–40. London (5 January).

POTTER, G. L., ELLSAESSER, H. W., MACCRACKEN, M. C. and LUTHER, F. M. (1975) Possible climatic impact of tropical deforestation. *Nature*, **258**, 697–8. London (25 December).

RAKIPOVA, L. R. (1962) Changing the climate by artificial action on the ice of the Arctic Basin, *Glav. Uprav. Gidromet. Služb., Met i Gidr.* **9**, 28–30. Leningrad. (In Russian: English translation available in U.K. Met. Office Library.)

RAKIPOVA, L. R. (1966) Variation in the zonal distribution of temperature as a result of climate modification, pp. 358–83 in *Sovremennye Problemy Klimatologii* (*Contemporary Problems of Climatology*) (ed. M. I. BUDYKO). Leningrad (Gidrometeoizdat). (In Russian: English translation by U.S. Joint Publ. Res. JPRS 43, 482. MGA 18.12–175.)

RASOOL, S. I. and SCHNEIDER, S. H. (1971) Atmospheric carbon dioxide and aerosols: effects of large increases on global climate, *Science*, **183**, 138–41. New York.

SAWYER, J. S. (1971) Possible effects of human activity on world climate, *Weather*, **26**, 251–62. London (Roy. Met. Soc.).

SCEP CONFERENCE (1970) *Man's Impact on the Global Environment* (Report of the Study of Critical Environmental Problems, SCEP). Cambridge, Mass. (M.I.T. Press). 319 pp.

SCHNEIDER, S. H. (1974) A new world climate norm? Implications for future world needs, *Bull. Amer. Acad. Arts and Sci.*, **28** (3), 20–35. Boston, Mass.

SCHNEIDER, S. H. and DENNETT, R. D. (1975) Climatic barriers to long-term energy growth, *Ambio*, **4**, (2), 65–74. Oslo.

SCHNEIDER, S. H. and MASS, C. (1975) Volcanic dust, sunspots and temperature trends: climatic theories in search of verification, *Science*, **190**, 741–6. Washington (21 November).

SELLERS, W. D. (1969) A global climatic model based on the energy balance of the Earth–atmosphere system, *J. Appl. Met.*, **8** (3), 392–400. Boston, Mass.

SIEMERLING, R. (1974) *The Negative Greenhouse Effect and Climatic Change*. Unpublished: Seminar held at Växtbiologiska Institutionen, Uppsala Univ., 29 March 1974.

SMIC CONFERENCE (1971) *Inadvertent Climate Modification* (report of the Study of Man's Impact on Climate, SMIC). Cambridge, Mass. (M.I.T. Press). 308 pp.

WASHINGTON, W. M. (1972) Numerical climatic-change experiments: the effect of Man's production of thermal energy, *J. Appl. Met.*, **11**, 768–72. Boston, Mass.

WEINBERG, A. M. and HAMMOND, R. P. (1970) Limits to the use of energy, *Amer. Scientist*, **58**, 412–18. New Haven, Conn.

WORLD METEOROLOGICAL ORGANIZATION (1972) *Climatic Fluctuation and the Problems of Foresight*. Unpublished Final Report of a Working Group of the Commission for Atmospheric Sciences, prepared by H. H. LAMB (Chairman), I. I. BORZENKOVA, A. BOUKLI-HACÈNE, H. FLOHN, H. J. KRIJNEN, J. M. MITCHELL and K. N. RAO. Geneva (W.M.O.) 157 pp.

Chapter 20   Approaches to the problem of forecasting

BRAY, J. R. (1971) Solar-climate relationships in the post-Pleistocene, *Science*, **171**, 1242–3. Washington.

BRYSON, R. A. (1973) World food prospects and climatic change. Testimony before a joint meeting of the Senate Subcommittee on Foreign Agricultural Policy and Subcommittee on Agricultural Production, Marketing and Stabilization of Prices. Washington, 18 October 1973. (Extracts published in *Fortune* magazine, February 1974.)

BRYSON, R. A. (1975) The lessons of climatic history, *Environmental Conservation*, **2** (3), 163–70.

BUCHINSKY, I. E. (1957) *The Past Climate of the Russian Plain*. Leningrad (Gidrometeoizdat), 2nd edn. (In Russian.)

BUSHBY, F. H. and TIMPSON, M. S. (1967) A 10-level atmospheric model and frontal rain, *Quart. J. Roy. Met. Soc.*, **93**, 1–17. London.

COLLIS, R. T. H. (1975) Weather and world food, *Bull. Amer. Met. Soc.*, **56** (10), 1078–83. Boston, Mass. (Amer. Met. Soc.).

COURT, A. (1967–8) Climatic normals as predictors, Parts 1–5. *Sci. Reps. 1–5 Contract AF19(628)-5176 Project* 8624. Bedford, Mass. (Air Force Cambridge Research Laboratories).

EBDON, R. A. (1975) The quasi-biennial oscillation and its association with tropospheric circulation patterns, *Met. Mag.*, **104**, 282–97. London.

ELLISON, T. H. (1970) Discussion meeting on long-range weather forecasting: introductory remarks, *Quart. J. Roy. Met. Soc.*, **96**, 326–8. London.

FENNER, J. A. (1975) The winter of 1974–5: the problem of extrapolation, *Weather*, **30** (8), 272–3. London.

FLOHN, H. (1974) Background of a geophysical model of the initiation of the next glaciation, *Quaternary Res.*, **4**, 385–404. Seattle.

GATES, W. L. (1975) Numerical modelling of climate change: a review of problems and prospects, pp. 343–54 in *Proc. WMO/IAMAP Symp. on Long-term Climatic Fluctuations, Norwich, 18–23 August 1975*. W.M.O. No. 421. Geneva (World Met. Org.). 503 pp.

JOSEPH, P. V. (1975) *Triennial Oscillation of Upper Tropospheric Westerlies and the Indian Summer Monsoon*. Scientific Report No. 223. New Delhi (India Met. Dept).

KUKLA, G. J. (1975) Missing link between Milankovitch and climate, *Nature*, **253**, 600–3. London (20 February).

KUKLA, G. J. and MATTHEWS, R. K. (1972) When will the present interglacial end? *Science*, **178**, 190–1. New York.

KUTZBACH, J. E. and BRYSON, R. A. (1974) *Variance Spectrum of Holocene Climatic Fluctuations in the North Atlantic Sector*. Unpublished memorandum. Madison (Dept Meteorol., Univ. Wisconsin). 15 pp.

LAMB, H. H. (1973) *The Seasonal Progression of the General Atmospheric Circulation Affecting the North Atlantic and Europe. Research Publication CRU RP1*. Norwich (Univ. of East Anglia, Climatic Research Unit). 131 pp.

LORENZ, E. N. (1969) The future of weather forecasting, *New Scientist*, pp. 290–1. London (8 May).

MCKAY, G. A. and BAIER, W. (1974) *Wheat Production*. Unpublished report presented at the Workshop on World Food Supply in a Changing Climate, Sterling Forest Conference Center, New York, 2–5 December 1974.

MCQUIGG, J. D. (1975) *Climate Change and World Food Production*. Unpublished Address to the Univ. of Florida Frontiers of Science Series, Gainesville, Florida, 23 April 1975. Presented also to the Conference on Climate Change, Food Production and Interstate Conflict, 4–8 June 1975, Bellagio, Italy (Rockefeller Foundation).

MAUNDER, W. J. (1970) *The Value of the Weather*. London (Methuen). 388 pp.

MITCHELL, J. M. (1972) The natural breakdown of the present interglacial and its possible intervention by human activities. *Quaternary Res.*, **2** (3), 436–45. Seattle.

NATIONAL ACADEMY OF SCIENCES, WASHINGTON (1975) *Understanding Climatic Change: A Program for Action*. Prepared by the Panel on Climatic Variation (Co-Chairmen W. L. GATES and Y. MINTZ) of the U.S. Committee for the Global Atmospheric Research Program. Washington, D.C. (National Research Council). 239 pp.

NATIONAL SCIENCE FOUNDATION, WASHINGTON (1974) *Report of the Ad Hoc Panel on the present Interglacial*. Prepared by the Interdepartmental Committee (D. M. HERSHFIELD, Chairman, J. O. FLETCHER, J. M. MITCHELL and J. S. PERRY) for Atmospheric Sciences. Washington D.C. (NSF, Science and Technology Policy Office). ICAS 18B-FY 75. 22 pp.

PISIAS, N. G., DAUPHIN, J. P. and SANCETTA, C. (1973) Spectral analysis of Late Pleistocene-Holocene sediments, *Quaternary Res.*, **3**, 3–9. Seattle.

PITTOCK, A. B. (1972) How important are climatic changes? *Weather*, **27**, 262–7. London.

RATCLIFFE, R. A. S. (1975) Can short-term climatic fluctuations be forecast?, *Abstracts in Advance: Australasian Conf. on Climate and Climatic Change, Monash Univ., 7–12 December 1975*. Clayton, Victoria (Roy. Met. Soc., Australian Branch).

SCHNEIDER, S. H. (1974) A new world climate norm? Implications for future world needs, *Bull. Amer. Acad. Arts & Sciences*, **28** (3), 20–35. Boston, Mass.

TEISSERENCE DE BORT, L. (1881) Étude sur l'hiver de 1879–1880 et recherches sur la position des centres d'action de l'atmosphère dans les hivers anormaux, *Ann. Bureau Centrale de Météorologie*, **4**, 17–62. Paris.

THOMPSON, L. M. (1975) Weather variability, climatic change, and grain production, *Science*, **188**, 535–41. New York.

WAUGH, F. V. (1967) Reserve stocks of farm products, in *Agricultural Policy: A Review of Programs and Needs, Vol. 5: Technical Papers of the National Advisory Commission on Food and Fiber*. Washington, D.C. (Government Printing Office).

WEST, R. G. (1968) *Pleistocene Geology and Biology*. London (Longmans), 377 pp.

WORLD METEOROLOGICAL ORGANIZATION (1972) Climatic fluctuation and the problems of foresight, *Final Report of the Working Group on Climatic Fluctuation of the Commission for Atmospheric Sciences* (H. H. LAMB, Chairman, I. I. BORZENKOVA, A. BOUKLI-HACÈNE, H. FLOHN, H. J. KRIJNEN, J. M. MITCHELL, and K. N. RAO). Unpublished (available from W.M.O.). 157 pp.

Subject Index

'Hanging valleys', 161

Hansa (Hanseatic League), 420, 449

HAROLD HARDRÅDE's Arctic voyage, 253

Harappan culture (Indus valley civilization, c.2500–1750 BC), 251, 389, 415

Harbour works, ancient, evidence of former sea levels, 257–8

Harvests (grain), 5, 7, 11, 16, 31, 451, 458, 461, 469, 471–3, 477, 540, 679–81, 689 (see also Agriculture, Cultivation, Wheat etc.)

Hazel (Corylus), 179, 197, 203, 205, 371 ff, 399, 416, 418

'Head', 164

Health, human and animal, affected by climatic vagaries, 12, 247, 258–68, 458 (see also Diseases, Plague etc.)

Heat, artificially generated, 674–6, 694, 698 (see also Urbanization and vol. 1 pp. 517–8)

Heat from the Earth's interior (geothermal heat), 675

'Heat island', urban, effects, 674 (see also vol. 1 pp. 517–8)

Heat (energy) storage in the atmosphere, 682–3

Heat (energy) storage in the oceans, 283, 285, 307, 662, 692

Heat transport in the atmosphere, 47, 283–4, 299, 323, 490, 516, 545, 548

Heat transport in the oceans, 283–4, 296, 299, 323, 364, 516–7, 545

Heat waves, 31, 465–6

Heather (Erica tetralix), 178, 372

Heather (Erica mediterranea), 179

Hedera helix, see Ivy

Helium in the atmosphere, 286

Helophorus beetle, 231, 233

Hemlock spruce (Tsuga), 201, 327, 392

Hengelo interstadial, 97, 337, 343

HERODOTUS, 4

Heron (Ardea cinerea), 186

Herring (Clupea harengus), 15, 183, 190

Hevea Brasiliensis (rubber tree), 181

Hickory tree (Carya), 392

Highland troubles/warfare in Scotland, 6, 459

HIPPARCHUS, recognition of climate's variability, 4

Hippopotamus, 176, 188

Historians of climate, 244, 246

Hittites, 421

Hokokam Indians (America), 249

Holly (Ilex aquifolium), 177, 178

Holocene (= postglacial times), 325, 363–422

Holstein interglacial, 325

Honey production, 16, 255, 679

Hopewell (American Indian) culture, 249

Hornbeam (Carpinus), 180, 205, 373, 392

Horses, 74, 157, 273, 310, 352

Horticulture, 4, 12, 182, 466, 679 (see also Cultivation limits, Vines etc.)

Hoxnian (Hoxne) interglacial, 197–9, 322, 325

Human activity, effects on climate, see Impact of Man's activities

Human (racial) adaptation to climate, 259

Human history, as evidence of past climate, 173, 183–4, 243–79

Human settlements and climatic changes, 134, 246, 388, 413 ff, 428, 432, 437–9, 452, 455 ff, 470–1 (see also Abandonment of villages, Migrations etc.)

Humboldt Current, 239

Humidity, 187

Hunger, see Famines

Hunting societies, 413–4

Hurricanes (Tropical cyclones q.v.), 16, 185, 192, 282, 504–5, 515, 640–1, 679

HUS, Hussite religious risings in Bohemia, 459

Hyaenas, 188

Hydroelectricity, 679

Hydrogen in the atmosphere, 286

Hylotrupes bajulus (house longhorn beetle), 187

Hymenoptera, see Ants

'Hypsithermal' period, 372, 378

Ibex, 352

Ice ages, 36, 39, 70, 88–9, 93–101, 106, 111, 115, 130–1, 150, 154, 159–65, 192, 207, 235, 237–43, 285, 296–7, 301, 304, 306, 308–62

Ice age geography (last climax and other stages), 344–8, 361

Ice age-interglacial cycles, 310–23, 687–8, 694–7

Ice age-interglacial cycles, causation, 312–23

Ice age, last, see Weichselian ice age

Ice age, next, 694–7, 707, 709

Ice age, onset of regime, 335–41

Ice age, Ordovician, 285, 296

Geographical Index

Milton Keynes UK
Ingram Content Group UK Ltd.
UKHW051857071024
449327UK00025B/1992